AF785296

Corrosion Basics: An Introduction

Corrosion Basics
An Introduction

Second Edition

Pierre R. Roberge

NACE INTERNATIONAL
1440 South Creek Drive
Houston, Texas 77084
http://www.nace.org

ISBN 1-57590-198-6

Library of Congress Control Number: 2005937891

Cover design by Michele Sandusky

Contents

Part III
Materials and How They Corrode

Part IV
Methods of Control

Abbreviations

ABS	Acrylonitrile-butadiene-styrene
AC	Alternating current
AE	Acoustic emission
AIWJ	Abrasive injected water jetting
ACS	American Chemical Society
ASM	American Society for Metals
ASTM	American Society for Testing and Materials
AWJ	Abrasive water jetting
AWWA	American Waterworks Association
BOD	Biological oxygen demand
CLIMAT	Classify Industrial and Marine ATmosphers
CCSRE	Copper/copper sulfate reference electrode
CCVT	Closed-cycle vapor turbogenerators
CDA	Corrosion-damage algorithm
CFA	Component failure analysis
CHC	Cyclohexylamine carbonate
CIPS	Close-interval potentials survey
COD	Chemical oxygen demand
CP	Cathodic protection
CPE	Constant-phase element
DC	Direct current
DCVG	Direct-current voltage gradient
DHN	Dicyclohexylamine nitrite
DO	Dissolved oxygen
DOT	U.S. Department of Transportation
DSA	Dimensionally stable anodes
ECM	Electrochemical machining
ECS	Electrochemical Society

EDM	Electical discharge
EFSM	Electrical-field signature method
EIS	Electrochemical impedance spectroscopy
EMF	Electromotive force
EN	Electrochemical noise
EPR	Electrochemical potentiodynamic reactivation
ER	Electrical resistance
EVS	Extreme value statistics
FN	Ferrite number
GCP	Galvanic cathodic protection
GFRP	Glass-fiber-reinforced polymer
GDP	Gross domestic product
GNP	Gross national product
HAZ	Heat-affected zone
HF	Hydrofluoric (acid)
HIC	Hydrogen-induced cracking
HPC	High-performance concrete
HSLA	High strength, low alloy
HT/HP	High temperature/high pressure
IAP	Ion-activity product
ICCP	Impressed current cathodic protection
IGSCC	Intergranular stress corrosion cracking
ISO	International Organization for Standardization
ISO CORRAG	International Organization for Standardization atmospheric-exposure program
KK	Kramers–Kronig transforms
LMC	Liquid-metal cracking
LPR	Linear polarization resistance
LSI	Langelier saturation index
MEK	Methyl ethyl ketone
MIC	Microbiologically influenced corrosion
MIL	Military
NACE	National Association of Corrosion Engineers
NDE	Nondestructive evaluation
NHE	Normal hydrogen electrode
OEM	Original equipment manufacturer
PB	Pilling–Bedworth ratio
PE	Polyethylene
PH	Precipitation hardening
PREN	Pitting-resistance equivalent number
PSI	Puckorius scaling index
PVC	Polyvinyl chloride
PWR	Pressurized-water reactors
RCA	Root-cause analyses
RCI	Root-cause investigation

RH	Relative humidity
%RH	Percent relative humidity
RSI	Ryzner stability index
SAE	Society of Automotive Engineers
SCC	Stress corrosion cracking
SCE	Saturated calomel electrode
SEM	Scanning electron microscopy
SHRP	U.S. Strategic Highway Research Program
SL	Saturation level
SRB	Sulfate-reducing bacteria
SSC	Sulfide stress cracking
SSCE	Saturated silver chloride electrode
SSPC	Society of Protective Coatings
SHE	Standard hydrogen electrode
TCA	Trans-cinnamaldehyde
TDS	Total dissolved solids
TFE	Polytetrafluoroethylene
TLA	Thin-layer activation
TOW	Time of wetness
UHP AB	Ultrahigh-pressure abrasive blasting
VCI	Volatile corrosion inhibitors
VPI	Vapor-phase inhibitors
WECS	Working Environmental Corrosion Standards
ZPC	Zero point of charge
ZRA	Zero-resistance ammeter

Preface to the First Edition

This book was originally issued as the Basic Corrosion Course in 1970. The utility of that publication can be appreciated when it is realized that 13 printings were necessary to supply the great number of persons interested in the subject.

As envisioned by the Editor in the Introduction of the original book, some revision is necessary with the passage of time. This has been accomplished with a minimum of change from the initial format and material content of the original. The authors of the first edition provided such an excellent coverage of the subject matter that only some rearrangement of the text and updating has been necessary.

Those responsible for the topical matter of the first edition were as follows:

Editor	A. deS. Brasunas
Scope of the Subject	A. deS. Brasunas and N. E. Hamner
Basics	F. L. LaQue and N. D. Greene
Metallurgy	R. F. Hochman
Materials	M. G. Fontana and J. H. Peacock
Localized Corrosion	H. P. Godard
Stress Corrosion Cracking	H. L. Logan
Cathodic Protection and Soils	A. W. Peabody and M. E. Parker
Inhibitors	N. Hackerman and E. S. Snavely
Atmospheric Corrosion	K. G. Compton
Coatings	N. E. Hamner
High Temperature Corrosion	A. deS. Brasunas and J. J. Moran, Jr.
Water Corrosion	W. E. Berry
Testing	B. W. Lifka and F. L. McGeary
Failure Analysis	E. D. Verink

A great amount of the material written by these authors has been retained. Changes have been made where it was believed a better continuity, less repetition, and more recent data would improve the development of the subject. Those responsible for the revisions are C. P. Dillon, J. S. Snodgrass, L. S. Van Delinder, and H. A. Webster.

This book provides a general coverage of the wide field of corrosion control. It is designed to be helpful to those being initiated into the work, and consequently attempts to present each corrosion process or control procedure in the most basic terms. A comprehensive discussion of the topics has not been attempted. Certainly the book is not represented as providing the latest or most erudite discussion of the subject. Yet for those knowledgeable in the field of corrosion control, we hope that each chapter presents some aspect of the work in terms that are stimulating and helpful. We solicit suggestions for improvement of the work.

To assist the reader, a bibliography is provided at the end of each chapter for those who may want to pursue a topic further. Many of the books that are listed in more than one chapter are obviously regarded as prime sources for corrosion information. The appendix to Chapter 1 contains a list of definitions of corrosion-related terms. A comprehensive and thoroughly cross referenced subject index is located in the back of the book and should provide a ready guide for all persons wishing to find specific items.

Education in corrosion control is our primary concern. We sincerely hope this new edition will contribute even more capably in attaining that goal. As with the preponderance of the first edition, a portion of the Introduction must be repeated: "We trust that . . . this book will be of considerable benefit to the nation, to numerous industries, and to individuals all over the world who will take advantage of the opportunity afforded them by this educational effort."

L. S. Van Delinder
Editor and Chairman of Subcommittee ETC-1
NACE Education and Training Committee
May, 1984

Preface to the Second Edition

As mentioned in the Preface of the first edition, the original material of the book was first published in 1970. Although corrosion principles have not changed much since the lecture notes that became the first edition were assembled in the late 1960s, there have been major advances and changes in the technologies used to combat corrosion damage.

When the NACE International Publication Committee became interested in updating this popular textbook, a few options were discussed. The decision was taken to prepare the new edition with a single author as the most expedient and efficient way of preserving this historical document. Hopefully, you will find for yourself that this objective has been realized. My main goals throughout this daunting project were to respectfully keep the spirit of the first edition and maintain the quality of the information it conveyed, while bringing all these interesting subjects into the 21st century. Education in corrosion control was the primary concern and motivation behind the production of the first lecture notes. It was again the primary focus during the production of the first edition. My 25 years of teaching corrosion and materials engineering, added to which are my 5 years as webmaster of the Corrosion Doctors Web Site, have helped me keep this focus.

Besides a complete reorganization of the order in which material is presented and the addition of new sections on electrochemistry and reinforced concrete, the readers will find that the new text uses SI units in compliance with NACE International policy. Appendix A of this book contains the official NACE glossary of corrosion related terms, and Appendix B presents a conversion table for many non-SI units, which some readers may find useful.

Finally, I would like to express my gratitude to all the reviewers who diligently helped to improve the book, and particularly to Ken Tator and J.P. Broomfield for their great help with the chapters on protective coatings and on reinforced concrete, respectively.

Pierre. R. Roberge
Editor, Second Edition
July, 2005

Part I

What Is Corrosion?

1

Chapter

Scope and Language of Corrosion

1.1 WHAT IS CORROSION?

Most people are familiar with corrosion in some form or another, particularly the rusting of an iron fence or a "tin" can, the degradation of steel pilings or boats and boat fixtures, or the rusting of common nails (Figure 1.1). However, invisible rusting often leads to surprises as exemplified by this picture of a city lamppost that fell on a windy day because of the corrosion of its steel base (Figure 1.2).

Almost everyone has seen bridges or tall, modern buildings erected with steel frameworks. These steel frames usually have a green, orange, or reddish color, which is the color of a protective coating placed on steel members even before they leave the steel mill. These coatings are applied to protect the frames from rust. Then, after installation (particularly on bridge construction), these surfaces are recoated with other materials that may lend a better appearance or added corrosion protection.

Piping is another major type of equipment subject to corrosion. This includes water pipes in the home (Figure 1.3), where corrosion usually attacks from the inside, and the underground water, gas, and oil pipe lines that crisscross our land. Thus, it would appear safe to say that almost everyone is at least somewhat familiar with corrosion, which is defined in general terms as the deterioration of a substance, usually a metal, or its properties because of a reaction with its environment.

Note that the words chemical or electrochemical are not used in this definition. This is because a few forms of corrosion do not involve these processes. Furthermore, this definition recognizes that materials other than metals or alloys may corrode. The deterioration of wood, ceramics, polymeric materials ("plastics"), and other materials must also be studied by the corrosion engineer and is included in the term corrosion.

This definition also indicates that properties, as well as the materials themselves, can and do deteriorate. In some forms of corrosion, there is no weight change or visible deterioration, yet properties change and the material may fail unexpectedly because of certain changes within the material. Such changes may defy ordinary visual examination

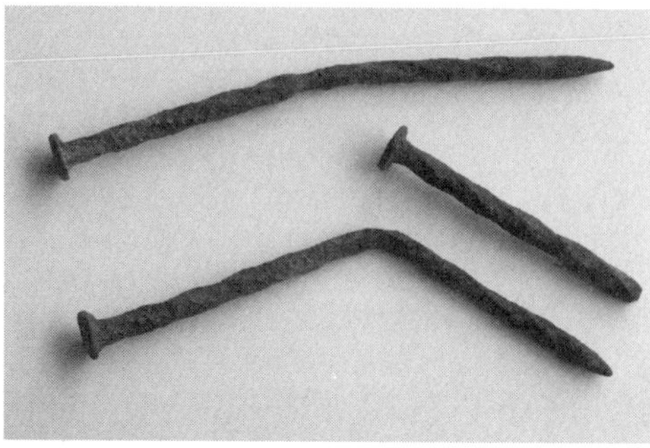

Figure 1.1 A prevalent example of corrosion is the rusting of an ordinary iron nail. Although commonplace, numerous technical articles can be devoted to this natural yet complex phenomenon.

or weight-change determinations. These are nevertheless important forms of corrosive action with which a corrosion engineer should be familiar.

1.2 PURPOSE OF THIS BOOK

As with the first edition of *Corrosion Basics*, this edition is intended to convey the scope of the field of corrosion prevention and control. It is important to realize the extent of the effort being made today to analyze and combat corrosion. Much of the experience and many of the workable solutions developed in one area of corrosion work can be used

Figure 1.2 This city lamppost fell because of corrosion of its base (courtesy of Bruce Hector, Rust Check Kingston).

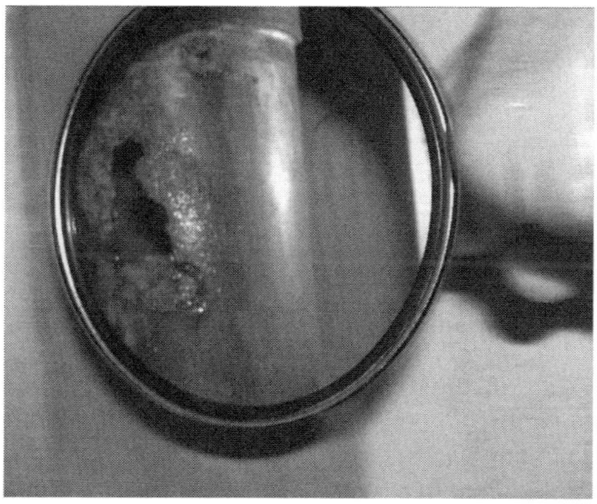

Figure 1.3 A common copper sink drain has suffered complete perforation after cleaning by a local plumber.

to improve the control procedures of another area. Although most people work in only one area of this discipline, there is always the possibility that a shift in responsibilities or interest may bring one to work in a completely different area of corrosion prevention and control.

This book is intended to acquaint individuals who have little or no corrosion knowledge with the fundamentals of corrosion and practices commonly used to control corrosion in the home, outdoors, and industry.

Many people become engaged in controlling or preventing corrosion by appointment rather than as the final step in a process of formal education that has this as its original goal. This book is therefore designed to be helpful to those people entering this field without the benefit of extensive training in the basic sciences related to corrosion, but who may be called upon from time to time to take at least the first steps in anticipating, diagnosing, and otherwise dealing with corrosion problems, either on their own or in collaboration with others.

The book also provides a good review of corrosion basics, which may serve as a refresher for those who have an education in the field.

Like any other technical discipline, corrosion work utilizes certain words, methods of presenting data, and shorthand symbols that are peculiar to the field. This book reveals this jargon in a manner that should, in the future, allow people to read or discuss corrosion topics with greater confidence and understanding. Although the book defines some terms as they appear, it would also be useful to develop the habit of referring to the glossary of corrosion-related terms that appears in Appendix A.

The best techniques available for detecting corrosion, determining the corrosion resistance of a material, and evaluating the efficacy of a control procedure serve as tools used daily to attack the problems faced by thousands of people engaged in corrosion work. This book will foster a better appreciation of these procedures.

Certainly anyone concerned with corrosion work is primarily interested in control procedures that allow the safe, economic use of a material in a specific environment. Thus, basic aspects of the most current control techniques are described. After learning the elements of each procedure, one can further investigate the possibility of using these approaches to solve specific problems. For instance, it might be found that cathodic protection (Chapter 11), which in many instances can halt corrosion with the throw of a switch, could be adapted to solve other specific problems.

A modern corrosion engineer or technician must find suitable materials for a great variety of corrosive environments and temperature ranges and must be familiar with the behavior of numerous alloys and possible substitute nonmetallic materials. However, because metals have and will undoubtedly continue to be the major engineering materials of concern, the focus of this book is mostly on the corrosion behavior of metals.

The only suitable answer to the question "In what environments does corrosion occur?" is "In just about any environment, depending on the material being used."

Another general question that often arises is "How many forms of corrosion are there?" Again, the answer may be unexpected. Some could say two or three, others propose eight, and still others more safely say, "Quite a few!" These forms of corrosion are discussed in detail in Chapter 9.

1.2.1 Economic Significance

Many people believe that corrosion is an inevitable foe that should be accepted as an inevitable process. Actually, something can and should be done to prolong the life of many metals exposed to corrosive environments. This is only a common-sense approach, considering corrosion involves the gradual destruction of a material that costs money to replace. Thus, corrosion control becomes a major factor in the very important area of dollars and cents.

Although corrosion processes form an interesting basis for scientific studies in chemistry and electrochemistry, by far the greatest interest in and concern for corrosion stems from its damaging effects and how they may be avoided.

As products and manufacturing processes have become more complex and the penalties of failures from corrosion, including safety hazards and interruptions in plant operations, have become more costly and more specifically recognized, the attention given to the control and prevention of corrosion has increased.

Since the first significant report by Uhlig in 1949 that the cost of corrosion to nations is indeed great,[1] the conclusion of all subsequent studies has been that corrosion represents a constant charge to a nation's gross national product (GNP). The annual cost of corrosion to the United States was estimated in Uhlig's report to be $5.5 billion or 2.1% of the 1949 GNP. This study attempted to measure the total costs associated with corroding components by summing up the cost for both the owner and operator (direct cost) and for the users (indirect cost).

Cost of corrosion studies of various forms and importance have been undertaken since then by several countries including the United States, the United Kingdom, Japan, Australia, Kuwait, Germany, Finland, Sweden, India, and China.[2] The common finding of these studies was that the annual corrosion costs ranged from approximately 1 to 5% of the GNP of each nation. Several studies separated the total corrosion costs into two parts:

1. the portion of the total corrosion cost that could be avoided if better corrosion-control practices were used and

2. costs where savings required new and advanced technology (currently unavoidable costs).

Estimates of avoidable corrosion costs varied widely with a range from 10 to 40% of the total cost. Most of the studies allocated corrosion costs to industrial sectors or to categories of corrosion-control products and services. All studies addressed direct costs. A common conclusion was that the indirect costs resulting from corrosion damage are often significantly greater than the direct costs. The indirect costs were more difficult to estimate.

Potential savings and recommendations of ways to realize the savings were included in most of the reports as formal results or as informal directions and discussion. Two of the most important and common findings were

1. better dissemination of the existing information through education and training, technical advisory and consulting services, and research and development activities; and
2. the opportunity for large savings through more cost-effective use of currently available means to reduce corrosion. Most studies addressed only the magnitude of possible savings without identifying the means of realizing such savings.

The most recent study resulted from discussions between NACE International representatives, members of the U.S. Congress, and the U.S. Department of Transportation (DOT). An amendment for the cost of corrosion was included in the Transportation Equity Act for the 21st Century, which was passed by the U.S. legislature in 1998. The amendment requested that a study be conducted in conjunction with an interdisciplinary team of experts from the fields of metallurgy, chemistry, economics, and others as appropriate.

Two different approaches were taken in the ensuing study to estimate the cost of corrosion:

- The first approach followed a method in which the cost was determined by summing the costs for corrosion-control methods and contract services. The costs of materials were obtained from various sources, such as the U.S. Department of Commerce Census Bureau, existing industrial surveys, trade organizations, industry groups, and individual companies. Data on corrosion-control services, such as engineering services, research and testing, and education and training, were obtained primarily from trade organizations, educational institutions, and individual experts. These services included only contract services and not service personnel within the owner/operator companies.
- The second approach followed a method where the cost of corrosion was first determined for specific industry sectors and then extrapolated to calculate a national, total corrosion cost. Data collection for the sector-specific analyses differed significantly from sector to sector, depending on the availability of data and the form in which the data were available. To determine the annual corrosion costs for the reference year of 1998, data were obtained for various years in the surrounding decade, mainly for the years 1996 to 1999.

Table 1.1 Summary of Estimated Direct Cost of Corrosion for Industry Sectors Analyzed in the 2001 Study

Category	Industry Sectors	Appendix	Estimated Direct Cost of Corrosion per Sector	
			$, billion	Percent
Infrastructure (16.4% of total)	Highway bridges	D	8.3	37
	Gas and liquid transmission pipe lines	E	7.0	27
	Waterways and ports	F	0.3	1
	Hazardous materials storage	G	7.0	31
	Airports	H	—	—
	Railroads	I	—	—
		Subtotal	$22.6	100%
Utilities (34.7% of total)	Gas distribution	J	5.0	10
	Drinking water and sewer systems	K	36.0	75
	Electrical utilities	L	6.9	14
	Telecommunications	M	—	—
		Subtotal	$47.9	100%
Transportation (21.5% of total)	Motor vehicles	N	23.4	79
	Ships	O	2.7	9
	Aircraft	P	2.2	7
	Railroad cars	Q	0.5	2
	Hazardous materials transport	R	0.9	3
		Subtotal	$29.7	100%
Production and Manufacturing (12.8% of total)	Oil and gas exploration and production	S	1.4	8
	Mining	T	0.1	1
	Petroleum refining	U	3.7	21
	Chemical, petrochemical, and pharmaceutical	V	1.7	10
	Pulp and paper	W	6.0	34
	Agricultural	X	1.1	6
	Food processing	Y	2.1	12
	Electronics	Z	—	—
	Home appliances	AA	1.5	9
		Subtotal	$17.6	100%
Government (14.6% of total)	Defense	BB	20.0	99.5
	Nuclear waste storage	CC	0.1	0.5
		Subtotal	$20.1	100%
		Total	$137.9	

The total cost of the impact of corrosion for the individual economic sectors was $137.9 billion per year (Table 1.1). A breakdown of these costs by individual sectors is shown in Figure 1.4. Because not all economic sectors were examined, the sum of the estimated costs for the analyzed sectors did not represent the total cost of corrosion for the entire U.S. economy. By estimating the percentage of U.S. gross domestic product (GDP) for the sectors for which corrosion costs were determined and by extrapolating the figures to the entire U.S. economy, a total cost of corrosion of $276 billion was estimated

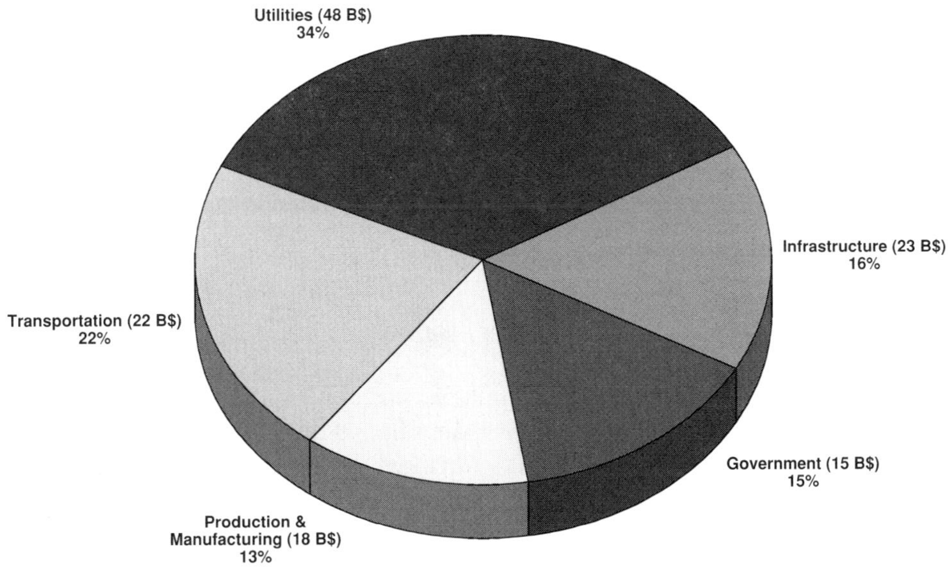

Figure 1.4 Percentage contribution to the total cost of corrosion for the five sector categories.

(Figure 1.5). This value shows that the effect of corrosion is approximately 3.1% of the nation's GNP (Figure 1.5). This cost is considered to be a conservative estimate because only well-documented costs were used in this latest study. Other costs of corrosion were discussed (not estimated) in the individual sectors but were left out because of lack of documentation.

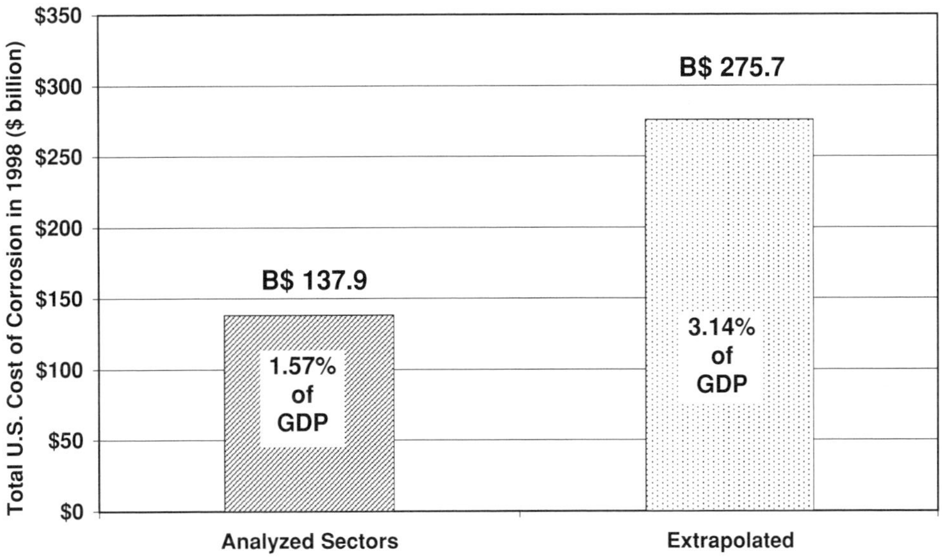

Figure 1.5 Total direct annual cost of corrosion.

The indirect cost of corrosion was conservatively estimated to be equal to the direct cost, giving a total direct plus indirect cost of $552 billion (i.e., 6% of the GDP). Evidence of the large indirect corrosion costs are

1. lost productivity because of outages, delays, failures, and litigation;
2. taxes and overhead on the cost of the corrosion portion of goods and services; and
3. indirect costs of non-owner/operator activities.

1.2.2 Making Choices Based on Economic Calculations

Economic calculations used to be laborious procedures that could only be properly performed by those who understood the relationship of all the factors involved. Calculations of the effects of the initial cost, life, rate of depreciation, taxes, and value of money to the owner/employer have been combined into easily used equations in a National Association of Corrosion Engineers (NACE) International standard that can be used without a great knowledge of economics.[3] This report advocates the use of the discounted-cash-flow method, which provides ready calculation of net present worth. Factors that need to be considered in calculating net present worth include

1. initial cost,
2. best estimate of expected life,
3. length of typical shutdown for emergency repair,
4. cost of planned maintenance during scheduled shutdowns, and
5. effect of failure on total plant operation.

When proposals are made with recommendations based on economic calculations, those controlling the purse strings are more inclined to listen. Calculated options provide them with an objective basis on which to make a decision regarding how and where to spend the money available and thus add more credibility to a proposal.

Notice that no number is given for the requirement of safety when selecting materials. If the safety of any individual or the public is endangered in some manner by the selection of a material or the use of a corrosion-control procedure, these safety considerations have to take precedence over purely economical calculations.

1.2.3 Corrosion Around Us

All metals and alloys can and generally do corrode. In looking about, it is easy to find corrosion occurring all around us as well as to find some metals that do not appear to corrode at all. The following are some examples of common objects and areas susceptible to corrosion.

1. The automobile is a veritable corrosion-testing device. Materials must be selected for use for atmospheric, high-temperature, closed-cycle-water wear, fatigue, and crevice exposures. Many different metals, polymeric materials

Figure 1.6 Photograph shows a dendrite on a circuit board after its exposure in a test chamber (courtesy of David A. Douthit LoCan LLC).

(plastics), paints, metallic coatings, and inhibitors are used to prolong the life of the parts. However, the body and certainly the muffler are generally seen in various stages of corrosion. In fact, when examining a car more than seven years old, it is often not necessary to look under the car; just looking at the car's body will probably reveal corrosion taking place in numerous locations.

2. According to the U.S. Census Bureau, the American electronic-manufacturing industries alone shipped $336 billion worth of electronic components in 1997. The trend toward miniaturization of technology has led to the development of small personal electronic devices that are now present everywhere, such as pagers, cellular phones, and palm-sized personal organizers and computers.[2] In the presence of moisture and an electric field, silver ions can migrate to a cathodically (negatively) charged surface and plate out, forming dendrites. The dendrites grow and eventually bridge the gap between the contacts, causing an electric short and possibly arcing and fire. Even a small volume of dissolved metal can result in formation of a relatively large dendrite (Figure 1.6).

3. In looking at the water pipes of an older home, particularly if they are made out of steel or even galvanized steel, some corrosion is usually evident, particularly at the joints where the galvanized coating might have been cut or at a junction where a brass valve has caused heavy corrosion in the adjacent area.
4. Barnyard roofs made of steel or galvanized iron rust through in time, whereas the properly chosen aluminum roofs look bright and shiny and appear to be free of corrosion much longer.
5. Kitchen appliances generally look bright and shiny, although their interiors are often stained below the water line and sometimes pitted.
6. The green patina on copper or brass roofs of churches and government buildings develops as a result of corrosion. Copper-bearing metals that have reacted with certain components of the atmosphere develop this patina, which is actually considered a protective layer that retards any further reaction and is also pleasing to the eye.
7. Various industries, particularly petroleum and chemical companies, have numerous corrosion problems unique to their own operations. The average person is probably totally unaware of them. Such companies generally maintain a corrosion-conscious staff to keep the plants out of trouble.
8. In the case of an aluminum canoe, corrosion seems almost negligible, although a close look often reveals tiny pits of corrosive attack that eventually can produce water leaks.
9. Metals exposed to seawater may corrode heavily. Ship hulls (inside and out), piling, oil-rig platforms, equipment for fresh-water recovery, and other items exposed to this highly corrosive environment must be protected in some manner. Luckily, most types of corrosion control may be used in seawater.
10. Concrete degradation of bridges, highways, buildings, and drainage ducts is often caused by reaction with the environment rather than by mechanical forces. When corrosion control is not exercised during initial construction, taxpayers' money is often required for the restoration of the failed facilities.
11. Many houses in the nation have cold-water lines, vent and drain ducts, moldings, and other construction components that are made of plastic materials. These have been successful, but how many people are aware of their limitations, proper use, and the differences in performance among the available materials?
12. A person may drive along a roadway and see a metal box at the fence line along the road right-of-way marked something like "XYZ Gas Co. Do Not Disturb." Although this may not be much of a clue to most people, it probably indicates a test-point installation that allows the corrosion-control personnel, by electrical means, to measure the corrosion protection being provided for a steel pipe many feet below the surface through a process called cathodic protection (Chapter 11).

Other common examples of corrosion, either occurring or under obvious control, are observed by everyone each day. This book should aid observation of these phenomena.

1.2.4 Choosing the Right Material

Of the 103 elements in the Periodic Table shown in Figure 1.7, approximately 80 are metals, each possessing different mechanical, chemical, and physical properties. Although all these metals can corrode, they do it in many different manners. Furthermore, many of these 80 metals have been alloyed to make tens of thousands of different alloys.

Obviously, one person cannot memorize the composition of such a large number of alloys. Reference texts are available to provide such information. Chapter 8 contains a general description of the mechanical properties and chemical composition as well as the corrosion characteristics of the most commonly used alloys. Chapter 8 also introduces many of the nonmetal materials an engineer has to consider for specific applications. In many situations these often newer materials combine gracefully with metals to provide strength, endurance, and light weight at an advantageous cost.

1.2.5 Defining the Environment

The fact that corrosion does occur should not be cause for surprise. Almost all materials should be expected to deteriorate with time when exposed to "the elements." Corrosion is a perfectly natural process, as natural as water flowing downhill. If water flowed uphill or remained stationary on a hillside, there may be cause for surprise, yet our human ingenuity can accomplish this by putting water in a closed container (pipe) and closing the bottom end, or by merely freezing it.

Similarly, if iron or steel was exposed to air and water, rust would be expected to develop within a matter of hours. In fact, it would be surprising if the exposed iron did not corrode or rust. Of course, if copper, brass, aluminum, or stainless steel was substituted for iron, a given degree of corrosion might take longer, but some corrosion would still be anticipated. Instead of forming rust (a form of iron oxide), some oxides of copper, aluminum, or chromium may form very slowly and coat the bare metal. This oxide coating, even if extremely thin, could form a partial barrier to continued attack and slow down the rate of corrosion almost to a standstill.

A surface-layer formation, whether it is oxide, carbonate, sulfate, or any other compound, is a major factor in corrosion resistance, particularly if the layer effectively separates the underlying metal from its environment. Such a naturally formed coating must be diffusion and moisture resistant to be effective. Ordinary iron does not naturally form an effective barrier; its rust permits oxygen and moisture to penetrate and continue rusting. Thus, unless precautions are taken, failure will eventually occur.

Precautions to prevent iron and its alloys from corroding constitute a major effort in corrosion control. It is here that we often resort to the use of protective coatings over the metallic surface to prolong its useful life. Other techniques that form the basis of modern corrosion control are discussed in later chapters, and particularly in Chapters 10–13 of the book.

Some metals, such as stainless steel, titanium, or aluminum, are frequently left unpainted. This is not because these metals are inert, but because oxygen in the air helps develop a protective oxide layer on the metallic surface. Although these oxide layers are so thin as to be invisible to the eye, they can be detected and their presence verified.

Key

Atomic number → 6 | +4 +2 } Combining capacity
Symbol → C
Carbon
12.0
↑
Atomic mass*

*Based on C^{12} = 12.00000

1	2	3	4	5	6	7	8	9	10	11	12	13	14	15	16	17	18
1 ±1 H Hydrogen 1.0																	2 He Helium 4.0
3 +1 Li Lithium 6.9	4 +2 Be Beryllium 9.0											5 +3 B Boron 10.8	6 +4 +2 C Carbon 12.0	7 −3 +5 N Nitrogen 14.0	8 −2 O Oxygen 16.0	9 −1 F Fluorine 19.0	10 Ne Neon 20.2
11 +1 Na Sodium 23.0	12 +2 Mg Magnesium 24.3											13 +3 Al Aluminum 27.0	14 +4 +2 Si Silicon 28.1	15 −3 P Phosphorus 31.0	16 −2 +4 +6 S Sulphur 32.1	17 −1 Cl Chlorine 35.5	18 Ar Argon 40.0
19 +1 K Potassium 39.1	20 +2 Ca Calcium 40.1	21 +3 Sc Scandium 45.0	22 +3 +4 Ti Titanium 47.9	23 +2 +5 V Vanadium 50.9	24 +2 +3 +6 Cr Chromium 52.0	25 +2 +7 Mn Manganese 54.9	26 +3 +2 Fe Iron 55.8	27 +2 +3 Co Cobalt 58.9	28 +2 +3 Ni Nickel 58.7	29 +1 +2 Cu Copper 63.5	30 +2 Zn Zinc 65.4	31 +3 Ga Gallium 69.7	32 +4 +2 Ge Germanium 72.6	33 −3 +5 As Arsenic 74.9	34 −2 +4 +6 Se Selenium 79.0	35 −1 Br Bromine 79.9	36 Kr Krypton 83.8
37 +1 Rb Rubidium 85.5	38 +2 Sr Strontium 87.6	39 +3 Y Yttrium 88.9	40 +4 Zr Zirconium 91.2	41 +5 +3 Nb Niobium 92.9	42 +6 +2 Mo Molybdenum 95.9	43 +7 +3 Tc Technetium (99)	44 +3 Ru Ruthenium 101.1	45 +3 Rh Rhodium 102.9	46 +2 +3 +4 Pd Palladium 106.4	47 +1 Ag Silver 107.9	48 +3 Cd Cadmium 112.4	49 +3 +1 In Indium 114.8	50 +2 +4 Sn Tin 118.7	51 −3 +5 Sb Antimony 121.8	52 −2 +4 +6 Te Tellurium 127.6	53 −1 I Iodine 126.9	54 Xe Xenon 131.3
55 +1 Cs Cesium 132.9	56 +2 Ba Barium 137.3	57 +3 La Lanthanum 138.9	72 +4 Hf Hafnium 178.5	73 +5 Ta Tantalum 180.9	74 +6 +2 W Tungsten 183.9	75 +7 +3 Re Rhenium 186.2	76 +2 +3 +4 Os Osmium 190.2	77 +2 +3 +4 Ir Iridium 192.2	78 +2 +4 Pt Platinum 195.1	79 +1 +3 Au Gold 197.0	80 +1 +2 Hg Mercury 200.6	81 +1 +3 Tl Thallium 204.4	82 +2 +4 Pb Lead 207.2	83 −3 +5 Bi Bismuth 209.0	84 ±2 +4 Po Polonium (209)	85 −1 At Astatine (210)	86 Rn Radon (222)
87 +1 Fr Francium (223)†	88 +2 Ra Radium (226)	89 +3 Ac Actinium (227)															

LANTHANIDE SERIES													
58 +3 +4 Ce Cerium 140.1	59 +3 +4 Pr Praseodymium 140.9	60 +3 Nd Neodymium 144.2	61 +3 Pm Promethium (145)	62 +2 +3 Sm Samarium 150.4	63 +2 +3 Eu Europium 152.0	64 +3 Gd Gadolinium 157.3	65 +3 +4 Tb Terbium 158.9	66 +3 Dy Dysprosium 162.5	67 +3 Ho Holmium 164.9	68 +3 Er Erbium 167.3	69 −2 −3 Tm Thulium 168.9	70 −2 −3 Yb Ytterbium 173.0	71 −3 Lu Lutetium 175.0

ACTINIDE SERIES													
90 +4 Th Thorium 232.0	91 +4 +5 Pa Protactinium (231)	92 +3 +4 +5 +6 U Uranium 238.0	93 +3 +4 +5 +6 Np Neptunium (244)	94 +3 +4 +5 +6 Pu Plutonium (244)	95 +3 +4 +5 +6 Am Americium (243)	96 +3 Cm Curium (247)	97 +3 +4 Bk Berkelium (247)	98 Cf Californium (251)	99 Es Einsteinium (252)	100 Fm Fermium (257)	101 Md Mendelevium (258)	102 No Nobelium (259)	103 Lr Lawrencium (260)

†Masses in parentheses are the mass numbers of the most stable isotope.

Figure 1.7 Elements in the Periodic Table are arranged in periods (rows) and groups (columns). Each of the seven periods is filled sequentially by atomic number. Groups include elements that have the same electron configuration in their outer shell, which results in group elements sharing similar chemical properties. The lanthanides and actinides are actually a part of periods 6 and 7, respectively.

Some environments are more corrosive than others. Although there are exceptions, the following statements are generally accepted as facts:

1. Moist air is more corrosive than dry air.
2. Hot air is more corrosive than cold air.
3. Hot water is more corrosive than cold water.
4. Polluted air is more corrosive than clean air.
5. Acids are more corrosive than bases (alkalies) to steels.
6. Salt water is more corrosive than fresh water.
7. Stainless steel will outlast ordinary steel.
8. No corrosion will occur in a vacuum, even at very high temperatures.

Although it may be a surprise to some, there are instances when all of these statements, including the last one, are incorrect. This would indicate that broad, categorical statements regarding corrosion should be considered to be suspect. There is essentially no statement regarding corrosion or the use of a material that does not have an exception.

1.3 HISTORICAL PERSPECTIVE

There have been organized studies of corrosion for a longer time than most people realize. The British Association for the Advancement of Science appropriated a sum of money for a series of corrosion experiments on the corrosion of cast and wrought iron in the 1830s. Experiments were conducted by Robert Mallet and reported by him in 1838, 1840, and 1843. Mallet's work was done in an era in which gifted scientists were investigating the properties of matter.

Many interesting references and reflections on historical aspects of corrosion can be found in the special symposium that was held to celebrate NACE International's fiftieth anniversary.[4] Table 1.2 presents some historical landmarks of discoveries related to the understanding and management of corrosion.

The science of corrosion has had two periods of rapid advancement. One, in the first half of the nineteenth century, was a result of intense and sustained scientific interest and activity aroused by the invention of the galvanic battery, and the controversy over the nature and source of the galvanic current. The other, in the first half of the twentieth century, was stimulated by the growing realization of the immense economic cost of corrosion in a rapidly developing industrial age. In the latter period, a number of theories and facts established in the earlier one were rediscovered or elaborated, or both. These include the electrochemical theory of corrosion, proposed by Wollaston in 1801, developed by de la Rive in 1830, confirmed by Ericson-Auren and Palmaer in 1901, and rediscovered by Whitney in 1903.[5]

As early as 1819, Hall demonstrated the necessity of dissolved oxygen for appreciable corrosion of iron in water at ordinary temperatures and Sir Humphrey Davy published results of his work on cathodic protection of copper bottoms for British naval vessels in 1824. These early experiments established a practical base for the application of cathodic protection, which led to the development of galvanized iron.

Table 1.2 Historical Landmarks of Discoveries Related to the Understanding and Management of Corrosion

Date	Landmark	Source
1500	Tin plate	
1675	Mechanical origin of corrosiveness and "corrodibility"	Boyle
1763	"Bimetallic corrosion"	HMS Alarm report
1788	Water becomes alkaline during corrosion of iron	Austin
1790	Passivity of iron	Keir
1791	Copper-iron electrolytic galvanic coupling	Galvani
1800	Galvanic battery	Volta
1801	Electrochemical theory of acid corrosion	Wollaston
1819	Insight into electrochemical nature of corrosion	Thenard
1824	Cathodic protection of Cu by Zn or Fe	Sir Humphrey Davy
1825	Differential-temperature cell	Walcker
1826	Metal/metal oxide and stress cells	Davy
1827	Differential solution concentration cell	Beckerel
1830	Microstructural aspect of corrosion (Zn)	de la Rive
1830	Differential-oxygen-concentration cell	Marianini
1834–40	Relations between chemical action and generation of electric currents	Faraday
1836	Cement-lined cast-iron pipe	Herschel
1837	Galvanized iron	Craufurd
1872	Organic inhibitors (essential and fixed oils)	Marangoni, Stefanelli
1873–78	Thermodynamics of corrosion	Gibbs
1899	Measurements of hydrogen overvoltage	Caspari
1903	Theory of corrosion of iron in water	Whitney
1904	Hydrogen overvoltage as a function of current	Tafel
1905	Carbonic and other acids are not essential for the corrosion of iron	Dunstan, Jowett Goulding, Tilden
1906	Monel alloy	Monell
1906	Phosphate coating	Coslett
1907	Chromate inhibitor	Cushman
1907	Oxygen action as cathodic stimulator	Walker, Cederholm
1908	Zinc-chromate pigment	Cushman
1908–10	Compilation of corrosion rates in different media	Heyn, Bauer
1909	Synthetic resin (Bakelite)	Baekeland
1910	Impressed current cathodic protection	Cumberland
1910	Microbiological corrosion	Gaines
1910	Inhibitive paint	Cushman, Gardner
1912	Austenitic chromium-nickel stainless steel	Maurer, Strauss
1913	Study of high-temperature oxidation kinetics of tungsten	Langmuir
1913	High silicon cast iron	Duriron
1916	Differential aeration currents	Aston
1917	Mechanism of corrosion fatigue	Haigh
1916	Chromium stainless steel	Brearley
1919	Mechanism of cavitation	Parsons, Cook
1920–23	"Season-cracking" of brass as intergranular corrosion	Moore, Beckinsale
1923	Anodized aluminum	Bengough, Stuart
1923	High-temperature formation of oxides	Pilling, Bedworth
1924	Galvanic corrosion	Whitman, Russell
1926	Chromium-nickel-molybdenum stainless steel	Strauss

Table 1.2 (Continued)

Date	Landmark	Source
1929	Inconel	Merica
1930–31	Subscaling or "internal corrosion"	Smith
1931–39	Quantitative electrochemical nature of corrosion	Evans
1936	Langelier index	Langelier
1938	Anodic and cathodic inhibitors	Chyzewski, Evans
1938	E-pH thermodynamic diagrams	Pourbaix
1939	Hastelloy C	McCurdy
1950	Autocatalytic nature of pitting	Uhlig
1956	Tafel extrapolation for measurement of kinetic parameters	Stern, Geary
1968	Electrochemical-noise signature of corrosion	Iverson
1970	Study of corrosion processes with electrochemical impedance spectroscopy (EIS)	Epelboin

In 1906, Committee U of the American Society for Testing Materials (ASTM) was formed to promote the development of corrosion tests. Shortly afterward, other organizations began to pay attention to corrosion and its control.

Among pioneers in studying the effects of corrosion was the American Committee on Electrolysis, which noted in 1921 that its preliminary report had been published in

Figure 1.8 An "electrolysis surveyor" using a McCollum Earth Current Meter from the early 1920s (courtesy of East Bay Municipal Utility District).

Figure 1.9 Pits in a 16-inch sheet iron water main in San Francisco's East Bay that failed in one month because of current leakage from electric streetcars (photo dated August 10, 1926, courtesy of East Bay Municipal Utility District).

October, 1916. This committee, composed of representatives of the American Institute of Electrical Engineers, American Electric Railway Association, American Railway Engineering Association, National Bureau of Standards, and others, concerned itself with the then-serious problem of stray current damage to underground metal structures, especially the protection of communication cables from electrified street and interurban railways (Figures 1.8 and 1.9).

In England, the Corrosion Committee of the Iron and Steel Institute issued its first report in 1931 and its sixth in 1959. An American Coordinating Committee on Corrosion was organized with representatives from 17 technical societies in 1938. This group, which aimed to coordinate the activities of societies to prevent duplicated work, was absorbed by the NACE in 1948 and was renamed the Inter-Society Committee on Corrosion Control. It functioned in a semi-autonomous manner until it was finally disbanded approximately 10 years later. This was largely because the growth of abstract publications and numerous other periodicals permitted easy interchange of most information.

Although Germany had a corrosion journal (*Korrosion und Metallschutz*) prior to World War II, which was interrupted during the war and reissued after its end under a new title (*Werkstoffe und Korrosion*), it was not until NACE began publishing the magazine *Corrosion* in 1945 that journals on corrosion control were started in other countries. Since then, one or more magazines about corrosion control have been started in most industrialized countries of the world.

In addition to the present NACE International association, many other scientific engineering, governmental, and trade organizations are active in corrosion-control work. Leaders among the scientific and engineering groups in North America are the American Society for Metals (ASM), ASTM, American Chemical Society (ACS), and the Electrochemical Society (ECS).

There are many other notable groups and associations dealing with some aspects of corrosion prevention and control in the Americas and in the rest of the world. A simple search of the World Wide Web reveals the busiest of these groups and associations. A summary of many of these and a compendium of information sources and databases for corrosion specialists can also be found in the latest revision of *ASM Metals Handbook* Volume 13.[6]

References

1. H.H. Uhlig, "The Cost of Corrosion in the United States," Chemical and Engineering News 27 (1949): p. 2764.
2. G.H. Koch, M.P.H. Brongers, N.G. Thompson, Y.P. Virmani, J.H. Payer, "Corrosion Costs and Preventive Strategies in the United States," National Technical Information Service, FHWA-RD-01-156, Springfield, VA, September 30, 2001.
3. NACE Publication 3C194, "Economics of Corrosion" (Houston, TX: NACE, 1994).
4. R.D. Gundry, Corrosion/93 Plenary & Keynote Lectures (Houston, TX: NACE International, 1993).
5. W. Lynes, "Some Historical Developments Relating to Corrosion," Journal of the Electrochemical Society 98 (1951): pp. 3C–10C.
6. S.D. Cramer, B.S. Covino, G.R. Holcomb, M. Ziomek-Moroz, "Information Sources and Databases for the Corrosionist," in vol. 13A, Corrosion: Fundamentals, Testing, and Protection, eds. D.S. Cramer, B.S. Covino (Metals Park, OH: ASM International, 2003), pp. 999–1001.

2

Chapter

Electrochemistry of Corrosion

2.1 INTRODUCTION

The information in this chapter is fundamental to an understanding of the remainder of the text. Thus, it is suggested that the reader proceed through this chapter carefully, concentrating particularly on learning the component parts of a corrosion cell and their relationships.

2.2 WHY METALS CORRODE

The driving force that causes metals to corrode is a natural consequence of their temporary existence in metallic form. To reach this metallic state from their occurrence in nature in the form of various chemical compounds (ores), it is necessary for them to absorb and store up the energy required to release the metals from their original compounds for later return by corrosion. The amount of energy required and stored varies from metal to metal. It is relatively high for metals such as magnesium, aluminum, and iron, and relatively low for metals such as copper, silver, and gold. Table 2.1 lists a few metals in order of diminishing amounts of energy required to convert them from their oxides to metal.

A typical cycle is illustrated by iron. The most common iron ore, hematite, is an oxide of iron (Fe_2O_3). The most common product of the corrosion of iron, rust, has a similar chemical composition. The energy required to convert iron ore to metallic iron is returned when the iron corrodes to form the original compound. Only the rate of energy change may be different.

The energy difference between metals and their ores can be expressed in electrical terms that are related to formation heats of the compounds. The difficulty of extracting metals from their ores in terms of the energy required, and the consequent tendency to

Table 2.1 Positions of Some Metals in the Order of Energy Required to Convert Their Oxides to Produce 1 kg of Metal

	Metal	Oxide	Energy (MJ kg^{-1})
Highest Energy	Li	Li_2O	40.94
	Al	Al_2O_3	29.44
	Mg	MgO	23.52
	Ti	TiO_2	18.66
	Cr	Cr_2O_3	10.24
	Na	Na_2O	8.32
	Fe	Fe_2O_3	6.71
	Zn	ZnO	4.93
	K	K_2O	4.17
	Ni	NiO	3.65
	Cu	Cu_2O	1.18
	Pb	PbO	0.92
	Pt	PtO_2	0.44
	Ag	Ag_2O	0.06
Lowest Energy	Au	Au_2O_3	−0.18

release this energy by corrosion, is reflected by the relative positions of pure metals in a list, which is discussed later as the electromotive series.

2.3 CORROSION FACTORS

Destruction by corrosion takes many forms, as described in chapter 9, and depends on the complex interaction of a multitude of factors such as the following:

- nature of the metal or alloy;
- presence of inclusions or other foreign matter at the surface;
- homogeneity of the metallic structure;
- nature of the corrosive environment;
- incidental environmental factors such as the presence of oxygen and its uniformity, temperature, and velocity of movement;
- other factors such as stress (residual or applied, steady or cyclic);
- oxide scales (continuous or broken);
- presence of porous or semiporous deposits on surfaces, built-in crevices;
- galvanic effects among dissimilar metals; and
- occasional presence of stray electrical currents from external sources.

Except in rare cases of a grossly improper choice of material for a particular service, or an unanticipated drastic change in the corrosive nature of the environment or complete misunderstanding of its nature, failures of metals by rapid general attack (wasting away) are not often encountered. Corrosion failures are more often subtle and a result of invisible localized effects in the form of pits, intergranular corrosion, or attack within crevices. These and several other forms of corrosion attack are discussed in Chapter 9.

2.4 CHEMISTRY OF CORROSION

2.4.1 The Atom

Matter, itself being made up of atoms, is also composed of those smaller particles that make up the atoms. These numerous particles arrange themselves so that those bearing positive charges or those that are neutral cluster together to form a nucleus around which negatively charged particles or electrons rotate in orbits or "shells." In a normal atom, the negative particles exactly balance the positive charges present in the atomic nucleus. The electrons occupy shells in an orderly fashion to balance the positive charge of the nucleus. The electrons in the outermost shell can be "stripped" from the atom, therefore drastically changing its properties. Thus, the charge of the nucleus is unbalanced and the atom displays a positive charge. This charged atom is called an ion and the process is called ionization.

Chemical shorthand exists to express these atomic states. For example, Fe is the chemical shorthand for a neutral atom of iron, whereas Fe^{2+} denotes an iron atom that has been stripped of two electrons and is called a ferrous ion. Similarly, Fe^{3+} denotes an iron atom stripped of three electrons and is called a ferric ion. The process of stripping electrons from atoms is referred to by electrochemists as oxidation. Note that the term oxidation is not necessarily associated with oxygen.

An opposite process can also occur in which extra electrons are added to the neutral atom to give it a net negative charge. Any increase in negative charge (or decrease in positive charge) of an atom or ion is called reduction.

Many chemical compounds are made up of two or more ions of opposite charge. When these are dissolved in water, they can readily split into two or more separate ions that display equal but opposite charges. This process is also called ionization. It is these particles that are responsible for the conduction of electric currents in aqueous solutions.

2.4.2 Acidity and Alkalinity (pH)

When discussing the ionic content of an aqueous medium, the question often arises as to the degree of acidity (or alkalinity) of the solution. Quite simply, this refers to whether there is an excess of H^+ (hydrogen) or OH^- (hydroxyl) ions present. The H^+ ion is acid whereas the hydroxyl ion is alkaline or basic. The other ionic portion of an acid or alkali added to water can increase its conductivity or change other properties of the liquid but does not increase or decrease its acidity. For instance, whether a given amount of H^+ ion is produced in water by introducing hydrochloric (HCl), sulfuric (H_2SO_4), or any other acid is immaterial. The pH of the solution will be the same for the same number of dissolved hydrogen atoms.

The pH may be measured with a meter or calculated if certain parameters are established. Water itself dissociates to a small extent to produce equal quantities of H^+ and OH^- ions displayed in the following equilibrium:

$$H_2O \rightleftarrows H^+ + OH^- \tag{2.1}$$

Originally defined by Danish biochemist Søren Peter Lauritz Sørensen in 1909, pH is a measure of the concentration of hydrogen ions. The term pH was derived from the manner in which the hydrogen-ion concentration is calculated; it is the negative logarithm of the hydrogen-ion (H^+) concentration:

$$pH = -\log_{10}(a_{H^+}) \tag{2.2}$$

where log is a base-10 logarithm and a_{H+} is the activity (related to concentration) of hydrogen ions. The "p" in Equation (2.2) stands for the German word for "power," potenz, so pH is an abbreviation for "power of hydrogen."

A higher pH means there are fewer free hydrogen ions, and that a change of one pH unit reflects a tenfold change in the concentrations of the hydrogen ion. For example, there are 10 times as many hydrogen ions available at pH 7 than at pH 8. The pH scale of normal environments ranges from 0 to 14 and a pH of 7 is considered to be neutral. Substances with a pH less that 7 are acidic and substances with pH greater than 7 are considered to be basic or alkaline. Thus, a pH of 2 is very acid and a pH of 12 is very alkaline.

Low-pH acid waters accelerate corrosion by supplying hydrogen ions. Although even absolutely pure water contains some free hydrogen ions, free carbon dioxide in the water can multiply the hydrogen-ion concentration many times. When carbon dioxide dissolves in water, it reacts with the water to form carbonic acid, a so-called weak acid but an effective source of acidity. Even more acidity is sometimes encountered in acid mine waters or in those contaminated with industrial wastes.

Many salts added to an aqueous system also have a direct effect on the pH of that mixture through the following process of hydrolysis shown here for the addition of ferric ions to water:

$$Fe^{3+} + 3H_2O \rightleftarrows Fe(OH)_3 + 3H^+ \tag{2.3}$$

In this particular example the equilibrium is established among ferric ions, water, ferric hydroxide or $Fe(OH)_3$, and the acidity of the water. This particular example is quite useful to explain the severity of a situation that can develop in confined areas such as crevices.

2.4.3 Corrosion as a Chemical Reaction

2.4.3.1 Corrosion in Acids

One of the common ways of generating hydrogen in a laboratory is to place zinc into a dilute acid, such as hydrochloric or sulfuric acid. When this is done, there is a rapid reaction in which the zinc is attacked or "dissolved" and hydrogen is evolved as a gas. This is shown in Equations (2.4)–(2.7):

$$Zn + 2HCl \rightarrow ZnCl_2 + H_2^{0}\uparrow \tag{2.4}$$

$$Zn + 2H^+ + 2Cl^- \rightarrow Zn^{2+} + 2Cl^- + H_2^{0}\uparrow \tag{2.5}$$

Equations (2.4) and (2.5) are the chemical shorthand for this statement: One zinc atom + two hydrochloric acid molecules dissociated as ions H^+ and Cl^- becomes one molecule of zinc chloride in Equation (2.4), written as a soluble salt in the form of Zn^{2+} and Cl^- ions in Equation (2.5) + one molecule of hydrogen gas, which is given off as indicated by the vertical arrow. It should be noted that the chloride ions do not participate directly in this reaction, although they could play an important role in real corrosion situations.

Similarly, zinc combines with sulfuric acid to form zinc sulfate (a salt) and hydrogen gas as shown in Equations (2.6) and (2.7):

$$Zn + H_2SO_4 \rightarrow ZnSO_4 + H_2^{0\uparrow} \tag{2.6}$$

$$Zn + 2H^+ + SO_4{}^{2-} \rightarrow Zn^{2+} + SO_4{}^{2-} + H_2^{0\uparrow} \tag{2.7}$$

Note that each atom of a substance that appears on the left-hand side of these equations must also appear on the right-hand side. There are also some rules that denote in what proportion different atoms combine with each other. As in the preceding reaction, the sulfate ions that are an integral part of sulfuric acid do not participate directly to the corrosion attack and therefore one could write both Equations (2.5) and (2.7) in a simpler form:

$$Zn + 2H^+ \rightarrow Zn^{2+} + H_2^{0\uparrow} \tag{2.8}$$

Many other metals are also corroded by acids, often yielding soluble salts and hydrogen gas as shown in Equations (2.9) and (2.10) for iron and aluminum, respectively:

$$Fe + 2H^+ \rightarrow Fe^{2+} + H_2^{0\uparrow} \tag{2.9}$$

$$2Al + 6H^+ \rightarrow 2Al^{3+} + 3H_2^{0\uparrow} \tag{2.10}$$

Note that zinc and iron reacted with two H^+ ions, whereas aluminum reacted with three. This is due to the fact that both zinc and iron, when corroding, each lose two electrons and display two positive charges in their ionic form. They are said to have a valence of 2, whereas aluminum loses three electrons when leaving an anodic surface and hence displays three positive charges and is said to have a valence of 3. Some metals have several common valences, others only one.

2.4.3.2 *Corrosion in Neutral and Alkaline Solutions*

The corrosion of metals can also occur in fresh water, seawater, salt solutions, and alkaline or basic media. In almost all of these systems, corrosion only occurs if dissolved oxygen is also present. Water solutions rapidly dissolve oxygen from the air, and this is the source of the oxygen required in the corrosion process. The most familiar corrosion of this type is the rusting of iron when exposed to a moist atmosphere or water:

$$4Fe + 6H_2O + 3O_2 \rightarrow 4Fe(OH)_{3\downarrow} \tag{2.11}$$

In Equation (2.11), we see that iron will combine with water and oxygen to produce an insoluble, reddish-brown corrosion product that falls out of the solution, as shown by the downward pointing arrow.

During rusting in the atmosphere, there is an opportunity for drying, and this ferric hydroxide dehydrates and forms the familiar red-brown ferric oxide (rust) or Fe_2O_3, as shown as follows:

$$2Fe(OH)_3 \rightarrow Fe_2O_3 + 3H_2O \tag{2.12}$$

Similar reactions occur when zinc is exposed to water or moist air followed by natural drying:

$$2Zn + 2H_2O + O_2 \rightarrow Zn(OH)_{2\downarrow} \tag{2.13}$$

$$Zn(OH)_2 \rightarrow ZnO + H_2O \tag{2.14}$$

The resulting zinc oxide is the whitish deposit seen on galvanized pails, rain gutters, and imperfectly chrome-plated bathroom faucets.

As discussed previously, the iron that took part in the reaction with hydrochloric acid in Equation (2.9) had a valence of 2, whereas the iron that takes part in the reaction shown in Equation (2.11) has a valence of 3. The clue to this lies in the examination of the equation for the corrosion product $Fe(OH)_3$. Note that water ionized into H^+ and OH^-. It is further known that a hydrogen-ion has a valence of 1 (it has only one electron to lose). It would require three hydrogen ions with the corresponding three positive charges to combine with the three OH^- ions held by the iron. It can thus be concluded that the iron ion must have been Fe^{3+} or a ferric ion.

Also note that there is no oxidation or reduction (electron transfer) during either reaction (2.12) or (2.14). In both cases the valences of the elements on the left of each reaction remain what it is on the right. The valences of iron, zinc, hydrogen, and oxygen elements remain unchanged throughout the course of these reactions, and it is consequently not possible to divide these reactions into individual oxidation and reduction reactions.

2.4.3.3 *Corrosion in Other Systems*

Metals can also be corrosively attacked in solutions containing neither oxygen nor acids. Typical environments that can cause this are solutions of oxidizing salts such as ferric and cupric compounds, such as in the following:

$$Zn + 2FeCl_{3(aq)} \rightarrow ZnCl_{2(aq)} + 2FeCl_{2(aq)} \tag{2.15}$$

$$Zn + CuSO_{4(aq)} \rightarrow ZnSO_{4(aq)} + Cu_{\downarrow} \tag{2.16}$$

Note that in one case ferric chloride is changed to ferrous chloride as it corrodes the zinc. The subscript (aq) in Equations (2.15) and (2.16) indicates that these chemical species are soluble and would therefore tend to dissociate in their respective ions. In the other case, zinc reacts with copper sulfate to yield soluble zinc sulfate plus a spongy mass of metallic copper deposited on the surface of the zinc. Equation (2.16) is often called a metal-replacement reaction. Once again, both the chloride and sulfate ions did not participate directly in these reactions, which could technically have been written without indicating their equal presence on either side of the reactions.

2.4.4 Corrosion Products

Corrosion products created during a corrosion reaction can be soluble, such as zinc chloride or zinc sulfate in the examples cited earlier, or insoluble compounds such as iron oxide or hydroxide. The presence of corrosion products is one way corrosion is detected (e.g., rust). However, it should be noted that insoluble corrosion products are not always visible. Upon exposure to air, aluminum forms an almost invisible oxide film that protects it from extensive atmospheric corrosion. It is invisible because it is so thin. This explains the widespread use of aluminum in storm windows, gutters, and automobile trim. The products of the corrosion processes frequently migrate through the environment to enter into further reactions that yield many of common visible corrosion products.

2.5 PRINCIPLES OF ELECTROCHEMISTRY APPLIED TO CORROSION

Although corrosion can take any one of several forms, the mechanism of attack in aqueous environments involves some aspect of electrochemistry. There is a flow of electricity from certain areas of a metal surface to other areas through a solution capable of conducting electricity, such as seawater or natural waters. The term anode is used to describe that portion of the metal surface that is corroded and the term cathode is used to describe the metal surface from which current leaves the solution and returns to the metal.

A solution capable of conducting electricity is called an electrolyte. Its ability to conduct electricity is due to the presence of ions. These are positively or negatively charged atoms or groups of atoms in solution. Pure water contains positively charged hydrogen ions (H^+) and negatively charged hydroxyl ions (OH^-) in equal concentration. The electrolyte forming a corrosive environment may be any solution, rain, or even moisture condensed from the air. It can range from fresh water or salt water to the strongest alkali or acid.

2.5.1 Electrochemical Reactions

An electrochemical reaction is defined as a chemical reaction involving the transfer of electrons. It is also a chemical reaction that involves oxidation and reduction. Because metallic corrosion is almost always an electrochemical process, it is important to understand the basic nature of electrochemical reactions. The discoveries that gradually evolved in modern corrosion science have, in fact, played an important role in the development of a multitude of technologies we are enjoying today.

During the 1780s, biologist Luigi Galvani performed experiments at the University of Bologna involving electric charges that were known in those days to make frog legs

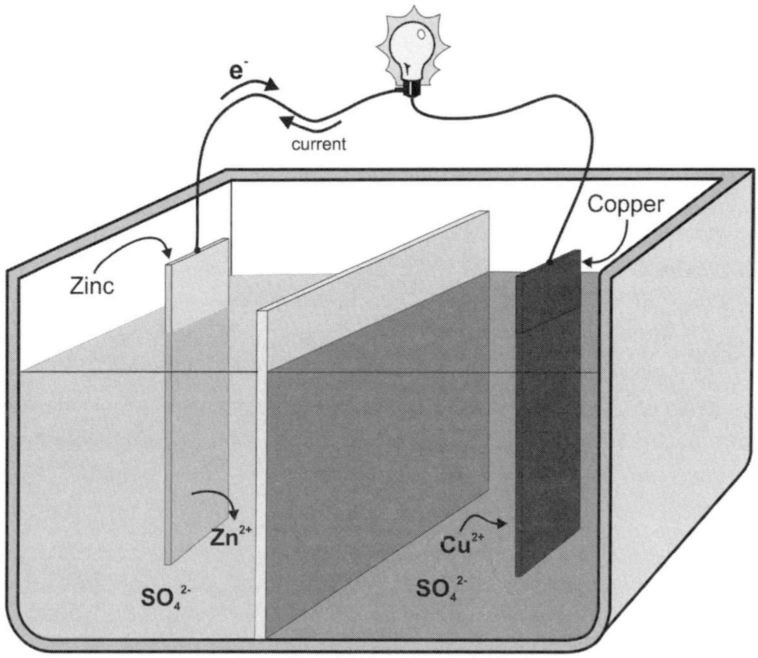

Figure 2.1 Schematic shows a Daniell cell.

jump even if the legs were no longer attached to a frog. While cutting a frog leg, Galvani's steel scalpel touched a brass hook that was holding the leg in place. The leg twitched. Further experiments confirmed this effect, and Galvani was convinced that he was seeing the effects of what he called animal electricity, the life force within the muscles of the frog. Not far from there, at the University of Pavia, Galvani's colleague Alessandro Volta was able to reproduce the results but was skeptical of Galvani's explanation.

In 1800, after extensive experimentation, Volta developed the voltaic pile. The original device consisted of a pile of zinc and silver discs with pieces of cardboard that had been soaked in salt water placed between alternate discs. A wire connecting the bottom zinc disc to the top silver disc could produce repeated sparks. No frogs were injured in the production of a voltaic pile. Unfortunately, these voltaic piles wore out quickly and after a short while, the current generated by the corrosion of the zinc died. Many scientists tried to solve this problem, but to no avail. Finally, in 1836, John Daniell invented a new battery that gave off a steady, lasting current.

This was no minor improvement. The "Daniell cell," as the invention was called, was the first truly practical and reliable electric battery. Without the Daniell cell, the telegraph could never have become a feasible tool for communication, and many other nineteenth-century electrical innovations would have been impossible.

Figure 2.1 illustrates the principle of a Daniell cell in which copper and zinc metals are immersed in solutions of their respective sulfates. In the process of the reaction, electrons can be transferred from the corroding zinc to the copper through an electrically conducting path as a useful electric current. Zinc loses electrons more readily than copper, so placing zinc and copper metal in solutions of their salts can

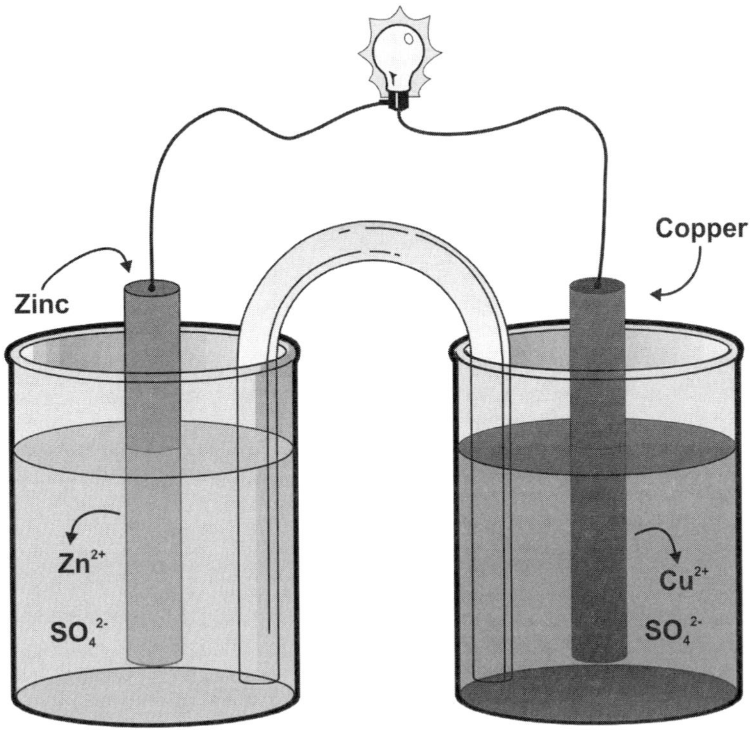

Figure 2.2 Schematic shows a Daniell cell with a salt bridge.

cause electrons to flow through an external wire that leads from the zinc to the copper.

$$\text{Zinc anode: } Zn_{(s)} \rightarrow Zn^{2+} + 2e^- \quad (2.17)$$

$$\text{Copper cathode: } Cu^{2+} + 2e^- \rightarrow Cu_{(s)} \quad (2.18)$$

The difference in the susceptibility of two metals to corrosion can often cause a situation that is called galvanic corrosion, named after the discoverer of the effect. The purpose of the separator shown in Figure 2.1 is to keep each metal in contact with its own soluble sulfates, a technical point that is critical to keep the voltage of a Daniell cell relatively constant. The same goal can be achieved by using a salt bridge between two different beakers as shown in Figure 2.2. The salt bridge, in that case, provides the electrolytic path that is necessary to complete an electrochemical cell circuit. This situation is common in natural corrosion cells where the environment serves as the electrolyte that completes the corrosion cell. The conductivity of an aqueous environment such as soils, concrete, or natural waters has often been related to its corrosivity.

The typical flashlight batteries or dry cells (Figure 2.3) that became so popular in the twentieth century also depend on similar galvanic corrosion to generate electrical power. In this case, the zinc (anode) is electrically connected to a graphite or carbon (cathode) in the presence of a corrosive electrolyte mixed with manganese dioxide. The reduction path of this material, which varies with details of the chemistry, can be summarized

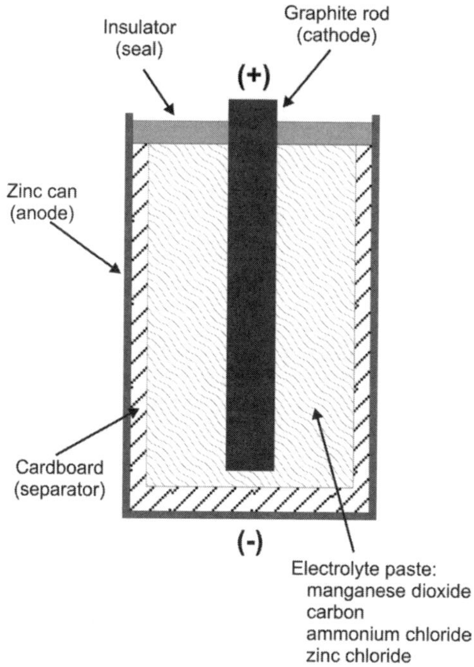

Figure 2.3 Schematic shows a dry Leclanché cell.

in Equation (2.19). When these electrodes are connected through a flashlight bulb or any other small load, electrical current flows between these two electrodes. This causes accelerated corrosion of zinc and produces a cathodic reaction at the graphite electrode. The battery is completely depleted of power when the zinc is completely corroded.

$$\text{Carbon cathode: } MnO_2 + H^+ + e^- \rightarrow MnOOH \tag{2.19}$$

An oxidation reaction, such as Equation (2.17), is indicated by an increase in valence or a production of electrons. In a similar fashion, a reduction reaction is indicated by a decrease in valence or the consumption of electrons, as shown in Equation (2.18). In corrosion terminology, an oxidation reaction is often called an anodic reaction, whereas reduction reactions are usually termed cathodic reactions. These terms are used interchangeably throughout this text.

The fact that corrosion consists of at least one oxidation and one reduction reaction is not always as obvious as it is in batteries. The two reactions are often combined on a single piece of metal as it is illustrated schematically in Figure 2.4. In this figure, a piece of zinc immersed in hydrochloric acid solution is undergoing corrosion. At some point on the surface, zinc is transformed to zinc ions, according to Equation (2.20). This reaction produces electrons and these pass through the solid conducting metal to other sites on the metal surface where hydrogen ions are reduced to hydrogen gas according to Equation (2.21).

$$\text{Anodic reaction: } Zn_{(s)} \rightarrow Zn^{2+} + 2e^- \tag{2.20}$$

$$\text{Cathodic reaction: } 2H^+ + 2e^- \rightarrow H_{2(g)}^{\uparrow} \tag{2.21}$$

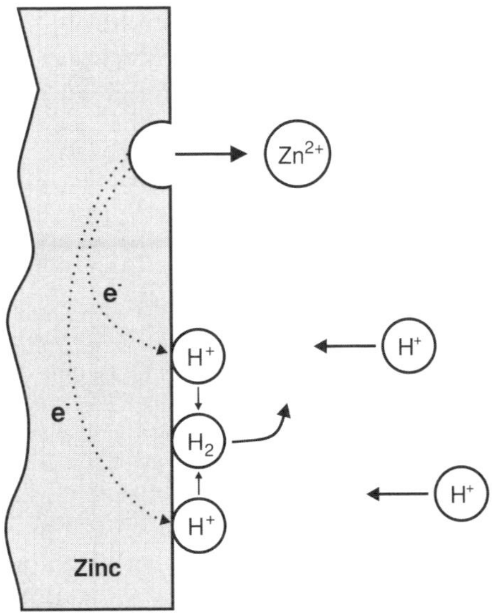

Figure 2.4 Electrochemical reactions occur during the corrosion of zinc in air-free hydrochloric acid.

Equations (2.20) and (2.21) illustrate the nature of an electrochemical reaction typically illustrated for zinc in Figure 2.4. During such a reaction, electrons are transferred, or, viewing it another way, an oxidation process occurs together with a reduction process. The overall corrosion processes are summarized in Equation (2.22):

$$\text{Overall corrosion reaction: } Zn + 2H^{+} \rightarrow Zn^{2+} + H_{2(g)}^{\uparrow} \tag{2.22}$$

Briefly, then, for corrosion to occur there must be a formation of ions and release of electrons at an anodic surface where oxidation or deterioration of the metal occurs. There must be a simultaneous acceptance at the cathodic surface of the electrons generated at the anode. This acceptance of electrons can take the form of neutralization of positive hydrogen ions or the formation of negative ions. The anodic and cathodic reactions must go on at the same time and at equivalent rates. However, corrosion occurs only at the areas that serve as anodes.

2.5.2 Anodic Processes

Let us consider in greater detail what takes place at the anode when corrosion occurs. For instance, reconsider Equations (2.8) or (2.22), (2.9), and (2.10), discussed earlier. All of these reactions involve the reduction of hydrogen ions to hydrogen gas, according to Equation (2.21), and the only difference among them is the nature of their anodic or oxidation processes. Thus, understanding corrosion by acids is greatly simplified, because in every case the cathodic reaction is simply the evolution of hydrogen as gas, as was previously shown in Equation (2.21). This hydrogen-evolution reaction occurs with a wide variety of metals and acids, including hydrochloric, sulfuric, perchloric, hydrofluoric, formic, and other strong acids.

By separating Equations (2.8) through (2.10) into anodic and cathodic reactions, the only difference that is found is in the oxidation reaction. Equation (2.8) involves the oxidation of zinc to its ions, whereas Equations (2.9) and (2.10) involve the oxidation of iron and aluminum to their ions. The individual anodic reactions for iron (Equation 2.23) and aluminum (Equation 2.24) are listed as follows:

$$\text{Iron anodic reaction: } Fe_{(s)} \rightarrow Fe^{2+} + 2e^- \quad (2.23)$$

$$\text{Aluminum anodic reaction: } Al_{(s)} \rightarrow Al^{3+} + 3e^- \quad (2.24)$$

Examining these equations shows that the anodic reaction occurring during corrosion can be written in the general form:

$$\text{General anodic reaction: } M_{(s)} \rightarrow M^{n+} + ne^- \quad (2.25)$$

That is, the corrosion of metal M results in the oxidation of metal M to an ion with a valence charge of $n+$ and the release of n electrons. The value of n, of course, depends primarily on the nature of the metal. Some metals, such as silver, are univalent, while others such as iron, titanium, and uranium are multivalent and possess positive charges as high as 6. Equation (2.25) is general and applies to all corrosion reactions. Just remember, the anode is where the action is.

2.5.3 Faraday's Law

If the current generated by one of the anodic reactions expressed earlier was known, it would be possible to convert this current to an equivalent mass loss or corrosion penetration rate with a very useful relation discovered by Michael Faraday, a nineteenth-century pioneer in electrochemistry. Faraday's empirical laws of electrolysis relate the current of an electrochemical reaction to the number of moles of the element being reacted. Supposing that the charge required for such a reaction was one electron per molecule, as is the case for the plating or the corrosion attack of silver described respectively in Equations (2.26) and (2.27):

$$Ag^+ + e^- \rightarrow Ag_{(s)} \quad (2.26)$$

$$Ag_{(s)} \rightarrow Ag^+ + e^- \quad (2.27)$$

According to Faraday's law, the reaction with 1 mol of silver would require 1 mol of electrons, or one Avogadro's number of electrons (6.022×10^{23}). The charge carried by 1 mol of electrons is known as one Faraday (F). The Faraday is related to other electrical units through the electronic charge, that is, electron charge is 1.6×10^{-19} coul. Multiplying the electronic charge by the Avogadro number means that 1 F equals 96,485 coul/(mol of electrons). Combining Faraday's principles with specific electrochemical reactions of known stoichiometry leads to Equation (2.28):

$$Q = \text{F} \times \Delta N \times n \quad (2.28)$$

Table 2.2 Conversion Among Current, Mass Loss, and Penetration Rates for All Metals[A]

	mA cm^{-2}	mm y^{-1}	mpy	g m^{-2} day^{-1}[B]
mA cm^{-2}	1	3.28 M/nd	129 M/nd	8.95 M/n
mm y^{-1}	0.306 nd/M	1	39.4	2.74 d
mpy	0.00777 nd/M	0.0254	1	0.0694 d
g m^{-2} day^{-1}	0.112 n/M	0.365 /d	14.4 /d	1

[A] Note: The table should be read from left to right, that is, 1 mA cm^{-2} = (3.28 M/nd) mm y^{-1} = (129 M/nd) mpy = (8.95 M/n) g m^{-2} day^{-1}.
[B] mpy = milli inch per year, n = number of electrons freed by the corrosion reaction, M = atomic mass, and d = density.

where

$$Q = \int_0^t I \times dt \tag{2.29}$$

and where N is the number of moles and ΔN the change in that amount, n is the number of electrons per molecule of the species being reacted, I is the total current in amperes (A), and t is the duration of the electrochemical process in seconds (s).

The corrosion current itself can be either estimated by using specialized electrochemical methods described later in the text or by using weight-loss data and a conversion chart (Table 2.2) based on Faraday's principle. Table 2.2 provides the conversion factors between commonly used corrosion rate units for all metals and Table 2.3 describes these conversion factors adapted to iron or steel (Fe) for which $n = 2$, $M = 55.85$ g/mol, and $d = 7.88$ g cm^{-3}.

2.5.4 Cathodic Processes

The electrons generated by corrosion and subsequent formation of metallic ions at the anodic part of a metal pass through the metal to the cathodic areas immersed in the electrolyte. Here, they restore the electrical balance of the system by reacting with the neutralizing positive ions, such as hydrogen ions, in the electrolyte. This particular point is quite important for understanding the corrosion processes and ways to mitigate them. The absolute equality between the anodic and cathodic currents expressed in Equation (2.30) does not mean that the current densities for these currents are equal.

$$I_{anodic} = I_{cathodic} \tag{2.30}$$

Table 2.3 Conversion Among Current, Mass Loss, and Penetration Rates for Steel[A]

	mA cm^{-2}	mm y^{-1}	mpy	$gm^{-2}day^{-1}$
mA cm^{-2}	1	11.6	456	249
mm y^{-1}	0.0863	1	39.4	21.6
mpy	0.00219	0.0254	1	0.547
g $m^{-2}day^{-1}$	0.00401	0.0463	1.83	1

[A] Note: The table should be read from left to right, that is, 1 mA cm^{-2} = 11.6 mm y^{-1} = 456 mpy = 249 g m^{-2} day^{-1}.

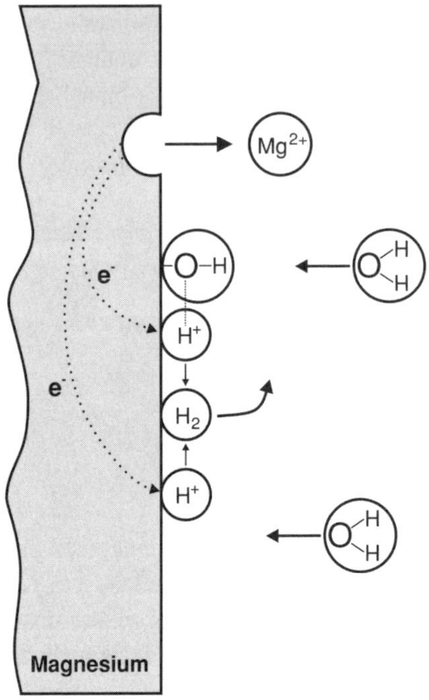

Figure 2.5 Electrochemical reactions occur during the corrosion of magnesium in neutral water.

If indeed Equation (2.30) is now expressed in terms of current densities in Equation (2.31), taking into account the relative anodic (S_a) and cathodic (S_c) surface areas and the associated current densities i_a and i_c, it becomes more clear that an imbalance in the surface area occupied by each reaction will have to be compensated by inequalities in the current densities as expressed in Equation (2.32).

$$I_{\text{anodic}} = i_a \times S_a = i_c \times S_c \tag{2.31}$$

$$i_a = i_c \frac{S_c}{S_a} \tag{2.32}$$

The implications of the surface area ratio S_c/S_a in Equation (2.32) is discussed later in the text, particularly in association with various forms of localized corrosion where a large surface-area ratio is a serious aggravating factor.

When hydrogen ions are reduced to their atomic form they often combine, as shown earlier, to produce hydrogen gas through reaction with electrons at a cathodic surface. This reduction of hydrogen ions at a cathodic surface will disturb the balance between the acidic hydrogen (H^+) ions and the alkaline hydroxyl (OH^-) ions and make the solution less acidic or more alkaline or basic in this region. In neutral waters the anodic corrosion of some metals such as aluminum, zinc, and magnesium develops enough energy to split water directly as illustrated in Figure 2.5 and Equation (2.33):

$$\text{Water-splitting cathodic reaction: } 2H_2O_{(l)} + 2e^- \rightarrow H_2^{\uparrow} + 2OH^- \tag{2.33}$$

The change in the concentration of hydrogen ions or increase in hydroxyl ions can be shown by the use of pH indicators that change color and thus can serve to demonstrate and locate the existence of surfaces on which the cathodic reactions in corrosion are taking place. There are several other cathodic reactions encountered during the corrosion of metals. These are listed as follows:

Oxygen reduction

$$\text{(acid solutions) } O_2 + 4H^+ + 4e^- \rightarrow 2H_2O \tag{2.34}$$

$$\text{(neutral or basic solutions) } O_2 + 2H_2O + 4e^- \rightarrow 4OH^- \tag{2.35}$$

$$\text{Hydrogen evolution: } 2H^+ + 2e^- \rightarrow H^{\uparrow}_{2(g)} \tag{2.21}$$

$$\text{Metal-ion reduction: } Fe^{3+} + e^- \rightarrow Fe^{2+} \tag{2.36}$$

$$\text{Metal deposition: } Cu^{2+} + 2e^- \rightarrow Cu_{(s)} \tag{2.37}$$

Hydrogen-ion reduction, or hydrogen evolution, has already been discussed. This is the cathodic reaction that occurs during corrosion in acids. Oxygen reduction (Equations [2.34] and [2.35]) is a very common cathodic reaction, because oxygen is present in the atmosphere and in solutions exposed to the atmosphere. Metal-ion reduction and metal deposition, although less common, can cause severe corrosion problems.

Note that all of these reactions are similar in one respect; they consume electrons. All corrosion reactions are simply combinations of one or more of these cathodic reactions, together with an anodic reaction similar to Equation (2.25). Thus, almost every case of aqueous corrosion can be reduced to these equations, either singly or in combination.

Consider the corrosion of zinc by water or moist air. By multiplying the zinc-oxidation reaction (2.20) by 2 and summing this with the oxygen-reduction reaction, the overall equation is a simplified form of that shown previously in Equation (2.13):

$$2Zn_{(s)} \rightarrow 2Zn^{2+} + 4e^- \text{ (oxidation)} \tag{2.38}$$

$$+O_2 + 2H_2O + 4e^- \rightarrow 4OH^- \text{ (reduction)} \tag{2.35}$$

$$2Zn + 2H_2O + O_2 \rightarrow 2Zn^{2+} + 4OH^- \rightarrow 2Zn(OH)_{2\downarrow} \tag{2.39}$$

The products of this reaction are Zn^{2+} and OH^-, which immediately react to form insoluble $Zn(OH)_2$. Likewise, the corrosion of zinc by copper sulfate (Equation [2.16]) is merely the summation of the oxidation reaction for zinc and the metal-deposition reaction involving cupric ions [Equation [2.37]).

$$Zn_{(s)} \rightarrow Zn^{2+} + 2e^- \text{ (oxidation)} \tag{2.20}$$

$$+Cu^{2+} + 2e^- \rightarrow Cu_{(s)} \text{ (reduction)} \tag{2.37}$$

$$Zn + Cu^{2+} \rightarrow Zn^{2+} + Cu_{\downarrow} \tag{2.40}$$

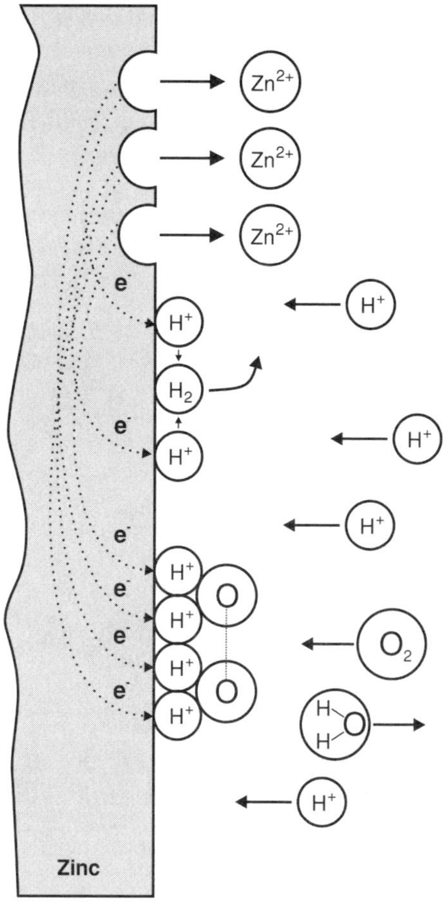

Figure 2.6 Electrochemical reactions occur during the corrosion of zinc in aerated hydrochloric acid.

A comparison of Equations (2.40) and (2.16) shows that they are essentially identical.

During corrosion, more than one oxidation and one reduction reaction may occur. For example, during the corrosion of an alloy, its component metal atoms go into solution as their respective ions. Thus, during the corrosion of a chromium-iron alloy, both chromium and iron are oxidized. Also, more than one cathodic reaction can occur on the surface of a metal.

Consider the corrosion of zinc in a hydrochloric acid solution containing dissolved oxygen. Two cathodic reactions are possible: the evolution of hydrogen and the reduction of oxygen (Figure 2.6). Because there are two cathodic reactions or processes that consume electrons, the overall corrosion rate of zinc is increased. Thus, acid solutions that either contain dissolved oxygen or are exposed to the air are generally more corrosive than air-free acids.

Therefore, removing oxygen from acid solutions will often render them less corrosive. This is a common method for reducing the corrosivity of many environments. Oxygen removal may be accomplished by either chemical or mechanical means.

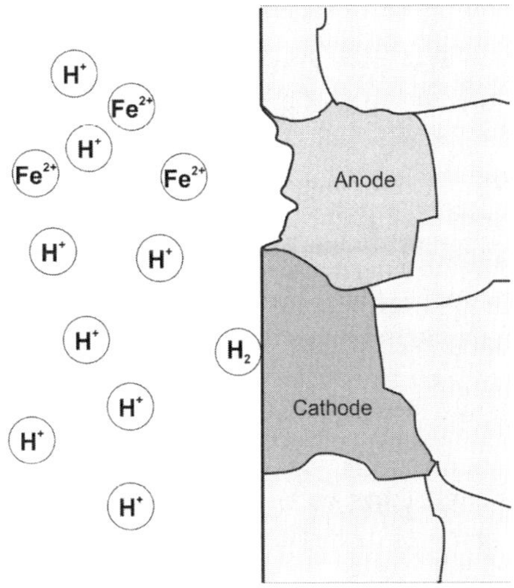

Figure 2.7 Ions from at an anodic area and hydrogen releases at a cathodic area in a local cell on an iron surface.

If a piece of mild steel is placed in a solution of hydrochloric acid, a vigorous formation of hydrogen bubbles is observed. Under such conditions, the metal corrodes very quickly. The dissolution of the metal occurs only at anodic surfaces. The hydrogen bubbles form only at the cathodic surfaces, even though it may appear they come from the entire surface of the metal rather than at well-defined cathodic areas. The anodic and cathodic areas may shift from time to time so as to give the appearance of uniform corrosion. If this action could be seen through a suitable microscope, many tiny anodic and cathodic areas would be observed shifting around on the surface of the metal. These areas, however, are often so small as to be invisible and so numerous as to be almost inseparable.

2.5.5 Combined Anodic and Cathodic Processes

In summary, then, if just one anode and one cathode could be seen in a magnified view of a piece of iron in an acid solution, electrons generated by the formation of ferrous ions would be observed flowing through the metal from an anodic area to a cathodic area (Figure 2.7). At the cathodic surface, the electrons would meet hydrogen ions from the solution. One hydrogen-ion would accept one electron and be converted into a hydrogen atom that could enter the metal and lead to hydrogen embrittlement, or, as in most cases, it could combine with another hydrogen atom and become molecular hydrogen gas, which would either cling to or be released as a bubble from the cathodic surface. As this process continues, oxidation (corrosion) of the iron occurs at the anodic surfaces and reduction of hydrogen ions occurs at the cathodes.

2.5.6 Polarization

As is the case with other chemical reactions, the driving force of a corrosion reaction is related to the difference in energy between an initial equilibrium that is higher in energy than the final equilibrium. As corrosion action proceeds, this difference in energy tends to decrease as a result of the effects of the products of anodic and cathodic reactions in the vicinity of the corrosion sites. The cathodic reaction, and with it the overall corrosion reaction, would slow down if, for example, the hydrogen product of the cathodic reaction were not removed by evolution as gas or some reaction involving oxygen. This slowing-down is said to be the result of cathodic polarization.

It is possible to measure this effect in terms of the potential of the metal on which the reaction is occurring. For example, if the potential of the surface of the more noble metal, the cathode, were to be measured before the flow of any galvanic current and subsequently after current flow had occurred for some time, it would be found that the potential measured would have changed to a value closer to that of the less noble metal in the couple.

Similarly, measurements of the potential of the anodic member of the couple would show a drift in potential closer to that of the cathodic member of the couple. This could be partly because of an increase in the concentration of the ions of the anodic metal in the immediate vicinity of the corroding metal surface.

There are three different types of polarization or ways that electrochemical reactions are retarded. These are activation polarization, concentration polarization, and the ohmic potential that can occur when the anodic and cathodic reactions are physically separated, as was the case with Daniell cells shown previously in Figures 2.1 and 2.2.

The subject of polarization is covered in greater details in a separate section of this chapter titled Corrosion Kinetics.

2.5.7 Area Effects

The intensity of corrosion effects is related not just to the total amount of current flow, but also to the current density or current flow per unit area. It is easy to understand that the effect of a certain amount of current concentrated on a small area of metal surface will be much greater than when the effect of the same amount of current is dissipated over a much larger area. This factor is expressed in Equation (2.32), which states that the ratio of cathodic to anodic surfaces has an amplification factor on the anodic current felt by a smaller anode that feeds in the electrons to a larger cathodic area.

$$i_a = i_c \frac{S_c}{S_a} \quad (2.32)$$

This area effect in terms of current density is illustrated by combinations of steel and copper as either plates or the fasteners used to join them and immersed in a corrosive solution. If steel rivets are used to join copper plates, the current density on the relatively large cathodic copper plates will be low, cathodic polarization of the copper will be slight, and the voltage of the galvanic couple will maintain a value close to the open circuit potential. At the same time, the current density on the small anodic steel rivets will be high and the consequent corrosion quite severe, giving rise to a particularly vicious form of corrosion called galvanic corrosion (Figure 2.8a).

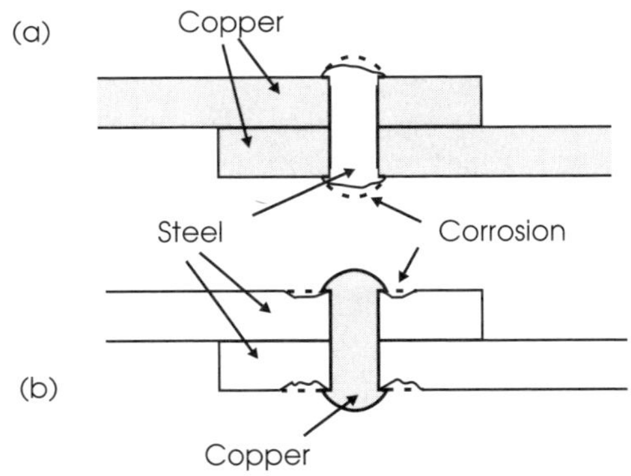

Figure 2.8 Galvanic coupling caused by riveting with dissimilar metals: a) steel rivets on copper plates and b) copper rivets on steel plates.

With the opposite arrangement of copper rivets joining steel plates, the current density on the copper cathodes will be high, with consequently considerable cathodic polarization of the copper reducing the open circuit potential below its initial value. The diminished anodic current will be spread over the relatively large steel plates and the undesirable galvanic effect will hardly be noticeable (Figure 2.8b).

Open-circuit potential measurements are grossly inadequate for predicting the magnitude of galvanic effects because they do not take into account area and polarization effects. They are reliable only for predicting the direction of such effects.

2.5.8 Importance of Oxygen

Oxygen is the most common of the cathodic depolarizers. The action of oxygen in increasing corrosion can be easily demonstrated by placing iron in two flasks filled with water. Oxygen is allowed to bubble through the water in one flask to supply it with oxygen whereas the water in the second flask is saturated with nitrogen to help eliminate dissolved oxygen. After the gases have bubbled for several hours, the iron in the oxygen-free solution remains bright, but the iron in the water saturated with oxygen already begins to rust.

The oxygen content of any solution ranks high on the list of factors influencing the corrosion of iron and numerous other metals. Elimination of oxygen by deaeration is a potent means of preventing corrosion, as in the case of steam boilers that are operated with deaerated feed water.

2.5.9 Differential Aeration Cells

Differential aeration cells can be caused by crevices, lap joints, dirt and debris, moist insulation, and so forth. Under these conditions, the oxygen-starved areas are anodic, whereas the areas with free access to oxygen are cathodic. Common terms for this type of corrosion include crevice corrosion, oxygen screening, and poultice action.

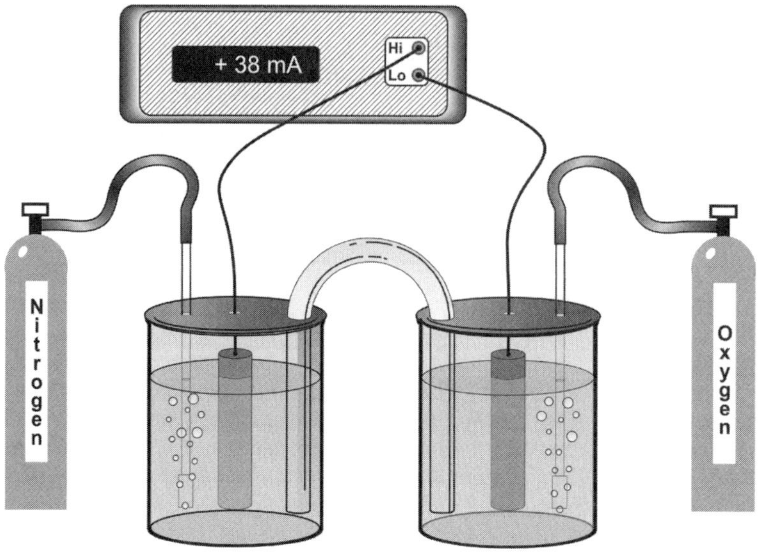

Figure 2.9 Experiment to demonstrate generation of a corrosion current by an oxygen concentration cell.

The role of oxygen in enabling a corrosion reaction to occur forms the basis for the fact that oxygen cannot only maintain a cathodic reaction but can also promote one. This occurs where there is a difference in the concentration of dissolved oxygen at one point on a metal surface as compared with another point. Because the direction of the reaction is toward equilibrium, the only way that equilibrium can be approached by corrosion will be to reduce the concentration of oxygen where it is highest. Such reduction can be done by consuming the oxygen as shown in Equation (2.35).

The result is that where there is a difference in the concentration of dissolved oxygen at two points on a metal surface, the surfaces in contact with the solution containing the higher concentration of dissolved oxygen will become cathodic to the surfaces in contact with the solution containing the lower concentration of dissolved oxygen. These surfaces exposed to the lower oxygen concentration will suffer accelerated corrosion as anodes in an oxygen concentration cell.

It is easy to demonstrate an oxygen concentration cell with an experimental setup using two containers (Figure 2.9). In this experiment, pieces of steel are connected and immersed in a sodium-chloride solution in the two containers. The solution in one container is saturated with oxygen and the solution in the other container is saturated with nitrogen. This establishes a large difference in the concentration of dissolved oxygen in contact with the two pieces of steel. The high concentration of dissolved oxygen in the solution in contact with the steel in one container makes this steel surface strongly cathodic to the steel in the other container. The potential that is measured is determined by the difference in oxygen concentration and the magnitude of the current by the areas of the metal surfaces and the resistance of the circuit.

Dissolved-oxygen-concentration differences can be established by velocity gradients and by crevices, but the location of anodes and cathodes from these sources is just

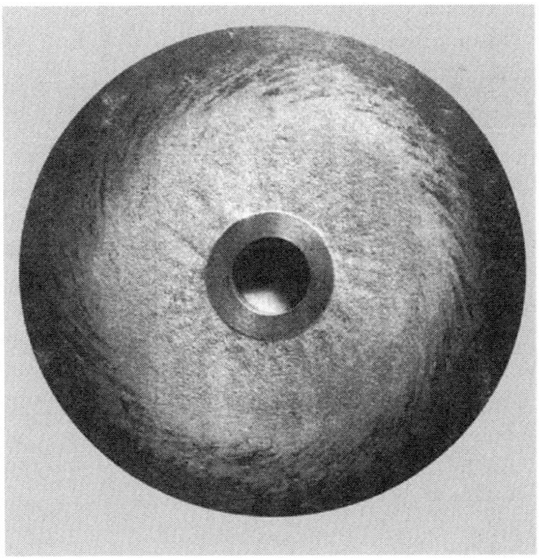

Figure 2.10 Corrosion by an oxygen concentration cell near the center of an iron disc rotating in seawater.

the opposite of that to be described for the metal-ion concentration cells. More oxygen is brought to the surfaces moving at the highest velocity so that these surfaces become cathodic to the anodic surfaces with lower oxygen availability because they are moving at the lower velocity. The surface near the center of a rotating disc, for example, will suffer accelerated corrosion as a result, as illustrated by Figure 2.10.

Similarly, because of the difference in the direction of oxygen concentration cells compared with metal-ion concentration cells, corrosion accelerated by an oxygen concentration cell will occur within a crevice or under a deposit rather than outside of it. Therefore, one type concentration cell may negate the effect of another type. This difference between the two types of concentration cells complicates the prediction of the intensity and location of corrosion resulting from concentration-cell action. It does, however, facilitate explanation after the fact.

As a general rule, those metals toward the top of the electromotive series (cf. galvanic corrosion in chapter 9), for example, iron, are usually more susceptible to acceleration of attack by oxygen concentration cells. Alloys made by combining metals near the top of the electromotive series, for example, iron and chromium (stainless steels), which exhibit a more noble potential than that of their constituents as a result of the passivating effect of a film based on a reaction with oxygen, will be particularly sensitive to oxygen availability and will, therefore, be particularly vulnerable to oxygen concentration cell action.

2.5.10 Metal-Ion-Concentration Cells

Under circumstances where there is a relatively low concentration of metal ions at one point on a metal surface and a higher concentration at another point, a difference in a potential between the two points will be established. The surface in contact with the lower concentration of metal ions will go into solution more easily, have the more

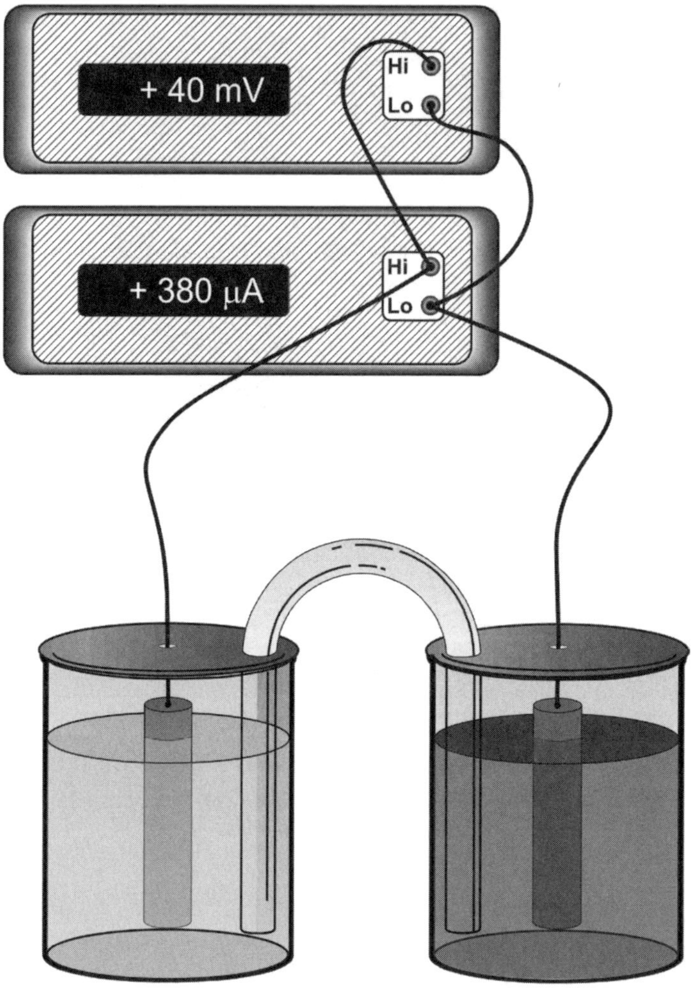

Figure 2.11 Experiment to demonstrate generation of a corrosion current by differences in concentration of a metal ion is shown; a potential of 40 mV and a current of 380 microamperes is obtained.

negative potential, and will thus act as the corroding anode in what is called a metal-ion concentration cell.

Such cells, like other chemical reactions, operate in a direction that will restore equilibrium. Corrosion occurs at the anodic surface where the metal-ion concentration is relatively low so that this concentration will increase. At the same time, metallic ions will plate out on the cathodic surface from the solution containing the higher concentration so as to decrease this concentration toward that of the originally lower concentration around the anode.

The action of a metal-ion concentration cell can be demonstrated by an experiment in which pieces of copper are immersed in two solutions of copper sulfate separated by a porous membrane that prevents the solutions from mixing, but which provides ionic conductance between the two solutions (Figure 2.11). The electrolytic cell established

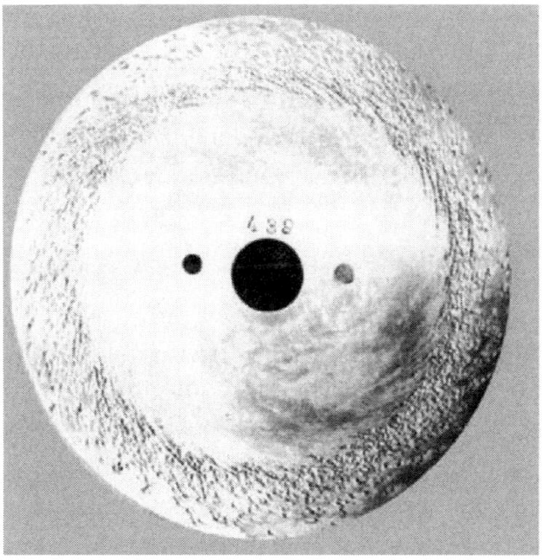

Figure 2.12 Corrosion by a metal-ion concentration cell near the periphery of a brass disc in seawater.

in this way generates a current at a voltage that depends on the difference in the concentrations of the copper ions in the two solutions. The magnitude of the current will be determined by the areas of the metal specimens and the resistance of the circuit.

In practice, differences in metal-ion concentrations that can give rise to corrosion cells of this type can also be aided by velocity gradients over a metal surface. Metal ions in corrosion products will be washed away faster where the velocity and turbulence are greatest and corrosion will be accelerated by the cell established in this way.

A velocity gradient over a metal surface can be established by spinning a disc immersed in a solution. The surfaces toward the periphery will move progressively faster than the surfaces nearer the center of the disc. A metal-ion concentration cell set up in this way on a copper alloy will cause accelerated corrosion of the faster-moving surfaces near the outer edge of the disc, as illustrated by Figure 2.12.

Metal ions can accumulate within crevices or under loosely adherent deposits so that the surfaces within such crevices can become cathodic to surfaces just outside of the crevices that will suffer accelerated attack as a result of the metal-ion concentration cell action.

2.6 CORROSION THERMODYNAMICS

As was mentioned earlier in this chapter, the energy difference between metals and their ores or their corrosion products can be expressed in electrical terms that are related to formation heats of the various compounds. One can use thermodynamic expressions or models, for example, Pourbaix or E-pH diagrams, derived from these formation heats to evaluate the theoretical stability of metals and compounds in given environments. However, for practical situations, it is important to realize that the environment is a variable that can change with time and conditions. It is also important to realize that the environment that actually affects a metal corresponds to the specific environmental conditions this metal really "sees", that is, the local environment at the surface of the metal.

This section introduces the basic thermodynamic information required to describe the stability of metals and provide a basis to understand the nature of corrosion potential, a variable that is frequently measured or monitored in many important applications.

2.6.1 Free Energy and Electrochemical Cells

Electrical work is the product of charge moved (Q) times the cell potential (E) through which it is moved. If the work done is that of an electrochemical cell in which the potential difference is E, and the charge is that of 1 mol of reaction in which n mol of electrons are transferred, then the electrical work ($-w$) done by the cell must be nE. In this relationship, the Faraday constant F is necessary to obtain coulombs from moles of electrons. In an electrochemical cell operating reversibly, no current flows and the energy change occurring in a reaction is expressed in Equation (2.41):

$$w = \Delta G = -n\mathrm{F}E \qquad (2.41)$$

Under standard conditions, the standard free energy of the cell reaction ΔG^0 is directly related to the standard potential difference across the cell, E^0:

$$\Delta G^0 = -n\mathrm{F}E^0 \qquad (2.42)$$

For solid or liquid compounds or elements, standard conditions are the pure compound or element; for gases they are 100 kPa pressure and for solutes they are the ideal 1 molar (mol/L) concentration.

Electrode potentials can be combined algebraically to give cell potential. For a galvanic cell, such as the Daniell cell shown previously, a positive cell voltage will be obtained if the difference is taken in the usual way, as Equation (2.43).

$$E_{\mathrm{cell}} = E_{\mathrm{cathode}} - E_{\mathrm{anode}} \qquad (2.43)$$

The free-energy change in a galvanic cell, or in a spontaneous cell reaction, is negative and the cell voltage positive.

Other thermodynamic quantities can be derived from electrochemical measurements. For example, the entropy change (ΔS) in the cell reaction is given by the temperature dependence of ΔG:

$$\Delta S = -\left(\frac{\partial \Delta G}{\partial T}\right)_{\mathrm{P}} \qquad (2.44)$$

hence

$$\Delta S = n\mathrm{F}\left(\frac{\partial E}{\partial T}\right)_{\mathrm{P}} \qquad (2.45)$$

and

$$\Delta H = \Delta G + T\Delta S = n\mathrm{F}\left[T\left(\frac{\partial E}{\partial T}\right)_{\mathrm{P}} - E\right] \tag{2.46}$$

where ΔH is the enthalpy change and T the absolute temperature (K).

The equilibrium constant ($\mathrm{K_{eq}}$) for the same reaction can be obtained with the following equation:

$$RT \ln \mathrm{K_{eq}} = -\Delta G^0 = n\mathrm{F}E^0 \tag{2.47}$$

2.6.2 Standard Electrode Potentials

There are several potential bench marks in common use, but all of them are related to a basic standard in that one-half of the cell that generates the potential that is measured is represented by a platinized platinum electrode over which hydrogen gas is bubbled while immersed in a solution that has a definite concentration of hydrogen ions (expressed as an activity of 1).

The potential difference across an electrochemical cell is the potential difference measured between two electronic conductors connected to the electrodes. In the external circuit, the electrons will flow from the most negative point to the most positive point. Because the electrode potential can be either positive or negative, the electrons in the external circuit can also be said to flow from the least positive electrode to the most positive electrode. A voltmeter can be used to measure the potential differences across electrochemical cells but cannot measure directly the actual potential of any single electrode. Nevertheless, it is convenient to assign part of the cell potential to one electrode and part to the other.

Tables of standard electrode potentials such as Tables 2.4 and 2.5 can be obtained if any one electrode, operated under standard conditions, is designated as the standard electrode or standard reference electrode with which other electrodes can be compared. An important electrochemical potential reference is called the standard hydrogen electrode (SHE). By definition, the equilibrium potential of this electrode is zero at any temperature. The SHE is also called normal hydrogen electrode (NHE), in reference to a solution containing one equivalent of protons. Strictly speaking, one must use unit activity rather than concentration of hydrogen ions and unit fugacity rather than unit pressure of hydrogen gas.

However, the SHE can be somewhat inconvenient to use because of the need to supply hydrogen gas. Therefore, other reference electrodes are much preferred for practical considerations.

The potential difference across a reversible cell made up of any electrode and a SHE is called the reversible potential of that electrode, E. If this other electrode is also being operated under standard conditions of pressure and concentration, then the reversible potential difference across the cell is the standard electrode potential E^0 of that electrode.

Table 2.4 Standard-State Reduction Half-Cell Potentials in Alphabetical Order

Half-Reaction	E^0_{red}
$Ag^+ + e^- \rightarrow Ag$	0.7996
$Al^{3+} + 3e^- \rightarrow Al$	−1.706
$Au^+ + e^- \rightarrow Au$	1.68
$Cl_2(g) + 2e^- \rightarrow 2Cl^-$	1.3583
$Co^{2+} + 2e^- \rightarrow Co$	−0.28
$Cr^{3+} + 3e^- \rightarrow Cr$	−0.74
$Cu^+ + e^- \rightarrow Cu$	0.522
$Cu^{2+} + 2e^- \rightarrow Cu$	0.3402
$Fe^{2+} + 2e^- \rightarrow Fe$	−0.409
$Fe^{3+} + 3e^- \rightarrow Fe$	−0.036
$2H^+ + 2e^- \rightarrow H_2$	0.0000
$H_2O_2 + 2e^- \rightarrow 2OH^-$	0.88
$H_2O_2 + 2H^+ + 2e^- \rightarrow 2H_2O$	1.776
$Hg^{2+} + 2e^- \rightarrow Hg$	0.851
$Hg_2{}^{2+} + 2e^- \rightarrow Hg$	0.7961
$K^+ + e^- \rightarrow K$	−2.924
$Mg^{2+} + 2e^- \rightarrow Mg$	−2.375
$Mn^{2+} + 2e^- \rightarrow Mn$	−1.04
$Na^+ + e^- \rightarrow Na$	−2.7109
$Ni^{2+} + 2e^- \rightarrow Ni$	−0.23
$O_2 + 2H_2O + 4e^- \rightarrow 4OH^-$	0.401
$O_2 + 4H^+ + 4e^- \rightarrow 2H_2O$	1.229
$O_3(g) + 2H^+ + 2e^- \rightarrow O_2(g) + H_2O$	2.07
$Pb^{2+} + 2e^- \rightarrow Pb$	−0.1263
$PbO_2 + 4H^+ + 2e^- \rightarrow Pb^{2+} + 2H_2O$	1.467
$Pt^{2+} + 2e^- \rightarrow Pt$	1.2
$Sn^{2+} + 2e^- \rightarrow Sn$	−0.1364
$Zn^{2+} + 2e^- \rightarrow Zn$	−0.7628

Because an electrochemical reaction can be written either as an oxidation or a reduction, causing confusion in relation to the sign of the potential of that reaction, a convention was adopted in Stockholm in 1953 to write the standard potential of a reaction in reference to its reduction (E^0_{red}) as shown again in Tables 2.4 and 2.5.

2.6.3 Introduction to Nernst Equation

The Nernst equation, named after the German chemist Walther Nernst, can be derived from the equation linking free-energy changes to the reaction quotient ($Q_{reaction}$):

$$\Delta G = \Delta G^0 + RT \ln Q_{reaction} \tag{2.48}$$

where, for a generalized equation of the form:

$$aA + bB + \cdots \rightarrow mM + nN + \cdots \tag{2.49}$$

Table 2.5 Standard-State Reduction Half-Cell Potentials by Decreasing Order of Potential

Half-Reaction	E^0_{red}
$O_3(g) + 2\,H^+ + 2e^- \rightarrow O_2(g) + H_2O$	2.07
$H_2O_2 + 2H^+ + 2e^- \rightarrow 2H_2O$	1.776
$Au^+ + e^- \rightarrow Au$	1.68
$PbO_2 + 4H^+ + 2e^- \rightarrow Pb^{2+} + 2H_2O$	1.467
$Cl_2(g) + 2e^- \rightarrow 2Cl^-$	1.3583
$O_2 + 4H^+ + 4e^- \rightarrow 2H_2O$	1.229
$Pt^{2+} + 2e^- \rightarrow Pt$	1.2
$H_2O_2 + 2e^- \rightarrow 2OH^-$	0.88
$Hg^{2+} + 2e^- \rightarrow Hg$	0.851
$Ag^+ + e^- \rightarrow Ag$	0.7996
$Hg_2^{2+} + 2e^- \rightarrow Hg$	0.7961
$Cu^+ + e^- \rightarrow Cu$	0.522
$O_2 + 2H_2O + 4e^- \rightarrow 4OH^-$	0.401
$Cu^{2+} + 2e^- \rightarrow Cu$	0.3402
$2H^+ + 2e^- \rightarrow H_2$	0.0000...
$Fe^{3+} + 3e^- \rightarrow Fe$	−0.036
$Pb^{2+} + 2e^- \rightarrow Pb$	−0.1263
$Sn^{2+} + 2e^- \rightarrow Sn$	−0.1364
$Ni^{2+} + 2e^- \rightarrow Ni$	−0.23
$Co^{2+} + 2e^- \rightarrow Co$	−0.28
$Fe^{2+} + 2e^- \rightarrow Fe$	−0.409
$Cr^{3+} + 3e^- \rightarrow Cr$	−0.74
$Zn^{2+} + 2e^- \rightarrow Zn$	−0.7628
$Mn^{2+} + 2e^- \rightarrow Mn$	−1.04
$Al^{3+} + 3e^- \rightarrow Al$	−1.706
$Mg^{2+} + 2e^- \rightarrow Mg$	−2.375
$Na^+ + e^- \rightarrow Na$	−2.7109
$K^+ + e^- \rightarrow K$	−2.924

The capital letters A, B, M and N in Equation (2.49) represent respectively the reactants and products of a given reaction while the small letters represent the coefficients required to balance the reaction.

$$Q_{\text{reaction}} = \frac{a_M^m \times a_N^n \cdots}{a_A^a \times a_B^b \cdots} \tag{2.50}$$

At equilibrium, $\Delta G = 0$ and Q_{reaction} corresponds to the equilibrium constant (K_{eq}) briefly described earlier (2.47).

In the case of an electrochemical reaction, substitution of the relationships $\Delta G = -nFE$ and $\Delta G^0 = -nFE^0$ into the expression of a reaction free energy and division of both sides by $-n$F gives the Nernst equation for an electrode reaction (Equation [2.51]):

$$E = E^0 - \frac{RT}{nF} \ln Q_{\text{reaction}} \tag{2.51}$$

Combining constants at 25°C (298.15 K) gives the simpler form of the Nernst equation for an electrode reaction at this standard temperature:

$$E = E^0 - \frac{0.059}{n} \log_{10} Q_{\text{reaction}} \tag{2.52}$$

In this equation, the electrode potential E is the actual potential difference across a cell in which this electrode and a standard hydrogen electrode are present.

Alternatively, the relationship in Equation (2.43) can be used to combine two Nernst equations corresponding to two half-cell reactions into the Nernst equation for a cell reaction:

$$E_{\text{cell}} = (E^0_{\text{cathode}} - E^0_{\text{anode}}) - \frac{0.059}{n} \log_{10} Q_{\text{reaction}} \tag{2.53}$$

Some of the species that take part in electrode reactions are pure solid compounds and pure liquid compounds. In dilute aqueous solutions, water can be treated as a pure liquid. For pure solid compounds or pure liquid compounds, activities are constant and their values are considered to be unity. The activities of gases are usually taken as their partial pressures and the activities of solutes such as ions are usually taken as their molar concentrations:

$$a_{\text{i}} = \gamma_{\text{i}}[i]_{\text{i}} \approx [i] \tag{2.54}$$

where $[i]$ and γ_i are respectively the molar concentration and the activity coefficient of species i.

2.6.4 Reference Electrodes (Half-Cells)

As mentioned earlier, the standard hydrogen half-cell is rather awkward to use under many circumstances in which potential measurements are to be made. The other half-cells most frequently used in corrosion studies, along with their potentials relative to the standard hydrogen half-cell, are listed in Table 2.6.

To illustrate conversion of values of potential measured with any one of these half-cells to values on the hydrogen-electrode scale, we can take the case of a measurement of the potential of a steel pipe buried in the ground, using a saturated copper/copper sulfate reference electrode (CCSRE). This might show a potential of −0.700 V measured in this way. To convert this potential to a value on the scale in which the hydrogen electrode has a potential of zero, it is necessary to add 0.318 V to the potential that was measured, making it −0.382 V vs. SHE.

The calomel half-cells are, in fact, mercury electrodes in contact with specific concentrations of mercury ions controlled by the concentration of potassium chloride (one-tenth normal to saturated) in a saturated solution of mercurous chloride. Calomel reference electrodes are used most frequently in laboratory experiments.

Most reference electrodes are used with a saturated solution and an excess of the salt crystals. The extra salt dissolves into the electrolyte as the ions diffuse out of the reference-cell body through the liquid junction in normal use. This extra buffer of salt

Table 2.6 Equilibrium Potential of the Main Reference Electrodes Used in Corrosion, at 25°C

Name	Equilibrium Reaction and Nernst Equation	Conditions	Potential (V vs. SHE)	T Coefficient (mV °C^{-1})
Standard hydrogen electrode (SHE)	$2H^+ + 2e^- = H_2$	pH = 0	0.00	
	$E^\circ - 0.059$ pH			
Silver chloride	$AgCl + e^- = Ag + Cl^-$	$a_{Cl^-} = 1$	0.2224	−0.6
	$E^\circ - 0.059 \log_{10} a_{Cl^-}$	0.1 M KCl	0.2881	—
		1.0 M KCl	0.235	—
		Saturated (KCl)	0.199	—
		Seawater	~0.250	—
Calomel	$Hg_2Cl_2 + 2e^- = 2Hg + 2Cl^-$	$a_{Cl^-} = 1$	0.268	
	$E^\circ - 0.059 \log_{10} a_{Cl^-}$	0.1 M KCl	0.3337	−0.06
		1.0 M KCl	0.280	−0.24
		Saturated	0.241	−0.65
Mercurous sulfate	$Hg_2SO_4 + 2e^- = 2Hg + SO_4^{-2}$	$a_{SO_4} = 1$	0.6151	
	$E^\circ - 0.0295 \log_{10} a_{SO_4^{-2}}$			
Mercuric oxide	$HgO + 2e^- + 2H^+ = Hg + H_2O$		0.926	
	$E^\circ - 0.059$ pH			
Copper sulfate	$Cu^{+2} + 2e^- = Cu$ (sulfate solution)	$a_{Cu^{2+}} = 1$	0.340	
	$E^\circ + 0.0295 \log_{10} a_{Cu^{2+}}$	Saturated	0.318	

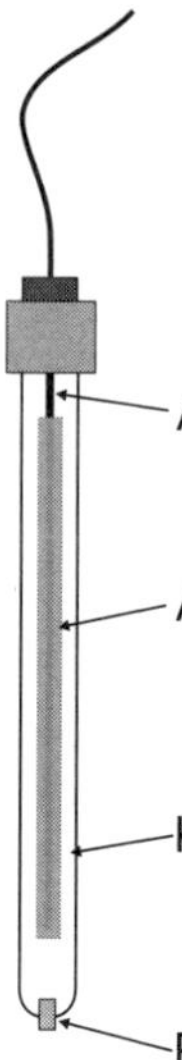

Figure 2.13 Schematic shows a silver/silver chloride reference electrode.

extends the time before the reference cell starts to drift because of the depletion of ions in the electrolyte.

2.6.4.1 Silver/Silver Chloride Reference Electrode

The silver/silver chloride reference electrode is a widely used reference electrode because it is simple, inexpensive, very stable, and nontoxic. It is mainly used with saturated potassium chloride electrolyte, but can be used with lower concentrations such as 1 M potassium chloride and even directly in seawater. Note that changing the electrolyte concentration changes the potential. Silver chloride is slightly soluble in strong potassium chloride solutions, so it is sometimes recommended that the potassium chloride solution be saturated with silver chloride to avoid stripping the silver chloride off the silver wire.

The electrode is a silver wire that is coated with a thin layer of silver chloride either by electroplating or by dipping the wire in molten silver chloride (Figure 2.13). When the electrode is placed in a saturated potassium chloride solution, it develops a potential of 199 mV vs. SHE. The potential of the half-cell reaction shown in Equation (2.55) is determined by the chloride concentration of the solution, as defined by the Nernst Equation (2.56).

$$AgCl_{(s)} + e^- \rightleftarrows Ag + Cl^- \quad E^0_{red} = 0.2224 \text{ vs. SHE} \tag{2.55}$$

$$E_{Ag/AgCl} = E^0_{Ag/AgCl} - 0.059 \log_{10} a_{Cl^-} \tag{2.56}$$

The potential of this electrode remains constant as long as the chloride concentration remains constant. The silver/silver chloride reference electrode develops a potential proportional to the chloride concentration, whether it is sodium chloride, potassium chloride, ammonium chloride, or some other chloride salt.

The silver/silver chloride electrode simplicity of fabrication and fundamental ruggedness makes it a good candidate for many industrial applications where the electrochemical potential has to been measured or controlled.

2.6.4.2 Copper/Copper Sulfate Reference Electrode

Copper/copper sulfate half-cells are favored on a traditional basis for measurements of potentials of systems buried in soils. What is often referred to as a pipe-to-soil potential is actually the potential measured between the pipe and the reference electrode used to make the measurement. The soil itself has no standard value of potential against which the potential of a pipe can be measured independently.

In principle, the half-cell potential of a CCSRE is dependent only upon the electrochemical equilibrium established between Cu and its ions in solution as shown in Equations (2.57) and (2.58):

$$Cu^{2+} + 2e^- \rightleftarrows Cu_{(s)} \quad E^0_{red} = 0.340 \text{ vs. SHE} \tag{2.57}$$

$$E_{Cu/Cu^{2+}} = E^0_{Cu/Cu^{2+}} + 0.059 \log_{10} a_{Cu^{2+}} \tag{2.58}$$

For Cu in a saturated solution of $CuSO_4$, this equilibrium is influenced modestly by temperature and not at all by other factors except for light. Therefore, this reference electrode has a constant half-cell potential, making it reliable for field potential measurements (Figure 2.14).

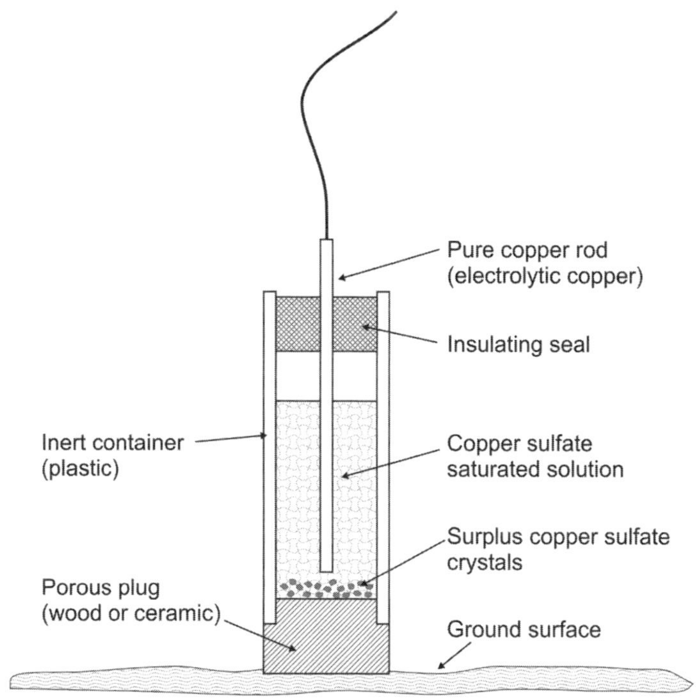

Figure 2.14 Schematic shows a copper/copper sulfate reference electrode.

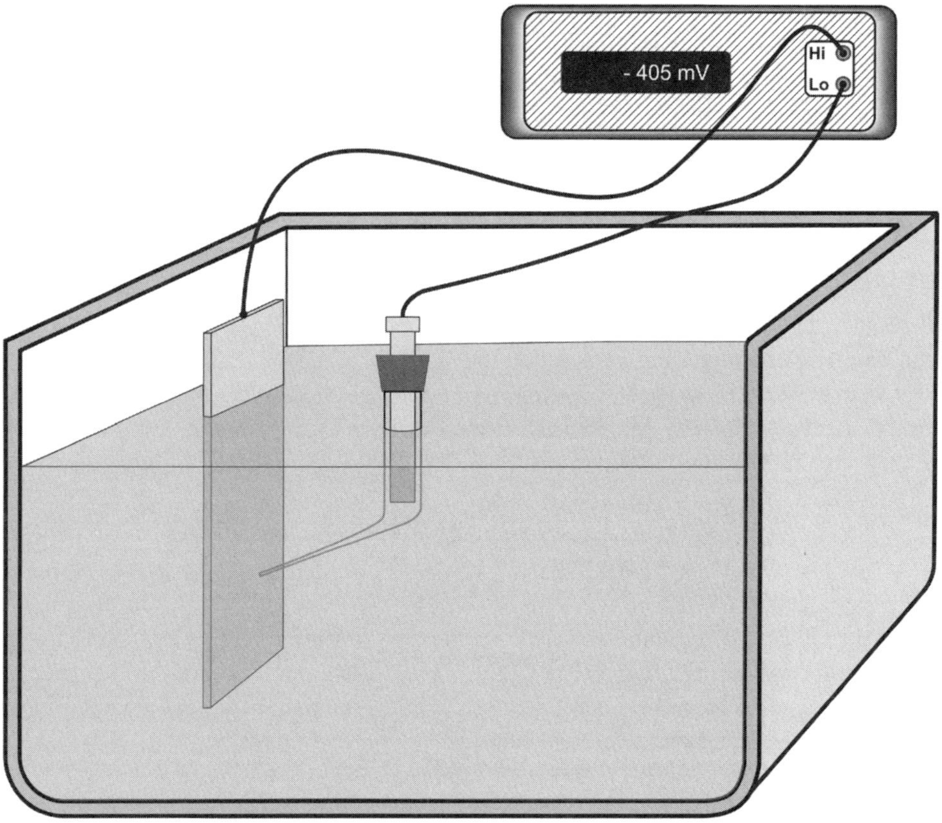

Figure 2.15 Experimental setup measures the corrosion potential of a specimen.

A saturated CCSRE can be fabricated with a solution of copper sulfate made with 40 g of $CuSO_4{\cdot}5H_2O$ in 25 mL of distilled water.[1] The saturated solution should contain approximately 260 g/L of $CuSO_4$ at 22°C.

2.6.5 Corrosion Potential

The potential of a corroding metal, often termed E_{corr}, is probably the single most useful variable measured in corrosion studies as well as during the corrosion monitoring of complex field situations. It is readily measured by determining the voltage difference between a metal immersed in a given environment and an appropriate reference electrode.

Figure 2.15 illustrates an experimental technique for measuring the corrosion potential of a metal M immersed in an electrolyte. This is accomplished by measuring the voltage difference between the reference electrode and the metal using a high-impedance voltmeter capable of accurately measuring small voltages without drawing any appreciable current. Note that in Figure 2.15 a capillary Luggin is used between the reference electrode and the corrosive solution. The capillary serves to prevent contamination of the reference electrode by the environment and to allow making voltage measurements very close to the active metal being monitored.

In measuring and reporting corrosion potentials, it is necessary to indicate the magnitude of the voltage and its sign. In the example shown in Figure 2.15, the corrosion potential of metal M is −0.405 V. The minus sign indicates that the metal is negative with respect to the reference electrode. However, if the metal was connected to the low point (Lo) and the reference electrode to the high point (Hi), the reading would be +0.405 V. It is customary to connect the reference electrode to the low point or the ground to avoid any confusion in reporting. Nonetheless, some manufacturers of electrochemical equipment have done the opposite.

In addition to recording the voltage and the plus or minus sign, it is also necessary to specify the kind of reference electrode used in making the corrosion-potential measurement. For example, if a saturated calomel electrode (SCE) is used, the experiment shown in Figure 2.15 would be reported as "−0.405 V vs. SCE."

The magnitude and sign of the corrosion potential is a complex function of the metal, the composition of the electrolyte, and the temperature and agitation of the electrolyte. However, experienced field personnel can readily recognize departures from normality of a system by monitoring its corrosion potential over time.

2.6.6 Potential-pH Diagram

Rates of general corrosion in aqueous environments depend on a multitude of factors, for example, chemistry of the aqueous solution, concentrations of components, temperature, presence of nonaqueous phases, hydrodynamic conditions, and metallurgical factors. Therefore, it is desirable to rationalize and predict the effects of these factors using mechanistic models. The first comprehensive model that summarized many of the basic variables underlying a corrosion situation became known as potential-pH (E-pH) diagrams, also called predominance or Pourbaix diagrams, which have been adopted universally since their introduction in late 1940s. E-pH diagrams are typically plotted for various equilibria on normal Cartesian coordinates with potential (E) as the ordinate (Y axis) and pH as the abscissa (X axis).[2]

The major uses of such diagrams, which can be constructed for all metals, are (1) predicting whether or not corrosion can occur; (2) estimating the composition of the corrosion products formed; and (3), predicting environmental changes that will prevent or reduce corrosion attack.

2.6.6.1 *E-pH Diagram of Water*

The following example illustrates how the predominance diagram of water can be constructed from its basic thermodynamic information. Equation (2.59) describes the equilibrium between hydrogen ions and hydrogen gas in an aqueous environment:

$$2H^+ + 2e^- \rightleftarrows H_{2(g)}^{\uparrow} \tag{2.59}$$

Adding sufficient OH^- to both sides of reaction (2.59) results in Equation (2.60) in neutral or alkaline solutions:

$$2H_2O + 2e^- \rightleftarrows H_{2(g)}^{\uparrow} + 2OH^- \tag{2.60}$$

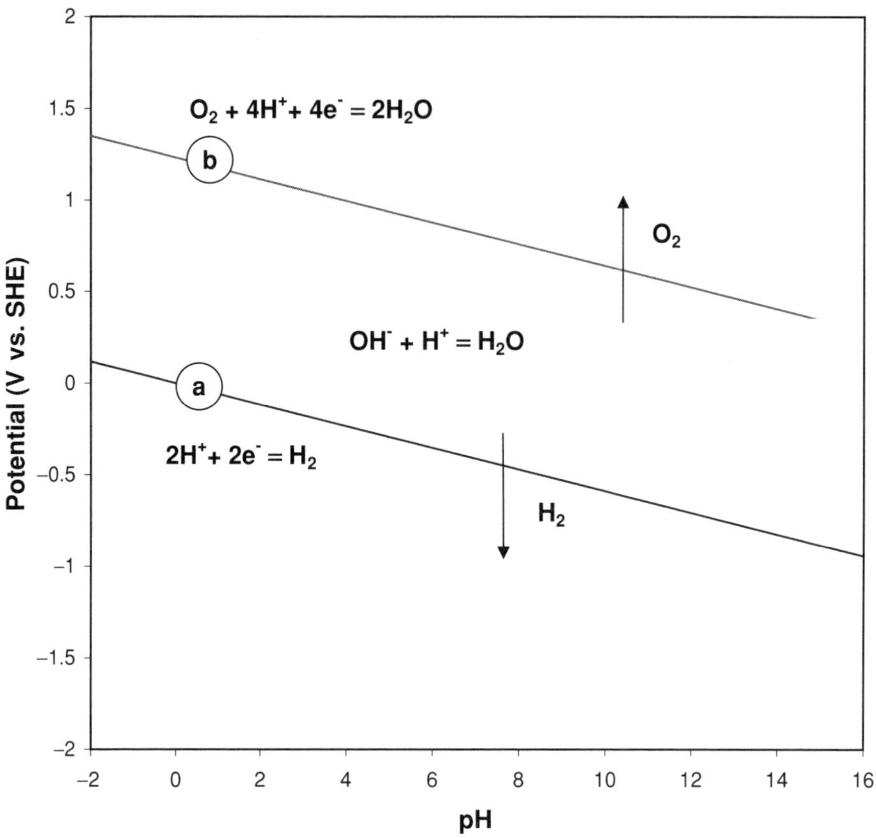

Figure 2.16 *E*-pH stability diagram of water at 25°C.

At higher pH than neutral, (2.60) is a more appropriate representation. However both representations signify the same reaction for which the thermodynamic behavior can be expressed by a Nernst equation (2.61):

$$E_{H^+/H_2} = E^0_{H^+/H_2} + \frac{RT}{n\mathrm{F}}\ln\frac{[H^+]^2}{p_{H_2}} \tag{2.61}$$

that becomes Equation (2.62) at 25°C and the partial pressure of hydrogen (p_{H_2}) of value unity:

$$E_{H^+/H_2} = E^0_{H^+/H_2} - 0.059\,\mathrm{pH} \tag{2.62}$$

Equation (2.59) and its alkaline or basic form, Equation (2.60), delineate the stability of water in a reducing environment and are represented in a graphical form by the sloping line (a) on the Pourbaix diagram in Figure 2.16. Below line (a) in this figure the equilibrium reaction indicates that the decomposition of H_2O into hydrogen is favored, whereas it is thermodynamically stable above that line. As potential becomes more positive or noble, water can be decomposed into its other constituent, oxygen, as illustrated in Equations (2.63) and (2.64) for the acidic form and neutral or basic form of the same process, respectively.

$$O_2 + 4H^+ + 4e^- \rightleftarrows 2H_2O \tag{2.63}$$

$$O_2 + 2H_2O + 4e^- \rightleftarrows 4OH^- \tag{2.64}$$

Again these equivalent equations can be used to develop a Nernst expression of the potential, that is, Equation (2.65) expressed as Equation (2.66) in standard conditions of temperature and oxygen partial pressure (p_{O_2} of value unity).

$$E_{O_2/H_2O} = E^0_{O_2/H_2O} + \frac{RT}{n\mathrm{F}} \ln \left(\mathrm{p}_{O_2} [H^+]^4\right) \tag{2.65}$$

$$E_{O_2/H_2O} = E^0_{O_2/H_2O} - 0.059\,\mathrm{pH} \tag{2.66}$$

The line labeled (b) in Figure 2.16 represents the behavior of E vs. pH for this last equation. Figure 2.16 is divided into three regions. In the upper one, water can be oxidized and form oxygen, whereas in the lower one it can be reduced to form hydrogen gas. In the intermediate region, water is thermodynamically stable. It is common practice to superimpose these two lines (a) and (b) on Pourbaix diagrams to mark the water stability boundaries.

2.6.6.2 E-pH of Diagrams of Metals

Aluminum provides one of the simplest cases for demonstrating the construction of E-pH diagrams. In the following discussion, only four species containing the aluminum element will be considered, that is, two solid species (Al and $Al_2O_3{\cdot}H_2O$) and two ionic species (Al^{3+} and AlO_2^-). The first equilibrium to consider examines the possible presence of either Al^{3+} or AlO_2^- expressed in Equation (2.67).

$$Al^{3+} + 2H_2O \rightleftarrows AlO_2^- + 4H^+ \tag{2.67}$$

Because there is no change in valence of the aluminum present in the two ionic species considered, the associated equilibrium is independent of the potential and the expression of that equilibrium can be derived to give an expression valid in standard conditions (Equation [2.68]):

$$RT \ln K_{eq} = RT \ln Q = -\Delta G^0_{\mathrm{reaction}} \tag{2.68}$$

where Q is expressed in Equation (2.69):

$$Q = \frac{a_{AlO_2^-} \times a^4_{H^+}}{a_{Al^{3+}} \times a^2_{H_2O}} \tag{2.69}$$

Assuming that the activity of H_2O is unity and that the activities of the two ionic species are equal, one can obtain a simpler expression of the equilibrium in Equation (2.68) based purely on the activity of H^+, Equations (2.70) and (2.71):

$$RT \ln [H^+]^4 = -\Delta G^0_{\mathrm{reaction}} \tag{2.70}$$

$$\log_{10}[H^+] = -\mathrm{pH} = \frac{-\Delta G^0_{\mathrm{reaction}}}{4 \times 2.303 \times RT} \tag{2.71}$$

Table 2.7 Possible Reactions in the Al-H_2O System

Equilibria Involving Aluminum Metal
$3\,e^- + Al^{3+} = Al$
$3\,e^- + Al(OH)_3 + 3\,H^+ = Al + 3\,H_2O$
$6e + Al_2O_3 \cdot H_2O + 6\,H^+ = 2\,Al + 4\,H_2O$
$3\,e^- + AlO_2^- + 4\,H^+ = Al + 2\,H_2O$
$3\,e^- + Al(OH)^{2+} + H^+ = Al + H_2O$
$3\,e^- + Al(OH)_2^+ + 2\,H^+ = Al + 2\,H_2O$
Equilibria Involving Solid Forms of Oxidized Aluminum
$Al(OH)_3 + H^+ = Al(OH)_2^+ + H_2O$
$Al_2O_3 \cdot H_2O + 2\,H^+ = 2\,Al(OH)_2^+$
$Al(OH)_3 + 2\,H^+ = Al(OH)^{2+} + 2\,H_2O$
$Al_2O_3 \cdot H_2O + 4\,H^+ = 2\,Al(OH)^{2+} + 2\,H_2O$
$Al_2O_3 \cdot H_2O + 6\,H^+ = 2\,Al^{3+} + 4\,H_2O$
$Al(OH)_3 + 3\,H^+ = Al^{3+} + 3\,H_2O$
$Al(OH)_3 = AlO_2^- + H^+ + H_2O$
$Al_2O_3 \cdot H_2O = 2\,AlO_2^- + 2\,H^+$
Equilibria Involving Only Soluble Forms of Oxidized Aluminum
$AlO_2^- + 4\,H^+ = Al^{3+} + 2\,H_2O$

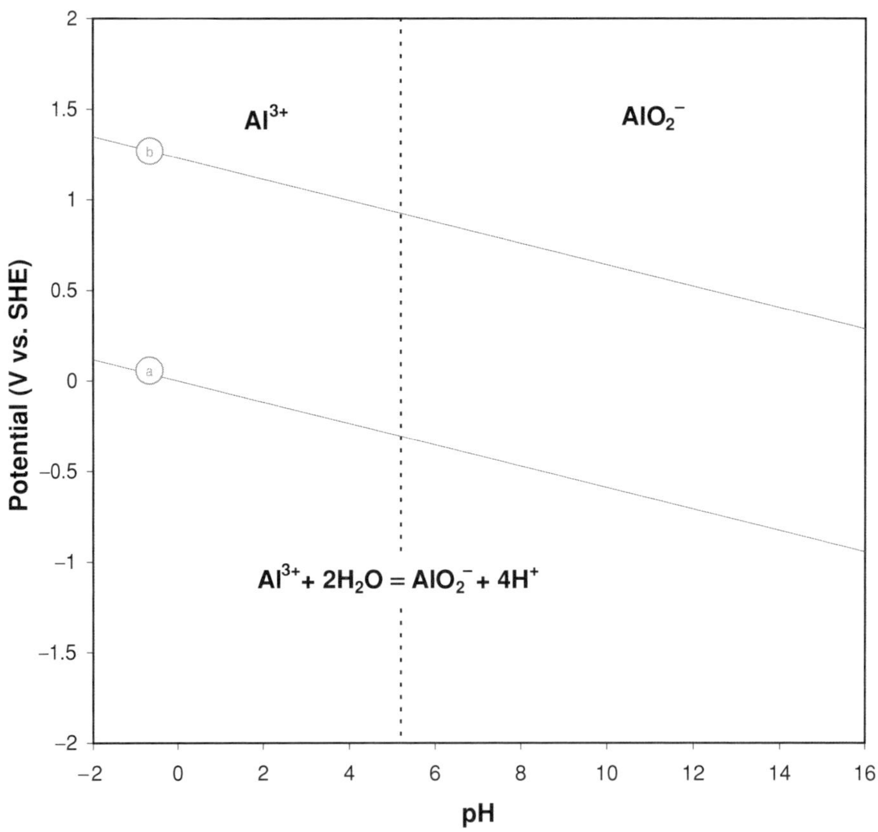

Figure 2.17 *E*-pH diagram shows the soluble species of aluminum in water at 25°C.

and if G^0 is expressed in Joules and the temperature is 25°C or 298 K Equation (2.71) is even further simplified (Eq [2.72]):

$$pH = 4.38 \times 10^{-5} \times \Delta G^0_{\text{reaction}} \tag{2.72}$$

By using the standard thermodynamic data from the literature,[2,3] it is possible to calculate that the free energy of reaction (2.67) is in fact equal to 120.44 kJ mol^{-1} (for either one [Al^{3+}] or [AlO_2^-]). Equation (2.72) then becomes Equation (2.73):

$$pH = 4.38 \times 10^{-5} \times 120{,}440 = 5.27 \tag{2.73}$$

This is represented, in the E-pH diagram shown in Figure 2.17, by a dotted vertical line separating the dominant presence of Al^{3+} at low pH from the dominant presence of AlO_2^- at the higher end of the pH scale.

The next phase for constructing the aluminum E-pH diagram is to consider all possible reactions between the four chemical species containing aluminum retained for this exercise, that is, Al, $Al_2O_3 \cdot H_2O$, Al^{3+}, and AlO_2^-. These reactions are summarized in Table 2.7. A computer program that would compare all possible interactions and rank the chemical species involved in terms of their thermodynamic stability for all conditions of pH and potential would typically carry out this work. Figure 2.18 illustrates

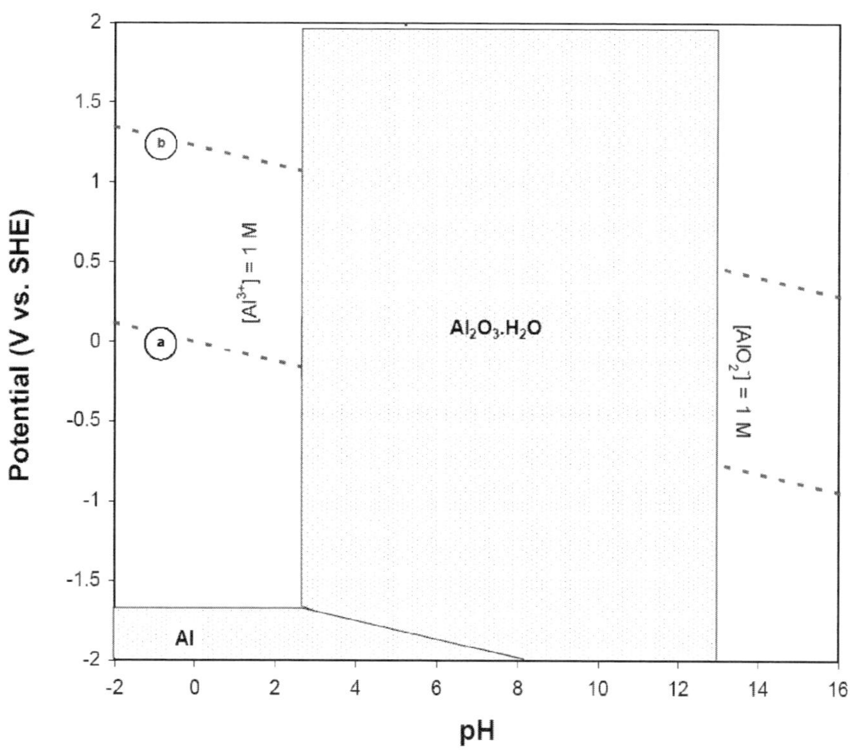

Figure 2.18 E-pH diagram shows solid species of aluminum when the soluble species are at 1 M concentration (25°C).

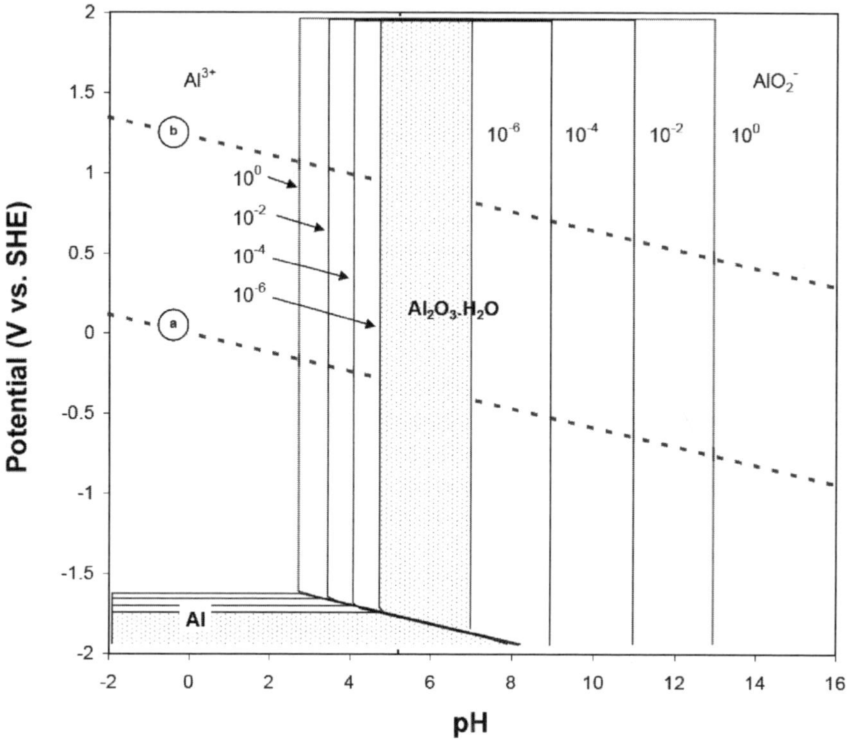

Figure 2.19 E-pH diagram shows aluminum with four concentrations of soluble species (25°C).

the results of such computation for aluminum in the presence of water and oxygen at 25°C when the activities of all species considered were set at value unity.

However, an additional consideration is necessary to make such diagrams useful for realistic conditions for which the presence of soluble species in the environment hardly ever reaches values of the order of one molar. Figure 2.19 illustrates the results that were computed by setting the concentrations of soluble species at decreasing values of one (10^0), one-hundredth (10^{-2}), one in ten thousand (10^{-4}), and one in a million (10^{-6}). The apparent stability of the solid species considered gradually recedes as lower values of soluble species are used in the calculations.

It is customary to use the lowest boundary (10^{-6}) as a practical indication of the corrosion stability of a metal and its solid products (Figure 2.20). The usefulness of this graphical representation of thermodynamic data for corrosion studies was discussed by Pourbaix, who showed three possible states of a metallic material.[2]

- Immunity region: In the conditions of potential and pH of that region, a metal would be totally immune from corrosion attack and safe to use. Cathodic protection is often used to bring the potential of a metal closer to that region by forcing a cathodic shift, as will be explained in much greater details in Chapter 11 of this book,

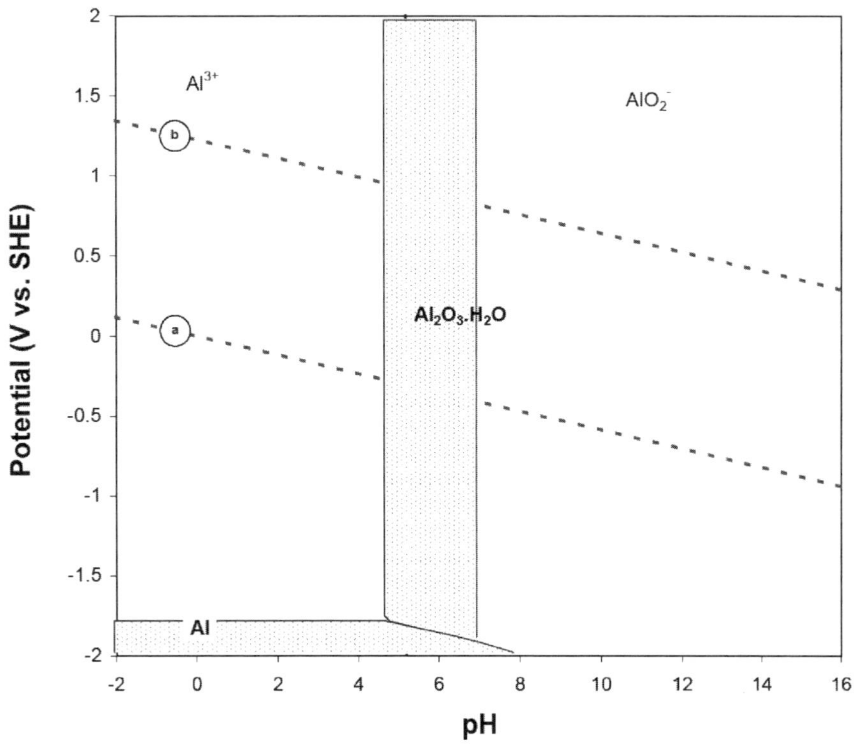

Figure 2.20 E-pH corrosion diagram shows aluminum at 25°C.

- Passivation region: In such region a metal tends to become coated with an oxide or hydroxide that can, according to the circumstances, form on the metal either as a nonporous film practically preventing all direct contact between the metal itself and the environment, or a porous deposit that only partially prevents contact between the metal and the environment,
- Corrosion region: Thermodynamic calculations indicate that, in such region of an E-pH diagram, a metal is stable as an ionic (soluble) product and therefore susceptible to corrosion attack. Experience is required to find out the extent and form of the corrosion attack that could occur in the corrosion region(s) of a Pourbaix diagram.

These three regions are indicated in Figure 2.20 where the chemical reactivity of pure aluminum is noticeable by the small size of immunity region. However, there is fortunately a band of stable oxy-hydroxide product at all potentials above the immunity region at neutral pH values. The presence of a protective oxide layer on aluminum is the main reason why aluminum alloys are so broadly used with success in indoor and outdoor environments provided they fall within the passivation potential/pH boundaries shown in Figure 2.20.

Although this oxide will, over time, form naturally a protective layer on aluminum, it is common practice to produce this oxide in a controlled process called anodization. The quality and properties of the protective oxide can thus be greatly enhanced.

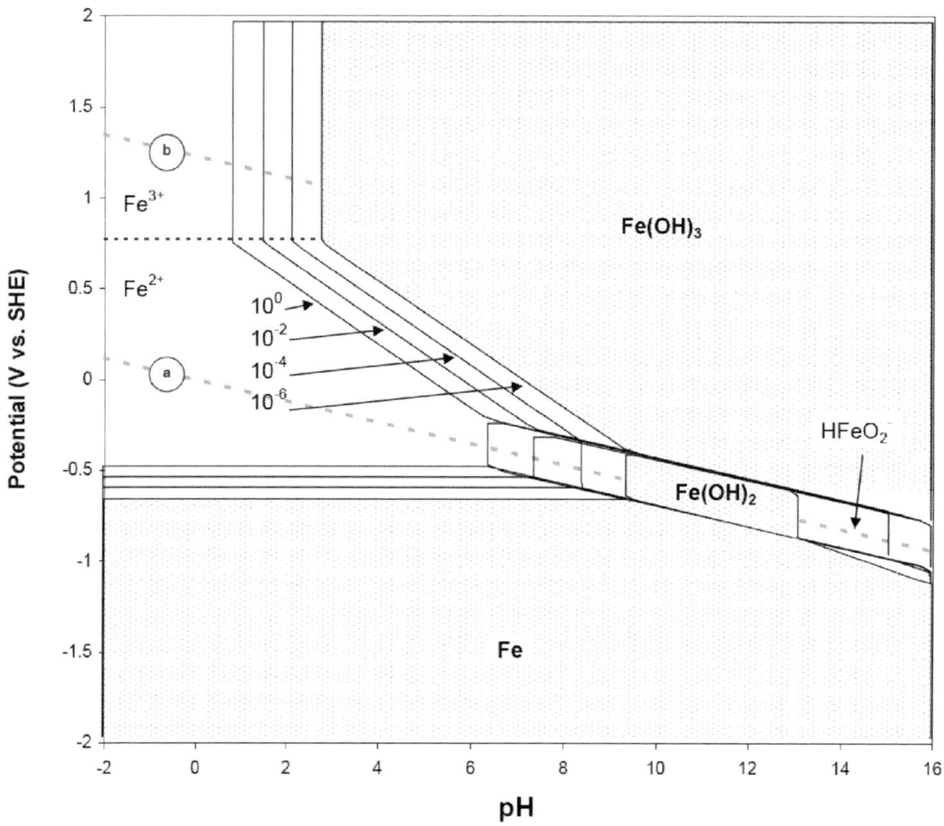

Figure 2.21 *E*-pH diagram shows iron or steel with four concentrations of soluble species (25°C).

Figure 2.21 illustrates the potential-pH diagram for iron in the presence of water or humid environments at 25°C, which was calculated by considering all possible reactions associated with iron in wet or aqueous conditions listed in Table 2.8, excluding therefore drier forms of corrosion products such as magnetite (Fe_3O_4) or iron (ferric) oxide (Fe_2O_3). The various stability regions are shown in Figure 2.22 where the predominant compounds and ions are also indicated. For example, at potentials more positive than -0.6 and at pH values below about 9, ferrous ion is the stable substance. This indicates that iron will be corroded under these conditions, yielding Fe^{2+} (Equation [2.23]). In other regions of this diagram, it can be seen that the corrosion of iron produces ferric ions (Fe^{3+}), ferric hydroxide [$Fe(OH)_3$], ferrous hydroxide [$Fe(OH)_2$], and at very alkaline conditions, complex $HFeO_2^-$ ions.

The relatively large immunity region in Figure 2.22 indicates that iron will not corrode under these potential and pH conditions. Thus, if the corrosion potential of iron is made sufficiently negative or shifted cathodically below approximately -1.2 V vs. SHE, iron will not corrode, ranging from very acidic to very basic conditions. In practice this is difficult to achieve in aqueous environments because such potential is cathodic to line (a) in Figure 2.22, indicating that such potential will tend to decompose water into hydrogen as indicated in Equations (2.59) and (2.60).

Table 2.8 Possible Reactions in the Fe-H_2O System Between the Species Most Stable in Wet Conditions

	Equilibria
1.	$2\,e^- + 2H^+ = 1H_2$
2.	$4\,e^- + 1O_2 + 4H^+ = 2H_2O$
3.	$2\,e^- + 1Fe(OH)_2 + 2H^+ = 1Fe + 2H_2O$
4.	$2\,e^- + 1Fe^{2+} = 1Fe$
5.	$2\,e^- + 1Fe(OH)_3^- + 3H^+ = 1Fe + 3H_2O$
6.	$1\,e^- + 1Fe(OH)_3 + 1H^+ = 1Fe(OH)_2 + 1H_2O$
7.	$1\,e^- + 1Fe(OH)_3 + 3H^+ = 1Fe^{2+} + 3H_2O$
8.	$1\,Fe(OH)_3^- + 1H^+ = 1Fe(OH)_2 + 1H_2O$
9.	$1\,e^- + 1Fe(OH)_3 = 1Fe(OH)_3^-$
10.	$1\,Fe^{3+} + 3H_2O = 1Fe(OH)_3 + 3H^+$
11.	$1\,Fe^{2+} + 2H_2O = 1Fe(OH)_2 + 2H^+$
12.	$1\,e^- + 1Fe^{3+} = 1Fe^{2+}$
13.	$1\,Fe^{2+} + 1H_2O = 1FeOH^+ + 1H^+$
14.	$1\,FeOH^+ + 1H_2O = 1Fe(OH)_{2(sln)} + 1H^+$
15.	$1\,Fe(OH)_{2(sln)} + 1H_2O = 1Fe(OH)_3^- + 1H^+$
16.	$1\,Fe^{3+} + 1H_2O = 1FeOH^{2+} + 1H^+$
17.	$1\,FeOH^{2+} + 1H_2O = 1Fe(OH)_2^+ + 1H^+$
18.	$1\,Fe(OH)_2^+ + 1H_2O = 1Fe(OH)_{3(sln)} + 1H^+$
19.	$1\,FeOH^{2+} + 1H^+ = 1Fe^{2+} + 1H_2O$
20.	$1\,e^- + 1Fe(OH)_2^+ + 2H^+ = 1Fe^{2+} + 2H_2O$
22.	$1\,e^- + 1Fe(OH)_{3(sln)} + 1H^+ = 1Fe(OH)_{2(sln)} + 1H_2O$
22.	$1\,e^- + 1Fe(OH)_{3(sln)} + 2H^+ = 1FeOH^+ + 2H_2O$
23.	$1\,e^- + 1Fe(OH)_{3(sln)} + 3H^+ = 1Fe^{2+} + 3H_2O$

2.7 CORROSION KINETICS

Thermodynamic principles can help explain a corrosion situation in terms of the stability of chemical species and reactions associated with corrosion processes. However, thermodynamic calculations cannot be used to predict corrosion currents or corrosion rates. The polarization effects mentioned earlier in the chapter control the balancing cathodic and anodic currents that are integrated components of corrosion processes.

When two or more complementary processes such as those illustrated in Figures 2.4, 2.5, and 2.6 occur over a single metallic surface, the corrosion potential described earlier that results from this situation is a compromise between the various equilibrium potentials of all the anodic and cathodic reactions involved. The difference between the resultant potential (E) and each individual reaction equilibrium potential (E_{eq}) is called polarization and quantified in terms of an overvoltage (η) as in Equation (2.74):

$$\eta = E - E_{eq} \tag{2.74}$$

The polarization is said to be either anodic, when the anodic processes on the electrode are accelerated by changing the specimen potential in the positive (noble) direction or cathodic, when the cathodic processes are accelerated by moving the potential in the

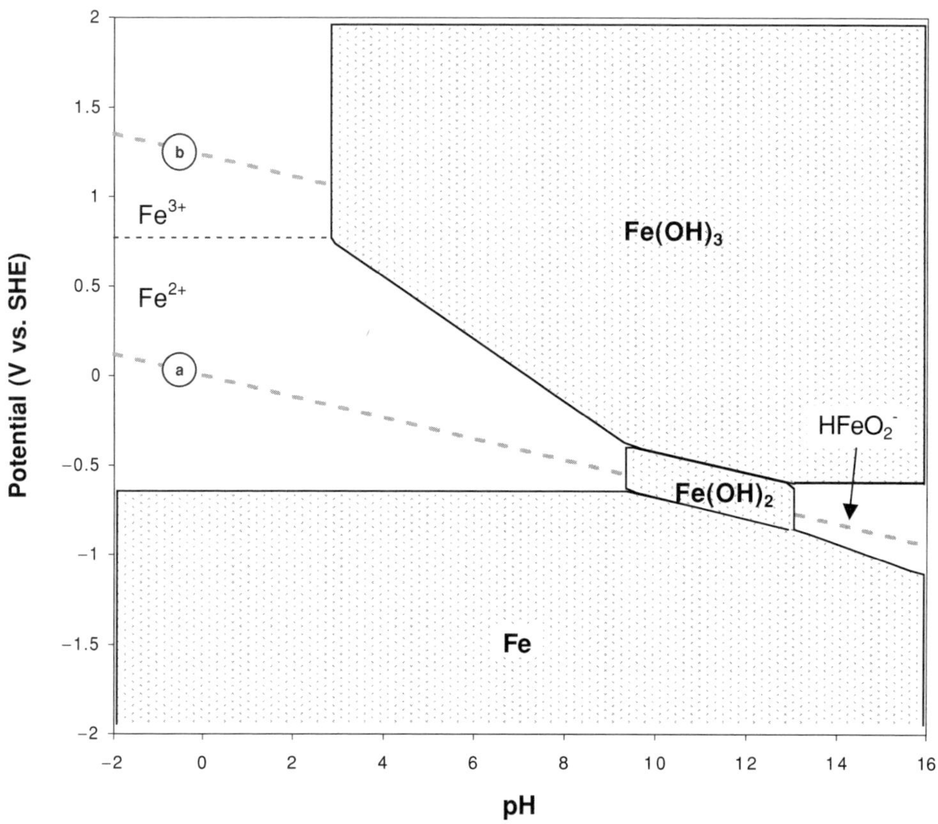

Figure 2.22 E-pH corrosion diagram shows iron or steel at 25°C.

negative (active) direction. There are three distinct types of polarization and these are additive, as expressed in Equation (2.75):

$$\eta_{total} = \eta_{act} + \eta_{conc} + iR \tag{2.75}$$

where:

- η_{act} is the activation overpotential, a complex function describing the charge transfer kinetics of an electrochemical reaction. η_{act} is predominant at small polarization currents or voltages;
- η_{conc} is the concentration overpotential, a function describing the mass-transport limitations associated with electrochemical processes. η_{conc} is predominant at larger polarization currents or voltages;
- iR is the ohmic drop. This is a simple function that takes into account the electrolytic resistivity of an environment when the anodic and cathodic elements of a corrosion reaction are separated by this environment while still electrically coupled.

Activation polarization is usually the controlling factor during corrosion in strong acids. Concentration polarization usually predominates when the concentration of the active species is low; for example, in dilute acids or in aerated waters.

Table 2.9 Approximate Exchange Current Density (i_0) for the Hydrogen Oxidation Reaction on Different Metals at 25°C

Metal	$\log_{10} i_0$ (A/cm^2)
Pb, Hg	−13
Zn	−11
Sn, Al, Be	−10
Ni, Ag, Cu, Cd	−7
Fe, Au, Mo	−6
W, Co, Ta	−5
Pd, Rh	−4
Pt	−2

Knowing the kind of polarization that is occurring can be very helpful, because it allows an assessment of the determining characteristics of a corroding system. For example, if corrosion is controlled by concentration polarization, then any change that increases the diffusion rate of the active species (e.g., oxygen) will increase the corrosion rate. In such a system, it would therefore be expected that agitating the liquid or stirring it would tend to increase the corrosion rate of the metal. However, if a corrosion reaction is activation controlled then stirring or agitation will have no effect on the corrosion rate.

2.7.1.1 Activation Polarization

The term activation polarization makes reference to retarding factors that are an inherent part of the kinetics of all electrochemical reactions. For example, consider the evolution of hydrogen gas illustrated previously in Equation (2.21) and Figure 2.4.

Although this reaction seems to be relatively simple, the rate at which hydrogen ions are transformed into hydrogen gas is in reality a function of several factors, including the rate of electron transfer from a metal to hydrogen ions. In fact, there is a wide variability in this transfer rate of electrons by various metals and, as a result, the rate of hydrogen evolution from different metal surfaces can vary greatly.

The exchange-current density (i_0) is surely the single most important variable that explains the large differences in the rate of hydrogen on metallic surfaces. Table 2.9 contains the approximate exchange-current density for the reduction of hydrogen ions on a range of materials. Note that the value for the exchange-current density of hydrogen evolution on platinum is approximately 10^{-2} A/cm^2 whereas on mercury it is 10^{-13} A/cm^2, 11 orders of magnitude for the rate of this particular reaction!

The stifling power of mercury on the production of hydrogen gas has been used for over more than a century in the production of battery material to increase the efficiency of active material transformation into power and prevent the hazardous accumulation of hydrogen in closed containers.

However, there is no simple means to estimate the exchange-current density for a specific system and it must be determined experimentally by scanning the potential with a laboratory setup such as shown in Figure 2.23. In this experimental arrangement

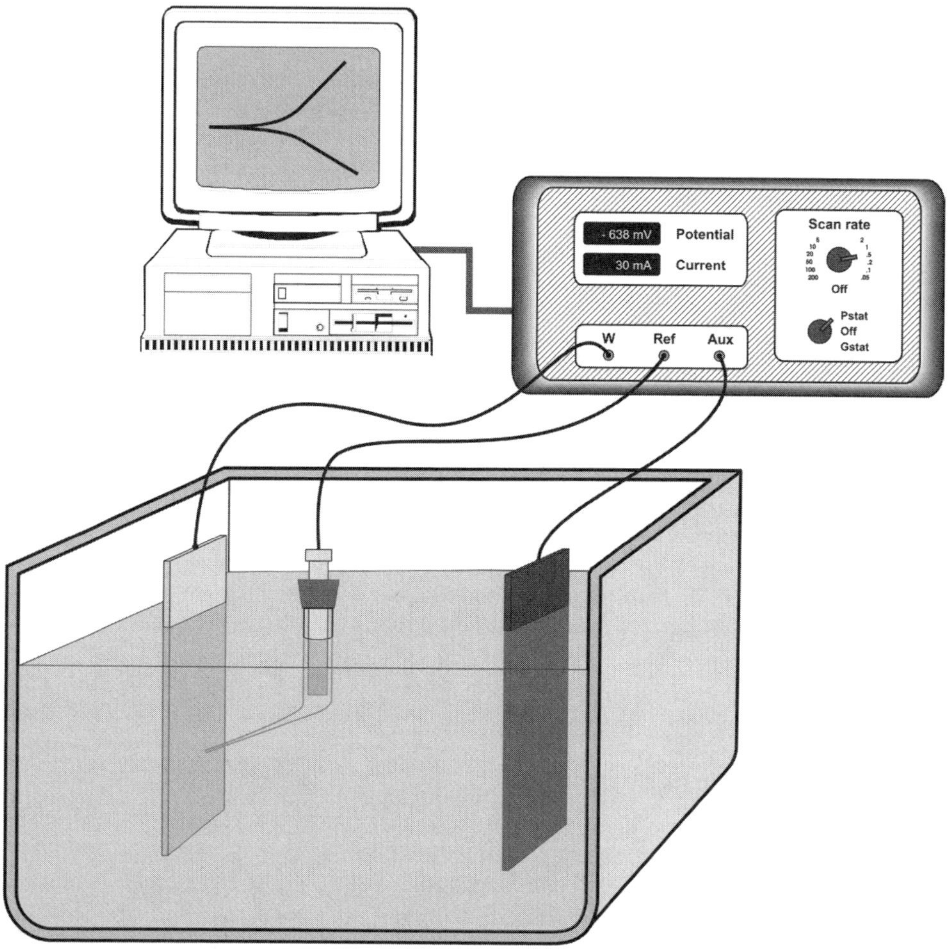

Figure 2.23 Electrochemical instrumentation to carry out potentiodynamic measurements in which a potentiostat/galvanostat power controller is used to pass current through the sample or working electrode (W) and an auxiliary electrode (AUX), while monitoring the potential of the working electrode with a reference electrode.

a potentiostat/galvanostat power controller is used to pass current through the sample, or working electrode (W), and an auxiliary electrode (AUX) immersed in solution while monitoring the potential of the working electrode with a reference electrode and a Luggin capillary.

The following theory explains the basic mathematics that can then be used to extract exchange-current density from the results obtained. Both the anodic and cathodic components of a reaction can be studied individually by using some well-established electrochemical methods where the response of a system to an applied polarization, current or voltage, is studied. A general representation of the polarization of an electrode supporting one specific reaction is given in the Butler–Volmer equation (2.76):

$$i_{\text{reaction}} = i_0 \left\{ \exp\left(-\beta \frac{n\text{F}}{RT} \eta_{\text{reaction}}\right) - \exp\left((1-\beta)\frac{n\text{F}}{RT}\eta_{\text{reaction}}\right) \right\} \tag{2.76}$$

where

- $i_{reaction}$ is the anodic or cathodic current;
- β = charge-transfer barrier or symmetry coefficient for the anodic or cathodic reaction, close to 0.5;
- n = number of participating electrons;
- R = gas constant, that is, 8.314 J mol^{-1} K^{-1};
- T = absolute temperature (K);
- F = 96,485 C/(mol of electrons).

When $\eta_{reaction}$ is cathodic, that is, negative, the second term in the Butler–Volmer equation becomes negligible and the cathodic current density (i_c) can be expressed by a simpler Equation [2.77]) and its logarithm (Equation [2.78]):

$$i_{reaction} = i_c = i_0 \exp\left(-\beta \frac{n\mathrm{F}}{RT} \eta_{reaction}\right) \tag{2.77}$$

$$\eta_{reaction} = \eta_c = b_c \times \log_{10}\left(\frac{i_c}{i_0}\right) \tag{2.78}$$

where b_c is the cathodic Tafel coefficient described in Equation (2.79) that can be obtained from the slope of a plot of η against log |i|, with the intercept yielding a value for i_0 as shown in Figure 2.24.

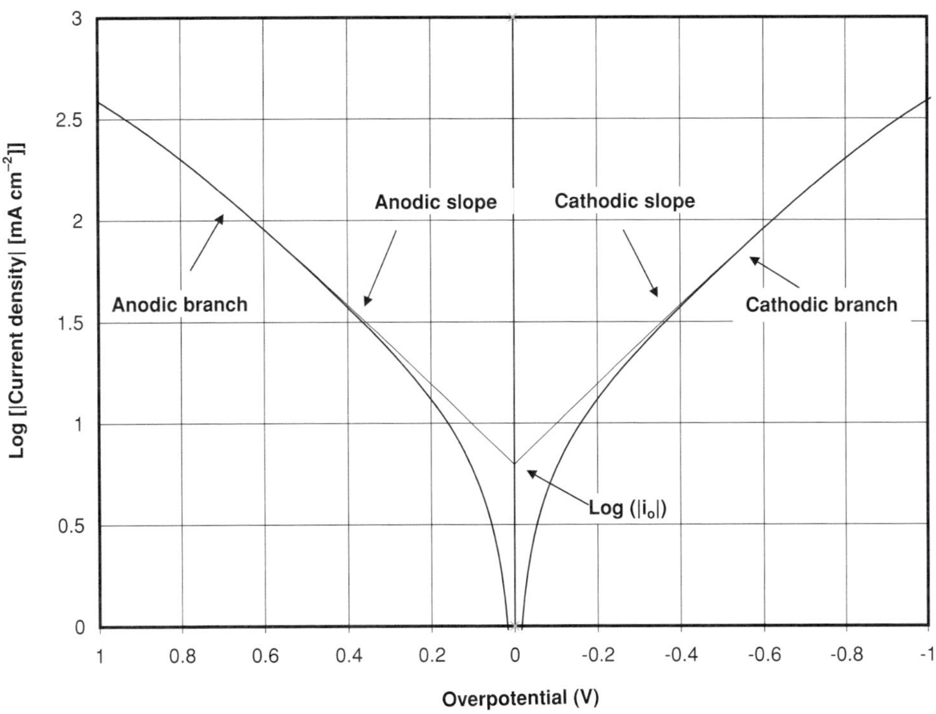

Figure 2.24 E of a plot of η against log $|i|$ or Tafel plot shows the exchange current density that can be obtained with the intercept.

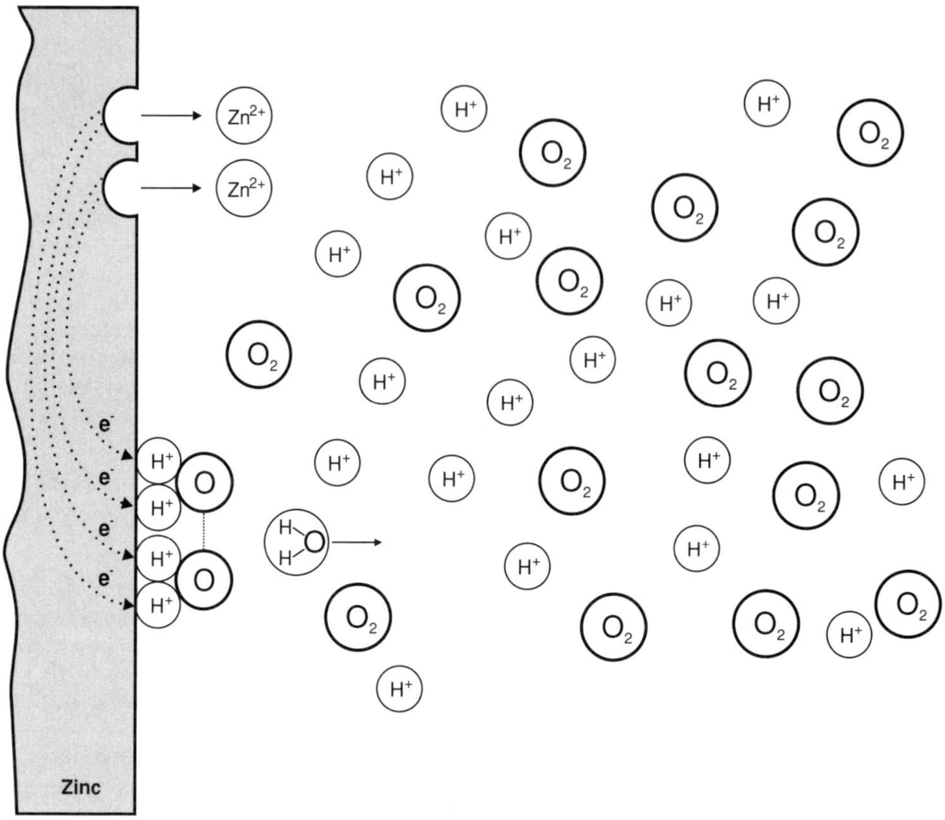

Figure 2.25 Concentration changes in the vicinity of an electrode cause a concentration polarization effect.

$$b_c = -2.303 \times \frac{RT}{\beta n \mathrm{F}} \tag{2.79}$$

Similarly, when η_{reaction} is anodic, that is, positive, the first term in the Butler–Volmer equation becomes negligible and the anodic current density (i_a) can be expressed by Equation (2.80) and its logarithm (Equation [2.81]), with b_a obtained by plotting η_{reaction} vs. log $|i|$ (Equation [2.82]):

$$i_{\mathrm{reaction}} = i_a = -i_0 \exp\left((1-\beta)\frac{n\mathrm{F}}{RT}\eta_{\mathrm{reaction}}\right) \tag{2.80}$$

$$\eta_a = b_a \times \log_{10}\left(\frac{|i_a|}{i_0}\right) \tag{2.81}$$

$$b_a = 2.303 \times \frac{RT}{\beta n \mathrm{F}} \tag{2.82}$$

2.7.1.2 Concentration Polarization

Concentration polarization refers to the polarization component of a metallic surface specific to concentration changes in the environment adjacent to the surface as illustrated

Table 2.10 Solubility of Oxygen in Air Saturated Water

Temperature (°C)	Volume (cm ^{3})$^{(A)}$	Concentration (ppm)	Concentration (M) (μ mol L^{-1})
0	10.2	14.58	455.5
5	8.9	12.72	397.4
10	7.9	11.29	352.8
15	7.0	10.00	312.6
20	6.4	9.15	285.8
25	5.8	8.29	259.0
30	5.3	7.57	236.7

$^{(A)}$ cm^3 at 0°C per kg of water.

in Figure 2.25. When the chemical species participating in a corrosion process is in short supply, the mass transport of that species to the corroding surface can become rate controlling. A frequent case of this type of limitation occurs when the cathodic processes depend on the reduction of dissolved oxygen because it is usually in low concentration, that is, in parts per million (ppm). Table 2.10 contains data related to the solubility of oxygen in air-saturated water at different temperatures, and Table 2.11 contains data on the solubility of oxygen in seawater of different salinity and chlorinity.[4]

As indicated in Figure 2.26, mass transport to a surface is governed by three forces: diffusion, migration, and convection. In the absence of an electrical field, the migration term is negligible whereas the convection force disappears in stagnant conditions.

For purely diffusion-controlled mass transport, the flux of a species O to a surface from the bulk is described with Fick's first law (Equation [2.83]):

$$J_O = -D_O\left(\frac{\delta C_O}{\delta x}\right) \tag{2.83}$$

Table 2.11 Oxygen Dissolved in Seawater in Equilibrium with a Normal Atmosphere

Chlorinity$^{(A)}$‰	0	5	10	15	20
Salinity$^{(B)}$‰	0	9.06	18.08	27.11	36.11
Temperature (°C)			Ppm		
0	14.58	13.70	12.78	11.89	11.00
5	12.79	12.02	11.24	10.49	9.74
10	11.32	10.66	10.01	9.37	8.72
15	10.16	9.67	9.02	8.46	7.92
20	9.19	8.70	8.21	7.77	7.23
25	8.39	7.93	7.48	7.04	6.57
30	7.67	7.25	6.80	6.41	5.37

$^{(A)}$ Chlorinity refers to the total halogen ion content as titrated by the addition of silver nitrate, expressed in parts per thousand, i.e., ‰.
$^{(B)}$ Salinity refers to the total proportion of salts in seawater, often estimated empirically as chlorinity × 1.80655, also expressed in parts per thousand, i.e., ‰.

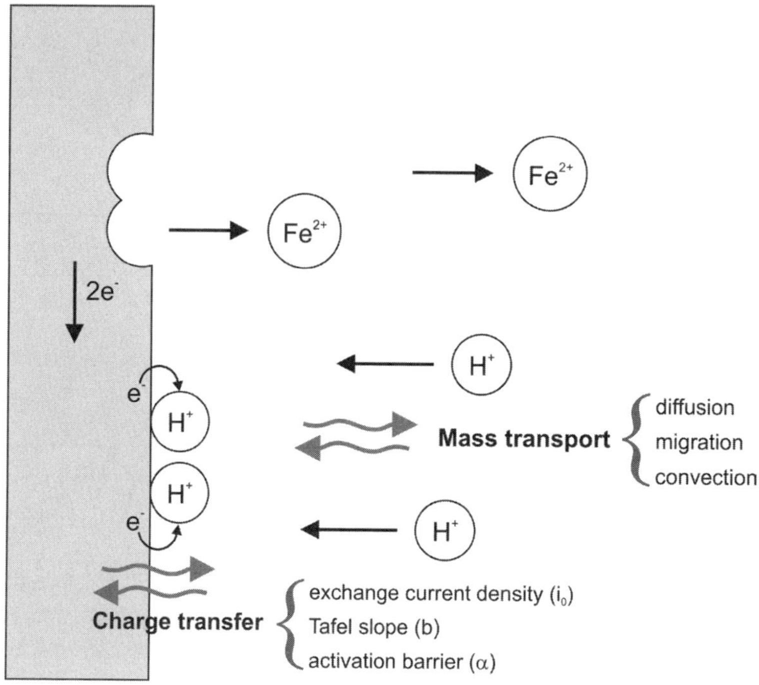

Figure 2.26 Graphical representation shows the processes occurring at an electrochemical interface.

where

- J_O is the flux of species O (mol s^{-1} cm^{-2}),
- D_O is the diffusion coefficient of species O (cm^2 s^{-1}),
- δC_O is the concentration gradient of species O across the interface between the metallic surface and the bulk environment (mol cm^{-3}), and
- δx is the thickness of the interface (cm).

Table 2.12 contains values for D_O of some common ions. For more practical situations the diffusion coefficient can be approximated with the help of Equation (2.84), which relates D_O to the viscosity of the solution (μ) and absolute temperature:

$$D_O = \frac{T \times A}{\mu} \tag{2.84}$$

where A is a constant for the system.

The region near the metallic surface where the concentration gradient occurs is also called the diffuse layer (δ). Because the concentration gradient ($\delta C_O/\delta x$) is greatest when the surface concentration of species O is completely depleted at the surface, that is, $C_O = 0$, it follows that the cathodic current is limited in that condition, as expressed by Equation (2.85):

$$i_c = i_L = -n \times \mathrm{F} \times D_O \times \frac{C_O^{\mathrm{bulk}}}{\delta} \tag{2.85}$$

Table 2.12 Diffusion Coefficients of Selected Ions at Infinite Dilution in Water at 25°C

Cation	$D \times 10^5$ ($cm^2\ s^{-1}$)	Anion	$D \times 10^5$ ($cm^2\ s^{-1}$)
H^+	9.30	OH^-	5.25
Li^+	1.03	F^-	1.47
Na^+	1.33	Cl^-	2.03
K^+	1.95	NO_3^-	1.90
Ca^{2+}	0.79	ClO_4^-	1.79
Cu^{2+}	0.71	SO_4^{2-}	1.06
Zn^{2+}	0.70	CO_3^{2-}	0.92
O_2	2.26	HSO_4^-	1.33
H_2O	2.44	HCO_3^{-1}	1.11

For intermediate cases, i.e. when the current is smaller than i_L, η_{conc} can be evaluated using an expression (Equation [2.86]) derived from Nernst Equation (2.51):

$$\eta_{conc} = \frac{2.303 \times R \times T}{n \cdot F} \log_{10}\left(1 - \frac{i}{i_L}\right) \tag{2.86}$$

where $2.303 \times R \times T/F = 0.059$ V when $T = 298.16$ K.

2.7.1.3 Ohmic Drop

The ohmic resistance of a cell can be measured by using a milliohmmeter with a high-frequency signal and a four-point technique. Table 2.13 lists some typical values of water conductivity.[4] Although the ohmic drop is an important parameter to consider when designing cathodic and anodic protection systems, it can be minimized, when carrying out electrochemical tests, by bringing the reference electrode in close proximity to the surface being monitored. For naturally occurring corrosion, the ohmic drop will limit the influence of an anodic or a cathodic site on adjacent metal areas to a certain distance, depending on the conductivity of the environment. For naturally occurring corrosion the anodic and cathodic sites are often adjacent grains or microconstituents and the distances involved are very small.

Table 2.13 Resistivity of Some Typical Waters

Water	ρ (Ω cm)
Pure water	20,000,000
Distilled water	500,000
Rain water	20,000
Tap water	1,000–5,000
River water (brackish)	200
Seawater (coastal)	30
Seawater (open sea)	20–25

2.7.2 Graphical Presentation of Kinetic Data

The use of polarization curves for the study of corrosion reactions can be traced back to the 1930s with the work of Wagner and Traud.[5] However, the representation of the mixed potential behavior is often associated with Evans, who popularized this representation of corrosion-polarization measurements.[6]

These polarization or diagrams are useful in describing and explaining many corrosion phenomena. According to the mixed-potential theory underlying these diagrams, any electrochemical reaction can be algebraically divided into separate oxidation and reduction reactions with no net accumulation of electrical charge. Under these circumstances the net measurable current is zero and the corroding metal is charge neutral, that is, all electrons produced by the corrosion of a metal have to be consumed by one or more cathodic processes.

To model a corrosion situation with mixed potential diagrams, one must first gather the information concerning the activation overpotential for each process that is potentially involved and any additional information for processes that could be affected by concentration overpotential. The following sections present some examples to illustrate how mixed potential models can be used to explain simple cases where corrosion processes are purely activation controlled or cases where concentration controls at least one of the corrosion processes.

2.7.2.1 Activation-Controlled Processes

For purely activation-controlled processes, each reaction can be described by a straight line on an E vs. log i plot, with positive Tafel slopes for anodic processes and negative Tafel slopes for cathodic processes.

The following example illustrates the polarization behavior of carbon steel in a deaerated solution maintained at 25°C and a pH of zero. The solid line in Figure 2.27 is the polarization plot itself and the dotted lines in this figure represent the anodic reaction(Equation [2.23]) and the cathodic reaction (Equation [2.21]) that explain the corrosion behavior of steel in these conditions. These lines are extrapolated from the linear sections of the plot on either the anodic or cathodic sides of the curve.

Although it is relatively easy to estimate the corrosion potential (E_{corr}) from the sharp peak observed at −0.221 V vs. SHE, when the current crosses zero (infinity on a log scale), the projected dotted lines are required to find the intercept indicating the specific point on the chart where the cathodic and anodic currents cancel each other. The corrosion current density (i_{corr}) can be obtained by dividing the anodic current by the surface area of the specimen, 1 cm^2 in the present case. According to the conversion table presented earlier (Table 2.3), the current density of 67 μA cm^{-2} evaluated in this example corresponds to a penetration rate of 0.8 mm y^{-1}.

The second example represents the carbon steel polarization behavior when exposed to a deaerated solution maintained at 25°C and pH of five. The mixed potential diagram of this system is shown in Figure 2.28. The E_{corr} in this example has shifted to a more negative value of −0.368 V vs. SHE and the modeled projected lines provides an

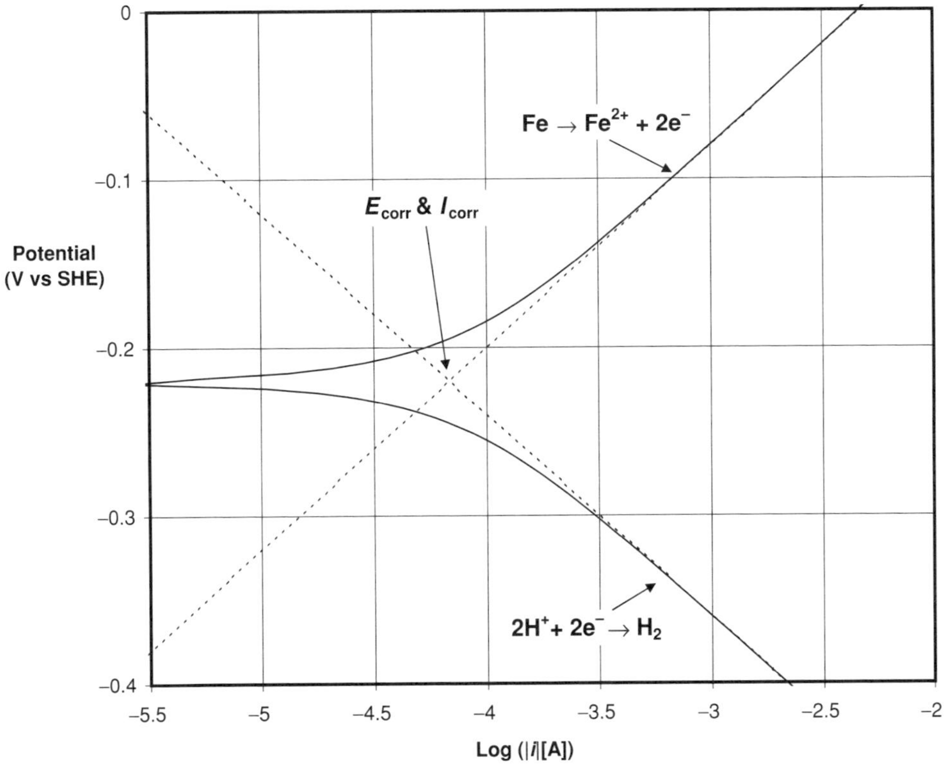

Figure 2.27 Polarization behavior of carbon steel in a deaerated solution maintained at 25°C and a pH of 0.

estimate of the corrosion current density of 4 μA cm^{-2} that translates into a penetration rate of 0.05 mm y^{-1}.

2.7.2.2 Concentration-Controlled Processes

When one of the reactions is limited by the rate of transport of the reactant to the metallic surface being corroded, the situation increases in complexity as illustrated in the polarization plot of the system in Figure 2.29. The system represented here is similar to the previous one, that is, pH of five at 25°C, with the exception that the environment is aerated and stagnant. In this situation the reduction of oxygen shown in Equation (2.34) now becomes possible as a second cathodic reaction.

To model the polarization plot, the total cathodic current corresponding to the sum of the currents of the hydrogen reaction and oxygen reduction has to be balanced by the single anodic current. The intercept where the opposing currents are balanced occurs at an E_{corr} of −0.33 V vs. SHE and, because the surface area is still 1 cm^2, an i_{corr} of 8.2 μA cm^{-2} or 0.1 mm y^{-1}. It should be noted that the current for the reduction of oxygen is constant across the potential range shown in Figure 2.29 and has the value of 6.3 μA cm^{-2}.

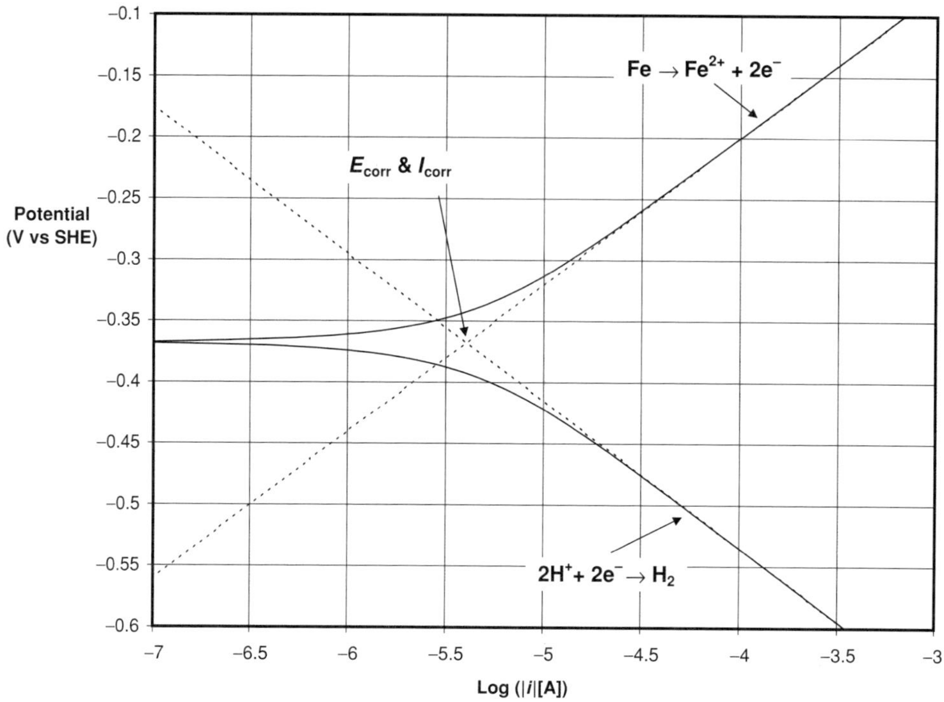

Figure 2.28 Polarization behavior of carbon steel in a deaerated solution maintained at 25°C and a pH of 5.

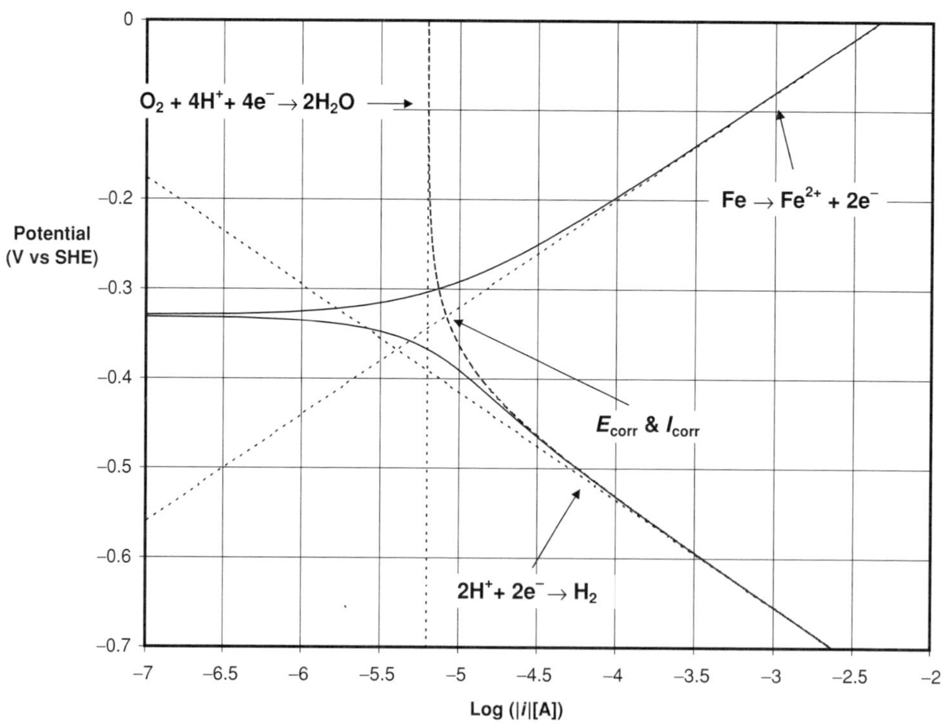

Figure 2.29 Polarization behavior of carbon steel in a stagnant aerated solution maintained at 25°C and a pH of 5.

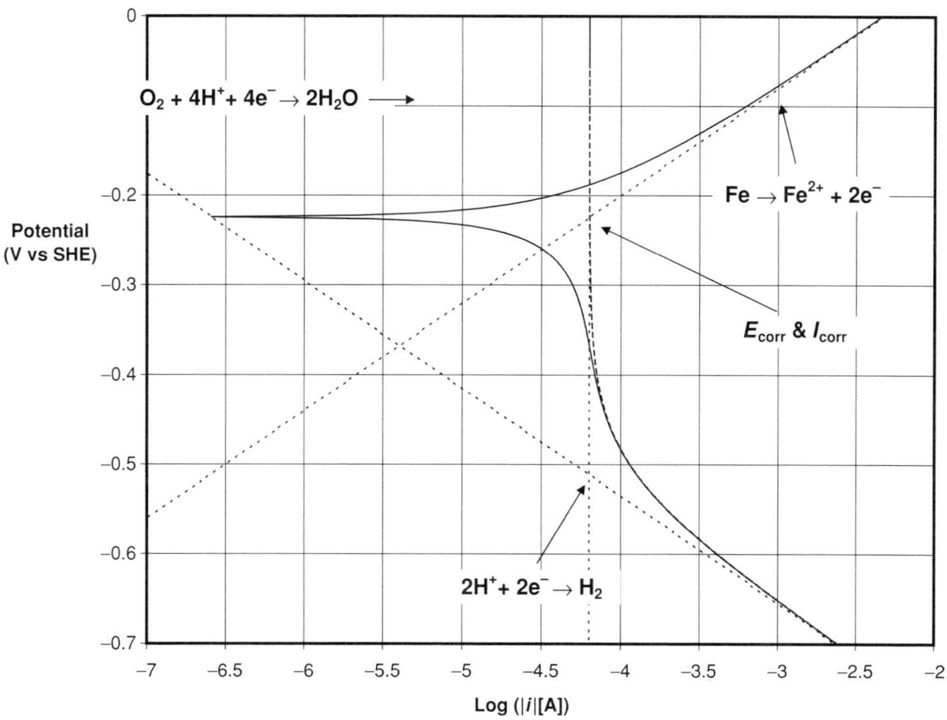

Figure 2.30 Polarization behavior of carbon steel in an agitated aerated solution maintained at 25°C and a pH of 5.

The reduction of oxygen depends, among other factors, on the level of agitation of the environment. If the limiting current of this reaction is now increased by a tenfold factor to reach the value of 63 μA cm^{-2}, a new situation emerges as depicted in Figure 2.30. The marked positive shift of E_{corr}, now -0.224 V vs. SHE, is accompanied by a marked increase in current density, now of 63 μA cm^{-2} or 0.8 mm y^{-1}.

2.7.3 Passivity and Protective Films

As mentioned previously, corrosion products and other surface films can have a profound effect on the corrosion behavior of metals. Oxide films that form naturally upon most metals when they are exposed to the air can provide substantial protection against further attack by many environments. If it were not for such films, many of the common metals near the bottom of the electromotive series (Table 2.5), would corrode rapidly in ordinary air and water. This is the case, for example, with chrome, nickel, and aluminum.

The passivation behavior of a metal is typically studied with a basic electrochemical testing setup shown in Figure 2.23. When the potential of a metallic component is controlled and shifted in the more anodic (positive) direction, the current required to cause that shift will vary. If the current required for the shift has the general polarization behavior illustrated in Figure 2.31, the metal is termed active–passive and can be

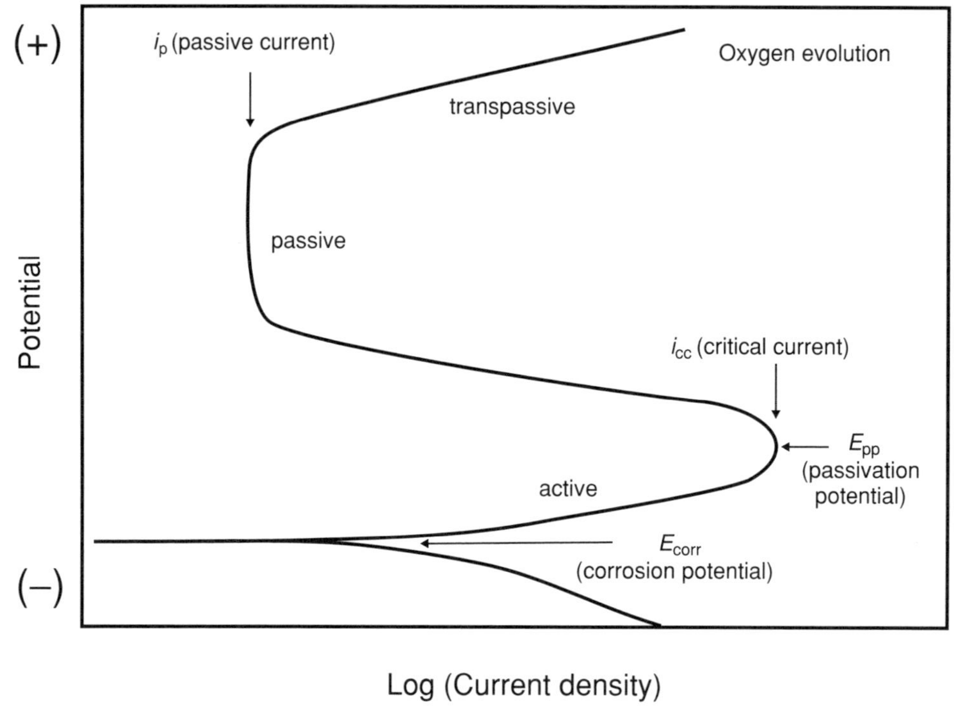

Figure 2.31 Generalized polarization diagram shows the various potential regions of a passivable metal.

anodically protected. Only a few systems exhibit this behavior in an appreciable and usable way. The corrosion rate of an active–passive metal can be significantly reduced by shifting the potential of the metal so that it is at a value in the passive range shown in Figure 2.31. The current required to shift the potential in the anodic direction from the corrosion potential E_{corr} can be several orders of magnitude greater than the current necessary to maintain the potential at a passive value. The current will peak at the passivation potential value shown as E_{pp}.

To produce passivation, the critical current density (i_{cc}) must be exceeded. The anodic potential must then be maintained in the passive region without allowing it to fall back in the active region or getting into the transpassive region, where the protective anodic film can be damaged and even break down completely. It follows that although a high current density may be required to cause passivation ($>i_{cc}$), only a small current density is required to maintain it, and that in the passive region the corrosion rate corresponds to the passive current density (i_p).

The relative tendency for passivation is strongly dependent on the interactions between a metal and its environment. Figure 2.32 illustrates how the sensitization of a S30400 stainless steel, for example, can affect its passivation behavior when exposed to sulfuric acid.[7] Among the parameters that are particularly affected by sensitization are i_p and i_{cc} as defined in Figure 2.31. In this example, the ability to sustain passivity increases as the current density to maintain passivity (i_p) decreases, and as the total film resistance increases, as indicated from measurements obtained with different metals

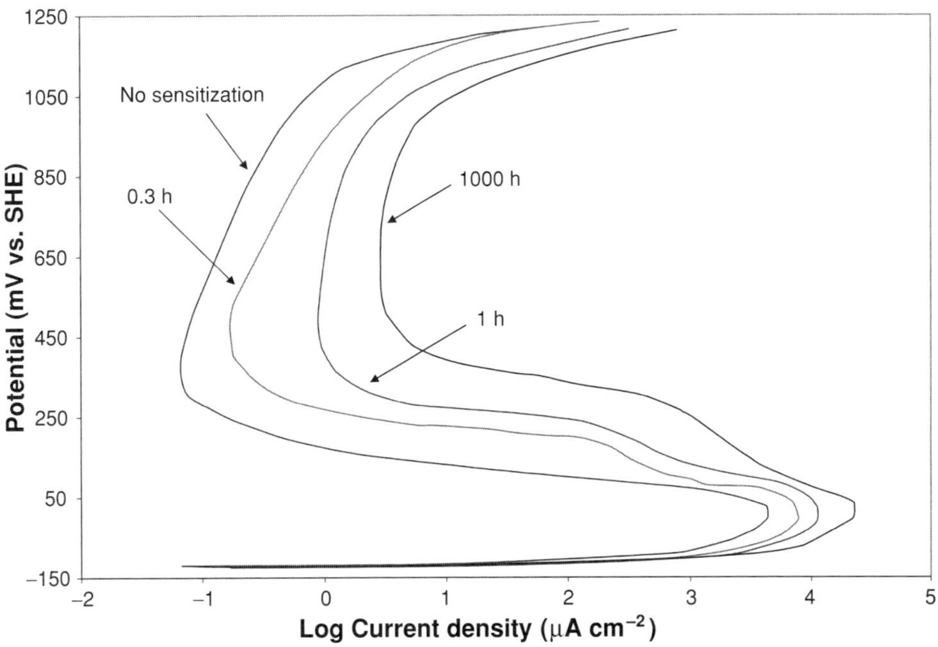

Figure 2.32 Anodic polarization curves of an S30400 stainless steel in a 1-M H_2SO_4 solution after sensitization for various periods.

exposed to 67% sulfuric acid (Table 2.14). The lower or more reducing the potential is at which a passive metal becomes active the greater the stability of passivity.

The depassivation potential corresponding to the passive–active transition, called the Flade potential, can differ appreciably from E_{pp} measured by going through the active-passive process of the same system. This technical distinction is important for the control aspect of anodic protection where E_{pp} is the potential to traverse to obtain passivation and the Flade potential the potential to avoid traversing back into active corrosion.

Passivity can also be readily produced in the absence of an externally applied passivating potential by using oxidants to control the potential of the environment. Apart from titanium and its alloys, most metals do not passivate in nonoxidizing environments.

Table 2.14 Current Density to Maintain Passivity and Film Resistance of Some Metals in 67% Sulfuric Acid

Metal or Alloy	i_p (μA cm $^{-2}$)	Film Resistance $M\Omega$ cm
Mild steel	150	0.026
S30400 steel	2.2	0.50
S31000 steel	0.5	2.1
S31600 steel	0.1	17.5
Titanium	0.08	1.75
N08020	0.03	4.6

Other corrosion-product films or scales are also protective. For example, insoluble films of lead sulfate are responsible for the resistance of lead to corrosion by sulfuric acid. The films that form on copper alloys in seawater contribute greatly to their durability. The extent to which these films are able to adhere, resist removal by turbulence effects, or be restored rapidly if broken largely determines the relative merits of the copper alloys in resisting velocity effects.

2.7.3.1 Active–Passive Cells

Under circumstances where there may be limited access by oxygen to a stainless-steel surface, passivity may be destroyed on such surfaces while the remainder of the stainless-steel surface will remain passive. A difference in potential between the active and passive stainless-steel surfaces can be as large as 700 mV. This will set up a powerful galvanic cell between the active and passive stainless-steel surfaces and result in serious corrosion of anodic areas (pitting).

The difference in potential between the active and passive states of stainless steels, and some other metals and alloys that develop passivity, will account for the dual location of these materials in positions representing either their active or passive states in the galvanic series shown in Tables 2.4.

Restoration of passivity within pits or crevices is frequently prevented by the acid nature of corrosion products when they accumulate within the pits or crevices. Consequently, circumstances such as stagnation and gravity effects that favor accumulation of corrosion products within pits or crevices will promote the most severe localized corrosion.

Because of the action of such active–passive cells, it is necessary in the use of stainless steel to eliminate designed-in crevices. It is also necessary to avoid any opportunities for the accumulation of loosely adherent deposits of any kind that could permit passivity to be destroyed within such crevices or under such deposits.

2.7.3.2 Effect of Alloying

Alloying with a more easily passivated metal normally increases the ease of passivation and lowers the passivation potential, as in the alloying of iron and chromium in 10% sulfuric acid (Table 2.15). Small additions of copper in carbon steels have been found to

Table 2.15 Effect on Critical Current Density and Passivation Potential of Chromium Content for Iron/Chromium Alloys in 10% Sulfuric acid

Chromium (%)	i_{cc} (mA cm^{-2})	E_{pp} (V vs. SHE)
0	1000	+0.58
2.8	360	+0.58
6.7	340	+0.35
9.5	27	+0.15
14.0	19	−0.03

Table 2.16 Effect on Critical Current Density and Passivation Potential on Alloying Nickel with Chromium in 0.5 M and 5 M H_2SO_4 Containing 0.25 M K_2SO_4

Ni	i_{cc} (mA cm $^{-2}$)		E_{PP} (V vs. SHE)	
(%)	0.5 M	5 M	0.5 M	5 M
100	100	23	+0.36	+0.47
91	0.95	3.9	+0.06	+0.14
77	0.11	0.82	+0.07	+0.08
49	0.020	0.20	+0.03	+0.06
27	0.012	0.041	+0.02	+0.05
10	0.0013	0.011	+0.04	+0.08
1	1.0	5.0	−0.32	−0.20
0	1.5	8.0	−0.30	−0.20

reduce i_p in sulfuric acid. Each alloy system has to be evaluated for its own passivating behavior as illustrated by the case Ni-Cr alloys where both the additions of nickel to chromium and chromium to nickel decrease the critical current density in a mixture of sulfuric acid and 0.25 M K_2SO_4 (Table 2.16).[8]

The effect of alloying is also illustrated in Figure 2.33, which schematically compares the potentiodynamic corrosion behavior of steel and chromium stainless steel in dilute sulfuric-acid solutions as a function of potential. It can be seen that the addition

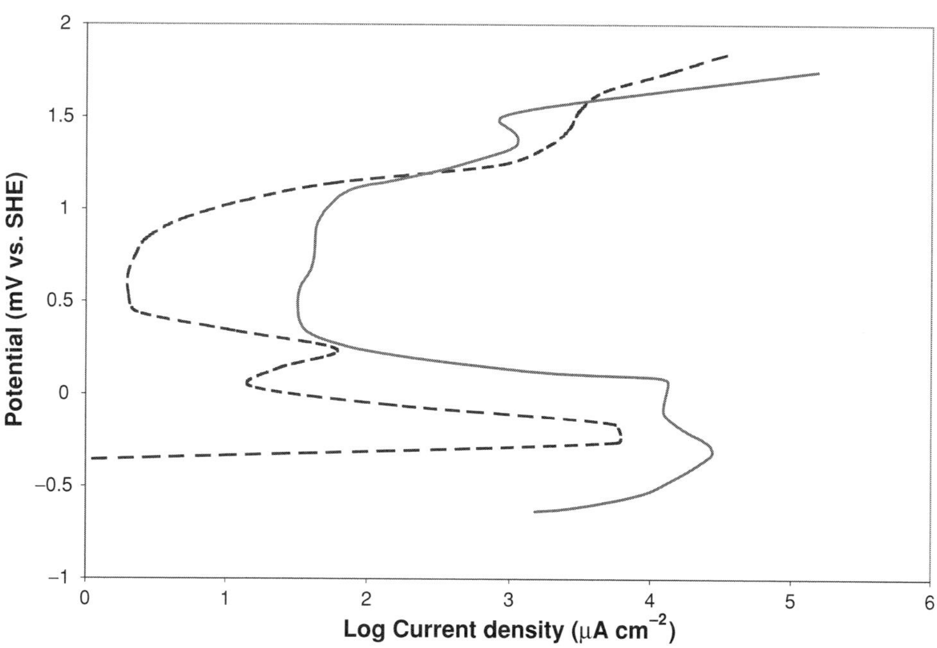

Figure 2.33 Anodic polarization curves of an S43000 stainless steel (dotted line) in an 0.5-M H_2SO_4 solution and mild carbon steel in a 1-M H_2SO_4 solution both at ambient temperature.

of 18% chromium to iron appreciably improves the passivation behavior of the steel and decreases the corrosion rate in the passive state.

2.8 CORROSION PREVENTION BY ELECTROCHEMICAL METHODS

2.8.1 Corrosion Testing

Electrochemical test methods for the rapid characterization of metallic materials have gained a tremendous level of acceptance since the introduction of potentiostat that followed the introduction of transistors in laboratory equipment. Chapter 14 provides a brief introduction to some of the techniques that are commonly used by many researchers around the world.

2.8.2 Cathodic Protection (CP)

CP, another electrochemical method for protecting metals, is widely used on small to extremely large structures by forcing the potential of a system toward less corrosive values. This can often be accomplished by applying current from an external source so that current will be made to flow to, instead of away from, the original anodic surface.

To accomplish this, the potential of the metallic system to be protected is reduced by using a protective current that is either generated by a power supply (impressed current cathodic protection, ICCP) or by another metal that corrodes more readily and therefore is sacrificed is the process. The amount of current that will be needed will depend on the requirement to support a cathodic reaction over the whole of the surface to be protected.

CP can be illustrated by a simple experiment using two iron nails and a piece of zinc. If one iron nail is immersed in water, but in contact with the zinc, it will not corrode. The nail by itself will corrode. This is why galvanized iron (zinc-coated steel) is so widely used.

A similar experiment can be conducted using iron nails with half their surfaces plated with copper and with one of these being connected to a piece of zinc. As would be expected, the unplated half of the first iron nail would become anodic to the copper-plated half and quickly corrode. In the case of the partially plated nail connected to the zinc, the galvanic corrosion of the iron half would be suppressed and a cathodic reaction would be made to occur along both the bare iron and copper-plated surfaces. Chapter 11 provides a more detailed discussion of this subject.

2.8.3 Anodic Protection

The term anodic protection refers to the corrosion protection achieved by maintaining an active–passive metal or alloy in the passive region by an externally applied anodic current. The basic principle for this type of protection is explained by the behavior shown in Figure 2.31.

In contrast to cathodic protection, anodic protection is relatively new. The feasibility of anodic protection was first demonstrated in 1954 and tested it on small-scale

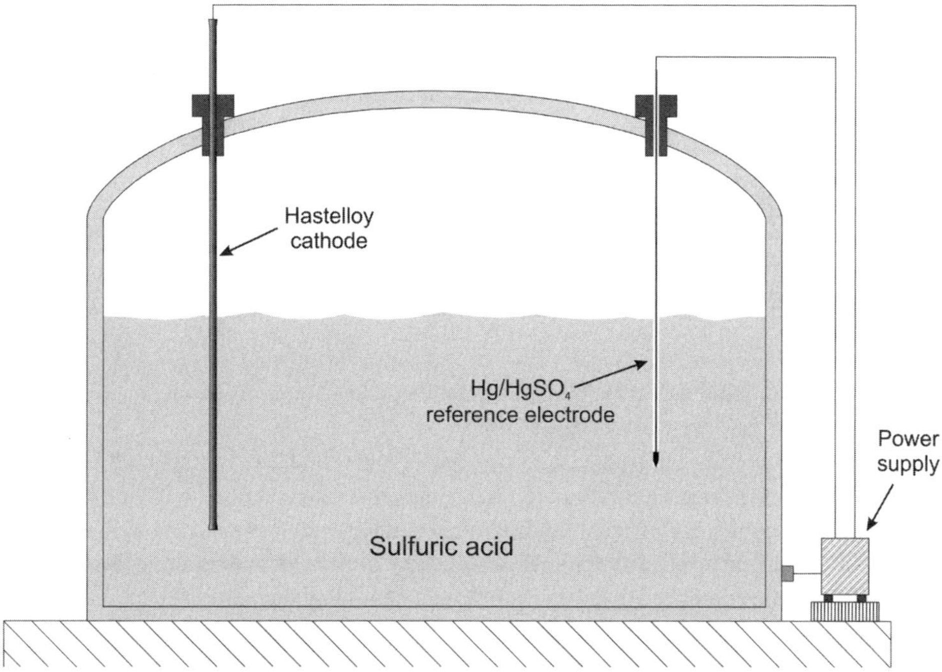

Figure 2.34 Schematic of an anodic protection system for a sulfuric-acid storage vessel.

stainless-steel boilers used for sulfuric-acid solutions.[3] Anodic protection possesses unique advantages. For example, the applied current is usually equal to the corrosion rate of the protected system. Thus, anodic protection not only protects but also offers a direct means for monitoring the corrosion rate of a system.

The main advantages of anodic protection are (1) low current requirements, (2) large reductions in corrosion rate (typically 10,000-fold or more), and (3) applicability to certain strong, hot acids and other highly corrosive media. It is important to emphasize that anodic protection can only be applied to metals and alloys possessing active-passive characteristics such as titanium, stainless steels, steel, and nickel-base alloys. Furthermore, it can only be utilized in certain environments, because electrolyte composition influences passivity, and under tightly controlled conditions in closed systems.

Because the potential must be maintained within the passive region, it is necessary to use a special device called a potential controller capable of maintaining a constant electrode potential by controlling the anodic current (Figure 2.34). The potential controller acts as a potentiostat with its three terminals connected to the proper electrodes.

The parameters defining and controlling the passivation domain of a system are directly related to the composition, concentration, purity, temperature, and agitation of the environment. This is illustrated with the current densities required to obtain passivity (i_{cc}), and to maintain passivity (i_p), for S30400 steel in different electrolytes. This information is presented in Table 2.17 where it can be seen that it is approximately 100,000 times easier to passivate large areas of this steel in contact with 115% phosphoric acid than in 20% sodium hydroxide. The concentration of the electrolyte is also important and for S31600 steel in sulfuric acid, although there is a maximum

Table 2.17 Critical Current Density and Current Density to Maintain Passivity of S30400 Stainless Steel in Various Electrolytes

Electrolyte	i_{cc} (mA cm^{-2})	ip (μ A cm^{-2})
20% NaOH	4.65	9.9
67% H_2SO_4 (24°C)	0.51	0.093
LiOH (pH = 9.5)	0.08	0.022
80% HNO_3 (24°C)	0.0025	0.031
115% H_3PO_4 (24°C)	0.000015	0.00015

corrosion rate at about 55%, the critical current density decreases progressively as the concentration of acid increases (Table 2.18).[8]

The presence in the environment of impurities that retard the formation of a passive film or accelerate its degradation is often detrimental. In this context, chloride ions can be particularly aggressive for steels and stainless steels. The addition of 3% hydrochloric acid to 67% sulfuric acid can thus raise the critical current density for the passivation of S31600 stainless steel from 0.7 to 40 mA cm^{-2} and the current density to maintain passivity from 0.1 to 60 μA cm^{-2}. Therefore, the use of chloride containing reference electrodes such as SCE or SSCE in anodic-protection systems is not recommended because of the possible leakage of chloride ions into the electrolyte, and metal/metal oxide and other electrodes are often preferred.[8]

2.9 SUMMARY

This chapter has examined the fundamental reactions of corrosion and found these to be primarily electrochemical reactions. It has explained how most corrosion processes involve electron transfer and can be greatly simplified by considering them as a series of anodic and cathodic reactions.

Electrochemical principles allow us to predict whether or not a pure metal will be corroded by a given environment. Furthermore, this chapter has shown that the

Table 2.18 Effect of Concentration of Sulfuric Acid at 24°C on Corrosion Rate and Critical Current Density of S31600 Stainless Steel

Sulfuric Acid (%)	Corrosion rate (mm y^{-1})	i_{cc} (mA cm^{-2})
0	0	4.7
40	2.2	1.6
45	5.6	1.4
55	8.9	1.0
65	7.8	0.7
75	6.7	0.4
105	0	0.1

electrochemical nature of corrosion often results in galvanic attack, which can be both understood and controlled by recognizing and understanding basic electrochemical principles. An understanding of electrochemistry has been shown to be useful because it can be utilized to reduce or prevent corrosion, and it can also be used to measure the corrosion rate of a metal.

References

1. S.J. Pawell, R.J. Lopez, E. Ondak, "Chemical and Environmental Influences on Copper/Copper Sulfate Reference Electrode Half Cell Potential," MP 38, 5 (1998): pp. 24–29.
2. M. Pourbaix, Atlas of Electrochemical Equilibria in Aqueous Solutions, 2nd ed. (Houston, TX: NACE International, 1974).
3. P.R. Roberge, Handbook of Corrosion Engineering (New York, NY: McGraw-Hill, 2000).
4. L.L. Shreir, R.A. Jarman, G.T. Burstein, Corrosion Control (Oxford, UK: Butterworths Heinemann, 1994).
5. C. Wagner, W. Traud, "Uber die deutung von korrosionvorgangen durch uberlagerung von elektrochemischen teilvorgangen und uber die potentialbildung an mischellektroden," Zeitschrift fur Elektrochemie, 44 (1938): pp. 391–402.
6. U.R. Evans, An Introduction to Metallic Corrosion (London, UK: Edward Arnold, 1948).
7. A.J. Sedriks, Corrosion of Stainless Steels (New York, NY: John Wiley & Sons, 1979).
8. R. Walker, "Anodic Protection," in Corrosion Control, eds. L.L. Shreir, R.A. Jarman, G.T. Burstein (Oxford, UK: Butterworths Heinemann, 1994), pp. 10:155–10:170.

Part II

Environments

3
Chapter

Atmospheric Corrosion

3.1 INTRODUCTION

The specific topic of corrosion in the atmosphere usually receives little attention in the writings of scientists and corrosion engineers who deal with the basic subject of corrosion. However, most corrosion damage to equipment and structures actually occurs in the atmosphere, as opposed to liquid immersion. The large segment of the paint industry committed to the manufacture and application of products for the protection of metals, as well as the large-scale operations of the galvanizing industry, attest to the importance of controlling atmospheric corrosion.

All of the general types of corrosion attack occur in the atmosphere (see Chapter 9 for details on types of corrosion). Because the corroding metal is not bathed in large quantities of electrolyte, most atmospheric corrosion operates in highly localized corrosion cells. Thus, calculation of the electrode potentials on the basis of ion concentration, the determination of polarization characteristics, and other electrochemical operations are not as simple in atmospheric corrosion as they are in liquid-immersion corrosion. However, all of the electrochemical factors that are significant in corrosion processes do operate in the atmosphere. Thus, an understanding of electrochemical factors is still vital to an understanding of the corrosion mechanism.

The economic losses caused by atmospheric corrosion are tremendous and therefore account for the disappearance of a significant portion of metal produced. Consider, for instance, agricultural machinery, steel structures, fences, exposed metals on buildings, automobile mufflers and bodies, and the myriad other metal items that are discarded when they become unusable as a result of corrosion. These constitute direct losses from corrosion.

Atmospheric corrosion has been reported to account for more failures in terms of cost and tonnage than any other type of material-degradation processes. This particular type of material degradation has recently received more attention, particularly by the aircraft industry, since the Aloha incident in 1988, when a Boeing 737 lost a major portion of the upper fuselage in full flight at 7,300 m.[1]

In one of the few detailed analyses of aircraft-corrosion costs ever performed, it was estimated that the direct corrosion costs in the U.S. Air Force aircraft exceeded $0.7 billion annually, with the oldest aircraft types accounting for approximately half these costs.[2] The total annual costs in the U.S. aircraft industry have been estimated to be approximately $4 billion. Atmospheric corrosion affects all structures exposed to potentially corrosive environments and many countries have programs dedicated to the understanding and monitoring of atmospheric corrosion.

3.2 TYPES OF CORROSIVE ATMOSPHERES

Although atmospheres can be classified into four basic types, most of them are mixed and present no clear lines of demarcation. Furthermore, the type of atmosphere may vary with the wind pattern, particularly where corrosive pollutants are concerned.

3.2.1 Industrial

An industrial atmosphere is characterized by pollution composed mainly of sulfur compounds and nitrogen oxides. Sulfur dioxide from burning coal or other fossil fuels is picked up by moisture on dust particles as sulfurous acid. This is oxidized by some catalytic process on the dust particles to sulfuric acid, which settles in microscopic droplets on exposed surfaces. Some sulfur dioxide and sulfurous acid are also present. The result is that contaminants in an industrial atmosphere, plus dew or fog, produce a highly corrosive, wet, acid film on exposed surfaces.

In addition to the normal industrial atmosphere in or near chemical plants, other corrosive pollutants may be present. These are usually various forms of chloride, which may be much more corrosive than the acid sulfates. The reactivity of acid chlorides with most metals is more pronounced than that of other pollutants such as phosphates and nitrates.

3.2.2 Marine

A marine atmosphere is laden with fine particles of sea-salt mist carried by the wind to settle on exposed surfaces. The quantity of salt contamination decreases with distance from the ocean and is greatly affected by wind currents. The marine atmosphere also includes the space above the sea surfaces where splashing and heavy sea spray are encountered. Some splash zones could be classified as intermittent immersion.

3.2.3 Rural

A rural atmosphere does not contain strong chemical contaminants but does contain organic and inorganic dusts. Its principal corrosive constituent is moisture and, of course, gaseous elements such as oxygen and carbon dioxide.

Arid or tropical atmospheres are special variations of the rural atmosphere. In arid climates there is little or no rainfall but there may be a high relative humidity and occasional condensation. This situation is encountered along the desert coast of northern Africa. In the tropics, in addition to the high average temperature, the daily cycle includes a high relative humidity, intense sunlight, and long periods of condensation during the night. In sheltered areas, the wetness from condensation may persist long after sunrise. Such conditions may produce a highly corrosive environment.

3.2.4 Indoor

The indoor atmosphere, although originally considered quite mild, actually can sometimes be quite severe. However, there is no typical contaminant or set of conditions associated with an indoor atmosphere. Any enclosed space that is not evacuated or filled with a liquid can be considered an indoor atmosphere. If not ventilated, it may contain fumes, which in the presence of condensation or high humidity could prove to be highly corrosive.

When considering atmospheres of any type, it must be understood that not only acidic contaminants are aggressive, but that alkaline materials (e.g., limestone, seashells, etc.) can also be corrosive when in contact with amphoteric metals (e.g., aluminum, zinc, and lead).

3.3 FACTORS AFFECTING ATMOSPHERIC CORROSION

The most important factor in atmospheric corrosion, overriding pollution or lack of it, is moisture, either in the form of rain, dew, condensation, or high relative humidity (RH). In the absence of moisture, most contaminants have little or no corrosive effect.

Rain also may have a beneficial effect in washing away atmospheric pollutants that have settled on exposure surfaces. This effect has been particularly noticeable in marine atmospheres. On the other hand, if the rain collects in pockets or crevices, it may accelerate corrosion by supplying continued wetness in such areas.

Dew and condensation are undesirable from a corrosion standpoint if not accompanied by frequent rain washing, which dilutes or eliminates contamination. A film of dew, saturated with sea salt or acid sulfates and acid chlorides of an industrial atmosphere, provides an aggressive electrolyte for the promotion of corrosion. Also, in the humid tropics, where nightly condensation appears on many surfaces, the stagnant moisture film either becomes alkaline from reaction with metal surfaces or picks up carbon dioxide and becomes aggressive as a dilute acid.

Temperature plays an important role in atmospheric corrosion in two ways. First, there is the normal increase in corrosion activity, which can theoretically double for each 10°C increase in temperature. Secondly, a little-recognized effect is the temperature lag of metallic objects, because of their heat capacity, behind changes in the ambient temperature.

As the ambient temperature drops during the evening, metallic surfaces tend to remain warmer than the humid air surrounding them and do not begin to collect condensation until some time after the dew point has been reached. As the temperature begins to rise in the surrounding air, the lagging temperature of the metal structures will tend to make them act as condensers, maintaining a film of moisture on their surfaces.

The period of wetness is often much longer than the time the ambient air is at or below the dew point and varies with the section thickness of the metal structure, air currents, RH, and direct radiation from the sun.

Cycling temperature has produced severe corrosion on metal objects in the tropics, in unheated warehouses, and on metal tools or other objects stored in plastic bags. Because the dew point of an atmosphere indicates the equilibrium condition of condensation and evaporation from a surface, it is advisable to maintain the temperature some 10 to 15°C above the dew point to ensure that no corrosion will occur by condensation on a surface that could be colder than the ambient environment.

3.3.1 Relative Humidity

RH is defined as the ratio of the quantity of water vapor present in the atmosphere to the saturation quantity at a given temperature, and it is expressed as a percentage. A fundamental requirement for atmospheric-corrosion processes is the presence of a thin-film electrolyte that can form on metallic surfaces when exposed to a critical level of humidity. Although this film is almost invisible, the corrosive contaminants it contains are known to reach relatively high concentrations, especially under conditions of alternate wetting and drying.

The critical humidity level is a variable that depends on the nature of the corroding material, the tendency of corrosion products and surface deposits to absorb moisture, and the presence of atmospheric pollutants.[3] It has been shown that, for example, this critical humidity level is 60% for iron if the environment is free of pollutants. In the presence of thin-film electrolytes, atmospheric corrosion proceeds by balanced anodic and cathodic reactions. The anodic-oxidation reaction involves the corrosion attack of the metal, whereas the cathodic reaction is naturally the oxygen-reduction reaction (Figure 3.1).

Marine environments typically have a high percent of relative humidity (%RH), as well as salt-rich aerosols. Studies have found that the thickness of the adsorbed layer of

Anode Reaction: $2Fe \rightarrow 2Fe^{2+} + 4e^-$

Cathode Reaction: $O_2 + 2H_20 + 4e^- \rightarrow 4OH^-$

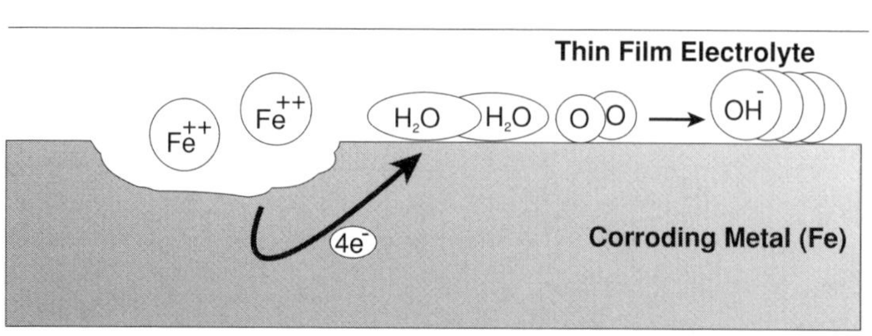

Figure 3.1 Schematic description of the atmospheric corrosion of iron.

water on a zinc surface increases with %RH and that corrosion rates increase with the thickness of the adsorbed layer. There also seems to be a finite thickness to the water layer that, when exceeded, can limit the corrosion reaction because of limited oxygen diffusion.[4] However, when metallic surfaces become contaminated with hygroscopic salts, their surface can be wetted at lower %RH. The presence of magnesium chloride ($MgCl_2$) on a metallic surface can make a surface apparently wet at 34% RH whereas sodium chloride (NaCl) on the same surface requires 77% RH to create the same effect.[5]

3.3.2 Deposition of Aerosol Particles

The behavior of aerosol particles in outdoor atmospheres can be explained by invoking the laws that govern their formation, movement, and capture. These particles are present throughout the planetary boundary layer and their concentrations depend upon a multitude of factors including location, time of day or year, atmospheric conditions, presence of local sources, altitude, and wind velocity. The highest concentrations are usually found in urban areas, reaching up to 10^8 and 10^9 particles per cm^3, with particle size ranging from around 100 μm to a few nm. Size is normally used to classify aerosol because it is the most readily measured property and other properties can be inferred from size information.[6] The highest mass fraction of particles in an aerosol is characterized by particles with a diameter in the range of 8 μm to 80 μm.[7] Some studies have also indicated that there is a strong correlation between wind speed and the deposition and capture of aerosols. In such a study of saline winds in Spain, a very good correlation was found between chloride deposition rates and wind speeds above a threshold of 3 m s^{-1} or 11 km h^{-1}.[8]

Aerosols can either be produced by ejection into the atmosphere or by physical and chemical processes within the atmosphere (called primary and secondary aerosol production respectively). Examples of primary aerosols are sea spray and wind-blown dust. Secondary aerosols are often produced by atmospheric gases reacting and condensing or by cooling vapor condensation (gas-to-particle conversion). Once an aerosol is suspended in the atmosphere, it can be altered, removed, or destroyed. It cannot stay in the atmosphere indefinitely, and average lifetimes are a few days to a week.

The lifetime of any particular particle depends on its size and location. Studies of the migration of aerosols inland of a sea coast have shown that typically the majority of the aerosol particles are deposited close to the shoreline (typically 400 to 600 m) and consist of large particles (>10-μm diameter), which have a short residence time and are controlled primarily by gravitational forces.[7,8] The aerosols that form also have mass and are subject to the influence of gravity, wind resistance, droplet dry-out, and possibilities of impingement on a solid surface as they progress inland.

3.3.3 Pollutants

Sulfur dioxide (SO_2), which is the gaseous product of the combustion of fuels that contain sulfur such as coal, diesel fuel, gasoline, and natural gas, has been identified as one of the most important air pollutants that contribute to the corrosion of metals.

Figure 3.2 Interlocking combs of gold and copper electrodes in a "Sereda" humidity sensor.

Less recognized as corrosion promoters are the nitrogen oxides (NO_X), which are also products of combustion. A major source of NO_X in urban areas is the exhaust fumes from vehicles. Sulfur dioxide, NO_X, and airborne aerosol particles can react with moisture and UV light to form new chemicals that can be transported as aerosols. A good example of this is the summertime haze over many large cities. Up to 50% of this haze is a combination of sulfuric and nitric acids.

3.4 MEASUREMENT OF ATMOSPHERIC FACTORS

Various methods have been developed for measuring many of the factors that influence atmospheric corrosion. The quantity and composition of pollutants in the atmosphere, the amount collected on surfaces under a variety of conditions, and the variation of these with time have been determined. Temperature, RH, wind direction and velocity, solar radiation, and amount of rainfall are easily recorded. Not so easily determined are dwelling time of wetness (TOW) and the quantity of sulfur dioxide and chloride contamination. However, methods for these determinations have been developed and are in use at various test stations. By monitoring these factors and relating them to corrosion rates, a better understanding of atmospheric corrosion can be obtained.

3.4.1 Time of Wetness

TOW is an estimated parameter based on the length of time when the relative humidity is greater than 80% at a temperature greater than 0°C. It can be expressed as the hours or days per year or the annual percentage of time.

A method of measuring the TOW has been developed by Sereda and correlated with the corrosion rates encountered in the atmosphere.[9] The moisture-sensing elements in this sensor are manufactured by plating and selective etching of thin films of appropriate anode (copper) and cathode (gold) materials in an interlaced pattern on a thin, nonconductive substrate (Figure 3.2). When moisture condenses on the sensor, it activates the cell, producing a small voltage (0 to 100 mV) across a 10^7-Ω resistor.

Thin sensing elements are preferred to preclude influencing the surface temperature to any extent. Although a sensor constructed using a 1.5-mm-thick glass-reinforced polyester base has been found to be satisfactory on plastic surfaces, this will not be the case with the same sensing element on a metal surface with a high thermal conductivity.[10] For metal surfaces, the sensing element should be appreciably thinner. Commercial epoxy sensor-backing products 1.5 mm thick or less are suitable for this purpose.

3.4.2 Pollutants

Sulfur dioxide is usually measured in terms of its concentration in air in units of $\mu g/m^3$. Precise methods are available to continuously monitor the amount of sulfur dioxide in a given volume of air. However, this is only indirectly related to the effect of sulfur dioxide on corrosion because only the actual amount of hydrated sulfur dioxide or sulfur trioxide deposited on metal surfaces is important. Corrosion-rate measurements of Classify Industrial and Marine ATmospheres(CLIMAT) coupons, described later in this chapter, at air-quality monitoring stations in a major city have continuously failed to correlate with the measured levels of NO_x, SO_2, and RH (Figure 3.3).

Because it is the SO_2 deposited on the metal surface that affects the corrosion, it is also often measured in terms of deposition rate on the surface in units of $mg/m^2/day$. The pollution levels can also be measured in terms of the concentration of the dissolved sulfate (SO_4^{2-}) in rain water. The collection of sulfur dioxide by the lead peroxide cylinder method has been employed and seems to correlate with corrosion rates when combined with the TOW.

3.4.3 Airborne Chlorides

Airborne salinity refers to the content of gaseous and suspended salt in the atmosphere. It is measured by the concentration in the air in units of $\mu g/m^3$. Because it is the salt that is deposited on the metal surface that affects the corrosion, it is often measured in terms of deposition rate in units of $mg/m^2/day$. The chloride levels can also be measured in terms of the concentration of the dissolved salt in rain water.

A number of methods have been employed for determining the contamination of the atmosphere by aerosol-transported chlorides, for example, sea salt and road-deicing salts. The "wet-candle method," for example, is relatively simple, but has the disadvantage that it also collects particles of dry salt that might not be deposited otherwise.[11] This technique uses a wet wick of a known diameter and surface area to measure aerosol deposition (Figure 3.4). The wick is kept wet using a reservoir of water or 40% glycol/water solution. Particles of salt or spray are trapped by the wet wick and retained. At intervals, a quantitative determination of the chloride collected by the wick is made and a new wick is exposed.

In reality, the wet-candle method gives an indication of the salinity of the atmosphere rather than the contamination of exposed metal surfaces. The technique is

Figure 3.3 An air quality sampling station near a very busy intersection in Montreal, Canada, with a CLIMAT coupon attached to the air intake (courtesy of Kingston Technical Software).

considered to measure the total amount of chloride arriving to a vertical surface and its results may not be truly significant for corrosivity estimates.

To model the chloride-deposition rates in a confined space such as a ventilated subfloor shown in Figure 3.5 where corrosion of fasteners can be quite severe (Figure 3.6) a special collecting box shown in Figure 3.7 was used in which the airborne chlorides were collected on horizontal and vertical filter papers positioned at different locations from the box openings (Figure 3.8).

For obvious reasons, the test chambers used in this study were thoroughly cleaned of chlorides before setting them outdoors. Rigid plastic-material panels stuck at predetermined positions with self-adhesive pads were used to hold the filter papers with

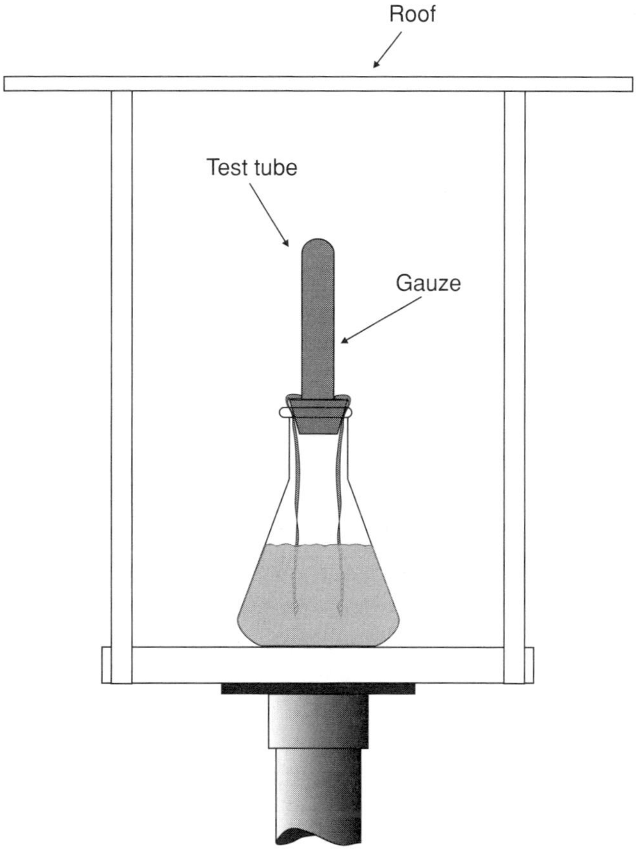

Figure 3.4 Schematic shows a wet-candle chloride apparatus.

chloride-free plastic paper clips. The roof protected the surface from direct rain but the filter papers were still exposed to deposition of the airborne chlorides and possibly some rain. The paper was removed every month for chemical analysis. A wet candle with its own roof cover was installed on each box (Figure 3.7) to provide a measure of the level of chlorides outside the boxes.

3.4.4 Atmospheric Corrosivity

The simplest form of direct atmospheric-corrosion measurement is by coupon exposures. Subsequent to their exposure, the coupons can be subjected to weight-loss measurements, pit-density and depth measurements, and to other types of examination. Flat panels exposed on exposure racks are a common coupon-type device for atmospheric corrosivity measurements. Various other specimen configurations have been used, including stressed U-bend or C-ring specimens for stress corrosion cracking (SCC) studies. The main drawback associated with conventional coupon measurements is that extremely long exposure times are usually required to obtain meaningful data, even on a relative scale. It is not uncommon for such programs to run for 20 years or longer.

Figure 3.5 A ventilated subfloor typical of construction in mild climate regions (courtesy of Branz, New Zealand).

Some variations of the basic coupon specimens can provide rapid material/corrosivity evaluations. The helical coil adopted in the ISO 9226 methodology is a high surface-area/weight ratio coupon that gives a higher sensitivity than panel coupons of the same material. The use of bimetallic specimens in which a helical wire is wrapped around a coarsely threaded bolt can provide additional sensitivity and forms the basis of the CLIMAT coupon for Classify Industrial and Marine Atmospheres.[12]

Figure 3.6 Severe corrosion of a subfloor fastener of a house built in coastal area (courtesy of Branz, New Zealand).

Figure 3.7 A typical subfloor simulation box is surmounted by a wet-candle chloride-measuring device that has been installed at several sites in Australia and New Zealand (courtesy of Branz, New Zealand).

An ASTM standard describes the construction of CLIMAT coupons. According to this standard CLIMAT coupons can be made from 1100 aluminum (UNS A91100) wire wrapped around threaded rods of nylon, 1010 mild steel (UNS G10100 or G10080), and CA110 copper (UNS C11000).[13] The mass loss of aluminum wire after 90 days of exposure is considered to be an indication of atmospheric corrosivity. However, the relative corrosivity of atmospheres could be quite different for the various combinations of materials.

The aluminum wire on copper bolts has been found by many to be the most sensitive of the three proposed arrangements in the ASTM standard. Although this arrangement is

Figure 3.8 Horizontal and vertical filter papers installed inside a subfloor simulation box (courtesy of Branz, New Zealand).

Figure 3.9 Aerial view of the NASA Kennedy Space Center beach corrosion test site, where atmospheric corrosivity is the highest of any test site in the continental United States.

the most sensitive, the use of triplicate coupons on a single holder additionally provides an indication of the reproducibility of the measurements. A CLIMAT coupon with three copper rods installed at the NASA Kennedy Space Center (KSC) beach corrosion test site (Figure 3.9) is shown immediately after it had been installed (Figure 3.10a), after thirty days (Figure 3.10b), and after sixty days (Figure 3.10c). KSC has the highest corrosivity of any test site in the continental United States.[14] The mass loss recorded after a shorter exposure than usual can be very high. In the present example it was already 16% of the original aluminum wire after 60 days.

The CLIMAT coupons' sensitivity to atmospheric corrosivity can be used to study fluctuations on a microenvironmental scale.[15] In the following example, a supporting panel (Figure 3.11) was placed about 2 m above the ground on a pedestrian bridge concrete support approximately 4 m from a moderately trafficked highway during the winter months when deicing salts are used. Six sets of CLIMAT units were deployed on the panel with each set in a different microenvironment produced by various baffle geometries. The overall atmospheric conditions, for example, temperature and RH, were therefore the same for each set except for the differences in the rate of aerosol deposition.

The average mass loss experienced by the CLIMATs exposed in different baffling geometries is also shown in Figure 3.11. One obvious conclusion from these

(a) (b) (c)

Figure 3.10 A CLIMAT coupon with three copper rods a) immediately after it was installed at the Kennedy Space Center beach corrosion test site, b) after 30 days, and c) after 60 days.

measurements is that shielding, whether from wind or direct precipitation, can dramatically reduce the corrosion rate to samples that are exposed to the same TOW factor. In fact, there was a 24-fold difference between the average mass loss in the boxed-in and boldly exposed coupons.

3.5 CLASSIFICATION SCHEMES

3.5.1 Environmental Severity Index

An environmental-corrosivity scale, based on atmospheric parameters, has been developed over the years for the maintenance management of structural aircraft systems.[16] A corrosion-damage algorithm (CDA) was proposed as a guide for anticipating the extent of corrosion damage and planning maintenance operations. This classification scheme was developed primarily for uncoated aluminum, steel, titanium, and magnesium alloys exposed to the external atmosphere at ground level.

The section of the CDA algorithm presented in Figure 3.12 considers distance to salt water, leading either to the very severe AA rating for close distance to seashore

Figure 3.11 Test panel and CLIMAT coupons shown exposed in different microenvironments adjacent to a moderately trafficked highway during the winter months when deicing salts are used.

or a consideration of moisture factors. Following the moisture factors, pollutant concentrations are compared with values of Working Environmental Corrosion Standards (WECS). The WECS values were adopted from the 50th-percentile median of a study aimed at determining ranges of environmental parameters in the United States and represent "average of averages." For example, a severe A rating would be given if any of the three pollutants considered in this scheme (sulfur dioxide, total suspended particles, and ozone levels) would exceed the WECS values in combination with a high moisture factor.

The environmental corrosivity, predicted from the CDA algorithm, of six marine air bases has been compared to the actual corrosion maintenance effort expended at each base. Considering the simplicity of the algorithms and simplifying assumptions in obtaining relevant environmental and maintenance data, the correlation obtained was considered to be reasonable. Further validation of the CDA algorithm approach was sought by comparison of the predicted corrosivity data to actual coupon-exposure results, and a good agreement was reported between the algorithm rankings and available experimental data.[16] However, subsequent attempts to enhance the PACER LIME

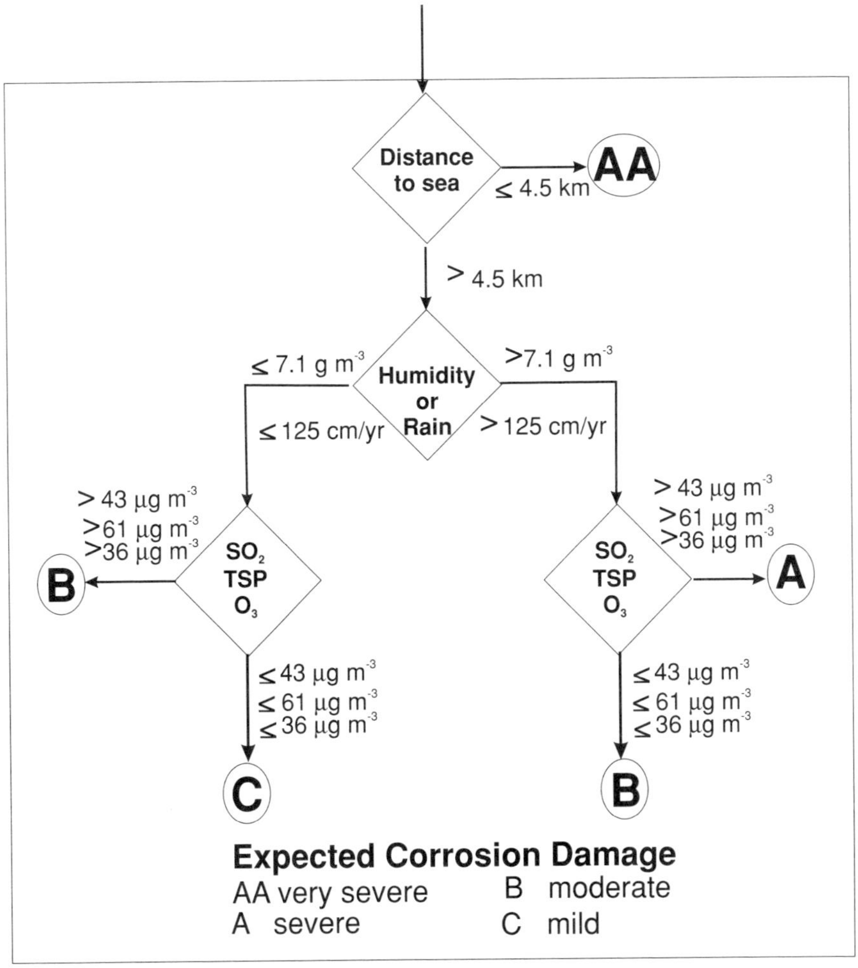

Figure 3.12 The PACER LIME algorithm for determining the corrosion severity for a given location.

algorithm by using the results obtained from broad-based corrosion-testing programs[17] have failed to provide enough differentiation between moderately corrosive environments and were not considered in the final classification scheme adopted.[18]

To remedy some of the deficiencies in this scheme, the Battelle Institute has been tasked to monitor the atmospheric corrosivity of air force and other sites worldwide.[19] The database describing the relative corrosive severity levels of different locations and actual corrosion rates of a variety of metals has now grown to more than 100 sites worldwide. The metals included in that study are three aluminum alloys (A92024, A96061, and A97075), copper, silver, and steel.

The data are collected for metals directly exposed to the outdoor environment in a standard sample-mounting configuration and test package. A typical plastic test rack with its metallic coupons is shown in Figure 3.13 beside a CLIMAT coupon

Figure 3.13 Metal coupons in a plastic test rack exposed at the Kennedy Space Center beach corrosion test site beside a CLIMAT coupon (courtesy of Battelle).

exposed at the KSC beach corrosion test site. Figure 3.14a shows a close-up view of the coupons before exposure. Once exposed to the environment for a given period of time, the corroded metal strips (Figure 3.14b) are sent back to the laboratory for mass-loss measurements following standard methods[20] and further analysis.

These efforts established a correlation between the corrosivity of various air-force-base environments to aluminum and the costs associated with the maintenance of various aircraft. Figure 3.15 summarizes three years of corrosion data compared to maintenance records.

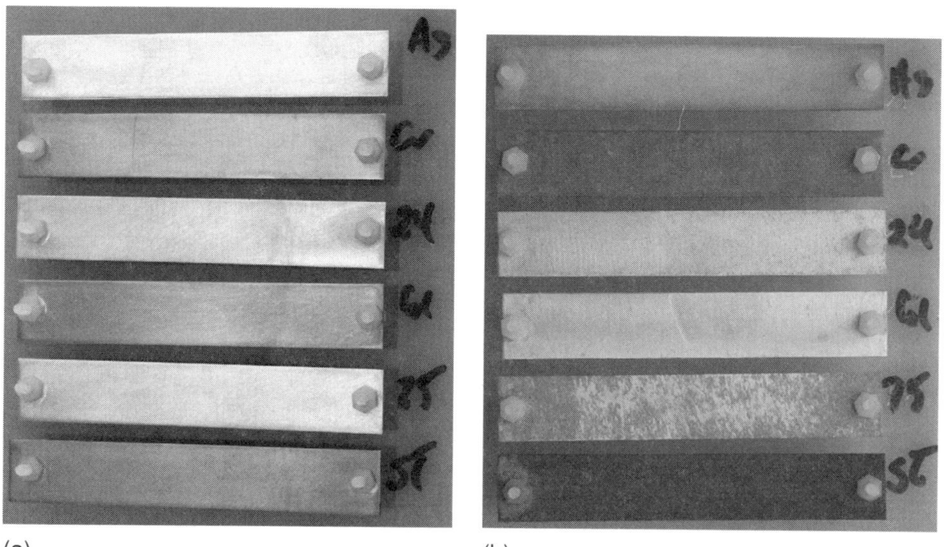

(a) (b)

Figure 3.14 Metal coupons a) before exposure to the environment and b) after a 3-month exposure in a rural environment (courtesy of Battelle).

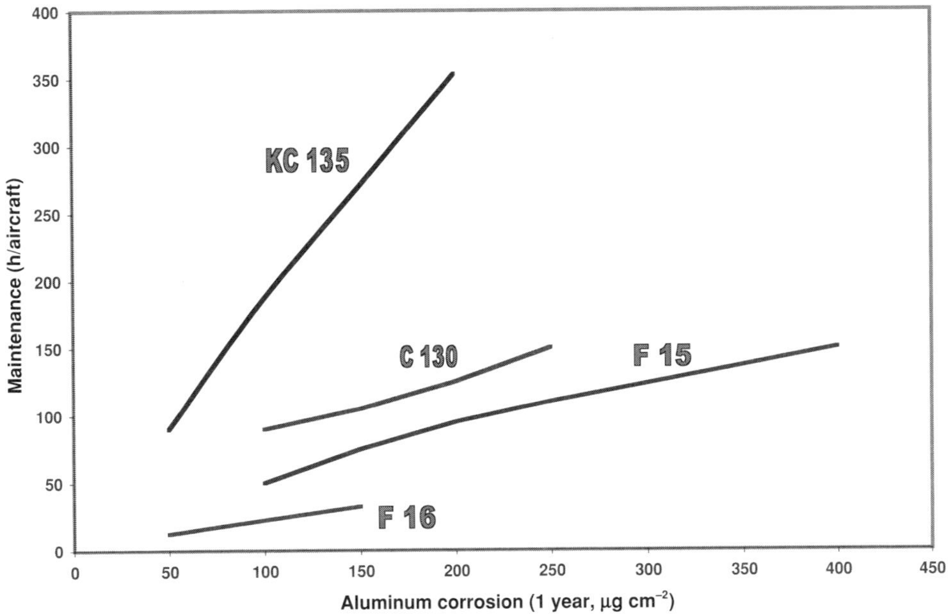

Figure 3.15 Maintenance costs for various aircraft as a function of the corrosivity of air-force-base environments on aluminum.

3.5.2 ISO Classification of Corrosivity of Atmospheres

The International Organization for Standardization (ISO) has implemented a worldwide atmospheric-exposure program known as ISO CORRAG. This program includes exposure of steel, copper, zinc, and aluminum for 1, 2, 4, and 8 years at sites located in 51 sites in 14 nations to generate the necessary data to produce a classification system for predicting atmospheric corrosivity from commonly available weather data.[21] The standard was obviously not intended for extreme-service atmospheres such as those within chemical- or metallurgical-processing facilities or where there is direct contact with salt spray.

The program consisted of six 1-year exposures of flat panels (100 × 150 × 2 mm) and helix specimens beginning every 6 months for 3 years. Triplicate specimens were

Table 3.1 ISO 9223 Corrosion Rates After One Year of Exposure Predicted for Different Corrosivity Classes

Corrosion Category	Steel (g/m^2year)	Copper (g/m^2year)	Aluminum (g/m^2year)	Zinc (g/m^2year)
C1	≤10	≤0.9	negligible	≤0.7
C2	11–200	0.9–5	≤0.6	0.7–5
C3	201–400	5–12	0.6–2	5–15
C4	401–650	12–25	2–5	15–30
C5	651–1500	25–50	5–10	30–60

Table 3.2 ISO 9223 Classification of Sulfur Dioxide and Chloride "Pollution" Levels

Sulfur Dioxide Category	Sulfur Dioxide Deposition Rate ($mg/m^2 day$)	Chloride Category	Chloride Deposition Rate ($mg/m^2 day$)
P0	≤10	S0	≤3
P1	11–35	S1	4–60
P2	36–80	S2	61–300
P3	81–200	S3	301–1500

used for each exposure. The metals selected were a low-carbon steel from a single supplier and commercially pure zinc, copper, and aluminum. These nonferrous metals were obtained from local sources in each of the participating nations. The program was initiated in 1986 and officially closed in 1998. At the conclusion of each exposure, the specimens were retrieved and sent to the laboratory that had done the initial weighing for cleaning and evaluation. Mass-loss values were obtained and converted to corrosion-thickness-loss values.[21] Based on these data, a simple classification scheme of five corrosivity classes was established for each metal (Table 3.1).

The environmental and weather data gathered in this program are based on SO_2, and Cl^- deposition rates (Table 3.2) combined with TOW measurements (Table 3.3) at each site to provide an estimated corrosivity class (Table 3.4). Having determined either a measured or an estimated corrosivity class, it is then possible to provide a prediction of the extent of corrosion damage of steel, copper, zinc, and aluminum.

Experience applying the ISO classification system has shown, however, that certain observations need further clarification. Substantial corrosion has also been measured on specimens exposed at temperatures well below 0°C, which is in direct contradiction to the ISO criterion.[22] It is proposed, on the basis of these results, that the TOW in these environments be estimated as the length of time that RH exceeds 50% and the ambient temperature exceeds –10°C.

3.5.3 Maps of Atmospheric Corrosivity

Maps are powerful tools for communicating information related to geographical landscapes. Corrosivity maps have been drawn to illustrate the corrosion severity of various regions of a country or a continent as it varies from coastal areas to deserts and from rural to heavily industrial locations. In the following example, many years of results

Table 3.3 ISO 9223 Classification of Time of Wetness (TOW)

Wetness Category	TOW (%)	TOW (hours per year)	Examples of Environments
T1	<0.1	<10	Indoor with climatic control
T2	0.1–3	10–250	Indoor without climatic control
T3	3–30	250–2500	Outdoor in dry, cold climates
T4	30–60	2500–5500	Outdoor in other climates
T5	>60	>5500	Damp climates

Table 3.4 ISO 9223 Corrosivity Categories of Atmosphere

TOW	Cl^-	SO_2	Steel	Cu & Zn	Al
T_1	S_0 or S_1	P_1	1	1	1
		P_2	1	1	1
		P_3	1–2	1	1
	S_2	P_1	1	1	2
		P_2	1	1	2
		P_3	1–2	1–2	2–3
	S_3	P_1	1–2	1	2
		P_2	1–2	1–2	2–3
		P_3	2	2	3
T_2	S_0 or S_1	P_1	1	1	1
		P_2	1–2	1–2	1–2
		P_3	2	2	3–4
	S_2	P_1	2	1–2	2–3
		P_2	2–3	2	3–4
		P_3	3	3	4
	S_3	P_1	3–4	3	4
		P_2	3–4	3	4
		P_3	4	3–4	4
T_3	S_0 or S_1	P_1	2–3	3	3
		P_2	3–4	3	3
		P_3	4	3	3–4
	S_2	P_1	3–4	3	3–4
		P_2	3–4	3-4	4
		P_3	4–5	3–4	4–5
	S_3	P_1	4	3–4	4
		P_2	4–5	4	4–5
		P_3	5	4	5
T_4	S_0 or S_1	P_1	3	3	3
		P_2	4	3–4	3–4
		P_3	5	4–5	4–5
	S_2	P_1	4	4	3–4
		P_2	4	4	4
		P_3	5	5	5
	S_3	P_1	5	5	5
		P_2	5	5	5
		P_3	5	5	5
T_5	S_0 or S_1	P_1	3–4	3–4	4
		P_2	4–5	4–5	4–5
		P_3	5	5	5
	S_2	P_1	5	5	5
		P_2	5	5	5
		P_3	5	5	5
	S_3	P_1	5	5	5
		P_2	5	5	5
		P_3	5	5	5

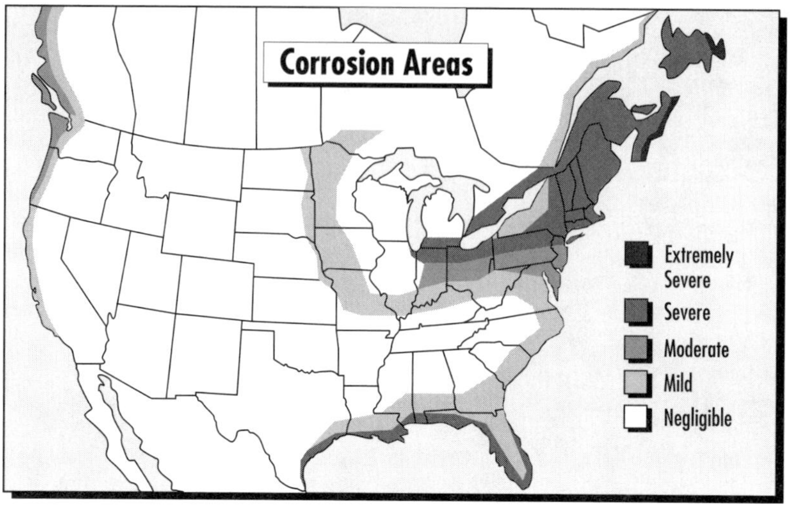

Figure 3.16 Corrosivity map of North America.

obtained by exposing bare-steel coupons attached to different vehicles in the northeastern United States and Canada have been used to produce the corrosivity map shown in Figure 3.16.[23] Periodically removed and measured for metal loss, the data from these coupons served to determine the most severe corrosion localities for captive fleet testing and future survey evaluations.

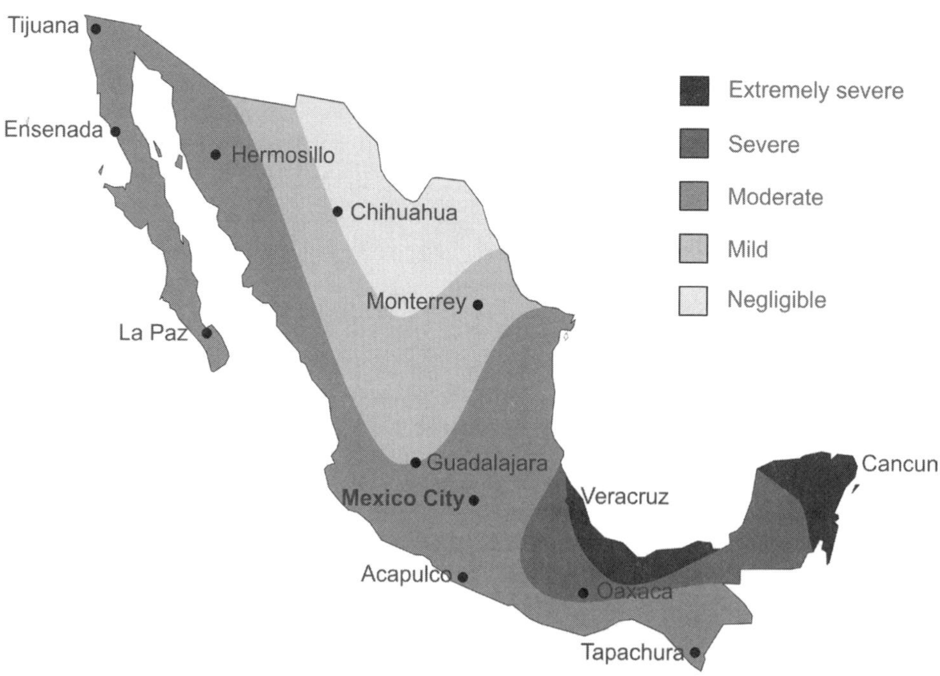

Figure 3.17 Corrosivity map of Mexico.

Figure 3.18 Corrosivity map of Japan.

Similar corrosivity maps have been created from data available in the literature such as the corrosivity map of Mexico and Japan shown respectively in Figure 3.17[24] an in (Figure 3.18).[25]

3.6 ATMOSPHERIC-CORROSION TESTS

For many years, different individuals and organizations have been putting specimens out in the atmosphere in all kinds of tests and shapes. There has been an attempt to standardize some of these tests, as well as the method of reporting the results. When initiating corrosion tests in the atmosphere, these standardized procedures should be consulted and used whenever possible. There tend to be three general types of specimens:

- panels,
- tensile specimens, and
- stress-corrosion-cracking specimens.

Figure 3.19 Atmospheric test rack exposed to a marine West Coast environment.

The panels are usually in the form of sheets 10 cm wide, 15 cm long, and about 0.2 cm thick; the tensile specimens in a machined "dumbbell" shape to fit tensile-testing machines; and the stress-corrosion specimens in jigs to suit the type of stress being considered.

3.6.1 Effect of Conditions of Exposure on Atmospheric Corrosion

In addition to the atmospheric conditions at the test site, the following factors are important aspects for the design and interpretation of atmospheric corrosion tests:

- shape of the specimen;
- direction it faces;
- amount of shelter, drip, or runoff from other specimens;
- elevation;
- shading; and
- unusual contamination.

Panel specimens are usually placed in racks at a 30-degree angle to the horizontal, facing the source of corrosive elements. They are electrically insulated from the racks on which they are mounted and are arranged so that drip from neighboring panels does not contaminate them (Figure 3.19). Cylindrical specimens are mounted horizontally, facing the same direction as the panels. They may be exposed fully to the weather or be partially sheltered, depending upon the requirements of the test. There is a great variety of test fixtures and shape of specimens used for environmental cracking tests. Figures 3.20 and 3.21 show respectively full-exposure and semisheltered test stations equipped with TOW galvanic sensors; temperature and SO_2 levels are recorded with a data logger. Figure 3.22 shows a custom-built test rack designed to be mounted on the full-exposure test station, shown in Figure 3.22, to reveal the atmospheric-corrosion resistance of aluminum and magnesium alloys to crevice corrosion.

Figure 3.20 Outdoor-exposure test station with specimen rack on top and data logger attached below at an army base in northern Australia (courtesy of DSTO Australia).

In most exposure tests, enough specimens are used so that removals may be made after periods of 1, 2, 7, and 20 years or 2, 5, 10, and 20 years. Such programs correct for materials that have changes in corrosion rate after the first one or two years. Very short-term tests usually can be misleading because the condition of the metal surfaces during the first few days of exposure may affect the initial corrosion rate and average weather conditions may not be encountered during the initial exposure period.

Some tests are continued until failure, as in SCC testing. However, in tests of protective coatings, as well as in others, periodic measurements of weight loss, pit depth, and change in tensile strength are made. All of these factors affect the duration of tests, but in most atmospheric exposures, a program covering a time span of 10 years or more is often considered.

Figure 3.21 Semisheltered test station with test coupons, wet-candle apparatus, and data logger attached at an army base in northern Australia (courtesy of DSTO Australia).

In many atmospheric-corrosion tests, the specimens are observed every year for changes in appearance. At predetermined intervals, samples are removed, the weight loss is determined, tensile tests are made, and if pitting is significant, the deepest pit is measured, as well as a number of other pits, to determine average depth. In a few instances, composition of corrosion products is determined.

3.6.2 Measurement of Corrosion

Results of corrosion tests are usually reported as

- loss in weight or reduction in thickness per unit of time;
- change in tensile strength;
- time to fracture (SCC);

Figure 3.22 Aluminum and magnesium test specimens equipped with crevice spacers and mounted on a rack to be exposed at the previous test station (courtesy of DSTO Australia).

- time to perforation;
- time to 10% increase in electrical resistance;
- time to initial rust, 10% rust, or complete rust; or
- in some cases, changes in ductility.

The reduction in thickness or penetration in microns per year (μm/y) or mils per year (mpy) is the most general and useful measurement of corrosion rate. A conversion between these penetration rates and other units is described in Tables 2.2 and 2.3 in Chapter 2.

3.6.3 Accelerated Tests

So far, reference has been made to natural exposures in the atmosphere. Many workers feel that evaluations are needed in times less than those required for on-site tests, and therefore have devised accelerated methods based on a previous determination of the dominant corrosion factors. The preferred practice is to design such tests to represent the most severe conditions for the corrosion mechanism involved.

However, it is useful to consider how realistically corrosion acceleration may be achieved. Raising the temperature can be useful but may cause changes in the form and nature of hydrous gels often important in the initial stages of corrosion.[26] Increasing the concentration or corrosive agents in the salt spray, for example, may not necessarily be appropriate during cyclic testing because even an initially dilute spray will, after a sufficient number of cycles, result in the solubility of ionic species being exceeded.[27]

Generally, corrosion products developed by synthetic environments such as those produced in the ASTM B117 test are substantially different from those produced during natural weathering or even by wet–dry mixed salt-spray tests.[26] For example, corrosion of aluminum or zinc specimens in ASTM B117 primarily produces soluble species such as $AlCl_3$ or $ZnCl_2$ with little corrosion product remaining on their surfaces. Exposures in a wet–dry test, in contrast, cause the formation of corrosion products on those metals more representative of those formed during natural exposure. On aluminum, for example, hydrated alumina containing chloride and amorphous material are produced in both the high-sulfate and high-chloride cyclic salt-spray tests.[26] The reality can be even more complex as illustrated in Table 3.5 where it can be seen that the products found on specimens exposed to real environments often consist of corrosion products mixed with various foreign materials.[28]

3.7 CORROSION BEHAVIOR AND RESISTANCE

The following sections contain some specific information on the atmospheric-corrosion performance of the most popular systems used in normal atmospheric conditions. For additional details on the alloys themselves, the reader should go to Chapter 8 where more details are provided.

Table 3.5 Results of X-Ray Diffraction of Products Found Specimens Exposed to Real Environments

Sample Description	Chemical or Mineral Name	Chemical Formula
Product formed on magnesium during 3-month immersion in tap water	Nesquehonite	$MgCO_3 \cdot 3H_2O$
	Calcium fluosilicate[(A)]	$CaSiF_6$
	Beta silicon carbide	β-SiC
	Sodium sulfide	*Na2S*
	Sodium fluoride	NaF
	Magnesium carbonate chloride hydroxide hydrate	$MgCl_2 \cdot 2MgCO_3 \cdot Mg(OH)_2 \cdot 6H_2O$
	Magnesium pyrophosphate	$Mg_2P_2O_7$
	Anorthoclase	$(Na,K)AlSi_3O_8$
	Alpha cristobalite	SiO_2
	Sodium hydroxide	$NaOH$
	Calcium aluminum oxide sulfate	$Ca_4Al_6O_{12}SO_4$
Substance found on heat exchanger	*Halite*	$NaCl$
Substance found beneath paint on metal surface	*Alpha quartz*	SiO_2
Product formed on auto mobile bumper support during 3-year service	Lepidocrocite	$\gamma-Fe_2O_3 \cdot H_2O$
	Goethite	$Fe_2O_3 \cdot 2H_2O$
Product from conversion unit in marine environment	Zinc ferrite	$ZnO \cdot Fe_2O_3$
	Cobalt ferrite	$CoO \cdot Fe_2O_3$
	Cobaltous ferrite	$CoO \cdot Fe_2O_3$
	Halite	$NaCl$
	Chromic oxide	Cr_2O_3
	Nickel, zinc ferrospinel	$(Ni,Zn)O \cdot Fe_2O_3$
	Sodium fluothorate	$Na_3Th_2F_{11}$
	Embolite	$Ag(Cl,Br)$
	Magnesioferrite	$MgFe_2O_4$
	Beryllium palladium	$BePd$
	Magnetite	Fe_3O_4
	Nickel titanium	NiTi
Product formed on copper during 3-month immersion in tap water	Botallackite	$CuCl_2 \cdot 3Cu(OH)_2 \cdot 3H_2O$
	Ilvaite	$Ca(Fe,Mn,Mg)_2(Fe,Al)(SiO_4)_2OH$
Product from Al-Cu alloy exposed to deep-sea environment	Ammonium copper fluoride dihydrate	$(NH_4)_2 \cdot CuF_4 \cdot 2H_2O$
	Potassium cyanide	KCN
	Chi alumina	Al_2O_3
	Calcium aluminate	$3CaO \cdot Al_2O_3$
	Alpha cadmium iodide	CdI_2
Product from Al-Zn-Mg-Cu alloy exposed to deep-sea environment	Chi alumina	Al_2O_3
	Alpha cadmium iodide	CdI_2
Product from Al-Mn alloy exposed to deep-sea environment	Ammonium copper fluoride dihydrate	$(NH_4)_2CuF_4 \cdot 2H_2O$
	Nobleite	$CaB_6O_{10} \cdot 4H_2O$

(A) Substances shown in italics are not corrosion products of the primary metals or alloys involved in the system.

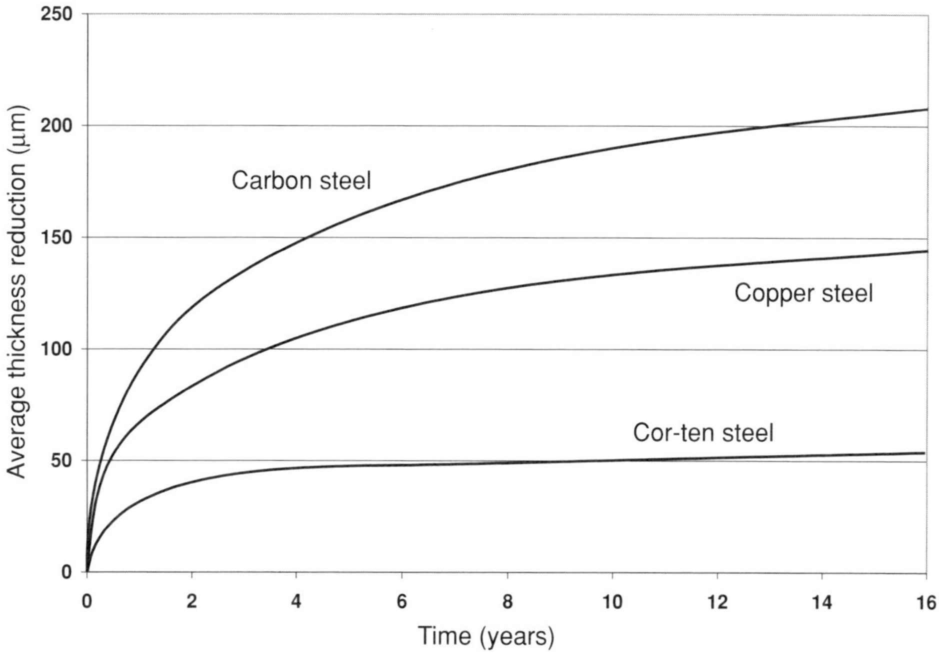

Figure 3.23 Time-corrosion curves of three steels in industrial atmosphere, Kearny, N.J.: (1) ordinary steel; (2) Custeel; and (3) Cor-ten.

3.7.1 Iron and Steel

Iron, in its various forms, is exposed to all kinds of environments. It tends to be highly reactive with most of them because of its natural tendency to form iron oxide. When it does resist corrosion it is due to the formation of a thin film of protective iron oxide on its surface by reaction with oxygen of the air. This film can prevent rusting in air at 99% RH, but a contaminant such as acid rain may destroy the effectiveness of the film and permit continued corrosion. Thicker films of iron oxide may act as protective coatings, and after the first year or so, could reduce the corrosion rate, as shown in Figure 3.23.

Although the corrosion rate of bare steel tends to decrease with time in most cases, the difference in corrosivity of different atmospheres for a particular product is tremendous. The relative corrosivity for open-hearth steel in atmospheres ranging from a desert to the spray zone on an ocean beach is shown in Table 3.6. Similar ranges in corrosivity were determined by the ISO 9223 corrosion rates for steel (Table 3.1). In a few cases, the corrosion rates of ferrous metals have been reported to increase with time, and careful analysis of the exposure conditions generally reveals that an accumulation of contaminating corrosive agents has occurred, thus changing the severity of the exposure.

It is generally conceded that steels containing low amounts of copper are particularly susceptible to severe atmospheric corrosion. In one test over a 3.5-year period in both a marine and an industrial atmosphere, a steel containing 0.01% copper corroded at a rate of 80 μm/y (3.1 mpy), whereas increasing the copper content by a factor of

Table 3.6 Relative Atmospheric Corrosion Susceptibility of Open-Hearth Steel

Location	Susceptibility
Khartoum, Egypt	1
Abisco, North Sweden	3
Singapore, Malaya	9
Daytona Beach, FL (inland)	11
State College, PA	25
South Bend, PA	29
Miraflores, Canal Zone, Panama	31
Kure Beach, NC, 800 ft. from ocean	38
Sandy Hook, NJ	50
Kearny, NJ	52
Vandegrift, PA	56
Pittsburgh, PA	65
Frodingham, U.K.	100
Daytona Beach, FL, near ocean	138
Kure Beach, NC, 80 ft. from ocean	475

five reduced the corrosion rate to only 35 μm/y (1.4 mpy). Further additions of small amounts of nickel and chromium reduced the corrosion rate to approximately 10 μm/y. This is illustrated in Figure 3.24 and Figure 3-25 for Ni and Cr, respectively.

Other tests comparing gray cast iron, malleable iron, and low-alloy steels indicated that their corrosion resistances were approximately the same.

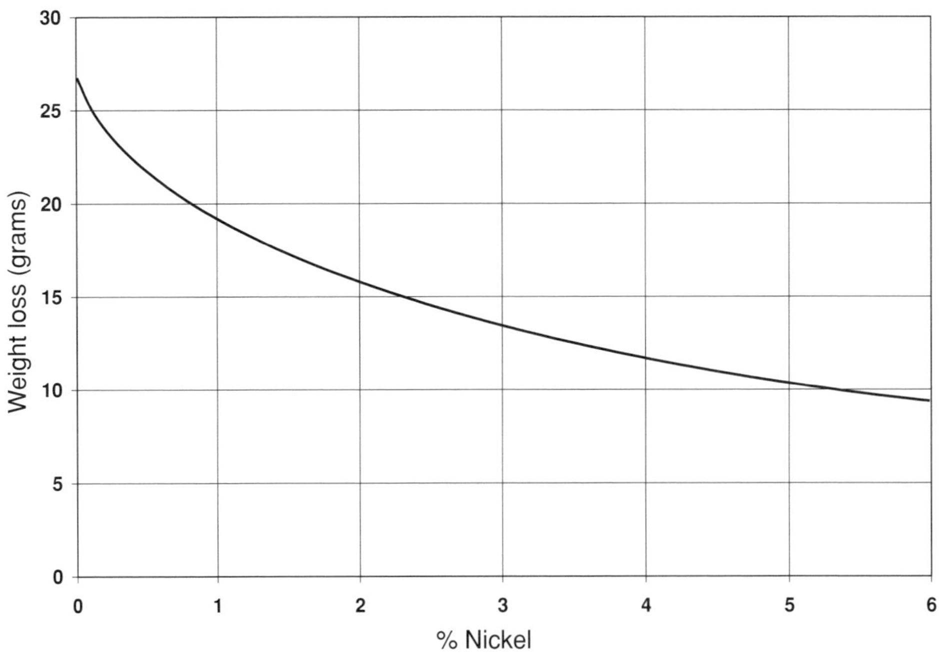

Figure 3.24 Effect of nickel content on the corrosion of steel in a marine atmosphere.

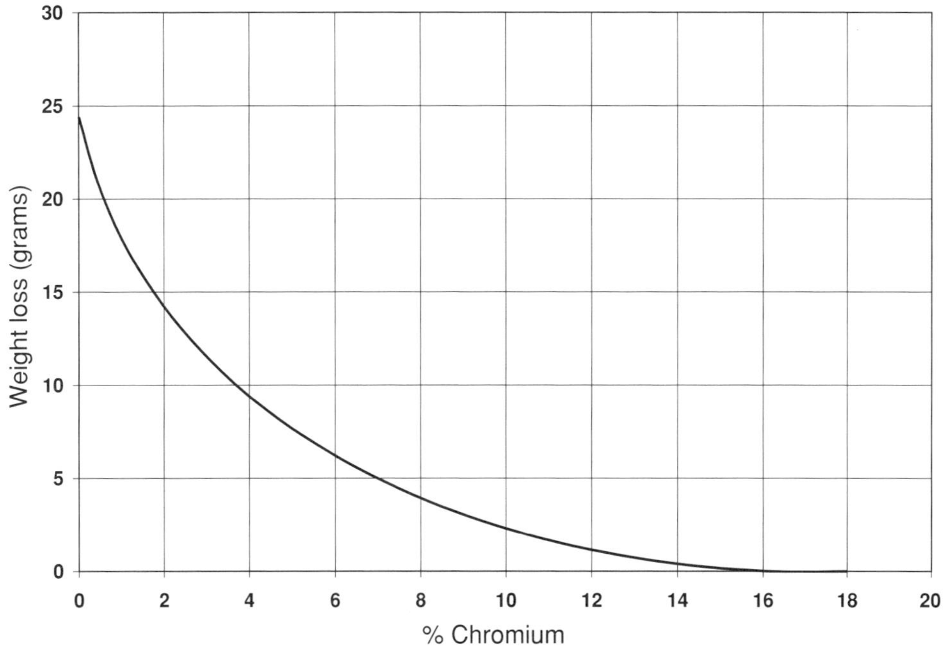

Figure 3.25 Effect of chromium content on the corrosion of steel in 44 months in a marine atmosphere 250 m from the ocean (specimens are 10×15 cm inches).

Plain-cast iron appears to have a corrosion rate about one half that of 0.2% copper steel in a marine atmosphere. One has to be careful in citing such differences to stipulate the composition of the carbon steel because corrosion behavior of carbon steels is influenced so markedly by small variations in copper and phosphorus content. In an industrial atmosphere, a structural carbon steel showed a penetration of about 20 μm (0.8 mil), a copper structural steel about 10 μm (0.4 mil), and a low-alloy steel about 4 μm (0.15 mil) after 5 years of exposure.

As shown in Table 3.1, it is impossible to give a corrosion rate for steel in the atmosphere without specifying the location, composition, and certain other factors. If one can relate exposure conditions to those described in the literature, a fairly good estimate can be made of the probable corrosion behavior of a selected material. However, all aspects of the exposure of the metal surface must be considered. A high-strength, low-alloy (HSLA) steel may show an advantage in corrosion resistance of 12:1 over carbon steel when freely exposed in a mild environment. As the severity or the physical conditions of exposure change, the HSLA steel will show less superiority, until in crevices or the backside of structural forms in a corrosive atmosphere, it will be no better than carbon steel.

Very little needs to be said about the behavior of stainless steels (Types 200 and 300), which contain high percentages of nickel and chromium, except that they can keep their shiny aspect without tarnishing for many decades as illustrated by "The Mudra," which was erected in the busy part of Toronto in 1974 (Figure 3.26). The steels containing only chromium (Type 400) as the principal alloying constituent tend to rust superficially, but

Figure 3.26 Toronto stainless-steel sculpture "the Mudra" by Ted Bieler was completed in 1974 (photo by Tom Skudra, courtesy of the Nickel Institute).

the others are relatively free from surface atmospheric corrosion. Many of them are susceptible to SCC in many common environments.

3.7.2 Copper and Copper Alloys

Copper and its alloys are not exposed to the atmosphere in great quantities when compared with steel. However, this material brings aesthetic value to building construction, in addition to excellent corrosion resistance. The black and then green patina formed on the surface provides an attractive decorative finish while sealing the metal from further corrosion. As a consequence, some copper has been used for roofs, gutters, and as flashings on wood or composition-shingled roofs.

Table 3.7 Corrosion of Copper and Copper Alloys in μm/Year

Material	New York/Newark	State College, PA	Kure Beach, La Jolla
T.P. copper	1.42	0.58	1.45
Phosphor bronze	1.91	0.56	2.51
Red brass	1.80	0.51	0.58
Yellow brass	2.16	0.58	0.46
90-10 Cu Ni	3.78	2.08	2.90
Tin brass	1.73	0.76	0.61
Admiralty metal	2.51	0.51	0.33
Manganese bronze	8.64	0.48	2.02
Si Al bronze	1.27	0.33	0.33
10% Ni Silver	1.96	0.64	0.58
18% Ni Silver	2.01	0.61	0.53
Advance	1.55	0.25	0.46
Be copper	1.75	0.76	1.02
12 different coppers	0.81	0.46	1.35

Extensive tests have been made of the corrosion resistance of copper and its alloys to various atmospheres. Various alloys were exposed in rural, industrial, and marine atmospheres for periods of up to 20 years. From data accumulated in these tests and the calibrations of relative corrosivity of the test sites, a fairly clear picture can be obtained of the corrosion behavior of copper.

In addition to the corrosion penetration rates (μm/y or mpy), one must be mindful of dezincification of brasses and selective attack on some bronzes, as well as SCC illustrated by season-cracking of brass. These types of corrosion contribute to the failure of the material in mechanical respects without significant weight changes or losses in thickness.

Where copper is used as flashing on roofs, corrosion has been encountered at the edge of the shingles as a continuous groove. This effect is more pronounced when the atmosphere contains both chlorides and sulfides, and with wood shingles as compared with roofs of other composition. Tests indicate that all-copper roofs 0.5 mm (20 mils) or more in thickness would last several centuries in an industrial atmosphere.

If the green patina on copper alloys is desired for aesthetic reasons, pretreatment of the surface with appropriate passivating solutions is recommended. If oxidation by sulfur compounds precedes the desired reaction, the surface will present only a dark brown color for many years. The behavior of copper and copper alloys in three typical atmospheres is summarized in Table 3.7.

3.7.3 Nickel and Nickel Alloys

Electrodeposited nickel and electroless nickel are widely used as a protective coating for atmospheric exposure, and some nickel alloys, although, selected for other reasons, are also exposed to atmospheric corrosion. The results shown in Table 3.8 were obtained for several representative alloys. As can be concluded from the data, nickel tends to be

Table 3.8 Corrosion in μm/Year of Nickel and Its Alloys in Various Atmospheres

Material	New York/Newark	State College, PA	Kure Beach, La Jolla
Nickel	0.15	3.66	0.22
	0.24	1.65	0.23
Monel	0.16	1.57	0.17
	3.56	0.81	0.21
Incoloy	0.06	0.02	0.02
Inconel	0.05	0.03	0.02
Nionel	0.03	0.02	0.02

passive in a marine atmosphere. The ratio between the corrosion rate for nickel exposed to the industrial atmosphere and that exposed to rural or marine atmospheres was 28:1.

3.7.4 Aluminum and Its Alloys

Aluminum in its many forms is exceeded only by steel in tonnage directly exposed to the elements. It is produced in the form of wrought products, extrusions, and castings with a large variety of alloying elements to impart desired mechanical properties. The atmospheric corrosion behavior of aluminum products fits into some fairly well-defined patterns that are related to composition.

Although pure aluminum has excellent atmospheric-corrosion resistance and is used extensively as a cladding materials for this very reason, alloys containing copper and silicon as the principal alloying constituents are susceptible and should be used with care.

In a rural atmosphere, the corrosion rate for most alloys is approximately 0.06 μm/y (2.5 microinches per year), with those containing large amounts of copper about double this low rate. Changes in tensile strength because of corrosion vary from 0 to less than 1% for sheet material.

In a marine environment, the differences between alloys appear as a 10-fold increase, from about 0.6 μm/y (25 microinches per year) for the less corrosion-resistant materials to about 0.7 μm (10 microinches) per year for the better materials. Pitting also is about 10 times greater in marine atmospheres. Corrosion can be much greater in a severe industrial atmosphere than in the marine atmosphere.

Some aluminum alloys develop severe pitting and a voluminous white corrosion product under some exposure conditions in a marine atmosphere. Aluminum roofs have been known to corrode severely at the overlaps. Some aluminum alloys also can be attacked intergranularly when exposed after certain metallurgical treatments (cold working or precipitation hardening). General intergranular attack or exfoliation can then occur (Chapter 9). The attack tends to start at sheared edges or punched holes but is not restricted to these areas. Aircraft manufacturers, in particular, must guard against this type of corrosion.

In designing aluminum equipment, care must be exercised to avoid dissimilar metal couples and the attendant galvanic corrosion. Copper and rusty steel are particularly bad in contact with aluminum. Because of the passive film on stainless steel, it can be used in contact with aluminum in the atmosphere with little expectation of accelerated corrosion, despite the differences in potential. In addition, designers should be aware of the possibility that some aluminum alloys may be sensitized to intergranular corrosion by heat treatment.

As would be expected, constant exposure to moisture with a limited supply of oxygen to the aluminum surface leads to the rapid corrosion of any aluminum apparatus or equipment component. This is due to the highly reactive nature of aluminum that leads to formation of oxides or hydroxides. In the presence of oxygen, a protective aluminum oxide film develops on any aluminum surface. This oxide film is substantially unreactive with the normal constituents of the atmosphere. If the film is removed by mechanical or chemical means and the aluminum exposed to water, a rapid reaction sets in and large quantities of the aluminum are converted to the hydroxide and subsequently to the oxide. Most of the 1000, 3000, and 6000 series of wrought alloys, magnesium, and the silicon/magnesium casting alloys are relatively immune to SCC in normal atmospheres. However, the other alloys may be susceptible under some conditions. This subject should be investigated carefully by the designer or user if stresses are high and the atmosphere is corrosive.

The usual corrosion behavior in the atmosphere involves pitting and roughening of the surface with a fairly large decrease in the corrosion rate after the first 1 to 3 years of exposure. In a marine atmosphere, the maximum pit depth might reach 150 μm (6 mils) within 2 years, yet not exceed 200 μm (8 mils) in 20 years. In an industrial atmosphere, initial pits might reach 100 μm (4 mils) in the first period and not exceed 125 μm (5 mils) at the end of 20 years. In terms of average corrosion rate, the initial rate would be about 0.8 μm/y (0.025 mpy), dropping to less than 0.3 μm/y (0.01 mpy) by 7 years.

Severe pitting has been encountered where aluminum surfaces were contaminated by either alkaline dust or coral dust containing chlorides, followed by condensation. On some of the South Pacific islands, dust collected from the surfaces of sheltered structures, such as those inside aircraft wings, contained 67% chloride by weight.

In the design of aluminum structures, the usual precautions of avoiding crevices or pockets and coupling with dissimilar metals must be observed. An example of this is in the overlapping encountered between aluminum roofs and siding. In some marine or industrial atmospheres, the aluminum perforated at the laps within a few months. In addition, stress concentrations should be avoided, such as those found in some riveted structures, in the vicinity of welds, and at notches or inside corners. Where it is impractical to avoid dissimilar metals, the aluminum should be electrically insulated from the more noble metal by means of washers, sleeves, and so forth. In some instances, covering the noble metal with an organic finish is sufficient to greatly reduce galvanic-coupling corrosion.

To illustrate the problems encountered in designing aluminum structures, an amusing and fairly well-known yet illustrative story comes to mind. A new processing plant was built with an uninsulated aluminum roof attached directly to steel support frames. This was known to have worked well in several instances. However, the high RH in the plant caused condensation on the cool roof during rainstorms, evenings, and cold spells in the winter.

Problem 1: Water dripped frequently, causing considerable annoyance to people working in the plant, and material was often damaged.

Partial Solution 1: Insulation of the cold roof seemed to be the obvious answer, so insulating material was sprayed on all interior surfaces. To ensure good adhesion of the insulation to the aluminum, an alkaline binder was used.

Problem 2: Moisture collected slowly and was retained in large quantities by the insulation. Now, chunks of dripping insulation "rained" down instead of water drops.

Partial Solution 2: The insulation was removed.

Problem 3: Galvanic corrosion aggravated by the alkaline environment resulted in numerous perforations of the aluminum roof along areas adjacent to the contacts with the steel supports.

Final Solution: The aluminum roof was not replaced, but the aluminum was electrically insulated from the steel supports. Nonalkaline insulation was applied and waterproofed by spraying a suitable compound to the exposed surfaces. This properly installed roof is now giving good service.

3.7.5 Zinc and Its Alloys

Zinc is exposed to the atmosphere in the form of galvanized sheet, as in flashings on roofs, as die castings, and as coatings on steel, either hot dipped or electroplated. Studies of the corrosion of zinc in the New York City atmosphere indicated that there was less than a 10% difference between the corrosion rate of galvanized iron, zinc die castings, and three grades of rolled zinc. It was also found that the corrosion rate tended to be a linear function with time. In some instances, where the rate changed after a period of time, it was concluded that the amount of contamination in the atmosphere had changed.

The general behavior of zinc metal and zinc coatings is described in the ISO tables presented earlier. Note the particularly low rates of attack on zinc as compared with steel in marine exposures where chloride deposition is important. Such excellent resistance is acquired by the hard, dense, protective products of corrosion in a chloride atmosphere. Similar results cannot be obtained in a sulfurous atmosphere where the products are soft, voluminous, and nonprotective.

Zinc-base die castings usually are not exposed boldly to the outside atmosphere without a protective coating. When breaks or pits occur in coatings such as nickel and chromium, the corrosion of the die casting may be accelerated because of the dissimilar metal contact and thus may give a false impression of the corrodibility of zinc. Many small parts of machinery, household appliances, and hardware are made of zinc-base die castings or "white metal" and are exposed to an indoor atmosphere where their corrosion behavior is very good. (White metal is typically 93 to 96% Zn, 4% Al, 0.05% Mg, and sometimes 1 to 3% Cu.) In these cases where severe corrosion is encountered in this relatively mild atmosphere, the cause may be improper alloy selection or the use of material containing too high a percentage of impurities.

Galvanized steel is the most important application of zinc. Worldwide, the use of zinc for galvanizing results in an annual consumption of more than 3 million tons of zinc, constituting nearly one-half of the world zinc production.[29] Most of these

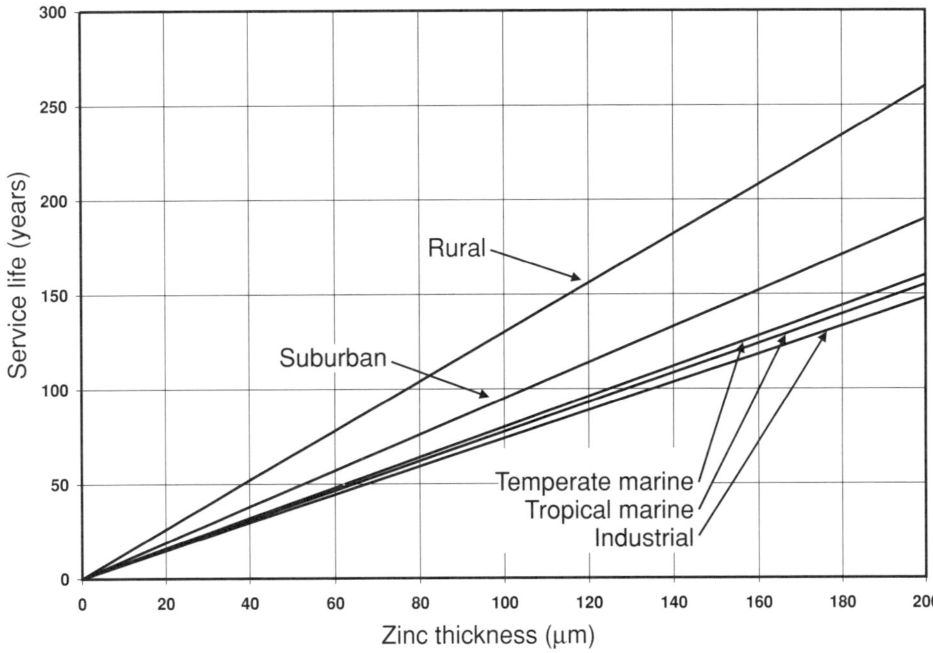

Figure 3.27 Service life of hot-dip-galvanized coatings as a function of zinc thickness and specific environments.

are hot-dipped galvanized coatings containing a small amount of aluminum. Thickness of electroplated coatings is considerably lower than those applied by the hot-dip process.

Galvanizing produces a zinc coating on the steel surface and is one of the most effective methods for corrosion protection of steel. This is attributed to the excellent corrosion resistance of zinc coatings, particularly in atmospheric environments.

Corrosion rates for zinc, like other metals, can vary by as much as two orders of magnitude in atmospheric environments depending on the specific environmental conditions. Therefore, it is important to know the specific corrosion rate in a given application environment to effectively use zinc-coated steels in outdoor structures.

The most commonly used method for corrosion life estimation of galvanized steels has been the use of generalized values for the different types of atmospheres (Figure 3.27).[30] This method uses a generalized value to represent the corrosion rates for five predetermined atmospheric environments as a function of zinc-coating thickness. Service life in Figure 3.27 is defined as the time to 5% rusting of the steel surface. It can be used to estimate the service life of a given coating thickness or to specify a coating for a given environment.

The method is applicable to zinc-coated steel produced by batch or continuous galvanizing. It is applicable to hot-dip, electrogalvanized, and thermal sprayed coatings. However, it does not apply to coatings containing more than 1% of alloying elements. The method assumes that the galvanized product is free of significant defects that could accelerate corrosion. Additionally the service-life prediction

Figure 3.28 Galvanized bolting assembly after 10 years of exposure to a deicing salt environment.

does not consider issues of water entrapment that can create severe crevice chemistry as explained in Chapter 9 and illustrated here for a lamppost bolting attachment (Figure 3.28).

3.7.6 Miscellaneous Alloys

Many other alloys are exposed to the atmosphere under a variety of circumstances. The resulting corrosion problems tend to be specific. Lead and tin alloys tend to be inert in the atmosphere unless some specific type of contamination exists. As far as exposure to the atmosphere is concerned, tin appears mostly as a coating on steel containers where it can accelerate the corrosion of the more anodic steel by pitting. Lead is sometimes used on roofs, but more often is found indoors in chemical plants.

Because of their inertness, silver and gold are used as alloys to confer corrosion resistance to more reactive metals. Silver reacts with sulfur compounds in the atmosphere to produce heavy coatings of black silver sulfide. This corrosion product is soft and rubs off of electrical contacts.

Titanium alloys have very good corrosion resistance in the atmosphere, as do the iron/nickel/chromium alloys. There are many alloys that hold a specific place in the metal family, such as pewter, speculum, constantan, and the like, which possess certain atmospheric corrosion-resistance properties, but are used so infrequently that they do not warrant general discussion.

3.7.7 Metallic Coatings

Any discussion of atmospheric corrosion should include consideration of the corrosion behavior of various metallic protective and decorative coatings. These are usually divided into two general categories:

- coatings that confer protection to the basic metal and
- coatings that are used for decorative purposes.

As protective coatings, a distinction must be made between those such as zinc, cadmium, and aluminum on steel that protect by sacrificial behavior, and those that must provide a substantially continuous protective envelope around the protected metal. In the latter category are metals such as nickel, tin, silver, brass, and chromium.

Sacrificial metals corrode as coatings in much the same manner as they do as solid metals until the base or protected metal is exposed at pores or bared areas. The galvanic couple effect then begins to accelerate the corrosion of the protective coating. This galvanic couple effect tends to protect the base metal at the pores or bared spots. General corrosion of the sacrificial type of coating tends to follow a linear function that is peculiar to the particular site.

This reaction is in contrast to what happens in the case of the more noble protective coatings that act as an envelope and resist the atmosphere because of their inertness, passivity, or protective films. As soon as a pore or bare spot appears, the corrosion of the base metal is accelerated. Die castings of a metal such as zinc covered with a copper/nickel/chromium coating will suffer severe corrosion at large pores or discontinuities in the coating.

Although most metal coatings are applied by electroplating, some are produced by flame or plasma spraying, hot dipping, electrostatic sputtering, or vapor deposition. The corrosion rate of a metal coating is largely independent of the method of application, except where impurities play a role. Powders of various metals have been attached to the surface of a base metal by either organic or inorganic binders. Where these metal particles are in mutual contact, and in contact with the base metal, they can perform much like a metallic coating (Chapter 10).

In noble types of coating, porosity is very important and tends to decrease rapidly as the coating thickness increases beyond 13 μm (0.5 mil). No clear-cut information is available on the factors that influence porosity in electrodeposited coatings.

3.7.8 Polymeric Materials

Essentially all polymers freely exposed to the elements will change in some manner. The active rays of the sun become potent agents of change in the organic materials. Further polymerization of the resin can occur to produce embrittlement. Other types of new bonding can be triggered to make polymers more crystalline. Any volatile component of the material, such as a plasticizer, can be evaporated. The polymer chains may be simply oxidized and broken up to destroy the product. Oxygen, ozone, and moisture act with the sunlight to degrade the plastics.

The external evidence of attack may be blushing (loss of gloss), chalking, or change in color of the product. This is often observed on the epoxy and polyester polymers when they have been boldly exposed to the environment. However, only mechanical tests will reveal the extent of degradation of either thermoplastic or thermosetting resins.

The effect of high atmospheric temperatures or heating from direct exposure to the sun can be particularly severe on thermoplastic polymers. Creep or distension of the polyvinyl chloride and polyethylene plastics will occur readily unless provision is made to prevent overheating or stressing of the materials. Certain other thermoplastics can be dimensionally stable under normal atmospheric temperatures. The strength of these thermosetting resins is not noticeably changed.

Polymeric materials should be thoroughly tested if they are to be exposed freely in the atmosphere. ASTM Recommended Practice D1435 describes the appropriate conditions for such a test exposure and suggests tests that might be used to evaluate changes in the materials. Changes in mechanical and physical properties of the polymers are determined for definitive results. Weight gains or losses can be of interest but do not provide substantive results.

Accelerated cabinet testing of the materials can be performed with essentially the same validity as when testing metallic materials. Poor materials can be eliminated, but extrapolation of the data to forecast the life of plastic parts in the atmosphere should not be attempted. ASTM D1499, D 2565, and G26 may be used to ensure the controlled cabinet testing of the plastics. ASTM D750 should be used for the evaluation of the elastomers.

The preventive measures used to ensure the integrity of polymeric materials are similar to those for metals, that is, the exterior should be covered. The exterior surfaces are painted or metal plated to provide a barrier between the material and the atmosphere. Alloying of the basic material with small amounts of other resins can often upgrade the stability of the material.

References

1. D. Miller, "Corrosion Control on Aging Aircraft: What Is Being Done?," MP 29 (1990): pp. 10–11.
2. V.S. Agarwala, P.K. Bhagat, G. L. Hardy, "Corrosion Detection and Monitoring of Aircraft Structures: An Overview," October 5–6, 1999 in AGARD Conference Proceedings 565, Corrosion Detection and Management of Advanced Airframe Materials, held 1994 (AGARD: Neuilly sloc Sein, France), pp. 19-1 to 19-6.
3. P.R. Roberge, Handbook of Corrosion Engineering (New York, NY: McGraw-Hill, 2000).
4. S.C. Chung, A.S. Lin, J.R. Chang, H.C. Shih, "EXAFS Study of Atmospheric Corrosion Products on Zinc at the Initial Stage," Corrosion Science 42 (2000): pp. 1599–610.
5. J.R. Duncan, J.A. Ballance, "Marine Salts' Contribution to Atmospheric Corrosion," in Degradation of Metals in the Atmosphere, eds. S.W. Dean, T.S. Lee (Philadelphia, PA: ASTM, 1988), pp. 316–326.
6. G.M. Hidy, Aerosols: An Industrial and Environmental Science (Orlando, FL: Academic Press, 1984).
7. S. Feliu, M. Morcillo, B. Chico, "Effect of Distance from Sea on Atmospheric Corrosion Rate," Corrosion 55 (1991): pp. 883–891.
8. M. Morcillo, B. Chico, L. Mariaca, E. Otero, "Salinity in Marine Atmospheric Corrosion: Its Dependence on the Wind Regime Existing in the Site," Corrosion Science 42 (2000): pp. 91–104.
9. P.J. Sereda, S.G. Croll, H.F. Slade, "Measurement of the Time-of-Wetness by Moisture Sensors and Their Calibration," in Atmospheric Corrosion of Metals, eds. S.W. Dean, E.C. Rhea (Philadelphia, PA: ASTM, 1982), p. 48.
10. ASTM G84-89, "Standard Practice for Measurement of Time-of-Wetness on Surfaces Exposed to Wetting Conditions as in Atmospheric Corrosion Testing" (West Conshohocken, PA: ASTM, 1999).

11. ASTM G140-02, "Standard Test Method for Determining Atmospheric Chloride Deposition Rate by Wet Candle Method" (Philadelphia, PA: ASTM, 2002).
12. D.P. Doyle, T.E. Wright, "Rapid Method for Determining Atmospheric Corrosivity and Corrosion Resistance," in Atmospheric Corrosion, ed. W.H. Ailor (New York: Wiley, 1982), pp. 227–243.
13. Standard Practice for Conducting Wire-on-Bolt Test for Atmospheric Galvanic Corrosion (West Conshohocken, PA: ASTM, 1999).
14. S. Coburn, "Atmospheric Corrosion," in Metals Handbook, vol. 1, 9th ed., Properties and Selection, Carbon Steels (Metals Park, OH: ASM, 1978), p. 720.
15. R.D. Klassen, P.R. Roberge, D.R. Lenard, G.N. Blenkinsop, Corrosivity Patterns near Sources of Salt Aerosols (West Conshohocken, PA: ASTM, 2002), pp. 19–33.
16. R. Summitt, F.T. Fink, "PACER LIME: An Environmental Corrosion Severity Classification System," AFWAL-TR-80-4102, USAF, June 1980.
17. S.W. Dean, "Corrosion Testing of Metals Under Natural Atmospheric Conditions," in Corrosion Testing and Evaluation: Silver Anniversary Volume, eds. R. Baboian, S.W. Dean (Philadelphia, PA: ASTM, 1990), pp. 163–176.
18. "Final Report for Air Force Base Environmental Corrosion Severity Ranking," Contract No. F09603-95-D-0053, NCI Information Systems, Fairborn, OH, November 1, 1998.
19. W.H. Abbott, R. Kinzie, "Corrosion Monitoring in Air Force Operating Environments," 2003 Tri-Service Corrosion Conference, held November 17, 2003 (Office of Naval Research Arlington, VA).
20. ASTM G1-03, "Standard Practice for Preparing, Cleaning, and Evaluating Corrosion Test Specimens" (West Conshohocken, PA: ASTM, 2003).
21. C. Leygraf, T.E. Graedel, Atmospheric Corrosion (New York, NY: Wiley 2000).
22. G. King, W. Ganther, J. Hughes, P. Grigioni, A. Pellegrini, "Studies in Antarctica Help to Better Define the Temperature Criterion for Atmospheric Corrosion" NACE International Northern Area Region Conference, held February 26–28, 2001 (Houston, TX: NACE, 2001).
23. R. F. Steinmayer, "Land Vehicle Management," AGARD Lecture Series No. 141 (Neuilly-sur-Seine, France: NATO, 1985).
24. D.C. Cook, A.C. Van Orden, J. Reyes, et al. "Atmospheric Corrosion in Marine Environments Along the Gulf of México," in Marine Corrosion in Tropical Environments, eds. S.W. Dean, G. Hernandez-Duque Delgadillo, J.B. Bushman (Philadelphia, PA: ASTM, 2000), p. 75–97.
25. K. Yamamota, K. Ima, Q & A on Corrosion and Protection for Metals (Tokyo, Japan: Japanese Society of Corrosion and Protection, 1988).
26. S.B. Lyon, G.E. Thompson, J.B. Johnson, "Materials Evaluation Using Wet–Dry Mixed Salt Spray Tests," in New Methods for Corrosion Testing of Aluminum Alloys, eds. V.S. Agarwala, G.M. Ugiansky (Philadelphia, PA: ASTM, 1992), pp. 20–31.
27. R.S. Treseder, "An Engineering View of Laboratory Corrosion Tests," in Laboratory Corrosion Tests and Standards, STP 866, eds. G.S. Haynes, R. Baboian, (Philadelphia, PA: ASTM, 1985), pp. 5–13.
28. R.S. Mapes, W.W. Berkey, "X-Ray Diffraction Methods for the Analysis of Corrosion Products," in Handbook on Corrosion Testing and Evaluation, ed. W.H. Ailor (New York, NY: Wiley, 1971), pp. 697–730.
29. "Zinc and Lead Statistics," International Lead Zinc Study Group, London, UK, 1996.
30. "Galvanizing for Corrosion Protection: A Special Guide," American Galvanizers Association, Centennial, CO, 1990.

4

Chapter

Corrosion by Water and Steam

4.1 INTRODUCTION

A major use of water in industry is the transfer of heat and the production of steam. There is extensive use of cooling water in almost every manufacturing process, in commercial air conditioning, and even a substantial percentage in domestic air conditioning. Water is used in the passive sense for potable and fire-control purposes. Fossil- and nuclear-fuel steam plants are encountered in the heating and power-generating fields.

According to the American Waterworks Association (AWWA) industry database, there is approximately 1,483,000 km of municipal water piping in the United States. This number is not exact, because most water utilities do not have complete records of their piping system. The sewer system consists of approximately 16,400 publicly owned treatment facilities that release some 155 million m^3 of waste water per day (1995). The total annual direct cost of corrosion for the nation's drinking water and sewer systems was estimated at $36 billion. The cost of replacing aging infrastructure, the cost of unaccounted water lost through leaks, the cost of corrosion inhibitors, the cost of internal mortar linings, and the cost of external coatings and CP all contributed to this cost.[1]

Water can be corrosive to most metals. Pure water, without dissolved gases (e.g., oxygen, carbon dioxide, and sulfur dioxide) does not cause undue corrosion attack on most metals and alloys at temperatures up to at least the boiling point of water. Even at temperatures of about 450°C, almost all of the common structural metals, except magnesium and aluminum, possess adequate corrosion resistance to high-purity water and steam. In summary, the factors influencing the corrosion of materials in water systems are the following:

1. the physical configuration of the system,
2. the chemistry of the water (hardness salts, chlorides, and dissolved gases being the most important),

3. the presence of solids in the water,
4. the flow rate,
5. the temperature of the water, and
6. the presence of bacteria.

This chapter discusses the corrosion problems associated with natural and treated waters at temperatures ranging from ambient up to about 540 to 600°C, the means of preventing or slowing down corrosion, and the proper choice of materials for each environment.

4.2 WATER CONSTITUENTS

The concentrations of various substances in water in dissolved, colloidal, or suspended form are typically low but vary considerably. A hardness value of up to 400 ppm of calcium carbonate, for example, is sometimes tolerated in public supplies, whereas 1 ppm of dissolved iron would be unacceptable. In treated water for high-pressure boilers or where radiation effects are important, as in some nuclear reactors, impurities are measured in very small units such as parts per billion (ppb). Water analysis for drinking water supplies is concerned mainly with pollution and bacteriological tests. For industrial supplies a mineral analysis is of more interest. The important constituents can be classified as follows[2]:

- dissolved gases (oxygen, nitrogen, carbon dioxide, ammonia, sulfurous gases);
- mineral constituents, including hardness salts, sodium salts (chloride, sulphate, nitrate, bicarbonate, etc.), salts of heavy metals, and silica;
- organic matter, including that of both animal and vegetable origin, oil, trade waste (including agricultural) constituents, and synthetic detergents; and
- microbiological forms, including various types of algae and slime-forming bacteria.

The pH of natural waters is rarely outside the fairly narrow range of 4.5 to 8.5. High values, at which corrosion of steel may be suppressed, and low values, at which gaseous hydrogen evolution occurs, are not often found in natural waters. Copper is affected to a marked extent by pH value. In acidic waters, slight corrosion occurs and the small amount of copper in solution causes green staining of fabrics and sanitary ware. In addition redeposition of copper on aluminum or galvanized surfaces sets up corrosion cells, resulting in severe pitting of the metals.

From a corrosion standpoint, the most significant contaminant is dissolved oxygen (DO) from ambient air. This is illustrated in Figure 4.1.[3] As described in Chapter 2, oxygen is a cathodic depolarizer that reacts with and removes hydrogen from the cathode during electrochemical corrosion, thereby permitting corrosion attack to continue.

The effect of oxygen on corrosion with increasing temperature is also shown in Figure 4.2, indicating the results in a closed versus an open container (the latter permitting natural deaeration by ebullition).[3] In a closed vessel, corrosion continues to increase with temperature, hence the requirement for removing DO from hot-water systems and boilers.

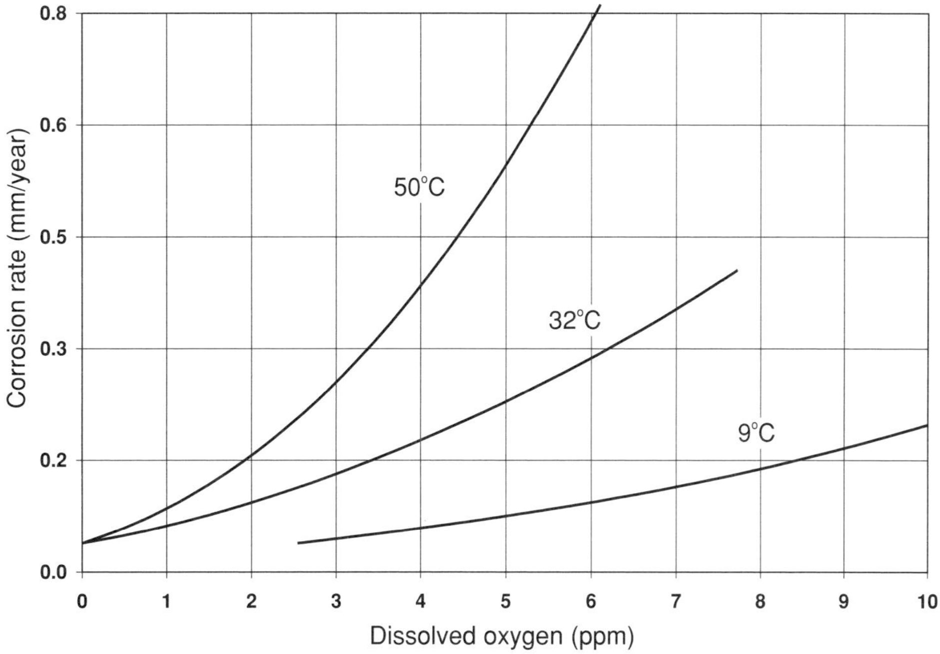

Figure 4.1 Effect of oxygen concentration on the corrosion of low-carbon steel in tap water at different temperatures.

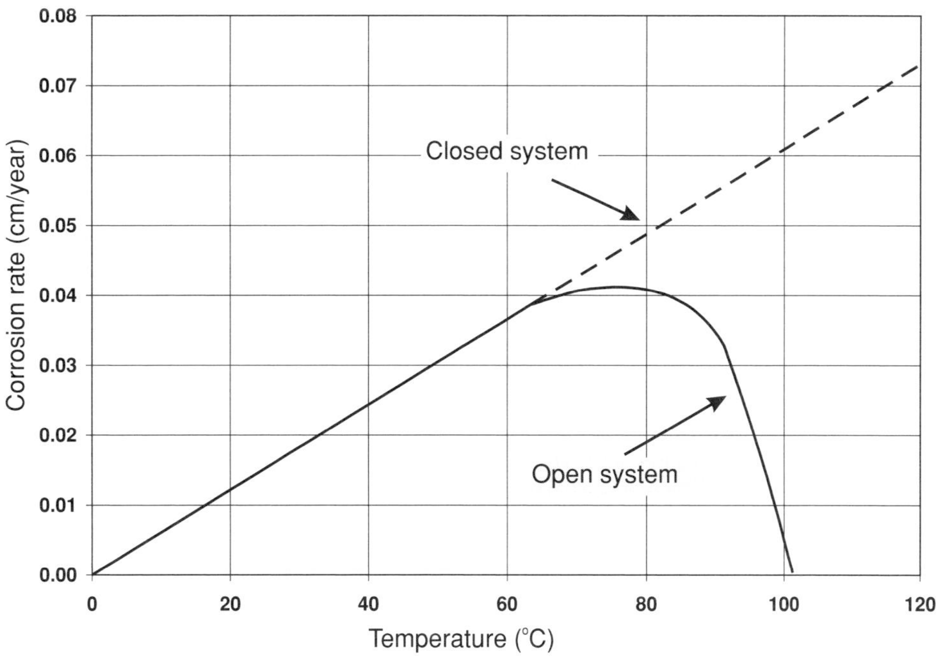

Figure 4.2 Effect of oxygen on corrosion of steel.

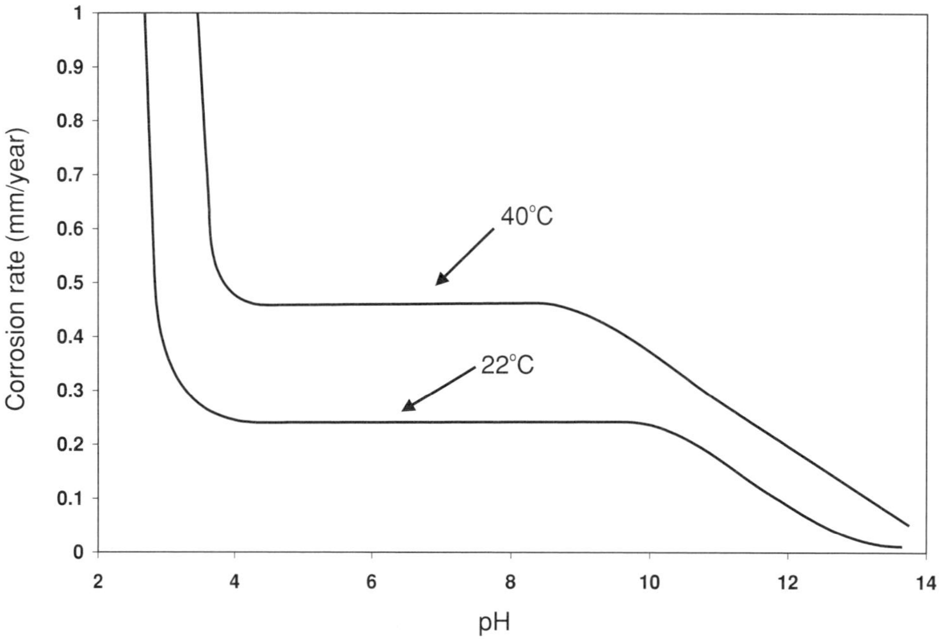

Figure 4.3 Corrosion of steel in water containing 5 ppm of dissolved oxygen at two different temperatures as a function of the water pH.

Other constituents that contribute to corrosion are chlorides and carbon dioxide, possibly calcium, as is explained later, and sulfides or ammonia from industrial or natural sources. Of course, many other man-made contaminants can be found in local areas if industries are permitted to discharge their waste products into water resources. As with other chemical reactions, corrosion increases with elevated temperature, unless stifled by insoluble scales, the removal of corrosive gases, or the addition of corrosion inhibitors.

Scales precipitated onto metal surfaces can provide excellent protection of the substrate or may accentuate pitting at pores, cracks, or other voids in the film. If the film attains any significant thickness, the loss of heat transfer through the metal and deposited scale can be a problem in certain applications. Thus, the development of scales on metal surfaces is an important consideration when using metals in waters.

The effect of oxygen and pH on the corrosion rate of steel at two temperatures is shown in Figure 4.3.[4] In a broad range of about pH 5 to 9, the corrosion rate can be expressed simply in terms of the amount of DO present (e.g., μm/y per ml, DO per liter of water). At about pH 4.5, acid corrosion is initiated, overwhelming the oxygen control. At about pH 9.5 and above, deposition of insoluble ferric hydroxide tends to stifle the corrosion attack.

4.3 TYPES OF WATER

In common terms, water is frequently described either in terms of its nature, usage, or origin. The implications in these descriptions range from being highly specific to so general as to be nondefinitive. Ground waters originate in subterranean locations (e.g., wells), whereas surface waters comprise the lakes, rivers, and seas.

4.3.1 Fresh Water

Fresh water may come from either a surface or ground source and typically contains less than 1% sodium chloride. It may be either "hard" or "soft," that is, either rich in calcium and magnesium salts and thus sometimes forming insoluble curds with ordinary soap or not. Actually, there are gradations of hardness, which can be estimated from the Langelier or Ryznar indexes or accurately determined by titration with standardized chelating-agent solutions (e.g., versenates).

4.3.2 Brackish Water

Brackish water contains between 1 and 2.5% sodium chloride, either from natural sources (e.g., salt deposits) around otherwise fresh water or by dilution of seawater (e.g., tidal rivers). Brackish water differs from open seawater in certain other respects. The biological activity, for example, can be significantly modified by higher concentrations of nutrients. Fouling is also likely to be more severe as a consequence of the greater availability of nutrients.

Within harbors, bays, and other estuaries, marked differences can exist in the amount and type of fouling present in the water. The main environmental factors responsible, singly or in combination, for these differences are the salinity, the degree of pollution, and the prevalence of silt. Moreover, the influence of these factors can be very specific to the type of organism involved. Apart from differences that can develop between different parts of the same estuary, there can also be differences between fouling in enclosed waters and on the open coast. In this respect the extent of offshore coastal fouling is strongly determined by the accessibility of a natural source of infection. Local currents, average temperature, seasonal effects, depth, and penetration of light are operative factors. The presence of pollutants can also be quite important and highly variable in coastal areas.

4.3.3 Seawater

Seawater typically contains about 3.5% sodium chloride, although the salinity may be weakened in some areas by dilution with fresh water or concentrated by solar evaporation in others. Seawater is normally more corrosive than fresh water because of the higher conductivity and the penetrating power of the chloride ion through surface films on a metal. The rate of corrosion is controlled by the chloride content, oxygen availability, and the temperature.

The 3.5% salt content of seawater produces the most corrosive chloride salt solution that can be obtained (Figure 4.4).[3] The combination of high conductivity and oxygen solubility is at a maximum at this point (oxygen solubility is reduced in more concentrated salt solutions). The corrosion of numerous metals in a wide range of saline waters is reported in Table 4.1.

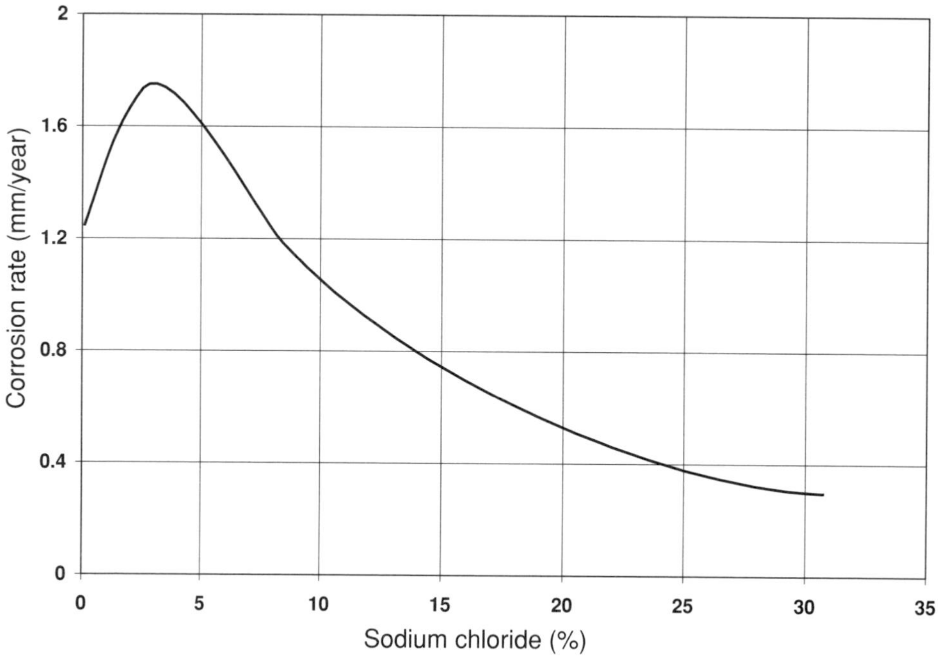

Figure 4.4 Corrosion of steel in various sodium chloride solutions. Note the peak in corrosion rate at approximately 3% salt.

The corrosion behavior of metals in seawater can be a complex function of many variables. Large areas of ocean have been surveyed at various depths regarding temperature, salt content, oxygen content, and other relevant factors. The varying corrosion rates to be expected at differing levels of a coastline have been generalized by Edwards (Figure 4.5).[5] Note the two maxima in the curve.

Table 4.1 Effect of Velocity on the Corrosion of Metals in Seawater

		Average Corrosion Rate (mm y^{-1})		
			Flowing Seawater	
Alloy	Deepest Pit (mm)	Quiet Seawater	8.2 m s^{-1}	35-42 m s^{-1}
Carbon steel	2.0	0.075	—	4.5
Grey cast iron (graphitized)	4.9	0.55	4.4	13.2
Admiralty gunmetal	0.25	0.027	0.9	1.07
85/5/5/5 Cu Zn Pb Zn	0.32	0.017	1.8	1.32
Ni resist cast iron type 1B	nil	0.02	0.2	0.97
Ni Al bronze	1.12	0.055	0.22	0.97
70/30 Cu Ni + Fe	0.25	< 0.02	0.12	1.47
Type 316 stainless steel	1.8	0.02	< 0.02	< 0.01
6% Mo stainess steel	nil	0.01	< 0.02	< 0.01
Ni-Cu alloy 40	1.3	0.02	< 0.01	0.01

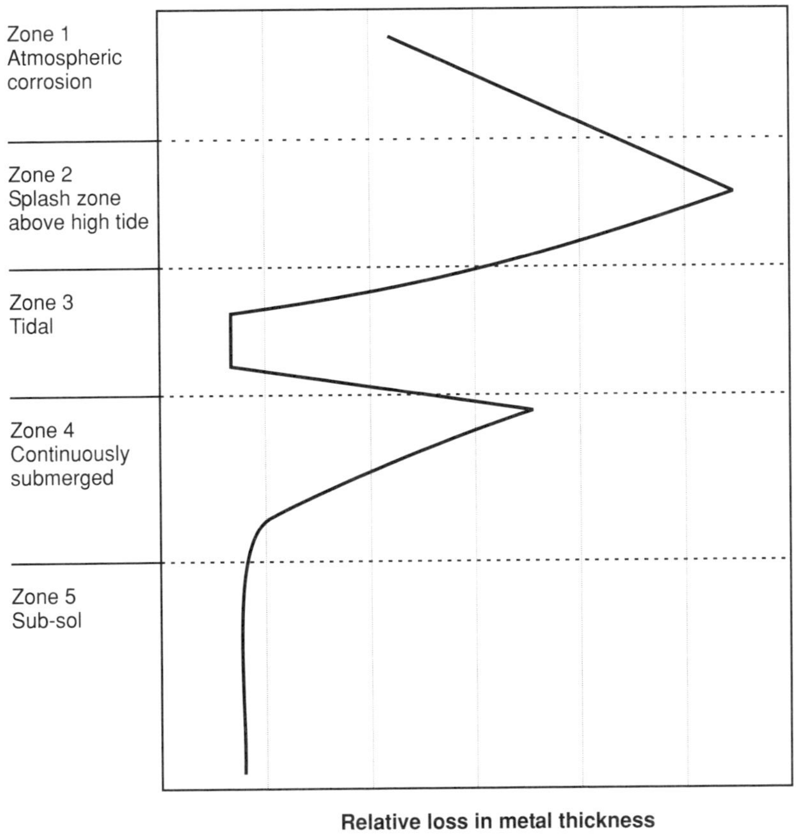

Figure 4.5 Corrosion profile of steel piling in seawater.

Essentially all of the corrosion mechanisms discussed in this book are accelerated in the seawater electrolyte. Greater pitting, more rapid dealloying, shorter fatigue life, and more severe galvanic problems are typical problems in seawater as compared with fresh-water exposures.

4.3.4 Distilled or Demineralized Water

The total mineral content of water can be removed by either distillation or mixed-bed ion exchange. In the first case, purity is described qualitatively in some cases (e.g., triple-distilled water), but is best expressed, for both distilled and demineralized water, in terms of specific conductivity. Water also can be demineralized by reverse osmosis or electrodialysis.

4.3.5 Steam Condensate

Water condensed from industrial steam is called steam condensate. It approaches distilled water in purity, except for contamination (as by DO or carbon dioxide) and the effect of deliberate additives (e.g., neutralizing or filming amines).

4.3.6 Boiler Feedwater Make-Up

The feedwater make-up for boilers is always softened and subsequently deaerated, as described in more detail later. It may vary in quality from fairly high dissolved solids (e.g., Zeolite-treated), to very pure demineralized feed for high-pressure boilers. It tends to be highly corrosive, because of its softness, until thoroughly deaerated. This term is more precise than "boiler feedwater," which may include recirculated steam condensate in various ratios to fresh make-up water.

4.3.7 Potable Water

Potable water is fresh water that is sanitized with oxidizing biocides such as chlorine or ozone to kill coliform and other bacteria and make it safe for drinking purposes. By definition, certain mineral constituents are also restricted. For example, the chlorinity will be not more than 250 ppm of chloride ions in the United States or 400 ppm on an international basis.

4.3.8 Process or Hydrotest Water/Firewater

These terms are essentially nondefinitive, because the water employed may be of almost any chemistry, ranging from demineralized water to quite saline fresh water or even seawater in some cases. "Produced water" is that which originates in oil and gas production, emanating from geological sources with the hydrocarbons.

4.3.9 Cooling Water

Cooling water is another undefined term, although it implies that any necessary treatment against excessive scaling or corrosion has been applied or corrosion-resistant material selected. This may include anything from fresh water to seawater and may comprise either an open- or closed-loop system or a once-through system, as is described later.

4.3.10 Waste Water

By definition, waste water is any water that is discarded after use. Sanitary waste from private or industrial applications is contaminated with fecal matter, soaps, detergents, and so forth, but is quite readily handled from a corrosion standpoint. Industrial wastes from chemical or petrochemical sources can contain strange and specific contaminants that greatly complicate materials selection, especially in the uses of plastics and elastomers.

4.4 SCALING INDICES

Encrustation of tubing, boilers, coils, jets, sprinklers, cooling towers, and heat exchangers arise wherever hard water is used. Scale formation can greatly affect heat-transfer performance. One-mm-thick scale, for example, can add 7.5% to energy costs, whereas 1.5 mm adds 15% and 7 mm can increase cost by more than 70%. Many factors can affect scaling. Scaling, which is basically the deposition of mineral solids on the interior surfaces of water lines and containers, most often occurs when water containing the carbonates or bicarbonates of calcium and magnesium is heated.

The saturation level (SL) of water in a mineral phase is a good indicator of the potential for scaling as a result of that specific scalant. SL is a ratio between the ion-activity product (IAP) and the thermodynamic solubility product (K_{sp}) of a specific compound in that water. For example, when calcium carbonate ($CaCO_3$) is the scalant, SL is defined as

$$SL = \frac{a_{Ca^{2+}} \times a_{CO_3^{2-}}}{K_{sp}} \tag{4.1}$$

where $a_{Ca^{2+}} \times a_{CO_3^{2-}}$ is the IAP of the two ions involved in the formation of $CaCO_3$, that is, Ca^{2+} and CO_3^{2-}.

K_{sp} is a measure of ionic concentration when dissolved ions and undissolved ions are in equilibrium. When a saturated solution of sparingly or slightly soluble salt is in contact with undissolved salt, equilibrium is established between the dissolved ions and the undissolved salt. In theory, this equilibrium condition is based upon an undisturbed water maintained at constant temperature and allowed to remain undisturbed for an infinite period of time.

In this example, water is said to be undersaturated (*SL* value of less than 1) if it can still dissolve calcium carbonate. When water is at equilibrium, SL will be 1.0 by definition. Supersaturated water (*SL* value of greater than 1) will precipitate calcium carbonate from water if allowed to rest. As the saturation level increases beyond 1.0, the driving force for the precipitation of calcium carbonate increases.

The following sections describe some indices that have gained wide acceptance in the corrosion community. However, it should be stated that these indices are designed to indicate the tendency of given waters to deposit scales on metal substrates and not to predict the absolute corrosivity of specific waters. Generally speaking, scales precipitated onto metal surfaces can provide protection of the substrate from general corrosion. If on the other hand the scales are defective and contain voids or cracks, they could lead to localized corrosion. The assumption that water below calcium-carbonate saturation is corrosive, although occasionally correct, is not reliable.

4.4.1 Langelier Saturation Index (LSI)

The LSI is probably the most widely used indicator of cooling-water-scale potential. This index indicates the driving force for scale formation and growth in terms of pH as a master variable. To calculate the LSI, it is necessary to know the alkalinity (mg L^{-1} as $CaCO_3$ or calcite), the calcium hardness (mg L^{-1} Ca^{2+} as $CaCO_3$), the total dissolved solids (mg L^{-1} TDS), the actual pH, and the temperature of the water (°C).

Table 4.2 Conversion Table Between the Conductivity of Natural Water and the Total Dissolved Solids (TDS) It Contains

Conductivity (micro-mho/cm)	TDS (mg/L as CaCO3)
1	0.42
10.6	4.2
21.2	8.5
42.4	17.0
63.7	25.5
84.8	34.0
106.0	42.5
127.3	51.0
148.5	59.5
169.6	68.0
190.8	76.5
212.0	85.0
410.0	170.0
610.0	255.0
812.0	340.0
1008.0	425.0

If TDS is unknown but conductivity is known, one can estimate mg L^{-1} TDS using a conversion table (Table 4.2). LSI is defined as

$$LSI = pH - pH_s \tag{4.2}$$

where pH is the measured water pH. pH_s is the pH at saturation in calcite or calcium carbonate and is defined as

$$pH_s = (9.3 + A + B) - (C + D) \tag{4.3}$$

$$A = \frac{(\log_{10}[TDS] - 1)}{10} \tag{4.4}$$

$$B = -13.12 \times \log_{10}(^{\circ}C + 273) + 34.55 \tag{4.5}$$

$$C = \log_{10}[Ca^{2+} \text{ as } CaCO_3] - 0.4 \tag{4.6}$$

$$D = \log_{10}[\text{alkalinity as } CaCO_3] \tag{4.7}$$

As for the SL reasoning described earlier, the LSI indicates three situations:

- If LSI is negative: No potential to scale, the water will dissolve $CaCO_3$.
- If LSI is positive: Scale can form and $CaCO_3$ precipitation may occur.
- If LSI is close to zero: Borderline scale potential. Water quality, changes in temperature, or evaporation could change the index.

As an example, suppose the drinking water supplied to animals has the following analysis:

pH = 7.5,
TDS = 320 mg/L,

calcium = 150 mg L^{-1} (or ppm) as $CaCO_3$, and
alkalinity = 34 mg L^{-1} (or ppm) as $CaCO_3$.

The LSI index is calculated at two temperatures, that is, 25°C (room temperature) and 82°C (cage wash cycle). The colder incoming water will warm to room temperature in the manifolds. Residual water in the rack manifold can be heated to 82°C when the rack is in the cage washer.

LSI formula:

$$LSI = pH - pHs,$$
$$pH_s = (9.3 + A + B) - (C + D), \text{ where} A = (\log_{10}[TDS] - 1)/10 = 0.15,$$
$$B = -13.12 \times \log_{10}(°C + 273) + 34.55 = 2.09 \text{ at } 25°C \text{ and } 1.09 \text{ at } 82°C,$$
$$C = \log_{10}[Ca^{2+} \text{ as } CaCO_3] - 0.4 = 1.78, \text{ and}$$
$$D = \log_{10}[\text{alkalinity as } CaCO_3] = 1.53.$$

Calculation at 25°C:

$$pH_s = (9.3 + 0.15 + 2.09) - (1.78 + 1.53) = 8.2,$$
$$LSI = 7.5 - 8.2 = -0.7,$$
hence no tendency to scale.

Calculation at 82°C:

$$pH_s = (9.3 + 0.15 + 1.09) - (1.78 + 1.53) = 7.2,$$
$$LSI = 7.5 - 7.2 = +0.3,$$
hence slight tendency to scale.

4.4.2 Ryznar Stability Index (RSI)

The RSI uses a correlation established between an empirical database of scale thickness observed in municipal water systems and associated water-chemistry data. Similar to the LSI, the RSI has its basis in the concept of saturation level. The Ryznar index takes the form $RSI = 2(pH_s) - pH$.

The empirical correlation of the RSI can be summarized as follows:

- RSI < 6: the scale tendency increases as the index decreases.
- RSI > 7: the calcium carbonate formation probably does not lead to a protective corrosion-inhibitor film.
- RSI > 8: mild steel corrosion becomes an increasing problem.

4.4.3 Puckorius Scaling Index (PSI)

The PSI is based on the buffering capacity of the water and the maximum quantity of precipitate that can form in bringing water to equilibrium. Water high in calcium but low in alkalinity and buffering capacity can have a high calcite-saturation level. The high calcium level increases the ion-activity product. Such water might have a high tendency to form scale because of the driving force, but scale formed might be of such a small quantity as to be unobservable. The water has the driving force but not the capacity and ability to maintain pH as precipitate matter forms.

The PSI index is calculated in a manner similar to the RSI. PSI uses an equilibrium pH rather than the actual system pH to account for the buffering effects:

$$PSI = 2\,(pH_s) - pH_{eq},$$

where

pH_s is still the pH at saturation in calcite or calcium carbonate,
$pH_{eq} = 1.465 \times \log_{10}[\text{Alkalinity}] + 4.54$, and
$\text{Alkalinity} = [HCO_3^-] + 2\,[CO_3^{2-}] + [OH^-]$

4.4.4 Larson–Skold Index

The Larson–Skold index is based upon evaluation of in situ corrosion of mild steel lines transporting Great Lakes waters. The index is the ratio of equivalents per million (epm) of sulfate (SO_4^{2-}) and chloride (Cl^-) to the epm of alkalinity in the form bicarbonate plus carbonate:

$$\text{Larson–Skold index} = \left(\text{epm } Cl^- + \text{epm } SO_4^{2-}\right) \,/\, \left(\text{epm } HCO_3^- + \text{epm } CO_3^{2-}\right) \tag{4.8}$$

Extrapolation to waters other than the Great Lakes, such as those of low alkalinity or extreme alkalinity, goes beyond the range of the original data. The index has proven to be a useful tool in predicting the aggressiveness of once-through cooling waters. The Larson–Skold index might be interpreted by the following guidelines:

- Index < 0.8: chlorides and sulfates probably will not interfere with natural film formation.
- $0.8 <$ index < 1.2: chlorides and sulfates may interfere with natural film formation. Higher than desired corrosion rates might be anticipated.
- Index > 1.2: the tendency toward high corrosion rates of a local type should be expected as the index increases.

4.4.5 Oddo–Tomson Index

The Oddo–Tomson index accounts for the effect of pressure and partial pressure of carbon dioxide on the pH of water and on the solubility of calcium carbonate. This empirical model also incorporates corrections for the presence of two or three phases (water, gas, and oil). Interpretation of the index is by the same scale as for the LSI and Stiff–Davis indices.[6]

4.5 COOLING SYSTEMS

Water treatment is done mainly with fresh water. Corrosion tendencies relate to the degree of aeration, hardness phenomena, and specific salts. Many municipal water plants adjust the relationship between pH and calcium salts so that the water is sufficiently scaling to minimize corrosion but not to plug up the piping. This type of control is obviously not suitable in cooling systems. If the supply is nonscaling, deposits will form in the hottest sections; if the hot water is adjusted, the cooler water is corrosive.

4.5.1 Once-Through Systems

Traditionally, industry has tended to develop in areas with an adequate supply of cooling water. Originally, it was sufficient to pipe water through the plant and discharge it back to its natural source. Only nominal attention was paid to control of water chemistry, and it is in fact economically ridiculous to attempt the chemical treatment of large volumes of once-through water.

Even such minor additions as the "threshold" treatment with 1 to 2 ppm sodium hexametaphosphate, for example, would be unacceptable in today's environmental awareness. Many modern countries, states, and provinces forbid the return to source even when the concentrations of natural constituents were higher in the intake water than in the return water. In some areas, thermal pollution is forbidden because the discharge of the same water at a higher temperature than the inlet temperature may be harmful to certain marine species (e.g., oyster beds).

The bottom line today for once-through cooling systems is that, where permitted, the materials of construction must be chosen to be resistant to the water, be it fresh water or seawater. All natural waters must be presumed to be corrosive to iron and steel by virtue of aeration. Historically, acceptable performance of bare steel in natural waters has been associated with biological or chemical oxygen demands (BOD or COD) that are today ecologically unacceptable. River waters that were once noncorrosive have become corrosive because they are now sufficiently clean to provide a suitable environment for game fish.

Biocidal control is likewise inimical. Treatments against bacteria (discussed later) are incompatible with environmental considerations, unless the effluent is treated for their removal.

4.5.2 Recirculated Systems

Recirculated systems are divided essentially into two types, closed and open.

4.5.2.1 Closed

Closed recirculated systems are characterized by an essentially permanent charge of water, greatly facilitating the selection of chemical control and permitting the use of relatively large amounts of chemical additives. Typical closed recirculated systems are automobile radiator-cooling systems, which are air cooled, and engine-jacket cooling systems, which may either be air cooled or have the sensible heat removed by exchange with another type of cooling-water system. In some locations, plants may be cooled with a closed loop of treated fresh water, and the fresh water in turn cooled in large exchangers carrying a once-through water coolant.

4.5.2.2 Open

Open recirculated cooling-water systems remove the heat picked up in the plant by evaporative cooling. This may be done by a spray pond, for example, combining air-conditioning needs with aesthetic considerations in industrial parks. The most common type of evaporative cooling, however, is effected in cooling towers of one type or another.

Cooling towers may operate on natural draft, as in the case of wind-cooled towers for small home air-conditioning systems or the large concrete hyperbolic towers used in power-generating stations (Figure 4.6). In process plants, the towers are more often aided by fans, either forced or induced-draft operations, to improve the cooling capacity (Figure 4.7).

Figure 4.6 Natural-draft cooling towers with their characteristic height and distinctive hourglass shape (courtesy of Russ Green, TMI).

Figure 4.7 Three-cell induced-draft counterflow cooling tower made of seamless double-wall polyethylene construction (courtesy of Delta Cooling Towers).

There are certain fundamental considerations that should be understood in relation to open recirculated systems. First is the concept of cycles of concentration. If three cups of boiling water in a tea kettle were allowed to boil away to one cup, the residual cup would contain a three-fold concentration of soluble water salts, assuming that only steam (i.e., pure H_2O) was driven off. The water would be said to be at three cycles of concentration.

To prevent this accumulation from becoming unacceptable from the standpoint of scale and corrosion, a small amount of blowdown (bleeding of the system) is maintained to control the number of cycles of concentration from evaporation. This means that make-up water must be added to equal the evaporation and blow-down losses, but this is a minor amount compared to the volume of the total system.

For example, if we needed 19,000 Lpm (5000 gpm) of cooling water in a system, the cost for treatment in a once-through design would be excessive. However, in a recirculating system, the make-up may only be 380 Lpm (100 gpm), of which only 95 Lpm (25 gpm) may need to be treated with inhibitors. This brings chemical treatment into the range of economic feasibility, as compared with a once-through system.

Not only are there tangible limits, imposed by water chlorinity and hardness, as to how far one can concentrate the water, but the savings effected by a recirculating system versus a once-through system are maximized at about 4 to 6 cycles of concentration. Below this range, treatment costs become prohibitive. At high cycles (e.g., 8 to 10), the additional water savings are not commensurate with the increased difficulty of effective treatment. If the blowdown is shut off entirely, there is still an effective upper limit of concentration dictated by water losses from drift or windage. The normal upper limits might be about 20 to 22 cycles of concentration for a mechanical-draft tower.

The advantages of water savings effected by the cooling tower impose certain inherent disadvantages as well. The water becomes air saturated, ensuring its full corrosion potential; its natural alkalinity tends to increase and aggravate scaling tendencies. The air-scrubbing action can contaminate the water with airborne materials, notably dust fines, which form silt in the tower basin, and spores of slime, algae, and fungi that can reproduce in the warm nutrient water of the system (Figure 4.8).

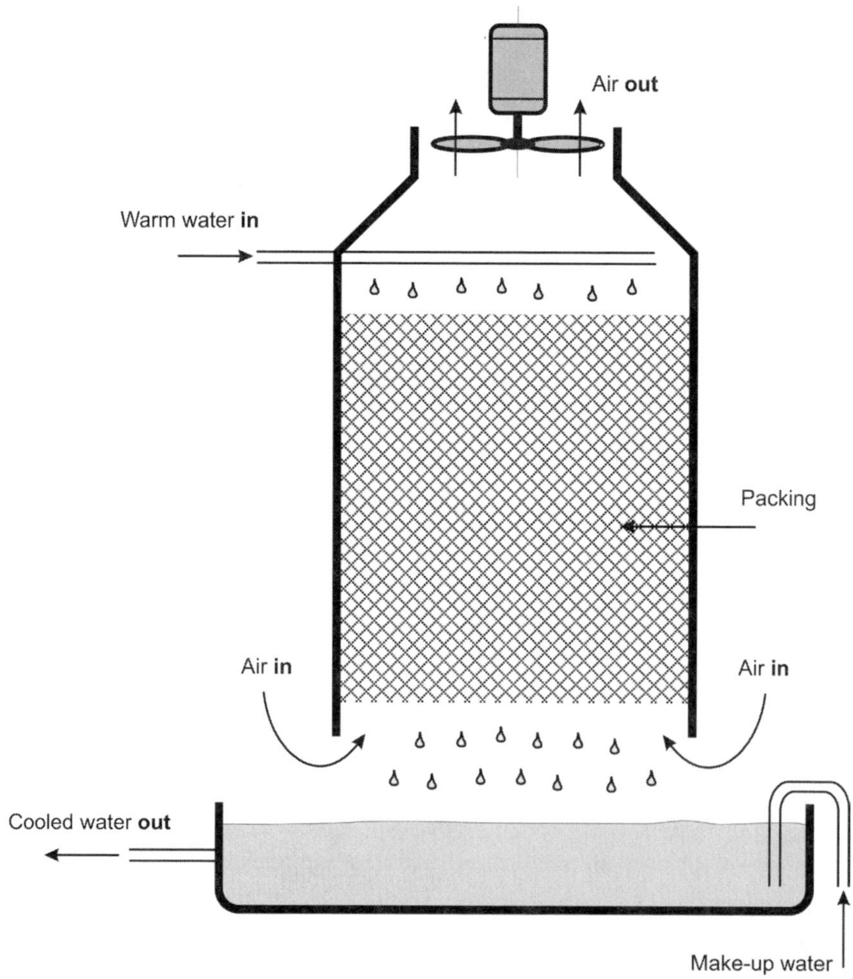

Figure 4.8 Schematic shows a forced-draft cooling tower.

4.5.3 Other Water Systems

Miscellaneous water systems, which bridge the cooling-water regimens and the high-temperature operations, range from cooling brines through tempered water to waste heat boilers.

4.5.3.1 *Brines*

Low-temperature brines are most often formulated at about 25% calcium chloride to maintain temperatures below freezing for the control of exothermic reactions. These solutions are corrosive to common materials of construction unless pH is adjusted to about 8.5 and inhibited with an appropriate corrosion-inhibitor system. Care must be taken to exclude entry of air, from which the absorption of carbon dioxide tends to lower the pH and encourage the deposition of calcium carbonate. The salt brines may also be rendered innocuous by completely removing oxygen and other oxidants from the system.

Methanol or glycol brines are sometimes employed. Organic water heat-exchanger agents are inhibited with organic, inorganic, and metalorganic compounds. Mercaptobenzothiazole is normally included as an inhibitor to protect any copper alloys in the system.

4.5.3.2 *Automotive Radiators (and Other Closed Glycol/Water Systems)*

"All-weather" ethylene glycol–type solutions require complex mixtures of additives, including corrosion inhibitors, stabilizers, and buffering agents. Without proper commercial inhibition, a 40% glycol solution at 70°C would corrode iron and steel at 250 to 500 μm/y (10 to 20 mpy) while also attacking the copper, brass, solder, and aluminum components at 25 to 50 μm/y (1 to 2 mpy).

4.5.3.3 *Hot-Water Systems*

Tempered-water systems are often employed to regulate chemical reactions at some moderately elevated temperature, typically 60°C to 80°C. Water is highly corrosive to ferrous metals at such temperatures, so the systems must be either strongly inhibited or deaerated. In domestic and industrial hot-water systems, corrosion tends to be self-arresting because the DO initially present is consumed by corrosion. However, if there should be regular ingress of oxygen for any reason, corrosion can be disastrous.

4.5.3.4 *Waste-Heat Boilers*

In many chemical or petrochemical processes, economy dictates that superfluous exotherms be utilized to generate steam as an energy-conservation measure. Examples are the cooling of a butane oxidation reaction or the condensation of hot sulfur vapors, where steam pressure is generated on the shell side of specially designed heat exchangers.

Corrosion problems in waste-heat boilers usually arise either from unusual materials of construction or from inattention to the required details of water treatment. Austenitic stainless steels may be required from the process-side corrosion aspects, yet be highly susceptible to SCC from boiler feedwater. The most frequent problem, however, is that the operating-department personnel, whose primary concern is the manufacture of marketable products, simply do not give adequate technical attention to the details of boiler feedwater chemistry. Many chemical converters, catalytic converters (oxycat) exchangers, or other waste-heat boilers have failed from steam-generating side corrosion as a result of such inattention.

4.6 WATER TREATMENT

4.6.1 Inhibition

Corrosion inhibition in recirculated cooling-water systems historically depended on the oxidizing inhibitors such as chromates and nitrites. Nitrites cannot be effectively used in open recirculated systems because they are highly conducive to organic growths and chromates have been phased out for environmental reasons.

The higher the chlorinity of the water, the more it seems to require a good dosage of corrosion inhibitors. Traditional glassy metaphosphates lose much of their effectiveness because they revert in time to orthophosphates. There is also an attendant possible danger of precipitation of tricalcium phosphate. This is a function of pH, calcium, orthophosphate, and solids concentration of the water. Silicates have also been used in water systems but tend to have an adverse effect on heat transfer because of the deposition of relatively heavy films.

Because of the ongoing search for effective nonoxidizing inhibitors over the past several decades, many new formulations have been developed, notably, polyphosphonate and organophosphate combinations. A careful study should be made of the behavior of specific cooling-water systems, because it is extremely difficult to predict the suitability of complex commercial formulations from water chemistry. Laboratory, pilot plant, and field tests are usually required to select the proper system. Even then, its effectiveness will depend upon the care taken in control of operating conditions.

There is no magic method of water treatment available. Beware of magnetic and electrostatic treatments that claim some beneficial change in the properties of the water.

4.6.2 Scale Control

There are three alternatives to prevent scale deposition:

1. pH control,
2. the addition of scale inhibitors, and
3. removal of scaling species.

One of the most effective methods for controlling crystallization fouling is by adding chemical inhibitors to potentially scaling water. Commonly used antiscalants are derived from chemicals such condensed polyphosphates, organophospates, and polyelectrolytes.

Scale control in open recirculated systems is effected primarily by limiting the concentration of the scale-forming species and related parameters, especially by controlled acid additions to effect a suitable pH range. Many of the commercial inhibitor formulations include scale-controlling additives (e.g., polyphosphates and chelating agents).

4.6.3 Microorganisms

Algae and slimes are found in cooling-water towers and spray ponds where sunlight and airborne contamination are to be expected. Slimes normally contain fungi, yeasts, bacteria, and entrapped quantities of inorganic and organic material. Some of these are described in more details in Chapter 13.

Conditions at the base of even thin slime films (biofilms) can be ideal for the growth of Sulfate Reducing Bacteria (SRB), with high organic-nutrient status, no oxygen, low redox potential, and protection from biocides. Figure 4.9 illustrates the steps in biofilm formation.[3] The SRB can produce active sulfide corrosion even in systems where the bulk liquid phase has a low nutrient status and a high oxygen concentration and will not support growth of anaerobic aerobic bacteria.[4]

Slime deposits can cause oxygen concentration cell–type corrosion and pitting. If sloughed off, they can contribute to fouling of filters and other equipment (Figure 4.10.)

Chlorination to a residual chlorine content of a few tenths of a ppm is the usual treatment for control of biological growths in open recirculated systems. To discourage the development of chlorine-resistant strains, it is advisable to surprise the microbes

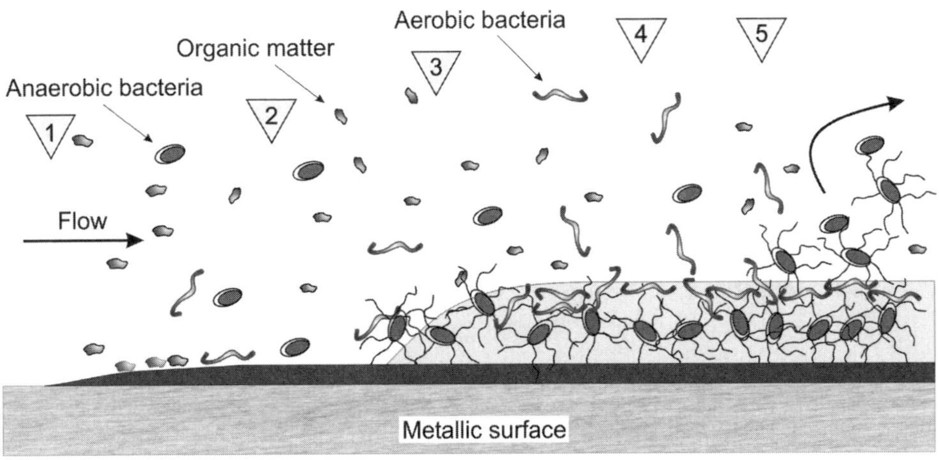

Figure 4.9 Biofilm formation is initiated when (a) small organic molecules become attached to an inert surface and (b) microbiological cells are adsorbed onto thee resulting layer. The cells (c) send out hair-like exopolymers to feed on organic matter, (d) adding to the coating. Flowing water (e) detaches some of the formation, producing an equilibrium layer.

Figure 4.10 Fouled chiller unit (courtesy of Kingston Technical Software).

at irregular intervals with other biocides (e.g., quaternary amines). Other oxidizing biocides (e.g., chlorine dioxide and ozone) used in municipal waters are normally too expensive for cooling-water systems.

Historically, copper sulfate had been used to control algae in some specific applications. It is effective at concentrations as low as 1 ppm but must be used at a sufficiently low pH to prevent precipitation of insoluble copper hydroxide. The use of copper sulfate as a biocide is incompatible with aluminum equipment because of deposition and subsequent galvanic corrosion. Similar copper plating would occur on zinc (as in galvanized-steel piping) and even steel.

Bacterial action (e.g., sporovibrio desulfuricans, the sulfate-reducing bacteria that generate hydrogen sulfide by metabolizing sulfates, and crenothrix, the "iron-eating" bacteria that metabolizes soluble iron and manganese, excreting the oxides) poses a further threat to such systems but usually is effectively controlled by the chlorination procedures mentioned earlier.

4.6.4 Silt Control

The accumulation of silt in the bottom of the tower sump is due to the scrubbing of large volumes of air, especially in a mechanical-draft tower. This can be controlled to a certain extent by the addition of polyelectrolytes (e.g., polyacrylate structures), which inhibit its aggregation from a dispersed form. In areas subject to large amounts of dust in the air, side-stream filtration of the recirculated water may be required.

4.7 STEAM GENERATION

The greatest use of high-temperature water and steam is in electrical-power generation. Historically, fossil fuels (i.e., wood, coal, gas, and oil) were used almost exclusively to heat water and make steam until the introduction of nuclear-power steam generators in the second part of the twentieth century. The two types of power plants have much in common but are sufficiently different to be discussed separately. Both, however, presuppose technically advanced water treatment and control for successful operation.

4.7.1 Treatment of Boiler Feedwater Make-Up

Boiler feedwater make-up and boiler feedwater must be softened and deaerated to an extent determined by specific requirements of boilers operating in different temperature and pressure ranges. Low-hardness salts to prevent scaling in the system and deaeration to prevent corrosion of steel are the goals.

A number of lime-softening treatments were used in the past, but these have given way to more sophisticated treatments. Probably the most common for boilers up to 2.8 to 4.0 MPa (400 to 600 psig) is Zeolite softening. In this treatment, a sodium salt of a long-chain polymeric organic molecule constitutes the ion-exchange bed. As the feedwater passes through the bed, sodium ions are exchanged for the calcium and magnesium ions that create the water hardness. Other cations are also exchanged. The water then exits from the Zeolite bed softened and at a higher pH than it entered. The beds are regenerated intermittently with sodium chloride solution to backwash the hardness salts and reform the sodium salt.

For purer water quality, the water may be totally demineralized by mixed beds of polymeric resins, which in turn exchange hydrogen ions for all cations and hydroxyl ions for all anions, effectively producing pure H_2O from a raw water stream. Such highly purified waters are required for boilers operating from about 6 MPa (900 psig) and for all nuclear boilers (to avoid radioactive half-life of water-borne salts).

The softened water is now at its most corrosive, being still saturated with DO and having no hardness that could even hope to slow down attack. Many plant problems arise from people stealing boiler feedwater, under the impression that it is "the best water in the plant." So it will be, but only after further treatment.

For reasons of economy, the removal of DO is usually first effected in part by thermomechanical deaeration. The boiler feedwater is preheated and then flashed in a deaerator to remove any free carbon dioxide and most of the dissolved oxygen. Then, the last traces of DO are chemically scavenged with sodium sulfite (Na_2SO_3) that reacts with oxygen to form sodium sulfate (Na_2SO_4):

$$2\ Na_2SO_3 + O_2 \rightarrow 2\ Na_2SO_4 \tag{4.9}$$

or, if a complete absence of solid precipitates is desired, with hydrazine (N_2H_2) that forms water and gaseous nitrogen in the presence of oxygen:

$$2\ N_2H_2 + O_2 \rightarrow 2\ H_2O + 2\ N_2\uparrow \tag{4.10}$$

Both sulfite and hydrazine are available in catalyzed form to promote more rapid reaction rates. A boiler designed to be operated with a catalyzed oxygen scavenger must never be operated on the uncatalyzed grades, or severe corrosion will be encountered in the economizers or even the steam drum.

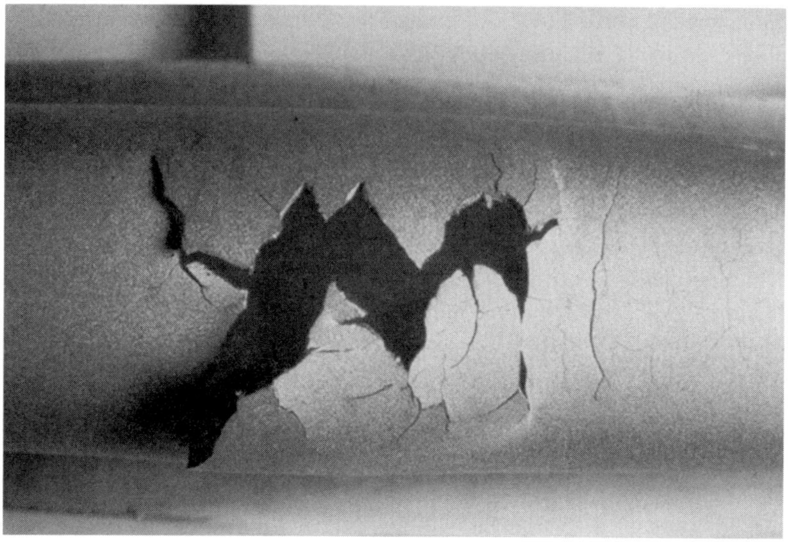

Figure 4.11 Type 321 stainless-steel expansion joint in 2.8 MPa (400 psi) steam service is cracked from caustic carryover in the steam.

A final step in boiler feedwater treatment consists of pH adjustment as a further aid to corrosion control. Usually the pH is adjusted to a range of 10 to 11 with trisodium phosphate (or combinations of caustic with sufficient mono- or di-sodium phosphate to form trisodium phosphate upon inadvertent evaporation of the water). This "coordinated phosphate" treatment is intended to preclude the environmental cracking of steel by free sodium hydroxide (caustic embrittlement), a catastrophic form of corrosion described in Chapter 9. Caustic carry-over with the steam can present severe corrosion problems (Figure 4.11).

Nuclear requirements are such that zero solids treatment is required, precluding the addition of sodium salts and necessitating the use of ammonium hydroxide for pH adjustment.

4.7.2 Fossil-Fuel Steam Plants

Conventional steam generation plants or more modern cogeneration steam plants (Figure 4.12) consist of many sections operating at different temperatures and pressures:

- feedwater heaters with water inside and steam outside the tubes;
- a boiler (water inside the tubes, hot combustion product gases outside), where water is heated to high temperatures under pressure and is sometimes flashed to steam;
- a steam drum, wherein steam is formed from water and water is separated from the steam (note: the steam drum is omitted in certain once-through systems);
- a superheater, where the steam is further heated to even higher temperatures;

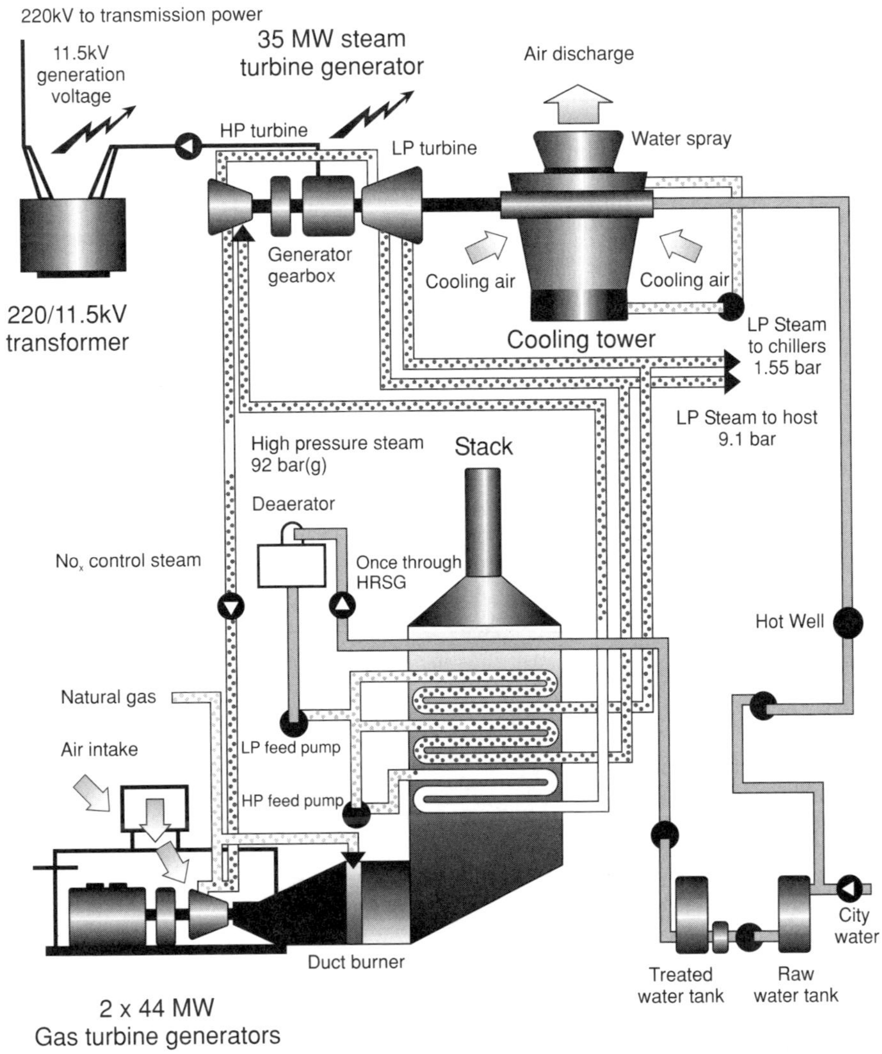

Figure 4.12 Schematic shows a 123-MW natural gas-fueled cogeneration plant that produces electrical power and industrial-use steam.

- a turbine, where the steam expands against the vanes of a wheel to drive the turbine that generates electricity; and
- a condenser, where the low-pressure steam is condensed to water and returned to the feedwater heaters.

In such plants, there are special corrosion problems in each of these sections. There are also special problems associated with exposure to hot combustion gases. The latter is discussed in Chapter 7.

The materials used in high-temperature steam and water include steels, stainless steels, and nickel-base alloys. Copper-base alloys are employed in the intermediate and low temperature ranges. Aluminum alloys generally are not used because of their poor resistance above about 200°C.

4.7.2.1 Boilers

The boiler proper is most often fabricated from ordinary carbon steel, which is useful under controlled conditions up to about 450°C. Alloy steels (e.g., 1.25 Cr–0.5 Mo, 2.25 Cr–1 Mo) are used at the 565°C maximum operating temperatures of most steam plants.

In a plant operated under properly degassed conditions, the steels develop a thin black protective film of magnetic iron oxide (magnetite, Fe_3O_4). If appreciable DO has entered the system, brick red Fe_2O_3 will also be found.

Steel may experience an occasional failure from on-load pitting during steady operations. This type of failure has been attributed to a variety of causes including copper deposits, thermogalvanic effects caused by temperature differences, incipient attack started during shutdown, and localized caustic concentration.

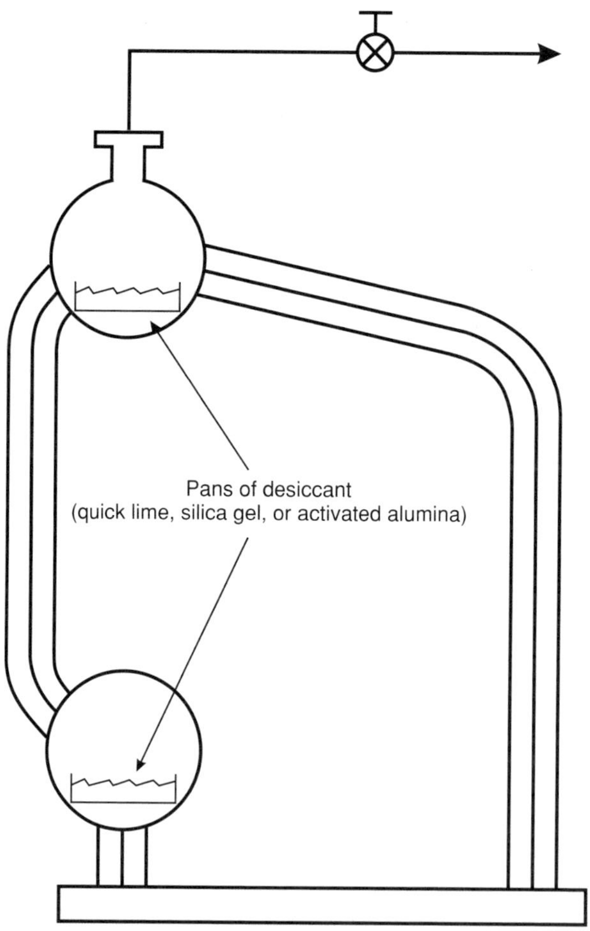

Figure 4.13 Diagram shows a boiler furnace in dry standby. Pans of desiccant, as indicated by the arrow, are placed in the unit, which is sealed airtight.

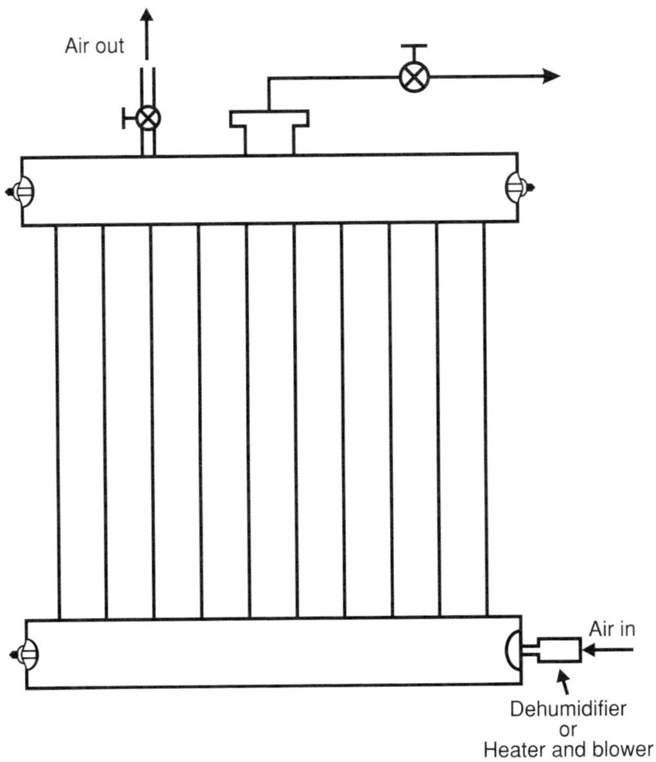

Figure 4.14 Diagram shows a boiler drum in dry standby with pans of desiccant as in Figure 4.13. Dry, heated air is circulated by blower.

There is no question that severe pitting can occur during shutdown if the boiler tubes are not completely dried out or completely filled with sulfite-treated water (Figures 4.13–4.18). Failure to do this during shutdown can result in the formation of air bubbles in the tubes and attendant air-bubble pitting.

Considerable quantities of copper may sometimes be found interspersed in the oxide scale in boilers. The source of the copper is usually the condenser and feedwater heater tubes and indicates inadequate DO control, often in conjunction with amine-type inhibitors used for condensate-corrosion control.

Copper-free corrosion products are often removed from the boiler by treatment with inhibited hydrochloric acid or other specialized acid formulations. If considerable copper is present, this will lead to deposition attack: the copper plating out on the boiler tubes. Copper may be removed from the boiler by first oxidizing it to cupric oxide by means of oxidants (e.g., chlorates, bromates, or persulfates) and then dissolving it in an ammoniacal solution, forming a typical royal-blue copper-ammonium complex.

4.7.2.2 Steam Drums

Corrosion problems in steam drums are similar to those in the boiler tubes. Localized concentration of boiler-water chemicals (a result of entrainment) and increased corrosion of the steel can occur at the water–steam interface or at steam separation plates.

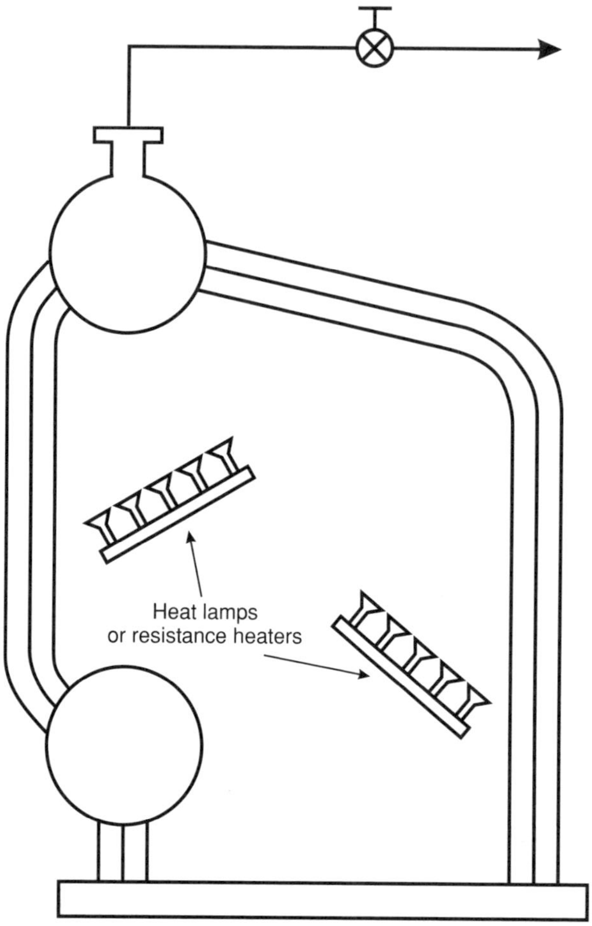

Figure 4.15 Diagram shows a boiler furnace in dry standby. Heating devices prevent condensation.

4.7.2.3 *Superheaters*

As indicated in Table 4.3, the 2.3% chromium steel is resistant to general corrosion by clean deaerated steam up to about 650°C. Accelerated attack can occur when the water from the boiler is not completely separated from the steam, and the chemicals in the entrained water become concentrated in localized areas. Concentration of caustic by this mechanism can produce pitting. The corrosion resistance of steel is optimum at pH 10 to 11 and decreases both above and below this range.

4.7.2.4 *Turbines*

In conventional steam stations, the areas of the turbine that are most susceptible to deterioration are the turbine blades because of erosion corrosion. Figure 4.19 shows the effects of flow-accelerated corrosion (erosion corrosion) on associated piping. To

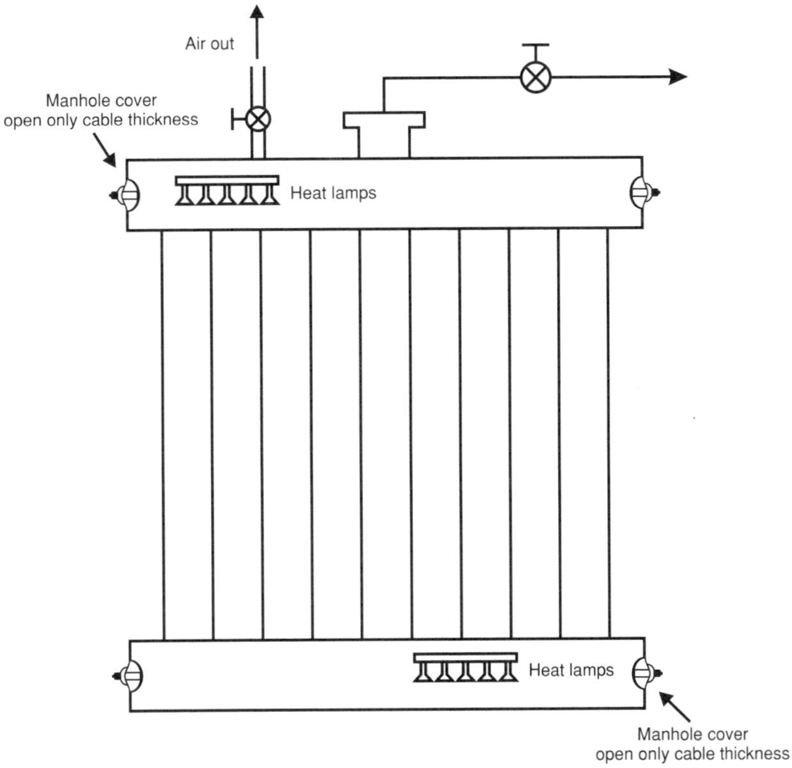

Figure 4.16 Diagram shows a boiler drum in dry standby. With some outside air circulating, heat lamps prevent condensation.

resist this type of attack, the turbine blades or blade tips are fabricated from hardened martensitic stainless steel (e.g., S41000).

Deposits on the turbine blades are also a problem if there is carry-over of entrained moisture, such as may occur when saturated steam directly from the steam drum is used to drive the turbine. Silica (SiO_2) carry-over and deposition may also be a problem unless the water chemistry is carefully controlled. At pressures above 4.0 MPa (600 psi), silica will carry and escape with the steam. It then deposits on low-pressure areas of the turbine. Silica may be controlled by eliminating it from the boiler feedwater, blowing down a portion of the steam, or by removing it from the steam by water washing.

Table 4.3 Maximum Recommended Operating Temperatures for Steels in Steam Service

Steel Composition	Maximum Operating Temperature (°C)
Carbon steel	450
1.25-Cr 0.5-Mo	590
2.25-Cr 1-Mo	650
3-Cr 1-Mo	650
9-Cr 1-Mo	650
18-Cr 8-Ni (stainless steel)	820

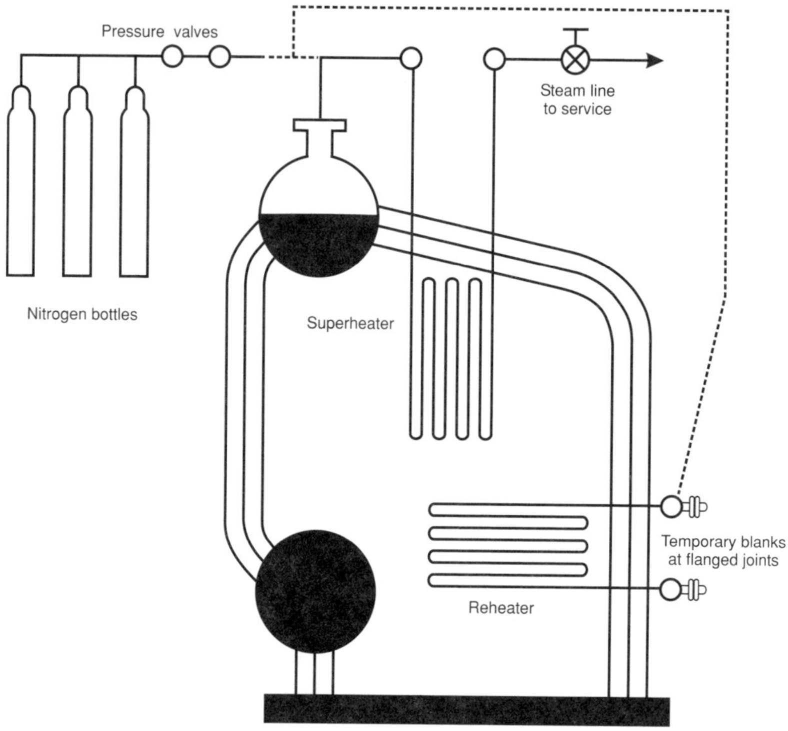

Figure 4.17 Diagram shows a boiler system in wet standby, using a nitrogen-blanketing technique. Water is maintained at or near operating level and nitrogen is introduced at a positive pressure of 5 to 10 psig.

4.7.2.5 Condensers

Condensers are shell-and-tube heat exchangers with steam condensing in the tubes and cooling water on the shell (Figures 4.20). Condenser tubes are frequently made of copper-base alloys, such as C44300 (arsenical admiralty) or C70600 (90–10 cupronickel). Selection of condenser-tube material is often dictated by the cooling water-which may be more corrosive than the condensing steam.

Corrosion by the condensed steam is usually the result of dissolved carbon dioxide. The CO_2 is released from carbonates in the boiler and, being volatile, passes through the turbine to contaminate the condensate, causing low pH (acidic) conditions. Grooving and thinning of the tubes will occur if protective measures are not taken. The usual remedy is to make the condensate alkaline with neutralizing amines (e.g., morpholine) to a pH of about 8.5 to 8.8, or to add filming amines such as octadecylamine. Higher pH values are not recommended because of the intrinsic danger to copper alloys in the event of ingress of oxygen to the condensate.

4.7.2.6 Feedwater Heaters

Feedwater heaters have also been tubed with copper-base alloys, usually cupronickels, although modern practice tends toward stainless steel or even titanium. Water at

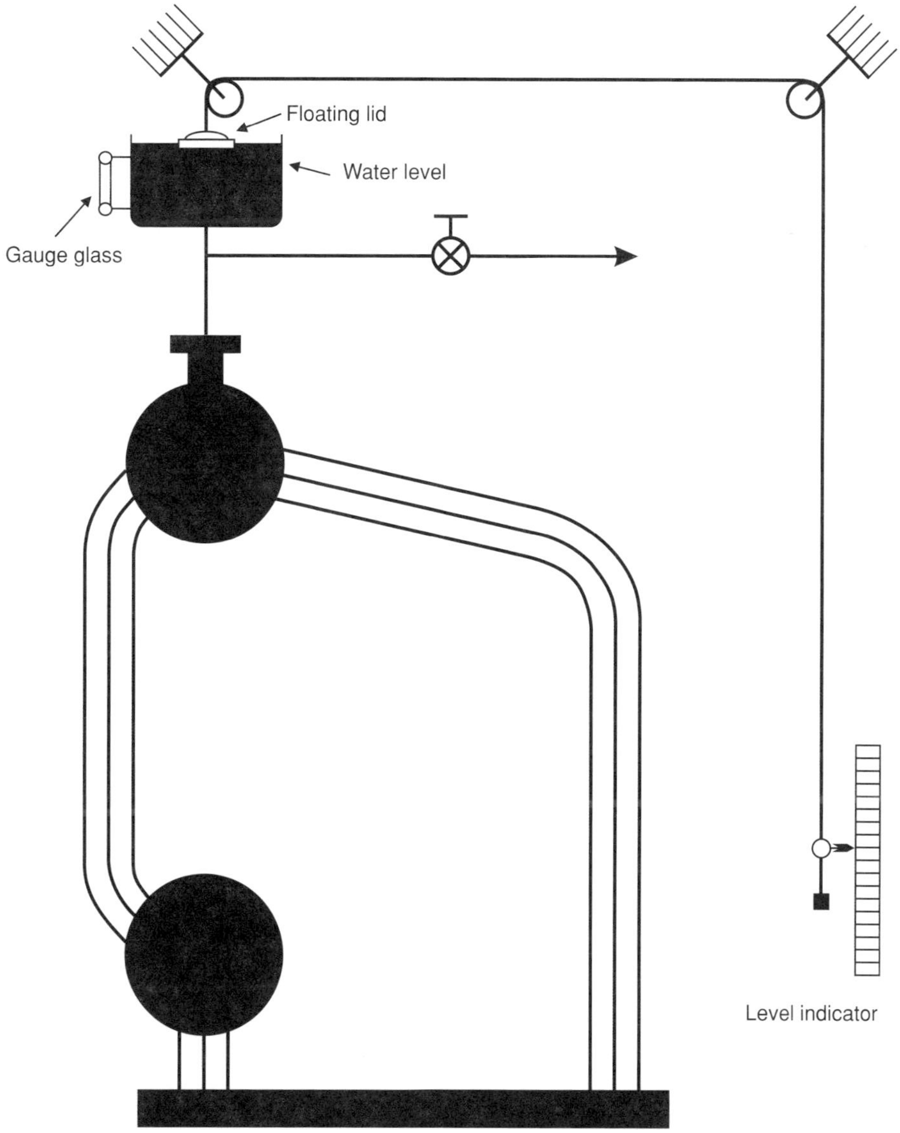

Figure 4.18 Diagram shows a boiler in wet standby using the complete flooding technique. This method is used when a system includes a drainable superheater. Nonvolatile materials such as caustic soda and catalyzed sodium sulfite are present for pH adjustment and oxygen scavenging.

about 260°C is inside the tubes, with steam outside the tubes. Corrosion problems are encountered on either or both sides.

Corrosion inside the tubes can be attributed mainly to DO introduced with make-up water or by air leaks to the condenser, which operates under a partial vacuum. Sodium sulfite additions may be required to remove the DO and prevent corrosion.

Corrosion outside the tubes occurs as layers of oxides (exfoliation) in installations that are operated in peak service (i.e., shut down overnight and weekends). This type of attack is severe on some cupronickels, but not so much with C70600 or Monel

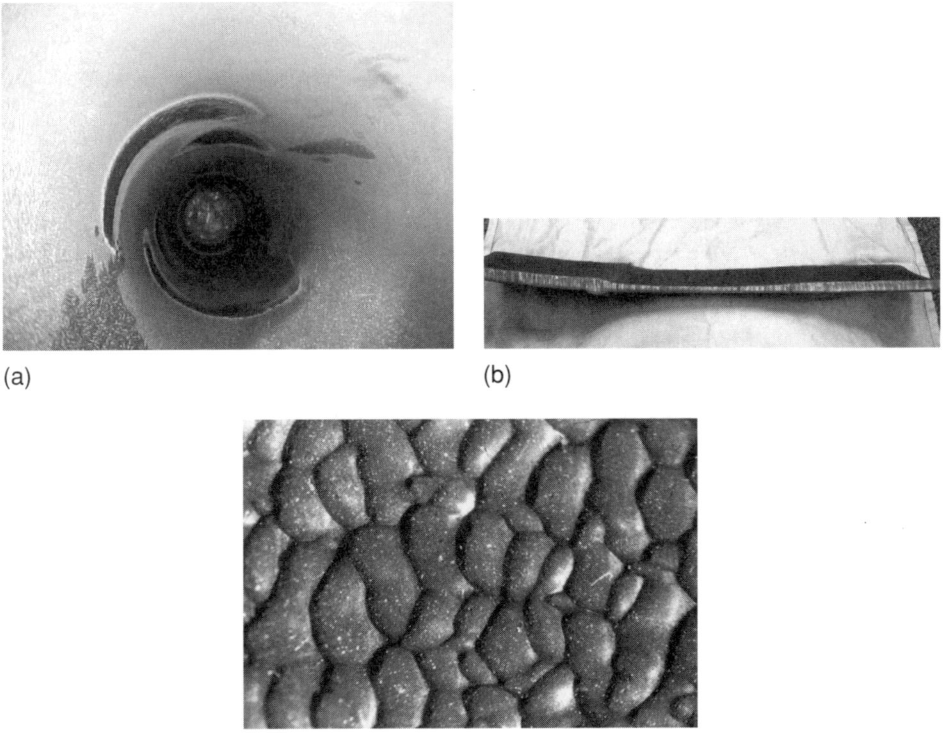

(a) (b) (c)

Figure 4.19 Flow-accelerated corrosion of (a) steam trap drain, (b) with thinning of the pipe wall, and (c) surface roughening (courtesy of Russ Green, TMI).

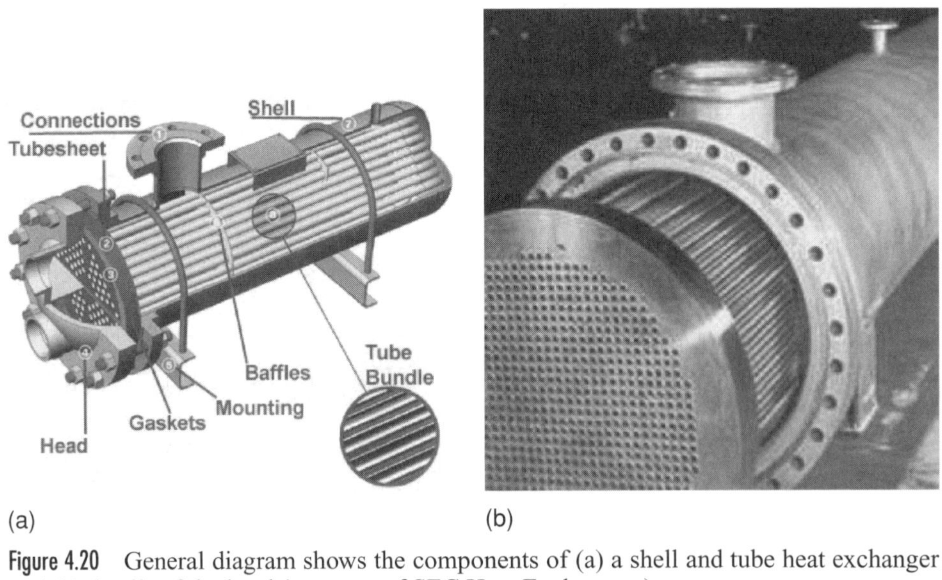

(a) (b)

Figure 4.20 General diagram shows the components of (a) a shell and tube heat exchanger and (b) details of the head (courtesy of SEC Heat Exchangers).

400 (N04400). Both oxygen, which enters the shell during shutdown, and the presence of contaminated condensate appear to be required for exfoliation to occur. The use of cupronickels containing 4 to 5% iron and manganese will largely eliminate the exfoliation problems in such instances.

4.7.3 Supercritical Steam Plants

Supercritical steam plants operating above the critical temperatures 375°C and critical pressure 22 MPa (3200 psi) of water have the same mechanical components as conventional steam plants. Major differences are greater wall thicknesses to withstand the pressure, more corrosion-resistant materials, and lower dissolved solids in the water and steam.

Almost all of the components in a supercritical steam plant are made of austenitic stainless steels of the 18-8 variety, for examples S30403 or 31603. These materials are employed to minimize corrosion products and their transport through the system. The steam temperature is no higher than in an ordinary superheater, but the pressures are such that many chemicals and corrosion products may show appreciable solubility in the steam.

Many chemicals exhibit an inverse solubility when the temperature of the steam is above the critical temperature and will therefore deposit out at these higher temperatures. In one supercritical plant, about 140 ppm of caustic was accidentally introduced into the plant. Within 30 minutes, caustic SCC occurred in that part of the plant where the temperature was about 425°C. This temperature corresponds to the minimum in the caustic solubility–temperature curve. The supercritical steam also undergoes a marked density increase above this temperature range, which could cause the deposition of chemicals.

For such reasons, the dissolved-solids content of the water must be kept at or near zero. A large fraction of the water is continuously cleaned up in a by-pass circuit containing ion-exchange (demineralizer) beds. At startup, 100% of the water may be passed through the clean-up beds. Chlorides and caustic are the most undesirable salts because they are known to cause SCC of the austenitic stainless steels (Chapter 8). DO may be reduced by the methods described to only a few parts per billion.

Copper deposition on turbine blades was an early problem with supercritical units. It was found that trace amounts of copper were dissolved from the condenser tubes and recirculated into the boiler section. Because of its excellent solvent properties, the superheated steam carried the dissolved copper into the turbine where, at the lower pressure, copper was deposited upon the turbine blades. This not only affected the efficiency but threatened to destroy the turbine by mechanical imbalance resulting from uneven deposition. The problem has been largely eliminated by using stainless steel or titanium in the condenser.

4.7.4 Nuclear-Fuel Steam Plants

There are several types of nuclear-fuel steam plants. In some designs, the heat from the nuclear reaction is removed by nonaqueous media, which then exchange heat with a water system to generate steam. Molten sodium, organic liquids, and helium gases have been used in this way. In other designs, an aqueous media (e.g., heavy water or pressurized light water) removes the heat of reaction, but the steam is generated in a heat exchanger. There also are designs (e.g., boiling-water reactors) in which the reactor cooling water is used to generate steam directly.

Nuclear and conventional plants have many components in common: the boiler, steam drum, turbine, condenser, and feedwater heaters. In the nuclear plant, however, most of these components, with the possible exception of the condenser, are made of austenitic stainless steel or some other nickel-chromium-iron alloy to minimize corrosion products in the system. Low corrosion is desired, because if corrosion products become radioactive by transport through the reactor core, they constitute a radiation hazard to personnel when deposited again in areas remote from the core.

When a water system surrounds the core, whether to generate steam directly or indirectly, specific problems arise that are peculiar to irradiated water.

4.7.4.1 Boiling-Water Reactors

In boiling-water reactors, the steam is generated directly in the pressure vessel and, after separation, goes directly to the turbine. This results in boiling occurring on the fuel elements, the possibility of solids deposition on the fuel elements, the stripping of volatiles from the water, and the production of 0.2 to 0.3 ppm oxygen in the water and up to 20 ppm in the steam by radiolytic decomposition of water (i.e., radiolysis).

High-purity neutral water is used because of the possible deposition of solid chemicals and the stripping of volatile components from any chemicals that would otherwise be used in water treatment.

Although the water and steam contain appreciable amounts of oxygen, carbon steels and stainless steels exhibit equivalent or superior corrosion resistance to that observed in pressurized water at the same temperatures. However, special problems do exist. Stainless-steel fuel cladding has failed by SCC in boiling-water reactors.

SCC has also occurred in sensitized stainless steels (i.e., those with chromium carbide precipitation at the grain boundaries) as a result of heating the steel in the 425 to 825°C range by welding or stress relieving. For this reason, Zircaloys are the preferred cladding materials. However, care must be exercised that alkaline components do not enter the water and become concentrated by boiling on the cladding surface because Zircaloys have poor corrosion resistance in high-temperature caustic solutions.

In a boiling-water reactor, the nuclear fuel boils the water and the steam goes directly to the turbine (Figure 4.21). Temperatures are approximately 230 to 290°C, and with saturated steam from 2.8 to 7.2 MPa (400 to 1050 psi). As described with conventional boilers, the silica and copper content of the water must be kept low to prevent their transport by steam and deposition in the turbine.

The boiling-water reactor design can be modified to accommodate a superheater to raise the steam temperature for improved efficiency, if desired. Experimental nuclear superheaters have been built to superheat the steam from the boiling-water reactor. Corrosion rates for stainless steels and nickel alloys under heat transfer in the 600°C temperature range are about 25 μm/y (1 mpy). However, experience has shown that

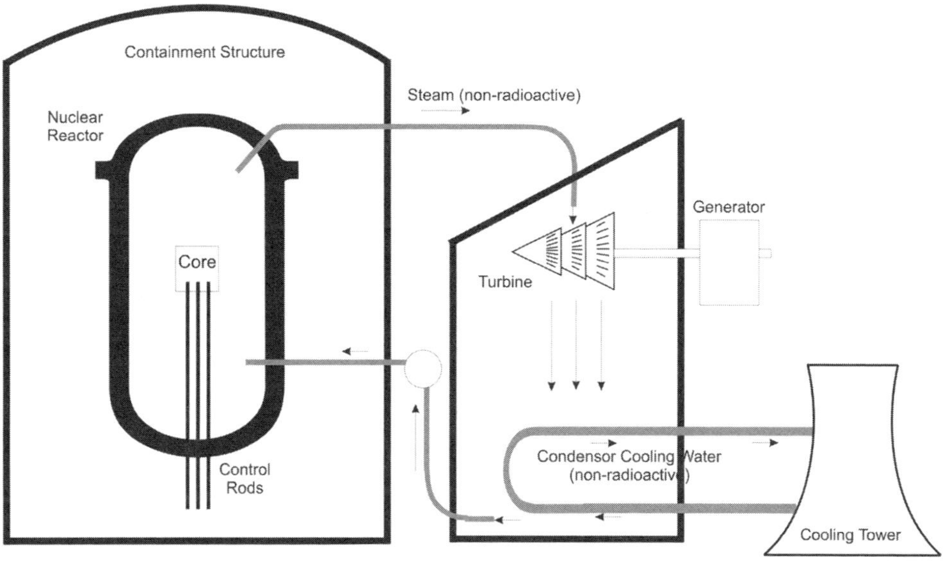

Figure 4.21 Boiling-water reactor.

even minute amounts of entrained moisture in the entering steam can carry sufficient chlorides that cracking of the stainless-steel cladding of the nuclear-fuel elements may ultimately ensue. Alloys with higher nickel contents (e.g., N08800, N06600, or N06690) have resisted SCC under these conditions.

4.7.4.2 *Pressurized-Water Reactors*

In the pressurized-water reactor (Figure 4.22), high-purity water is pumped past nuclear-fuel elements where it becomes heated. The heated water then goes to the tube side of

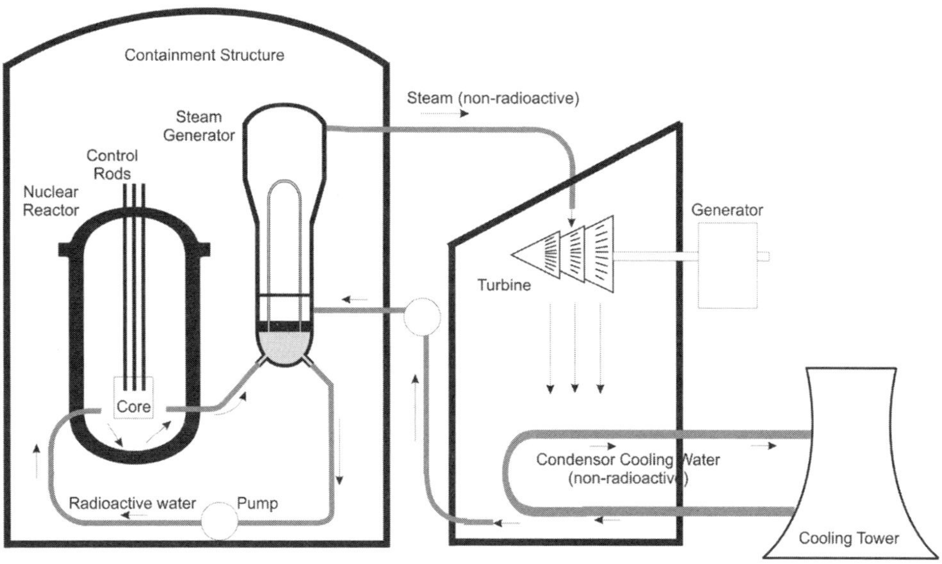

Figure 4.22 Pressurized-water reactor.

a heat exchanger, giving up its heat to boil water on the shell side. Steam is generated from the boiler water.

The water adjacent to the core is called primary water. It is usually treated with hydrazine to remove oxygen, with lithium hydroxide or ammonium hydroxide to pH 10 to 11 to minimize corrosion-product transport, and with 25 to 50 cc of hydrogen per kilogram of water to suppress radiolytic decomposition of water (thereby minimizing corrosion-product transport).

Fuel and Fuel Cladding The nuclear fuel must be clad with a corrosion-resistant material to prevent the release of radioactive gases and fission products to the primary water. The fuel itself is usually uranium oxide, which is quite resistant to the primary water. The fuel is clad with austenitic stainless steel or Zircaloy-2 (R60802) or the extra-low nickel Zircaloy-4 (R60804).

The Zircaloys are preferred because they do not "poison" the nuclear reaction as much as the stainless steels. The Zircaloys perform well in the 285 to 315°C temperature of the primary water for the two- to three-year expected life of a nuclear core. They develop a shiny, adherent black oxide film, which is protective and has excellent heat-transfer properties. However, after long exposure times (two to four years) or shorter times at higher temperatures (e.g., 40 to 120 days at 360 to 400°C), the corrosion rate of the Zircaloys increases and a white, relatively nonadherent insulating film develops.

The stainless steels, on the other hand, develop a relatively thick tarnish film with a thin, powdery surface film. Both films have a nominal composition of M_3O_4 where M represents iron, nickel, or chromium. The corrosion rate of stainless steel is not greatly affected by temperature in the range of 260 to 400°C, and it does not exhibit the marked effect of temperature on corrosion rate that is characteristic of Zircaloys.

Control-Rod Mechanisms The drives and bearing surfaces for inserting and withdrawing the control rods in the core are made of hardenable alloys. AISI Type 410 (S41000) and 17-4 PH (S 17400) stainless steels have exhibited SCC or hydrogen-induced cracking (Chapter 8) when heat treated to hardness in excess of Rockwell C 35-40. Lowering the strength level (lower hardness) has eliminated the problem. Chromium plating has been ineffective. Hard cobalt-base alloys have been used in bearing surfaces but tend to contribute the long-lived radioactive isotope, cobalt 60, as a contaminant in the water.

The control material itself (neutron absorbers such as boron or cadmium) is clad with stainless steel for corrosion protection. Hafnium is sometimes used bare or unclad as a control material because its corrosion resistance is superior to the Zircaloys.

Piping and Steam Generator The piping, steam generator, and pressure vessel constitute much of the total exposed area in the primary section of a pressurized-water reactor. The internal surface of the pressure vessel is clad with austenitic stainless steel. The piping and steam-generator components are primarily austenitic stainless steel or Alloy 600 (N06600). Carbon steel has been used in a few systems.

The corrosion rates of austenitic stainless steels and Alloy 600 are about the same, approximately 1.5 μm/y (0.06 mpy). The corrosion rate of carbon steels is 5 to 10 times higher, to a maximum of about 13 μm/y (0.5 mpy). These rates are acceptable from a structural standpoint. However, as much as one-half of this oxide may not remain on the surface but becomes a radioactive "crud" in its passage through the reactor core. Its subsequent deposition constitutes a personnel hazard.

The isotopes in the corrosion products that contribute most to radioactivity are Co-60, Co-58, Fe-59, Mn-54, and Cr-51. These isotopes are formed from the elements in stainless and nickel-base alloys. Their half-lives (i.e., the period over which one-half of the initial activity decays) range from 27 days to more than five years. The Co-60 has a long half-life, and this is the reason for minimizing cobalt content in nuclear-grade alloys and for using cobalt-base alloys sparingly.

It is generally agreed that pH in the range of 6 to 10, oxygen up to 5 ppm, and irradiation have little or no effect on the corrosion rate of stainless steels. However, these factors do affect the amount of crud released into the water; the amount increases with decreasing pH, increasing oxygen content, and irradiation.

Because the primary water is pressurized, it does not boil. There is, therefore, no opportunity for traces of chloride contamination or caustic from the water treatment to concentrate and cause SCC. The dissolved chloride content is usually restricted to less than 0.1 ppm. Flow rates are of the order of 7 to 10 m s^{-1} (20 to 30 fps), and the temperature is about 260 to 290°C, with water treatment as described for boiling-water reactors.

Most structural components are also fabricated from stainless steel with corrosion characteristics as described for the piping.

Secondary (Boiler) Water There is no radioactivity on the secondary (shell) side of the heat exchanger. Crud transfer is therefore no problem. This secondary water is treated like any conventional boiler water.

The usual conditions are 230 to 260°C, pH 10 to 11, 100 to 300 ppm phosphate, less than 0.1 ppm chloride, and sodium sulfite treatment to an oxygen content of less than 0.1 ppm.

Under these conditions, the oxide film that forms on the stainless steel of the steam generator and on the carbon steel of the steam drum is the protective black magnetic M_3O_4. Any evidence of the brick-red Fe_2O_3 indicates that there have been traces of DO in the water.

Both caustic and chloride SCC have been observed in the stainless-steel-tube heat exchanger (steam generators) in pressurized-water reactors. The cracks have initiated from the secondary water side (i.e., outside the tubes). The cracks have been found in the crevices where the tubes pass through the baffle plates or where the tubes are rolled into the tube sheet. Cracking has also been found where steam blanketing and localized boiling has occurred on the tubes.

Just a few ppm of chlorides or free caustic in the boiler water can become sufficiently concentrated at these areas to cause cracking. DO in the water contributes to chloride cracking. Analyses of corrosion products in the vicinity of chloride stress corrosion cracks have revealed the presence of several thousand ppm of chloride. Caustic, on the other hand, although it may become concentrated in such areas, does not remain in the corrosion product.

Double tube sheets have been suggested as a remedy for this problem but are not necessarily effective. Experience has shown that some leakage occurs past the expanded tubes in the inner tube sheet and into the drain area between the two tube sheets. This water flashes to steam, and the resulting concentration of corrosive species can still cause SCC.

4.7.4.3 Supercritical Reactors

Many of the same problems exist with nuclear supercritical power plants as with the fossil-fueled variety: 1) the supercritical steam is an excellent solvent, so solids in the water must be kept to a minimum to avoid deposition in the turbine, and 2) changes in density can result in localized buildup of any chlorides or caustic, with the attendant possibility of eventual SCC by these ions.

References

1. G.H. Koch, M.P.H. Brongers, N.G. Thompson, Y.P. Virmani, J.H. Payer, "Corrosion Costs and Preventive Strategies in the United States," FHWA-RD-01-156, National Technical Information Service, Springfield, VA, 2001.
2. P.R. Roberge, Handbook of Corrosion Engineering (New York, NY: McGraw-Hill, 2000).
3. W.E. Berry, "Water Corrosion," in Corrosion Basics, eds. L.S. Van Delinder, A.D. Brasunas (Houston, TX: NACE, 1984), pp. 149–176.
4. H.H. Uhlig, "Iron and Steel," in The Corrosion Handbook, ed. H.H. Uhlig (New York, NY: Wiley, 1948), pp. 125–143.
5. F.L. LaQue, Marine Corrosion: Causes and Prevention (New York, NY: Wiley 1975).
6. J.E. Oddo, M.B. Tomson, "Scale Control, Prediction and Treatment or How Companies Evaluate a Scaling Problem and What They Do Wrong," Corrosion 1992, paper no. 34 (Houston, TX: NACE, 1992).

5

Chapter

Corrosion in Soils

5.1 INTRODUCTION

It has long been observed that an iron pipe buried in a dry soil suffers much less corrosion than one buried in wet soil. Because of rain, natural springs, and rivers, soils are rarely dry, especially beneath the surface, and can contain a variety of mineral matter.

An explanation of corrosion behavior under a variety of conditions is not easy to give; however, one important factor to consider is the electric conductivity of the medium that surrounds the metal. Pure water is a very poor electric conductor, but as substances dissolve in it, particularly those that ionize, conductivity rises sharply. This simple principle is the foundation of the corrosivity scale that will be described later in this chapter.

5.2 HOW DO METALS CORRODE?

Any two or more of the factors described in the following sections can operate at the same time. Sometimes they will work in synergy, sometimes they will be opposed. For example, in a well casing, there is usually an oxygen cell and a temperature cell, both tending to make the deep pipe act as an anode. At the same time, there may be two or more ordinary concentration cells, because the casing passes through different strata of soil.

The corrosion of guyed tower anchors such as those shown in Figures 5.1 and 5.2, which is becoming an issue of critical importance, is used in the next sections to illustrate some of these corrosion cells. Soil type can vary even within a few centimeters and this variability can cause multiple corrosion cells on the same structure. In many cases, the loss of one anchor to corrosion (Figure 5.3) has resulted in the catastrophic loss of a supported communication tower (Figure 5.4).

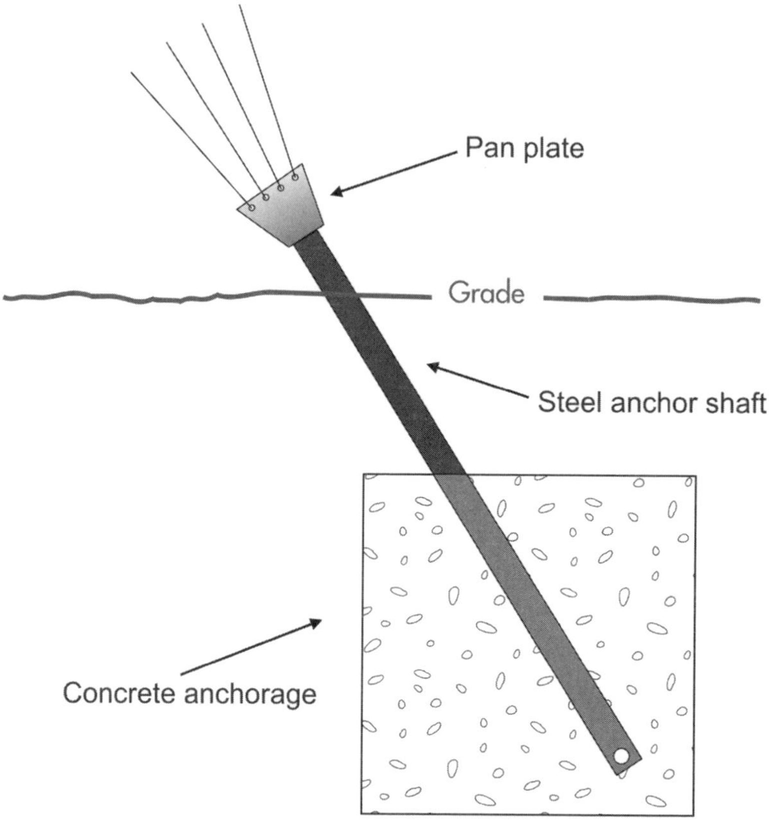

Figure 5.1 Basic anchor-support design (courtesy of Anchor Guard).

Figure 5.2 Typical anchor support with a copper-rod ground connection (courtesy of Anchor Guard).

Figure 5.3 Anchor assembly was cut by corrosion at the concrete anchorage (courtesy of Anchor Guard).

Figure 5.4 Mangled communication tower fell because of loss of the anchor in Figure 5.3 (courtesy of Anchor Guard).

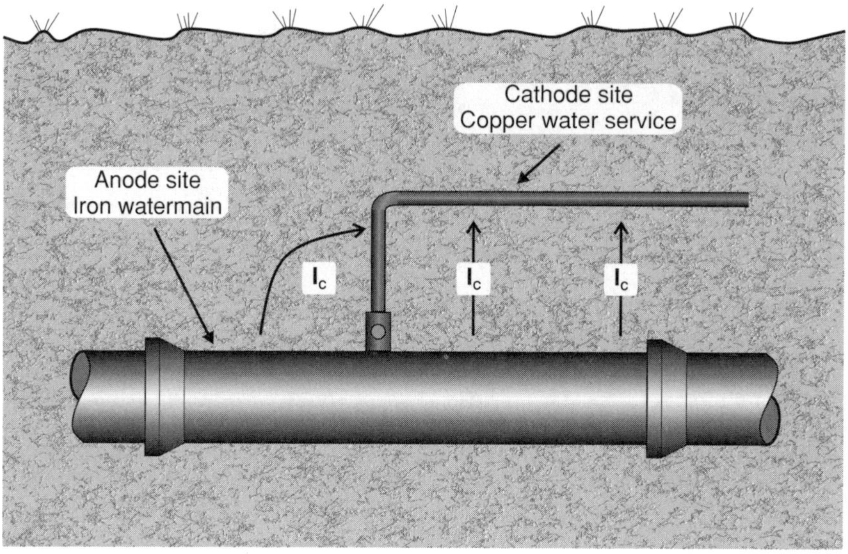

Figure 5.5 Underground-corrosion cell involving connection of dissimilar metals.

5.2.1 Galvanic Corrosion

Any two metals can be used to make a galvanic cell. The subject of galvanic corrosion and galvanic cells is covered in great detail in Chapter 9. Whether a metal will behave as an anode or a cathode in combination with another metal in the soil can usually be determined by its relative position on the galvanic series in soils (Table 9.4, Chapter 9). The metals higher on the list (more negative) will, in general, be anodes and thus will corrode. The metals lower on the list (more positive) will be cathodes and thus will not corrode. Of course, this galvanic action will not take place under open-circuit conditions; there must be a connecting circuit.

Do galvanic cells occur often in soils? Are they important as sources of corrosion? The answer to both questions is yes, although galvanic cells are not the most important kind of corroding cell. Whenever a copper-pipe service line is directly connected to a cast-iron gas or water main, a galvanic cell is formed (Figure 5.5) . The soil is the electrolyte, the copper service line is the cathode, the iron (or steel) main is the anode, and the connecting circuit is completed by attaching the line to the main. Normally, such cells do not do any great amount of damage, because the anode (corroding metal) is so much larger than the cathode that the attack is spread out over a large area (refer to Area Effects in Chapter 2, Section 2.5.7).

However, the situation can lead to the corrosion of steel when the surface area of the cathode (copper) approaches that of the anode (steel), such as shown in Figure 5.6, or when the surrounding soil is more conductive or corrosive in the region where the galvanic coupling exists Figure 5.7. In this example the anodic corrosion of the anchor is partly due to the galvanic corrosion cell formed by the connection of ground copper rods. One way to minimize the dissimilar metal interaction that causes the corrosion of the structural-steel anchor is to break the electrical path between these two components by adding insulators as shown in Figure 5.8. However, this measure alone has proven

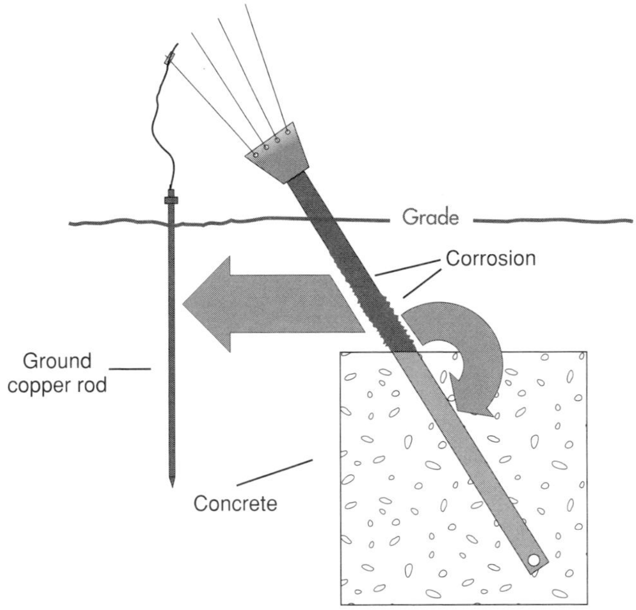

Figure 5.6 Schematic shows the corrosion cells acting on an anchor support with a ground connection (courtesy of Anchor Guard).

to be only a partial solution to this type of problem because another corrosion cell is still active, as is described in the next section.

5.2.2 Concentration Cells

This type of corrosion cell is formed when a metal is exposed to an environment containing varying levels of electrolytes, either of different substances or of the same

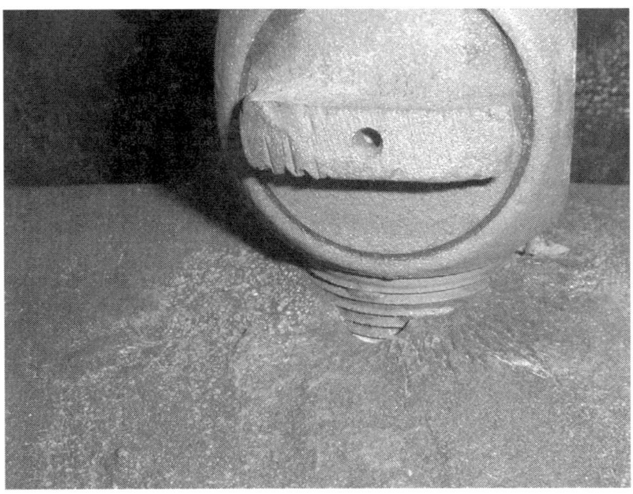

Figure 5.7 Corrosion of a water main: ductile iron adjacent to a copper fitting (courtesy of Drinking Water Services, City of Ottawa).

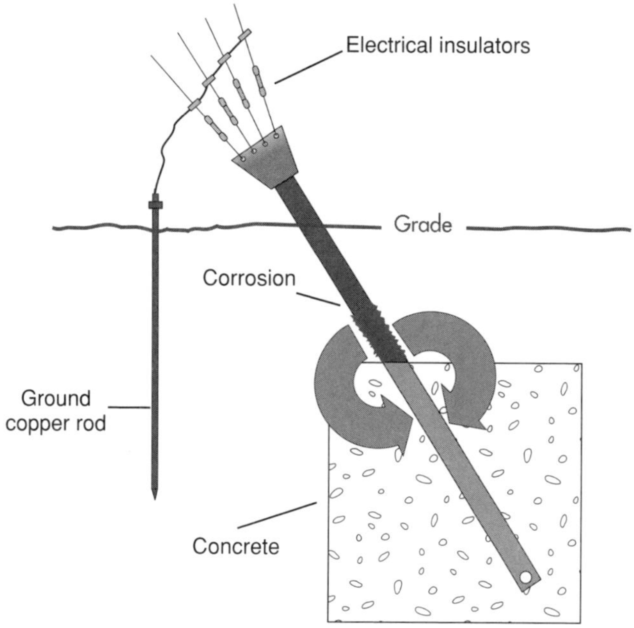

Figure 5.8 The addition of insulators between the ground rod and the steel anchor stops the galvanic action (courtesy of Anchor Guard).

substance in different amounts (see Principles of Electrochemistry Applied to Corrosion in Section 2.5). For example, if one electrolyte is a dilute salt solution and the other a concentrated salt solution, a concentration cell may be formed. This takes place because one of the factors that determine an electrode potential is the electrolyte concentration.

The dissolved salts can be quite complex mixtures. They usually include compounds of aluminum, calcium, magnesium, and other metals, and they may be sulfates, chlorides, hydroxides, or any one of a variety of other compounds. The role played by these salts can also vary greatly. Chloride ions, for example, can be quite aggressive toward steel, and sulfate ions can serve as nutrients to SRB, which themselves can be extremely damaging to most buried metals.

In the example shown in Figure 5.8, the main difference in the environment surrounding the steel anchor is due to the variation of pH between the soil, which can vary greatly depending on the type of soil but is often acidic, and the pH in the concrete anchorage, which is always basic and higher than 10. Because this higher pH is protective to steel, it naturally follows that the steel, in concrete becomes the cathode, drawing an anodic current form the adjacent steel in the soil.

5.2.3 Oxygen-Concentration Cells

This is a specific type of concentration cell that has a difference in the amount of oxygen reaching different parts of a metallic structure. A potential difference can develop between them, as it was explained in Chapter 2.

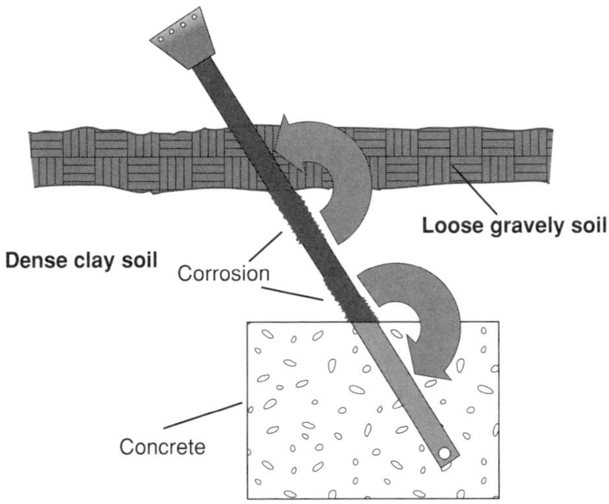

Figure 5.9 Differences in the porosity of the soil can lead to an oxygen-concentration corrosion cell (courtesy of Anchor Guard).

In the following example the concentration cell is caused by differences in the electrolyte around a buried anchor rod. Figure 5.9 illustrates how various soil types can create a corrosion cell on a tower-anchor shaft. In this illustration, we find the upper soil layer a looser, somewhat gravelly soil and below that more dense clay-type soil. The portion of the shaft in contact with the clay-type soil acts as an anode to the portion of the shaft in contact with the looser gravely soil, which is consequently the cathode. Again, we have a corrosion cell where the shaft deteriorates in the anodic areas.

Such differential aeration cells are very common on buried pipes. For example, a pipe usually rests on undisturbed soil at the bottom of the ditch. Around the sides and on top of the pipe is relatively loose backfill that has been replaced in the ditch. Because the backfill is more permeable to oxygen (and the path is shorter) diffusing down from the surface, a cell is formed. The anode is the bottom surface of the pipe and the cathode is the rest of the surface. The electrolyte is the soil, and the connecting circuit is the pipe metal itself. This explains why most corrosive attacks on pipe lines are on the bottom quarter of the pipe.

When a pipe or cable passes under paving, such as an airport runway, parking lot, or street (Figure 5.10), the portion under the paving has less access to oxygen than does the area lying under unpaved soil. Thus, a cell is formed:

- the anode is the pipe under the paving,
- the cathode is the pipe outside the paving,
- the electrolyte is the soil, and
- the connecting circuit is the pipe or cable.

In this particular example, although the entire length of pipe under the paving is anodic, most of the attack will take place not far from the edge. Because the path through the electrolyte is shorter to this part, most of the current takes this low-resistance path.

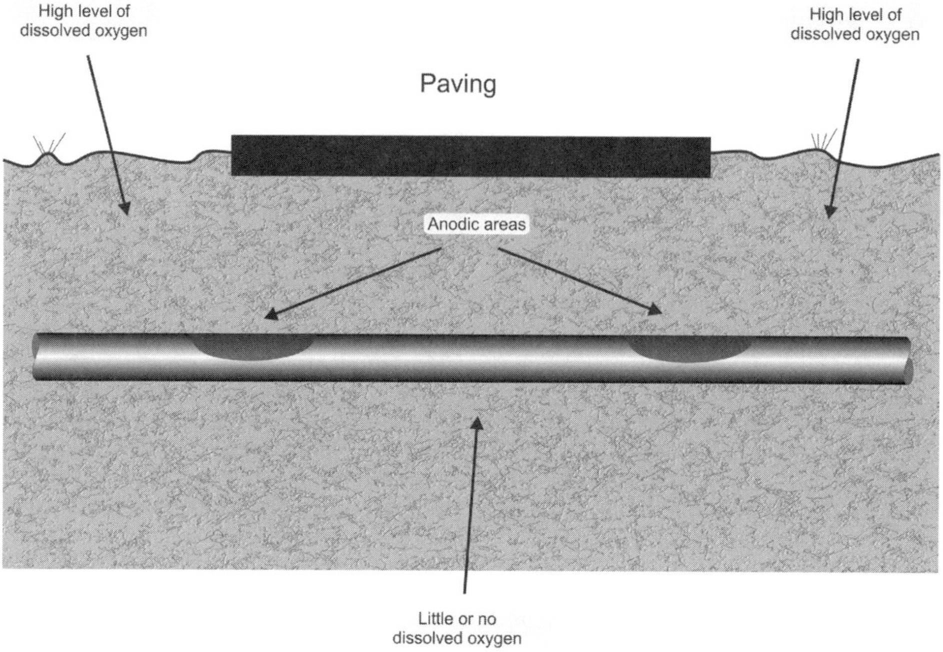

Figure 5.10 Oxygen differential cell is a result of burial under paving, producing an oxygen-concentration cell.

The casing of an oil or gas well usually is connected to a network of surface piping, that is, a gathering system that lies just below the surface. This can form an oxygen cell:

- the casing, at depth, forms the anode;
- the surface piping, with greater access to oxygen, becomes the cathode;
- again the soil is the electrolyte; and
- the piping and casing together make up the connecting circuit.

The action of this cell can be controlled by insulating the surface piping from the casing by means of a special insulating union or other fitting. Note, however, that this will not stop all attacks, because of the concentration cells described in the preceding section.

Oxygen concentration cells are responsible for much of the corrosion near the water line on piling driven into the bed of the sea or in fresh-water bodies. Wave action maintains a constant supply of oxygen to the metal just below the surface, whereas at greater depths there is little oxygen penetration.

5.2.4 Temperature Cells

The two cells just described, concentration and oxygen, are responsible for perhaps 90% of the corrosion in soils and much of that in natural waters. There is still a great

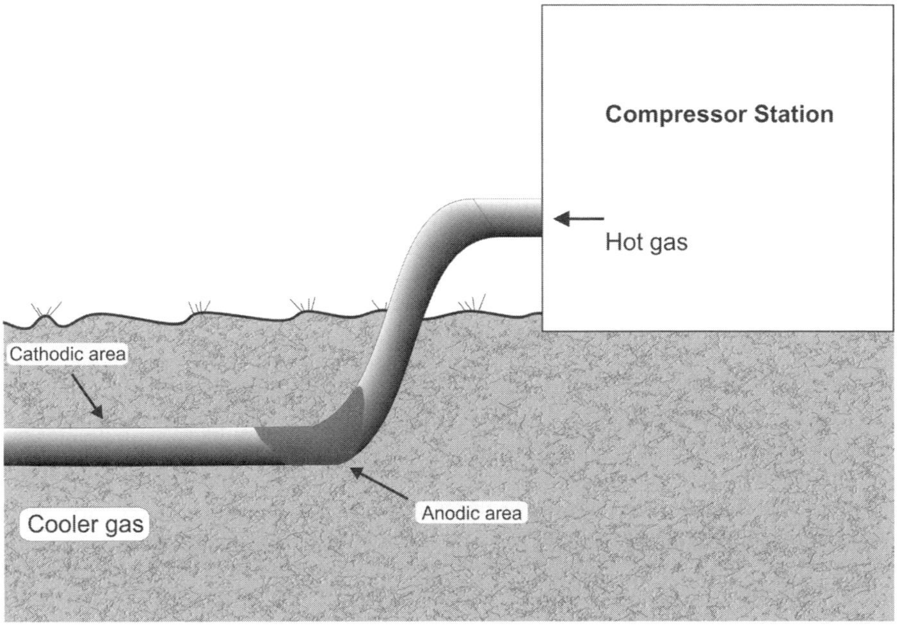

Figure 5.11 Concentration cell resulting from heat differential.

deal of corrosion from other types of cells, however, so additional types of cells cannot be ignored.

One such cell is the temperature cell shown in Figure 5.11, in which the two electrodes are of the same metal, but one is maintained at a higher temperature than the other by some external means. In most cases, the electrode at the higher temperature becomes the anode, causing thermogalvanic corrosion.

A gas transmission line emerging from a compressor station serves as an example. The compressed gas coming out of the station is hot, and, as it travels down the line, it loses heat by transfer to the surrounding soil (as well as losing some by expansion). The result creates a cell:

- the hot pipe near the compressor is the anode,
- the cooler pipe down the line is the cathode,
- the soil is the electrolyte, and
- the pipe itself is the connecting circuit.

This is a particularly unfortunate type of cell because the high temperature just outside the compressor tends to damage the coating, so that the part of the line that is the anode is also likely to have the poorest coating.

External corrosion of underslab, copper, hot-water tubing has resulted in a great deal of construction-defect litigation because of this type of corrosion.[1] Oil and gas-well casings also experience similar cell attack. The pipe at some depth below the surface is at a higher temperature (this is a natural phenomenon; temperature increases with depth) and thus becomes the anode, the cooler pipe near the surface and the surface piping make up the cathode, the soil is the electrolyte, and the pipe is the connection. Note that this cell operates in the same direction as the oxygen cell described earlier; both tend to concentrate the corrosion on the casing at depth.

5.2.5 Stray Currents

In all cells described so far, the source of the energy that makes the cell active has been within the cell. The source has been the solution potential of the anode (the amount by which it is greater than that of the cathode), some distribution of energy in the metal, or a concentration gradient in the electrolyte.

It is possible to have a cell in which the energy is supplied in the form of electrical current from an outside source. An electroplating cell is one example. In this case, energy comes from a generator supplying direct current. A car battery that is being charged is another example. The power in this instance comes from the car's generator, which actually forces current through the cell in a direction opposite to the driving voltage of the cell itself (the battery is really a set of three or six cells).

In a cell of this kind, the anode is still the electrode, from which current flows to the electrolyte, and the cathode is the one to which current flows from the electrolyte. In a storage battery, the "corrosion" that takes place at the anode is a reversible reaction: When the battery is discharging, this electrode becomes the cathode, and the chemical action is the reverse of that taking place during charging.

Impressed current cells in which the soil is the electrolyte are of two kinds: accidental and deliberate. The deliberately impressed current system is the one that supplies cathodic protection to a threatened structure. Accidental systems, however, may exist under a variety of circumstances. Any direct current flowing in the soil from any source whatsoever can, if it finds a pipe line or other metal object in its path, collect on the pipe in one area and discharge from it in another. The area where it collects becomes a cathode whereas the area where it discharges is an anode, and thus corrosion occurs.

The source of energy for such cells may be a distant generator, a direct-current transmission line, a CP rectifier on some other line, a street-car system, or an electric railway. Such a cell can result in troublesome problems if left uncontrolled, as was found out early in the twentieth century (see Chapters 1 and 11 for some examples).

5.2.6 Stress Cells

If an ordinary nail is dropped into a vessel of salt water, it will be attacked, and, after some time, rust will be observed discoloring the water. The initial points of attack will be, in almost every case, the point and the head. This is an example of a type of cell known as a stress cell. Figure 5.12 shows how dramatic this effect can be on the heads and nuts used to hold fire hydrants on their base in the soil. In these cases the two electrodes are of the same metal and the electrolyte is uniform. The difference lies in that one electrode is more mechanically stressed than the other. The area of high stress is always the anode of the cell because of the extra energy supplied by the stress itself.

Figure 5.12 Severe corrosion of fire-hydrant bolts and nuts because of stress cells (courtesy of Drinking Water Services, City of Ottawa).

Stress cells can take on two basic forms. One, like the nail just described, has its anode established by residual internal stress or stress left because of something that has happened to the metal. In the case of the nail, the "happening" was the cold-forming of the head and point. If these stresses are relieved by heating the nail and letting it cool slowly, the stresses will disappear and this type of stress cell will be eliminated.

In the other type of stress cell, the metal is part of some kind of structure that is under stress. The most highly stressed part of the metal becomes the anode, with the less stressed or unstressed metal acting as the cathode.

Pipe that has been subjected to cold bending can set up such a cell. Other pipes, especially in manifolds and similar locations, that happen to have been forced into alignment and thus remain under large stresses, can also set up this type of cell. An improperly designed manifold may be relatively free from stress when built, but then develops some highly concentrated stresses because of operating pressures, temperatures, or both while in service. In addition to the hazard of simple mechanical failure, such a situation can intensify into an, SCC problem, as discussed in Chapter 9.

5.2.7 Surface-Film Cells

Some chemical or electrochemical reactions can take place on a metal surface, making it behave quite differently from a piece of clean metal, a situation closely related to dissimilar-metal corrosion. A film may be formed that is invisible, actually only a few molecules in thickness, but that may have a potential as much as 0.3 V different from the unfilmed metal. Naturally, such a potential difference is enough to create an active corrosion cell. Steels in soil have a tendency to film with time. "Old" steel, that is, steel that has been in the ground for several years, is cathodic with respect to "new" steel, even when the two are identical in composition. It is thus strictly a surface phenomenon.

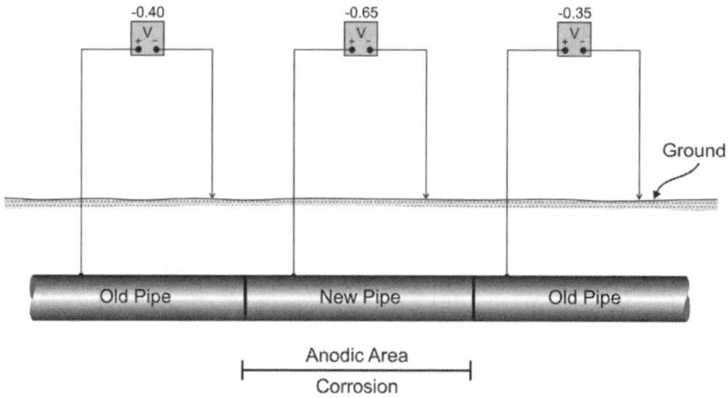

Figure 5.13 Schematic shows a differential corrosion cell created by the replacement of a section of pipe.

For example, suppose that a gas line had been in operation for ten years. During that time it has developed a number of leaks, and the rate of pit development is increasing. Before the leaking becomes critical, however, the gas demand increases to the point that the capacity of the line is not adequate. Accordingly, a new line is laid, parallel to the old one on the same right-of way. Also, in order to give maximum flexibility in operation, crossovers are installed approximately every sixteen km.

This situation forms an active corroding cell. The new line is the anode, the old line is the cathode, the soil is the electrolyte, and the crossovers form the connecting circuit.

The first effect is that corrosion is halted on the old line. Although it had developed a leak rate that was beginning to be troublesome, now it has no new leaks whatsoever. This seemingly happy state of affairs continues for several months, perhaps even a year or two. Then, suddenly, and to the great surprise and alarm of everyone concerned, the new line starts developing leaks, and at a greater rate than the old line had developed them. If the forces at work in this situation were not fully understood, it could well lead to the erroneous belief that the new steel was not as good as the old steel.

The same condition occurs when new steel pipe is intermixed with old steel pipe (Figure 5.13)[2]. This has often been found in older distribution-piping systems where a section of pipe has been replaced because of corrosion damage. The new piece of pipe, exposed to the same corrosion conditions, logically would be expected to last as long as the original section. However, the new section will usually fail sooner than expected unless it is electrically insulated from the remainder of the system.

5.3 PIPE-LINE CORROSION

By far the greater portion of the metal threatened by corrosion in soil and water is that which makes up pipe lines. When a new line is laid in the earth, a vast number of cells come into being:

- the line passes through a variety of soils, giving rise to concentration cells;
- there are variations in oxygen access, setting up oxygen cells;

- there are impurities and differences on the surface of the pipe, creating galvanic cells;
- cold bends and welding stresses set up stress cells; and
- in the great complexity of the soil, there are the opportunities for other, less-known types of cells.

These cells are of various sizes and shapes. In some cases, anodes and cathodes are separated by only a few centimeters; in others, they may be kilometers apart. The cells can also greatly vary in strength, from a few millivolts to sometimes half a volt. Anodic areas are of all sizes, tiny to large, as are the cathodes.

As soon as these areas become active, they begin changing. On cathodic areas, hydrogen may begin to form, as described in Chapter 2. Changes affect both cell resistance and potential. Anodes begin corroding, which puts new ions into solution at the surface. As these react with various components in the environment, concentrations change. Cell potentials also change; some increase, whereas others decrease.

Some areas that were initially anodic become cathodic. The reverse also may take place, although not as often. Generally, as time passes, the total anodic area on a line gets smaller, although total activity does not decrease at the same rate. The result is that the rate of attack at the worst locations tends to increase. Finally, among all of the various cells along the entire line, the most active anode loses enough metal by corrosion to penetrate the pipe wall completely. The line thus has its first leak and its corrosion history begins.

5.3.1 Auxiliary Effects of Cell Operation

When a CP system is in full operation, there is a high possibility for oxygen to be produced at the anode, and in nearly all cells, hydrogen is formed at the cathode. If chloride ions are present, chlorine gas may be formed at the anode. This generation of gas, either oxygen or chlorine, at the anode is not nearly as likely to occur in a natural corroding cell as it is in a CP system, especially where inert anodes are used.

5.3.1.1 Hydrogen

The formation of hydrogen at the cathode, however, can occur during both normal corrosion and CP. Initially, it may be formed as nascent hydrogen, that is, single atoms are produced. These may then combine with oxygen to form water or combine with some other ion in the environment to form some other compound. It may dissolve in the metal (cathode area) or combine in pairs to form ordinary gaseous hydrogen.

Under special circumstances, hydrogen formation can be damaging. Atomic hydrogen formed at the cathode may diffuse into the steel, and, at some depth below the metal surface, later recombine to form the ordinary molecular gaseous hydrogen. Unlike the single atoms, these hydrogen molecules cannot migrate through the structure of the metal. Consequently, there is an accumulation of hydrogen gas within the metal, which can generate enough pressure to blister or split the solid steel. Hydrogen can also react with the metal, causing hydrogen-induced cracking HIC. Chapter 9 provides more details on this particular type of corrosion damage.

5.3.1.2 Electroendosmosis

Another side effect of the passage of electrical current through a porous medium such as soil is electroendosmosis. Water is "carried along" with the current. Normally this effect is significant only with higher current densities than natural corrosion cells would produce. Like hydrogen formation, it is more likely to be associated with cathodic protection. Its effect there is to carry water to the cathode and away from the anode, which can result in an unwanted increase in anode resistance if the anode is above or not far below the water table.

5.3.1.3 Cathode Scale

A third effect is the deposition of compounds from the soil on the pipe. This effect is extremely pronounced in seawater, where a cathode scale can be built up over a period of time that is like a hard, glossy enamel, and which can build up to more than a couple of centimeters. This material is, when deposited in seawater, a mixture of calcium, magnesium oxides, and hydroxides. It is a fairly complex substance, and its composition depends upon current density, among other factors.

In fresh or brackish water, cathode-scale composition is even more variable than that formed in the sea because its formation is based on highly variable ion concentrations. This is also true in soils. Some soils will not form a visible scale, although there is enough calcium present in most soils for some scale formation. Often, when a line has been under cathodic protection, no scale will be visible until the pipe surface dries out. It then appears as a whitish discoloration. On a line that has not been under protection, cathode scale is frequently found in an irregular mottled pattern; this makes the actually active cathodic areas visible to the eye.

5.3.1.4 Pitting

When a pipe line that has been in corrosive soil without adequate protection for some time is examined, it is usually found that the greater part of the area by far is unaffected. Where corrosion has taken place, it is in the form of pits, which are relatively small areas where the attack has been deep (Figure 5.14).

In general, there always will be one spot on any pipe line where all of the conditions combine to give the highest rate of penetration. This is the spot where the first leak will occur. Because the underground pipe is invisible to the observer, it also will be the first place where corrosion will make its presence known.

The effectiveness of CP to reduce leak rates on a pipe was demonstrated more than 50 years ago when CP was applied to an old piping network that had been developing leaks at a rapidly increasing rate. Reduction in the number of leaks immediately after

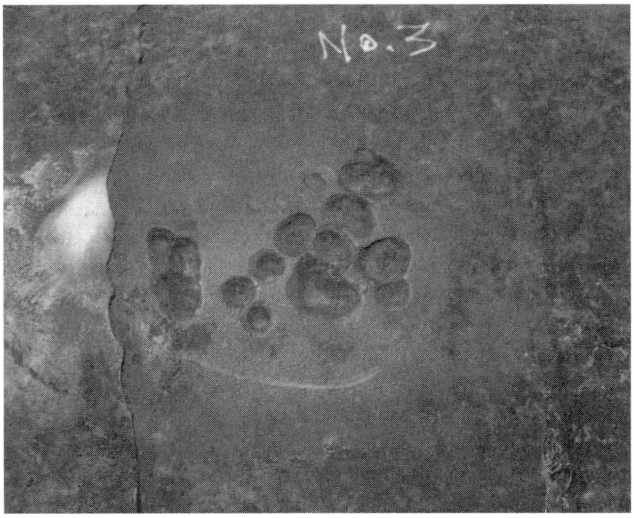

Figure 5.14 Typically deep, sharp-sided pits are usually found under disbonded coatings or hard accumulations on a pipe surface. When disturbed, the area may smell of hydrogen sulfide, a telling sign of microbial activity (courtesy of *MACAW's Pipeline Defects*, published by Yellow Pencil Marketing Co.).

the CP installation was impressive, as shown in Figure 5.15. This study was made on mains in corrosive soil, where leaks were becoming so numerous that abandonment was seriously considered prior to the decision to apply CP.[3] A similarly impressive reduction in the frequency of leaks on a cast-iron water main was reported in 1980 after application of CP [4] and more recently with additional details as shown in Figure 5.16.

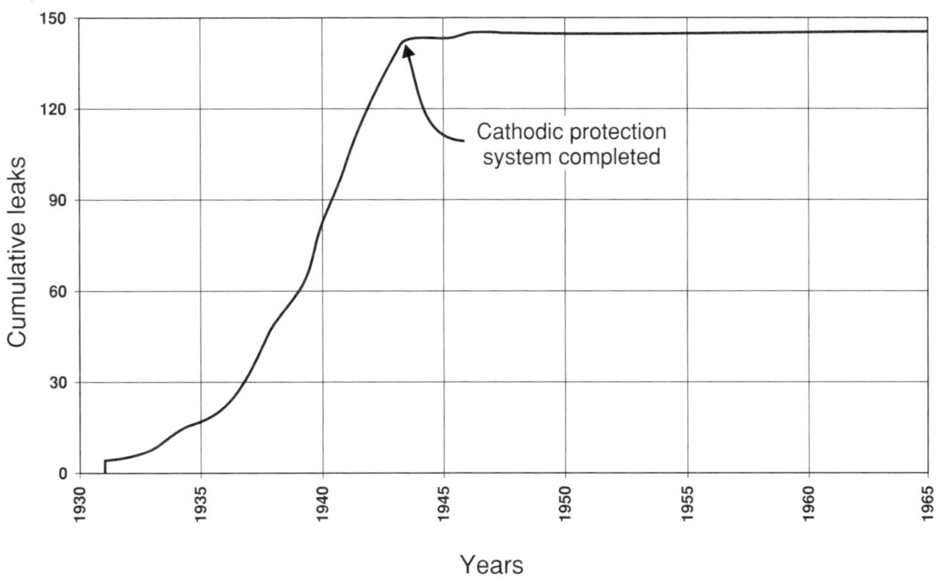

Figure 5.15 Effectiveness of CP in stopping the development of leaks.

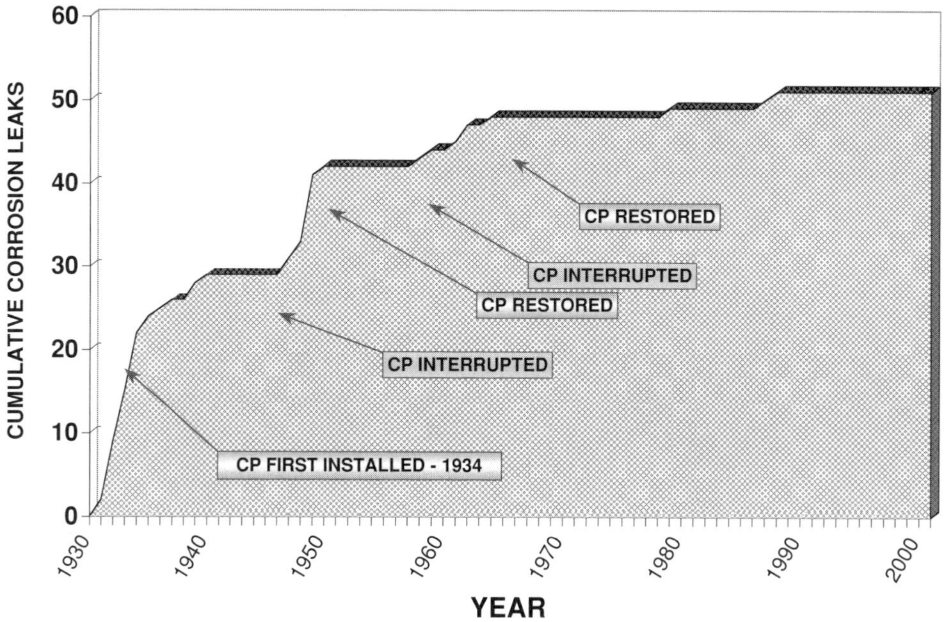

Figure 5.16 Leak history and CP on Aqueduct No. 1 (East Bay Municipal Utility District, Oakland, California).

5.3.2 Electrolyte Resistivity

The high leak rate of an unprotected line depends on a few factors. The greatest difference is in the soils in which lines lie. An easy way to get an approximate measure of the quantity of dissolved salts in soil is by measuring the electrical resistivity of the soil.

Probably the second largest factor, which can easily be argued to be the first, is the coating. This point is discussed in Chapter 10. Pipe-wall thickness, obviously, will also have an effect on leak rates. Other things being equal, a thin-walled pipe will develop a leak at a given location in less time than one with a thicker wall.

Soils come in many varieties and have many different properties that can be measured. Soils may be classified by texture, organic content, dissolved chemicals, color, origin, and so forth, depending upon in what one is interested. In the control of pipe-line corrosion, resistivity measurements are valuable in two respects: they help predict the probable corrosiveness of soils and waters, and they are needed for the proper design of CP systems.

Because ionic-current flow is associated with soil-corrosion reactions, high soil resistivity will arguably slow down corrosion reactions. Soil resistivity generally decreases with increasing water content and the concentration of ionic species. Sandy soils are high up on the resistivity scale and therefore are considered the least corrosive. Clay soils, especially those contaminated with saline water, are on the opposite end of the spectrum. The generally adopted corrosion-severity ratings are shown in Table 5.1.

Table 5.1 Corrosivity Ratings Based on Soil Resistivity

Soil Resistivity (ohm cm)	Corrosivity Rating
> 20, 000	Essentially noncorrosive
10,000 to 20,000	Mildly corrosive
5,000 to 10,000	Moderately corrosive
3,000 to 5,000	Corrosive
1,000 to 3,000	Highly corrosive
< 1, 000	Extremely corrosive

Soil resistivity is by no means the only parameter affecting the risk of corrosion damage. A high soil resistivity alone will not guarantee absence of serious corrosion. The AWWA has developed a numerical soil-corrosivity scale, applicable to cast-iron alloys (Table 5.2).[5] The severity ranking is based on assigning points for different variables. When the point total of a soil in the AWWA scale equals 10 (or higher), corrosion-protective measures (e.g., CP) should be considered or recommended.

Table 5.2 Point System for Predicting Soil Corrosivity According to the AWWA C-105 Standard

Soil Parameter	Assigned Points
Resistivity (ohm cm)	
< 700	10
700–1000	8
1000–1200	5
1200–1500	2
1500–2000	1
> 2000	0
pH	
0–2	5
2–4	3
4–6.5	0
6.5–7.5	0
7.5–8.5	0
> 8.5	3
Redox Potential (mV)	
> 100	0
50–100	3.5
0–50	4
< 0	5
Sulfides	
Positive	3.5
Trace	2
Negative	0
Moisture	
Poor drainage, continuously wet	2
Fair drainage, generally moist	1
Good drainage, generally dry	0

5.4 EFFECTS OF COATING ON CORROSION AND ON CP

An obvious method of controlling corrosion is that of interposing a barrier between the threatened metal surface and the corrosive medium, that is, some kind of coating. Because corrosion always requires the presence of an electrolyte (moisture) in contact with the metal, if a metal could be coated with a material that was absolutely waterproof and absolutely free from holes, all attack would be stopped. The coating, it should be noted, would not need these two properties only when it was applied: the two properties would have to be permanent.

Unfortunately, a perfect coating does not exist. There are, however, coatings that approach this degree of perfection, available at a high cost. We already know that corrosion can be controlled on a pipe line and on many other structures with a combination of a reasonably good coating and CP, and we know how much this combination costs. Thus, we are forced to acknowledge that no matter how good a new, improved coating might be, we cannot afford to pay more for it than we do for the older combination. Many questions therefore arise about the use of imperfect coatings:

- how good they are when used alone,
- what effect they have on the progress of corrosion, and
- how they affect the need for CP.

5.4.1 Without CP

Compared with a bare pipe line in the same location, a coated line will, in the long run, have fewer leaks. It may, however, have its first leak within a shorter time simply because the coating, with its relatively few holes, may happen to concentrate corrosion activity at one or more of the small imperfections in it, called holidays.

Under special circumstances, this effect can be even more pronounced. For example, the tendency of new steel to be anodic with respect to older steel can lead to a perplexing situation. Suppose that a bare line has been in operation for some time. At the worst spot, a leak develops. Soon others develop in the same general area and repairs are needed. So, a section of line is removed and replaced, thereby introducing new steel into the circuit.

At the same time, it is also decided to coat the new section. Because the coating is not perfect, there will be some holidays, or small areas not completely coated. This will have the effect of creating a cell with a small anode (the bare metal in the holidays) and a large cathode (the old steel pipe on either side of the new area). The potential of the cell will be as great as if the entire line were bare, but total activity will be concentrated on the small area of the holidays, where an early failure is to be expected.

Any other circumstance in which the anodic section of a line is coated while the cathodic area remains bare may produce similar results. The total amount of corrosion may be reduced by the coating, but the speed of penetration (and hence the rapidity of appearance of the first leak) is increased.

In a stray-current exposure area, it is extremely important that a good coating be applied to the line in the area of current pickup (which is cathodic and not subject to attack) to increase electrical resistance and thus minimize the amount of current that will have to be disposed of elsewhere.

5.4.2 With CP

The effect that coating has on CP requirements is simple and can be stated briefly: Application of a coating greatly reduces the amount of current required to obtain protection. The reduction may run from as much as 99.8% (for an extremely good coating) to as low as 50% (for a very poor, old, damaged coating).

5.5 EFFECTS OF CORROSION AND CP ON A COATING

The opposing effects are also worth considering; that is, both corrosion and CP can adversely affect a coating. As stated earlier, it is essential for a coating, aside from being affordable, to be waterproof and free from holes or imperfections, not just at the time of installation, but permanently.

To achieve such a long-term performance goal, it is necessary that a coating bond firmly to the metal surface. If the coating were perfect in all other respects, this would not be important. However, because the coating does have imperfections, it is necessary to keep moisture from penetrating indefinitely at each break. Realistically, then, a good bond is a definite requirement for any coating for pipe lines.

It is also necessary, or at least extremely helpful, if the coating has a high electrical resistance. This is so that CP can be applied as cheaply as possible. The more conductive the coating, the larger the amount of current needed for CP. Thus, high electrical resistance is another coating requirement.

It is often said that a good pipe line coating should have high dielectric strength, which is the ability to withstand high voltage without failure. All of the pipe-line coatings that have proven useful do, in fact, have high dielectric strength. Therefore, any pipe line coating material with low dielectric strength is likely to be ineffective. A high dielectric strength also makes it possible to use a high-voltage holiday detector in the inspection of the finished coating: if the detection sparks through, then there is either a bubble, a piece of foreign matter, or a thin spot in the coating.

Two cathodic effects have already been described: the formation of cathode scale and the release of hydrogen. These can take place at the cathode of any type of cell, although the latter effect, the actual generation of hydrogen gas, takes place only at fairly high negative-potential values not likely to be found in a corrosion cell but common enough in CP systems or in stray-current exposures.

Either of these effects is capable of damaging a coating. Given a small defect in the coating, the deposition of scale or of hydrogen under the edge of a sound coating can cause a progressive enlargement of the unbonded area. Under rare conditions, cathodic current can literally strip off a coating in a matter of hours. This can be one of the limitations of a CP-system design.

5.6 CORROSION SURVEYS

5.6.1 Resistivity

A corrosion survey may be conducted to determine these factors:

1. how severely a pipe line or other structure is attacked in a proposed location;
2. how severely an existing structure is being attacked;
3. how much damage has already been done, or,
4. What steps can be taken to control the corrosive exposure.

A survey of the resistivity along an existing or proposed route will give good answers to questions (1), (2), and (4), and will permit some fairly good estimations of question (3).

Except under what can only be described as research conditions, it is seldom profitable to conduct an extensive survey of resistivity along the route of a coated line, either existing or planned. There is really only one decision to make about a coated line, and that is whether to place it under CP. If the answer is negative, then the resistivity data serve no useful purpose. If it is affirmative, then all that is needed is information about possible anode-bed sites, and these should be taken at a greater depth than those for the pipe-line corrosion survey.

On a bare line, however, a resistivity survey makes a great deal of sense. It assembles data concerning two important problems: it tells where the line is subject to attack (hot spots), and it furnishes data needed for the selection of galvanic anodes of the proper sizes or the amount of impressed current that is likely to be needed.

There are many adherents to the belief that any soil that is sufficiently corrosive to justify coating a line also justifies CP. In other words, no line should be coated unless it is also protected. Acceptance of this conclusion, of course, covers the first decision referred to previously, thereby eliminating any justification (except as a research study) for a detailed resistivity survey along a coated line. Instead, a much more limited survey is conducted in conjunction with current-requirement tests to obtain the data needed for the design of the CP system.

5.6.2 Potential Surveys

Once the line has been placed under CP, potential surveys, with readings usually at wide spacing using permanent test leads, are needed to ensure complete and continued protection. If trouble occurs and some sections of the line are found to be less than fully protected, then close potential measurements, for example, 8-m spacing, may be useful in tracking down the difficulty.

A profile of the potential along the surface of the soil immediately above the pipe line can be run by any of several methods. Any sharp break in that profile must have an explanation:

- it may be an uninsulated lateral,
- it may be an accidentally shorted crossing line,
- it may be a shorted casing,

- it may be just a sharp change in soil resistivity, or
- it may be something else.

However, it must have some reason for existence, and whatever is causing it may be what is robbing the line of protection.

5.7 OTHER MEANS OF PROTECTING STEEL

The preceding discussion has cited the use of CP as the prime method of protecting steel buried in soil. Other measures, however, also may be adopted in special cases.

When economically and mechanically attractive, a steel line or tank in the ground may be covered with an alkaline material such as limestone. Leachings from the stone will neutralize any acidic constituents in the soil and surround the steel with an alkaline environment, which is noncorrosive. Calcium salts will eventually precipitate from the alkaline water to coat the steel. Any material with a reserve alkalinity could be used to encase the steel for this purpose. There are also proprietary products on the market that claim to provide a protective sheath around buried steel in a wet soil.

Casing the exterior of a pipe underground is often required for mechanical and corrosion reasons. The working pipe is placed on insulating spacers inside a pipe of larger diameter. Only the exterior pipe is in contact with the soil and must be protected in the usual manner. The working pipe is now immune from soil corrosion. The annulus between the two pipes must be considered, however. Condensation of moisture, and perhaps corrosive gases, occurs in this annulus. An inhibited grease, oil, formulated aqueous mixture, or amine should be pumped into this annulus and maintained properly to ensure full protection of the working pipe line.

5.8 STRUCTURES OTHER THAN PIPE LINES

5.8.1 Distribution Systems

Most of the previous discussion has been in direct reference to pipe lines, for reasons given earlier. Other structures may be considered in terms of how they may differ from pipe lines.

Distribution systems generally are composed of pipe of many different sizes and ages. The latter difference, in particular, can cause problems. Distribution systems are much more subject to accidental contacts with other lines, typically water lines. It is also common for distribution systems to be made of a greater variety of materials. For example, copper service lines may be used. Because of these differences, the approach to the protection of such systems is likely to differ in that more emphasis is placed on the use of small units and galvanic anodes, even when a coated pipe is involved.

5.8.2 Gathering Systems

Gathering systems do not ordinarily differ much from pipe lines. They offer more variety and often have more accidental contacts. Protection systems may encounter more interference problems in both directions. One common problem is that gathering lines, particularly in the early development of a field, may be laid in great haste, with little or no thought to the corrosion-control problem. Although this is not, strictly speaking, a difference in corrosion behavior, it does account for some serious problems.

5.8.3 Plant Piping

The corrosion activity in a complex of piping in a refinery or other industrial plant has several unique features:

- combinations of aboveground and underground piping;
- a considerable variety of sizes, materials, and functions; and
- almost invariably, a great variety of coatings applied in original construction, creating a range from bare to very well-coated pipe.

Another factor present in almost every plant is the large amount of bare copper in the ground system, typically a collection of ground rods that are connected by a network of large-diameter bare-copper cable. This can cause much difficulty, both in causing corrosion when no protective measures are taken and again as a complicating factor when protection is applied. Further problems may arise because of the presence of many paved and unpaved areas which sets, up troublesome aeration cells. Furthermore, the usual construction activity ultimately results in badly mixed soils. Soil-resistivity maps, covering entire plant areas, are useful both in anticipating trouble areas and in planning control measures.

5.8.4 Well Casings

Although a well casing may be considered as merely a vertical pipe line, it also has several unique features. It costs much more per unit length, so that leak repair is enormously expensive, if at all possible. The corrosive exposure is complicated by the variety of soil strata through which it passes. Normally, these are much more varied than would be found on an equal length of horizontal pipe line. Casing is usually bare (by necessity or custom). Any investigation is severely hampered by the fact that only one end of the pipe is accessible, so that most conventional surveys cannot be conducted.

5.8.5 Underground Tanks

Underground tanks usually are large enough to contact more than one stratum of soil, so they may be subject to concentration-cell action. They are almost always subject to oxygen-cell attack, even when under pavement. Often, there are fittings of different metals, and seldom is the coating as good as that used on pipe lines. One of the major problems in dealing with these structures is that it is difficult to justify an engineering study for each tank, yet they cannot all be treated alike.

5.8.6 Steel Piling

An important difference with steel piling is that a few pits or even holes have little effect on its structural strength. Consequently, much more corrosion can be tolerated than with pipe lines. Piling is almost always bare, vertical, and hence subject to the same kinds of cells that attack oil-well casings. Bonding often may be a problem because individual piles may not be interconnected electrically, a condition that makes both investigation and protection a problem.

5.8.7 Transmission Tower and Pole Footings, Guys, and Grounds

Electrical lines often have built-in problems because of the common use of extensive copper ground wire. The galvanic cells thus created can be ruinous. Although this problem can be avoided by using insulates between grounded components, there is still a problem with the corrosion of a large number of small units, which makes both study and protection difficult.

5.9 CORROSION OF MATERIALS OTHER THAN STEEL

5.9.1 Cast Iron

Pure iron, wrought iron, or mild steel have similar corrosion rates in underground situations. However, cast iron may be removed from the metal, leaving a network of carbon particles by the dealloying phenomenon, termed "graphitic corrosion" (Chapter 9). The residual carbon retains the form of the object and, unless the weak structure is fractured, the graphitized form will convey liquids. Some cast-iron water mains that are almost one century old are in this condition and continue to adequately serve their purpose.

Once the cast iron is graphitized, the exterior becomes an extremely noble electrode in any galvanic couple. Thus, care should be used in attaching uncoated or unprotected cast iron to other metals in the soils.

Cast iron can be also subject to pitting and, depending on soil corrosivity and the presence of bacteria, can lead to complete perforation of the pipe wall (Figures 5.17 and 5.18).

Coating of the exterior of cast-iron pipe with bituminous or lower oil coatings is often used to provide a measure of protection. More recently tape wrapping with self-adherent tapes have been found to be a very cost-economical method of providing an efficient barrier between the metal and a corrosive soil Figure (5.19). CP can then be applied if the mechanical joints used are bonded to provide one continuous electrical conductor.

5.9.2 Aluminum

Aluminum has been used underground with good results under certain circumstances. However, the metal is subject to attack by acids, alkalies, heavy-metal salts, chlorides,

Figure 5.17 A corrosion pit that led to the perforation and leak of a water main (courtesy of Drinking Water Services, City of Ottawa).

and galvanic action. If attack does occur, either from soil corrosion, galvanic effect, or electrolysis, the area affected is normally small but the pitting depth can be severe.

As discussed previously, the metal may be cathodically protected in either the coated or bare conditions with appropriate precautions. The dense oxide film on the metal surface acts as a coating to a significant degree when protection is applied.

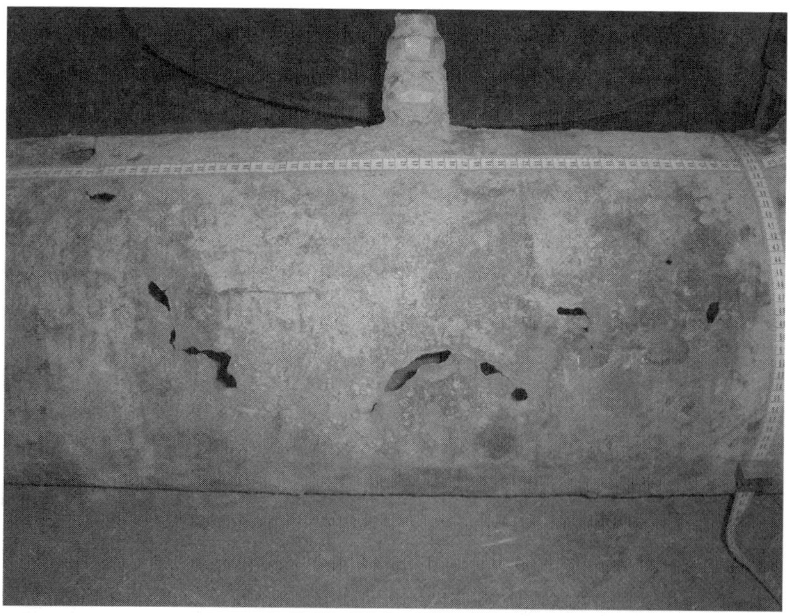

Figure 5.18 Patterns in corrosion pits that are probably a result of the galvanic situation shown in Figure 5.7 (courtesy of Drinking Water Services, City of Ottawa).

Figure 5.19 Wrapping a water-main pipe with adhesive tape provides a barrier against soil corrosion (courtesy of Drinking Water Services, City of Ottawa).

5.9.3 Zinc

Galvanized steel is not normally installed underground. The thin zinc coating is quickly corroded away by galvanic action with any exposed steel. As discussed previously, zinc is sometimes used as a reference half-cell in soils.

5.9.4 Lead

At one time, lead was used extensively as sheathing for telephone cables but has been supplanted by plastics in most areas. Although the metal is amphoteric (attacked by either acids or alkalies), a reasonable corrosion resistance was shown in most soils.

If CP of the metal is attempted, close control of the protection potentials is necessary to avoid alkaline attack on the lead.

5.9.5 Stainless Steels

It is sometimes necessary to use the austenitic stainless steels as underground piping. Very light-gauge pipe is used, and any attack on the exterior can be serious. In many soils, the passive film on the metal remains continuous and no attack occurs. However, pitting can develop in heavy soils and in chloride environments. When in doubt, such as with the lack of prior experience, the metal should be protected in the usual manner.

5.9.6 Copper

It is well known that copper serves well in underground exposure, unless high sulfide concentrations exist, for example, SRB. Burying copper in cinders, however, can cause severe pitting.

The brasses can degrade by dezincification in many soil exposures. Other copper alloys are resistant to essentially all natural soils.

5.9.7 Concrete

Concrete is extensively used underground as footings, piers, tanks, piping, and so forth. The material is normally used with a reinforcing agent such as steel bar of thin steel pipe (see Chapter 6), as in cylinder pipe. In most areas the concrete will be stable, but serious corrosion can occur from a number of sources. When appraising the potential corrosivity of a soil toward concrete, both the soil chemistry and the effect of ground water must be considered. The soil must be judged aggressive if

1. the pH is less than 6;
2. the pH is neutral, but hydrogen ion's are available by exchange;
3. the sulfate or sulfide content is high; or
4. the magnesium content is high.

Ground water percolating into the soil to extract the aggressive ions is responsible for the attack. Dry soil is innocuous. Acid attack on the alkaline concrete is to be expected. Thus, any acid spills in the vicinity of concrete structures should be neutralized and not allowed to penetrate deeply into the ground.

Sulfate and magnesium ions can be corrosive to the concrete in neutral solutions. Reactions occur with the calcium salts in the concrete to destroy the cohesion and produce a soft, porous mass. A Type V portland concrete (see Chapter 8) will resist the sulfate more capably, but aluminate cements are used to combat magnesium-salt attack.

Organic compounds, particularly esters, can rapidly degrade the usual concrete. Detergents can accelerate the rate of degradation. The freezing of absorbed water in the concrete can cause spalling or cracking.

Thus, when proposing the use of concrete underground one should be attentive to water levels in the ground, the selection of the proper concrete (there are many types), the proper cure of the material, the density of the finished product, the cleanliness of the water and sand used, the depth of coverage of the reinforcing metal, the possible need to seal the exterior, and the possible need to maintain the reinforcing metal as an electrically continuous structure.

The surface can be treated chemically to densify and harden the exterior (fluoride treatments or sodium-silicate washes) as a method of corrosion protection. Bituminous coatings are often applied to seal the exterior. Epoxy coatings are most compatible with the concrete as a coating, patching compound, or adhesive.

5.9.8 Polymeric Materials

Polymeric materials (plastics) have revolutionized many of the underground-materials applications. Although certainly not immune to failure, the use of plastics negates the pitting, galvanic action, and other forms of localized attack experienced when using steel underground. The smooth interior will normally remain clean for the easy conveyance of liquids or gases. In diameters of 15 cm or less, the laying of pipe in the ground can be accomplished at the pace of a slow walk, which is an important economic advantage when combined with the savings from unneeded external protection.

The three major polymeric materials used underground are thermoplastics: polyvinyl chloride (PVC), acrylonitrile-butadiene-styrene (ABS), and polyethylene (PE). These materials are not attacked by the concentration of acids, alkalies, or solvents encountered underground. Small amounts of moisture are absorbed by the polymeric materials but are not harmful. Certain specific organic compounds can cause SCC of the polymeric materials. Of these, only the cracking of stressed PE by detergents in a contaminated soil could be envisioned as a potential problem.

Mechanical support of the materials by soil at a constant temperature overcomes two of the major problems associated with the use of polymeric materials: the low modulus and strength, and the large thermal-expansion property. When installed at sufficient depth to prevent collapse from heavy top-soil loadings (trucks and tractors), the materials will provide many years of trouble-free service.

When piping installations of these polymeric materials are made, care should be used in making up the joints to ensure their integrity, notches should not be made in the outer surface (notch sensitivity is common), and the pipe should be laid on a uniform base free of rocks or other hard, irregular pieces.

Other polymeric materials, such as polypropylene, polybutylene, and glass-reinforced polyester or epoxy, are also used for specific services at greater expense. The glass-reinforced thermosetting plastics have been particularly successful as tank materials or underground storage for a wide range of products (from water to gasoline). Long exposures in wet soils show only a superficial attack on the exterior of such thermoset polymers.

References

1. G.E.C. Bell, M.J. Schiff, D.F. Wilson, "Field Observations and Laboratory Investigations of Thermogalvanic Corrosion of Copper Tubing," CORROSION/97, paper no. 568 (Houston, TX: NACE, 1997).
2. J.A. Beavers, "Fundamentals of Corrosion," in Peabody's Control of Pipeline Corrosion, eds. A.W. Peabody, R.L. Bianchetti (Houston, TX: NACE, 2001), pp. 297–317.
3. J.A. Beavers, "Cathodic Protection—How It Works," in Peabody's Control of Pipeline Corrosion, eds. A.W. Peabody, R.L. Bianchetti (Houston, TX: NACE, 2001), pp. 21–47.
4. F.E. Stetler, "Accelerating Leak Rate in Cast-Iron Water Mains Yields to CP," MP 19 (1980): pp. 15–20.
5. J.D. Palmer, "Environmental Characteristics Controlling the Soil Corrosion of Ferrous Piping," in Effects of Soil Characteristics on Corrosion, eds. V. Chaker, J.D. Palmer (Philadelphia, PA: ASTM, 1989), pp. 5–17.

6

Chapter

Reinforced Concrete

6.1 INTRODUCTION

Concrete is surely the most-used material on earth. The world consumption of concrete is estimated at more than 2.5 tons per person per year. The use of cement, a key ingredient of concrete, dates back more than 3500 years. In the construction of the Egyptian pyramids, an early form of mortar was used as a structural binding agent. The Roman Coliseum is another example of a historic landmark utilizing cement mortar as a construction material.

To understand corrosion damage in concrete, a basic understanding of the nature of concrete is required. The fundamental ingredients to make concrete are cement clinker, water, fine aggregate, and coarse aggregate. Cement clinker is essentially a mixture of several anhydrous oxides. For example, standard portland cement consists mainly of the following compounds, in order of decreasing weight percent:

- $3CaO{\cdot}SiO_2$,
- $2CaO{\cdot}SiO_2$,
- $3CaO{\cdot}Al_2O_3$, and
- $4CaO{\cdot}Al_2O_3{\cdot}Fe_2O_3$.

The reaction of the cement and water to form the cement paste is actually a series of complex hydration reactions that produce a multiphase cement paste. One example of a specific hydration reaction is the following:

$$2(3CaO{\cdot}SiO_2) + 6H_2O \rightarrow 3Ca(OH)_2 + 3CaO{\cdot}2SiO_2{\cdot}3H_2O \tag{6.1}$$

Following the addition of water, the cement paste develops a microstructure with microscopic spaces that are filled with a highly alkaline ionic solution. Apart from calcium hydroxide, sodium, and potassium hydroxides are also formed, resulting in a pH of the aqueous phase in concrete typically between 12.5 and 13.6. Under such alkaline conditions, reinforcing steel tends to display a completely passive behavior. Ordinary carbon-steel reinforcing has thus an excellent corrosion resistance in absence of corrosive species penetrating into the concrete. For additional information on concrete as a material see Chapter 8.

6.2 CORROSION DAMAGE IN REINFORCED CONCRETE

The main causes of corrosion of steel in concrete are chloride attack and carbonation. These two mechanisms are unusual in that they do not attack the integrity of the concrete. Instead, aggressive chemical species pass through the pores in the concrete and attack the steel. This is unlike normal deterioration processes resulting from chemical attack on concrete. Other acids and aggressive ions such as sulfate destroy the integrity of the concrete before the steel is affected. Most forms of chemical attack are therefore concrete problems before they are corrosion problems. Carbon dioxide and chloride ions are very unusual in penetrating the concrete without significantly damaging it.[1]

Forty percent of the bridges in North America are 40 years old and concrete is the primary materials of construction.[2] It was recognized by the mid-1970s that the deterioration of concrete-bridge structures was caused by the corrosion of the reinforcing steel in the concrete, which, in turn, was induced by the intrusion of chloride from the deicing salts into the concrete. According to a 1997 report, of the 581,862 bridges in and off the U.S. national bridge inventory, about 40% were either functionally or structurally deficient. Most of these bridges were severely deteriorated with extensive loss of serviceability and reduced safety, with some of the bridges load-posted so that overweight trucks would be required to take a longer alternative route.[3]

The estimated cost to eliminate all backlog bridge deficiencies (both structural and functional) in Canada was estimated at $10 billion[2] and in the United States to be $78 billion, increasing to as much as $112 billion, depending on the number of years it would take to meet the objective. Although corrosion of the reinforcing steel was not the sole cause of all structural deficiencies, it was a significant contributor and has therefore become a matter of major concern.

Reinforced-concrete bridge decks constitute the weakest link in North America's infrastructure network. An extensive area of these decks is severely damaged through the exposure to deicing salts and traffic. It is estimated that one-third to one-half of projected direct costs will be allocated to bridge decks. However, even though the cost of maintaining bridge decks is becoming prohibitively expensive, the benefits provided by deicing salts are too great in terms of reducing vehicular accidents and minimizing traffic disruption.[4] Therefore, its use is not likely to decrease in the future. In fact, the use of road-deicing salts, which are extremely corrosive because of the disruptive effects of its chloride ions on protective films on metals, has actually increased in the first half of the 1990s after leveling off during the 1980s.

Figure 6.1 Growth of the corrosion-oxide products can lead to cracking and spalling of the concrete cover.

6.2.1 Corrosion Damage

Most problems associated with corrosion of steel in concrete are not due to the direct loss of steel but to the growth of the corrosion oxide products, which eventually can lead to cracking and spalling of the concrete cover (Figure 6.1). Structural collapses of reinforced-concrete structures as a result of corrosion are rare. The most common problem caused by corrosion is spalling of concrete covers, which has caused a few deaths and accidents in recent history. It has been suggested that less than 100 μm of steel section loss is needed to start cracking and spalling the concrete. The actual amount needed will depend on the geometry in terms of cover, proximity to corners, rebar spacing, bar diameter, and the rate of corrosion.[1]

The importance of vertical cracks in accelerating corrosion by allowing access of corrosion agents to the steel surface is well established. If reinforcing steel is doing its job in areas of tension in the structure, small cracks can occur as the tensile load exceeds the tensile strength of the steel. Most of these are small cracks (that is, less than 0.5 mm) that intersect the reinforcing steel at right angles. These cracks do not usually lead to corrosion of the steel because any local ingress of chlorides, moisture, and carbonation is limited and contained by the local alkalinity. Larger cracks can potentially stay open more easily and accelerate corrosion processes.[5]

Corrosion can also cause horizontal cracking along the plane of the rebar and the comer cracking around the end rebar. This leads to the loss of concrete cover as shown in Figure 6.2. This is the main consequence of reinforcement corrosion with its subsequent risk of falling concrete and unacceptable appearance.

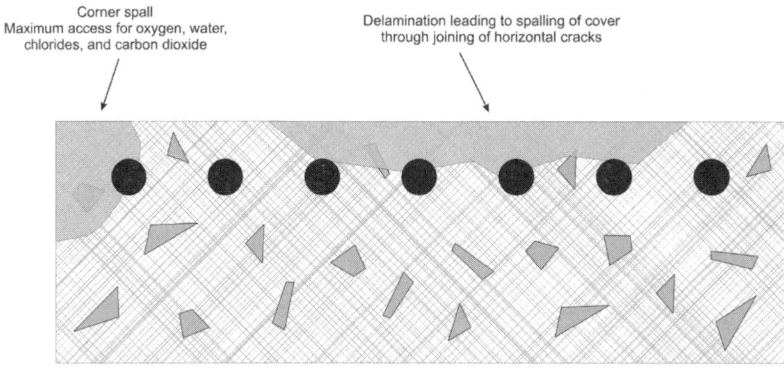

Figure 6.2 Corrosion-induced cracking and spalling.

6.2.2 Chloride Attack

6.2.2.1 Sources of Chlorides

Chlorides can come from several sources. They can be cast into the concrete or they can diffuse in from the outside. Chlorides cast into concrete can be due to these factors[1]:

- deliberate addition of chloride set accelerators (calcium chloride was widely used until the mid-1970s),
- use of seawater in the mix, and
- contaminated aggregates (usually sea-dredged aggregates that were unwashed or inadequately washed).

An important source of chlorides from the external environment, as mentioned in the previous section, is the widespread use of deicing salts on road surfaces in cold climates. Around 10 million tons of deicing salt is used annually in the United States, whereas the Canadian figure is about 3 million tons.[6] The actual annual tonnage used fluctuates with the severity of particular winter seasons. The main purpose of deicing-salt application is to keep roadways safe and passable in winter and to minimize the disruption of economic activity. The application of salt to ice and snow results in the formation of brine, with a reduced freezing point.

Salt, primarily in the form of rock salt (halite or NaCl), is the most widely used deicing agent in North America because of its low cost, general availability, and ease of storage and handling. Other chloride compounds in use for deicing purposes are calcium chloride ($CaCl_2$) and magnesium chloride ($MgCl_2$). Alternate drying and wetting cycles promote the build-up of chlorides on surfaces, which can thus be well in excess of those of the bulk environment.

General exposure to sea salt and particularly in tidal and splash zones is a constant source of salt-rich environment. Marine mist carrying salt-laden aerosols is also quite an important source of damage in concrete constructions in coastal areas.

Salt water is rapidly absorbed by dry concrete. The capillary movement of the salt-laden water through the pores is then followed by "true" diffusion. There are other opposing mechanisms that slow the chlorides' movement. These include chemical reactions to form chloroaluminates and adsorption onto the pore surfaces. An important problem when trying to predict the chloride penetration rate is defining the initial concentration, because chloride diffusion produces a concentration gradient that is quite different than the clear-cut "front" produced during carbonation.

6.2.2.2 Chloride Attack Mechanism and Chloride Threshold

Chlorides act as catalysts to corrosion when there is sufficient concentration at the rebar surface to break down the passive layer. These chlorides are not consumed in the corrosion processes. Chlorides in concrete exist as either free chlorides or as bound chlorides. The former are mobile chlorides dissolved in the pore solution, whereas the latter type represents relatively immobile chloride ions that can react by chemical binding or by adsorption with the cement paste.

The determination of a critical chloride level, above which serious rebar-corrosion damage is expected, is appealing for design and maintenance planning purposes. Such a "chloride threshold" has been defined as a function of the chloride/hydroxyl ratio required before corrosion initiates. This threshold has been estimated in laboratory tests with calcium hydroxide solutions during which corrosion was observed when the chloride concentration exceeded 0.6 of the hydroxyl concentration, which translates to a concentration of 0.4% chloride by weight of cement if chlorides are cast into concrete and 0.2% if they diffuse in the concrete later.

In North America a commonly quoted threshold is 0.455 kg of chloride per 0.76 cubic meters of concrete.[7] This empirical threshold is based on experimental evidence and practical observations of real structures. However, any threshold can only be an approximation because of these factors.

1. Concrete pH varies with the type of cement and the concrete mix. A very small change in pH, basically a logarithmic function of the inverse concentration of acid ions, can represent a massive change in absolute hydroxyl ion (OH^-) concentration.
2. Chlorides can be bound chemically (by aluminates in the concrete) and physically (by adsorption on the pore walls). This removes them temporarily or even permanently from the corrosion reaction. Sulfate-resisting cements have low aluminate content, for example, tricalcium aluminate (C_3A), which leads to more rapid diffusion and lower chloride thresholds.
3. In very dry concrete, corrosion may not occur even at very high chloride concentration because the water is missing from the corrosion reaction.
4. In sealed or polymer-impregnated concrete, corrosion may not occur even at a very high Cl^- concentration if no oxygen or moisture is present to fuel the corrosion reaction.
5. Corrosion can be suppressed when there is total water saturation as a result of oxygen starvation, but if some oxygen gets in, then pitting corrosion rapidly prevails.

For example, corrosion has been observed at a threshold level of 0.2% chloride by weight of cement when the concrete quality was poor and water and oxygen were readily available. In other cases no corrosion has been observed at 1.0% chloride or more when oxygen and water were excluded.

6.2.2.3 *Macrocell Formation*

In the case of chloride attack, the corrosion processes are often well separated, with areas of rusting adjacent to areas of "clean" steel. This is known as the macrocell phenomenon. Chloride-induced corrosion is particularly prone to macrocell formation when high levels of water are present to carry chlorides into the concrete. The presence of water in the pores increases the electrical conductivity of the concrete. The higher conductivity allows the separation of anode and cathode because ions can move more freely through the water-impregnated pores.

In comparison, carbonation occurs when the concrete is much drier. Otherwise the carbon dioxide (CO_2) would not be able to penetrate the concrete. This other type of corrosion attack is therefore on a "microcell" level with apparently continuous corrosion observed along the rebars.

6.2.3 Carbonation-Induced Corrosion

Carbon dioxide present in the atmosphere can react with calcium hydroxide ($Ca[OH]_2$) and other hydroxides to produce insoluble carbonate ($CaCO_3$) and water in the concrete as follows:

$$Ca(OH)_2 + CO_2 \rightarrow CaCO_3 + H_2O \tag{6.2}$$

Because the carbonation process neutralizes the alkaline material in the concrete it can also significantly reduce the pH of the pore solution in the outer layers of the concrete. This type of attack often appears as a well-defined "front," parallel to the external surface. This front can conveniently be made visible by applying a phenoplthalein-indicator solution to freshly exposed concrete surfaces. Phenolphthalein changes from colorless at low pH (carbonated zone) to pink at high pH (uncarbonated concrete). Behind the front, where all the calcium hydroxide has been depleted, the pH is around 8, whereas ahead of the front, the pH remains in excess of 12.5.[8]

Care must be taken to prevent dust or water from contaminating the surface to be measured, but the test, with the indicator sprayed on to the freshly exposed concrete, is cheap and simple. Measurements can also be taken on concrete cores and fragments, and down drilled holes.

There is much more calcium hydroxide in the concrete pores than what can be dissolved in the pore water. This helps maintain the pH between 12 or 13 as the carbonation reaction occurs. However, eventually all the locally available calcium hydroxide reacts, precipitating the calcium carbonate and allowing the pH to fall to a level where steel can corrode.

At the carbonation front there is a sharp drop in alkalinity from pH 11–13 down to less than pH 8 as illustrated in Figure 6.3.[1] At that level the passive layer on the steel is

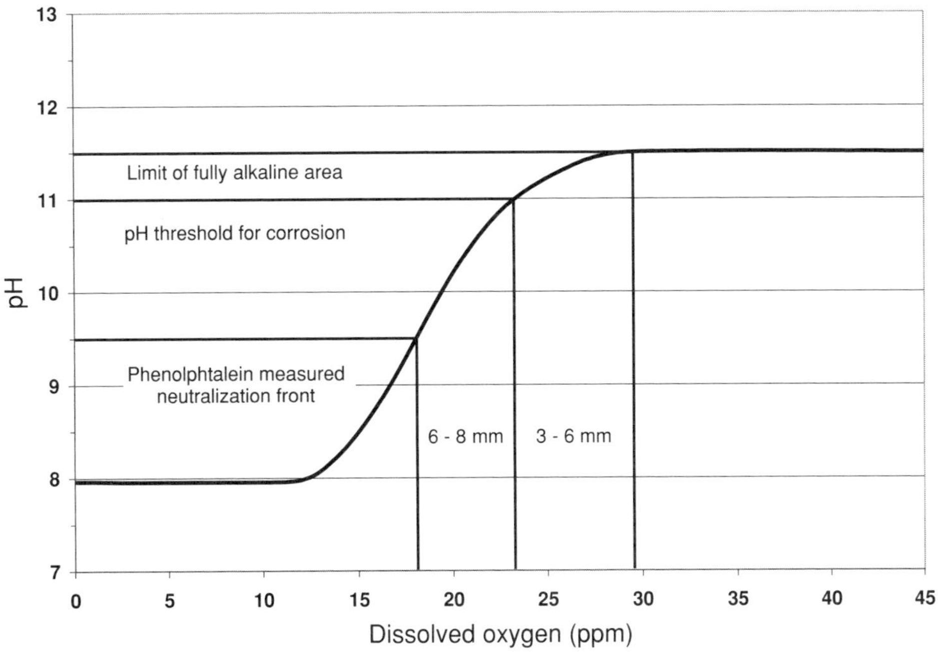

Figure 6.3 Carbonation front and its relationship to the corrosion threshold and the color change for phenolphthalein.

no longer sustained, so corrosion proceeds by the general corrosion mechanism. Carbon dioxide diffuses through the concrete and the rate of movement of the carbonation front approximates to Fick's second law of diffusion. The following equation derived from this law is a useful expression that relates the carbonation depth (x) as a function of time (t):

$$x = At^n \tag{6.3}$$

where A and n are constants describing the specific behavior of a concrete component, with n being very close to 0.5.

However, because the carbonation process modifies the concrete-pore structure as it proceeds, this is only an approximation. Cracks and changes in concrete composition and moisture levels with depth will also lead to deviation from the perfect diffusion equation.

Several variables affect the rate of carbonation. In general low-permeability concrete is more resistant. The diffusion process is made easier if the concrete has an open-pore structure. On the macroscopic scale this means that there should be good compaction. On a microscopic scale well-cured concrete has small pores and lower connectivity of pores so the carbon dioxide movement is slowed down. Microsilica and other additives can block pores or reduce pore sizes. The rate of carbonation-induced corrosion is also an inverse function of cover thickness, so good cover is essential to resist carbonation. Because the process is one of neutralizing the alkalinity of the concrete, good reserves of alkali are needed, that is, a high cement content.

Carbonation tends to proceed most rapidly at RH levels between 50 and 75%. At lower humidity levels, carbon dioxide can penetrate into the concrete relatively rapidly but little calcium hydroxide is available in the dissolved state to react with it. At higher humidity levels, the water-filled pores offer a more effective barrier to the ingress of carbon dioxide. Environmental cycles of alternately dry and wet conditions have been associated with rapid carbonation damage.

Carbonation is common in old structures, badly built structures (particularly buildings), and reconstituted stone elements containing reinforcement that often have a low cement content and are very porous. Carbonation is rare on modern highway bridges and other civil-engineering structures where water/cement ratios are low, cement contents are high with good compaction and curing, and there is enough cover to prevent the carbonation front advancing into the concrete to the depth of the steel within the lifetime of the structure.

6.2.4 Synergy Between Chlorides and Carbonation Attack

In many practical situations carbonation- and chloride-induced corrosion can act in synergy with the resultant corrosion attack being greater than the sum of the individual chlorides- and carbon-dioxide-induced damage. Research studies have demonstrated that corrosion caused by carbonation was intensified with increasing chloride-ion concentration.[8] This particular situation has been found responsible for major corrosion problems in hot coastal areas.

As stated earlier, chlorides can be bound by the concrete constituents to form chloroaluminates, for example. This binding process is strongest for chlorides cast into concrete, which explains why it has been considered acceptable to use seawater to make concrete for many years.

The extent and effectiveness of this binding is not well understood. However, it is known that a reduction in pH as caused by carbonation will break down the chloroaluminates and leads to a wave of chlorides moving in front of the carbonation front. This phenomenon makes structures containing built-in chlorides particularly vulnerable to the synergistic attack by combined carbon dioxide and chlorides.

6.3 REMEDIAL MEASURES

A number of fundamental measures can be taken to address the problem of reinforcing-steel corrosion. However, an important distinction has to be made in the applicability of remedial measures to new and existing structures. Unfortunately, the options in the following list of remedial measures are more limited for fixing existing, aging structures than for designing new constructions:[6]

- repairing the damaged concrete;
- modifying the external environment, for example, choosing a different deicing agent or changing deicing-salt practices (however, this will not remove salt already in the concrete);

- modifying the internal concrete environment, for example, realkalization of concrete, chloride extraction, addition of corrosion inhibitors, and so forth;
- creating a barrier between the concrete and the external environment;
- creating a barrier between the rebar steel and the internal concrete environment, for example, epoxy coating and galvanization;
- applying cathodic protection to the rebar structure;
- alternative, more corrosion-resistant rebar materials, for example, stainless steel; and
- alternative methods of reinforcement, for example, fiber-reinforced polymer composites.

For new structures, it is believed that much progress will be made toward effective corrosion control as the implementation of life-cycle costing strategies are adopted, as opposed to awarding contracts on the basis of lowest initial capital-cost outlay. With such a vision, using a more corrosion-resistant stainless-steel rebar material, for example, could prove to be a cost-effective route even if it increases the initial cost of rebar by an appreciable number.

In areas of strategic importance such as the highway belts of most modern cities, the total cost of repairs is greatly amplified by adding the indirect costs of traffic disruptions.

6.3.1 Repair Techniques

The major scale of the concrete-infrastructure deterioration by corrosion processes prevents many organizations from being proactive in the matter, simply because these organizations are busy maintaining the functionality of existing structures. Essentially these procedures involve the removal of loose, spalled concrete, followed by further systematic removal of the concrete surrounding the corroded rebar. Finally, the rebar and concrete surfaces are cleaned and primed before the new repair concrete is applied. The repair procedures thus create three different material zones that interact with the reinforcing steel:

1. old chloride-contaminated or carbonated concrete,
2. new concrete, and
3. the interface between the old and new concrete.

The interface may represent a zone of weakness with respect to further ingress of corrosive species. It is important to remove the damaged concrete to a depth well below the corroded reinforcing bars. Failure to do so can easily produce a detrimental galvanic-corrosion cell in the repaired area, as shown in Figure 6.4.

An undesirable galvanic-corrosion cell involving the new and existing concrete can also be created, despite this precaution, as shown in Figure 6.5. To avoid the recurrence of rebar corrosion in the existing concrete in this situation, it is advisable to remove as much of the old chloride-contaminated concrete as necessary. Even then the "ring

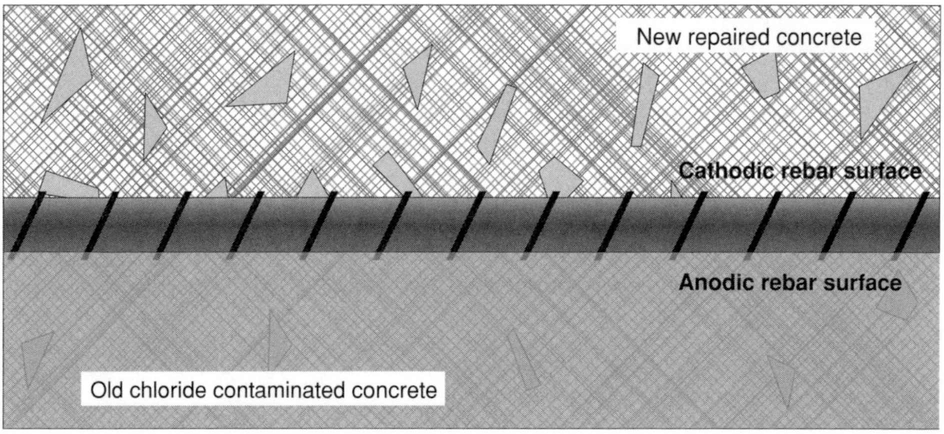

Figure 6.4 Schematic shows a galvanic corrosion cell in concrete repair.

anode" or "incipient anode" effect means that the previously anodic area becomes cathodic, creating new anodes around it and leading to corrosion around new patch repairs.

Cathodic protection of the rebar or chloride extraction could also be considered as part of the repair specifications because of the ring anode effect. The latest option is to install proprietary zinc anodes that are encapsulated in a special lithium-based mortar that keeps the zinc active and absorbs the zinc corrosion products without damaging the concrete (Figure 6.6 and Figure 6.7).

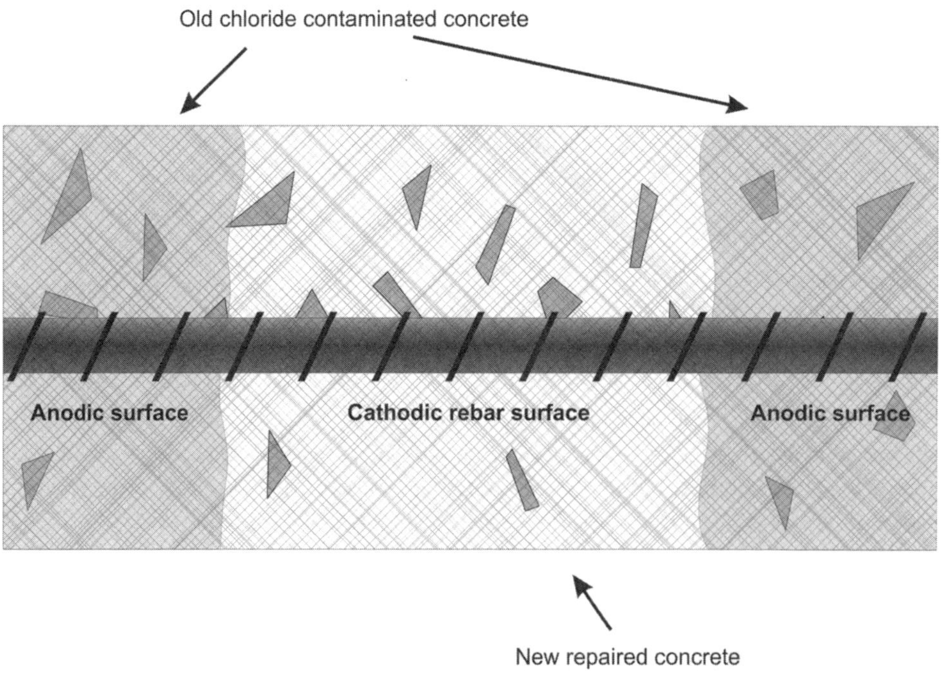

Figure 6.5 Another view of the corrosion cell in concrete repair.

Figure 6.6 Zinc anodes are connected to the rebar structure to control ongoing corrosion and prevent the initiation of new corrosion activity in concrete structures (courtesy of Vector Corrosion Technologies).

Figure 6.7 Embedded zinc anodes are used in concrete rehabilitation to prevent the formation of new corrosion sites adjacent to completed patch repairs (courtesy of Vector Corrosion Technologies).

Figure 6.8 ICCP is retrofitted on an old bridge structure; note vent tubes and drill holes anodes (courtesy of Vector Corrosion Technologies).

Other approaches have been proposed and tried. The following approaches based on the use of corrosion inhibitors, for example, have been extensively tried without meeting with much success[9]:

- corrosion-inhibiting admixtures in conjunction with polymer-modified cementitious coatings applied to the rebar,
- other cement-based barrier coatings in combination with corrosion inhibitors, and
- migrating corrosion inhibitors.

6.3.2 Electrochemical Techniques

6.3.2.1 Impressed-Current Cathodic Protection (ICCP)

CP is one of the few techniques that can be applied to control corrosion on existing structures (Figure 6.8). ICCP of conventional rebar, as opposed to pre- and posttensioned systems, is well established, with applications dating back more than twenty-five-years.

The first reinforced-concrete ICCP system was an experimental system installed on a bridge support beam in 1959.[10] A more advanced system was subsequently installed on a bridge deck in 1972.[11] The anode system used in both applications was based on a conventional ICCP system for pipe lines, but spread out over a bridge deck. The conventional pipe-line system was corrosion-resistant silicon iron "primary" anode in a backfill of conductive carbon coke breeze.

The principles and theory of cathodic protection are the subject of a separate chapter (Chapter 11). In the present context, ICCP consists of polarizing the rebar to a cathodic potential where anodic dissolution of the rebar is minimized. A DC current source (rectifier) is usually employed to establish the rebar as the cathode of an electrochemical cell, whereas a separate anode is required to complete the electric circuit. Three basic methods are available for controlling the output of a rectifier:

- In ***constant-current mode*** the rectifier maintains a constant-current output. The output voltage then varies with changes in the circuit resistance. The potential of the reinforcing steel can be measured with a reference cell as a function of the applied current to ensure that certain protection criteria are met.
- In ***constant-voltage mode*** a constant-output voltage is maintained by the rectifier. The applied current then changes with variations in circuit resistance. Low concrete resistance, often associated with increased risk for corrosion damage, can result in increased current output. It should be noted that in this mode, the rebar potential is not necessarily constant. It can again be monitored with a reference cell.
- In ***constant-rebar-potential mode***, a method that can be used for immersed concrete where potentials are relatively uniform and stable, the current output is adjusted continuously to provide a constant (preselected) rebar potential. The rebar potential, measured continuously with reference electrodes, is fed back to the rectifier unit. Successful operation in this mode depends on minimizing the IR drop error in the rebar-potential measurements and on the accuracy and stability of the reference electrodes over time.

An important issue in ICCP of reinforcing steel is how much current should be impressed between the reinforcing steel and the anode. Too little current will result in inadequate corrosion protection of the rebar, whereas excessive current can result in problems such as hydrogen embrittlement and concrete degradation. Hydrogen embrittlement of high-strength steel is a particularly severe problem for pre- and posttensioned prestressed-concrete systems.

Uniformity in the current distribution is obviously a desirable feature. Unfortunately, the current requirement cannot be measured directly and various indirect criteria have been proposed (see Table 6.1). The ICCP current requirements are often expressed in terms of the potential of the reinforcing steel relative to a reference electrode, or alternatively as a shift in this potential when the ICCP system is activated or deactivated.

These reference electrodes can be located externally, in contact with the outside concrete surface, or embedded directly in the concrete with the rebar. The current densities involved in meeting commonly used protection criteria are typically around 10 mA per square meter of rebar surface.

Adequate anode lifetime is also an important factor that is directly related to the magnitude and uniformity of current flow. A variety of anode systems have evolved for ICCP of reinforcing steel, each with certain advantages and limitations. Continuous surface anodes have been based on conductive bituminous overlays and conductive

Table 6.1 CP Criteria for Steel in Concrete

Criterion	Details	Comments
Potential shift	100-mV shift of rebar potential in the positive direction when system is depolarized	Depolarization occurs when CP current is switched off. Time period required for rebar to depolarize is debatable. The potential reading before interrupting the CP current should be IR corrected.
Potential shift	300-mV shift of rebar potential in the negative direction because of application of CP current	The potential reading with CP current should be IR corrected. The method relies on astable rebar potential before the application of CP current.
E-log (i) curve	The decrease in corrosion rate resulting from the application of CP current can be determined provided the relationship between rebar potential (E) and current (i) can be measured and modeled. A simple model is Tafel behavior with a linear relationship between E and log(i).	This methodology is structure-specific and the measurements involved are relatively complex and require specialist interpretation. Ideal Tafel behavior is rarely observed for steel in concrete.
Current density	Application of 10 mA/m^2 of rebar surface area	Empirical approach based on limited experience. Does not consider individual characteristics of structures and environments.

surface coatings. The former are only suited to horizontal surfaces. In general, good current distribution is achievable with such systems. Discrete anodes have been used without overlays and with cementitious overlays.

For horizontal surfaces, anodes without overlay can be recessed in the concrete surface. Nonuniform current distribution is a fundamental concern in these systems. Anodes in the form of a titanium mesh, with proprietary surface coatings of precious metals, are commonly used in concrete structures, in conjunction with cementitious overlays. These systems are applicable to both horizontal and vertical surfaces and generally provide uniform current distribution.

However, adhesion of the overlay to the original concrete problem can be a problem, particularly on vertical and soffit surfaces where the overlay is of sprayed concrete.

RP0290-2000[12] and DIN/BS/EN 12696[13] provide guidelines for ICCP of reinforcing steel in atmospherically exposed concrete structures. These documents provide performance requirements for cathodic protection of steel in atmospherically exposed concrete, in both new and existing structures, including normal reinforcement and prestressed reinforcement embedded in the concrete. However, they do not apply to buried or submerged elements of buildings and structures.

6.3.2.2 *Galvanic Cathodic Protection (GCP)*

One alternative to ICCP is galvanic or sacrificial-anode cathodic protection. This technique uses a metal more susceptible to corrosion than steel as an anode that corrodes and generates the current needed to protect the steel. For steel in concrete this is generally zinc or a zinc-based alloy. Although ICCP systems have been widely used since the 1980s, GCP systems have only become available around the end of the 1990s. Various configurations exist, including the following:

- thermal sprayed zinc, a few hundred microns in thickness, applied directly to the concrete surface. A "humectant" may be added to increase moisture, reducing concrete resistance and increasing the current level. Alternatively a high current-output, proprietary aluminium/zinc/indium alloy can be used.
- Thin sheets of zinc with a conductive adhesive backing applied to the concrete surface.
- Expanded zinc mesh in a glass-reinforced, plastic permanent form filled with a proprietary cementitious grout, usually applied to marine exposed piles in the splash and tidal zone.
- Zinc disks in a proprietary, high-alkaline mortar jacket embedded in patch repairs to prevent the ring-anode or incipient-anode effect discussed in the previous section on repair techniques.
- Zinc plugs in a cylinder of a proprietary, high-alkaline mortar for installation in cored holes in the concrete.

The advantage of GCP over ICCP is that it has no power supply control system and therefore low monitoring requirements. Its disadvantage is that it cannot be controlled to ensure complete corrosion control, the anode types are more limited, and anode lives are more limited and less understood.

6.3.2.3 Electrochemical Chloride Extraction

Another electrochemical technique that is applicable to existing concrete structures contaminated with chlorides involves the electrochemical removal of these harmful ions. The hardware involved is similar to that involved in cathodic protection. Electrochemical extraction of chloride ions is achieved by positioning an anode in contact with a caustic electrolyte spread on the external concrete surface and impressing a direct current between the anode and the reinforcing steel acting as the cathode (Figure 6.9). Under the application of this electrical field, chloride ions migrate away from the negatively charged steel and toward the positively charged external anode.

Chloride extraction has been recommended for structures that do not contain pre- or posttensioned steel and have little damage to the concrete itself. The current densities and applied voltages involved are significantly higher than those used in CP, therefore increasing the risk of hydrogen evolution and subsequent hydrogen embrittlement (Chapter 9). The unsuitability of the technique to prestressed concrete is thus not surprising. Additional requirements are a high degree of rebar electrical continuity and preferably low concrete resistance. Because several days or even weeks are involved in the extraction processes using suitable current densities, the technique is more realistically applicable to highway substructures rather than bridge decks, where it could cause serious traffic disruptions as illustrated in Figure 6.10.

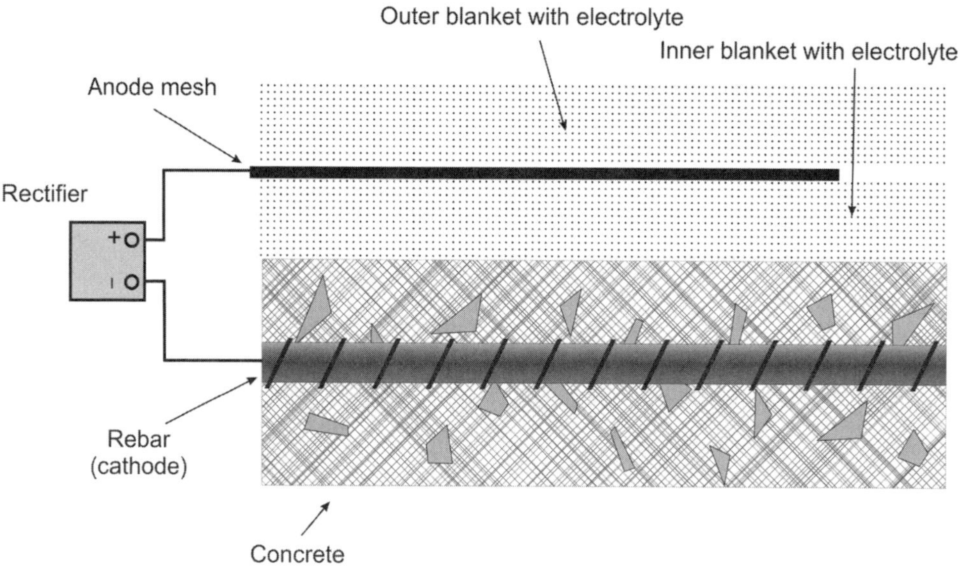

Figure 6.9 Schematic shows electrochemical chloride extraction and realkalization treatments.

In practice, the chloride-extraction process does not remove the chloride ions from the concrete completely. Rather, a certain percentage is removed and the balance is redistributed away from the reinforcing bars. There is also increasing evidence that the protective layer around the reinforcing steel is considerably enhanced, elevating the threshold for corrosion.

Figure 6.10 Bridge deck is energized for electrochemical chloride extraction (courtesy of Vector Corrosion Technologies).

As with ICCP, the applied current density has to be controlled. Under excessive current magnitude several problems can arise, such as reduction in bond strength between rebar and concrete, softening of the cement paste around the rebar steel, and cracking of the concrete. Concrete containing alkali-reactive aggregates is not considered a suitable candidate for the process because the expansive reactions leading to cracking and spalling associated with these aggregates tend to be aggravated.[14,15]

6.3.2.4 *Realkalization*

The realkalization treatment is applied to existing structures to restore alkalinity around reinforcing bars in previously carbonated concrete. The electrochemical principle and hardware is similar to that of electrochemical chloride extraction. Direct current is applied between the cathodic rebar and external anodes positioned at the external concrete surface and surrounded by electrolyte (Figure 6.9). Compared to ICCP, the current densities in realkalization are again significantly higher. Typically, the process is applied for several days to restore alkalinity in carbonated concrete.

The external electrolyte used in realkalization is usually a sodium or potassium carbonate solution with a caustic pH. Apart from the hydroxyl (OH^-) ions generated at the cathode and their migration away from the rebar under the electrical field, other mechanisms can account for formation of alkaline solution in the concrete. Initially, diffusion effects may arise as a result of concentration gradients in the concrete. In later stages, bulk flow of external solution into the concrete may occur by either direct absorption or by electro-osmosis. In dry concrete, absorption effects can occur to a depth of several centimeters in a matter of one day.

The potential disadvantages of realkalization are similar to those of chloride extraction, namely risk of reduced bond strength between rebar and concrete, hydrogen embrittlement of the reinforcing steel, alkali–aggregate reaction, and other microstructural changes in the concrete.[16]

6.3.3 New Construction—Rebar Options

6.3.3.1 *Epoxy-Coated Reinforcing Steel*

Epoxy coatings represent an inert physical barrier that isolates the reinforcing steel from the corrosive environment. In North America, the use of epoxy-coated rebar dates back nearly three decades and at present represents a commonly used alternative rebar material to standard reinforcing steel. Standards covering these materials include ASTM A 775 and BS 7295-1. Besides the type of epoxy material used, variables such as surface cleanliness and preparation, coating thickness, coating adhesion to the rebar, coating continuity, and coating thickness have to be considered for optimal corrosion resistance.

Although epoxy coatings have reportedly performed satisfactorily in many applications such as bridge decks, incidents of severe corrosion have been observed in the substructure of four bridges in the Florida Keys (USA) after only 6 to 10 years of exposure.[17] The marine environment in these cases was particularly corrosive.

Figure 6.11 Ajax overpass, Hwy 401, Ontario, was built in 1998 and used 245 tons of S31603 for the upper bridge deck and carbon steel for the lower deck (courtesy of Nickel Institute).

Because the epoxy coating functions as a corrosion barrier, the coating continuity is obviously quite important. Although it may be possible to control coating defects (holidays) within tight limits in the manufacturing plant, the risk of coating damage during transportation, off-loading, storage, installation on site, and concrete pouring and vibration is considerably greater. Efforts have been directed at repairing visible damage on site, prior to placement in concrete.[18]

6.3.3.2 Stainless-Steel Rebar

The results from a number of research projects have concluded the corrosion performance of stainless alloys is superior to uncoated "black" carbon steel.[19] Several advantages could lead to a growing interest in stainless rebars for better concrete reinforcement:

- corrosion resistance is integral to the material (this does not imply that the material is always immune to corrosive attack);
- no coatings are involved that could chip, crack, or degrade;
- it is able to withstand shipping, handling, and bending;
- there are no exposed ends to cover or coat;
- it has good ductility, strength, and weldability;
- it is magnetic or nonmagnetic, depending on grade.

There are numerous examples of modern construction in which different stainless steels have been used as the reinforcing material. In the example shown in Figure 6.11,

Figure 6.12 Two piers in Yucatan, Mexico: in the foreground is the ruins of the pier built with carbon-reinforcing bar 30 years before the picture was taken in 1998 and in the background is the pier built between 1937 and 1941 with stainless-steel rebars (courtesy of Nickel Institute).

245 tons of S31603 were used for the upper bridge deck with carbon steel for the lower deck. However, the earliest use of this technology goes back to the 2100-m concrete pier in Progreso de Castro in Yucatan, Mexico, which was constructed between 1937 and 1941 by the Danish contractor Christiani & Nielsen (Figure 6.12). It has 175 spans of 12 m and consists of massive columns and arches. Because of the severe marine exposure of the pier and the use of concrete with a relatively high porosity (use of local limestone aggregate), the pier was designed for a minimum use of reinforcement (compressive stresses in the arches). The pier has not undergone any major repair work during its lifetime and no significant routine maintenance activities have been carried out.[20]

A range of stainless steels is available for rebar applications. The final selection depends on mechanical-design requirements, expected corrosivity, and cost considerations. For rebar, the austenitic and duplex grades have received the most attention (see Chapter 8 for more details on stainless steels).

Although the initial cost of a structure with stainless-steel rebar may seem to be higher than conventional structures, the overall construction costs are actually only modestly increased. The case for stainless-steel rebar can be strengthened when a life-cycle cost analysis is performed to compare total cost options over the lifetime of a structure. The cost analysis should include all direct costs, for example, costs associated with the frequency and cost of future maintenance and replacement work plus any loss of revenue for toll facilities, and all indirect costs associated with various losses in productivity resulting from traffic or usage disruptions.

In such an analysis, the cost–benefit crossover of using austenitic stainless steels over carbon steel in a bridge-construction project was shown to be between 18 and 23 years.[21]

6.3.3.3 *Galvanized Rebars*

Advocates of the corrosion protection of rebars by galvanizing have invoked three reasons for using this protection scheme:

- zinc-coated rebars remain free of corrosion at somewhat lower pH levels in concrete than ordinary steel;
- the zinc coating represents a sacrificial material that will protect the steel cathodically;
- zinc-corrosion products have a smaller volume than steel-corrosion products, therefore reducing the expansive stresses produced during the corrosion processes; and
- galvanized rebars are easier to handle than epoxy-coated bars, with lower risk of damage and fewer consequences to corrosion protection if damaged.

Despite these considerations, the performance of galvanized reinforcing steel in actual service has had somewhat mixed reviews. One explanation provided for low performance levels has, in fact, been attributed to the formation of particularly voluminous corrosion products with a resulting expansion greater than those produced during the corrosion of plain carbon steel.

6.3.4 Corrosion Inhibitors

The addition of corrosion inhibitors to a concrete mix can improve the corrosion performance of good-quality reinforced concrete. However, it is surely not the role of these inhibitors to make good concrete out of poor quality concrete.[22]

One of the more well-known corrosion inhibitors for concrete is calcium nitrite, $Ca(NO_2)_2$. The mechanism of inhibition involves nitrite ions (NO_2^-) that compete with chloride ions in their reaction with ferrous ions (Fe^{2+}) produced during the corrosion of steel rebar.[22] Essentially, the nitrite ions limit the formation of unstable iron-chloride complexes and promote the formation of stable compounds, for example, Fe_2O_3 or FeOOH, that passivate the rebar surface. The following reactions have been proposed:

$$2Fe^{2+} + 2OH^- + 2NO_2^- \rightarrow 2NO^{\uparrow}_{(g)} + Fe_2O_3 + H_2O \tag{6.4}$$

or

$$Fe^{2+} + OH^- + NO_2^- \rightarrow NO^{\uparrow}_{(g)} + FeOOH \tag{6.5}$$

Both these reactions produce a volatile nitric-oxide gas (NO). Because nitrite ions compete with chloride ions for interacting with ferrous ions, the ratio of nitrite/chloride ions is an important variable for specifying an adequate corrosion-protection scheme.

A serious consideration when choosing a corrosion-inhibiting admixture is its potential effect on the properties of concrete, such as workability, curing time, and strength. The details of commercial rebar corrosion-inhibitor reactions have remained relatively obscure, because of the proprietary nature of their formulations.

6.3.5 Concrete Cover and Mix Design

Because chlorides and other corrosive species diffuse through the concrete to the reinforcing steel, one obvious method of mitigating corrosion damage is to increase the concrete cover. An often-quoted rule of thumb is that a two-fold increase in the concrete cover produces a four-fold life extension of the rebar, and that a three-fold cover increase results in a nine-fold benefit, and so on. This relationship is based on the simplistic model described by Fick's second law of diffusion (6.3). The beneficial effect of increased cover is also applicable to cracked concrete but is less significant in exposure to chloride-laden environments involving alternate wetting and drying cycles.[23]

The diffusion rate of chlorides into concrete increases distinctly with increasing porosity and the porosity itself increases as the water-to-cement ratio increases or as the cement content decreases. In the case of portland cement, the penetration of chlorides by diffusion decreases with increasing content of the $3CaO \cdot Al_2O_3$ (C_3A) cement component. The following recommendations have therefore been made for portland cement concrete for improving its resistance to chloride ingress:[23]

- water/cement ratio < 0.45,
- cement content > 400 kg/m^3, and
- C_3A content in cement $> 11\%$ (by weight).

An important option to consider for the reduction of the permeability of concrete is to perform an adequate curing (2 to 4 weeks of hydration) to favor the development of a dense internal texture with low porosity. Another beneficial practice in concrete-mix design has been the addition of supplementary cementitious materials such as

- **pozzolans**, for example, fly ash and silica fume, that react with the cement-hydration products, notably calcium hydroxide; and
- **hydraulic materials**, for example, granulated blast-furnace slag, that undergo direct hydration reactions.

An obvious advantage for blending these cementitious materials in concrete mixes is the low cost of these additives, because they are often considered environmentally critical waste products. As an added bonus, cementitious materials can also improve the strength and durability of concrete, provided that the concrete is cured adequately.[8]

Concrete mixes described as high-performance concrete (HPC) have received considerable attention in recent years. HPC refers to various mixes with enhanced attributes resulting form advances in the fields of admixtures and new cementitious materials. Examples of recent major HPC construction projects in highly corrosive environments include the Hibernia offshore oil platform and the 13-km bridge structure linking Prince Edward Island to New Brunswick in Canada.

HPC mixes with particularly low water-to-cement ratios, resulting in high compressive strengths, have been achieved with super plasticizer additives. These additives obviously play a key role in ensuring satisfactory workability for the low water contents. High density and low permeability typically characterize mixes that can be expected to represent an effective barrier to the ingress of corrosive species.

Table 6.2 Conventional and New Methods for Reinforced-Concrete Structures

Property	Test Methodology
Existing techniques in SHRP bridge assessment guide	
Depth of concrete cover	Magnetic flux devices
Concrete strength from test cylinders	ASTM C 39 for test cylinders
Concrete strength from core samples	ASTM C 42
Concrete strength from pull-out tests	ASTM C 900
Concrete strength/quality from rebound hammer test	ASTM C 805
Concrete strength/quality from penetration tests	ASTM C 803
Air-void system characterization in hardened-concrete	ASTM C 457
Microscopic evaluation of hardened-concrete quality	ASTM C 856 petrographic examination
Alkali–silica reactivity	
Delamination detection	ASTM D 4580 sounding
Damage assessment by pulse velocity	ASTM C 597
Cracking damage	ACI 224.1R
Probability of active rebar corrosion	ASTM C 876 based on corrosion potential (note no corrosion rate is determined)
New techniques in SHRP bridge-assessment guide	
Instantaneous corrosion-rate measurement	Electrochemical measurements, applicable to uncoated steel
Condition assessment of asphalt-covered decks with pulsed radar	
Condition of pre-formed membranes on decks using pulse velocity	
Relative effectiveness of penetrating concrete sealers with electrical resistance	
Evaluating penetrating concrete sealers by water absorption	
Chloride content in concrete by specific ion sensor	
Relative concrete permeability by surface air flow	
Other	
Chloride-ion content by titration	AASHTO T-260
Rebar location	X-ray and radar
pH and depth of carbonation	Phenolphthalein solution or pH electrode in extracted pore solution
Concrete permeability with respect to chloride ions	ASTM C 1202 and ASTM C 642
Delamination, voids, and other hidden defects	Impact echo, infrared thermography, pulse echo, and radar
Material properties	ASTM C 642 (density), moisture content (ASTM C 642), shrinkage (ASTM C 596, C 426), dynamic modulus (ASTM C 215), modulus of elasticity (ASTM C 464)

Table 6.3 Initial Evaluation Survey of Reinforced Bridge Components

Assessment Procedure	Comments
Air void and petrographic core samples	Applicable to structures up to 15 years old
Alkali–silica reactivity test	Applicable to structures that are between 1 and 15 years old
Concrete strength	Applicable at all ages
Relative permeability	Applicable at all ages
Rebar cover	Applicable at all ages

Note: Test and sampling details may vary, depending on the age of the structure.

6.4 CONDITION ASSESSMENT OF REINFORCED-CONCRETE STRUCTURES

Techniques that can provide an early warning of critical corrosion problems can be particularly helpful. Once rebar corrosion has proceeded to an advanced state that visually apparent on external surfaces, it is usually too late to implement effective corrosion-control measures and high repair or replacement costs are inevitable. Tools for assessing the effectiveness of remedial measures in short, practical time frames are also greatly required.

A systematic comparison of condition-assessment techniques has been developed for concrete bridge components under a contract of the U.S. Strategic Highway Research Program (SHRP). The comparison scheme developed in this program (SHRP product 2032) compares 13 conventional, well-established test methods and 7 new methodologies (refer to Table 6.2) divided in the following survey types:[24]

- The initial (baseline) evaluation surveys focus on parameters that undergo relatively little change over time. The test methods recommended for this baseline survey essentially represent tests that should form part of the acceptance testing of new concrete bridge components (Table 6.3).
- The goal of the second type of surveys in the SHRP scheme is to provide subsequent evaluation during the life of a structure. The initial step in these surveys is visual inspection. The nature of subsequent inspection techniques depends on the morphology of damage observed in the visual-inspection phase. The emphasis on reinforcing-steel-corrosion damage is placed under concrete-spalling phenomena. The recommended assessment procedures for this form of damage are presented in Figure 6.13.
- The third survey category describes special surveys, applicable to asphalt-covered decks, pretensioned and posttensioned concrete members, and rigid deck overlays.

Classical methods such as visual inspection, core sampling for compressive strength tests and petrographic analysis, and chain drag sounding have formed the basis of traditional condition assessments. However, as is apparent from Table 6.2, a host of new nondestructive-evaluation (NDE) methods and corrosion-monitoring techniques are available in modern engineering practice. Some of these techniques are described in the following sections.

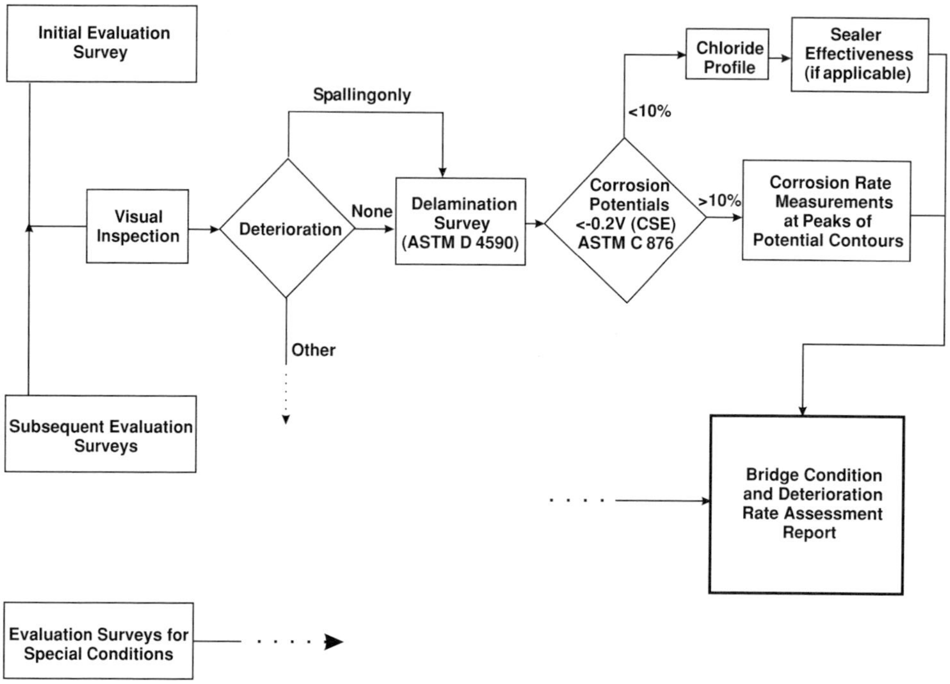

Figure 6.13 Part of the SHRP guide to assessing concrete-bridge components (applicable to uncoated-steel rebar).

6.4.1 Electrochemical-Corrosion Measurements

There are three different types of measuring device:

- Half-cell potential measurements.[25] These can be used to map the anodic areas across the concrete surface.
- Corrosion-rate measurement by polarization-resistance measurement.[26,27] These can be used to measure the corrosion rate of the bars nearest to the sensor.
- Measurement of galvanic-current flow between embedded steel sensors. These are usually embedded prior to construction. The corrosion sensors are essentially small sections of rebar steel, with shielded electrical leads attached for potential and current measurements. Preferably these sensors are embedded at different depths of cover and their use must obviously be defined in the design stage of the structure.

These measurements can be performed nondestructively on the actual reinforcing steel or on sensors that are embedded in the concrete structure. However, it is necessary to expose a reinforcing bar to make an electrical connection for the portable sensors. In general, electrochemical techniques are highly sensitive and therefore can detect corrosion damage at a very early stage.

6.4.2 Chloride Content

Samples for determining the chloride level in concrete are collected in the form of powder produced by drilling or by the extraction of cores, sections of which are subsequently crushed. This latter method can provide an accurate chloride-concentration depth profile. The chloride-ion concentration is subsequently determined by potentiometric titration.

Two distinctions are made in chloride-ion concentration testing: acid-soluble chloride content (ASTM C114) refers to the total chloride-ion content, whereas the water-soluble content represents a lower value.

6.4.3 Petrographic Examination

Petrographic examination is a microscopic analysis of concrete, performed on core samples removed from the structure according to the ASTM C 856 standard. This technique yields information such as the depth of carbonation, density of the cement paste, air content, freeze–thaw damage, and direct attack of the concrete.

6.4.4 Permeability Tests

Permeability tests are either based on ponding core samples in chloride solution with subsequent chloride-content analysis or on forced migration of chloride ions under the influence of an external electric field. The application of the electrical field accelerates chloride-ion migration and hence reduces the testing time.

6.5 OTHER FORMS OF CONCRETE DEGRADATION

Apart from rebar corrosion-induced damage, there are three other commonly cited forms of concrete degradation, namely alkali–aggregate reaction, freeze–thaw damage, and sulfate attack.

6.5.1 Alkali–Aggregate Reaction

Alkali–aggregate reaction refers to chemical reactions between certain reactive aggregates and the highly alkaline concrete-pore solution. Reactive silica is known for such reactions and the term *alkali–silica* reaction is often used. The damage is associated with an internal volume increase, producing cracking and spalling of the concrete. The expansion of aggregate particles and the formation of hygroscopic gels that swell are thought to produce internal stresses. The cracking and spalling of alkali–aggregate reaction damage can make the underlying steel more susceptible to further corrosion attack.

For this type of damage the adage "prevention is better than cure" certainly holds true. Screening tests, for example, ASTM C 289 and C 227, are available to identify problematic aggregates. Methodologies are under development to improve the reliability and reduce the testing time of existing screening tests. The addition of lithium salts to concrete mixes may arrest the undesirable expansive effects. Drying of concrete and sealing to minimize ingress of moisture have been suggested to limit the damage in existing structures. These methods are obviously not always easy to implement.

6.5.2 Freeze–Thaw Damage

Scaling of concrete surfaces is closely related to freeze–thaw damage. Repetitive freeze and thaw cycles can cause concrete surfaces to scale, leading to a pitted surface morphology. Freeze–thaw damages are related to the porous nature of concrete. If the solution trapped in the pores freezes, a volume expansion can occur when the solution freezes and can result in increased tensile stresses followed by cracking and spalling of the concrete. Concrete with high moisture content is most susceptible to this damage mechanism.

In cold climates, where this form of damage is a problem, the use of air-entrained concrete is a viable solution. Such concrete has demonstrated an ability to provide durable long-term service through the formation of air voids in the concrete. The freezing pore solution can then expand into the interconnected air voids. Usually the air content of this type of concrete is between 3 and 8 volume percent. However, the total air content alone is not necessarily sufficient to insure complete resistance to freeze–thaw damage. The distribution and size of the entrained air voids are also of major importance. A trade-off exists between air content and strength.

6.5.3 Sulfate Attack

Soluble sulfate species can cause deterioration of concrete because of expansion during the reaction between sulfate and calcium aluminates in the cement paste. Sulfate ions are ubiquitous; they are found in soils, seawater, ground water, and effluent solutions. Use of cement with low tricalcium aluminate content is beneficial for reducing the severity of this type of attack. Alternatively the use of cement replacement material such as ground, granulated blast-furnace slag can produce a high-quality chloride- and sulfate-resistant concrete ideal for marine tidal- and splash-zone applications.

References

1. J.P. Broomfield, Corrosion of Steel in Concrete (London, UK: E & FN Spon, 1997).
2. Z. Lounis, "Maintenance Management of Aging Bridges: Economical and Technological Challenges," Canadian Civil Engineer 19 (2002): pp. 20–23.
3. "Corrosion Protection: Concrete Bridges," U.S. Department of Transportation, FHWA-RD-98-088, Washington, DC, 1998.
4. "Highway Deicing: Comparing Salt and Calcium Magnesium Acetate," National Research Council, Transportation Research Board, Special Report 235, Washington, DC, 1991.
5. "The Relevance of Cracking in Concrete to Corrosion of Reinforcement," Concrete Society, Technical Report CSTR 44, Camberly, UK, 1995.
6. P.R. Roberge, Handbook of Corrosion Engineering (New York, NY: McGraw-Hill, 2000).

7. "Protection of Metals in Concrete Against Corrosion," American Concrete Institute (ACI), Report 222R-01 2001.
8. C.M. Hansson, "Concrete: The Advanced Industrial Material of the 21st Century," Metallurgical and Materials Transactions A 26A (Farmington Hill, 1995): pp. 1321–1341.
9. A.A. Sohanghpurwala, "Long-Term Effectiveness of Corrosion Inhibitors Used in Repair of Reinforced-Concrete Bridge Components," CORROSION/2003, paper no. 286 (Houston, TX: NACE, 2003).
10. R.F. Stratfull, "Progress Report on Inhibiting the Corrosion of Steel in a Reinforced Concrete Bridge," Corrosion 15 (1959): pp. 65–69.
11. R.F. Stratfull, "Experimental Cathodic Protection of a Bridge Deck," Tranportation Research Board, Transportation Research Record 500, Washington, DC, 1974.
12. RP0290-2000, "Impressed Current Cathodic Protection of Reinforcing Steel in Atmospherically Exposed Concrete Structures" (Houston, TX: NACE, 2000).
13. "Cathodic Protection of Steel in Concrete," DIN/BS/EN 12696, British Standards Institute, London, UK, March 2000.
14. "Electrochemical Chloride Extraction from Concrete Bridge Components," Canadian Strategic Highway Research Program (C-SHRP), Technical Brief 2, Toronto, Canada, June 1995.
15. "Electrochemical Chloride Extraction from Steel-Reinforced Concrete—A State of the Art Report," NACE, Houston, TX, 2001.
16. "Electrochemical Realkalization of Steel-Reinforced Concrete—A State of the Art Report," NACE, Houston, TX, 2004.
17. A. Sagues, H.M. Perez-Duran, R.G. Powers, "Corrosion Performance of Epoxy-Coated Reinforcing Steel in Marine Substructure Service," Corrosion 47 (1991): pp. 884–893.
18. D.G. Manning, "Corrosion Performance of Epoxy-Coated Reinforcing Steel North American Experience," Construction and Building Materials Journal 10 (1996): pp. 349–365.
19. U. Nürnberger, "Stainless Steel in Concrete — State of the Art Report," The Institute of Materials, European Federation of Corrosion, London, UK, 1996.
20. P. Castro-Borges, O.T. de Rincon, E.I. Moreno, A.A. Torres-Acosta, M. Martinez-Madrid, A. Knudsen, "Performance of a 60-Year-Old Concrete Pier with Stainless-Steel Reinforcement," MP 41 (2002): pp. 50–55.
21. D.D. MacDonald, M.R. Sherman, D.W. Pfeifer, Y.P. Virmani, "Stainless-Steel Reinforcing as Corrosion Protection," Concrete International 17 (1995): pp. 65–70.
22. B. El-Jazairi, N.S. Berke, "The Use of Calcium Nitrite as a Corrosion-Inhibiting Admixture to Steel Reinforcement in Concrete," in Corrosion of Reinforcement in Concrete, eds. C.L. Page, K.W.J. Treadaway, P.B.E. Bamforth (London, UK: Applied Science, 1990), pp. 571–585.
23. U. Nürnberger, "Chloride Corrosion of Steel in Concrete, Fundamental Relationships—Practical Experience, Part 1 and Part 2," Betonwek und Fertigteil-Technik (1984): pp. 601–704.
24. R. Kessler, "Concrete Deterioration Manual," National Research Council, Strategic Highway Research Program, SHRP 2032, Washington, DC, 1991.
25. ASTM [C 976-91], "Standard Test Method for Half-Cell Potentials of Uncoated Reinforcing Steel in Concrete," (West Conshohocken, PA: ASTM, 1997).
26. J. Flis, H.W. Pickering, K. Oseo-Asare, "Assessment of Data from Three Electrochemical Instruments for Evaluation of Reinforcement Corrosion Rates in Concrete Bridge Components," Corrosion 51 (1995): pp. 602–609.
27. J. Wenzlick, "An Electrochemical Method for Detecting Ongoing Corrosion of Steel in a Concrete Structure with CP Applied," National Research Council, Strategic Highway Research Program, SHRP 4003, Washington DC, 1991.

7
Chapter

High-Temperature Corrosion

7.1 INTRODUCTION

High-temperature corrosion is a form of corrosion that does not require the presence of a liquid electrolyte. Sometimes, this type of damage is called "dry corrosion" or "scaling." The first quantitative approach to oxidation behavior was made in the early 1920s with the postulation of the parabolic-rate theory of oxidation by Tammann and, independently, by Pilling and Bedworth.

Although temperatures greater than approximately 90 or 150°C are sometimes considered "high temperature," this chapter is concerned primarily with temperatures greater than the "red-hot range," primarily 650°C and greater.

Alloys often rely upon the oxidation reaction to develop a protective scale to resist corrosion attack's such as sulfidation, carburization, and other forms of high-temperature attack. In general, the names of the corrosion mechanisms are determined by the most abundant dominant corrosion products. For example, oxidation implies oxides, sulfidation implies sulfides, sulfidation/oxidation implies sulfides plus oxides, and carburization implies carbides.[1]

Oxidizing environments refer to high-oxygen activities with excess oxygen. Reducing environments are characterized by low-oxygen activities, with no excess oxygen available. Clearly, oxide-scale formation is more limited under such reducing conditions. It is for this reason that reducing industrial environments are generally considered to be more corrosive than the oxidizing variety.

The properties of high-temperature oxide films, such as their thermodynamic stability, ionic-defect structure, and detailed morphology, play a crucial role in determining the oxidation resistance of a metal or alloy in a specific environment. High-temperature corrosion is a widespread problem in various industries such as

- power generation (nuclear and fossil fuel),
- aerospace and gas turbine,

- heat treating,
- mineral and metallurgical processing,
- chemical processing,
- refining and petrochemical,
- automotive,
- pulp and paper, and
- waste incineration.

The need for a careful study of the properties of a heat-resistant alloy and its behavior in the anticipated environment is of considerable importance in the selection of a suitable alloy for a particular service application. New alloys and nonmetallic materials that are continually being made available to industry are making it possible to make better selections and to establish safe working limits within which the material can be expected to give satisfactory performance over a reasonable length of time.

7.2 MATERIALS

The most common alloys used for exposure at high temperatures are the iron/nickel/chromium alloys, sometimes referred to as the stainless steels. The compositions of the more commonly used stainless steels and their applications are described in Chapter 8. The materials used may be cast or wrought. For most furnace operations, the cast metals are used.

Ordinary iron has been used successfully from room temperature up to temperatures of about 480°C, and for short periods of time up to 590°C. At temperatures beyond this, ordinary steels tend to corrode rather heavily. However, it is possible to resort to cementation (surface-diffusion treatments) on carbon and low-alloy steels to render them more resistant to corrosion. Depending upon the corrosiveness of the medium involved and the exposure temperature, one or several of the following may be used: siliconizing, aluminizing, chromizing, or a somewhat different method of surface treatment such as plating, plasma spraying, cladding, and so forth.

If medium- or high-alloy steels are used, chromium-bearing stainless steels are generally recommended. Both the straight chromium steels (400 series) and the chromium/nickel/steels (300 series) are used successfully up to approximately 870°C. At still higher temperatures, the higher nickel and higher chromium alloys of the 300 and 400 series are used, and at temperatures in excess of 1100°C, only the alloys with more than 20% Cr can be used with relative safety and particularly for short or intermediate periods of time.

During long-term use at high temperatures, a phenomenon known as creep is experienced, during which relatively low loads may cause a metal to deform very slowly and possibly rupture. Although this is not a corrosion phenomenon, it is nevertheless a factor that must be considered by anyone considering or recommending alloys for use at very high temperatures. It is on the basis of the creep (percent strain per unit time) to be expected at the operating temperature that furnace parts are designed to withstand the mechanical stress.

Table 7.1 Melting Point of Various Metals

Metal	°C	°F
Aluminum	659	1218
Beryllium	1280	2340
Carbon	3600	6512
Chromium	1890	3430
Cobalt	1495	2723
Copper	1083	1981
Gold	1063	1945
Iridium	2454	4449
Iron	1539	2082
Lead	327	621
Magnesium	650	1202
Manganese	1245	2273
Molybdenum	2625	4760
Nickel	1455	2651
Niobium	2468	4475
Osmium	2700	4892
Platinum	1773	3224
Silver	961	1761
Rhenium	3180	5756
Rhodium	1966	3571
Silicon	1430	2605
Tantalum	2996	5425
Tin	232	450
Titanium	1668	3034
Tungsten	3000	5432
Zinc	419	787

At still higher temperatures, it may be necessary to use refractory metals that have good high-temperature mechanical properties, plus melting temperatures greater than 1650°C, as shown in Table 7.1. Alloys of chromium, niobium, molybdenum, tantalum, rhenium, or tungsten will therefore be considered. Although titanium and zirconium melt above 1650°C, their high-temperature mechanical properties are rather unsatisfactory, so they are not generally considered in the category of refractory metals. Similarly, vanadium is not a popular contender for high-temperature applications.

7.3 BASIS OF CORROSION RESISTANCE

Corrosion resistance at high temperatures stems from one of two basic factors: one deals with thermodynamics, the other with kinetics.

7.3.1 Thermodynamics

In a corrosion context, the thermodynamic properties of a metallic material describe the energy that is lost (or gained) in the form of heat as the metal reacts to form oxides or other compounds. The stability of the various oxides also differ.

Two important points are involved. First, the more energy released when forming the compound, such as an oxide, the more stable the compound. Second, as the temperature is raised, more of the oxidizing species is required to maintain the oxide or other film.

As an illustration, let us consider a bar of copper and a bar of iron at 980°C in an atmosphere made up of 20% H_2 and 80% H_2O. This atmosphere is partly oxidizing and partly reducing. Under these conditions, copper will not corrode and will remain bright. Iron, on the other hand, will become coated with a layer of iron oxide.

In the case of copper, the reaction with water that produces copper oxide and hydrogen is thermodynamically not favored:

$$2Cu + H_2O\ (80\%) \leftarrow Cu_2O + H_2\ (20\%) \tag{7.1}$$

In fact, the reverse reaction would occur if some copper oxide were on the copper; that is, the copper oxide would be reduced by the hydrogen to form H_2O and metallic copper.

In the case of iron, however, the reaction with water to form iron oxide and hydrogen in the same conditions is thermodynamically possible and would occur:

$$Fe + H_2O\ (80\%) \rightarrow FeO + H_2\ (20\%) \tag{7.2}$$

The 80% H_2O–20% H_2 mixture would not permit H_2 to reduce FeO to form Fe and H_2O, as in the case of copper. Thus, we may say that copper is corrosion resistant under these circumstances, or is more noble than iron. Iron, however, is thermodynamically subject to corrosive attack.

It is possible to use plots of the free energy of formation of metal oxides vs. temperature to predict the temperatures at which a metal is stable and the temperatures at which it will spontaneously oxidize. For temperatures at which the free energy of formation of the oxide is positive, the reverse reaction is favored and the oxide will spontaneously decompose to the metal.

Plots of the standard free energy of reaction ($\Delta G°$) as a function of temperature, commonly called Ellingham diagrams, can help to visualize the relative stability of metals and their oxidized products. Figure 7.1 shows an Ellingham diagram for many simple oxides. The $\Delta G°$ values on an Ellingham diagram are expressed as kJ per mole O_2 to normalize the energy scale and compare the stability of these oxides directly; that is, the lower the position of the line on the diagram the more stable is the oxide.[2] For a given reaction, expressed in generic terms in Equation (7.3), Equation (7.4), or its logarithmic form, Equation (7.5), may be used to express the oxygen partial pressure (p_{O_2}) at which the metal and oxide are at equilibrium:

$$M + O_2 \rightleftarrows MO_2 \tag{7.3}$$

$$p_{O_2} = e^{\frac{\Delta G°}{RT}} \tag{7.4}$$

$$\text{Log}(p_{O_2}) = \frac{\Delta G°}{2.303 \cdot RT} \tag{7.5}$$

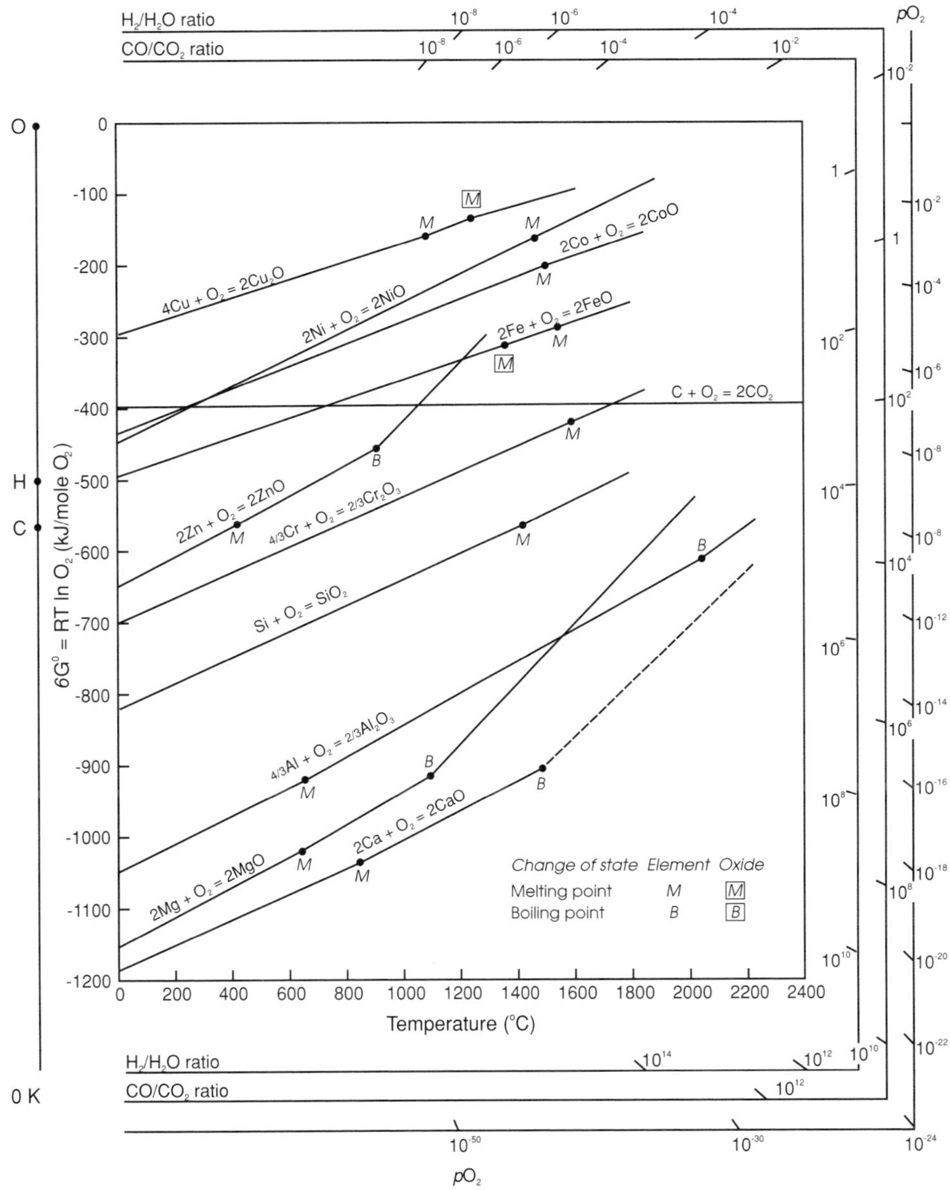

Figure 7.1 Standard free energy of formation of selected oxides as a function of temperature.

The oxygen partial pressures may be obtained directly from the Ellingham diagram by drawing a straight line from the origin marked **O** through the free energy line at the temperature of interest and reading the oxygen pressure from its intersection with the scale at the right side labeled p_{O_2}. Values for the pressure ratio H_2/H_2O for equilibrium between a given metal and oxide may be obtained by drawing a similar line from the point marked **H** to the scale labeled H_2/H_2O ratio and values for the equilibrium CO/CO_2 ratio may be obtained by drawing a line from point **C** to the scale CO/CO_2 ratio.

Table 7.2 Oxidation-Reduction Conditions Existing at High Temperature

Oxidizing Environment	Reducing Environment
Oxygen	Hydrogen
Sulfur dioxide	Ammonia
Sulfur trioxide	Hydrogen sulfide
Water (steam)	Sulfur
Chlorine	Carbon
Carbon dioxide	Carbon monoxide
Molten Salts	

Note that environments other than oxygen, water, or CO_2 can produce metallic oxides because of their oxidizing capacity. Table 7.2 lists a number of common gases encountered in high-temperature works, and the type of reaction to be expected when they predominate.

Referring back to our original statements regarding copper and iron at 980°C in a 20% H_2–80% H_2O gas atmosphere, Figure 7.1 indicates that although pure H_2O would oxidize copper (use point **H** on left vertical axis as a fulcrum), as little as about 0.1% H_2 (H_2/H_2O ratio of 10^{-3}) would be adequate to suppress copper corrosion.

This is important to remember and is in accord with the relatively high nobility of copper. In the case of iron, however, about 70% H_2 would be necessary to suppress iron corrosion at 980°C. It can be determined from Figure 7.1 that the $2\ Fe + O_2 = 2FeO$ line at 980°C indicates an equilibrium gas mixture of about 2 : 1 for H_2/H_2O ratio. Because a gas made up of 2 parts H_2 and 1 part H_2O is a 67% H_2 gas, 70% H_2 or more would favor oxide reduction at 980°C, and iron would remain bright in this atmosphere. If the iron were to be cooled in this atmosphere, it would oxidize because drier hydrogen is necessary at lower temperatures.

By using the heavy line shown from fulcrum **O**, we can determine that 10^{-7} or more volume percent of oxygen also will sustain the oxide on iron at 980°C.

By using Figure 7.1 in an appropriate fashion, one could determine the partial pressure of oxygen that would cause any of the listed metals to corrode, as well as the CO/CO_2 ratios or the H_2/H_2O ratios to produce a similar oxidation. Of course, if a mixture of all these gases were present simultaneously, the situation would become more complicated and its discussion would be beyond the intended scope of this chapter.

One can further add to the complexity of a situation by introducing other gases such as sulfur dioxide, hydrogen sulfide, chlorine gas, or other "contaminants" that, in addition to oxides, tend to form sulfides, chlorides, and so forth.

7.3.2 Kinetics

In the case of kinetics, whether a metal is considered corrosion resistant is dependent not on whether it corrodes but on how fast it corrodes. Practically all real situations come under this category. That is, more than 99% of cases involve situations where corrosion is thermodynamically possible. The determination of corrosion resistance lies simply in how fast corrosion occurs. If it is extremely slow, it can be concluded that some form of corrosion resistance has developed.

Conversely, if corrosion is rapid or intermediate in speed, it can be concluded that the metal in question does not have enough corrosion resistance and that a corrosion failure can soon occur. The meaning of "rapid" or "slow" is usually dependent on the expected life of the object in question.

As to what factors determine whether a high-temperature corrosion reaction will occur slowly or rapidly, we must consider the characteristics of the corrosion-product layer that forms.

7.4 CORROSION PRODUCTS

When a metal is oxidized at elevated temperatures, a stable oxide or other compound often covers and "coats" the exposed-metal surfaces. Thus, the corrosion-product layer may act as a barrier between the underlying metal and the corrosive environment (air, flue gas, molten salt, or any other corrosive agent). The compound may be a solid, liquid, or gas. For instance, if chromium, vanadium, and molybdenum were heated in air at 1100°C,

1. chromium would form solid Cr_2O_3,
2. vanadium would form liquid V_2O_5, and
3. molybdenum would form gaseous MoO_3.

Needless to say, only chromium would not be severely corroded. If the oxide formed serves as an effective barrier, corrosion can be strongly retarded and only a thin oxide layer will form. On the other hand, if the barrier is not very effective, corrosion can and will continue at a moderate rate. It could then be concluded that the oxide scale was only slightly protective and that the metal was not corrosion resistant under these conditions of exposure.

The most common observation on iron-base alloys is a multiple-layer oxide scale similar to that shown in Figure 7.2. With ordinary iron, the innermost layer is wustite, FeO; the intermediate layer is magnetite, a spinel form of oxide, Fe_3O_4; and the outermost layer is hematite, Fe_2O_3. There are cases where only one or two oxide layers will form, whereas in other instances as many as seven layers have been observed.

Another unusual observation of representative nonprotective oxide formation is a corrosion product with a low melting point. This would be expected to be accompanied by severe corrosive attack. An illustration of this phenomenon showing partial melting of the scale layer is shown in Figure 7.3.

7.4.1 How Corrosion Proceeds

Once a corrosion-product layer is formed, continued reaction must involve the diffusion of at least one of the reactants through the corrosion-product layer so that the corrosion-reaction will continue. This will be true regardless of whether single or multiple scale layers exist.

Figure 7.2 Three oxide layers often are observed on low-alloy iron above 570°C. In this case, the inner layer was 55% of total layer thickness; the outer layer was 30% (× 2.7 magnification).

Let us assume, for example, that nickel is to be exposed to air at high temperature. Corrosion can theoretically continue through the nickel-oxide layer by means of diffusion in either direction, alone or by countercurrent diffusion, as illustrated in Figure 7.4. From the atomic viewpoint, diffusion through the oxide occurs by atomic movement, which is made easier by existing atomic vacancies, as illustrated in Figure 7.5 and

Figure 7.3 Molten phase forms sometimes as it has on the high-nickel-alloy specimen after exposure to reducing flue gas that contains less than 1% hydrogen sulfide (×2 magnification).

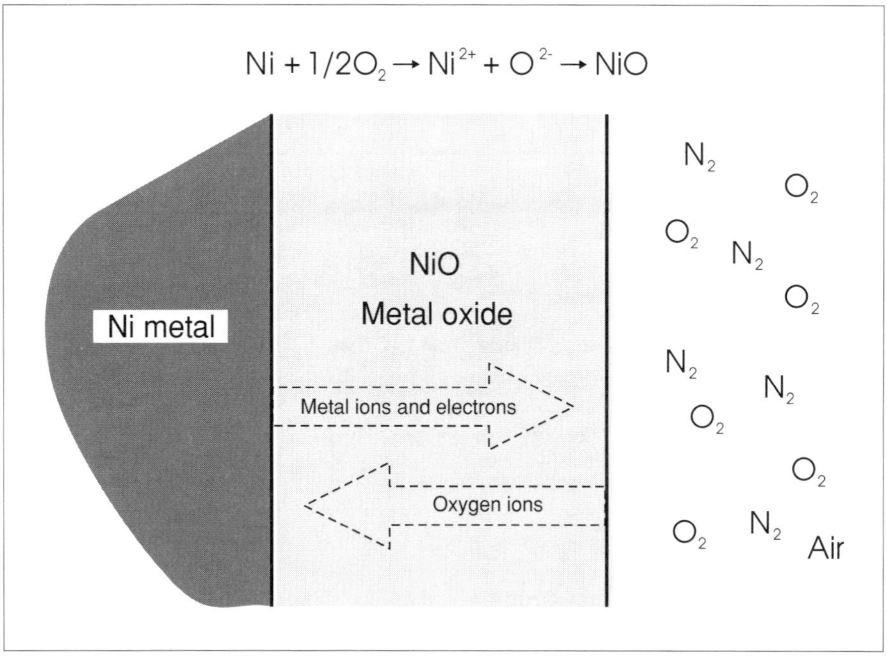

Figure 7.4 Schematic shows countercurrent diffusion, that is, diffusion of oxygen inward and metal ions plus electrons outward. Thus, the anode is at the metal/metal-oxide interface A, where oxidation ($M \rightarrow M^{2+} + 2e^{-}$) occurs. The cathode is at C, the oxide–air interface where reduction ($1/2O_2 + 2e^{-} \rightarrow O^{2-}$) occurs.

Figure 7.6. The schematic diagram in Figure 7.5 represents the type of oxide that forms on nickel (and on a number of other metals, including copper, cobalt, manganese, and chromium). Nickel forms only one oxide, NiO, but, as the figure shows, it is a metal-deficit oxide; that is, a slight excess of oxygen atoms over nickel atoms will be found under normal conditions, and the charge balance (the scale must remain electrically neutral) will be maintained by a few of the nickel atoms, assuming a higher valence state.

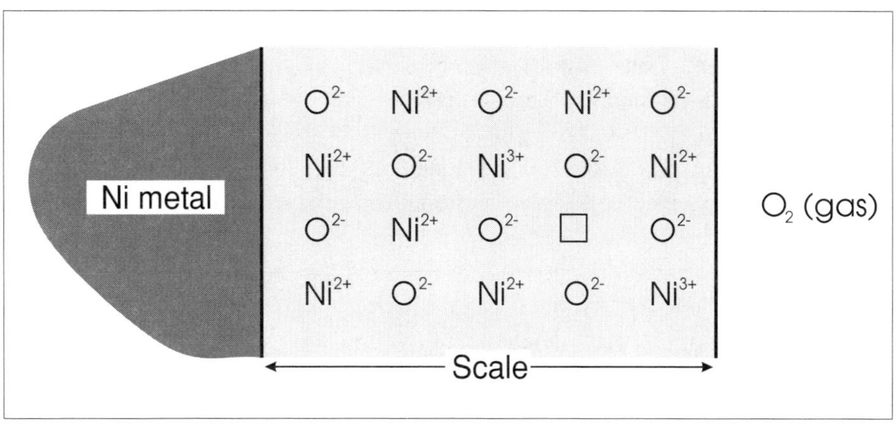

Figure 7.5 Nickel oxide is an example of metal-deficit scale. Schematic of NiO lattice shows a few Ni^{3+} sites among the Ni^{2+}. For every two Ni^{3+} sites, there is a vacant Ni^{2+} site.

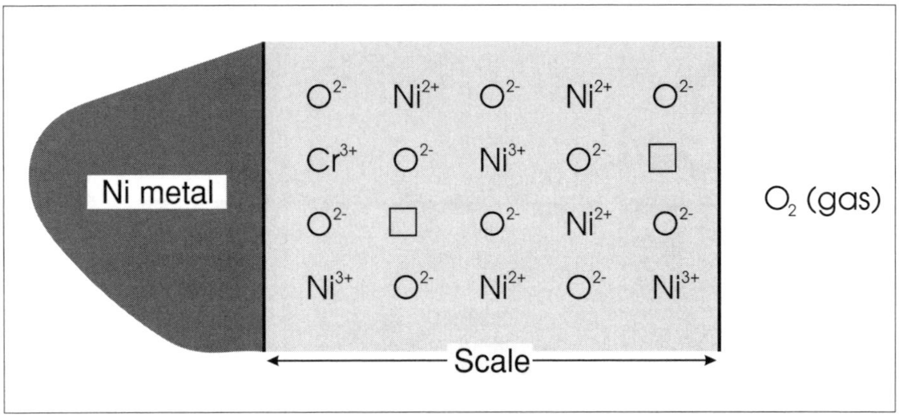

Figure 7.6 Nickel oxide containing small amounts of chromic oxide. Note that this increases the density of vacant Ni^{2+} sites.

For this type of oxide, the controlling step in increasing scale thickness is the ability of the nickel ions to migrate outward toward the surface. If a small quantity of a higher valence ion (Cr^{3+}, for example) is added to the nickel, as in Figure 7.6, charge balance will require a greater number of nickel vacancies and hence an easier, quicker path for nickel ions to migrate to the surface. Conversely, if a lower-valence metal ion (Li^{+} is the best-known example) is added, charge-balance requirements will reduce the number of defects or vacancies, and the oxidation rate, conforming to the nickel-ion diffusion rate, will decrease.

The opposite situation, a metal-excess scale, is also possible (zinc is an example). In this case, extra metal ions are believed to occupy intermediate positions in the scale. The introduction of a higher-valence metal ion will reduce the number of zinc ions in the scale and thereby lower the oxidation rate. A lower-valence metal ion will, of course, have the opposite effect, causing the metal-excess scale to respond to the addition of other metal ions in a manner just the reverse of that in metal-deficit scales.

Some metals form more than one oxide. In the case of iron, for example, Fe_2O_3, Fe_3O_4, and FeO can form, depending upon the availability of oxygen, usually expressed as a partial pressure. Generally, all three types will form concurrently on the metal surface. Fe_2O_3, which requires the most oxygen, will be found on the outside, and FeO, which requires the least, on the inside, with Fe_3O_4 between them. The boundaries between the various oxides usually are quite well defined and, provided the several layers remain intact, each will grow according to the parabolic-rate law (as discussed in a later section).

In most instances, the predominant diffusion is that of metal ions plus electrons outward to the cathode (i.e., to the oxide–air interface shown at "C" in Figure 7.4).

The migration of electrons should not be overlooked. If electron mobility was low, as in the case of oxides that behave as electrical insulators (Al_2O_3, BeO, Cr_2O_3, etc.), the metals developing these compounds as surface films could be expected to have good corrosion resistance.

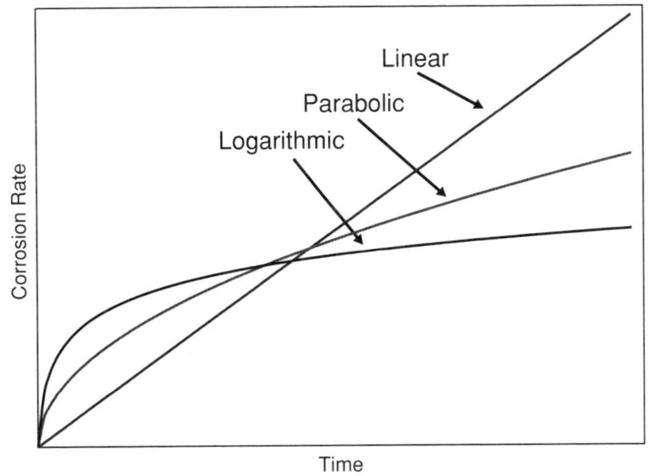

Figure 7.7 Linear, parabolic, and logarithmic relationships between time and extent of corrosion may be measured as depth of penetration or weight change (gain or loss).

Because nickel at high temperatures does not form an oxide layer with a good diffusion-resistant coating, alloying elements are generally added to provide added imperviousness to the oxide layer. Chromium is an excellent alloying element, and the 80 Ni–20 Cr alloy is one of several compositions commercially used for heating elements in high-quality domestic electric toasters or electric irons. This alloy generally develops a suitable NiO-Cr_2O_3 spinel (oxide) on the surface, but under suitable conditions Cr_2O_3 forms an even more diffusion-resistant barrier, the spinel.

7.4.2 Linear Behavior

If the oxide film or scale cracks or is porous (i.e., if the corrosive gas can continue to penetrate readily and react with the base metal), no protection will be afforded and attack will proceed at a rate determined essentially by the availability of the corrosive gas. In this case, the rate will not sensibly change with time, and, as is apparent from Figure 7.7, the weight change or depth of penetration from oxidation is a straight line or linear function of time and may be expressed as

$$y = kt \tag{7.6}$$

where y = scale thickness or weight change, t = time, and k = a constant, which will, however, be dependent upon temperature, the material being tested, and other conditions of the test.

7.4.3 Logarithmic Behavior

The logarithmic-rate behavior follows an empirical relationship, which has no fundamental underlying mechanism. This behavior, also shown in Figure 7.7, is mainly applicable to thin oxide layers formed at relatively low temperatures and therefore is rarely applicable to high-temperature engineering problems:

$$y = k' \log(ct + 1) \tag{7.7}$$

where k' is the rate constant and c is a constant.

7.4.4 Parabolic Behavior

If, conversely, the scale formed is continuous, adherent, and prevents easy access of the corrosive gas to the underlying base metal, a considerable measure of protection may occur, and the extent of protection will increase as the scale thickens. In this case, the availability of the corrosive gas will not determine the reaction rate.

Diffusion through the scale will be the slowest and, hence, the rate-controlling step. The diffusing element that controls the reaction rate may be either the oxygen moving inward or the metal moving outward. The latter is more likely in alloys possessing good corrosion resistance. As the scale thickens, everything else remaining the same, the diffusion rate will decrease, slowing proportionally the rate of scale growth. A mathematical expression of the process is a parabola (Figure 7.7) having the equation:

$$y = k'' \sqrt{t} \tag{7.8}$$

After an initial period of time, the rate of scale growth will decrease to a low level, that is, a considerable measure of oxidation resistance will be obtained and, provided the scale remains intact, will be maintained for a very long time.

Under special conditions, particularly at intermediate temperatures, the oxidation process may essentially stop after the initial scale formation. In such cases, no change in scale thickness or weight would be observed after very long intervals. Presumably this occurs because when the scale reaches some minimum thickness, the diffusion rate is so low as to be essentially zero, thus creating a condition under which the metal should last almost indefinitely without further deterioration. Some investigators have chosen to call this type of behavior tarnishing and thereby to distinguish it from the higher-temperature phenomenon of scaling. The latter is of greater concern in the remainder of this discussion.

The parabolic-rate law, although obviously an oversimplification of the actual state of affairs during oxidation, nevertheless provides a relatively accurate approximation of the behavior of the majority of metals and alloys practically used at high temperatures.

7.4.5 Characteristics of Scales

Thus far, principally oxide-scale formation was considered, and it was assumed that once formed, the scale would remain intact and protective. Scale thickness may be an important factor. In early stages, the scale is thin and reasonably elastic. It also will have little intrinsic strength and usually will remain tightly fixed on the base metal. As it thickens, however, its own less-desirable properties may become manifest. Heavy scales tend to be brittle and therefore to rupture and spall from the surface.

Figure 7.8 presents data on a stainless steel that reflects this type of behavior. At the lower temperatures, the scale thickens slowly and the parabolic-rate law is followed.

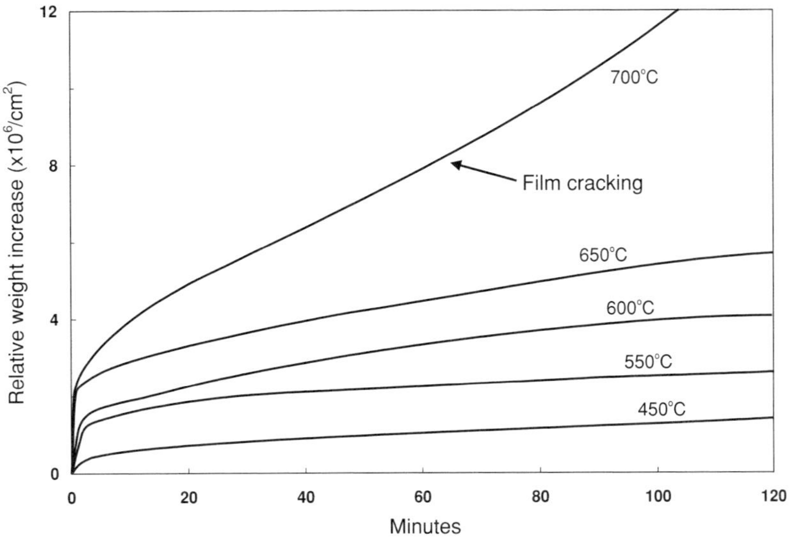

Figure 7.8 Oxidation of stainless steel at various temperatures in 0.1 atm oxygen.

At the highest temperature, the rate also is initially parabolic, but at some point the scale breaks away locally and there is an immediate increase in the scaling rate as a new protective scale is formed.

The volume of the oxide formed, relative to the volume of the metal consumed, is an important parameter in predicting the degree of protection provided by the oxide scale. If the oxide volume is relatively low, tensile stresses will tend to crack the oxide layers. Oxides, which are essentially brittle ceramics, are particularly susceptible to fracture and cracking under such tensile stresses. If the oxide volume is very high, stresses will tend to lead to a break in the adhesion between the metal and oxide. For a high degree of protection, the oxide formed should be similar to that of the metal consumed. This argument is the basis for the Pilling–Bedworth ratio (PB):

$$PB = \frac{\text{Volume of oxide}}{\text{Volume of metal consumed}} = \frac{Wd}{nDw} \tag{7.9}$$

where W = molecular weight of oxide, D = density of the oxide, n = number of metal atoms in the oxide molecule, d = density of the metal, and w = atomic weight of the metal.

PB ratios slightly greater than 1 could be expected to indicate "optimal" protection, with modest compressive stresses generated in the oxide layer. Table 7.3 provides the PB ratio of a few metal/oxide systems.[2] In practice, it has been found that PB ratios are generally poor predictors of the actual protective properties of scales. Some of the reasons advanced for deviations from the PB rule include these:[3]

- some oxides actually grow at the oxide–air interface, as opposed to the metal–oxide interface;

Table 7.3 Oxide-Metal Volume Ratios of Some Common Metals

Oxide	Oxide/Metal Volume Ratio
K_2O	0.45
MgO	0.81
Na_2O	0.97
Al_2O_3	1.28
ThO_2	1.30
ZrO_2	1.56
Cu_2O	1.64
NiO	1.65
FeO (on α-Fe)	1.68
TiO_2	1.70–1.78
CoO	1.86
Cr_2O_3	2.07
Fe_3O_4 (on α-Fe)	2.10
Fe_2O_3 (on α-Fe)	2.14
Ta_2O_5	2.50
Nb_2O_5	2.68
V_2O_5	3.19
WoO_3	3.30

- specimen and component geometries can affect the stress distribution in the oxide films;
- continuous oxide films are observed even if $PB < 1$;
- cracks and fissures in oxide layers can be "self healing" as oxidation progresses;
- oxide porosity is not accurately predicted by the PB parameter; and
- oxides may be highly volatile at high temperatures, leading to nonprotective properties, even if predicted otherwise by the PB parameter.

Figure 7.9 represents, in a schematic fashion, typical ways in which scales have been observed to rupture. If temperature variations occur, a weakly adherent scale is likely to separate locally from the base metal, a phenomenon sometimes termed blistering, and, because it has little intrinsic tensile strength, eventually rupture and spall. The relative thermal expansivity between scale and metal also has been cited as an influential factor when temperature fluctuations are involved. Probably of greater significance is the nature of the interface between the metal and the scale. If the interface is planar, it is easy to conceive of a shear crack initiating at the interface and propagating rapidly across the surface.

If, on the other hand, the interface is rough, where incipient grain-boundary attack beneath the scale is occurring, a keying action may prevent rapid crack propagation and thereby improve scale adherence. It is thought by some that the major effect of reactive rare-earth additions to heat-resistant alloys is not to enter the scale and lower diffusion rates, but to accumulate in grain-boundaries and thus ensure a small amount of grain-boundary attack to provide the keying action.

A considerable effort is often made in industry to produce an initial scale on the interior that has the optimum properties. Centrifugally cast tubing is often bored to

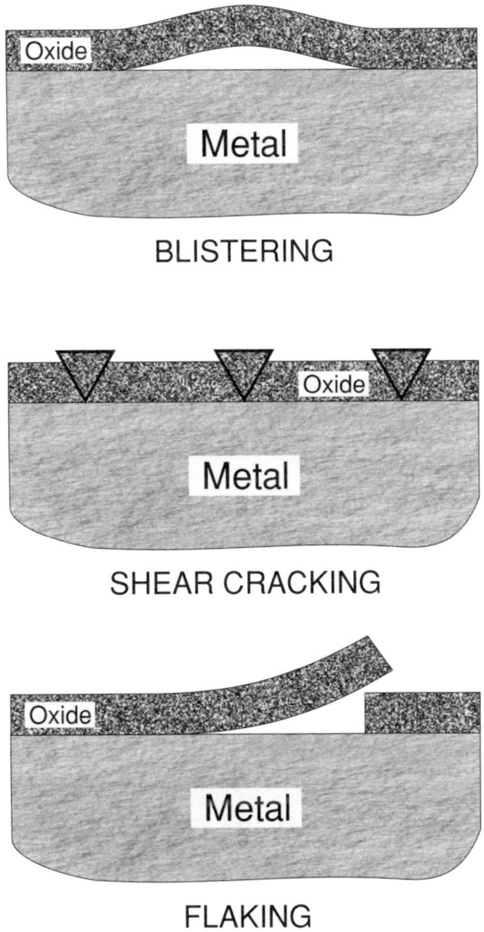

Figure 7.9 Schematic shows some of the hypothetical ways the integrity of film or scale is impaired.

remove porosity near the inner surface, honed to provide a smooth surface profile, and then steamed heavily to produce a thin, continuous, dense, resistant oxide layer on the interior. Such a procedure has remarkably increased the life of tubing sections in some services.

7.4.6 Intergranular Corrosion

Intergranular attack has been shown to be severe in some cases, thereby reducing the effective metallic cross-sectional area. This results in a weak structural member. Furthermore, intergranular attack tends to act as a stress raiser in stressed members. The detrimental effects of notches and cracks on mechanical properties are well recognized, and the danger of deep corrosion penetration via the cracks cannot be overemphasized.

There are several ways of measuring the extent of high-temperature corrosion attack. Measurement of weight change per unit area in a given time has been a popular procedure. However the weight change/area information is not directly related to

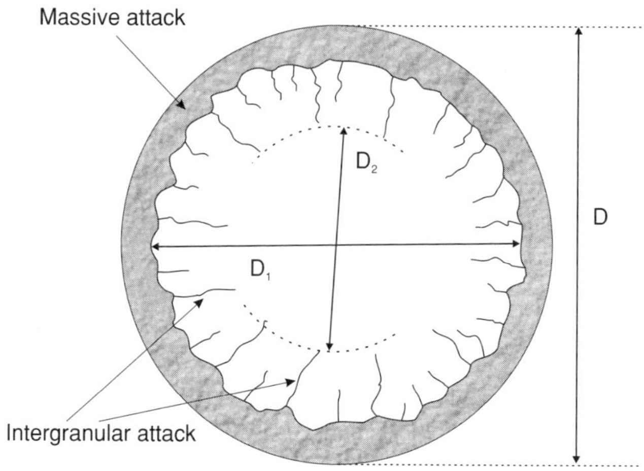

D = original diameter
D_1 = diameter of apparently useful metal
D_2 = diameter of metal unaffected by intergranular attack

Figure 7.10 Metallographic method of measuring hot corrosion attack.[4]

the thickness (penetration) of corroded metal, which is often needed in assessing the strength of equipment components. Corrosion is best reported in penetration units, which indicate the sound-metal loss. A metallographic technique to determine with relative precision the extent of damage is illustrated in Figure 7.10.[4] The parameters shown in Figure 7.10 relate to cylindrical specimens and provide information about the load-bearing section (metal loss) and on the extent of grain-boundary attack that can also affect structural integrity.

7.4.7 Internal Oxidation

The development of external scales and their effect upon continuing oxidation is only one facet of high-temperature corrosion. It is very possible for oxidation to proceed without an external scale, provided certain conditions exist. It is most important that the alloy contain, as a minor ingredient, an element that is considerably more reactive than the major constituent. Silicon in copper is one example, chromium in nickel is another. It is also essential that the rate at which the minor element diffuses toward the surface be less than that at which oxygen, sulfur, or nitrogen diffuses inward. It is possible, however, to stabilize the minor element by formation of a compound, for example, chromium carbide.

Internal oxidation is controlled by the concentration of oxygen in surface layers. This quantity must be insufficient to oxidize the nobler major constituent but sufficient to oxidize the reactive minor constituent. External scale will form only when the effective pressure of oxygen at the metal–gas interface exceeds this range. Conversely, failure of surface scale to form indicates that the oxygen partial pressure at the interface is at such a level that it will not react with the major constituent.

An early example of destructive internal oxidation was found in failures of the chromel segments of chromel–alumel thermocouples used to measure the temperature of a hydrogen-rich gas. The hydrogen was able to penetrate the protection tube and set up an atmosphere within the tube of 90% nickel–10% chromium chromel alloy, which was reducing to the nickel but oxidizing to the chromium. No surface scale formed, but the fracture surfaces showed the characteristic green color of chromium oxide. One method to correct situations of this sort is to include in the thermocouple assembly an active getter for oxygen, such as a strip of titanium, which, by removing the oxygen, would leave an atmosphere that is reducing to both constituents of the alloy.

7.5 CORROSION IN AIR

Of all gaseous, high-temperature environments, air is no doubt the most common. Although we consider air to be roughly 80% nitrogen (N_2) and 20% oxygen (O_2), it may also contain small but sometimes significant amounts of water vapor (H_2O), carbon dioxide (CO_2), carbon monoxide (CO), sulfur dioxide (SO_2), hydrogen sulfide (H_2S), and various other chemical fumes and rare gases. The following discussion of air corrosion is limited to iron-base alloys, especially those containing chromium and nickel (stainless steels and heat-resistant alloys).

Adequate resistance to corrosion by air at elevated temperatures, for example, greater than 650°C, implies the use of appropriate alloying elements. These elements generally include chromium and may also contain one or several of the following beneficial alloying elements: nickel, silicon, aluminum, and possibly titanium.

The relative amounts of alloying elements are determined generally by alloy composition, temperatures, and environment to be encountered; degree of corrosion resistance required; and anticipated life, as well as the physical or mechanical properties required. However, the discussion to follow is restricted to a consideration of the influence of various alloy additions on corrosion resistance.

The influence of alloy additions on corrosion resistance is largely dependent on the composition of the metal prior to the addition. That is, the effect of a certain alloy addition would be entirely different in degree, and sometimes in nature, for a carbon steel when compared with the effect of the same addition to a high-alloy steel.

Nickel and chromium, the two main alloying elements in heat-resistant alloys, will be treated in greater detail than the others. Silicon and aluminum are also quite effective but must be used carefully because they can have a harmful effect on mechanical properties and other alloy characteristics. Still other elements also are sometimes added to the ironnickel chromium base to produce specific characteristics in fabrication or performance requirements.

7.5.1 Effects of Chromium

Chromium (Cr) is considered the most highly desirable addition to heat-resistant steels. In addition to its ability to provide effective corrosion resistance, chromium also improves certain high-temperature mechanical properties.

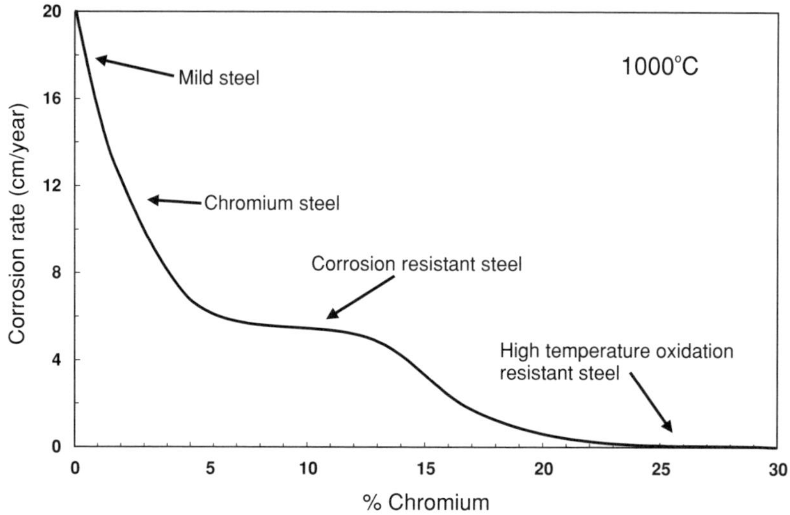

Figure 7.11 Effect of chromium on reducing corrosion in air at 1000°C in Fe-Cr alloys.

Although additions of chromium to iron are described here, it seems logical that chromium additions to other metals will also have a similar beneficial effect. It is the formation of a thin chromium-rich oxide layer on the metal surface that lends resistance to further corrosion.

Figure 7.11 shows a sharp decrease in iron corrosion (oxidation) by air at a temperature of 1000°C, produced by additions of Cr. Note that additions of chromium beyond 20% have little effect, particularly at lower temperatures. As is described later, these iron/chromium alloys are ferritic (i.e., they have a body-centered cubic structure) or have notoriously poor high-temperature strength. Nickel additions to iron/chromium alloys not only improve air-corrosion resistance, as described in the following section, but mechanical strength as well.

7.5.2 Effects of Nickel

Nickel is not alloyed to ordinary iron for the development of improved high-temperature corrosion properties. It is effective, however, as illustrated in Figure 7.12, in alloys having 11% Cr. Note that its beneficial effects, like chromium, tend to fade after reaching some specific amount. However, the amount of nickel needed to achieve a low corrosion rate is greater than that of chromium, and the needed amount also increases for higher exposure temperatures.

Thus, from the corrosion viewpoint, nickel of up to about 20% is effective at 870°C; up to about 30% is effective at 980°C; and up to about 50% would be desirable at 1200°C.

At higher chromium levels, the optimum nickel level for maximum air-corrosion resistance decreases. This effect can be shown clearly by means of isocorrosion lines (compositions with the same weight losses) on a ternary diagram; that is, one that shows

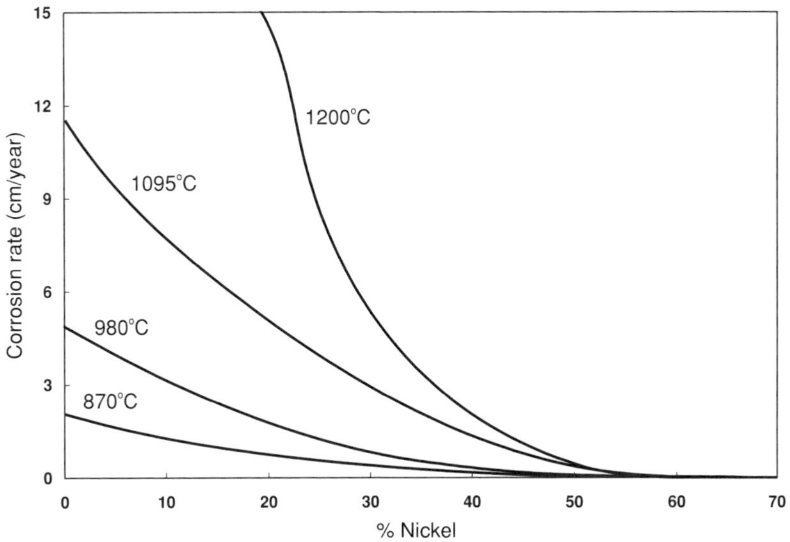

Figure 7.12 Effect of adding nickel to a Fe 11–Cr alloy in air at 870, 980, 1095, and 1200°C.

the changing composition of an alloy made up of three alloying elements, in this case iron, nickel, and chromium.

Figure 7.13 shows the relative amounts of nickel and chromium necessary as alloying additions to iron to reduce very heavy corrosion to lesser amounts. Note that at 1200°C in air, the addition of about 28% Cr is about equivalent to the combined addition of about 50% nickel plus 10% Cr, because all compositions on a given isocorrosion line yield the same corrosion rate.

The combined effect of nickel and chromium additions to iron in reducing corrosion by air at 980°C is more clearly portrayed in the space diagram (three dimensional), shown in Figure 7.14. This diagram has been drawn from the voluminous data reported from corrosion studies at Battelle.

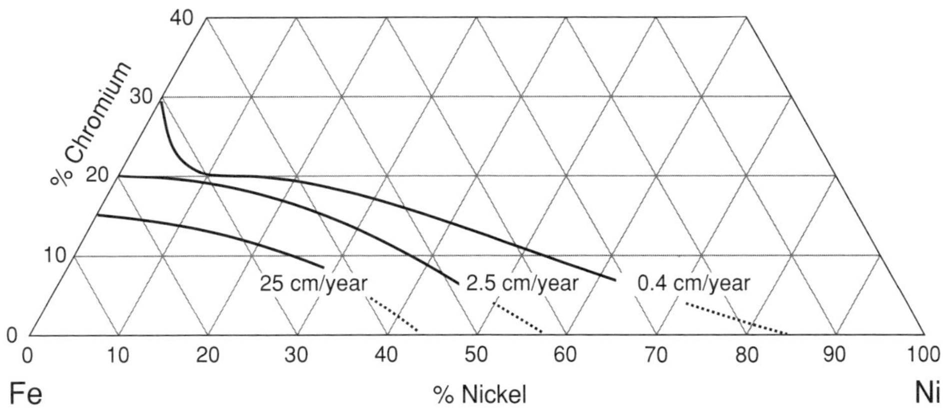

Figure 7.13 Ternary diagram shows alloy composition areas of relatively heavy and light corrosion in air at 1200°C. Based on 100-h tests.

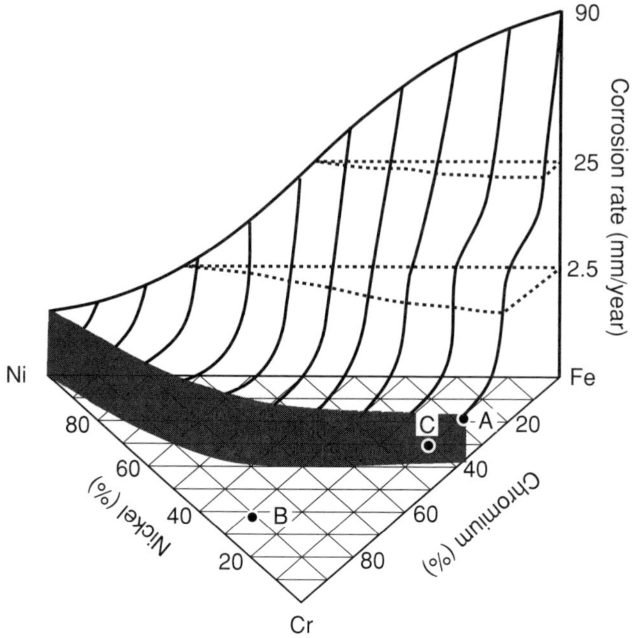

Figure 7.14 Space diagram illustrates relative effects of nickel and chromium contents on intensity of iron-alloy corrosion in air at 980°C: A) 18 Cr–8 Ni–74 Fe, B) 60 Cr–30 Ni–10 Fe, and C) 30 Cr–10 Ni–60 Fe.

Figure 7.14 should emphasize that the data reported to this point are based on 100-h tests, and because of the deceleration of rates encountered in most high-temperature tests, the reported corrosion rates are conservative, as far as long-term projections are concerned. In other words, the average corrosion rate actually should be less in a 200-h or 1000-h exposure test.

7.5.3 Effects of Aluminum

The beneficial effect of aluminum additions in suppressing air corrosion of iron alloys is surprisingly strong. When amounts of up to 12% Al are added, a considerable reduction in high-temperature corrosion is achieved. However, mechanical properties suffer quite strongly, thus making this approach to corrosion resistance less attractive.

The marked beneficial effects of aluminum additions are clearly shown in Figure 7.15. Where low mechanical properties can be tolerated, the use of limited amounts of aluminum should be considered.

One approach to the aluminum-rich use with relatively little effect on mechanical properties is surface impregnation or aluminizing. This high-temperature cementation process is the diffusion of aluminum into iron and has been known for many years to produce an effective surface for exposure in air or sulfur dioxide fumes to temperatures in the neighborhood of 1095°C. The commercial process of calorizing or aluminizing has been in use for almost a century and has definite advantages for certain high-temperature applications.

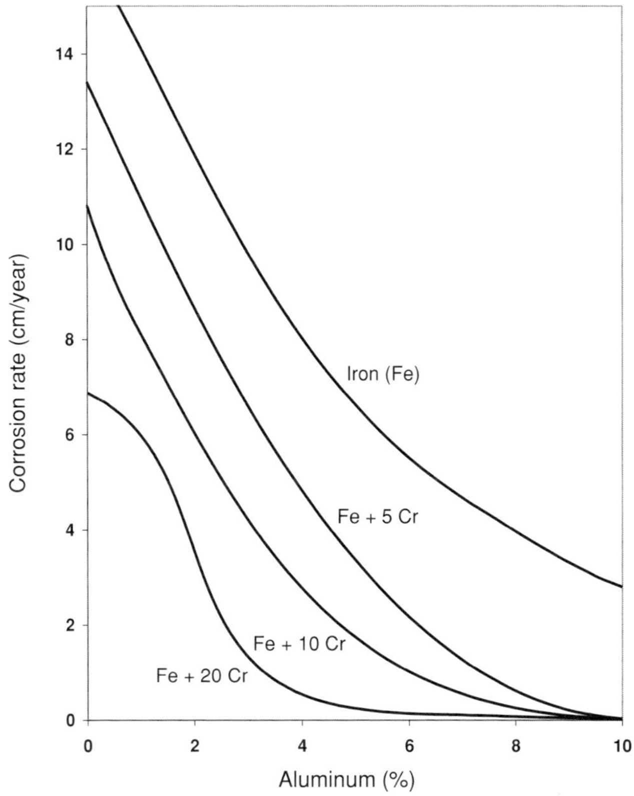

Figure 7.15 Effect of adding aluminum to iron and to Fe-Cr alloys on air corrosion at 1100°C.

7.5.4 Effects of Silicon

Although silicon additions to steels are beneficial both in high and ordinary temperatures, the use of silicon is restricted because of marked embrittlement of the alloy when the amount exceeds 2 or 3%. Thus, the actual use of silicon is generally limited to about 2%.

When used in combination with other beneficial alloying elements, silicon can be quite effective. This is illustrated in Figure 7.16. The commercial steel containing 5% Cr and 0.5% Mo can have its corrosion resistance improved by raising the silicon content from 0.2% to about 1%. There appears to be no further advantage to increasing silicon beyond 1% for applications limited to 815°C in air environments.

For instance, a test made with a 5% Cr steel, with and without Si and Al additions, gave the results shown in Table 7.4 in a 1-day test in air at 732°C and at 1024°C.

7.5.5 Effects of Other Elements

A simple and direct statement concerning the effects of other alloying elements on high-temperature corrosion cannot be reliably made. When very strong effects result,

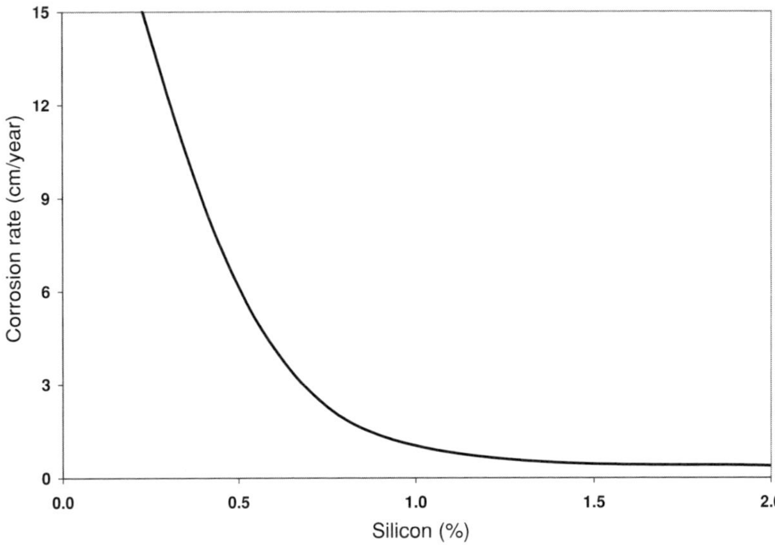

Figure 7.16 Effect of increasing silicon content in a 5 Cr–0.5 Mo steel on resistance to air corrosion at 815°C.

as from Cr, Ni, Si, and Al, mentioned previously, clear-cut statements can be made. However, if the effects are weaker, they are difficult to describe and may be masked by small variations in other factors. The following statements should therefore only be considered with a certain degree of caution.

Manganese (Mn) Small additions of manganese have very little effect. Although few data are available, Mn is not known to have a pronounced effect on high-temperature corrosion rates. In several actual tests, a mild detrimental effect has been observed. Significant amounts of Mn are used in 200-series steels to replace part of the nickel in the analogous 300 steels.

Cobalt (Co) Cobalt additions have very little effect. Cobalt additions beyond 50% have been made at 850 to 1100°C, and no significant effects have been noted in air or flue-gas atmospheres on alloys containing 22 to 29% Cr.

Table 7.4 Effects of Silicon and Aluminum on Performance of 5-Cr Steel up to 1025°C

	Weight Loss (mg/cm^2)	
Composition (%)	730°C	1025°C
5 Cr (plain)	56.0	164.1
5 Cr + 2 Al	23.2	146.1
5 Cr+ 2 Si	0.4	68.1
5 Cr + 2 Si + 2 Al	0.2	4.0

Table 7.5 Influence of Copper on 0.45-C 15.3-Cr 35-Ni Alloys. (Corrosion Rate in mm/year)

	Air Corrosion		Reducing Flue Gas at 980°C	
Cu %	980°C	1100°C	Low Sulfur	High Sulfur
0.00	0.46	1.22	0.46	1.30
1.38	0.99	1.37	1.09	6.9

Molybdenum (Mo) Molybdenum additions are generally considered detrimental. Although Mo additions to stainless steels can be highly beneficial in ambient- temperature environments, their use at high temperatures is inadvisable. Small amounts (less than 4%) may have little effect on high-alloy materials. Larger additions to high alloys or smaller amounts in lower alloy steels can result in catastrophic attack, which is described later.

Copper (Cu) Copper additions can be somewhat harmful at high temperature. Ordinarily, copper is not added to alloys for high-temperature service, although it is sometimes advantageous in stainless steels used to resist chemical attack. As shown in Table 7.5, two heat-resistant alloys were prepared: one with 0.0% Cu, the other with 1.38% Cu (both alloys also contained 0.45 C, 15.3 Cr, 35 Ni). Corrosion rates for the exposures indicated how copper can be harmful in heat-resistant alloys.

Arsenic (As) Additions of arsenic have very little effect as shown in Table 7.6. There is evidence that arsenic is beneficial to atmospheric-corrosion resistance of mild steel, but this effect does not appear to exist at elevated temperatures.

Vanadium (V) All evidence points to vanadium as a very harmful alloying element to heat-resistant alloys, as far as corrosion resistance is concerned. This is attributable to the very low melting point of its oxide, V_2O_5, which is well known to promote more rapid corrosion, even to the extent of catastrophic rates.

Carbon (C) Carbon is recognized as being highly undesirable in 12% Cr steels and austenitic 18-8 stainless steels. Carbon's deleterious influence is two-fold. When present in steels in excess of 0.1%, a substantial amount of carbides is generally formed,

Table 7.6 Influence of Arsenic on 0.45-C 15.3-Cr 35-Ni Alloys (Corrosion Rate in mm/year)

	Air Corrosion		Reducing Flue Gas at 980°C	
As %	980°C	1100°C	Low Sulfur	High Sulfur
0.00	0.28	0.79	0.18	0.43
0.086	0.28	0.99	0.28	0.30

especially when the metal is in an aged or sensitized condition. This is the usual condition of the heat-resistant steels, because their application is generally at elevated temperatures. In forming a carbide, some of the chromium that ordinarily would be present in the matrix becomes associated with the carbon particles as chromium carbide. Hence, the matrix is depleted of some of its chromium, and the resulting steel could have the corrosion characteristics of chromium-bearing steel containing less chromium than a chemical analysis would indicate.

Because the approximate chemical formula for chromium carbide is Cr_4C, and recognizing that the atomic weights for chromium and carbon are 52.01 and 12.0, respectively, we can calculate that for every unit of weight of carbon, 17.3 weight units of chromium are required according to the following calculations:

$$\frac{4\,\mathrm{Cr}}{\mathrm{C}} = \frac{4(52.01)}{12.0} = \frac{208.04}{12} = 17.3 \tag{7.10}$$

However, because some carbon is soluble in the metal matrix, the ratio 16:1 is more suitable. Thus, in calculating the probable chromium content of the matrix (background metal) of the low (0.30% C) and high (1.0% C) carbon 26 Cr–20 Ni alloy of Figure 7.17, the following information is obtainable:

$$\text{Matrix chromium} = \text{Total Cr} - 16\text{C} \tag{7.11}$$

For the 0.3% carbon alloy,

$$\text{Matrix chromium} = 26 - 16(0.30) \tag{7.12}$$

For the 1.01% C alloy,

$$\text{Matrix chromium} = 26 - 16(1.01) \tag{7.13}$$

This would seem to imply a marked susceptibility of the high-carbon alloy to corrosion. However, high-temperature air-corrosion tests made on these alloys indicate that the chromium gets into the oxide layer to supply corrosion protection equally well when obtained from the carbide phase as when derived from the matrix.

One method by which carbon may act as a detrimental element is its formation of a continual network of interdendritic carbides (also referred to as primary or eutectic carbides) in cast alloys. These particles are often extremely susceptible to attack. This phenomenon is a serious problem in the liquid corrosion field and, to a lesser extent, to the hot-gas-corrosion field. Not all iron/chromium carbides are equally susceptible. As the composition of the matrix varies according to the general composition of the steel, the composition of the carbide is also altered. Therefore, a different behavior should be anticipated.

Carbide-forming elements, such as titanium and niobium, are often added to alloys, particularly to those of the 18-8 austenitic type. These elements act as carbide stabilizers, as is described in Chapter 8. These elements have a greater affinity for carbon than chromium does, and consequently they form carbides of their own, leaving all the chromium as a matrix constituent. The carbides formed by the stabilizing elements are found as very small particles scattered throughout the grains themselves, rather than in grain boundaries, thereby appreciably reducing the prevalence of intergranular corrosion.

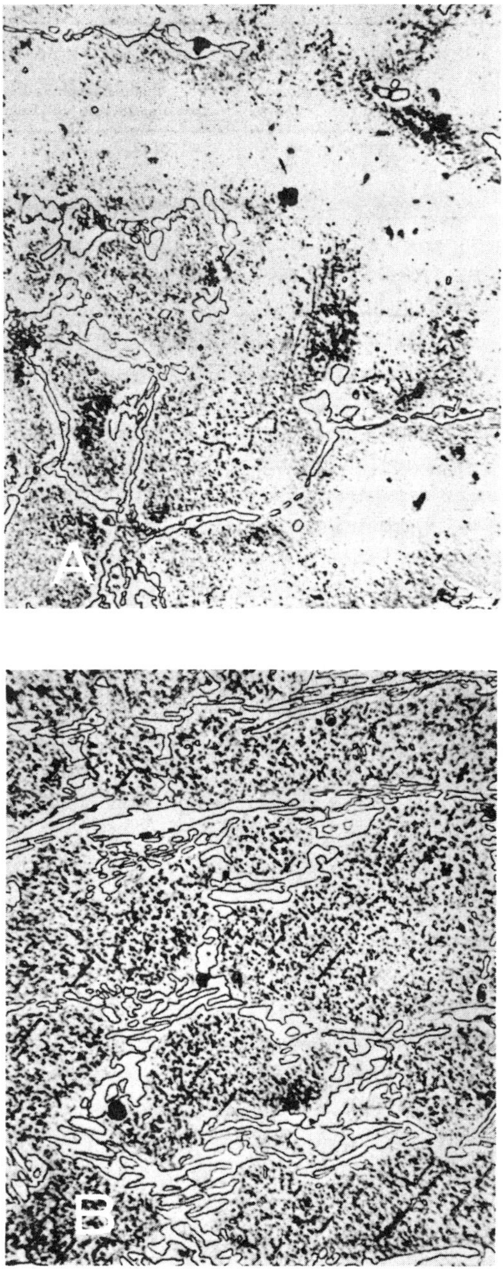

Figure 7.17 Microstructure of aged medium (A = 0.32% C) and high (B = 1.0% C) carbon 26 Cr–20 Ni alloys at ×250 magnification.

The influence of moderate carbon variations (0.3% C) on the high-temperature corrosion resistance of alloys containing substantial amounts of alloying materials is essentially nil. Although the depletion of approximately 5% matrix chromium from a 12% Cr steel or an 18-8 steel might be a serious matter, with richer alloys of the 25 Cr–20 Ni type, the effect of removing 5% chromium would be almost imperceptible.

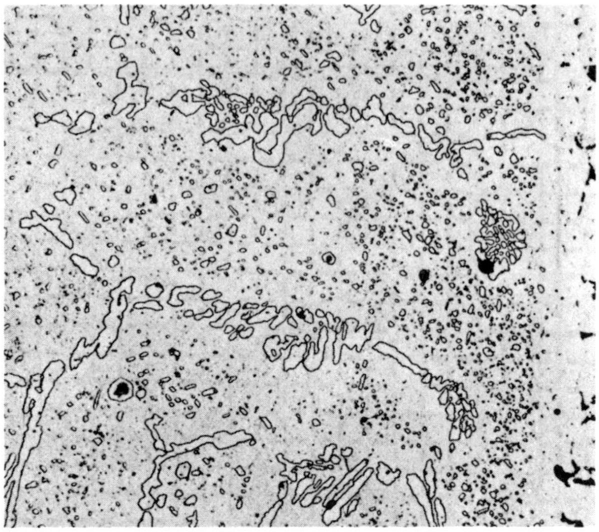

Figure 7.18 Photomicrograph shows a cross section of heat-resistant alloy after mild corrosive attack. Although primary and secondary carbides appear throughout, they do not appear in the surface zone because of decarburization.

In all likelihood, because decarburization (Figure 7.18) generally occurs at the metal surface preceding oxide formation, it releases chromium to the matrix. Thus, chromium can serve equally well in low- or high-carbon alloys at elevated temperatures:

$$Cr_4C + O_2 \rightarrow CO_2 + 4\,Cr \tag{7.14}$$

$$4\,Cr + 3O_2 \rightarrow 2\,Cr_2O_3 \tag{7.15}$$

Other Alloying Elements Specimens of the 35 Ni–15 Cr type were also prepared with 2 and 4% additions of various other alloying elements. These included tungsten (W), niobium (Nb), tantalum (Ta), titanium (Ti), zirconium (Zr), beryllium (Be), as well as vanadium and molybdenum, which were discussed earlier.

In a long-term laboratory test of 1000 h, specimens were exposed to air at 1100°C, cooled, weighed, and handled lightly to remove loose oxide scale daily, then reexposed for additional 24-h periods until a total of 1000 h (42 days) had elapsed. Results are summarized in Table 7.7.

7.6 OTHER GASEOUS MEDIA

Oxygen reactions were discussed at great length for several reasons. It is, of course, the most common high-temperature corrosion reaction. Consequently, it has been more widely studied and is better understood than less common forms of high-temperature corrosion.

A more compelling reason, however, is that what one learns about oxygen is directly applicable to most other gas–metal, high-temperature reactions in which a scale is formed. In fact, all of these reactions should be classified as oxidation, whether the

Table 7.7 Effect of Alloy additions on Air Corrosion of 35Ni–15Cr Alloy at 1100°C

Behavior	Alloy Additions 35-Ni-15Cr (%)	Influence
Slightly beneficial	2 and 4 Mo 2 and 4 Zr 2 and 4 Be 2 W	Reduces corrosion
Slightly harmful	4 W 2 Ti	~1.5×
Harmful	4 Ti 2 and 4 Nb 2 and 4 Ta	2× or 3×
Extremely harmful	2 and 4 V	~20×

scale is an oxide, a sulfide, or a halide, because the valence charge of the metal has been increased. The main differences between the reactions under consideration and those already discussed is discussed in the following sections.

7.6.1 Corrosion by Flue Gases

Corrosion by flue gases has many elements in common with air corrosion at high temperatures. Oxygen is the major element to consider, largely in the form of uncombusted gaseous oxygen (O_2) or its combination in the form of carbon dioxide (CO_2), water vapor (H_2O), or possibly even sulfur dioxide (SO_2). These gases, in addition to nitrogen (N_2), are the major constituents in oxidizing flue gases, which contain an excess of O_2.

If the oxygen level was insufficient for complete fuel combustion, these oxidizing gas constituents would be present in smaller amounts and would be partially replaced by carbon monoxide (CO), hydrogen (H_2), and hydrogen sulfide (H_2S), respectively. Such gases are referred to as reducing flue gases.

Of course, there sometimes exists an in-between condition when the appropriate amount of air (or oxygen) is used for fuel combustion and the flue gas cannot be appropriately referred to as oxidizing or reducing. In such cases, the term neutral may be used.

7.6.1.1 Sulfur Content

The major constituent in flue-gas corrosion that differentiates it from air corrosion is the sulfur content. Although it originally may be present in the form of complex organic compounds, after combustion it occurs as SO_2 or H_2S, depending on the oxygen availability, because either can be converted to the other by either of the two following equations:

$$H_2S + O_2 \rightarrow H_2 + SO_2 \tag{7.16}$$

$$H_2S + 2\ CO_2 \rightarrow H_2 + SO_2 + 2\ CO \tag{7.17}$$

Table 7.8 Sulfur Content in Some Fuels and Their Combustion Products

	Sulfur Content of Gas	
Fuel	Before Combustion	After Combustion
Good natural gas	Trace	Trace
Coke-oven gas		
Purified	30 gS[1]	6 gS
Unpurified	600 gS	125 gS
Fuel Oil		
Low S	0.7% S	25 gS
High S	2.5% S	100 gS
Bituminous coal		
Low S	1% S	55 gS
High S	5% S	300 gS

[1] One grain of sulfur (1 gS) per 100 cubic feet of gaseous fuel equals 15.8 ppmv.

As mentioned earlier, reducing flue gases tend to have significant amounts of H_2, H_2S, and CO, whereas oxidizing flue gases are relatively rich in H_2O, SO_2, and CO_2 (Table 7.2). One should keep in mind that no flue gas is completely or totally reducing or oxidizing. When CO_2 is largely present, some CO can also be found; when H_2 is present, some H_2O will be there, and so forth.

Some fuels are known to be quite high in sulfur, whereas others can be low. As a rule, bituminous coal is quite heavy (Table 7.8) and is a major "villain" in today's city air-pollution problems. Fuel oils vary somewhat in sulfur, and sometimes also contain trace amounts of organic compounds of vanadium that can be much more corrosive, as is mentioned in the next section dealing with catasrophic corrosion.

Among the lowest sulfur fuels are the natural gases. Some have significant traces of sulfur, but many are extremely low. Economics and availability usually dictate which fuel a given industry will use, and the corrosion engineer is expected to keep the plant relatively free of serious operating problems.

Two principal methods are used to report sulfur content of gases. One is by volume percent and the other by referring to its weight in 100 cubic feet of gas. The former method is easier; that is, a 2 vol% SO_2 content means that for every 100 gas molecules, 2 are likely to be SO_2 molecules. This system is less suitable when trace amounts are present, and references must be made to very small amounts, such as 0.00164 vol% SO_2. In such cases, the weight in grains of sulfur per 100 ft^3 of gas is used.

A grain is one seven-thousandth of a pound. This is an ancient unit of weight that represents the average weight of a grain of wheat. The above amount, 0.00164 vol% SO_2, represents one grain of sulfur per 100 ft^3 of gas, and is abbreviated in this section of the text as 1 gS. One grain per 100 ft^3 of gaseous fuel equals to 15.8 ppmv.

Thus, one can compare fuels more easily in regard to sulfur content. A coal with a 4% sulfur content would generate a flue gas under ideal combustion with the proper amount of air, containing about 250 gS. Similarly, other typical flue gas comparisons are shown in Table 7.8.

Table 7.9 Differences Between Metal and Sulfide Eutectic Melting Points

Element	Melting point (°C)	
	Metal	Metal–Sulfide Eutectic
Co	1495	877
Cr	1850	1350
Cu	1083	1070
Fe	1539	985
Mn	1260	—
Ni	1455	645

7.6.1.2 Sulfidation

In general, oxidation by sulfur or sulfidation is a considerably more destructive form of high-temperature corrosion than oxidation by oxygen. Sulfide scales tend to crack and spall more readily than oxides. In some cases, depending on the form in which sulfur is present in the atmosphere, continuous sulfide scales cannot form, so attack will proceed linearly; that is, the scale will afford no protection. The melting points of metallic sulfides usually are lower than those of the corresponding oxides.

Table 7.9 compares the melting points among a few metal sulfide/metal eutectics with the base-metal melting points. Note that most sulfide eutectics are molten at temperatures well below the melting points of the base metals. Oxide eutectics usually melt at temperatures much closer to the melting points of the metals on which they form. Thus, the sulfide-eutectic melting temperature frequently will constitute a limit to the upper service temperature of the metal or alloy, whereas only rarely does this occur with an oxide. As is discussed later, the presence of a molten phase generally will cause a vast increase in the corrosion rate of any metal.

Nevertheless, the basic mechanism of sulfidation is closely akin to air oxidation and, to a considerable degree, the same alloying elements are useful in providing resistance to attack. As shown in Figure 7.19, the behavior of chromium follows the same pattern in sulfidation as in oxidation. It is particularly interesting to note that, when sufficient chromium is present, the relative amount of nickel or iron is of secondary importance, despite the fact that pure nickel is much more susceptible to sulfur attack than pure iron. Chromium, however, is much more attracted to sulfur than either iron or nickel, and the scale formed is essentially a chromium sulfide, when sufficient chromium is present in the alloy, whatever the balance of the composition. Eventually, of course, when the chromium is depleted, the remaining base metal will be attacked rapidly.

The form in which sulfur is present will affect the rate of attack. When oxygen is also present, as would be the case with SO_2 or SO_3, a mixed oxide/sulfide scale frequently forms and offers a greater degree of protection than the sulfide scale produced by H_2S or organics (which in most cases decompose to give H_2S) and, of course, sulfur vapor. Figure 7.20 shows some data on the relative performance of a nickel/chromium alloy exposed to either pure SO_2 or pure H_2S.[5] Note that parabolic behavior is indicated, although nonprotective scale is evident at an early stage. Figure 7.21 shows the results

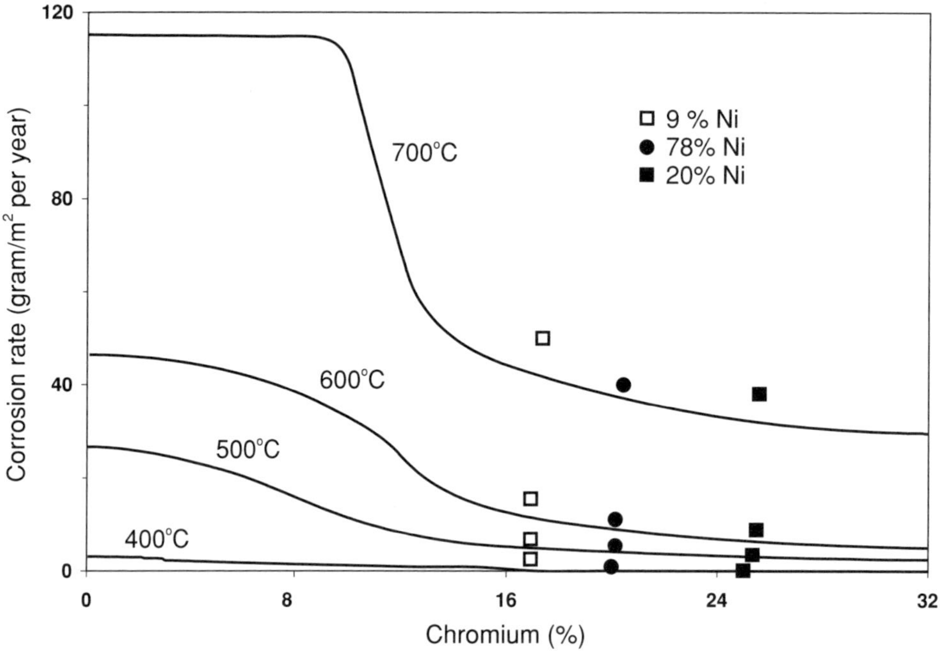

Figure 7.19 Effect of Cr content of steels on corrosion resistance in hydrogen sulfide is shown. Curves were drawn through open circles, representing binary Cr-Fe steels. Data for Cr-Fe-Ni alloys were plotted as solid circles and were found close to curves for binary alloys, despite the great differences in Fe-Ni content.

of exposing a Ti-Al codefusion coating on N08800 after 1000-h testing at 700°C in a metal-dusting environment containing 7% hydrogen sulfide. Some small iron sulfides can be observed in Figure 7.21b between Ti oxides, Ti sulfides, and Al oxides.

As with oxygen attack, aluminum may be a useful addition to improve sulfidation resistance. Table 7.10 gives some data on the performance of a number of iron/nickel/chromium alloys with and without aluminum. In every case, the presence of aluminum provided an additional degree of resistance. Similar improvement could be expected when cementation techniques are used to develop siliconized, chromized, or aluminized coatings on ferrous alloys.

Table 7.10 Influence of Aluminum on Sulfidation Resistance of Alloys[(1)] (Corrosion Rate in mm/year)

	Aluminum Addition		
Material	0%	3%	4%
18-Cr 8-Ni	1.13	0.43	0.08
20-Cr 32-Ni	0.41	0.18	—
16-Cr 78-Ni	2.51	0.31	—

[(1)] Test conditions: 530°C, 3.1 MPa (480 psi) H_2S in H_2. Vol% from 1.4 to 1.7

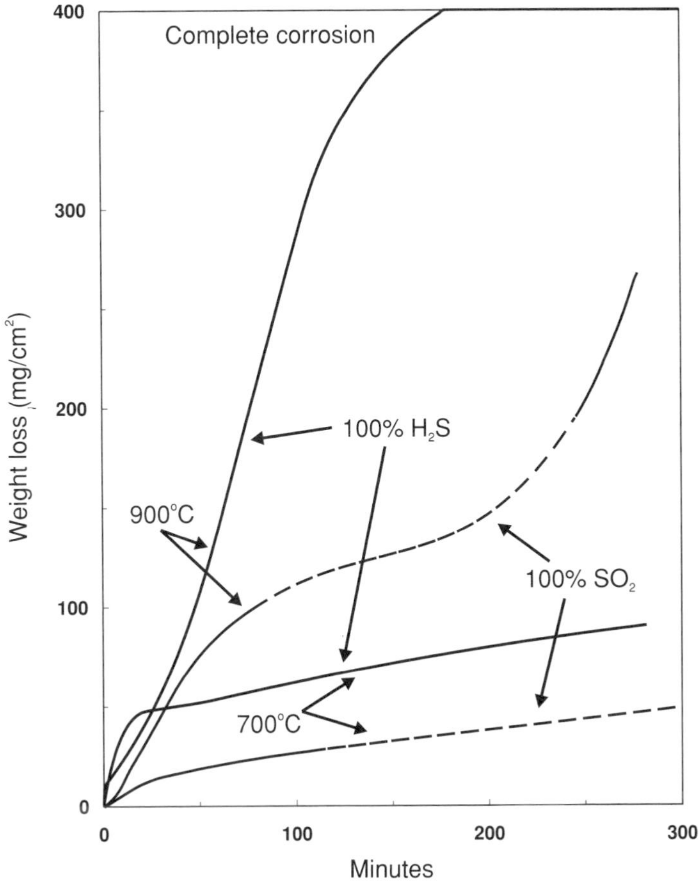

Figure 7.20 Corrosion of Ni-Cr alloy in sulfur-bearing gases.

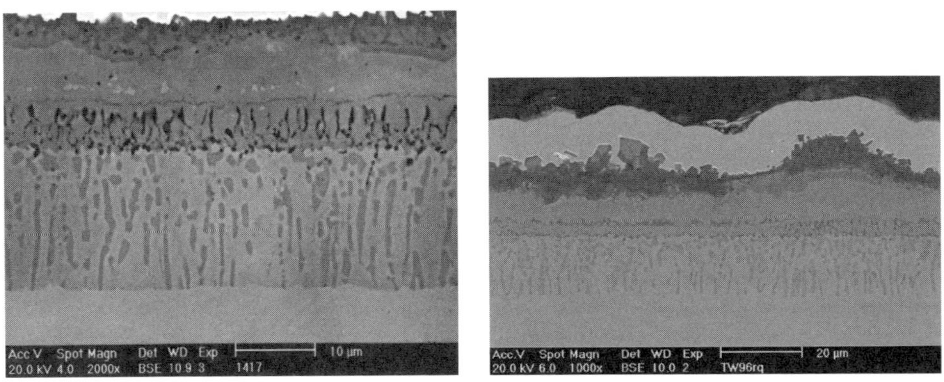

(a) (b)

Figure 7.21 Ti-Al-codiffusion coating on alloy 800 (N08800) a) before testing and after b) 1000 h testing at 700°C is shown. The test atmosphere was CH_4 79%, CO_2 1, CO 1, H_2 10, H_2S 7 (courtesy of Karl-Winnacker-Institut, Dechema, e.V.).

7.6.1.3 Rate of Attack

The attack on iron/nickel/chromium alloys by oxidizing and reducing flue gases has shown strong similarities to air corrosion, particularly to the oxidizing gases. Two major differences are as follows:

1. Oxidizing flue gases are less oxidizing than air; that is, they contain less available oxygen, and although they also contain CO_2, H_2O, and SO_2, the oxygen supply is smaller. In reducing flue gases, the available oxygen supply is still smaller.
2. They contain sulfur-bearing gases that may not attack high-chromium, nickel-free alloys excessively, but that will be corrosive to high-nickel alloys because of the instability and low melting points of the nickel/sulfur corrosion products (Figure 7.3).

Reducing flue gases, however, are very sensitive to the sulfur content. If the sulfur content is very low (below 5 grains per 100 ft^3 of gas), corrosion is generally less severe than in air. This is attributed to the relatively small oxidizing power of the gas, and may be likened to air that is somewhat depleted in oxygen. However, when the sulfur content is appreciable, exceeding 100 grains per 100 ft^3 of gas, corrosion resulting from intergranular sulfide penetration can be quite severe. Weight-change indications may not properly describe the severity of metal damage in such instances. High-nickel alloys (greater than 35% Ni) with less than 15% Cr are especially susceptible to this form of corrosion.

Whenever low-grade (Bunker C-type) fuel oils are used for heating boilers and so forth, entrained tiny particles of vanadium pentoxide and other impurity particles can be deposited on hot metal surfaces and result in extremely severe corrosion attack. This subject is discussed under a later section concerning molten phases.

7.6.2 Nitriding

Nitriding has been employed for years to provide hard, wear-resistant surfaces on certain low-alloy steels, and probably the first corrosion problems involving nitriding related to the retorts used to produce nitrided surfaces. Ammonia is the source of active nitrogen in the hard-surfacing process and is the principal source of nitrogen as a corrosive agent.

The nitrogen molecule is relatively stable and, except at very high temperatures over very long periods, has not created serious corrosion problems. Active nitrogen, readily produced by the decomposition of ammonia, is another matter, and rapid nitride formation will occur at temperatures below 540°C. Because ammonia is an important component in many chemical processes and is essential in the fertilizer industry, the need to cope with nitriding is readily evident.

If active nitrogen is available at the surface of the metal, its passage into the metal can be followed by the etching response of the alloy; a smooth front parallel to the surface will be found, evidence that preferential grain-boundary diffusion is not a significant mode of entry. This is illustrated in Figure 7.22. If, however, a limited supply of nitrogen is available, the continuous diffusion front will not be observed. Nitride needles frequently can be seen not only near grain boundaries but within the grains as well.

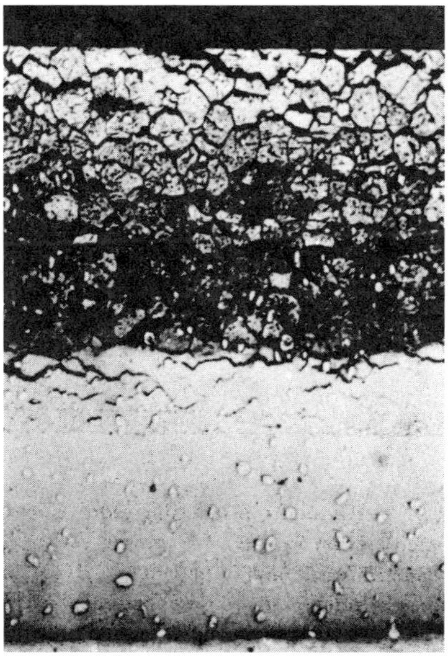

Figure 7.22 Type 446 steel (28 % Cr) after 1540 h at 260°C in anhydrous ammonia. Note cracking in upper areas ×100.

Metals such as iron, chromium, aluminum, and titanium readily form nitrides and are not generally suitable if the nitriding potential is high. Nickel and copper, on the other hand, do not form stable nitrides at elevated temperatures and might be expected to confer some resistance to nitriding. In fact, both nickel and copper are successfully used to "stop off" portions of steel parts during commercial nitriding when only a partial hard surface is desired. Neither element is entirely satisfactory for prolonged high-temperature service, because they are relatively sensitive to contaminants, such as sulfur, which may be present, nor do they have outstanding oxidation resistance.

7.6.3 Carburization

Carburization is not directly a form of high-temperature corrosion. Scale formation and metal waste are not direct products of the carburizing reaction, although loss of ductility can occur. Nevertheless, it is an important gas–metal reaction and is so intimately connected with high-temperature corrosion that it must be studied and understood by anyone seriously interested in the factors affecting the behavior of metals at elevated temperatures.

Like nitriding, carburization is employed commercially to develop hard, wear-resistant surfaces on certain low-alloy steels. Steels, when exposed to a suitable carbonaceous gas or liquid at high temperatures, may absorb some portion of the carbon atoms, which reach the metal surface. The higher carbon surface will respond better to heat treatment and develop a higher surface hardness, while the lower carbon interior will be softer but tougher, a desirable characteristic for many applications.

Again, some of the earlier difficulties with carburization undoubtedly originated because of the materials used to contain the reaction for carburizing steels. Currently, because of the enormous use of high-temperature treatment to alter or crack the molecular

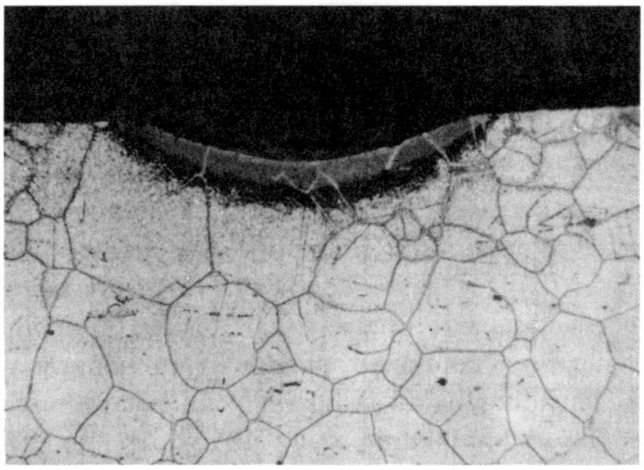

Figure 7.23 Photomicrograph shown cross section of a very small pit. Intense carburization is clearly evident in zones being corroded (×130 magnification).

structure of hydrocarbons derived from petroleum products, the problem of carburization and the selection of materials to withstand it have great industrial significance.

As already mentioned, carburization is largely a gas–metal reaction. Although under some very critical conditions it may be possible to introduce carbon into a metal via a solid-state reaction (that is, directly from carbon or graphite into the metal), the incidence of such an event is so infrequent as to be of no importance. Generally, a gas containing carbon monoxide or hydrocarbons such as methane, propane, and so forth will be the source of the carbon.

At high temperatures, carbon will be released at the surface of the metal and, if the conditions are right, will diffuse inward. As in the case of nitriding, the diffusion occurs uniformly across the grain, in most instances, and is not intergranular (Figure 7.23).

Most metals have a significant solubility for carbon; that is, they can absorb a quantity without undergoing any drastic physical change, although some properties (such as response to heat treatment) may be altered measurably. Eventually, if carbon continues to be driven into the metal, its solubility can be exceeded and more dramatic changes will occur. In pure nickel, which forms no stable carbide at high temperatures, embrittlement from graphite precipitation may occur, and in the more commonly used heat-resistant alloys of the iron/nickel/chromium system, precipitation of chromium-rich carbides will occur.

Initially, such carbides will have little effect on the chemical or physical behavior of the alloy. In fact, most commercial, chromium-containing alloys such as stainless steels have a small quantity of chromium carbides present. However, as carburization continues and the quantity of carbides increases, the situation will change. Because chromium forms carbides more readily than the other constituents of stainless steels, the matrix of the alloy will be depleted of chromium.

As discussed previously, chromium is the element that contributes most to a resistance to oxidation and sulfidation. The chromium-depleted matrix will be unable to develop adequate protective scales and will become quite susceptible to these and other types of attack (Equation [7.10]).

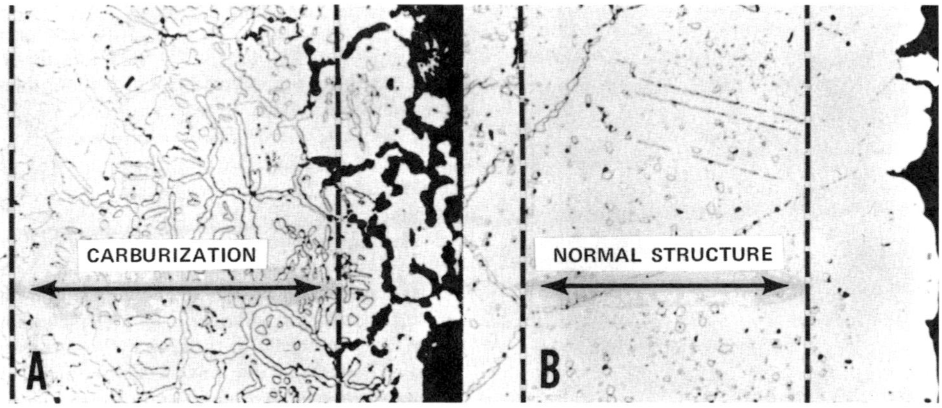

Figure 7.24 Effect of variations of silicon content on carbon pickup after 7340 h at 915°C in a carburizing atmosphere: A) 0.47 Si–0.96 C (significant carbon pickup) and B) 1 Si–0.14 C (no carbon pickup). Atmosphere is hydrogen 34%; CO 14, CH_4 12.4, CO_2 0.4, N_2 balance (×250 magnification).

Furthermore, the presence of continuous grain-boundary carbides, as illustrated in Figure 7.24, will both increase the rate of intergranular attack (because the carbides are more reactive than the matrix) and harmfully affect the mechanical properties because cracks can propagate more readily along the brittle cracks. In chemical reactions of the type involved in carburization, for example,

$$2\,CO \rightarrow C + CO_2 \tag{7.18}$$

where C indicates carbon in solution in the metal; there is some equilibrium quantity of carbon in solution in the metal for each ratio of CO : CO_2. The precise amount will be strongly influenced by alloying elements. If the ratio of CO : CO_2 in Equation (7.18) is high, the driving force to cause carbon to enter the metal is great. Conversely, if the ratio is low, the driving force is reduced.

If reduced sufficiently, carbon may actually be removed from the metal; that is, not carburization but decarburization would be the result. At some intermediate ratio, of course, the carbon content of the metal would not change because the quantity already present in the alloy would be in equilibrium with the amount available from the gas.

Carburization may then be minimized by selecting an alloy that has a carbon content close to the equilibrium amount for the gas in question. Such a selection will reduce the driving force, and even if perfect equilibrium is not attained, harmful changes in the alloy should proceed at a relatively slow rate. For example, consider the photomicrographs in Figure 7.24. These alloys of the same base composition were exposed in the same carburizing environment, yet one was carburized a great deal and the other almost not at all. It is evident that at fixed chromium and nickel contents, increases in silicon improve the carburization resistance of iron/nickel/chromium alloys.

Another ultimate effect of carburization can be metal dusting. This phenomenon occurs in process operations where oxidizing and reducing conditions alternate. When the environment is on the reducing side (CO predominant), carburization of the metal to a shallow depth can occur at breaks in the protective oxide film. When the exposure then

(a) (b)

Figure 7.25 Two alloys tested in the same H_2-CO-H_2O metal-dusting atmosphere: a) 2.25 Cr 1-Mo at 620°C, 1600 h showing general attack; and b) 9 Cr–1 Mo (P91 steel) at 620°C, 1008 h, showing metal-dusting pitting attack.

changes to oxidizing, the high-carbon area of the metal is burned out and the metal reacted to the oxide. A depression is left in the metal surface where the carburized area existed (Figure 7.25), and the metal oxide is swept downstream in the process as metal dust.

7.6.4 Hydrogen Attack

High-temperature hydrogen attack should be clearly distinguished from hydrogen embrittlement, which occurs at temperatures near ambient and is usually associated with aqueous corrosion as a source of the hydrogen. At higher temperatures, above the dew point, high-pressure hydrogen is frequently a component of some process gas and can be a source of hydrogen attack. Ammonia production, mentioned previously in connection with nitriding, is an industrially important example. To join nitrogen with hydrogen to form ammonia (NH_3), the two gases must be brought together in the presence of a catalyst under very high pressure. The resistance to hydrogen attack is an important consideration in selection of materials to contain this reaction.

The attack relates to the reaction of hydrogen with readily reducible carbides, or in some cases oxides, within the alloy to form methane or steam. These gaseous products, under high pressure, can cause small local ruptures that would impair the structural integrity of the metal and lead to early failure. In some cases, if structural discontinuities are present, hydrogen under pressure will accumulate at these points and cause blistering when the external pressure is reduced, as during a shutdown. In such cases, the internal pressure in the defect will be maintained at a high level and will attempt to relieve itself by expanding against the wall, now no longer supported by the pressure of the contained gases.

The main recourse is to make certain that the carbides formed are sufficiently stable to resist reaction with hydrogen. Elements such as chromium and molybdenum form carbides considerably more stable than iron carbides, so steels containing additions of these elements have been quite successful in resisting hydrogen attack. Note, however, that the problem of blistering results from defects such as seams and inclusions and will not be eliminated merely by selecting a stabilized alloy.

Table 7.11 Melting Points, Temperatures at which Chloride Vapor Pressure Reaches 10^{-4} atm, and Boiling Points of Various Chlorides

Chloride	Melting Point (°C)	Temperature at 10^{-4}atm (°C)	Boiling Point (°C)
$FeCl_2$	676	536	1026
$FeCl_3$	303	167	319
$NiCl_2$	1030	607	987
$CoCl_2$	740	587	1025
$CrCl_2$	820	741	1300
$CrCl_3$	1150	611	945
CrO_2Cl_2	−95	—	117
CuCl	430	387	1690
$MoCl_5$	194	58	268
WCl_5	240	72	—
WCl_6	280	11	337
$TiCl_2$	1025	921	—
$TiCl_3$	730	454	750
$TiCl_4$	−23	−38	137
$AlCl_3$	193	76	—
$SiCl_4$	−70	−87	58
$MnCl_2$	652	607	1190
$ZrCl_4$	483	146	—
$NbCl_5$	205	—	250
$NbCl_4$	—	239	455
$TaCl_5$	216	80	240
$HfCl_4$	434	132	—
CCl_4	−24	−80	77
NaCl	801	742	1465
KCl	772	706	1407
LiCl	610	665	1382
$MgCl_2$	714	663	1418
$CaCl_2$	772	1039	2000
$BaCl_2$	962	—	1830
$ZnCl_2$	318	349	732
$PbCl_2$	498	484	954

7.6.5 Halide Attack

The corrosive effect of halogen on passivating alloys is well known in aqueous media. Chlorides and fluorides also contribute to high-temperature corrosion by interfering with the formation of protective oxides or breaking them down if already formed. The main reason for the reduced corrosion resistance in the presence of halogens is the formation of volatile corrosion products that are nonprotective. The melting points, boiling points, and the temperature at which the vapor pressure reaches 10^{-4} atm of selected metal chlorides is presented in Table 7.11.[6] The high volatility and relatively low melting points of these chlorides should be noted. Clearly these properties are not conducive to establishing an effective diffusion barrier on the corroding alloy surfaces.

In refining operations, chlorides most commonly enter the process operations as salt water or brine. Organic chlorides find their way into crude feed. These are not removed in the desalters but are generally removed in the distillation process. Chlorides can enter the downstream processes. Fluoride contamination is usually the result of blending streams from an alkylation operation. This downstream contamination now can spill over to petrochemical facilities that take feed from these sources.

Halogen contamination during shipment and storage is also of concern to petrochemical facilities.[7] Chlorination processes are used to produce certain metals, as well as in nickel extraction. Calcining operations used in the production of certain rare-earth metals and for producing ceramic ferrites for permanent magnets are also associated with high-temperature chlorine-containing environments.

In high-temperature chlorine atmospheres chromium and nickel additions to iron are both regarded as beneficial. Stainless steels are therefore more resistant than the lower alloyed steels. Austenitic stainless steels tend to outperform the ferritic grades (at equivalent chromium levels). Nickel and nickel-based alloys are widely used under such conditions. The high-nickel alloys are significantly more resistant than the stainless steels to chlorine but not fluorine, which is more soluble in nickel than chlorine is.

When both chlorine and oxygen are present in the environment, essentially a competing situation arises between the formation of stable oxide and volatile chloride species. Therefore, the degradation rate can fluctuate between parabolic, linear, and hybrid behavior. Molybdenum- and tungsten-alloying additions are considered highly undesirable in such service environments because of the formation of highly volatile oxy-chlorides (Table 7.11). Aluminum additions are regarded as beneficial because of the formation of a protective alumina scale at high temperatures.

7.7 MOLTEN PHASES

There are many situations where the corrosion product itself may melt, or contaminants may lower the melting point into the service temperature range. Moreover, important industrial processes are based upon the ability to contain static or flowing molten salts and liquid metals for long times without harmful corrosion.

The contaminants that can interfere with the normal development of a protective scale during high-temperature service are virtually impossible to enumerate. Much effort has been spent on problems resulting from the combustion of residual oils, because the ash contains, among other things, vanadium pentoxide, a compound which in the pure form melts at 690°C, and whose melting temperature may be lowered further by the presence of other compounds.

The small amount of sea salt that may enter a reactor or gas turbine with the combustion air can also initiate a sequence of events that can culminate in catastrophically high corrosion rates. Welding slag, improperly or incompletely removed, will also cause difficulty and frequently premature failures.

These harmful contaminants or reaction products have one thing in common: whether they themselves are solid, liquid, or gaseous, they react with the metal or corrosion products to form a molten phase. This phase will destroy the integrity of the protective scale by fluxing or dissolving it and leave the underlying metal available for further corrosion. Frequently, the molten phase will attract and concentrate potent corrosive agents present in the atmosphere at exactly the location where they do the maximum amount of damage, that is, where the scale is disrupted.

Ideally, the contaminating phase would be removed and the problem eliminated. However, this is frequently not possible. Molybdenum, for example, is an important and necessary component of many heat-resistant alloys, and the cost of removing vanadium from residual oil and removing salt from sea air is prohibitive.

In some cases, careful alloy selection can improve performance. However, other necessary characteristics often limit this approach. Alloys of 50 to 60% chromium and 30 to 40% nickel, respectively, have excellent resistance to residual oil ash. However, their rather poor high-temperature mechanical properties limit their range of applications, although not their usefulness when they are mechanically suitable.

Protective coatings are used frequently and appear to be effective where small, readily inspected parts such as turbine blades are involved. In some cases, additives which prevent the molten phase from forming offer a solution. For example, the addition of magnesium sulfate (common Epsom salts) to fuels has permitted gas turbines to operate on residual oils without excessively high corrosion rates.

7.7.1 Hydroxides

When molten hydroxides are present, similar results may be expected. However, in cases where the air atmosphere is excluded, reaction with oxygen can still occur because oxygen is a component of hydroxides. In the case of a metal in sodium hydroxide (NaOH), the following can occur:

$$\text{Metal} + 2\,\text{NaOH} \rightarrow \text{Metal Oxide} + \text{Na}_2\text{O} + \text{H}_2^{\uparrow} \quad (7.19)$$

Sometimes another reaction can also form, depending on the metal involved and other conditions of temperature, pressure, and metallic sodium (Na), as in the following equation:

$$\text{Metal} + 2\,\text{NaOH} \rightarrow \text{Metal Oxide} + 2\,\text{Na} + \text{H}_2^{\uparrow} \quad (7.20)$$

7.7.2 Cyanides

Molten sodium cyanide is used commercially both to carburize and nitride steels. High-temperature containers for molten cyanide are generally made of high-alloy materials to resist this reaction. Tests at 843°C in a mixture of sodium carbonate, cyanide, and chloride have shown that carburization resistance increases with higher chromium, nickel, or silicon content. However, although Cr up to 30% was found to be desirable, and specimens with nickel contents beyond 50% were among the best, silicon content seemed to be optimum between 1.5 and 2.0%. Lesser amounts of Si caused higher carbon pickup, and larger amounts resulted in oxidation phenomena, similar to what was described under chloride attack when Cr was oxidized.

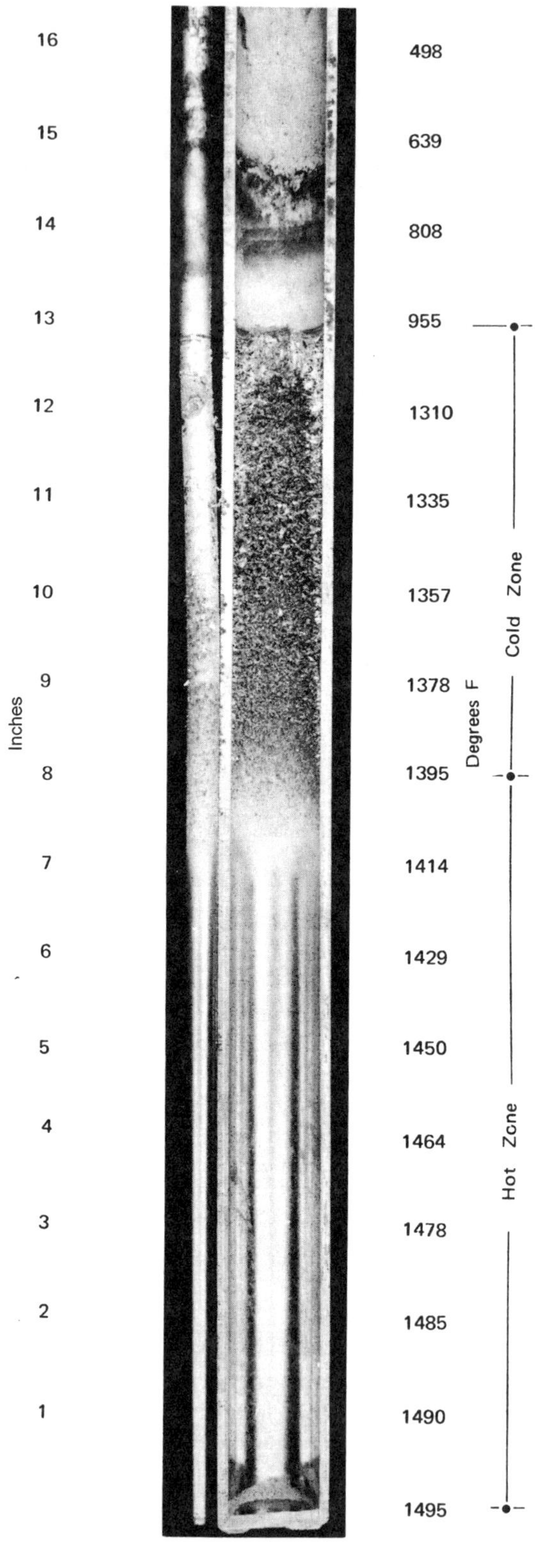

Figure 7.26 A standing thermal-convection liquid-metal corrosion test. The thermal gradient between the 1- and 13-in. levels causes natural convection and mass transfer from the hot zone (polished surface identifies dissolution zone from which metal dissolved) to the cold zone where solid metal crystals build up. There was no overall weight loss or gain during this eight-day test ($\times 0.5$ magnification).

7.8 CORROSION IN LIQUID METALS

Solid metals in contact with liquid metals are frequently encountered in high heat-flux heat exchangers, mercury boilers, hot dip tinning or galvanizing operations, die casting, and similar processes.

A certain degree of potential interaction can be predicted from binary-phase diagrams, and these interactions may be considered chemical in nature. That is, intermetallic phase formation would be indicated and the extent of mutual solubility could be noted. However, very low solubility is no indication of negligible interaction. The mercury/iron phase diagram shows practically no solubility of one for the other (less than one part per million), yet damaging interaction has occurred because of solution deposition or mass transfer caused by thermal gradients.

Corrosion in liquid metals can result from direct interaction of the metals, or, if alloys are involved, selective dissolution of certain constituents by the liquid metal. Furthermore, if impurities are present in the solid or liquid metal, even in trace amounts, the situation can become more complex.

Tests for compatibility have included static and dynamic tests. Usually numerous liquid metal/solid metal combinations are screened in the isothermal static test, and fewer of the more promising materials are tested in the expensive and time-consuming thermal-gradient dynamic tests. Some of the numerous varieties of dynamic tests are as follows;

1. forced circulation loops using electromagnetic or mechanical pumps,
2. thermal-convection loops of various designs,
3. see-saw-type tests,
4. gyrated toroid loops, and
5. standing thermal-convection pipe (Figure 7.26).

In Figure 7.26, molten metal is introduced up to the 13-inch level and placed vertically in a furnace to a depth of 8 inches. The upper portion was cooler, but well above the melting point of the molten metal contained. A temperature difference of about 93°C was measured in this particular test.

Assume that the solid metal, A (container), and liquid metal, B, have a negligible or very slight mutual solubility, and therefore show practically no corrosion in this thermal-gradient test. There would obviously be only a very slight difference in solubility of metal A in metal B at 815°C vs. 705°C , as illustrated in Figure 7.27. Therefore, metal A would dissolve slightly at 815°C up to the saturation point. Because of thermal convection, this hot, saturated liquid then rises to the top, where at 705°C it becomes supersaturated and precipitates a metal on the cooler wall in the form of dendrite-like A crystals. Then, the cooled liquid metal sinks to the bottom to dissolve more solid metal. A close-up of the cold-wall crystals is shown in Figure 7.28.

Continuation of this simple process can cause a problem at both hot and cold ends of a closed-cycle system. At the hot end, the metal wall gradually becomes thinner and could perforate. At the cold end, crystals could build up to the point where liquid metal could no longer flow through.

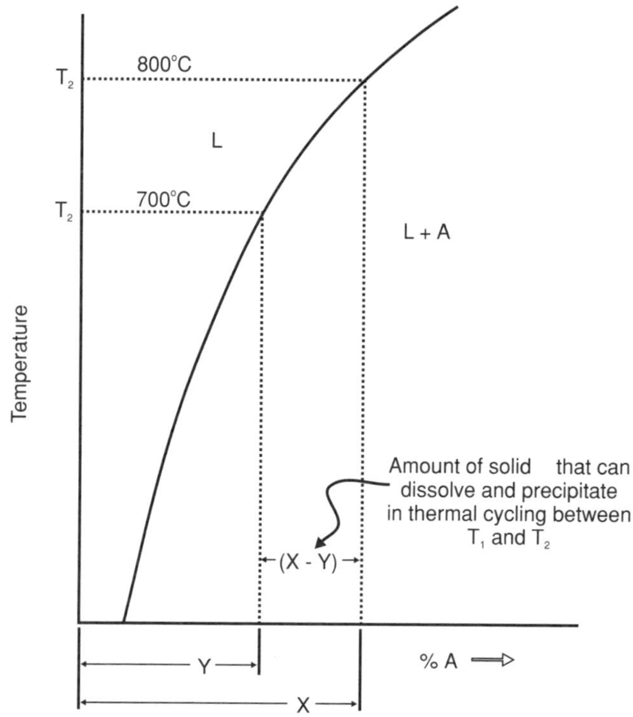

Figure 7.27 Solubility of solid metal A in liquid metal B as a function of temperature. (This line is known in phase-diagram studies as a liquidus.) An amount, X, is soluble at T_2 (800°C), and Y is the amount of A soluble at T_1 (700°C). Thus, X-Y may be precipitated during a thermal cycle between T_2 and T_1.

7.9 CORROSION IN VACUUM

A perfect vacuum is a space devoid of gaseous particles. Yet, one of the best man-made vacuums (about 10^{-9}-mm Hg pressure) still contains about 40 million molecules per cubic centimeter (640 million per cubic inch).

Figure 7.28 Cold-zone crystals that have precipitated during a solution-deposition reaction caused by thermal cycling (×24 magnification).

The more noble metals would not react with the remaining "few" gaseous molecules, for example, O_2, CO_2, and H_2O, to form any oxide, but the more active metals (Such as Cr, Ti, Al, Be, or Mg) would react, probably superficially, to form a thin oxide layer, or in some cases, a nitride layer, depending on the metal and nature of residual gases. This is why purging a "vacuum" with hydrogen or argon can be very helpful in making the residual molecules more inert.

Perhaps a more important form of high-temperature deterioration in vacuum has to do with evaporation or volatilization. All substances have a vapor pressure, which may be extremely low even at elevated temperature. However, changes can occur over a period of time that should be given some passing attention.

Brass, an alloy of copper and zinc, can undergo a form of corrosion (even in perfect vacuum), which can be termed dezincification. Zinc has a relatively high vapor pressure, whereas that of copper is much lower. Thus, zinc would vaporize preferentially, leaving a copper-rich surface. Whereas in aqueous corrosion, zinc often goes into solution as ions, in a vacuum it goes into space as atoms. The effect on the remaining metal is much the same.

7.10 MECHANICAL BEHAVIOR

A serious problem facing the engineer or designer in the realm of high-temperature applications is due to the mechanical properties of alloys at high temperatures. Loss of strength and ductility, poor creep characteristics at elevated temperatures, and phase transformations under service conditions can cause more failures than corrosion. Therefore, a consideration of these properties and the effects of corrosion on mechanical properties is essential to the corrosion engineer who is concerned with metals at temperatures above 540°C.

7.10.1 Mechanical-Property Testing at Elevated Temperatures

Tensile tests at elevated temperatures have been made to determine the extent of decrease in strength as influenced by an increase in temperature. These tests are generally made in the conventional manner, that is, by increasing the load at a given rate until fracture occurs, which generally requires a few minutes. These short-term tensile tests may be of fundamental interest but have no direct correlation with actual service conditions where a constant and low load is usually encountered and creep of the alloy occurs.

Creep is the gradual stretch of an alloy under low load at high temperature. Creep strength is the present-day basis to determine the relative merits of various heat-resistant alloys for high-temperature service. Creep-strength tests may be likened to low constant-load tensile-strength tests at elevated temperatures. In this type of test, nothing seems to happen under low loads, but accurate measurements show that something is going on all the time. It may produce fracture in time periods up to 10,000 h and provide engineering-design data for high-temperature applications.

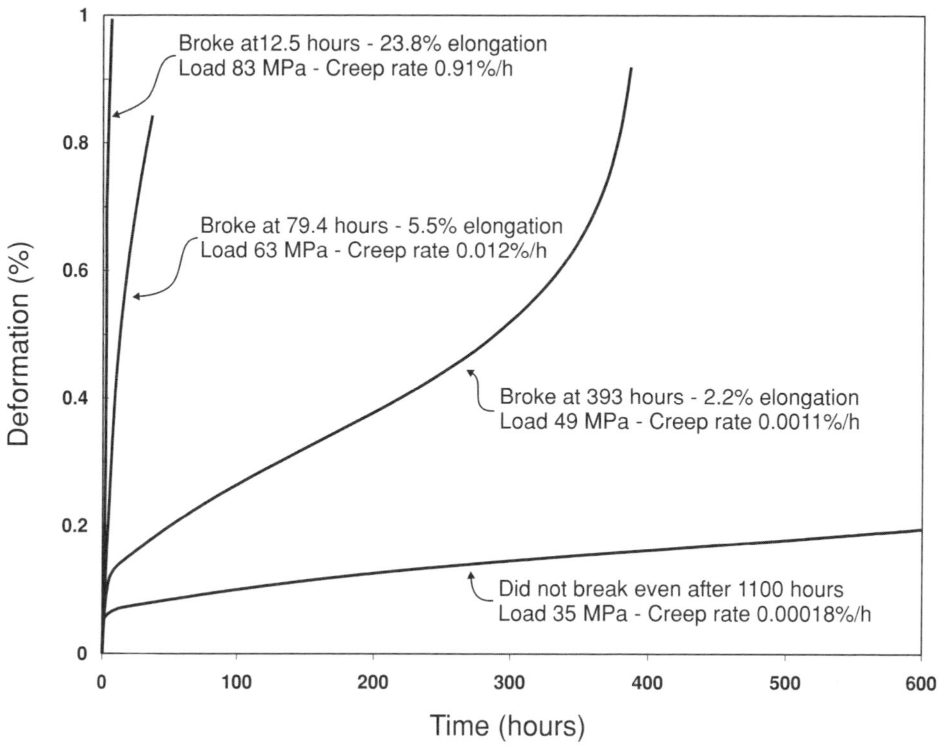

Figure 7.29 Time-deformation curves for HT-type cast alloy for four loads at 870°C. Three important mechanical properties are assessable from creep–rupture tests: creep rate, rupture time, and percent elongation.

This type of test closely duplicates actual service conditions. To illustrate this method of testing, consider the following example:

A common high-temperature alloy is one containing about 35 Ni, 15 Cr, 0.5 C, and 1% each of Mn and Si. If creep-tested at 870°C, it would slowly deform under load. With greater tensile loads, of course, the stretching rate would increase. Figure 7.29 illustrates four different loads ranging from 83 MPa (12,000 psi) down to 35 MPa (5000 psi). The creep (stretching) rate at the high load was 5000 times greater than the low load. Fracture occurred in 12.5 h at 83 MPa but did not occur in 50 days with the lower load, at which time the test was suspended. Note that ductility decreases as rupture time increases; this is typical in creep-rupture testing.

The four creep curves shown in Figure 7.29 would be similar for increasing temperature as well as for increasing load; that is, for a given load, the higher the temperature, the higher the creep rate.

A practical method of designing for creep involves generating creep-test results at relatively short times, high stresses, and high temperatures. The test data can then be extrapolated to the time–temperature–stress region of interest using a parametric relationship that describes the creep behavior in terms of the variables of interest. Several parametric relationships for creep have been suggested. The Larson–Miller parameter is among the most well-known of these parametric relationships used for the creep characterization of materials.[8]

Use of the Larson–Miller parameter is based on the observation that creep is a thermally activated process and that the creep rate can be described by an Arrhenius-type equation of the following form:

$$\frac{1}{t} = \mathrm{A}e^{\frac{-E_\mathrm{a}}{RT}} \tag{7.21}$$

where t = time to a specified creep strain or to final rupture, E_a = the activation energy for the creep process, R = the universal gas constant, T = the absolute temperature (K), and A = an empirical constant.

Rearrangement of this equation results in the following relationship:

$$\frac{E_\mathrm{a}}{\mathrm{R}} = T(\mathrm{C} + \log t) \tag{7.22}$$

or

$$P_\mathrm{LM} = T(\mathrm{C} + \log t) \tag{7.23}$$

where C is a constant related to the constant A in Equation (7.21). E_a/R is often written as P_LM, as in Equation (7.23), and is termed the Larson–Miller parameter. Larson and Miller made the assumptions that P_LM is a function of stress but independent of temperature, and that the constant C is independent of stress and temperature. Empirical evidence has shown that these assumptions are generally valid for many commercial alloys if the creep-deformation mechanism remains constant and there are no phase transformations between the test temperature and the temperature of interest. It has also been shown that P_LM is a function of creep strain. Values of P_LM and C may be obtained by plotting log time vs. $1/T$ and fitting the curve by using a linear-regression process.

The Larson–Miller parameter can also be used as a figure of merit for the creep resistance of materials with the same value of C. The larger the Larson–Miller parameter, the greater is the creep resistance of the material. Experimentally, as composition varies within normal chemical-specification allowances, differences in strength properties should be expected. Furthermore, if other alloying elements are introduced, changes can be expected. The nature of change will depend on base composition, temperature, and environment.

7.10.2 Effect of Composition Variations on Strength

The following statements summarize the general effects of various alloy elements on the strength behavior of steels at high temperature:

1. Carbon is added to strengthen steel, and it does actually increase the creep strength below 480°C. Above 600°C, the effect diminishes, and the influence of carbon on steel is definitely weak above this temperature.
2. Titanium is beneficial below 815°C, and is of questionable advantage above this temperature.

3. Molybdenum definitely strengthens low- and high-alloy steels.
4. Chromium is a feeble strengthener and is used mainly to impart corrosion resistance.
5. Manganese has been found to increase strength when added in amounts up to 1.3%. Additional amounts of manganese cause a very slight decrease in strength.
6. Silicon has a very weak effect and is added chiefly for corrosion resistance in amounts up to 1.5%. It is a ferrite promoter, being about three times as effective as chromium in this respect. It has been known to facilitate the transformation of alpha to sigma in high-chromium alloys at about 740°C. Sigma is a very brittle phase.
7. Aluminum generally weakens metals at high temperatures and is added, other than in deoxidation amounts, primarily for corrosion resistance or to control grain size.
8. Nickel has no perceptible effect when added in amounts up to a few percent. Greater amounts increase the strength somewhat.
9. Cobalt is a feeble strengthener for iron, but in the case of certain highly alloyed materials, it appears to be a definite strengthener.
10. Tungsten is positively a strengthener, but only half as effective as molybdenum. Tungsten is a ferrite promoter, increases the hot-hardness of heat-resistant steels, improves creep strength, and tends to reduce ductility.
11. Vanadium, when added in amounts of 0.5 to 1.5% , is quite effective in strengthening iron, especially in the 650°C range. It is a renowned carbide and nitride former.
12. Zirconium (Zr), titanium (Ti), and niobium (Nb) are elements that form stable carbides and nitrides and are considered to be ferrite promoters. The addition of small amounts of these elements are known to render low-carbon, 12% Cr steels non hardenable.
13. Phosphorus, surprisingly enough, in amounts of 0.2 and 0.4% does strengthen iron but greater amounts embrittle it.

7.10.3 Effects of Microstructure

Ferritic heat-resistant stainless steels are very stable when subjected to aging or cyclic heating through a wide range of temperatures. These and the martensitic stainless steels all have relatively low creep strengths at high temperatures. These alloys are strongly magnetic at room temperatures.

Austenitic stainless steels are characterized by their superior creep strengths and retained ductility over long periods of time. Fully austenitic alloys retain their stability under any thermal treatment. A slight microstructural change is effected, however, by the agglomeration of secondary carbides from the supersaturated austenite, resulting in a slight decrease in strength and ductility. Austenitic alloys can be nonmagnetic, weakly magnetic, or strongly magnetic, depending upon their composition and prior mechanical treatment.

Alloys containing more than one phase are rather susceptible to wide variations in properties and characteristics, depending upon the thermal history of the alloy. Possible high-temperature phases (other than carbide) are ferrite, nitride, austenite, and sigma.

Partially austenitic alloys are considerably inferior to the fully austenitic alloy in strength characteristics. Therefore, care should be exercised in the selection of alloy types for high-temperature service, or in balancing the composition of marginal alloys in the multiphase vicinity (such as 18 Cr–8 Ni, 25 Cr–12 Ni, and 25 Cr–20 Ni) by proper control of carbon, nitrogen, nickel, manganese, and other austenite-promoting alloying elements.

References

1. R.C. John, "Compilation and Use of Corrosion Data for Alloys in Various High-Temperature Gases," CORROSION 99, paper no. 99073 (Houston, TX: NACE, 1999).
2. N. Birks, G.H. Meier, Introduction to High Temperature Oxidation of Metals (London, UK: Edward Arnold, 1983).
3. D.A. Jones, Principles and Prevention of Corrosion, 2nd ed. (Upper Saddle River, NJ: Prentice Hall, 1996).
4. A.J. Sedriks, Corrosion of Stainless Steels (New York, NY: Wiley 1979).
5. E.J. Bradbury, P. Hancock, H. Lewis, "The Corrosion of Nickel-Base Material in Gas-Turbine and Boiler Atmospheres," Metallurgia 67 (1963): pp. 3–14.
6. G.Y. Lai, "High-Temperature Corrosion: Issues in Alloy Selection," Journal of Materials (1991): pp. 54–60.
7. D.J. Tillack, J.E. Guthrie, "Wrought and Cast Heat-Resistant Stainless Steels and Nickel Alloys for the Refining and Petrochemical Industries," NiDI Technical Series 10071, Nickel Development Institute, Toronto, Canada, 1998.
8. F.R. Larson, J. Miller, "A Time–Temperature Relationship for Rupture and Creep Stresses," Trans. ASME 74 (1957): pp. 765–775.

Part III

Materials and How They Corrode

8

Chapter

Engineering Materials

8.1 INTRODUCTION

The corrosion literature is filled with data on the performance of various materials in myriads of chemical environments. Engineers must constantly be on guard when considering such information to be certain not only that the chemical environment is adequately defined, but also that both the particular alloy (and heat treatment) and the character of attack are fully described.

For example, corrosion rates for "aluminum" often are given without further alloy designation. There are approximately 200 commercial compositions of aluminum alloys available. Their level of corrosion resistance varies widely with composition. The wrought alloys of the 1000, 3000, 5000, and 6000 series are roughly similar in their corrosion behavior and have far better corrosion resistance in most chemical environments than have alloys of the heat-treatable 2000 and 7000 series. The corrosion behavior of the 2000 and 7000 series alloys is strongly influenced by heat-treatment practices. Merely by varying heat treatment, the character of attack on the 7075 (A97075) alloy in a 3.5 NaCl solution can be changed from intergranular to pitting.

Stainless steels likewise tend to be treated in the literature as a class and not as individual alloys. Naturally, this leads to much confusion and some misapplications. It is therefore necessary to differentiate between the austenitic and ferritic steels, as well as between stabilized and nonstabilized grades of austenitic stainless steels.

The technical literature is far more reliable in deciding which materials cannot be used than in deciding which can be used. Conversely, the literature can narrow the choice of materials to a manageable number of alternatives from which a final choice can be made more quickly. Major materials suppliers often maintain staffs of consultants to advise customers on the proper use of their materials. These suppliers recognize that a bad application can result in damaging publicity. Failures of materials in service usually are traceable to misapplications resulting from the following factors:

- choice of the wrong material,
- improper treatment or fabrication of the material,
- an inadequately controlled (or defined) environment, or
- improper design.

8.2 MATERIALS SELECTION

A large variety of materials, ranging from platinum to concrete, is used by the engineer to construct bridges, vehicles, process-plant equipment, pipe lines, and power plants.[1] Although the primary focus of a corrosion engineer is on the chemical stability and corrosion resistance of these materials, a general knowledge of mechanical, physical, and other properties is required to ensure that the desired materials performance is maintained. The properties of engineering materials depend upon their physical structure as well as their basic chemical composition.

For most materials, there exists a standard that normally identifies the analysis, form, properties, and other details of their characteristics. The use of such specifications provides a recognizable basis for the supplier to furnish the specific material desired. The various fabrication and use codes are based, with few exceptions, on these standardized materials. Many of the nonmetallic materials may also be identified by existing standards. The use of these consensus standards is recommended whenever possible.

Most metals can be generally identified by a common name (e.g., yellow brass), by a group number (e.g., Copper Development Association No. 268), by the newer UNS (unified numbering system) number (e.g., C26800), or by the ASTM specification (e.g., B36 or B587). The UNS is a coordinated effort devised by the ASTM and the SAE (Society of Automotive Engineers) to provide a consistent mode of identification for all alloys. A description of the alloy identifiers is given in ASTM E 527. Throughout this text, the UNS number is used in parentheses following the older designator for an alloy to familiarize the reader with the UNS system.

8.2.1 Mechanical Properties

Mechanical properties are related to behavior under load or stress in tension, compression, or shear. Properties are determined by engineering tests under appropriate conditions. Commonly determined mechanical properties are tensile strength, stress rupture, fatigue, ductility, impact strength (toughness and brittleness), hardness, and modulus of elasticity. Strain may be elastic (present only during stressing) or plastic (permanent) deformation. These properties are helpful in determining whether a part can be produced in the desired shape and also resist anticipated mechanical forces.

Many unexpected catastrophic failures have occurred because of stress environment interactions. Knowledge of the relation among stress, mechanical properties, and SCC resistance is critical to the proper application of many materials in engineering design. An understanding or at least an awareness of this problem cannot be overemphasized.

Unfortunately, the ability of a metal or alloy to withstand mechanical loading often is used as the sole criterion in material selection. Only if the structure is protected from all environmental effects is the mechanical property the most important consideration. Too often, however, the relation of the metal to stress plus environment is neglected. It is necessary to relate environmental effects to the loads on a structure and to knowledge of the mechanical properties of the metal.

8.2.2 Other Properties

The corrosion engineer is often required to consider one or more properties in addition to corrosion resistance and strength when selecting a material. These include density or specific gravity (needed to calculate corrosion rates); fluidity or castability; formability; thermal, electrical, optical, acoustical, and magnetic properties; and resistance to atomic radiation. The requirements for a particular part may, for example, include the ability to be cast into an intricate shape, good heat-transfer characteristics, and nondegradation by atomic radiation. In another case, equipment requirements may include being a good insulator, heat reflector, and of low unit weight.

Although cost is not an intrinsic property of a material, keep in mind that it may be the overriding factor in selecting of material for engineering use.

8.3 PRINCIPLES OF METALLURGY

Because metals are the principal materials that suffer corrosive deterioration, it is important to develop a background in the principles of metallurgy to fully understand corrosion.

Nearly all metals and alloys exhibit a crystalline structure. The atoms that make up a crystal exist in an orderly three-dimensional array. Figure 8.1 is a schematic representation of the unit cells of the most common crystal structures found in metals and alloys. The unit cell is the smallest portion of the crystal structure that contains all of the geometric characteristics of the crystal. The crystals, or grains, of a metal are made up of these unit cells repeated in a three-dimensional array.

The crystalline nature of metals is not readily obvious because the metal surface usually conforms to the shape in which it has been cast or formed. In some instances, the crystal structure can be observed naturally. Figure 8.2 is a photograph of the surface of a hot-dip-galvanized post with characteristic spangle patterns that are in fact large grains visible without any etching. Brass doorknobs are normally bright and shiny; however, after some time the corrosive perspiration from hands etches the crystalline features of the alloy on the surface as shown in this second example (Figure 8.3).

Controlled etching with selected chemical solutions will normally show the granular characteristics of metals and alloys. The grains visible with the naked eye in Figures 8.2 and 8.3 are usually revealed by chemical etching to promote the controlled corrosion of grain boundaries. The grains shown in these figures are extremely large for most metal crystals. Normally, metal grains are so small they can only be satisfactorily observed under a microscope. The general range of grain size usually runs from 25 to 250 μm in diameter.

To determine the grain size or microstructure of a metal or alloy, it is usually necessary to prepare a sample for microscopic study by special grinding and polishing

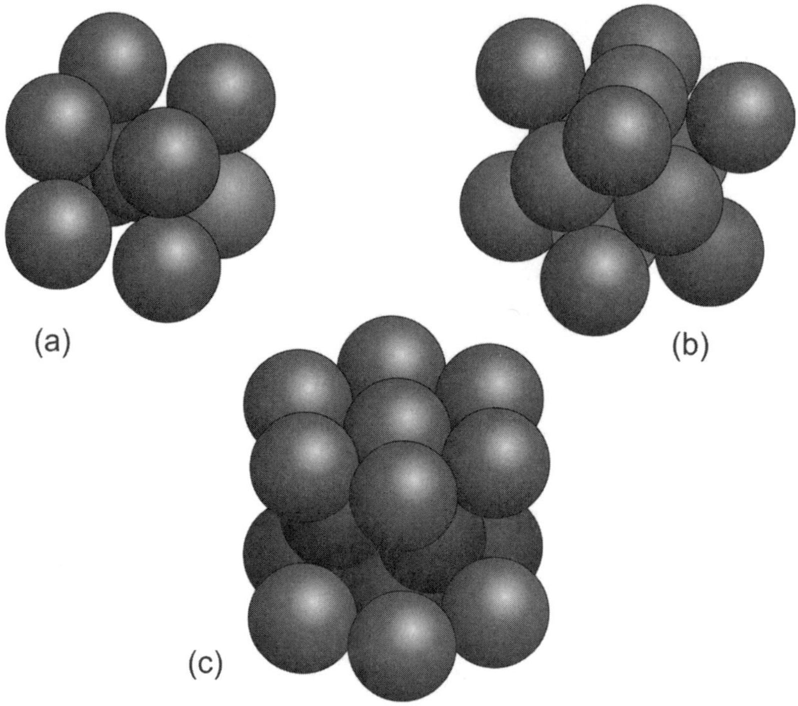

Figure 8.1 Schematic representation shows the unit cells of the most common crystal structures found in metals: a) body-centered cubic, b) face-centered cubic, and c) hexagonal close packed.

Figure 8.2 Hot dip-galvanized steel with spangle patterns that are a form of crystallized grains.

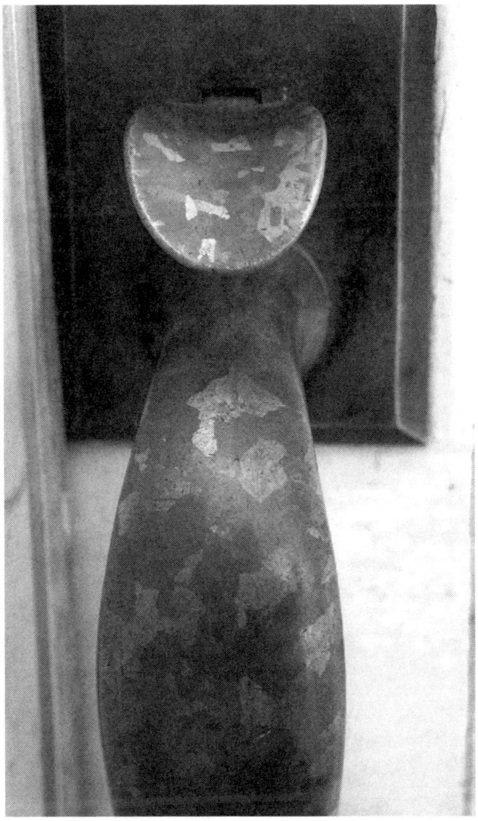

Figure 8.3 Brass doorknob shows evidence of crystal structure etched by corrosive perspiration (courtesy of Kingston Technical Software).

techniques of the specimen surface. The polished surface is then reacted with a suitable etching reagent. This reveals the grain boundaries and the distinguishing features of the microstructure.

Figure 8.4 shows micrographs (photographs taken through a microscope) of annealed pure iron and pure copper at a magnification of 100 times. Even though grains are very small, each is still made up of thousands upon thousands of unit cells. The contrast in the grains in these photographs is due to the difference in orientation of the crystal with respect to the surface being examined.

Many features in the microstructure of metals can be related to the properties of the material. For example, Figure 8.5a is a relatively pure iron with sulfide inclusions. Sulfide inclusions have a marked tendency to react in even mild corrosive environments. Figure 8.5b, also of iron, shows a sample in which the grains have been heavily deformed by plastic deformation (permanent deformation from the original shape). Metals are frequently plastically deformed in fabricating. Terms often used for metals deformed at room temperature are cold worked, cold rolled, cold drawn, and so forth, depending upon the method of fabrication.

In highly deformed metals, the grains are deformed and the grain structure is completely disrupted. Normally, in this condition, the material is somewhat more reactive

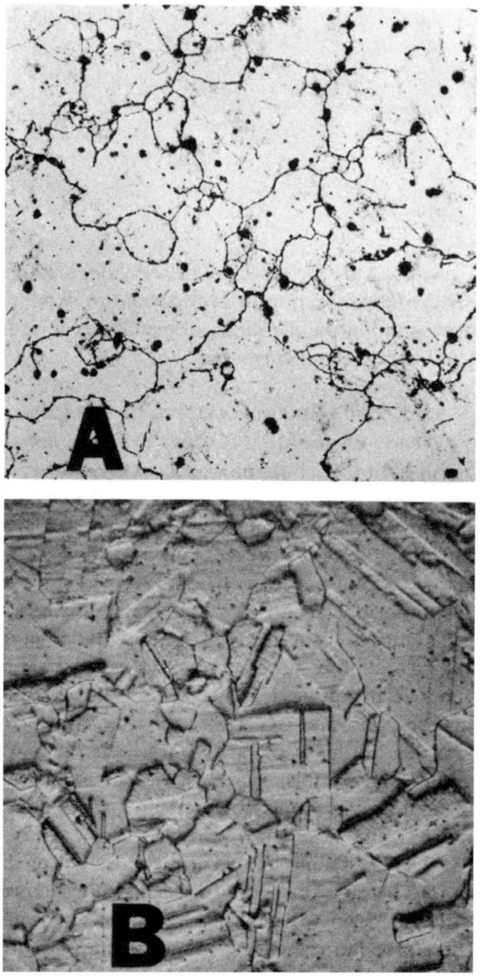

Figure 8.4 Photomicrographs of a) annealed pure iron and b) pure copper ($\times 100$ magnification).

in most environments. Relatively pure metals may show definite increases in electrochemical reactivity resulting from impurities or mechanical deformation.

In addition to the influence of impurities, inclusions, and cold work, grain boundaries and differences in grain orientation also may result in significantly different electrochemical reactivity in many metals and alloys. For example, the grain boundaries of the samples shown in Figures 8.4a and b have been selectively attacked by the etching reagent. A marked difference in grain contrast in these figures has occurred because of the variation in etching characteristics of different grain orientations.

8.3.1 Imperfections and Defects

On a smaller scale, differences in submicroscopic characteristics of a metal also must be considered. The atomic crystal structure presented earlier encounters an increasing

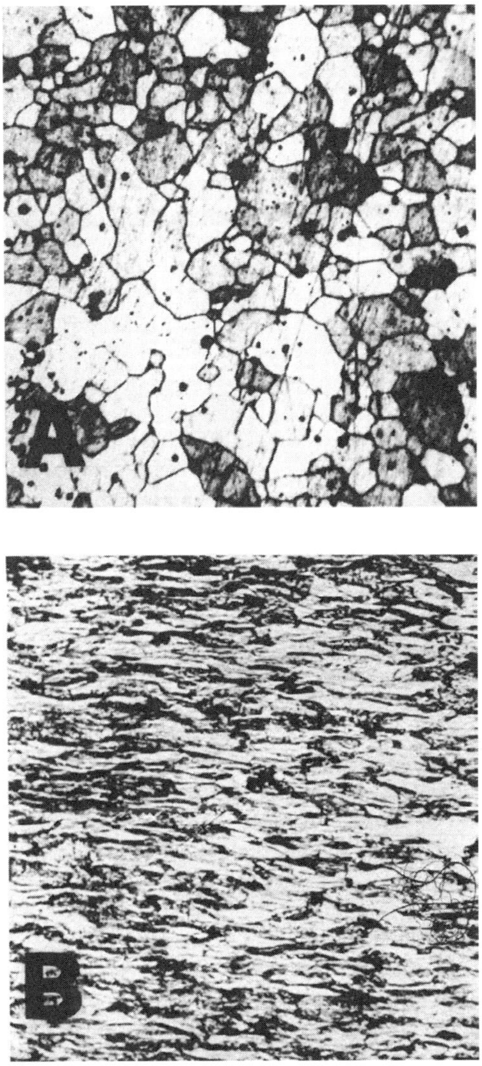

Figure 8.5 a) Relatively pure iron with sulfide inclusions and b) iron in which grains have been heavily deformed.

level of variations as the structure increases in size because of a variety of crystalline defects. These defects may be vacancies caused by the absence of atoms in the crystal, impurity atoms of different sizes, interstitial atoms (small atoms in the space between large atoms), and large lattice disturbances, called dislocations.

Each of these imperfections can produce highly localized differences in the electrochemical behavior of the metal. However, the vacancy, the impurity atom, and the interstitial atom are point defects affecting a much greater volume of the crystal. In etching or corroding environments, these areas are usually more anodic than the surrounding matrix.

Examples of pits formed at intersections of dislocations with the surface of a copper sample are shown in Figure 8.6. The large number of triangular etch pits are a result

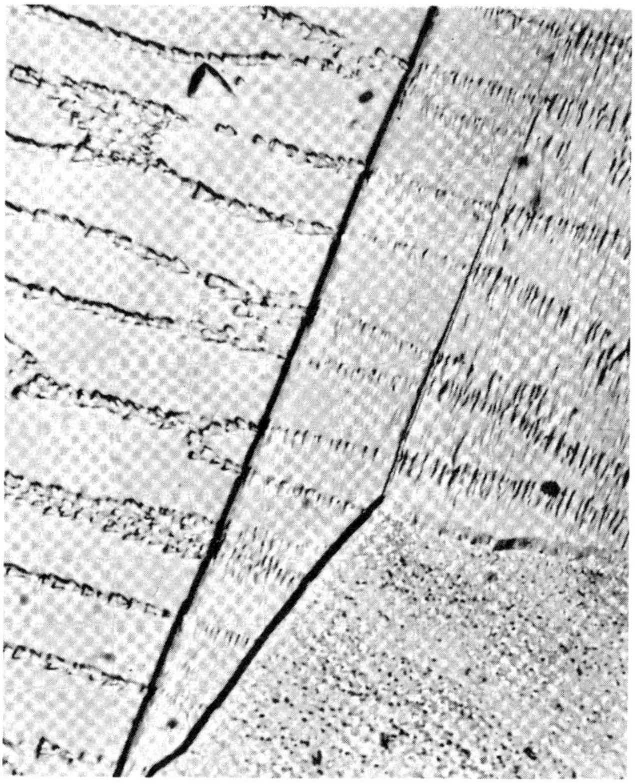

Figure 8.6 Pits form at the intersection of dislocations with the surface of a copper sample.

of selective electrochemical attack caused by the stress field around a dislocation. The shape of the etch pit is related to the orientation of the grain to the etched surface.

Metals listed as pure or commercially pure actually contain a variety of impurities and imperfections. These impurities and imperfections are inherent causes of corrosion in an aggressive environment. It has been shown that as purity increases, the tendency for a metal to react in electrochemical environments reduces proportionately. Very high purity metals can be produced by a special technique called zone refining. Final purities of 99.99999% or better are possible, depending on the metal and nature of impurity. Such materials often exhibit corrosion resistance much better than their commercial counterparts, which are only 99 to 99.9% pure.

However, high-purity metals generally have lower mechanical strength than their alloys and are rarely used in engineering applications with the important exception of copper and aluminum when used for the transport of electricity for which purity of the metal is an asset. It is necessary, therefore, to work with metallic materials that are stronger and are usually formed from a combination of several elemental metals. One metal usually serves as a major or base metal to which other metals or nonmetallic constituents are added. These metallic mixtures are properly called alloys and are a major class of engineering materials.

Although alloys can exist in an almost unlimited number of combinations, only a small fraction of these will possess the manufacturing and overall qualities required to be produced on an industrial scale. The common alloys are those that have a good combination of mechanical, physical, and fabrication qualities that tend to make them structurally, as well as economically, useful. Although an alloy's resistance to corrosion is also very important, this factor is often sadly neglected in the aim to improve mechanical strength.

Alloying by itself has produced a number of materials that have improved corrosion resistance compared to high-purity metals. Designations of many of these so-called corrosion-resistant alloys may be misleading because they are not truly corrosion-resistant under all conditions. In fact, they may be subject to catastrophic failure as with, for example, intergranular attack of sensitized stainless steel and stress-corrosion cracking of brass, stainless steel, and high-strength steel. To understand the corrosion characteristics of alloys, it is important to examine some of the fundamental characteristics of alloying.

8.3.2 Alloying

When an alloying element is added to a base metal, it is possible that the crystal structure will remain essentially stable and produce a simple solid solution. For example, copper is face-centered cubic. The addition of nickel, which is also face-centered cubic, does not alter the crystal structure. This ability of nickel to dissolve in copper, or copper to dissolve in nickel, without a change of crystal structure, results in what is termed a solid solution or a single phase.

Basically, there are two common solid solutions. In the substitutional solid solution, the atoms of the alloying element take a position on the lattice of the base metal. For substitutional solid solutions to exist, a number of limiting conditions related to atomic size and electronic structure must be met. Another type of solid solution is possible in which small atoms actually fit into void (interstitial) areas between the atoms in the crystal lattice of a metal with a larger atomic size. Such alloy phases are called interstitial solid solutions. The most notable of these is carbon, which dissolves interstitially in iron to produce steel.

Although there is a wide range of solid solutions, in many cases it is impossible to dissolve a large amount of one metal into another. When this occurs in an alloy, it results in the formation of two or more phases, depending on the number and type of components in the alloy.

The properties of steel, or any multiphase material, depend greatly on the relative physical and structural characteristics (amount, distribution, size, shape, and strength) of the various phases in the alloy. In many cases, multiphase materials present a problem from a corrosion standpoint because the two phases may have marked differences in electrochemical characteristics. It is possible that one phase will be selectively attacked in a corrosive environment. However, some two-phase alloys have small electrochemical differences and excellent corrosion resistance, and others develop a protective film, which results in improved corrosion properties for both phases. The existence of more than one phase in an alloy usually results in poorer corrosion resistance than for equivalent single-phase materials. There are, however, many notable exceptions.

8.3.2.1 Phase Diagrams

Graphs can be plotted showing the equilibrium relation between various compositions of one metal in another as a function of temperature. Although pressure also can be considered a variable, as a rule it has little influence on the metallurgy of metals. These graphs of phase stability as a function of temperature and composition are called phase diagrams. The phase diagram often is called the equilibrium diagram, because it is based on equilibrium conditions in the alloy, with the most stable phases represented at each temperature and composition on the diagram.

A diagram consisting of two metallic components is called a binary phase diagram. Binary systems can range from the very simple to the very complex. The copper/nickel system (Figure 8.7), which was briefly discussed previously, is one of the simplest because there is complete solubility of the metallic components in all compositions in the liquid and solid state. The vapor state is ignored for most metals because it normally exists well above the temperatures of exposure. However, in simple binary systems, rates of heating and cooling can greatly influence the characteristics of certain alloys. For example, rapid cooling may not allow diffusion of atoms to occur, and a nonequilibrium phase, or phases, may result at room temperature.

Figure 8.8 is an example of a phase diagram that is extremely important to stainless steel technology, the iron/chromium system. Although this diagram is for the Fe-Cr

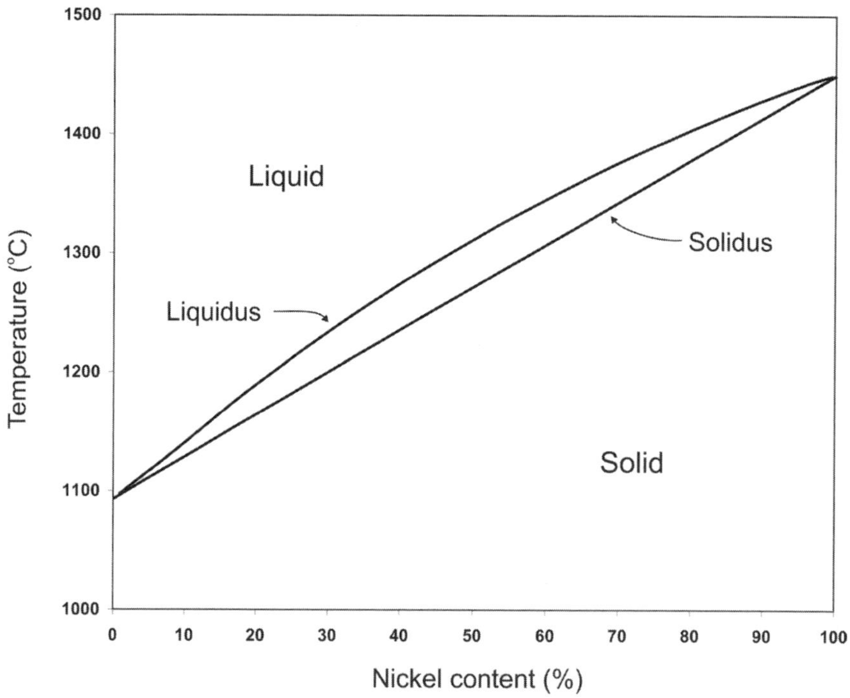

Figure 8.7 Copper/nickel phase diagram.

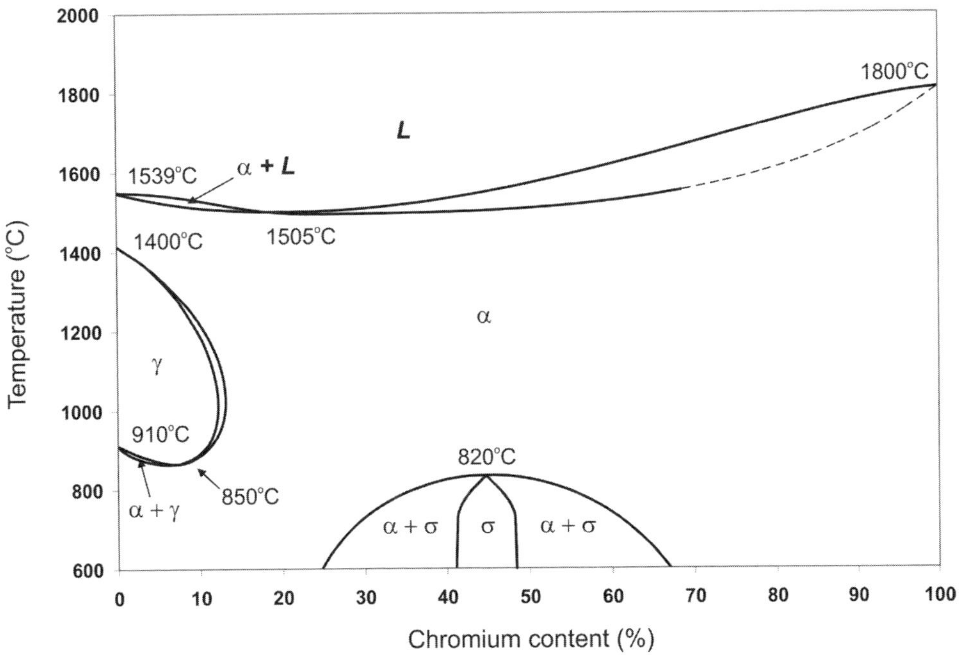

Figure 8.8 Iron/chromium phase diagram.

binary system, the influence of a third important alloying element, carbon, will cause an enlargement of the gamma (γ, face-centered cubic) region with increasing carbon content. The sigma (σ) phase is also noted in this diagram. This phase is important because its formation usually results in a reduction in the mechanical properties and corrosion resistance of the alloy.

A ratio of 1 chromium atom to 7 iron atoms results in a naturally occurring surface oxide that greatly improves corrosion resistance. On a weight basis, this represents slightly more than 10% chromium, and this portion of the iron/chromium system is the basis for the 400 series of stainless steels (S40000).

The 400 series of stainless steels (S40000) is divided into two types: 1) the ferritic type, represented by alloys that remain alpha (α, body-centered cubic iron) at room temperature and 2) the martensitic types that result from the formation of a nonequilibrium phase by quenching from the gamma (γ) or austenite region. The latter usually have higher carbon content to enlarge the gamma region. Quenching from the gamma region (face-centered cubic iron) produces the extremely hard, strong, and often brittle, nonequilibrium phase called martensite.

Figure 8.9 represents the general features of the iron-rich end for a constant 18% chromium in the iron/nickel/chromium system. A dotted line represents the composition, 18–8, which is the most common composition for austenitic stainless steels. Austenitic Fe-Cr-Ni, 300-series stainless steels (S30000), are generally more corrosion-resistant than the Fe-Cr, 400 series (S40000). Of course, there are some exceptions because of environment, particularly in environments that contain the chloride (Cl^-) ion. It is obvious from the diagram that nickel has a strong influence in changing the

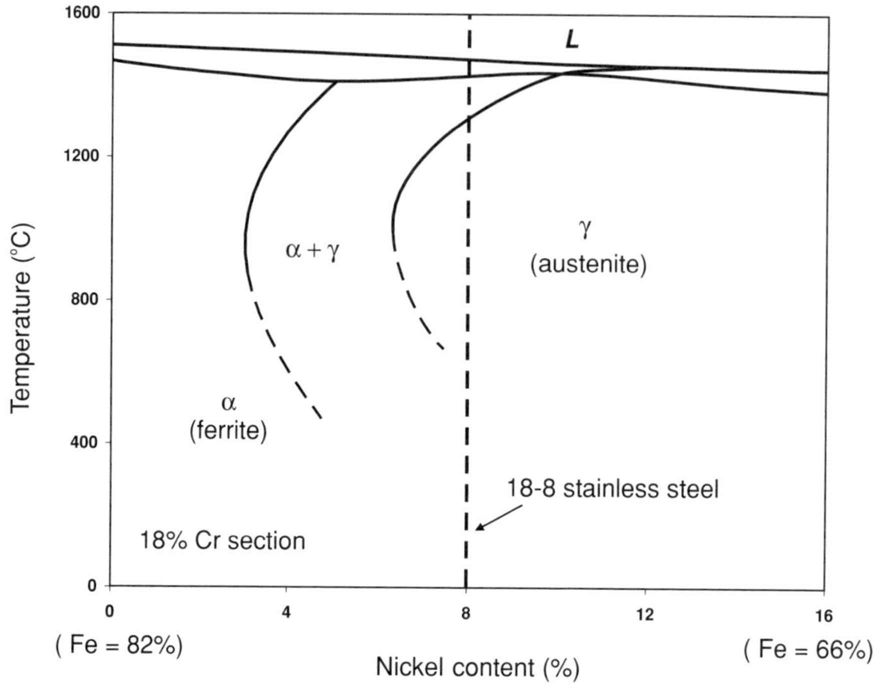

Figure 8.9 Binary phase diagram for stainless steel with a constant level of chromium (18%).

structure of the alloy from body-centered cubic alpha (ferrite), to face-centered-cubic gamma (austenite). This section shows that austenite is essentially "stable" at room temperature for composition above 8% nickel.

From the foregoing discussion it should be obvious that although phase diagrams are complex, they also are extremely important to the understanding of the application, fabrication, and even prediction of nonequilibrium states of the most commonly used alloys. Reference to equilibrium diagrams for heat treatment, annealing, and fabrication information is made later in this chapter.

8.3.2.2 *Castings*

Probably the best example of how the phase diagram can be applied to predicting a nonequilibrium condition is found in the prediction of inhomogeneous chemical composition as a result of nonequilibrium solidification. This inhomogeneity is very important in determining the corrosion characteristics of many cast alloys.

When an alloy cools through the solid-liquid, two-phase region, solid and liquid phases will exist with widely different compositions (Figure 8.10). Under slow equilibrium cooling, these variations disappear in the solid state because of rapid diffusion of the atoms in the solid near its melting point. The result is a homogeneous solid once the alloy has completely solidified. But what happens if cooling is rapid? Under these conditions, sufficient diffusion cannot occur and nonequilibrium variations result.

Consider a single grain of material rapidly solidified from a liquid. A variation in composition will exist from the interior to the surface of the grain because diffusion is

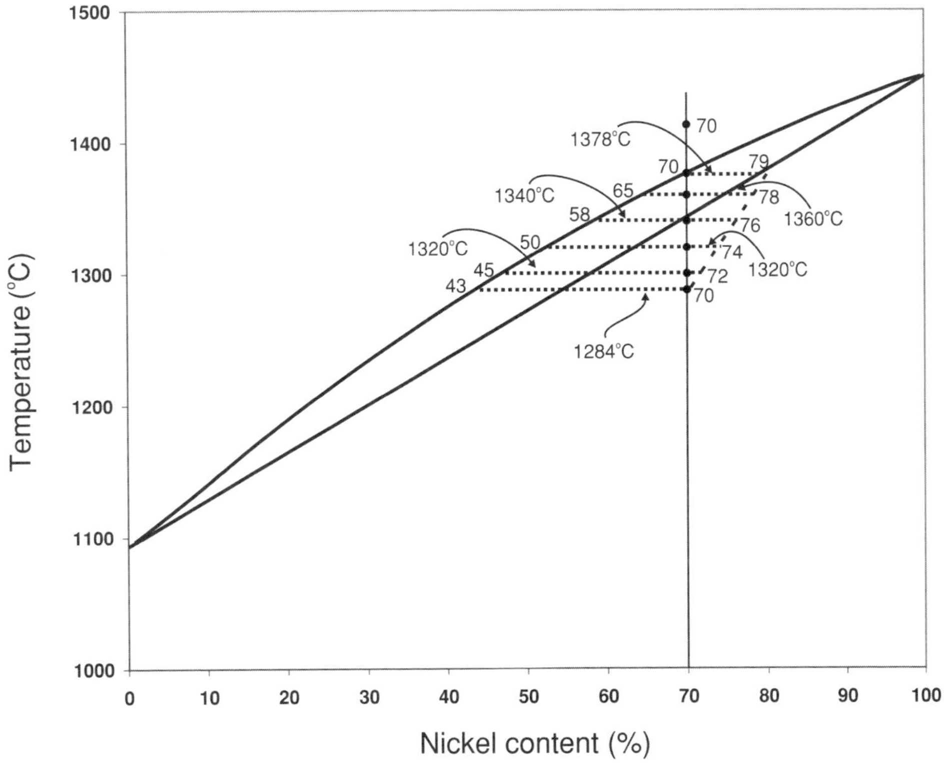

Figure 8.10 Copper/nickel phase diagram with nonequilibrium solidification of a 70 Ni–30 Cu alloy.

not fast enough to chemically homogenize the grain as it solidifies and grows. When grains of a rapidly cooled alloy grow together, the grain boundaries have a definitely different composition than their interiors. This variation in chemical composition from the interior to the exterior of a grain is referred to as microsegregation or coring. The dendrites quite visible in Figure 8.11a are typical of the microsegregation associated with nonequilibrium solidification. The same alloy, that is, 70 Ni–30 Cu (N04400) also known as Monel 400, would normally have in its hot-rolled form the crystalline grain structure revealed in Figure 8.11b).

In some cases, it is possible to reduce major chemical differences in a cored structure by reheating the alloy and holding it for a relatively long time at a temperature just below the solidus line. This allows diffusion to occur more rapidly and assists in homogenization of the alloy. Cored structures generally have widely different corrosion characteristics and actually have a built-in electrochemical cell.

The possibility of severe coring in a cast alloy can be predicted by examining the phase diagram. It is possible also to determine the approximate temperatures for homogenization annealing. The importance of chemical homogeneity in effecting corrosion resistance for a cast, corrosion-resistant alloy cannot be overstressed. It is important also to note that rapid cooling through two-phase regions in the solid state also can result in inhomogeneous phases. Many cases of intergranular corrosion result from this type of chemical inhomogeneity.

(a) (b)

Figure 8.11 Surface corrosion of respectively the same 70 Ni–30 Cu alloy (N04400): a) when cast and left to cool and b) after hot rolling.

These are examples of the important metallurgical and corrosion data contained in alloy equilibrium diagrams. This information, if used wisely, can be a tremendous help in obtaining the maximum corrosion resistance from alloys.

8.3.3 Heat Treatment of Metals

Many of the final mechanical and corrosion-resistance properties of a metallic material can be related to its heat treatment. The following section defines and discusses the most common heat treatments and their relation to mechanical strength and corrosion resistance of many metals and alloys.

8.3.3.1 Annealing

The annealing of metals is usually done to produce one of two effects:

1. homogenization of cast alloys that may exhibit chemical inhomogeneity (coring) and
2. removal of residual stress and cold work in deformed metals.

Variations in chemical homogeneity of alloys usually are the result of rapid cooling of the cast metal. (The way in which this occurs was covered in the preceding section on phase diagrams.) The homogenization anneal is designed to eliminate or reduce this type of chemical inhomogeneity.

A metal that has been cold worked (plastically deformed to a different permanent shape) has experienced distortion and fragmentation of its crystalline grain structure that has produced high residual stresses. This often can be detrimental if the material is subject to SCC. The second form of annealing reduces or eliminates this cold-worked condition by stress relief and recrystallization of the metal.

8.3.3.1.1 Homogenization Anneal. The importance of chemical homogeneity of an alloy in relation to corrosion resistance cannot be overstressed. In the as-cast condition,

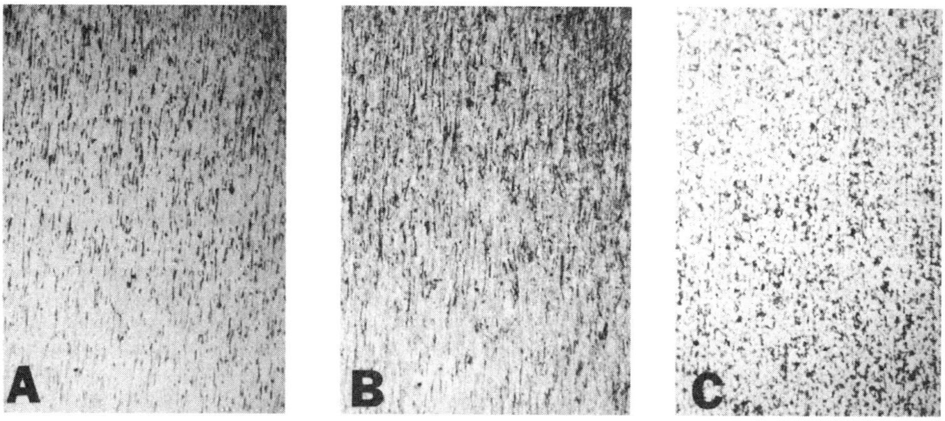

Figure 8.12 Typical structure of low-carbon steel after a) 30% cold work, b) 60% cold work, and c) hot rolling.

coring or chemical segregation often exists. Appreciable reduction of the mechanical properties, as well as reduced corrosion resistance, often occur because of this inhomogeneity.

The major purpose of the homogenization anneal is to produce chemical homogeneity by diffusion (movement of atoms in the solid state). To obtain the maximum diffusion rates, temperatures as close as possible to the melting point are usually used. For example, for most nonferrous alloys (alloys not based on iron), temperatures not far below the solidus temperature (Figure 8.7) are usually used for annealing. In ferrous alloys (based on iron), homogenization may be realized by heating to some intermediate temperature above the gamma to alpha transformation.

Length of time for a homogenization anneal can vary from a few minutes to several days, depending upon the material and the inhomogeneity of the structure, as well as the economic factors involved. Alloys that have been previously cold worked usually homogenize more rapidly than castings. Rearrangement of the disrupted crystal lattice at elevated temperatures aids in the redistribution of the various elements in an alloy.

8.3.3.1.2 Annealing of Cold-Worked Material. Because of the plasticity of metals (their ability to deform without failure), it is possible to produce large changes in shape and cross section. Thus, hot and cold forging, rolling, drawing, and so forth results in many useful fabricated shapes. Figures 8.12 a and b show the typical changes in grain structure that occur under the effect of cold rolling mild (low-carbon) steel. The structure after cold working shows elongated and highly deformed grains. Figure 8.12c illustrates the grain structure of a metal after hot rolling where the grains have rapidly reformed by diffusion, are stress free, and have uniform size.

When a metal or alloy is deformed by cold working, hardness and strength increase, but the ability to withstand more deformation decreases. Figure 8.13 shows the effect of cold working and the resulting change in properties. It is often important to be able to remove the effects of cold work. This can be done by annealing at a temperature at which the cold-worked structure can be completely eliminated by reforming of stress-free grains by diffusion.

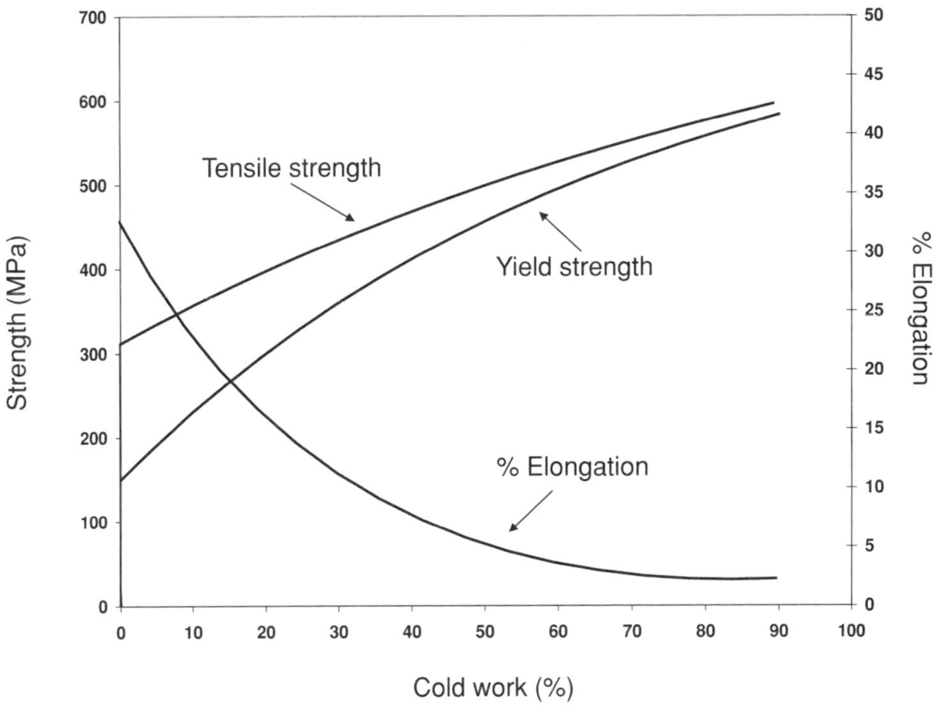

Figure 8.13 The effect of cold work on the mechanical properties of copper.

Annealing of deformed material involves three stages: 1) recovery, 2) recrystallization, and 3) grain growth. These stages are noted in Figure 8.14, which also contains a series of photomicrographs showing the typical microstructure changes that occur in each stage.

In the first stage, recovery, the cold-worked metal recovers a slight amount of its lost ductility while losing little or none of its strength characteristics. Because higher strength properties in the cold-worked condition often are desired, a heat treatment in the recovery portion of the annealing curve is often used. As a result, the hardness and tensile properties remain relatively high, and improvement often is noted in fatigue life, toughness, and corrosion resistance. This type of partial or preanneal is called stress relief.

The second stage of annealing of a cold-worked material is recrystallization. In this stage, new stress-free grains are initiated at the grain boundaries of the cold-worked metal. These grains grow by diffusion until new stress-free grains replace all the heavily deformed metal. Upon recrystallization, large changes in mechanical properties occur in which ductility, hardness, and other strength properties return to their pre–cold-worked level.

The last step in an annealing heat treatment results in grain growth. In this step, some recrystallized grains grow at the expense of others. This further increases ductility and lowers the strength and hardness of the metal. It is important to control grain size because a large grain size usually leads to difficulties in further fabrication processes and fatigue life.

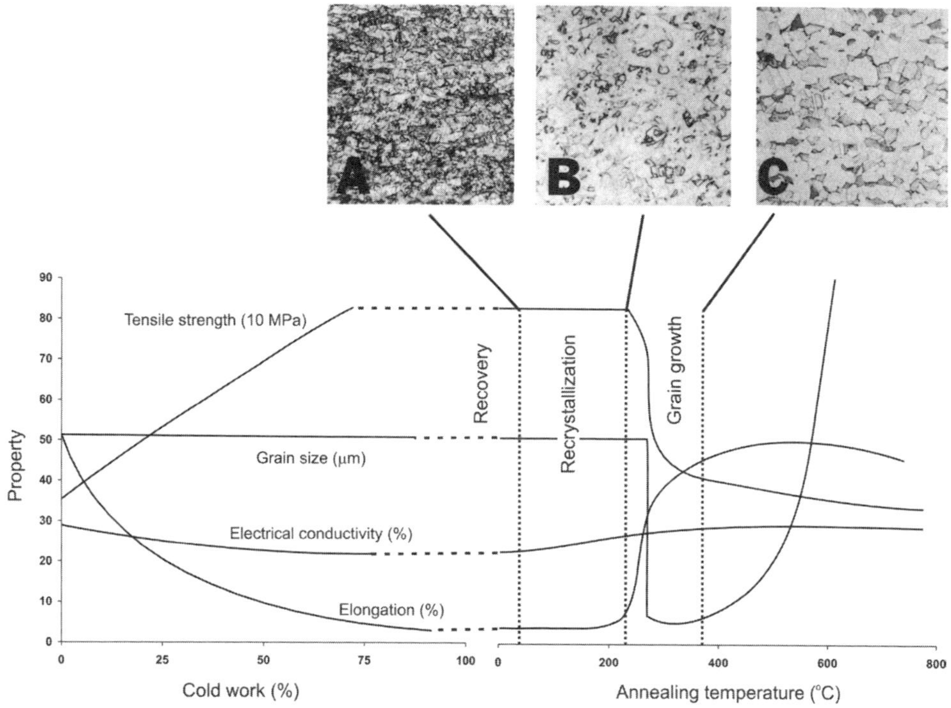

Figure 8.14 The effect of cold work on the properties of a Cu-35% Zn alloy and the effect of annealing temperature on the properties of a Cu-35% Zn alloy that is cold worked 75%: a) recovery, b) recrystallization, and c) grain growth.

8.3.3.1.3 Relation of Cold Work to Corrosion Resistance. In general, cold working can reduce resistance to general corrosion. In many alloys, it results in a condition more prone to SCC. However, evidence in recent years indicates that cold working of ultrahigh-purity metals has very little effect on their overall corrosion resistance. It appears that impurities and alloying elements, in addition to cold work, are responsible for diminished corrosion resistance in plastically deformed metals. However, the cost and low-strength properties of very high-purity metals rule them out for most applications.

Deformation decreases corrosion resistance basically because it increases dislocations in the cold-deformed metal. As shown in Figure 8.6, these areas are usually subject to pitting. Impurities or atoms of alloying metals often migrate to these imperfections, causing an even greater change in the electrochemical characteristics of these defects. Thus, annealing a cold-worked metal may result in a decrease in dislocations and important improvements in corrosion resistance. Just as a homogenization anneal is required to produce a more uniform chemical composition, full annealing tends to produce a more uniform crystal structure with fewer defects.

8.3.3.2 Hardening Heat Treatments

Certain alloys may be subjected to heat treatments which result in improved mechanical properties. This hardening may be produced in some alloys by what is termed age hardening or precipitation hardening. In other alloys, usually steels, quenching can produce martensite, a nonequilibrium phase with extremely high mechanical strength. The next two sections cover the essential features of both types of these hardening reactions.

8.3.3.2.1 Precipitation Hardening. Aluminum, magnesium, nickel, copper, and some forms of stainless steel, when alloyed with certain elements, can be strengthened by precipitation hardening. A number of these alloys have achieved a high degree of importance in alloy technology. Aluminum and magnesium age-hardenable alloys have become particularly important as structural components in aircraft because of their combination of lightness and strength.

Historically, Duralumin was the first precipitation-hardened alloy. Duralumin is the trade name for the original German alloy developed and patented early in the twentieth century. This alloy contains approximately 4% copper, which is the important element in the age-hardening process of this particular aluminum alloy.

Aluminum/copper alloys (2000 series) that have resulted from this earlier work have poor corrosion resistance. Basically, they are more noble than pure aluminum. However, it is well known that pure aluminum forms a thin film of aluminum oxide, which provides a very effective barrier to environmental attack. This explains the extremely good corrosion resistance of aluminum, which is otherwise a rather active element.

The corrosion resistance of these high-strength, age-hardened aluminum alloys can be improved by using a composite aluminum product known as Alclad aluminum. The composite consists of a thin layer (usually 5 to 10% of the total composite thickness) of either pure aluminum or a special aluminum alloy that is metallurgically bonded to one or both sides of the high-strength aluminum alloy core. The cladding component is selected to be anodic to the high-strength core. As a result, the cladding will provide CP to the core.

It is possible that deterioration of the resistance to corrosion of clad materials can occur because of improper heat treatment. If the solution anneal is overextended, or several reanneals are used, diffusion of copper into the aluminum cladding will result. This occurs rapidly at the grain boundaries and reduces the ability of clad material to retain its original oxide protection where the grain boundaries intersect the surface. Thus, with an overextended annealing time or too high a solution-anneal temperature, it is possible to reduce the corrosion resistance of a clad alloy.

In the precipitation-hardened aluminum alloys, the precipitating phase may cause what is referred to as a denuded zone adjacent to the grain boundaries that is more subject to corrosion attack. In various corrosive environments, this results in intergranular corrosion adjacent to and at the grain boundaries.

8.3.3.2.2 Martensite Transformation. In contrast to the quenching of age-hardenable alloys, which results in a soft material in the as-as quenched condition, steels quenched properly from the gamma region will form a hard nonequilibrium phase before reaching room temperature. This phase is termed martensite, after Martens, an early German metallurgist who recognized the basic characteristics of the reaction. The extent of this transformation depends upon the composition of the steel, the size and shape of the work piece, and the quenching ability of the cooling medium.

The quenching of steel from its face-centered cubic gamma region results in a highly supersaturated condition, because solubility of carbon in alpha iron may be more than 100 times less than it is in gamma iron. Thus, the intermediate martensite phase is formed as the result of the enormously supersaturated condition. The martensite that forms has a highly stressed lattice, which results in exceptional hardening of the steel.

In general, as-quenched martensitic steel is extremely brittle. Many steels may crack or fracture during rapid quenching as the result of martensite formation. However, although the martensite structure is hard and brittle at room temperature, it can be softened by reheating to some intermediate temperature. This causes the metastable martensite to begin to transform into a more stable structure. This process, called tempering, decreases brittleness, and although the total strength is lowered, ductility is increased and the best combination of strength and toughness results. Thus, steels are nearly always tempered following quenching.

Generally, an increase in strength and hardness because of heat treatment is accompanied by decreased corrosion resistance. If possible, hardened steels are protected in a corrosive environment by some form of surface treatment ranging from simply painting or lubricating to special plating or coating procedures.

A good example of the problem encountered when using a heat-treated, unprotected, high-strength steel is the so-called sucker rods used in oil-well pumping. When salt water or hydrogen sulfide solutions are encountered, SCC is very likely to occur. Consequently, special alloys and extreme care in heat treatment must be used to prevent these failures.

8.3.3.3 *Sensitization of Austenitic Stainless Steel*

Stainless steels, particularly the 300 series (S30000) are subject to a heat-treating effect called sensitization. These steels, when heated in the 427 to 760°C range, form chromium carbide. During short exposures at these temperatures, the chromium carbide forms only at the grain boundary. (This is analogous to the $CuAl_2$ precipitates referred to in aluminum-alloy aging treatments.) Thus, the chromium near the grain boundaries is tied up as carbide and no longer can act as a deterrent to corrosion. The grain boundaries are susceptible to intergranular attack and are anodic to the surrounding grains.

The importance of care in the heat treatment not only of stainless steels, but also other metals and alloys, cannot be overstressed. The principles of heat treatment applicable to each material must be considered to produce the most effective combination of mechanical properties and environmental stability.

8.3.3.4 *Welding*

Through the application of heat, welding may

1. induce phase transformations,
2. cause secondary precipitation, and
3. produce high stresses in and adjacent to welds, which can greatly reduce corrosion resistance in these zones.

For example, welding can cause intergranular sensitization of stainless steels in the heat-affected zone (HAZ) adjacent to welds.

During welding, a large variation of thermal expansion occurs between the solidifying molten metal pool and the base metal. Upon solidification, this often can result in high residual tensile stresses. Under these circumstances, the highly stressed areas can be susceptible to SCC when exposed in a corrosive environment. It is important to stress-relieve welds exposed to environments that could cause SCC. Design changes can often eliminate a weld zone that might induce stresses that could lead to such failure.

The other basic metallurgical considerations discussed in this chapter should be considered in welding. For example, for optimum corrosion resistance, it is important to maintain homogeneity between the weld and base metals. Weld filler metals should be used that are chemically and electrochemically (neither strongly anodic nor cathodic to the base metal) similar to the base metals. In addition, any phase transformations that could occur in welding, especially in the HAZ, must be considered in estimating the corrosion stability of welded structures.

8.3.4 Metallurgical Principles of Corrosion Prevention

The following brief sections illustrate the importance of some metallurgical principles in corrosion control.

8.3.4.1 High-Purity Metals

A high-purity metal has better general corrosion resistance and a reduced tendency to pitting corrosion compared to its commercial counterpart. Of course, the limits on their mechanical properties greatly reduce their application possibilities. One of the best examples of improvement of corrosion characteristics of a material is found in results achieved by reducing the sulfur content of plain carbon steels. Corrosion attack on a steel may be greatly reduced when its sulfur level is low.

Another example where an impurity plays an extremely large role in the corrosion characteristics of a material is the effect of iron in magnesium. The solid solubility of iron in magnesium is 0.003%. If the iron is below this amount, it has little or no effect on the corrosion tendencies, but above this amount (0.003% Fe), an iron-rich phase precipitates. Because of the extremely large potential difference between the magnesium and the iron phase, the magnesium corrodes rapidly. This results in severe pitting in areas containing the iron-rich precipitate.

Lead in zinc die-cast alloys has a marked effect on the corrosion characteristics of the material. If the lead concentration exceeds 0.002%, it precipitates at the grain boundaries and produces an increased tendency for intergranular corrosion of the casting alloy.

The low-carbon 300-series (S30000) stainless steels designated as 300L (S30003) have a maximum limit of 0.03% carbon. This reduces the possibility of sensitization as a result of welding and heat treatment. There are many other similar examples of instances when minor concentrations of alloying elements have a great effect upon corrosion resistance of a metal.

The importance of keeping elements of widely different electrochemical potentials in solid solution cannot be minimized. In fact, very small amounts of impurities in solid solution can be detrimental to corrosion resistance. The amount of carbon and nitrogen in the conventional stainless steel is very small. Yet when these are reduced further in quantity by vacuum melting, an increased corrosion resistance is noted.

8.3.4.2 Alloy Additions

It is possible to improve the corrosion resistance of certain alloys by specific alloying additions. The most notable, discussed previously, is the addition of chromium to iron in amounts of 12% or more. At or above this concentration, a passive film of chromium/iron oxide, which is the basis of stainless-steel corrosion resistance, is formed on the surface.

It is also possible by controlled additions of much smaller amounts of selected alloying additions to greatly improve stability of the surface films that form on an alloy surface. The best example of this is the weathering of certain low-alloy architectural steels, which form a stable protective rust after a short exposure.

Corrosion resistance also may be improved by changing the electrochemical potential of the second phase in an alloy. For aluminum alloys, this is accomplished by adding chromium or manganese. In these alloys, the intermetallic compound, $FeAl_3$, greatly affects pitting and corrosion resistance. The addition of manganese or chromium changes the $FeAl_3$ to a complex Al-Fe-Mn or Al-Fe-Cr compound whose potential now approaches that of the aluminum itself. Consequently, the pitting tendency is greatly reduced.

The intergranular sensitization of stainless steels resulting from the formation of chromium carbide at grain boundaries was discussed previously, but alloying for prevention of this effect was not. In these alloys, it is possible to add columbium (S34700) or titanium (S32100), which selectively tie up the carbon as carbides, eliminating chromium-carbide formation and the sensitization of the alloy.

A number of alloying elements, most notably chromium and aluminum, improve the corrosion resistance of alloys by forming or assisting in the formation of passive surface films that are resistant to corroding environments. It is possible to add aluminum or manganese to copper alloys to get passive surface films that resist corrosive attack. The addition of magnesium to aluminum, even though it may lower the corrosion potential of the alloy, improves its corrosion resistance because magnesium stabilizes and thickens the natural oxide film on aluminum.

8.3.4.3 Heat Treatment

It has been shown that it is possible to modify the structure of metals and alloys in many ways through heat treatment. The improvement of cast alloys by homogenization annealing has been discussed already. Age-hardening heat treatments also have a great effect on the corrosion resistance of alloys. Many alloys have a temperature range for aging in which they are not susceptible to intergranular corrosion. However, the aging temperature for optimum mechanical properties of an alloy may produce susceptibility to intergranular corrosion.

Often, it is important to sacrifice some mechanical properties to improve corrosion resistance. This is particularly true of alloys susceptible to SCC. It has been shown that stress relief greatly improves the resistance of 300-series (S30000) stainless steels to SCC. High-zinc brasses, susceptible to SCC in the cold-worked state, also have been protected effectively by stress relief.

Heat treatment of the surface to increase surface hardness or improve the stability of surface films is important in resisting all types of corrosion but is particularly useful in improving fretting and erosion-corrosion resistance.

Usually, the producers or users of a specific alloy have carefully evaluated the various heat-treating procedures to produce the maximum stability for a particular corrosion environment. This information should be examined carefully before heat treating an alloy for a specific application.

8.3.4.4 Metallurgical History and Corrosion

Nearly all forms of metal deterioration are dependent upon the metallurgical history of the material. Impurities retained from the original extractive processes, the inclusions and imperfections introduced in casting and forming, plus structure variations caused by heat treatment, all alter the corrosion stability of the metal or alloy. Thus, a thorough knowledge of the background of the material is important. Here are some examples:

1. When steels were made in the acid open hearth, higher sulfur resulted, and, of course, corrosion resistance was lower.
2. Hot rolling a steel results in scale formation, which, if not properly removed by pickling, serves to produce corrosion-pit initiation sites.
3. Slow cooling of regular the S30000 stainless steels through the 760–427°C temperature range sensitizes it to intergranular corrosion susceptibility. Rapid cooling or quenching through this range prevents sensitization.
4. Residual stress resulting from cold working of an alloy or unequal cooling of welded structures can lead to SCC.

8.4 CORROSION BEHAVIOR OF METALS AND ALLOYS

8.4.1 Cast Irons

Cast iron is a generic term that applies to high-carbon/iron alloys containing silicon. The common ones are designated as gray cast iron, white cast iron, malleable cast iron, and ductile or nodular cast iron. Gray cast irons contain about 2 to 4% carbon and 1 to 3% silicon. These are the least expensive of the engineering metals. The dull or grayish fracture is due to the free graphite flakes in the microstructure. Gray cast irons can be readily cast into intricate shapes because of their excellent fluidity and relatively low melting points. They can also be alloyed for improvement of corrosion resistance and strength.

In most cases, gray cast iron is about twice as resistant to natural waters as steel because of the bonding of rust products by the graphite flakes. Soft or low pH waters may cause graphitic corrosion. In modern engineering practice, most cast-iron piping receives an internal cementatious coating for improved water-side corrosion resistance.

8.4.1.1 *Malleable Cast Irons*

Malleable cast irons are produced by high-temperature heat treatment of white irons of suitable composition. The graphite forms as rosettes or clusters instead of flakes, and the material shows good ductility (hence the name malleable).

8.4.1.2 *White Cast Irons*

White cast irons have practically all of their carbon in the form of iron carbide. These are extremely hard and brittle. Silicon content is low because this element promotes graphitization. Graphite formation is related to rate of cooling from the melt, so chilling can produce a white iron from one that would be normally gray.

8.4.1.3 *Ductile Cast Irons*

These materials, also known as nodular cast irons, exhibit ductility in the as-cast form. The graphite is present as nodules or spheroids as a result of special ladle additions to the molten metal. The mechanical properties of ductile cast irons can be altered by heat treatment similar to that used for ordinary steels.

8.4.1.4 *High-Silicon Cast Irons*

When the silicon content of gray cast iron is increased to more then 14%, it becomes extremely corrosion resistant in many environments. The notable exception is hydrofluoric acid. In fact, these high-silicon irons are the most universally resistant of the commercial (nonprecious) metals and alloys. Their inherent hardness makes them resistant to erosion–corrosion. A straight high-silicon iron, such as Duriron, contains about 14.5% silicon and 0.95% carbon. This composition must be closely controlled within narrow limits to provide the best combination of corrosion resistance and mechanical strength. The excellent corrosion resistance of high silicon irons is due to of the formation of an inert SiO_2 surface layer, which forms during exposure to the environment.

Durichlor 51 exhibits increased resistance to hydrochloric acid, chlorides, bleaches, chlorinated compounds, and pitting and has replaced the former molybdenum-containing alloy, Durichlor. Durichlor 51 contains chromium in addition to some molybdenum, which imparts improvcd resistance to oxidizing conditions. For instance, the presence of ferric or cupric chlorides in hydrochloric acid inhibits corrosion instead of causing severe selective attack as it does with most metals and alloys.

Duriron and Durichlor 51 exhibit tensile strengths around 138 MPa (20,000 psi) and hardness of Rockwell C53. The specific gravity is 7.0. Both alloys are brittle and can be machined only by grinding. Welding of the alloy is very difficult, and although simple shapes such as pipe can be welded when proper precautions are observed, it is not practical to weld complex shapes.

High-silicon cast irons are available only in cast form for drain lines, pumps, valves, and other process equipment. They have also found extensive use as anodes for ICCP (Figure 8.15).

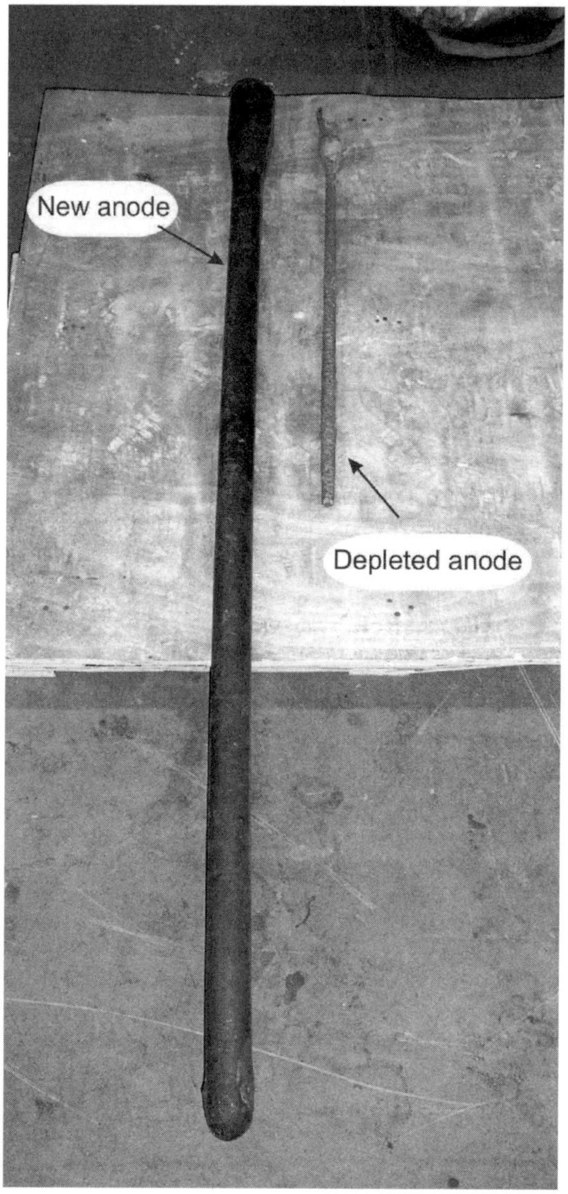

Figure 8.15 ICCP anode made of high-silicon cast iron. A new anode is on the left and a depleted anode is on the right (courtesy of Dean Rookes, West Coast Corrosion Prevention Ltd.).

8.4.1.5 Other Alloy Cast Irons

In addition to silicon and molybdenum, nickel, chromium, and copper are added to cast irons for improved corrosion, abrasion, heat resistance, and mechanical properties. Copper additions impart better resistance to sulfuric acid and atmospheric corrosion.

The high-nickel/chromium cast irons, with and without copper (up to 7%), are the most widely used of this group. These austenitic alloys, known as Ni-Resist, are the toughest of the gray cast irons. Seven varieties of Ni-Resist contain from 14 to 32% nickel and from 1.75 to 5.5% chromium and possess tensile strength from 170 to 310 MPa (25 to 45,000 psi). They are also produced as ductile cast irons with tensile strengths up to 480 MPa (70,000 psi) and elongations up to 40%. One variety contains 35% nickel and is used where low thermal expansion is required.

Ni-Hard is a white cast iron containing about 4% nickel and 2% chromium. It is very hard, with a Rockwell hardness of C55 to C65. Ni-Hard has found wide application where erosion-corrosion resistance is needed in near-neutral and alkaline solutions or slurries.

8.4.2 Carbon Steels and Irons

A wide range of carbon steels is available to take advantage of the great diversity of properties obtainable from this most common material of construction. Industrial carbon and low-alloy steels are generally designated by four numerical digits (Table 8.1). The first two refer to the alloy type and the next indicate the carbon content.

Commercially pure irons are ingot iron and Armco iron. These are soft, weak, and not used where strength is a major requirement. However, the softness of the metal can be used to advantage in gaskets and other applications where deformation is required.

The lower-carbon steels (G10050–G10100) may be severely cold worked without cracking. Consequently, shipping cans and drums, or other items requiring extensive cold bending operations, are made from such low-strength alloys.

Table 8.1 Composition of Steels and Other Iron-Base All: Carbon and Low-Alloy Steels

Designation	Description
10XX	Pain carbon steels
11XX	Cabon steels with higher than normal sulfur content for easier machining
13XX	1.75% Mn (manganese)
23XX	3% Ni (nickel)
25XX	5% Ni
31XX	1% Ni and some chromium
33XX	3% Ni and some chromium
40XX	0.25% Mo (molybdenum)
41XX	0.25% Mo and 1% Cr
43XX	0.25% molybdenum, 0.8 chromium, and 1.8% nickel
50XX	0.3% Cr
51XX	1.0% Cr
52XX	1.5% Cr
61XX	1% Cr and 0.15% V (vanadium)
86XX	0.5% Cr, 0.5i, and 0.2% Mo
92XX	2% Si

XX refers to percent carbon; thus a 4340 steel contains 0.40% C, 0.25% Mo, 0.8% Cr, 1.8% Ni, 0.7% Mn, 0.04% (max) S, 0.04% (max) P, and 0.3% Si.

Carbon steel is severely corroded by natural waters containing oxygen. It is much more corrosion resistant in soft, anaerobic waters. With the exception of specifically treated boiler feedwater make-up, duly deaerated water, or water specifically inhibited for cooling purposes, one should assume that ordinary surface or ground waters will be corrosive to steel. Rates may be as low as 130 μm/y (5 mpy) at room temperature but escalate rapidly with increasing temperature. Pitting rates will be about 10 times the general rate of corrosion. Deposition corrosion can occur if copper ions are present.

Corrosion resistance of the various carbon steels is essentially the same. If any difference were noted, it would be a somewhat higher corrosion rate for the higher carbon materials in certain media (e.g., strong sulfuric acid at ambient temperatures). Phase changes developed by welding or heat treatment produce greater changes in corrosion resistance than any difference in the annealed metals.

Iron and steel corrode in water to form various compounds, depending on the temperature and other environmental conditions. The first corrosion product formed on iron and steel is ferrous hydroxide (see Chapter 2 for details). However, even extremely small quantities of dissolved oxygen in the water will produce ferric ions, which color the ferrous hydroxide green, brown, or black, depending on the quantity of the chromophoric ferric ions.

The oxides normally found include lepidocrocite (gamma $FeO{\cdot}OH$), a yellow to orange rust, such as that found on well-used train rails after a rain, and ordinary rust ($Fe_2O_3{\cdot}3H_2O$). At higher temperatures, one may encounter magnetite (Fe_3O_4), a black magnetic film strongly cathodic to steel and therefore conducive to localized attack in aerated water. Other high-temperature forms include gamma Fe_2O_3 (a brown, protective, slightly magnetic product) and alpha Fe_2O_3 (a brick-red oxide). Its presence in steam-generating equipment indicates oxygen contamination.

8.4.2.1 SCC of Steels

The most common types of failure in low-carbon structural steels were those occurring during the first part of the twentieth century in riveted steam boilers and in welded vats used to contain caustics at high temperatures.

Failures in boilers have been known to occur as the result of fine leaks at riveted joints where deposits from evaporated water could build up and where residual stresses were present in the steel as the result of a mismatch of rivet holes or of extensive caulking to prevent leaking. This particular form of SCC, also often referred to as caustic embrittlement, is believed to have triggered many boiler explosions.

The replacement of steam locomotives with diesels and of riveted stationary boilers by welded pressure vessels has substantially eliminated this type of failure. However, failures of welded steel tanks containing hot concentrated caustics for deenameling steel and for treating bauxite ores have been reported.

SCC of structural-steel specimens is easily produced in the laboratory in boiling concentrated nitrate solutions. Although some authors have set 0.20% carbon as the upper limit of carbon content of steels susceptible to SCC in nitrates, Logan has produced SCC in a steel containing 0.03% carbon. On the other hand, he has also produced failures in a decarburized steel containing less than 0.01% carbon at the surface[1].

One special case of failure in low to medium-carbon steels should be mentioned; namely, the failures of tanks used for storage and transportation of agricultural ammonia. A number of catastrophic failures triggered by SCC in containers for this service had been reported. After an extensive investigation, the following procedures were recommended to prevent failures of these tanks:

1. Tanks (more than 1 m in diameter) are to be stress-relief annealed following fabrication to remove residual stresses.
2. Tanks are to be purged to remove air before they are filled.
3. Water content is to be maintained above 0.2% .

Elimination of air most probably removes one of the sources of the corrosive process, and the addition of water acts to inhibit the corrosion process in the event of accidental ingress of air to the ammonia.

Steels are also known to suffer environmental cracking in mixtures of carbon monoxide, carbon dioxide, and water (as steam or vapor). The specific mixture is required; the individual components or binary mixtures have no effect. Steels can additionally suffer cracking in the presence of carbonates or bicarbonates under some conditions. In some cases, there is obviously reversion to free caustic (as at elevated temperatures). In other cases, as in soils of certain compositions, there may be HIC effects predominating.

The problem of combating SCC in low-carbon steels has been solved in several ways. First, boiler water in locomotives was treated with sodium nitrate, phosphates, tannic acid, and a variety of other materials. Second, where welded steel vessels are used to contain hot caustic liquids, many authorities recommend annealing the welded structure at about 650°C to remove residual stresses. Coatings also have been applied to serve as a barrier between the hot caustic and the stressed steel. Where welded structures are subject to gaseous environments, the welds should be annealed, if possible, or a protective coating should be considered. Another possibility is to use a different material, one not susceptible to SCC in the particular environment anticipated.

8.4.3 High-Strength Low-Alloy Steels

Carbon steel can be alloyed, singly or in combination, with chromium, nickel, copper, molybdenum, phosphorus, and vanadium in the range of a few percent or less to produce HSLA steels. Typically, HSLA steels are low-carbon steels with up to 1.5% manganese, strengthened by small additions of elements, such as columbium, copper, vanadium or titanium, and sometimes by special rolling and cooling techniques. Improved-formability HSLA steels contain additions such as zirconium, calcium, or rare-earth elements for sulfide-inclusion shape control. The higher alloy additions usually have better mechanical properties and hardenability.

Because HSLA alloys are stronger, they can be used in thinner sections, making them particularly attractive for transportation-equipment components where weight reduction is important. The lower range of about 2% total maximum is of greater interest from the corrosion standpoint. Strengths are appreciably higher than those of plain carbon steel, but the most important attribute is a better resistance to atmospheric corrosion when freely exposed.

Alloy steels of the chromium/molybdenum family are employed in parts of boilers because of their improved high-temperature properties; the steam and water have already been rendered substantially noncorrosive.

Steels with very high strengths are of particular interest for aerospace applications where a high strength–weight ratio is the overriding factor. A good example is H-11 steel (5 Cr–1.5 Mo–0.4 V–0.35 C), which can be heat-treated to tensile strengths greater than 2060 MPa (300,000 psi).

8.4.4 Stainless Steels

The main reason for the existence of the stainless steels is their resistance to corrosion. Chromium is the main alloying element, and the steel should contain at least 11%. Chromium is a reactive element, but it and its alloys passivate and exhibit excellent resistance to many environments. A large number of stainless steels are available. Their corrosion resistance, mechanical properties, and cost vary over a broad range. For this reason, it is important to specify the exact stainless steel desired for a given application.

Stainless steels can be divided in six groups that are briefly described in the following sections. The nominal compositions for these alloys are described in Tables 8.2 to 8.7 and key mechanical properties for most of these alloys are described in Tables 8.8 to 8.12.

Note the wide range of properties available from these materials. The high-strength materials exhibit good strength/weight ratios for aircraft and missile applications. High hardness is desirable for wear and some applications where resistance to erosion corrosion is required (e.g., trim for high-pressure steam valves).

The 200- and 300-series stainless steels all exhibit roughly the same mechanical properties in the annealed condition. The major exception involves cast alloys with duplex microstructures. One of the main uses for stainless steels in the chemical industry is for handling concentrated acids acid.

A common form of representing the corrosion resistance of materials is what is known as iso-corrosion diagrams. The use of the prefix "iso" refers to lines (or regions) of constant corrosion behavior across variations in concentration and temperature. The diagram shown in Figure 8.16 illustrates how some stainless steels compare to other metallic materials in industrially important nitric or sulfuric acids and their mixtures[2].

For most environments, the corrosion resistances of cast and wrought structures are considered equivalent. A cast part is not rolled or deformed after casting into shape, so the microstructure can be varied considerably with regard to ferrite content for "duplex" alloys. High ferrite content in an austenitic matrix substantially increases strength. The amount and extent of ferrite phase present can be controlled by raising the percentage of ferrite formers (Cr and Mo) and by decreasing the percentage of austenite formers and stabilizers. (Ni, N, and C). (Ordinary 18-8 has enough nickel added to balance the alloy, i.e., make it completely austenitic.) Producers of wrought products cannot take advantage of this situation because mixed austenite/ferrite structures cause difficulty in rolling.

Table 8.2 Nominal Chemical Composition (%) of Martensitic Stainless Steels (Max. Unless Noted Otherwise)

UNS	Type	C	Mn	P	S	Si	Cr	Ni	Mo	Other
S40300	403	0.15	1.00	0040	0.030	0.50	11.50/13.00			
S41000	410	0.15	1.00	0.040	0.030	1.00	11.50/13.50			
S41400	414	0.15	1.00	0 040	0.030	1.00	11.50/13.50	1.25/2.50		
S41600	416	0.15	1.25	0 060	0.15 min	1.00	12.00/14.00		0.60	
S41623	416Se	0.15	1.25	0060	0.060	1.00	12.00/14.00			0.15Se min
S42000	420	0.15 min	1.00	0.040	0.030	1.00	12.00/14.00			
S42020	420F	0.15 min	1.25	0.060	0.15 min	1.00	12.00/14.00		0.60	
S42200	422	0.20/0.25	1.00	0.025	0.025	0.75	11.00/13.00	0.50/1.00	0.75/1.25	0.15/0.30 V, 0.75/1.25 W
S43100	431	0.20	1.00	0.040	0.030	1.00	15.00/17.00	1.25/2.50		
S44002	440A	0.60/0.75	1.00	0.040	0.030	1.00	16.00/18.00	0.75		
S44004	440B	0.75/0.95	1.00	0.040	0.030	1.00	16.30/18.00	0.75		
S44004	440C	0.95/1.20	1.00	0.040	0.030	1.00	16.30/18.00	0.75		

Table 8.3 Nominal Chemical Composition (%) of Ferritic Stainless Steels (Max. Unless Noted Otherwise)

UNS	Type	C	Mn	P	S	Si	Cr	Ni	Mo	Other
S40500	405	0.08	1.00	0.040	0.030	1.00	11.50/14.50	0.60		0.10/0.30 Al
S40900	409	0.08	1.00	0.045	0.045	1.00	10.50/11.75	0.50		6 × C/0.75 Ti
S42900	429	0.12	1.00	0.040	0.030	1.00	14.00/16.00	0.75		
S43000	430	0.12	1.00	0.040	0.030	1.00	16.00/18.00	0.75		
S43020	430F	0.12	1.25	0.060	0.15 (min)	1.00	16.00/18.00		0.60	
S43023	430FSe	0.12	1.25	0.060	0.060	1.00	16.00/18.00			0.15 Se (min)
S43400	434	0.12	1.00	0.040	0.030	1.00	16.00/18.00		0.75/1.25	
S43600	436	0.12	1.00	0.040	0.030	1.00	16.00/18.00		0.75/1.25	5 × C/0.70 Cb+Ta
S44200	442	0.20	1.00	0.040	0.030	1.00	18.00/23.00	0.60		
S44600	446	0.20	1.50	0.040	0.030	1.00	23.00/27.00	0.75		0.25N
S44635	Monit	0.25	1.00	0.040	0.030	0.75	24.50/26.00	4.00	3.5/4.5	0.035N, 4 × (C+N) Ti + Nb

Table 8.4 Standard Designations for Austenitic Stainless Steels (Composition as Maximum in Percentage Unless Indicated as Range or Minimum)

UNS	Type	C	Mn	P	S	Si	Cr	Ni	Mo	Other
S20100	201	0.15	5.5/7.5	0.060	0.030	1.00	16.00/18.00	3.50/5.50	—	0.25N
S20200	202	0.15	7.5/10.0	0.060	0.030	1.00	17.00/19.00	4.00/6.00	—	0.25N
S20500	205	0.2	14.0/15.5	0.030	0.030	0.50	16.50,18.00	1.00/1.75	—	0.32/0.40N
S30100	301	0.15	2.00	0.045	0.030	1.00	16.00/18.00	6.00/8.00	—	—
S30200	302	0.15	2.00	0.045	0.030	1.00	17.00/19.00	8.00/10.00	—	—
S30215	302B	0.15	2.00	0.045	0.030	2.00/3.00	17.00/19.00	8.00/10.00	—	—
S30300	303	0.15	2.00	0.20	0.15 min	1.00	17.00/19.00	8.00/10.00	0.60	—
S30323	303Se	0.15	2.00	0.20	0.060	1.00	17.00/19.00	8.00/10.00	—	0.15Se min
S30400	304	0.08	2.00	0.045	0.030	1.00	18.00/20.00	8.00,10.50	—	—
S30403	304L	0.030	2.00	0.045	0.030	1.00	18.00/20.00	8.00/12.00	—	—
S30430	18-9-LW	0.08	2.00	0.045	0.030	1.00	17.00/19.00	8.00/10.00	—	3.00/4.00Cu
S30451	304N	0.08	2.00	0.045	0.030	1.00	18.00/20.00	8.00/10.50	—	0.10/0.16N
S30500	305	0.12	2.00	0.045	0.030	1.00	17.00/19.00	10.50/13.00	—	—
S30800	308	0.08	2.00	0.045	0 030	1.00	19.00/21.00	10.00/12.00	—	—
S30900	309	0.20	2.00	0.045	0.030	1.00	22.00/24.00	12.00/15.00	—	—
S30908	309S	0.08	2.00	0.045	0.030	1.00	22.00/24.00	12.00/15.00	—	—
S31000	310	0.25	2.00	0.045	0.030	1.50	24.00/26.00	19.00/22.00	—	—
S31008	310S	0.08	2.00	0.045	0.030	1.50	24.00/26.00	19.00/22.00	—	—
S31400	314	0.25	2.00	0.045	0.030	0.50/3.00	23.00/26.00	19.00/22.00	—	—

(Continued)

Table 8.4 Continued

UNS	Type	C	Mn	P	S	Si	Cr	Ni	Mo	Other
S31600	316	0.08	2.00	0.045	0.030	1.00	16.00/18.00	10.00/14.00	—	—
S31620	316F	0.08	2.00	0.20	0.10 min	1.00	16.00/18.00	10 00/14.00	1.75/2.5	—
S31603	316L	0.030	2.00	0.045	0.030	1.00	16.00/18.00	10.00/14.00	2.0/3.0	—
S31651	316N	0.08	2.00	0.045	0.030	1.00	16.00/18.00	10.00/14.00	2.0/3.0	0.10/0.16N
S31700	317	0.08	2.00	0.045	0.030	1.00	18.00/20.00	11.00/15.00	3.0/4.0	—
S31703	317L	0.030	2.00	0.045	0.030	1.00	18.00/20.00	11.00/15.00	3.0/4.0	—
—	317LMN	0.030	2.00	0.045	0.030	0.75	17.00/20.00	13.50/17.50	4.0/5.0	0.10/0.20N
S32100	321	0.08	2.00	0.045	0.030	1.00	17.00/19.00	9.00/12.00	—	5xC Ti min
N08830	330	0.08	2.00	0.040	0.030	0.75/1.50	17.00/20.00	34.00/37.00	—	0.10 Ta 0.20Cb
S34700	347	0.08	2.00	0.045	0.030	1.00	17.00/19.00	9.00/13.00	—	Cb+Ta 10 × C (min) Ta 0.10 max Co 0.20 max
S34800	348	0.08	2.00	0.045	0.030	1.00	17.00/19.00	9.00/13.00	—	—
S38400	384	0 08	2.00	0 045	0 030	1.00	15.00/17.00	17.00/19.00	—	—
S31254	254 MO	0.01	2.00	0.045	0.030	1.00	20	18	6.1	0.20N, Cu
S32654	654 MO	0.01	3.50	0.045	0.030	1.00	24	22	7.3	0.50N, Cu
S30815	253 MA	0.09	2.00	0.045	0.030	1.7	21	11	—	Ce
N08904	904L	0.01	2.00	0.045	0.030	1.00	20	25	4.5	0.06N, 1.5Cu
N08020	20Cb-3	0.07	2.00	0.045	0.030	1.00	20	34	2.2	Cb
N08367	AL 6 × N	0.03	2.00	0.045	0.030	1.00	20	25	6	0.18–0.25N

Table 8.5 Compositions of Precipitation-Hardening (PH) Stainless Steels

UNS	Alloy	C	Mn	Si	Cr	Ni	Mo	P	S	Other
Martenitic										
S13800	PH13-8 Mo	0.05	0.10	0.10	12.25–13.25	7.5–8.5	2.0–2.5	0.01	0.008	0.90–1.35 Al, 0.01 N
S15500	15-5 PH	0.07	1.00	1.00	14.0–15.5	3.5–5.5	—	0.04	0.03	2.5–4.5 Cu, 0.15–0.45 Nb
S17400	17-4 PH	0.07	1.00	1.00	15.0–17.5	3.0–5.0	—	0.04	0.03	3.0–5.0 Cu, 0.15–0.45 Nb
S45000	Custom 450	0.05	1.00	1.00	14.0–16.0	5.0–7.0	0.5–1.0	0.03	0.03	1.25-1.75 Cu, 8 × %C min Nb
S45500	Custom 455	0.05	0.50	0.50	11.0–12.5	7.5–9.5	0.50	0.04	0.03	1.5–2.5 Cu, 0.8-1.4 Ti, 0.1–0.5 Nb
Semiaustenitic										
S15700	PH 15-7 Mo	0.09	1.00	1.00	14.5–16.0	6.5–7.75	2.0–3.0	0.04	0.04	0.75–1.50 Al
S17700	17-7 PH	0.09	1.00	1.00	16.0–18.0	6.50–7.75	—	0.04	0.04	0.75–150 Al
S35000	AM-350	0.07–0.11	0.50–1.25	0.50	16.0-17.0	4.5–5.0	2.50–3.25	0.04	0.03	0.07–0.13 N
S35500	AM-355	0.10–0.15	0.50–1.25	0.50	15.0–16.0	4.0–5.0	2.50-3.25	0.04	0.03	0.07–0.13 N
Austenitic										
S66286	A-286	0.08	2.00	1.00	13.5–16.0	24.0–27.0	1.0–1.5	0.025	0.025	1.90–2.35 Ti, 0.35 max Al, 0.10–0.50 V, 0.003–0.010 B
—	JBK-73	0.015	0.05	0.02	14.5	29.5	1.25	0.006	0.002	2.15 Ti, 0.25 Al, 0.27 V, 0.0015 B

Table 8.6 Nominal Chemical Composition (%) of Cast Heat Resistant Stainless Steels

UNS	Type	C	Cr	Ni	Fe
—	HA	0.2	8–10	—	Rem
—	HB	0.3	18–22	2	Rem
J92605	HC	0.5	26–30	4	Rem
J93005	HD	0.5	26–30	4–7	Rem
J93403	HE	0.2–0.5	26–30	8–11	Rem
J92603	HF	0.2–0.4	19–23	9–12	Rem
J93503	HH	0.2–0.5	24–28	11–14	Rem
J94003	HI	0.2–0.5	26–30	14–18	Rem
J94224	HK	0.2–0.6	24–28	18–22	Rem
J94604	HL	0.2–0.6	28–32	18–22	Rem
J94213	HN	0.2–0.5	19–23	23–27	Rem
J94605	HT	0.35–0.75	13–17	33–37	Rem
J95405	HU	0.35–0.75	17–21	37–41	Rem
N08001	HW	0.35–0.75	10–14	58–62	Rem
N06006	HX	0.35–0.75	15–19	64–68	Rem
J95705	HP	0.4	25	35	Rem

Table 8.7 Nominal Compositions of First- and Second-Generation Duplex Stainless Steels

UNS	Type	Cr	Mo	Ni	Cu	C	N	Other
First Generation								
S32900	Type 329	26	1.5	4.5	—	0.08	—	—
J93370	CD-4MCu	25	2	5	3	0 04	—	—
Second Generation								
S31200	44LN	25	1.7	6	—	0.030	0.14–0.20	—
S31260	DP-3	25	3	7	0.5	0.030	0.10–0.30	0.3 W
S31500	3RE60	18.5	2.7	4.9	—	0.030	0.05–0.1	1.7 Si
S31803	2205	22	3	5	—	0.030	0.08–0.20	—
S32304	SAF 2304	23	—	4	—	0.030	0.05–0.20	—
S32550	Ferralium 255	25	3	6	2	0.04	0.1–0.25	—
S32750	SAF 2507	25	4	7	—	0.030	0.24–0.32	—
S32760	Zeron 100	25	4	7	0.8	0.03	0.2–0.3	1.0 Si, 0.8 W
S32950	7-Mo PLUS	26.5	1.5	4.8	—	0.03	0.15–0.35	—

Table 8.8 Mechanical Properties of Martensitic Stainless Steels

UNS	Type	Tensile Strength (Mpa)	Yield Strength (0.2%) (Mpa)	Elongation (50 mm) (%)	Hardness (Rockwell)
S40300	403	483	310	25	B80
S41000	410	483	310	25	B80
S41400	414	827	724	15	B98
S41600	416	517	276	30	B82
S41623	416Se	517	276	30	B82
S42000	420	655	345	25	B92
S42020	420F	655	379	22	B97
S42200	422	1000	862	18	C23
S43100	431	862	655	20	C24
S44002	440A	724	414	20	B95
S44004	440B	738	427	18	B96
S44004	440C	758	448	14	B97

Table 8.9 Mechanical Properties of Ferritic Stainless Steels

UNS	Type	Tensile Strength (Mpa)	Yield Strength (0.2%) (Mpa)	Elongation (%)	Hardness (Rockwell)
S40500	405	448	276	25	B75
S40900	409	446	241	25	B75
S42900	429	483	276	30	B80
S43000	430	517	345	25	B85
S43020	430F	655	586	10	B92
S43023	430FSe	655	586	10	B92
S43400	434	531	365	23	B83
S43600	436	531	365	23	B83
S44200	442	552	310	20	B90
S44600	446	552	345	20	B83
S44635	Monit	620	515	20	C28

Table 8.10 Nominal Mechanical Properties of Austenitic Stainless Steels

UNS	Type	Tensile (Mpa)	Yield (0.2%) (Mpa)	Elongation (%)	Hardness (Rockwell B)
S20100	201	655	310	40	90
S20200	202	612	310	40	90
S20500	205	831	476	58	98
S30100	301	758	276	60	85
S30200	302	612	276	50	85
S30215	302	655	276	55	85
S30300	303	621	241	50	84
S30323	303Se	621	241	50	96
S30400	304	579	290	55	80
S30403	304L	558	269	55	79
S30430	302HQ	503	214	70	70
S30451	304N	621	331	50	85
S30500	305	586	262	50	80
S30800	308	793	552	40	80
S30815	253 MA	600	310	40	90
N08904	904L	460	200	25	95
S30900	309	621	310	45	85
S30908	309S	621	310	45	85
S31000	310	655	310	45	85
S31008	310S	655	310	45	85
S31254	254 MO	550	260	35	95
S31400	314	689	345	40	85
S31600	316	579	290	50	79
S31620	316F	586	262	60	85
S31603	316L	558	290	50	79
S31651	316N	621	331	48	85
S31700	317	621	276	45	85
S31703	317L	593	262	55	85
S31726	317LMN	662	373	49	88
S32100	321	621	241	45	80
S32654	654 MO	620	340	35	98
N08830	330	552	262	40	80
S34700	347	655	276	45	85
S34800	348	655	276	45	85
S38400	384	517	241	55	70
N08020	20C-3	550	240	30	96
N08367	AL 6XN	760	360	50	88

Table 8.11 Minimum Mechanical Properties of Precipitation-Hardening (PH) Stainless Steels

UNS	Alloy	Yield Strength (0.2%) (MPa)	Tensile Strength (MPa)	Elongation (%)	Hardness (Brinell)
Martensitic					
S13800	PH13-8Mo	1410	1520	10	430
S15500	15-5PH	1275	1380	14	388
S17400	17-4PH	1170	1310	10	388
S45000	Custom 450	1170	1240	10	363
S45500	Custom 455	1520	1620	8	444
Semiaustenitic					
S15700	PH15-7 Mo	1210	1380	7	415
S17700	17-7PH	1035	1275	6	388
S35000	AM-350	1170	1380	15	380
S35500	AM-355	1070	1170	12	341
Austenitic					
S66286	A-286	690	1005	25	320

An aid in determining which structural constituents can occur is the Schaeffler–de Long diagram. With knowledge of the properties of different phases, it is possible to judge the extent to which they affect the service life of the weldment. The diagram indicates the structure obtained after rapid cooling to room temperature from 1050°C and is not an equilibrium diagram. It was originally established to provide a rough estimate of the weldability of different austenitic steels. In creating the diagram, the alloying elements commonly used for making stainless steels are categorized as either austenite or ferrite stabilizers.[3] In this diagram the ferrite number (FN) is an international measure of the delta or solidification ferrite content of the weld metal at room temperature. The Cr and Ni equivalents that form the two axes of the Schaeffler diagram in Figure 8.17

Table 8.12 Minimum Mechanical Properties of Duplex Stainless Steels

UNS	Type	Yield Strength (0.2%) (MPa)	Tensile Strength (MPa)	Elongation (%)	Hardness (Brinell)
First					
S32900	329	485	620	15	269
J93370	CD-4MCu	490	696	12	287
Second					
S31200	44LN	450	690	25	—
S31260	DP-3	450	690	25	270
S31500	3RE60	440	630	30	290
S31803/S32205	2205	450	620	25	293
S32304	SAF2304	400	600	25	290
S32550	Ferralium 255	550	760	15	302
S32750	SAF 2507	550	795	15	310
S32760	Zeron 100	550	750	25	270
S32950	7-Mo PLUS.	485	690	15	—

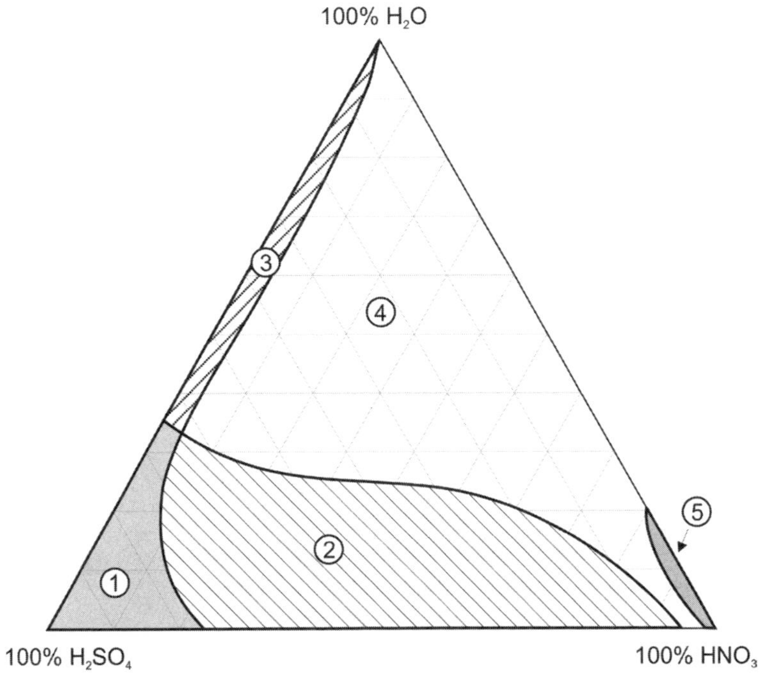

Materials in shaded zones have reported corrosion rates of <0.50 mm/year

Zone 1	Zone 2	Zone3	Zone4	Zone 5
20Cr 30Ni	18Cr 8Ni	20Cr 30Ni	18Cr 8Ni	18Cr 8Ni
Gold	20Cr 30Ni	Gold	20Cr 30Ni	20Cr 30Ni
Lead	Cast Iron	Platinum	Gold	Aluminum
Platinum	Gold	Silicon Iron	Platinum	Gold
Silicon Iron	Lead	Tantalum	Silicon iron	Platinum
Steel	Platinum		Tantalum	Silicon Iron
Tantalum	Silicon Iron			Tantalum
	Tantalum			

Figure 8.16 Mixed acid isocorrosion diagram.

can be estimated with the following relations[4]:

$$\%\,\text{Cr equivalent} = 1.5\ \text{Si} + \text{Cr} + \text{Mo} + 2\ \text{Ti} + 0.5\ \text{Nb} \qquad (8.1)$$

$$\%\,\text{Ni equivalent} = 30\,(\text{C} + \text{N}) + 0.5\ \text{Mn} + \text{Ni} + 0.5(\text{Cu} + \text{Co}) \qquad (8.2)$$

8.4.4.1 Martensitic Stainless Steels

Martensitic stainless steels are called such because they can be hardened by similar heat treatments to carbon steels. Strength increases and ductility decreases with increasing hardness. Corrosion resistance is usually less than for ferritic or austenitic stainless steels. Martensitic steels can be heat treated to obtain high tensile strengths. Corrosion resistance is generally better in the hardened condition than in the annealed or soft condition. They are used in applications requiring moderate corrosion resistance plus high strength or hardness.

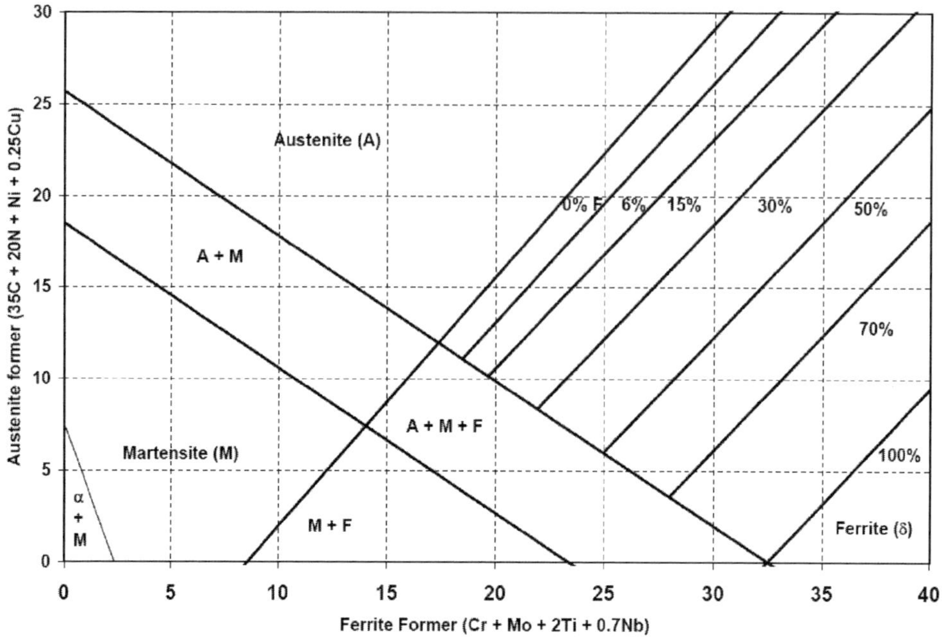

Figure 8.17 Schaeffler diagram.

The composition of the most common martensitic stainless steels is shown in Table 8.2 with key mechanical properties in Table 8.8. Martensitic stainless steels of the 11 to 13% chromium variety (e.g., S41000) are corrosion resistant to water, but very prone to severe localized chloride pitting. In deaerated boiler feedwater make-up, the cast varieties (e.g., CA6NM) are finding successful application. Others are valve parts, ball bearings (S44002), and surgical instruments (S42000). These steels are not normally used in process equipment such as tanks and pipe lines.

Type 416 (S41600) is easier to cut and is used for valve stems, nuts, bolts, and other parts to reduce machining costs.

8.4.4.2 *Ferritic Stainless Steels*

Ferritic stainless steels are nonhardenable steels because they do not undergo a high-temperature phase transformation and therefore cannot be hardened by heat treatment.

The composition of the most common ferritic stainless steels is shown in Table 8.3 and key mechanical properties in Table 8.9. Type 430 (543000) can be formed readily and has good corrosion resistance to the atmosphere. This is one reason why its most extensive use is for automobile trim. It is used also in ammonia oxidation plants for making nitric acid and in tank cars and tanks for the storage of nitric acid. The first chemical-plant application of stainless steel was a Type-430 tank car for shipping nitric acid. However, in these chemical applications, it has been largely displaced by 18-8 because of its ease of welding, better ductility, and better corrosion resistance.

The ferritic grades (e.g., S43000) are more reliable than the lower chromium grades (e.g., S41000, S40500, and S40900), but still are prone to pitting in chloride-bearing waters. The low interstitial superferritics (e.g., S44625), however, are extremely resistant to both pitting and SCC in otherwise aggressive waters, whereas the 29-4 varieties (e.g., S44700, S44800, and S44735) will even cope with hot seawater.

Types 442 and 446 (544200 and S44600) find application where heat resistance is required, such as in furnace parts and heat-treating equipment. They possess good resistance to high-temperature oxidation and sulfur gas attack because of their high chromium content. These materials do not possess very good structural stability or high-temperature strength and should be selected with care.

One of the most interesting aspects of the martensitic stainless steels is their resistance to SCC. They do quite well in many cases where the austenitic types fail, particularly in chloride-containing waters.

8.4.4.3 *Austenitic Stainless Steels*

Austenitic stainless steels are essentially nonmagnetic and cannot be hardened by heat treatment. Like the ferritic steels, they are hardenable only by cold working.

The composition of the most common austenitic stainless steels is shown in Table 8.4 and key mechanical properties in Table 8.10. Most of these steels contain nickel as the principal austenite former, but the relatively new ones such as Types 201 and 202 (S20100 and S20200) contain less nickel and therefore more substantial amounts of manganese.

The austenitic steels retain good ductility and impact resistance at very low temperatures and are used for handling liquid oxygen and nitrogen. The only method available for hardening the austenitic stainless steels is cold working. This usually decreases corrosion resistance only slightly, but in certain critical environments, a galvanic cell could form between cold-worked and annealed material. Type 301 (S30100) is utilized mostly in the cold-worked condition for such applications as train and truck bodies. The austenitic steels can be cold rolled to strengths in the neighborhood of 2000 MPa (300,000 psi) in wire form. Types 301 and 302 (S30100 and S30200) are not used for severe corrosion applications, the former because of its lower Cr and Ni and the latter because of higher carbon.

The 18-8 stainless steels containing molybdenum (e.g., S31603 and J92800) may be successfully used in seawater, but their performance is contingent upon high-velocity (4 m s^{-1} or more), aeration (or total deaeration) and freedom from deposits of scale or marine organisms. The regular carbon grades, unless stabilized, may be subject to intergranular corrosion in natural waters under some conditions (e.g., high chlorinity, bacterial action, oxygen cells, etc.).

Perhaps the major loss of austenitic stainless steels has been by exposure to heating or cooling waters when used as tubing in tube-and-shell heat exchangers. Figure 8.18 shows the actual problem area. If the tube sheet is sufficiently hot, salts can accumulate on the tube in the crevice or immediately behind the sheet as water is evaporated. Chlorides contained in these salts will crack the stressed metal in this area.

When using the austenitic stainless steels in tube-and-shell heat exchangers, the design factors become most important. Preventing the accumulation of silt or evaporated salt deposits on the metal surface may, for example, require running the water on the tube

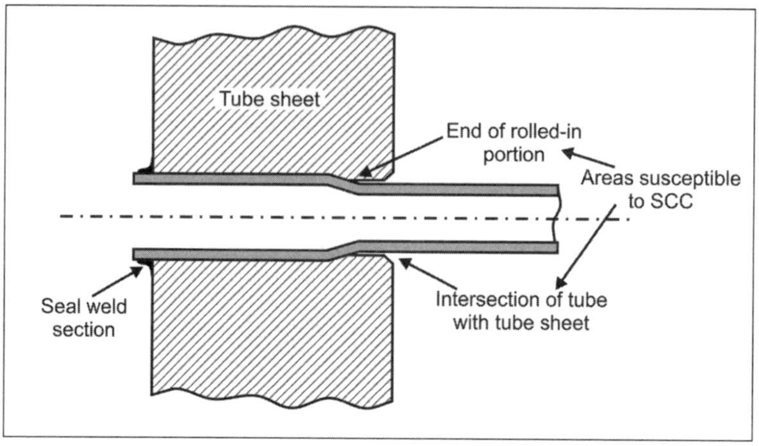

Figure 8.18 Stainless-steel tube-to-tube sheet joint (exaggerated).

side and the process stream on the shell side of condensers. With water on the shell side, void spaces above the cooling water may be eliminated by placing the exchanger in a horizontal rather than vertical position. Other approaches have been used to circumvent this problem.

The superstainless steels exemplified by the 20 Cr–25 Ni–4 Mo–2 Cu alloys (e.g., N08904) are quite resistant to aggressive waters, as are the alloys intermediate between high stainless and low-nickel specialty materials. Alloy 825 (N80825) will withstand high heat flux in aerated waters of quite high chlorinity (e.g., at least 1500 ppm Cl^-).

The austenitic steels possess better corrosion resistance than the straight chromium steels (martensitic and ferritic). For this reason, austenitic steels are widely specified for the more severe corrosion conditions such as those encountered in the process industries. They are rust resistant in the atmosphere and find wide use for architectural purposes, in the kitchen, in food manufacture and dispensing, and for applications where contamination (rust) is undesirable.

An empirical ranking scheme has been developed over the years to describe the corrosion pitting susceptibility of austenitic and duplex stainless steels. The pitting-resistance equivalent number (PREN) predicts the pitting resistance of these alloys as a function of their content in chromium, molybdenum, tungsten, and nitrogen:

$$\text{PREN} = \text{Cr} + 3.3(\text{Mo} + 0.5\,\text{W}) + 16\text{N} \tag{8.3}$$

8.4.4.4 Age- or Precipitation-Hardened Stainless Steels

Precipitation-hardened stainless steels are hardened and strengthened by solution quenching, followed by heating for substantial times at temperature in the approximate range of 427 to 538°C.

The composition of the most common precipitation-hardened stainless steels is shown in Table 8.5 and key mechanical properties in Table 8.11. Tensile strengths up to around 1375 MPa (200,000 psi) can be obtained. The martensitic precipitation-hardened steels find their widest application in the aircraft and missile industry under relatively mild corrosion conditions. Corrosion resistance to severe environments is generally less than that of the austenitic stainless steels.

The higher hardness of precipitation-hardened stainless steels reduces the tendency for seizing and galling of rubbing parts such as valve seats and disks.

8.4.4.5 Cast Stainless Steels

Cast stainless steels and alloys usually have somewhat different chemical compositions than their wrought counterparts. One difference is higher silicon content, which increases castability. Silicon content is usually 1% maximum in wrought stainless steels. This is necessary particularly for casting thin sections or small parts. The composition of the most common precipitation-hardened stainless steels is shown in Table 8.6.

8.4.4.6 Duplex Stainless Steels

The composition of first- and second-generation stainless steels is shown in Table 8.7 and key mechanical properties in Table 8.12. This is a stainless-steel alloy group, with two distinct microstructure phases—ferrite and austenite. The duplex alloys have greater resistance to chloride SCC and higher strength than the other austenitic or ferritic grades. Duplex grades are the newest of the stainless steels. These materials are a combination of austenitic and ferritic material. Modern duplex stainless steels have been developed to take advantage of the high strength and hardness, erosion, fatigue and SCC resistance, high thermal conductivity, and low thermal expansion produced by the ferrite/austenite microstructure. These steels have high chromium content (18–26%), low amounts of nickel (4–8%), and generally contain molybdenum. They are moderately magnetic, not hardenable by heat treatment, and readily weldable in all section thicknesses. Duplex stainless steels are less notch sensitive than ferritic types but suffer loss of impact strength if held for extended periods of high temperature greater than 300°C. Duplex stainless steels thus combine some of the features of the two major classes. They are resistant to SCC, albeit not quite as resistant as the ferritic steels, and their toughness is superior to that of the ferritic steels but inferior to that of the austenitic steels. Duplex steels yield strength is appreciably greater than that of the annealed austenitic steels, by a factor of about two. Table 8.10 describes basic mechanical properties for many commercial austenitic stainless steels.

8.4.4.7 SCC of Stainless Steels

Although martensitic and ferritic stainless steels can be stress-corrosion cracked by hot caustic (and the ferritic stainless steels by chlorides under some special conditions), the prominent SCC problems arise with the austenitic grades.

The 18-8 grades (e.g., S30400, S31600, and their low-carbon and niobium- or titanium-stabilized variants) are highly susceptible to SCC, particularly by chloride ions. A common problem is SCC of stainless-steel heat-exchanger tubes in chloride-bearing cooling waters when the skin temperature is above about 60°C. Crevices where chlorides can accumulate, for example, where the tubes are rolled into the tube sheet (Figure 8.18), are particularly susceptible. Calcareous or other films or deposits aggravate the situation by occluding and concentrating chlorides at a level higher than that in the bulk solution.

A particularly obnoxious form of SCC occurs from the outside when 18-8 equipment becomes contaminated under insulation with chlorides from water leaks or storm-driven water salts. Chloride cracking is usually transgranular (even when the 18-8 is sensitized) in natural waters. There appears to be little difference between conventional 18-8s and their manganese-substituted counterparts (e.g., S20400, S21600) in this type of environment.

The most common form of SCC for 18-8 austenitic grades is from chloride contamination. The ubiquitous chloride is sodium chloride, as found in water and atmospheric exposures. NaCl usually causes SCC only between about 50°C and 200°C. However, SCC can occur at room temperature and even at cryogenic temperatures in the presence of other halides (e.g., sulfuric/NaCl mixtures, aqueous solutions of hydrogen sulfide/NaCl, HCl). For boldly exposed surfaces, there are rough correlations between pH and chloride concentration at which SCC may be anticipated. There is no minimum chloride concentration below which SCC will not occur if there is a possibility of concentration either by evaporation or by occlusion or adsorption in films adhering to the stainless surface (e.g., mill scale, heat tints, welding slag, rust deposits, calcareous deposits, biomasses).[5]

External SCC occurs when chlorides from the atmosphere, water leaks, or insulation concentrate on the surface of an 18-8-type stainless steel. Ferric ions from rusted steel flanges will aggravate the situation. External SCC is best prevented by coating vessels and piping larger than 10 cm with a zinc-free and chloride-free paint system.

The morphology of SCC in austenitic stainless steels is characteristically transcrystalline in both chloride and caustic environments, except for the SCC specific for sensitized 18-8 grades in refinery applications from polythionic acids and the DO phenomenon in the supercritical water in nuclear-power generation.

Higher temperatures are not required to produce SCC of the austenitic stainless steels. Highly stressed parts, such as cold-drawn wire cable, have cracked at atmospheric temperatures. Austenitic stainless steels of the 18-8 variety are also susceptible to SCC by hot caustic solutions or even by traces of caustic entrained in high-pressure steam.

In heavy-water plants, it has been observed that the presence of hydrogen sulfide weakens the passive oxide film on 18-8-type stainless, and chloride SCC occurs at much lower chloride concentrations and temperatures than would otherwise occur in a chloride-bearing natural water. If 18-8 stainless steels are sensitized, they becomes subject to SCC in an intergranular mode in very high-temperature water, as in nuclear reactors, a form of attack known as oxygen SCC.

Likewise, sensitized stainless steels suffer intergranular stress corrosion cracking (IGSCC) in refineries during down time, a mode of attack ascribed to polythionic acid. These polythionic acids are formed by the interaction of hydrogen sulfide or metallic sulfides with sulfur dioxide and water when there is ingress of atmospheric air to the system. Stabilized grades are reportedly immune, but low-carbon grades are unreliable because of the possibility of long-time sensitization at normal service temperatures. This type of attack is combated by alkaline or ammoniacal washes plus nitrogen blanketing of the equipment.

New superferritic stainless steels, special proprietary grades, and super stainless steels have been developed specifically to resist chloride SCC. In general, those alloys containing more than 29% nickel are immune to chloride cracking. Resistance to caustic and other environments is a more nebulous topic, requiring investigation.

8.4.5 Aluminum and Its Alloys

Pure aluminum is soft and weak, but it can be alloyed and heat-treated to a broad range of mechanical properties. Strengthening usually decreases corrosion resistance, particularly resistance to SCC, and for this reason the structural alloys are clad with a thin skin of pure aluminum (Alclad).

Aluminum alloys are divided into two broad classes, castings and wrought, or mechanically worked, products. The latter is subdivided into heat-treatable and non-heat-treatable alloys, and into various forms produced by mechanical working. Casting compositions are described by a three-digit system followed by a decimal value. This decimal indicates the product form, that is, 0 for casting and 1 and 2 for ingots. For wrought alloys, a four-digit system is used to produce a list of wrought composition families.

Table 8.13 describes the designation system for cast and wrought aluminum alloys.[6] The 3000-, 5000-, and 6000- series alloys are widely used in the process industries. Alloys are sandcast, die cast, or cast into permanent molds. This table also lists mechanical properties of some aluminum alloys. This table illustrates the wide range

Table 8.13 Designation System for Cast and Wrought Aluminum Alloys

Wrought Alloys	Composition Family	Heat Treatment
lxxx[(a)]	Commercially pure (> 99%)	Not age-hardenable
2xxx	Al-Cu and Al-Cu-Li	Age-hardenable
3xxx	AI-Mn	Not age-hardenable
4xxx	Al-Si and AI-Mg-Si	Age-hardenable if magnesium is present
5xxx	Al-Mg	Not age-hardenable
6xxx	AI-Mg-Si	Age-hardenable
7xxx	Al-Mg-Zn	Age-hardenable
8xxx	Al-Li, Sn, Zr, or B	Age-hardenable
9xxx	Not currently used	
Casting alloys		
lxx.x.[(b)]	Commercially pure Al	Not age-hardenable
2xx.x.	Al-Cu	Age-hardenable
3xx.x.	AI-Si-Cu or Al-Mg-Si	Some are age-hardenable
4xx.x.	AI-Si	Not age-hardenable
5xx.x.	Al-Mg	Not age-hardenable
7xx.x.	Al-Mg-Zn	Age-hardenable
8xx.x.	AI-Sn	Age-hardenable
9xx.x.	Not currently used	

(a) The first digit shows the main alloying element, the second digit shows modification, and the last two digits shows the decimal percentage of the Al concentration (e.g., 1060 will be 99.6% Al alloy).
(b) Last digit indicates product form, 1 or 2 is ingot (depends upon purity) and 0 is for casting.

of properties available. For example, the tensile strength of annealed, commercially pure aluminum is 90 MPa (13,000 psi) and 600 MPa (88,000 psi) for heat-treated 7178 (A97178) alloy. Alloy 5052 (A95052) exhibits the highest strength of non-heat-treatable alloys. Alloy 7178 is one of the highest strength heat-treatable aluminum alloys utilized in aircraft and aerospace applications.

The corrosion resistance of aluminum is dependent upon a protective oxide film. This film is stable in aqueous environments when the pH is between about 4.0 and 8.5. The oxide film is naturally self-renewing and accidental abrasion or other mechanical damage of the surface film is rapidly repaired. The conditions that promote corrosion of aluminum and its alloys, therefore, must be those that continuously abrade the film mechanically or promote conditions that locally degrade the protective oxide film and minimize the availability of oxygen to rebuild it.

The high-copper alloys are used mainly for structural purposes. When using these alloys, considerations must be given to their protection even in a normal atmosphere. The copper-free or low-copper alloys are used in the process industries or where better corrosion resistance is required.

The acidity or alkalinity of the environment significantly affects the corrosion behavior of aluminum alloys. At lower and higher pH, aluminum is more likely to corrode but by no means always does so. For example, aluminum is quite resistant to concentrated nitric acid. When aluminum is exposed to alkaline conditions, corrosion may occur, and when the oxide film is perforated locally, accelerated attack occurs because aluminum is attacked more rapidly than its oxide under alkaline conditions. The result is pitting. In acidic conditions, the oxide is more rapidly attacked than aluminum, and more general attack should result.

Unalloyed aluminum and those alloys containing combinations of magnesium, zinc, and silicon are most resistant to water. Aluminum alloys containing appreciable percentages of copper may be severely corroded. Aluminum and its alloys often pit in seawater and brackish waters, especially because of oxygen-concentration cells under deposits.

In both fresh and salt water, extreme care must be exercised to avoid galvanic couples of aluminum to more noble metals (e.g., steel or copper). The copper/aluminum couple is particularly bad. Copper or iron ions in the system can cause attack even when not directly connected; the heavy-metal ions plate out on the aluminum by reduction effects and cause pitting or other types of localized corrosion such as deposition corrosion.

Aluminum corrosion products in water are white to translucent. At low temperatures, an amorphous (i.e., noncrystalline) hydrated alumina ($Al[OH]_3$) film forms at first. This may revert to boehmite ($Al_2O_3{\cdot}3H_2O$ or $AlO{\cdot}OH$) and finally bayerite ($Al_2O_3{\cdot}3H_2O$). At higher temperatures (greater than the boiling point of water), the oxide film consists of an inner layer of amorphous alumina and randomly oriented boehmite, and an outer layer of highly oriented boehmite.

Aluminum is a reactive metal, but it develops an aluminum-oxide coating or film that protects it from corrosion in many environments. This film is quite stable in neutral and many acid solutions, and it can be artificially increased by passage of electric current. This process is called anodizing. Aluminum is not used in water hotter than 200°C.

SCC of aluminum alloys is confined primarily to the higher-strength alloys, such as those of the 2000 (A92000) and 7000 (A97000) series, whose strength properties are due to precipitation-strengthening heat treatment rather than cold rolling. However, SCC may also occur in aluminum/magnesium alloys containing more than 3.5% magnesium, and failures have occurred in certain high-strength casting alloys.

Failures have occurred most commonly in marine atmospheres but are no longer confined to coastal areas because of the general use of salt for ice and snow removal. Nor, apparently, is a large amount of salt or chloride necessary. Failures of highly stressed castings have occurred in only mildly corrosive industrial atmospheres.

The susceptibility of many alloys to SCC is usually greater if they are stressed perpendicular to the rolling direction instead of being stressed in the rolling direction. This condition is especially pronounced in aluminum alloys. The high-strength alloys generally are quite resistant to SCC if they are stressed in the direction of rolling or extrusion, but most susceptible if stressed in the short transverse direction.

The grains of an aluminum alloy are very much elongated in the direction of rolling or extrusion (longitudinal direction). The boundaries of grains normal to a stress applied in the longitudinal direction are very short and mostly do not link up with that part of the boundary of the adjacent grain that is attacked. Hence, although there is attack on boundaries of individual grains, the regions attacked do not link up to form a continuous corroded path, and the material, therefore, is relatively immune to SCC if stressed in this direction.

On the other hand, the boundaries are very much elongated normal to a stress applied in the short transverse direction, so a crack formed on the elongated boundary of one grain frequently can join a crack formed at the boundary of the next grain. The continuous crack networks thus formed cause an alloy stressed in the short transverse direction to fail under a much lower stress than is required to cause a failure from stress applied in the longitudinal direction with respect to the rolling or extruded direction. Figure 8.19 shows how cracks can join up to produce a continuous crack in a material stressed in the transverse direction.

Specifically, the yield strength of the 7075-T6 (A97075) alloy, tested in air, is reported to decrease from 550 MPa (80,000 psi) in the longitudinal direction to 410 MPa (60,000 psi) in the short transverse direction. However, the threshold stress (maximum stress that the material can withstand without failure when subjected to cyclic alternate immersion for 10 min in a 3.5% NaCl solution, 50 min in air) decreases from approximately 410 MPa (60,000 psi) (three-fourths of the yield strength) for material tested in the longitudinal direction to 82 MPa (12,000 psi) (one-fifth of the yield strength) for material tested in the short transverse direction.

Figure 8.20 gives a generalized picture of the grain structure to be expected in a thick aluminum alloy plate or extrusion. Figure 8.21 shows how a fraction of material is often sacrificed to alleviate the impact on the susceptibility to SCC of the short transverse sections of a component. Corrosion products building up along these grain boundaries exert pressure between the grains and the end result is a lifting or leafing effect. The damage often initiates at end grains encountered in machined edges, holes, or grooves and can subsequently progress through an entire section.

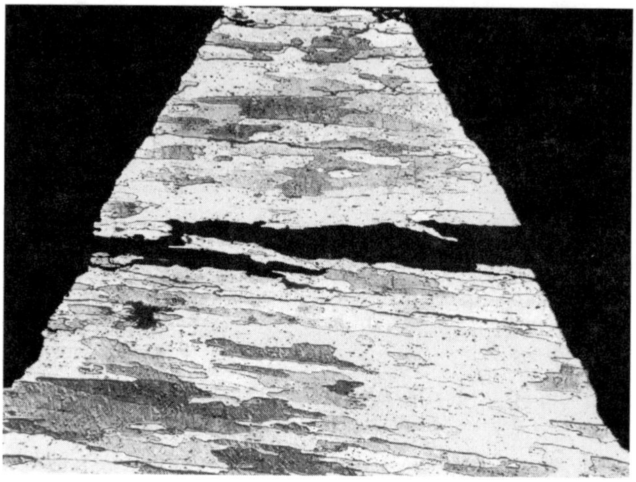

Figure 8.19 Intergranular SCC in a nut machined from a 2024-aluminum-alloy extrusion. Stresses were in the transverse direction. Note how cracks follow a series of boundaries. Etched 2.5% HNO_3, 1.5% HC1, 1% HF ($\times$100 magnification).

8.4.6 Magnesium and Its Alloys

Magnesium is one of the lightest commercial metals, with a specific gravity of 1.74 g cm^{-3}. It is utilized in trucks, automobile engines, ladders, portable saws, luggage, aircraft, and missiles because of its light weight and also its good strength when alloyed. However, it is one of the least corrosion-resistant and is accordingly used as sacrificial anodes for cathodic protection and seawater-activated batteries. Magnesium parts can be severely damaged in many environments unless galvanic couples are avoided by proper design or surface protection. Unalloyed magnesium is not extensively used for structural purposes. Consequently, the corrosion resistance of magnesium alloys is of primary concern.

Magnesium exhibits good resistance to ordinary atmospheres because of the formation of a protective oxide film. This protection tends to break down, forming corrosion pits in air contaminated with salt, so protective measures are required. These include

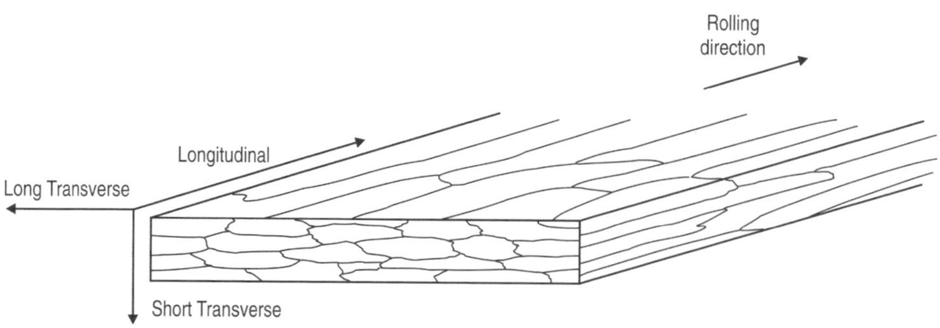

Figure 8.20 Schematic shows the three-dimensional grain structure typical of wrought high-strength aluminum alloys.

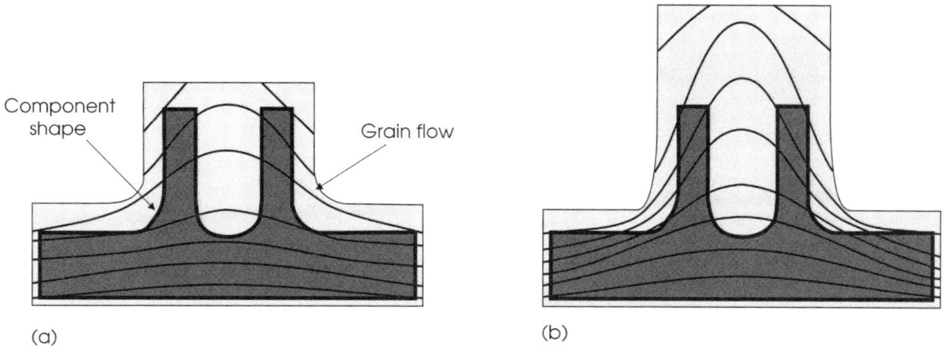

Figure 8.21 Machining for neutralizing the effects of grain flow on corrosion resistance: a) saving on material and losing on lifetime and b) losing on material for increased lifetime.

coatings and "chrome" pickling, which also provides a good base for coatings. Corrosion resistance generally decreases with impurities and alloying. Alloys are quite susceptible to SCC and must be protected. The presence of dissolved oxygen in water has no significant effect on corrosion. The metal is susceptible to erosion corrosion. Magnesium is much more resistant than aluminum to alkalies. It is attacked by most acids except chromic and hydrofluoric (HF). The corrosion product in HF acts as a protective film.

Magnesium and its alloys are available in a variety of wrought forms and die castings. Tensile strengths in the approximate range of 100 to 340 MPa (15,000 to 50,000 psi) are obtainable. Two major magnesium alloy systems are available to the designer.[7]

The first includes alloys containing 2 to 10% Al, combined with minor additions of zinc and manganese. These alloys are widely available at moderate costs and their mechanical properties are good up to 95–120°C.

The second group consists of magnesium alloyed with various elements (e.g., rare earth, zinc, thorium, silver with the exception of aluminum), all containing a small but effective zirconium content that imparts a fine grain structure and thus improved mechanical properties. These alloys generally possess much better properties at elevated temperatures, but their more costly alloying additions, combined with the specialized manufacturing technology required, result in significantly higher costs.

A clean, unprotected magnesium-alloy surface exposed to indoor or outdoor atmospheres free from salt spray will develop a gray film that protects the metal from corrosion while causing only negligible losses in mechanical properties. Chlorides, sulfates, and foreign materials that hold moisture on the surface can promote corrosion and pitting of some alloys unless the metal is protected by properly applied coatings.[7]

Magnesium is rapidly attacked by all mineral acids except HF acid and chromic acid. HF acid does not attack magnesium to an appreciable extent because it forms an insoluble, protective magnesium-fluoride film on the magnesium. However, pitting corrosion can develop at low acid concentrations.

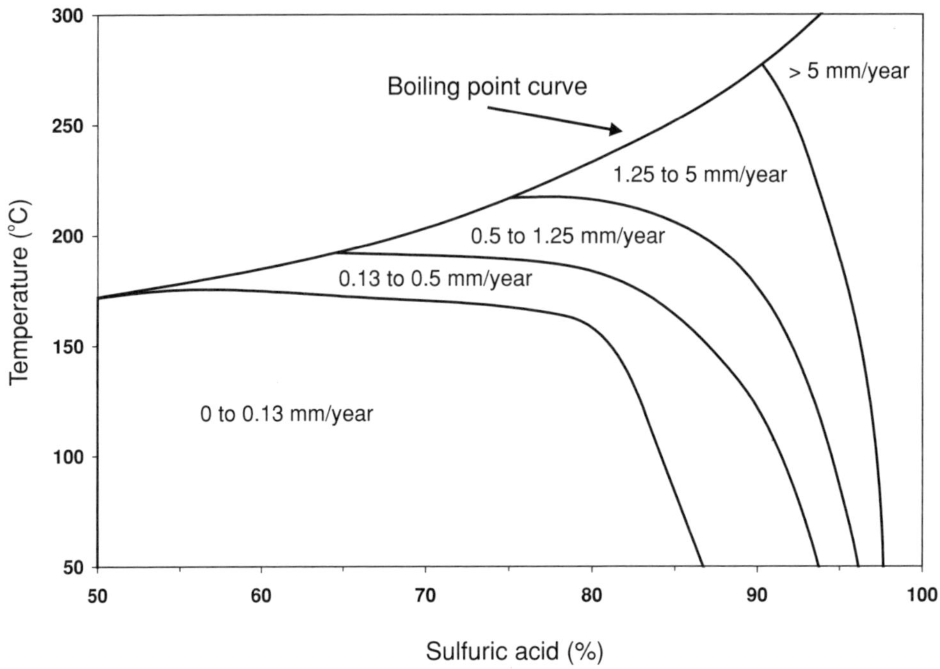

Figure 8.22 Isocorrosion chart for chemical lead in stagnant sulfuric acid shows effects of temperature and concentration.

Certain magnesium alloys are susceptible to SCC by laboratory corrodents. Because cracking can be produced easily and quickly, they have been used in mechanistic studies. There have been no recent reports of service failures of these alloys attributed to SCC.

8.4.7 Lead and Its Alloys

Lead is one of the most stable metals and one of the oldest metals used by humans. It was used for water piping during the time of the Roman Empire, and some of it is still in operation. Lead ornaments and coins were utilized several thousand years ago. Lead forms protective films consisting of corrosion products such as sulfates, oxides, and phosphates. Lead can be used in sulfuric-acid service (Figure 8.22). Lead is still used as a roofing material in the United Kingdom, particularly on historic buildings. Upon exposure to the outdoor environment, lead develops a gray patina, resulting in the metal's characteristic appearance. This patina may be nonadherent and so may result in "run-off" staining of adjacent areas, thus spoiling the aesthetic appeal of a lead roof. Lead is soft, easily formed, and has a low melting point. Lead-lined steel has often been made by "burning on" the lead. It is subject to erosion corrosion because of its softness.

The processes used in its production are essentially the same as those used for zinc. Lead-sulfide ore is converted into lead oxide, which is reduced to metallic lead by a continuous process in a blast furnace. Refining of blast-furnace bullion yields 99.99% pure lead. The relatively low cost, easy workability, and good environmental resistance of lead have made it in the past an obvious choice for sheathing purposes. However, in more recent years its high toxicity, high density, and its low creep strength have led to substitution by polymeric materials and metals such as aluminum, and its use is becoming restricted to applications where its special properties offer unique advantages to the end product.

When protection against corrosion is required for process equipment, chemical lead containing about 0.06% copper is specified, particularly for sulfuric acid. This lead is resistant to sulfuric, chromic, HF, and phosphoric acids in certain strengths; neutral solutions; seawater; and soils. It is rapidly attacked by acetic acid and generally is not used in nitric, hydrochloric, or organic acids.

Chemical lead exhibits a tensile strength of about 16 MPa (2300 psi) at room temperature. Hard leads, containing 3 to 18% antimony, doubles this strength. However, the strength of both materials drops rapidly as temperature increases, and they show about the same strength around 110°C. Design strength at appreciably higher temperatures drops to zero.

8.4.8 Copper and Its Alloys

Copper occurs naturally with elements such as lead, nickel, silver, and zinc. It is widely used in industry both as a pure metal and as an alloying material. The copper industry is composed of two segments: producers (mining, smelting, and refining industries) and fabricators (wire mills, brass mills, foundries, and powder plants). The end products of copper producers, the most important of which are refined cathode copper and wire rod, are sold almost entirely to copper fabricators. The end products of copper fabricators can be generally described as mill products and foundry products, and they consist of wire and cable, sheet, strip, plate, rod, bar, mechanical wire, tubing, forgings, extrusions, castings, and powder metallurgy shapes. These products are sold to a wide variety of industrial users.

The resistance of all grades of copper to atmospheric corrosion is good, hence their usage for roofing and for contact with most waters. The metal develops adherent protective coatings, initially of oxide but subsequently thickening to give a familiar green patina on roofs and the dark brownish color of bronze statuary. Because copper is largely unaffected by potable water, it is widely used for tubes carrying domestic and industrial water. In the following broad classifications, copper and copper alloys have demonstrated superior corrosion performance:

- atmospheric exposure such as roofing and other architectural applications;
- plumbing systems with superior corrosion resistance to both potable waters and soils;
- marine applications involving supply lines, heat exchangers, and hardware where resistance to seawater and biofouling are mandatory; and
- industrial- and chemical-plant process equipment involving exposure to a wide variety of organic and inorganic chemicals.

The corrosion product formed on copper alloys in low-temperature water is reddish cuprous oxide (Cu_2O). It may be present as a very thin film in the form of an iridescent tarnish. Continued exposure to oxygenated water at low temperatures, or to high-temperature water, can produce black cupric oxide (CuO). Usually, CuO forms on the surface of Cu_2O to produce a grayish-black, powdery appearance. The corrosion resistance of copper and its alloys depends on the retention of such protective films, which may be removed by acidic dissolution or by velocity effects, or may suffer chemical conversion because of contamination (e.g., by traces of hydrogen sulfide).

As the oxygen content of the water is increased and the chlorinity increases, increasingly resistant alloys are demanded. Although copper and red brass are used extensively in fresh water (e.g., potable water), the more-resistant, inhibited yellow brasses (e.g., admiralty, C44300) are used to withstand aerated, moving, cooling waters. In seawater, copper may be used up to about 8 m s^{-1} under wholly anaerobic conditions. In conventional heat exchangers, however, at least arsenical aluminum brass (C68700) is employed. This is largely being replaced in current seawater applications requiring copper alloys for the process side with the iron-modified 90–10 cupronickel (C70600) or the related 70–30 alloy (C71500). The tin-modified aluminum bronze (C61300) is also used in severe seawater applications.

Waters in which copper alloys perform poorly are those contaminated with carbon dioxide, sulfide, or ammonia.

Soft, aerated waters containing free carbon dioxide often result from decomposition of organic products such as in reservoirs and certain rivers. The settling treatment used in treating reservoir waters may release copious quantities of carbon dioxide. Such potable waters may cause green staining of copper plumbing fixtures.

Sulfides may be found in certain well, surface, and mine waters and may be produced by sulfate-reducing bacteria in many waters. Sulfides interfere with the otherwise protective film on copper alloys. Even 90–10 cupronickel may suffer pitting if seawater becomes contaminated with as little as 10 to 15 ppm hydrogen sulfide. Fresh waters contaminated with sulfides are better resisted by brasses than by copper.

Polluted seawater and brackish water containing sulfides and nitrogenous material (e.g., ammonia) are particularly corrosive to copper alloys. Improved resistance to polluted water has been reported for brass alloys containing more than about 4% tin.

Copper is different from other metals in that it combines corrosion resistance with high electrical and heat conductivity, formability, machinability, and strength for use in urban, marine, and industrial atmospheres and waters. Copper is a relatively noble metal, and hydrogen evolution is not usually part of the corrosion process. For this reason it is not corroded by acids unless oxygen or other oxidizing agents (e.g., HNO_3) are present. Copper-base alloys are resistant to neutral and slightly alkaline solutions with the exception of those containing ammonia, which cause SCC and sometimes rapid general attack. In strongly reducing conditions at 300 to 400°C, copper alloys are often superior to stainless steels and stainless alloys.

Table 8.14 provides a generic classification of wrought and cast copper alloys and some common copper-base materials. Hundreds of compositions with a wide variety of mechanical properties are commercially available. For example, a range of tensile strength from about 205 to 1350 MPa (30,000 to 200,000 psi) is exhibited by pure copper and copper alloyed with 2% beryllium. The most common alloys are brasses (Cu-Zn), bronzes (Sn, Al, or Si additions to Cu), and cupronickels (Cu-Ni). Nickel silver (an alloy of Cu, Ni, and Zn) does not contain silver, but resembles it in appearance. Everdur,

Table 8.14 Generic Classification of Wrought and Cast Copper Alloys

	UNS Number	Composition
Wrought alloys		
Coppers		
Coppers	C10100-C15760	>99% Cu
High-copper alloys	C16200-C19600	>96% Cu
Brasses		
Brasses	C20500-28580	Cu-Zn
Leaded brasses	C31200-C38590	Cu-Zn-Pb
Tin brasses	C40400-C49080	Cu-Zn-Sn-Pb
Bronzes		
Phosphor bronzes	C50100-C52400	Cu-Sn-P
Leaded phosphor bronzes	C53200-C54800	Cu-Sn-Pb-P
Copper/phosphorus and copper/ silver/phosphorus alloys	C55180-CS5284	Cu- P-Ag
Aluminum bronzes	C60600-C64400	Cu-AI-Ni-Fe-Si-Sn
Silicon bronzes	C64700-C66100	Cu-Si-Sn
Others		
Other copper/zinc alloys	C66400-C69900	
Copper/nickels	C70000-C72950	Cu-Ni-Fe
Nickel silvers	C73200-C79900	Cu-Ni-Zn
Wrought alloys		
Coppers		
Coppers	C80100-C81100	>99% Cu
High-copper alloys	C81300-C82800	>94% Cu
Brasses and Bronzes		
Red and leaded red brasses	C83300-C85800	Cu-Zn-Sn-Pb (75–89% Cu)
Yellow and leaded yellow brasses	C85200-C85800	Cu-Zn-Sn-Pb (57–74% Cu)
Manganese and leaded manganese bronzes	C86100-C86800	Cu-Zn-Mn-Fe-Pb
Silicon bronzes, silicon brasses	C87300-C87900	Cu-Zn-Si
Tin bronzes and leaded tin bronzes	C90200-C94500	Cu-Sn-Zn-Pb
Nickel/tin bronzes	C94700-C94900	Cu-Ni-Sn-Zn-Pb
Aluminum bronzes	C95200-C95810	Cu-AI-Fe-Ni
Others		
Copper/nickels	C96200-C96800	Cu-Ni-Fe
Nickel silvers	C97300-C97800	Cu-Ni-Zn-Pb-Sn
Leaded coppers	C98200-C98800	Cu-Pb
Miscellaneous alloys	C99300-C99750	

a silicon bronze (C65100, C65500, or C66100), possesses relatively high strength, and this property is utilized in hardware such as nuts, bolts, and valve stems for copper-base equipment, thus minimizing two-metal corrosion.

Copper and brasses are subject to erosion–corrosion or impingement attack. The bronzes and aluminum brass are much better in this respect. The bronzes are stronger and harder. The cupronickels with small iron additions are also superior in erosion-corrosion resistance.

Copper and copper alloys are available as duplex tubing or plate (one metal inside, another outside) in combination with steel, aluminum, and stainless steels. This construction solves many heat-exchanger materials problems, for example, steel tubing carrying ammonia and immersed in brackish water (admiralty metal).

Ammonia, or amines that may break down to give ammonia, generally are considered to be environments active in the cracking of brass and bronze. The resistance of brass to SCC is considered to increase as its copper content increases. Hence, a red brass (copper content 85%) generally would be more resistant than a yellow brass (copper content approximately 65%).

However, under certain conditions, even commercial copper tubing can fail by SCC. Two cases of failure have been reported in tubing used to connect homes with gas mains in the street. The copper tubing (deoxidized with and containing 0.033% phosphorus) had been supplied in rolls, and the stresses were believed to result from straightening out the rolled material or from rerolling the tubing after too much had been unrolled for use at a time. The corrodent would come from decayed vegetable matter in the soil. Fortunately, failures in installations such as these are rare (e.g., only four or five failures have occurred in more than 50,000 installations).

Susceptibility of cartridge brass (C26000) to SCC was studied extensively during World Wars I and II. Small-arms cartridge cases made from this alloy are deep drawn in a series of steps with intermediate anneals to remove residual stresses. These anneals are expensive but seem to be necessary to prevent the metal from cracking from excessive deformation and to reduce susceptibility to SCC.

8.4.9 Nickel and Its Alloys

An important group of materials for corrosion applications is based on nickel. Nickel and its alloys are resistant to many corrosives and are a natural choice for handling alkaline solutions. Most tough corrosion problems involving caustic solutions are handled with nickel alloys. In fact, the corrosion resistance of alloys to sodium hydroxide is roughly proportional to their nickel content. For example, 2% Ni cast iron is much superior to unalloyed cast iron.

Commercially pure nickel (N02200) is rarely used in service waters whereas Monel 400 (N04400) has been widely used in both fresh water and seawater. It is less resistant to fouling by marine organisms than the copper-base alloys. As with many other materials, crevices must be avoided to prevent pitting in water containing dissolved oxygen. However, Monel has been used for salt-water evaporators, with the cupronickels used for piping and heat-exchanger tubing.

Nickel generally shows good resistance to neutral and slightly acid solutions. It is widely used in the food industry. It is not resistant to strongly oxidizing solutions, for example, nitric acid and ammonia. Nickel forms a good base for alloys requiring high strength at high temperatures. However, nickel and its alloys are attacked and embrittled by sulfur-bearing gases at elevated temperatures.

In the chromium-free alloys, the oxide formed on nickel is NiO. In low-temperature waters, a thin, barely visible yellow tint develops on nickel. At boiler-water temperatures, the oxide film on nickel is dark gray and slightly powdery.

In chromium containing nickel alloys, the spinel oxides (M_3O_4), where M is a metal atom such as Cr, Ni, or Fe, are formed in high-temperature water. These alloys without molybdenum, such as Inconel (N06600) and Incoloy (N08800), are prone to pitting in chloride-bearing waters. The molybdenum-bearing grades such as Incoloy 825 (N08825) are a substantial improvement, and the high-molybdenum grades such as Alloy C-276 (N10276) and Alloy 625 (N06625) are resistant even to hot seawater.

Table 8.15 provides a generic classification and composition of high-performance nickel, nickel-iron, and cobalt-base alloys. A wide range of compositions and mechanical properties is available. Aged Duranickel (N03301) possesses very good mechanical properties and also good resistance to many environments. Monel is a natural for HF acid. Chlorimet 3 (N30107) and Hastelloy C (N10002) are two of the most generally corrosion-resistant alloys commercially available. Chlorimet 2 (N30007) and Hastelloy B (N10001) are very good in many cases where oxidizing conditions do not exist. Nichrome (N06003) is used for electrical resistors (heating elements). Nickel, high-nickel alloys, and alloys containing substantial amounts of nickel (8%) are the most common materials used for most of the more severe corrosion problems.

Alloy 200 (N02200) and Monel 400 (N04400) can suffer SCC when exposed to very hot, concentrated caustic (e.g., 50% caustic at more than 200°C) but are still an order of magnitude more resistant than the austenitic stainless steels. Nickel is even more resistant than Inconel. For temperatures above about 450°C, a low-carbon grade of nickel (less than 0.02% carbon) is recommended. No instances of SCC of Hastelloy B have been reported.

Monel 400 is also susceptible to SCC by the vapors of hydrofluoric acid in the simultaneous presence of oxygen. Apparently, cracking does not occur when the metal is fully immersed. A blanketing nitrogen atmosphere may be employed to prevent ingress of atmospheric moisture and oxygen and militate against SCC. However, moderate general corrosion and copper deposition may occur simultaneously with the SCC. Another important attribute is the large and rapid increase in SCC resistance as the nickel content of stainless alloys exceeds 30%. For example, Inconel shows excellent SCC resistance.

8.4.10 Zinc and Its Alloys

About three-fourths of zinc produced is consumed as metal, mainly as a coating to protect iron and steel from corrosion (galvanized metal), as an alloying metal to make bronze and brass, as zinc-based die-casting alloy, and as rolled zinc. The remaining one-fourth is consumed as zinc compounds mainly by the rubber, chemical, paint, and agricultural industries. Zinc is also a necessary element for proper growth and development of humans, animals, and plants; it is the second-most-common trace metal, after iron, naturally found in the human body.

Zinc is employed primarily in the form of hot-dip coating (i.e. galvanizing) or electroplate. Useful in cold-water applications, its potential application for hot-water systems should be carefully evaluated. In waters of a particular chemical makeup and at temperatures of approximately 80°C, zinc may become cathodic to steel and cause accelerated corrosion. Dissolved oxygen and carbon dioxide aggravate the corrosion of zinc, as do extremes of pH. Zinc is amphoteric, that is, attacked by both free acid and free alkali.

Table 8.15 Generic Classification and Composition of High-Performance Nickel, Nickel-Iron, and Cobalt-Base Alloys

Alloy	UNS	C	Cr	Ni	Co	Fe	Mo	W	Other
263	N07041	0.06	20	Rem	20	—	5.8	—	A10.5, Ti 2.2
20Cb-3	N08020	0.02	20	33	—	Rem	2.2	—	Cu 3.3, Cb 0.5
20Mo-4	N08024	0.02	23.5	37	—	Rem	3.8	—	Cu 1.0, Cb 0.25
20Mo-6	N08026	0.02	24	36	—	Rem	5.6	—	Cu 3.0
625 Plus	N07716	0.02	20	Rem	—	5	9	—	Cb 3.1, Al 0.2, Ti 1.3
Alloy 150(UMCo-50)	—	0.06	27	—	Rem	18	—	—	—
Alloy 188	R30188	0.1	22	22	Rem	3	—	14	La 0.04
Alloy 214	N07214	0.04	16	Rem	—	3	—	—	Al 4.5, Y
Alloy 230	N06230	0.1	22	Rem	—	3	2	14	La 0.02, B O.Ol5
Alloy 242	—	0.03	8	Rem	—	—	25	—	—
Alloy 25 (L-605)	R30605	0.1	20	10	Rem	3	—	15	—
Alloy 556	R30556	0.1	22	20	18	Rem	3	2.5	T a 0.6, L a 0.02, N 0.2, Z r 0.02
Alloy 6B	R30016	1.2	30	—	Rem	1.5	4.5	—	—
Alloy HR- 120	—	0.05	25	37	—	Rem	—	—	Cb 0.7, N 0.2
Alloy HR-160	—	0.05	28	Rem	29	1.5	—	—	Si 2.75
AR 213	—	0.17	19	—	Rem	—	—	4.5	Al 3.5, Ta 6.5, Zr 0.15, Y 0.10
Astroloy	—	0.06	15	Rem	17	—	5.3	—	Al 4.0, Ti 3.5, B 0.03
Chromel D	—	—	18.5	36	—	Rem	—	—	Si 1.5
Cupro 107	—	—	—	Rem	—	0.8	—	—	Cu 68.0, Mn 1.1
D-979	N09979	0.05	15	Rem	—	27	4	4	A1 1.0, Ti 3.0
Discaloy	K66220	0.06	14	26	—	Rem	3	—	Al 0.25, Ti 1.7
Fecralloy A	—	0.03	15.8	—	—	Rem	—	—	Al 4.8, Y 0.3
Ferry alloy	—	—	—	Rem	—	—	—	—	Cu 55.0
Hastelloy B	N10001	0.05	—	Rem	2.5	5	28	—	V 0.03
Hastelloy B-2	N 10665	0.01	—	Rem	—	2	28	—	—
Hastelloy C	N10002	0.08	15.5	Rem	2.5	6	17	4	—

Hastelloy C-22	N06022	0.01	22	Rem	2.5	3	13	3	—
Hastelloy C-276	N10276	0.01	15.5	Rem	2.5	5.5	16	4	—
Hastelloy C-4	N06455	0.01	16	Rem	2	3	15.5	—	—
Hastelloy G	N06007	0.05	22	Rem	—	19.5	6.5	—	Cb + Ta 2.0, Cu 2.0
Hastelloy G-3	N06985	0.015	22	Rem	5	19.5	7	1.5	Cb + Ta 0.3, Cu 2.0
Hastelloy G-30	—	0.03	29.5	Rem	5	15	5	2.5	Cu 2.0
Hastelloy H-9M	—	0.03	22	Rem	5	19	9	2	—
Hastelloy N	N 10003	0.06	7	Rem	—	5	16.5	-	—
Hastelloy S	N06635	0.02	15.5	Rem	—	3	14.5	..	La 0.05, B O.Ol5
Hastelloy W	N10004	0.12	5	Rem	2.5	6	24	—	—
Hastelloy X	N06002	0.1	22	Rem	1.5	18.5	9	0.6	—
IN 100	N13100	0.15	10	Rem	15	—	3	—	Al 5.5, Ti 4.7, Zr 0.06, V 1.0, B 0.015
IN 100 Gatorize	—	0.07	12.4	Rem	18.5	—	3.2	—	Al 5.0, Ti 4.3, Zr 0.06, B0.02, V 0.8
IN 102	N06102	0.06	15	Rem	—	7	3	3	Cb 3.0, Al 0.4, Ti 0.6, Mg0.02, Zr 0.03
IN 587	—	0.05	28.5	Rem	20	—	—	—	Cb 0.7, Al 1.2, Ti 2.3, Zr 0.05
IN 597	—	0.05	24.5	Rem	20	—	1.5	—	Cb 1.0, Al 1.5, Ti 3.0, Zr 0.05
Incoloy 800	N08800	0.05	21	32.5	—	Rem	—	—	A 1 0.3, Ti 0.3
Incoloy 800H	N08810	0.08	21	32.5	—	Rem	—	—	Al 0.4, Ti 0.4
Incoloy 800HT	N08811	0.08	21	32.5	—	Rem	—	—	Al + Ti 1.0
Incoloy 801	—	0.05	20.5	32	—	Rem	—	—	Ti 1.1
Incoloy 802	N08802	0.4	21	32.5	—	Rem	—	—	—
Incoloy 803	—	0.5	27	35	—	Rem	—	—	Ti 0.5
Incoloy 825	N08825	0.03	21.5	Rem	—	30	3	—	Cu 2.2
Incoloy 903	N19903	—	—	38	15	Rem	—	—	Ti 1.4, A 1 0.9, C b 3.0
Incoloy 904	N19904	—	—	32.5	14.5	Rem	—	—	Ti 1.6
Incoloy 907	N19907	—	—	38	13	Rem	—	—	Ti 1.5, Cb 4.7, Si 0.15

(Continued)

Table 8.15 Continued

Alloy	UNS	C	Cr	Ni	Co	Fe	Mo	W	Other
Incoloy 909	N19909	—	—	38	13	Rem	—	—	Ti 1.5, Cb 4.7, Si 0.4
Incoloy 925	N09925	0.01	21	Rem	—	28	3	—	Cu 1.8, Ti 2.1, A1 0.3
Incoloy DS	—	0.06	17	35	—	Rem	—	—	Si 2.3
Inconel 600	N06600	0.08	15.5	Rem	—	8	—	—	—
Inconel 601	N06601	0.1	23	Rem	—	14.4	—	—	Al 1.4
Inconel 617	N06617	0.07	22	Rem	12.5	1.5	9	—	Al 1.2
Inconel 625	N06625	0.1	21.5	Rem	—	2.5	9	—	Cb 3.6
Inconel 671	—	0.05	48	Rem	—	—	—	—	Ti 0.35
Inconel 690	N06690	0.02	29	Rem	—	9	—	—	—
Inconel 706	N09706	0.03	16	Rem	—	37	—	—	Ti 1.8, Al 0.2, Cb 2.9
Inconel 718	N07718	0.04	18	Rem	—	18.5	3	—	Cb 5.1
Inconel 751	N07751	0.05	15	Rem	—	7	—	—	Ti 2.5, Al 1.1, Cb 1.0
Inconel X-750	N07750	0.04	15.5	Rem	—	7	—	—	Ti 2.5, Al 0.7, Cb 1.0
Kanthal AF	—	—	22	—	—	Rem	—	—	Al 5.3, Y
Kanthal Al	K92500	—	22	—	—	Rem	—	—	Al 5.8
M 252	N07252	0.15	20	Rem	10	—	10	—	Al 1.0, Ti 2.6, B 0.005
MAR-M 918	—	0.05	20	20	Rem	—	—	—	Ta 7.5, Zr 0.10
Monel 400	N04400	—	—	Rem	—	1.2	—	—	Cu 31.5, Mn 1.1
Monel 401	N04401	—	—	Rem	—	0.3	—	—	Cu 55.5, Mn 1.63
Monel 450	C71500	—	—	Rem	—	0.7	—	..	Cu 68.0, Mn 0.7
Monel K-500	N05500	—	—	Rem	—	1	—	—	Cu 29.5, Ti 0.6, Al 2.7
Monel R-405	N04405	—	—	Rem	—	1.2		—	Cu 31.5, Mn 1.1
MP 159	—	—	19	25.5	Rem	9	7	—	Cb 0.6, A1 0.2, Ti 3.0
MP 35N	R30035	—	20	35	Rem	—	10	—	—
Multimet (N-155)	R30155	0.1	21	20	20	Rem	3	2.5	Cb + Ta 1.0, N 0.15
NA 224	—	0.5	27	Rem	—	18.5	—	6	—
Ni 200	N02200	0.08	—	99.6	—	—	—	—	—

Ni 201	N02201	0.02	—	99.6	—	—	—	—	—
Ni 270	N02270	0.01	—	99.98	—	—	—	—	—
Nichrome 80	—	—	20	Rem	—	—	—	—	Si 1.0
Nimonic 105	—	0.08	15	Rem	20	—	5	—	Al 4.7, Ti 1.3, B 0.005
Nimonic 115	—	0.15	15	Rem	lS.0	—	4	—	A15.0, Ti 4.0
Nimonic 70	—	—	20	Rem	—	25	—	—	Al l.O, Ti 1.25, C b 1.5
Nimonic 75	—	0.1	19.5	Rem	—	—	—	—	—
Nimonic 80A	N07080	0.06	19.5	Rem	—	—	—	—	A1 1.4, Ti 2.4
Nimonic 81	—	0.03	30	Rem	—	—	—	—	A10.9, Ti 1.8
Nimonic 86	N07090	—	25	Rem	—	—	10	—	C e 0.03
Nimonic 90	—	0.07	19.5	Rem	16.5	—	—	—	Al l.S, Ti 2.5
Nimonic 901	—	—	12.5	Rem	—	36	5.8	—	Ti 2.9
Nimonic 91	—	—	28.5	Rem	20	—	—	—	Al 1.2, Ti 2.3
Nimonic AP 1	—	—	15	Rem	17	—	5	—	Al 4.0, Ti 3.5
Nimonic PE 11	—	0.05	18	Rem	—	34	5.2	—	Al 0.8, Ti 2.3
Nimonic PE 16	—	0.05	16.5	Rem	—	34	3.3	—	A 1 1.2, Ti 1.2
Nimonic PK 31	—	—	20	Rem	14	—	4.5	—	A 1 0.4, Ti 2.35, C b 5.0
Nimonic PK 33	—	0.04	18	Rem	14	—	7	—	A 1 2.1, Ti 2.4
Nimonic PK 37	N07001	—	19.5	Rem	16.5	—	—	—	A 1 1.5, Ti 2.5
Nimonic PK 50	—	—	19.5	Rem	13.5	—	4.25	—	A 1 1.4, Ti 3.0
Pyromet 31	N07031	0.04	22.5	Rem	—	15	2	—	Al 1.4, Ti 2.3, Cu 0.9, B 0.005
Pyromet 860	—	0.05	13	Rem	4	28.9	6	—	Al 1.0, Ti 3.0, B 0.01
Pyromet CTX- I	—	0.03	—	37.7	16	Rem	—	—	Cb 3.0, Al 1.0, Ti 1.7
RA 330	N08330	0.05	19	35	—	Rem	—	—	Si 1.2
RA 330HC	—	0.4	19	35	—	Rem	—	—	Si 1.2
RA 333	N06333	0.05	25	Rem	3	18	3	3	—
Refractory 26	—	0.03	18	Rem	20	16	3.2	—	Al 0.2, Ti 2.6, B 0.015

(Continued)

Table 8.15 Continued

Alloy	UNS	C	Cr	Ni	Co	Fe	Mo	W	Other
René 100	—	0.16	9.5	Rem	15	—	3	—	A15.5, Ti 4.2, Z r 0.06, B 0.015
René 41	—	0.09	19	Rem	11	5	10	—	Al 1.5, Ti 3.0, B 0.006
René 95	—	0.15	14	Rem	8	—	3.5	3.5	C b 3.5, Al 3.5, Ti 2.5, Zr 0.05
S-816	R30816	0.38	20	20	Rem	4	4	4	Cb 4.0
Sanicro 28	N08028	0.01	27	31	—	Rem	3.5	—	Cu 1.0
Udimet 400	—	0.06	17.5	Rem	14	—	4	—	Cb 0.5, Al l.S, Ti 2.5, Z r 0.06, B 0.008
Udimet 500	—	0.08	18	Rem	18.5	—	4	—	— Al 2.9, Ti 2.9, Z r 0.05, B 0.006
Udimet 520	—	0.05	19	Rem	12	—	6	1	Al 2.0, Ti 3.0, B 0.005
Udimet 630	—	0.03	18	Rem	—	18	3	3	C b 6.5, A 1 0.5, Ti 1.0
Udimet 700	—	0.03	15	Rem	18.5	—	5.2	—	A 1 5.3, Ti 3.5, B 0.03
Udimet 710	—	0.07	18	Rem	15	—	3	1.5	Al 2.5, Ti 5.0
Udimet 720	—	0.03	17.9	Rem	14.7	—	3	1.3	A 1 2 5, Ti 5.0, Z r 0.03, B
Ultimet	—	0.06	26	9	Rem	3	5	2	N 0.08
Unitemp AF2-1DA	—	0.35	12	Rem	10	—	3	6	Ta 1.5, Al 4.6, Ti 3.5, Zr 0.10
Unitemp AF2-1DA6	—	0.04	12	Rem	10	—	2.7	6.5	Ta 1.5, Al 4.0, Ti 2.8, Zr 0.1, B 0.015
V-57	—	0.08	14.8	27	—	Rem	1.25	—	Al 0.25, Ti 3.0, V 0.5, B 0.01
W-545	K66545	0.08	13.5	26	—	Rem	1.5	—	Al 0.2, Ti 2.85, B 0.05
Waspaloy	—	0.08	19	Rem	14	—	4.3	—	A 11.5, Ti 3.0, Z r 0.05, B 0.006

Zinc alloy parts are made by die casting because of their low melting points. Many automobile components such as grilles and door handles are die cast, but are usually plated with corrosion-resistant metals.

In a normal atmosphere, zinc forms a basic zinc-carbonate film that greatly retards its corrosion rate, which is similar to the aluminum-oxide film that forms on aluminum and which accounts for its low corrosion rate. Zinc, when connected with metals below it in the electrochemical series, will sacrificially protect that metal, which accounts for its wide usage in the galvanized-steel industry. It is also utilized in the form of bars or slabs as sacrificial anodes to protect ship hulls, pipe lines, and other structures.

8.4.11 Tin and Tinplate

More then half of the tin produced goes into coating other metals, primarily steel to make the "tin" can. In addition to offering corrosion resistance, tin-coated steel is easily formed and soldered, provides a good base for organic coatings, is nontoxic, and has a pleasant appearance. Tin is a weak, soft, and ductile metal. Tensile strength at room temperature is around 17 MPa (2500 psi), and this drops rapidly with increasing temperature.

Tin coating can be applied by dipping, but most of it is electroplated. Two advantages of the latter method are controlled thinness and the fact that steel can be plated with a different thickness on each side of a sheet, usually thicker on the inside of the can than on the outside. Tin cans are used for food products, beverages, petroleum products, and paints. Alloys of tin with zinc, nickel, cadmium, or copper can be deposited as electroplates.

Tin is cathodic to iron in most common environments, but the potential reverses in most sealed cans containing food products; tin acts as a sacrificial coating, thus protecting the steel. Complex ion formation apparently causes this reversal. Tin is relatively inert, but in the presence of oxygen or other oxidizing agents, it is attacked.

Tin shows excellent resistance to relatively pure water. Solid tin pipe and sheet and also tinned copper are utilized for producing and handling distilled water. Collapsible tubes have been used for dentifrices and medicants. Solders are another important outlet for tin in which tin and lead contents vary from 30 to 70% . Tin babbits are used for bearings.

Tin shows good resistance to atmospheric corrosion, dilute mineral acids in the absence of air, and many organic acids but is corroded by strong mineral acids. It is generally not used for handling alkalies.

8.4.12 Cadmium

Cadmium is a relatively weak metal with a tensile strength around 70 MPa (10,000 psi) and elongation of 50%. Except for low-melting alloys and some bearings, cadmium is used almost exclusively as an electroplated coating. Its bright appearance and ease of soldering accounts for its use in electronic equipment and hardware. Cadmium is less electronegative than zinc and not as effective as a sacrificial anode in many cases.

It is somewhat less protective than zinc for steel in industrial atmospheres. Cadmium is more expensive than zinc and its salts are toxic. Unlike zinc, cadmium has some measure of resistance to alkalies.

Cadmium plating is utilized on high-strength steels in aircraft because of improved resistance to corrosion fatigue. However, hydrogen embrittlement can be a problem. Proper plating techniques and baking out the hydrogen after plating have solved this problem. At temperatures in the neighborhood of the melting point (321°C), steel can be attacked by cadmium.

8.4.13 Titanium and Its Alloys

Titanium is a relative newcomer in that it was first used as a structural metal in 1952. It is strong and has a specific gravity of 4.5 g cm^{-3}, about halfway between aluminum and steel. Its strength to-weight ratio makes it an excellent choice for aircraft and ordnance, where it was first introduced as a structural material. Titanium is now widely utilized in the space and chemical process industries.

Titanium is a reactive metal and depends on a protective film (TiO_2) for corrosion resistance. Melting and welding must be done in inert environments or the metal becomes brittle because of absorbed gases. Titanium has three outstanding characteristics that account for many of its applications in corrosive services, such as

- seawater and other chloride-salt solutions,
- hypochlorite and wet chlorine, and
- nitric acid, including fuming acids.

Salts such as $FeCl_3$ and $CuCl_2$, which tend to pit most other metals and alloys, actually inhibit corrosion of titanium. Titanium is not resistant to relatively pure sulfuric and hydrochloric acids, but does a good job in many of these acids when they are heavily contaminated with heavy-metal ions such as ferric and cupric.

Titanium is highly resistant to aggressive waters, even more so than zirconium. It develops a tarnish film of TiO_2 only after long-time exposure to steam at temperatures of approximately 400°C. Titanium is an excellent choice for heat-exchanger tubing in seawater-cooled exchangers, provided that the process side is also compatible. Titanium is catastrophically attacked in red fuming nitric acid with high NO_2 and low water content and also in dry halogen gases.

Alloying with about 30% Mo greatly increases resistance to hydrochloric acid. Small amounts of tin reduce scaling losses during hot rolling. Small additions of palladium, platinum, and other noble metals increase resistance to moderately reducing acids. One such commercial titanium alloy contains about 0.15% palladium. Other commercial alloys contain aluminum, chromium, iron, manganese, molybdenum, tin, vanadium, and zirconium.

Table 8.16 lists mechanical properties of titanium and a few commercial alloys. The alloys fall into three metallurgical types, as indicated in the table. Titanium has a low modulus of elasticity (115 GPa or 16,800,000 psi) as compared with iron (200 GPa or 28,500,000 psi).

Table 8.16 Mechanical Properties of Titanium Alloys

UNS	ASTM	Trade Name	Tensile Strength (MPa)	Yield Strength (MPa)	Elastic Modulus (GPa)
R50250	1	Unalloyed Ti	241	172	103
R50400	2	Unalloyed Ti	345	276	103
R50550	3	Unalloyed Ti	448	379	103
R60700	4	Unalloyed Ti	552	483	103
R56400	5	Ti-6Al-4V	896	827	113
	6	Ti-5Al-2.5Sn	827	793	110
R52400	7	Ti-0.15Pd	345	276	103
R56320	9	Ti-3Al-2.5V	620	483	90
	10	Ti-11.5Mo-6Zr-4.5Sn	689	620	103
R52250	11	Ti-0.15Pd	241	172	103
R53400	12	Ti-0.3-Mo-0.8Ni	483	345	103
	13	Ti-0.5Ni-0.05Ru	276	172	103
	14	Ti-0.5Ni-0.05Ru	414	276	103
	15	Ti-0.5Ni-0.05Ru	483	379	103
R52402	16	Ti-0.05Pd	345	276	103
R52252	17	Ti-0.05Pd	241	172	103
R56322	18	Ti-3Al-2.5V-0.05Pd	620	483	105
R58640	19	Ti-3Al-8V-6Cr-4Zr-4Mo	793	758	103
R58645	20	Ti-3Al-8V-6Cr-4Zr-4Mo-0.05Pd	793	758	103
R58210	21	Ti-15Mo-2.7Nb-3Al-0.25Si	793	758	103
	23	Ti-6Al-4V ELI[a]	793	758	112
	24	Ti-6Al-4V-0.05Pd	896	827	113
	25	Ti-6Al-4V-0.5Ni-0.05Pd	896	827	113
	26	Ti-0.1Ru	345	276	103
	27	Ti-0.1Ru	241	172	103
	28	Ti-3Al-2.5V-0.1Ru	620	483	90
	29	Ti-6Al-4V-0.1Ru	827	758	112

[a] Extra-low interstitial.

The first SCC problem developed as the result of contact of one titanium alloy with an organic chloride at a temperature above 290°C. The chloride was believed to have broken down to form HCl. It was found subsequently that SCC developed in highly stressed alloy specimens coated with sea salt, chemically pure sodium chloride, or with a number of other chlorides (barium, cesium, lithium, or strontium) and heated to 400 to 430°C in a laboratory furnace.

To avoid the possibility of failure from high-temperature chloride cracking, it was found that titanium-alloy parts should be cleaned after fabrication and prior to heat treatment in chloride-free solvents. It has even been suggested that perspiration or even fingerprints have caused failures of material subsequently heat treated.

SCC has also been reported in tanks containing liquid nitrogen/tetroxide (N_2O_4) or methyl alcohol. Small amounts of water have inhibited SCC of titanium alloys in the methyl alcohol. SCC has occurred in a much shorter time in specimens of commercially pure titanium and in an alloy exposed to alcohol vapor than in those completely submerged in the liquid. Methyl alcohol is much more potent in producing cracking than the higher alcohols such as ethyl alcohol.

The mechanisms of SCC of titanium alloys in hot salt, N_2O_4, or methyl alcohol are complex and difficult to explain. Cracking of the alloy in chlorides and methanol at room temperature probably can be considered an example of the mechanism of hydrogen-induced cracking.

8.4.14 Refractory Metals

Refractory metals are characterized by very high melting points as compared with iron and steel. The jet engine and outer-space programs have provided the impetus in making these metals commercially available. Unfortunately, they show poor resistance to high-temperature oxidation, and protective coatings are needed. Niobium, molybdenum, tungsten, and zirconium are relative newcomers in the field of corrosion by aqueous solutions. Tantalum has long been used for severely corrosive conditions. Some typical properties of these metals are shown in Table 8.17.

8.4.14.1 Niobium

Niobium exhibits good corrosion resistance to organic and inorganic acids except HF and hot, concentrated sulfuric and hydrochloric acids. Apparently the formation of a Nb_2O_5 film results in protection. Niobium is poor in alkaline solutions.

8.4.14.2 Molybdenum

Molybdenum shows good resistance to HF, hydrochloric, and sulfuric acids, but oxidizing agents such as nitric acid cause rapid attack. It is good in aqueous alkaline solutions. This metal forms a volatile oxide (MoO_3) in air at temperatures above about 700°C.

One great advantage of molybdenum from the mechanical standpoint is its high modulus of elasticity, which is 345 GPa (50,000,000 psi). This means it is much stiffer and will deflect much less than steel under a given load and cross section.

8.4.14.3 Tantalum

Tantalum has been used for many years because of its superior resistance to most environments. A few exceptions include alkalies and HF and hot, concentrated sulfuric acid. It is used in handling chemically pure solutions of such corrosives as hydrochloric acid. Because of its excellent corrosion resistance across a wide spectrum of environments, tantalum is utilized in repair of glass-lined equipment.

Any evolution of hydrogen (corrosion reaction or otherwise) near tantalum will result in absorption and embrittlement of this metal. Desorption of hydrogen to restore ductility is not practical because of the high temperature and high vacuum required. Tantalum sheet is strong, so cost is reduced through the use of thin sections. Tantalum is also used for surgical implants.

Table 8.17 Typical Properties of Molybdenum, Niobium, Tantalum, and Tungsten

	Unit	Mo	Nb	Ta	W
Property					
Atomic number		42	41	73	74
Atomic weight	(g mol^{-1})	95.95	92.91	180.95	183.86
Atomic radius	(nm)	0.1363	0.1426	0.143	0.1371
Lattice type		bcc	bcc	bcc	bcc
Lattice constant; 20°C	(nm)	0.31468	0.3294	0.33026	0.31585
Mass					
Density at 20°C	(g cm^{-3})	10.2	8.57	16.6	19.3
Thermal properties					
Melting point,	(°C)	2610	2468	2996	3410
Boiling point, °C	(°C)	5560	4927	6100	5900
Linear coefficient of expansion	per °C	4.9×10^{-6}	7.1×10^{-6}	6.5×10^{-6}	4.3×10^{-6}
Thermal conductivity at 20°C.	W m^{-1} K^{-1}	147	219	54	167
Specific Heat, 20°C	(J kg^{-1} K^{-1})	255	525	151	134
Electrical properties					
Conductivity	% IACS (Cu)	30%	13.2%	13%	31%
Resistivity 20°C	$\mu\Omega$-cm	5.7	15	13.5	5.5
Coefficient of resistivity	per °C (0–100°C)	0.0046		0.0038	0.0046
Mechanical properties					
Tensile strength, 20°C	(MPa)	700–1,400	195	240–500	700–3,500
500°C	(MPa)	240–450		170–310	500–1,400
1000°C	(MPa)	140–210		90–120	350–500
Young's modulus, 20°C	(GPa)	320	103	190	410
500°C	(GPa)	280		170	380
1000°C	(GPa)	270		150	340
Working temperature	(°C)	1600		Room	1700
Recrystallizing temperature	(°C)	900–1200	800–1100	1000–1250	1200–1400
Stress-relieving temperature	(°C)	800		850	1100

8.4.14.4 Tungsten

As indicated in Table 8.17, tungsten has the highest melting point of all metals. Its chief use involves strength at high temperatures. A common example is filaments in light bulbs. Tungsten shows good resistance to acids and alkalies but it is not commonly used for handling aqueous solutions. Tungsten exhibits a tensile strength of 138 MPa (20,000 psi) at 1649°C.

8.4.14.5 Zirconium

Zirconium is occasionally specified for seawater-cooled exchangers where the process side is incompatible with titanium or Alloy C-276 (N10276). Zirconium possesses a tensile strength of about 110 MPa (16,000 psi) at 427°C and 550 MPa (80,000 psi) at room temperature. Its modulus of elasticity is 95 GPa (13,700,000 psi).

Zirconium exhibits good corrosion resistance to alkalies and acids (including hydriodic and hydrobromic acids), except for HF acid and hot, concentrated hydrochloric and sulfuric acids. Cupric and ferric chlorides cause pitting. Zirconium has found some application in hydrochloric-acid service. Its corrosion resistance is affected by impurities such as nitrogen, aluminum, iron, and carbon.

Zirconium alloyed with small amounts of tin, iron, chromium, and nickel (Zircaloys) shows improved resistance to high-temperature water. Zirconium and its alloys in high-temperature water first show a decreasing corrosion rate that may be followed by a rapid linear rate of attack (termed breakaway). They also tend to pick up hydrogen from the corrosion reaction and become brittle.

Zirconium has found increased use following the advent of nuclear energy, primarily because it has a low thermal neutron cross section and resists high-temperature water and steam. The greatest use of zirconium is as the tin-bearing alloy Zircaloy-4 (R60704), particularly in demand in the nuclear industry for pressurized-water reactors (PWRs). A bright luster is maintained at temperatures well above boiling. At higher temperatures of about 260°C, a black, shiny film of zirconium oxide develops. The oxide-to-zirconium ratio is slightly less than 2:1, and the film has some metallic properties (e.g., luster and reasonably good heat-transfer properties). After long-time exposure of several years at 260°C or several days at 425°C, the black oxide changes to white, powdery ZrO_2. This film behaves as an insulator, but it less corrosion protective because of its powdery nature.

8.4.15 Noble Metals

Noble materials are characterized by high free-corrosion potentials relative to other metals, excellent corrosion resistance, and high cost. The latter accounts for the term precious metals. In most cases, formation of protective films (passivity) is not required. The noble metals are gold, silver, platinum, and the other five "platinum group" metals: iridium, osmium, palladium, rhodium, and ruthenium. Gold, silver, platinum, and palladium are available in many commercial forms. The first three are used extensively in industry. The other noble metals are used chiefly as alloying elements. Table 8.18 describes the physical and mechanical properties for the platinum group of metals.[8]

The use of noble metals for jewelry is well known. In spite of their high cost, the noble metals are the most economical for numerous corrosion applications. Rhodium and osmium electroplate well and are used for critical valve parts and other applications requiring total resistance to an aggressive environment.

8.4.15.1 Gold

Gold is one of the oldest metals utilized by humans because it is found in nature in its native metallic state. Jewelry and coinage represent its first applications, with jewelry accounting for most of its modern consumption. Gold corrosion performance is excellent in dilute nitric acid and hot, strong sulfuric acid. It is attacked by aqua regia, concentrated nitric acid, chlorine, bromine, mercury, and alkaline cyanides.

Table 8.18 Physical and Mechanical Properties for the Platinum Group of Metals

	Platinum	Iridium	Osmium	Palladium	Rhodium	Ruthenium	Gold	Silver
Chemical symbols	Pt	Ir	Os	Pd	Rh	Ru	Au	Ag
Atomic number	78	77	76	46	45	44	79	47
Atomic weight	195.09	192.22	190.2	106.4	102.9055	101.07	196.967	107.87
Crystal structure[a]	FCC	FCC	HCP	FCC	FCC	HCP	FCC	FCC
Density g/cm^3	21.45	22.65	22.61	12.02	12.41	12.45	19.3	10.5
Melting point C	1769	2443	3050	1554	1960	2310	1064.43	961.93
Vickers hardness (annealed)	41	220	—	41	101	—	20	26
Electrical resistivity at 0 C microohm·cm	9.85	4.71	8.12	9.93	4.33	6.80	2.4	1.6
Thermal conductivity Watts/m/C	73	148	87	76	150	105	293	419
Tensile strength (annealed) tons per square inch	9	71	—	11	45	36	7	9

[a]FCC, face-centered cubic; HCP, hexagonal close-packed.

For jewelry applications, gold is alloyed with copper because of the extreme softness of pure gold. A karat is 1/24 part, so pure gold is 24 karats and 12 karat would be 50% gold. In addition, gold and its alloys are used for dental inlays, electrical contacts, plating, tableware, process equipment (e.g., condensers and stills), printed circuits, gold leaf, and surgical and body implants, and for decorative purposes in signs and displays. A thin gold underplate (topped with cadmium) prevents hydrogen embrittlement of high-strength steels.

8.4.15.2 *Platinum*

Platinum is highly resistant to many oxidizing environments and particularly to air at high temperatures. Platinum exhibits good mechanical properties at high temperatures, as shown in Table 8.18. Gold and silver, however, are poor in this respect.

Platinum is also used for thermocouples (Pt-PtRh), tanks, spinnerets for molten glass, crucibles for chemical analytical work, windings for electrical-resistance furnaces (alloys up to 1760°C), and combustion or reaction chambers for extremely corrosive environments at temperatures greater than 982°C. Sheet linings on a base alloy substrate are used for the latter. An applied oxide layer (e.g., Al_2O_3) prevents alloying or reaction with the substrate material. It has replaced fused quartz in several chemical applications of this type. Incidentally, contact with silica at very high temperature embrittles the platinum. The inertness of platinum is attested to by its extensive service as a catalyst.

Other uses for platinum and its alloys are spinnerets (70 Au, 30 Pt) for rayon, safety or frangible disks up to 482°C (silver is good only up to 149°C and gold to 71°C, sulfuric-acid absorbers, electroplating anodes, impressed-current anodes for CP, other chemical equipment, and high-quality jewelry. It is not attacked by mercury.

Platinum is attacked by aqua regia, hydriodic and hydrobromic acids, ferric chloride, chlorine, and bromine.

8.4.15.3 *Silver*

Silver is best known for its use as coinage and tableware in solid and plated form. Sterling silver contains not more than 7.5% copper for hardness. Silver loses its "nobility" in contact with sulfur compounds and is evident by tarnishing. It serves as electrical contacts, electrical bus bars (even at red heat), brazes, solders, and dental alloys with mercury. Silver is widely used in the chemical industry as solid silver and also as loose, clad, or brazed linings. Applications include stills, heating coils and condensers for pure HF acid, evaporating pans for production of chemically pure anhydrous sodium hydroxide, autoclaves for production of urea, and all kinds of equipment for production of foods and drugs where purity of the product is paramount. It is highly resistant to organic acids.

Silver is attacked by nitric acid, hot hydrochloric acid, hydriodic and hydrobromic acids, mercury, and alkaline cyanides, and may be corroded by reducing acids if oxidizing agents are present.

8.5 POLYMERIC MATERIALS

Production and utilization of polymeric materials have increased tremendously during the past 50 years. One of the early incentives for development of these materials was the need for an inexpensive substitute for ivory billiard balls. These materials are now available in a great variety of forms and can be produced by a multitude of processes such as casting, molding, extrusion, and calendering. They are available as solid parts, linings, coatings, foams, fibers, and films.

When compared with metals, polymeric materials are generally weaker, softer, more resistant to chloride ions and hydrochloric acid, less resistant to concentrated sulfuric and oxidizing acids such as nitric, less resistant to solvents, and most have definitely lower temperature limitations. Cold flow, or creep, at ambient temperatures is a problem, particularly with thermoplastics.

Polymeric materials do not generally dissolve like metals. Degradation or corrosion occurs from three basic sources. Some active species from an environment can be absorbed into the polymer and swell or react internally with the polymer chains. Softening and distortion normally develop, although actual loss of weight from the polymer can occur. Oxidation of the resinous molecule can occur in the atmosphere or other oxidizing conditions. This often results in hardening and cracking of the polymer. Continued polymerization of the resin can also occur with certain resinous components, resulting in hardening, shrinkage, and cracking of the material.

The important aspect of the corrosion mechanism is that degradation is not a surface effect but occurs internally in a polymer. The intrinsic problem with polymers is their capability of selectively absorbing parts per million (ppm) of organic species from industrial waste water to their ultimate detriment. Traces of chlorinated solvents or aromatic hydrocarbons will ultimately attack many elastomers; polypropylene has been destroyed by accumulations of sorbic acid. In industrial wastes, the polymeric material selected should be resistant to a 100% concentration of any contained organic solvents at ambient temperature.

Polymeric materials are inherently resistant to natural waters up to a minimum of about 80°C, or within their inherent temperature and pressure limitations. The selection of materials may start with solid polymeric material (e.g., polypropylene, polyethylene, and PVC), then move to the reinforced polymers (e.g., glass-fiber-reinforced polymer or GFRP) and thence to the polymer-lined steel (e.g., polypropylene or Saran-lined steel) pipe.

The selection of polymeric materials, included elastomeric materials, should be based on a thorough evaluation of the functional requirements for the specific application. Dependent upon application, properties to be documented and included in the evaluation are as follows:

- thermal stability and aging resistance at specified service temperature and environment,
- physical and mechanical properties,
- thermal expansion,
- swelling and shrinking by gas and by liquid absorption,

- gas and liquid diffusion,
- decompression resistance in high-pressure oil/gas systems,
- chemical resistance, and
- control of manufacturing process.

Necessary documentation for all important properties relevant for the design, area and type of application and design life should be thoroughly documented, including results from relevant and independently verified tests, and/or confirmed successful experience in similar design, operational, and environmental situations.

The chemistry and compounding of polymeric materials are quite complex and information is not readily available to consumers. The best procedure is to discuss the situation at hand with suppliers to obtain the proper material for a given problem. If corrosion testing is involved, it is imperative that the specimens be representative of the actual installation. Furthermore, if lining is involved, the test specimens should be covered, as bond failure is sometimes the primary cause of unsatisfactory performance.

8.5.1 Natural and Synthetic Rubbers

The outstanding characteristic of rubber and other elastomers is their resilience in association with a low modulus of elasticity. Flexibility accounts for most applications such as tubing, belting, gasketing, and automobile tires. However, chemical and abrasion resistance and good insulating qualities result in many corrosion applications. Rubber and hydrochloric acid form a natural combination in that rubber-lined steel pipes and tanks have been used for this service for many years.

Generally speaking, the natural rubbers have better mechanical properties (resilience and resistance to cuts and their propagation) than the synthetic or artificial rubbers, but the synthetics have better corrosion resistance.

8.5.1.1 Natural Rubber

Organically speaking, rubber is a long-chain molecule of isoprene (polyisoprene). It comes from trees as a liquid latex. The coil-like structure of these molecules accounts for the high elasticity (100 to 1000% elongation). Soft rubber has a temperature limitation of around 70°C. This temperature limit can be raised to about 82°C by hardening through compounding. Adding sulfur and heating makes the rubber harder and more brittle. Charles Goodyear first discovered this process, called vulcanization, in 1839.

The addition of 50% sulfur results in a hard product, known as ebonite, which is used to make bowling balls. Semihard and hard rubbers are used for tires, tank linings, and many other items. Corrosion resistance usually increases with hardness. In the case of lining pipes and tanks, the rubber is usually applied soft and then cured in place (for large items) or in autoclaves. Modulus of elasticity varies roughly from 3 to 3500 MPa (500 to 500,000 psi) for soft and hard rubber, respectively.

8.5.1.2 Synthetic Rubbers

Neoprene was developed in the early 1930s by DuPont, but it was in small supply at the beginning of World War II, when the main sources of natural rubber became inaccessible. Neoprene was to become one of the five strategic materials in World War II; the other four were metals.

A wide variety of synthetic rubbers is available, including combinations with polymers. Plasticizers, fillers, and hardeners are compounded to obtain a large range of properties, including elasticity and temperature and corrosion resistance. Table 8.19 summarizes the general chemical resistance of plastics and elastomers used in pipe-line construction.[9] Note the variations obtainable in hardness, elongation, tensile strength, and elasticity, and temperature, tear, and corrosion resistance. Neoprene and nitrile rubber possess resistance to oils and gasoline. One of the first extensive applications of neoprene was, and still is, gasoline hoses. Neoprene and nitrile-rubber-lined vessels can also handle pure strong sodium hydroxide.

An outstanding characteristic of butyl rubber is impermeability to gases. This accounts for its use as inner tubes and process equipment such as seals for floating-top storage tanks. Butyl rubber exhibits better resistance to oxidizing environments such as air and dilute nitric acid. The maximum permissible temperatures shown in Table 8.19 are for constant and short-term use in water. Note the value of 300°C for Teflon. The temperature limits listed in Table 8.19 can be considerably reduced in some corrosives. The temperature limit of natural rubber, for example, would be down to room temperature in 70% sulfuric acid.

Soft rubbers are best for their abrasion and erosion-corrosion resistance, a counterintuitive statement for metals-oriented thinkers. Linings also may consist of hard and soft layers. Another high-performing elastomer is a perfluoro compound known as Kalrez. Other fluorinated products were often wrongly considered to be comparable to the polytetrafluoroethylene (TFE) Teflon polymer in chemical stability when first introduced. However, the Kalrez product approaches this ideal and is the first elastomer to provide such a wide range of resistance.

8.5.2 Thermoplastics

Thermoplastic polymers are composed of long chains produced by joining together small molecules, or monomers, that behave in a plastic, flexible manner. These polymers soften and are formed by viscous flow when heated to elevated temperatures. Thermoplastic polymers may be recycled, will soften with increasing temperature, and return to their original hardness when cooled. Most are meltable. Nylon, for example, is extruded into fibers or filaments from the molten state.

Table 8.20 list some important engineering properties of several high-performance polymeric materials.[10] A wide range of properties is available. These properties can be changed considerably through the addition of plasticizers, fillers, and hardeners.

8.5.2.1 ABS

ABS resin is used extensively in the production of pipe for the distribution of gas and water. Many irrigation and sprinkling systems are installed in ABS. The material is available in a high-impact grade that is used widely for telephones, tool housings, automotive parts, and toys.

Table 8.19 Chemical Resistance of Plastics and Elastomers Used in Pipe-Line Construction

Material	Abbreviation	Remarks	Maximum permissible Temperature (Water)°C	
			Constant	Short Term
Polyvinyl chloride	PVC	Resistant to most solutions of acids, alkalis, and salts and to organic compounds miscible with water. Not resistant to aromatic and chlorinated hydrocarbons.	60°	60°
Chlorinated polyvinyl chloride	CPVC	Can be used similarly to PVC but at higher temperatures. Consult factory for specific applications.	90°	110°
High-density polyethylene	PE 50	Resistant to hydrous solutions of acids, alkalis, and salts as well as to a large number of organic solvents. Unsuitable for concentrated oxidizing acids.	60°	80°
Polypropylene, heat stabilized	PP	Chemical resistance similar to that of PE but suitable for higher temperatures.	90°	110°
Polyvinylidene fluoride	PVDF (SYGEF®)	Resistant to acids, solutions of salts, aliphatic, aromatic and chlorinated hydrocarbons, alcohols, and halogens. Conditionally suitable for ketones, esters, organic bases, and alkaline solutions.	140°	150°
Polybutylene-1	PB	Similar to PE 50, but can be used up to 90°C	90°	100°
Polyoxymethylene	POM	Resistant to most solvents and hydrous alkalis. Unsuitable for acids.	60°	80°
Polytetrafluoroethylene (e.g., Teflon®)	PTFE	Resistant to most chemicals	250°	300°
Nitrile rubber	NBR	Good resistance to oil and gasoline. Unsuitable for oxidizing media.	90°	120°
Butyl rubber ethylene propylene rubber	IIR EPDM	Good resistance to ozone and weather. Especially suitable for aggressive chemicals. Unsuitable for oils and fats.	90°	120°
Chloroprene rubber (e.g. Neoprene®)	CR	Chemical resistance very similar to that of PVC-U and between that of nitrile and butyl rubber.	80°	110°
Fluorine Rubber (e.g., Viton®)	FPM	Has best chemical resistance to solvents of all elastomers.	150°	200°
Chlorine sulphonyl polyethylene (e.g., Hypalon®)	CSM	Chemical resistance similar to that of EPDM.	100°	140°

®Registered trademark of DuPont Dow Elastomers.

Table 8.20 Engineering Properties of High-Performance Polymeric Materials

Chemical Name	Commercial Name	SPG[1]	Tens[1]	TEB[1]	Flex[1]	CTE[1]	HDT[1]	Tmax[1]	Tcon[1]	UL94[1]	Dist[1]	Cost
Acrylonitrile-butadiene-Styrene	ABS	1.04	45	25	76	9.5	102	60	—	H-B	—	$$
PolyOxy-methylene	Acetal (copolymer)	1.41	65	30	83	9.7	104	82	0.23	H-B	165	$$
Polymethyl-methacrylate	Acrylic	1.18	62	2	97	12.6	71	82	0.56	—	158	$
Chlorinated poly vinyl chloride	CPVC	1.52	57	27	103	6.7	103	93	0.14	V-0	493	$$$
Polybenzlmidazole	Celazole PBI	1.3	159	3	221	2.3	427	343	0.40	V-0	217	$$$$$
Polyoxy-methylene	Delrin (acetal homopolymer)	1.41	76	30	90	8.5	121	82	0.36	H-B	177	$$
Polyimide	Duratron XP	1.4	110	4	138	4.9	360	304	0.22	V-0	276	$$$$$
Polyethylene terephthalate polyester	Ertalyte PET-P	1.41	85	20	124	5.9	116	99	0.29	H-B	152	$$$
Filled (mica) PTFE	Fluorosint 500 PTFE	2.32	8	10	15	4.3	132	260	0.76	V-0	108	$$$$
Ethylene-chlorotrifluoroethylene	Halar ECTFE	1.68	48	200	—	10.1	116	149	0.16	V-0	197	$$$$
Polybutylene teraphalate polyester	Hydex 4101 PBT-P	1.31	65	50	80	8.1	93	105	—	H-B	162	$$$
Nylon with Kevlar® aramid fiber	Hydlar Z	1.16	110	4	159	2.9	243	149	—	—	—	$$$
Polyvinylidene fluoride	Kynar PVDF	1.77	43	50	67	11.9	110	149	0.17	V-0	670	$$$$
Polyamide	MC901 Nylon 6 (cast)	1.15	83	20	110	6.3	93	127	0.25	H-B	197	$$

(Continued)

Table 8.20 Continued

Chemical Name	Commercial Name	SPG[1]	Tens[1]	TEB[1]	Flex[1]	CTE[1]	HDT[1]	Tmax[1]	Tcon[1]	UL94[1]	Dist[1]	Cost
Polyphenylene oxide-styrene	Noryl PPO (30% gf)	1.36	123	—	138	2.5	104	104	—	V-0	209	$$$$
	Noryl PPO (unfilled)	1.08	66	30	93	5.9	129	135	—	V-0	197	$$$
Extruded nylon	Nylatron GS (moly-lube)	1.16	86	25	117	7.2	93	104	0.25	V-2	138	$$
	Nylatron GSM (moly-lube)	1.16	72	30	110	6.3	93	93	—	H-B	158	$$
	Nylatron GSM Blue (oil & moly)	1.16	69	35	103	10.6	—	93	—	—	—	$$
	Nylatron NSM (solid-lube)	1.15	76	20	110	9.0	93	93	—	H-B	158	$$$
Cast nylon	Nyloil (cast)	1.14	72	50	103	6.3	191	110	0.25	—	217	$$
	Nylon (30% gf)	1.35	186	3	269	2.2	93	99	0.25	H-B	209	$$$
	Nylon 6 (cast)	1.15	83	20	110	6.3	93	93	0.25	H-B	197	$$
	Nylon 6/6 (extruded)	1.15	79	50	103	9.9	93	99	0.25	V-2	158	$$
Polychlorotrifluoroethylene	PCTFE (Kel-F)	2.1	39	150	59	12.6	75	177	0.14	—	197	$$$$
Polyetheretherketone	PEEK (15% gf)	1.4	121	3.2	177	2.3	249	249	0.43	V-0	—	$$$$
	PEEK (20% gf)	1.45	130	2.2	190	2.3	304	249	0.43	V-0	—	$$$$
	PEEK (30% cf)	1.41	179	1	262	1.8	288	260	0.92	V-0	13	$$$$
	PEEK (30% gf)	1.54	157	1.6	233	2.2	316	249	0.43	V-0	69	$$$$
	PEEK (bearing)	1.44	76	3	190	3.1	196	252	0.25	V-0	—	$$$$
	PEEK (unfilled)	1.31	110	20	172	4.7	160	249	0.26	V-0	189	$$$$
Polyettrafluorotehylene	PTFE Teflon (25% cf)	2.08	13	75	16	10.8	66	260	0.65	V-0	—	$$$
	PTFE Teflon (25% gf)	2.25	14	270	13	11.5	66	260	0.45	V-0	—	$$$
	PTFE Teflon (unfilled)	2.16	27	300	—	13.5	56	260	0.25	V-0	112	$$$
Chlorinated polyvinyl chloride	PVC	1.41	48	29	83	11.0	80	60	0.13	V-0	214	$
	Phenolic CE Canvas	1.37	83	—	131	2.0	—	125	0.29	H-B	217	$$
	Phenolic G-10	1.8	310	—	517	1.1	—	141	0.29	H-B	315	$$

	Phenolic G-11	1.8	296	—	552	1.3	—	166	0.29	H-B	355	$$$
	Phenolic G-5/G-9	1.85	303	—	345	1.4	—	141	0.29	V-0	118	$$
	Phenolic G-7	1.8	138	—	207	1.3	—	221	0.29	H-B	138	$$$
	Phenolic LE linen	1.34	90	—	152	1.8	—	141	0.29	H-B	246	$$
	Phenolic XX paper	1.35	138	—	138	1.4	—	125	0.29	H-B	296	$$
	Polycarbonate (30% gf)	1.43	131	10_	159	2.2	146	132	0.19	H-B	185	$$$
	Polycarbonate	1.2	65	60_	103	7.0	132	121	0.19	H-B	154	$$
	Polyethylene - HDPE	0.95	32	900_	—	20.3	71	82	—	—	—	$
	Polyethylene - LDPE	0.92	14	600	—	20.7	43	71	—	—	—	$
	Polyethylene - UHMW	0.93	21	1200	—	20.0	96	82	0.42	H-B	355	$
	Polypropylene (copolymer)	0.9	33	12	48	11.2	99	82	0.12	H-B	227	$
	Polypropylene (homopolymer)	0.89	33	23	37	11.9	78	77	0.12	—	187	$
	Polystyrene (gen purp)	1.05	52	47	42	7.2	93	66	—	H-B	24	$
	Polystyrene (high–impact)	1.04	28	55	60	7.6	91	60	—	H-B	18	$
PAS polyarylsulfone	Radel A PES	1.37	84	15	111	4.9	204	160	0.16	V-0	150	$$$
	Radel R PAES	1.3	76	30	107	5.6	207	149	0.35	V-0	142	$$$
Cross-linked polystyrene	Rexolite 1422 (unfilled)	1.05	62	—	79	6.8	—	100	0.14	—	197	$$$
Filled PTFE	Rulon J (gold)	1.95	14	200	—	varies	—	260	0.25	—	79	$$$
	Rulon LR (maroon)	2.27	10	200	—	varies	—	260	0.33	—	177	$$$
Polyphenylene uslfide	Ryton PPS (40% gf)	1.7	90	2	159	4.5	254	232	0.30	V-0	152	$$$
	Ryton PPS (bearing)	1.55	69	1	103	2.2	254	232	0.32	V-0	—	$$$
Polyteherimide	Semitron 225 (acetal)	1.33	42	10	41	16.7	107	82	—	H-B	—	$$$
	Semitron 410C (PEI)	1.41	62	2	83	3.2	210	171	—	V-0	—	$$$$
	Semitron 420 (PSU)	1.45	65	2	100	5.8	216	177	0.22	V-0	—	$$$$
	Semitron 500 (PTFE)	2.3	8	10	15	10.3	132	260	0.76	V-0	154	$$$
	Semitron 520HR (PAI)	1.58	83	3	138	2.7	271	260	—	V-0	—	$$$$
Polymethylpentene	TPX Polymethylpentene	0.83	28	10	42	3.1	100	93	0.17	H-B	8	$$$

(Continued)

Table 8.20 Continued

Chemical Name	Commercial Name	SPG[1]	Tens[1]	TEB[1]	Flex[1]	CTE[1]	HDT[1]	Tmax[1]	Tcon[1]	UL94[1]	Dist[1]	Cost
Polyphenylene sulfide	Techtron PPS (unfilled)	1.35	93	15	14	5.0	121	218	0.29	V-0	213	$$$$
Ethylene Tetrafluoroethylene	Tefzel ETFE	1.7	42	300	—	13.3	—	155	—	V-0	709	$$$
Polyamide-imide	Torlon 4203 (unfilled)	1.41	124	10	165	3.1	278	260	0.26	V-0	229	$$$$
	Torlon 4301 (bearing)	1.45	83	3	159	2.5	279	260	0.53	V-0	—	$$$$
	Torlon 5530 (30% gf)	1.61	103	3	138	4.7	271	260	0.36	V-0	276	$$$$
Acetal (polyoxy-meothylene)	Turcite TA (acetal)	1.49	52	15	76	9.4	96	82	—	—	—	$$$
	Turcite TX (acetal)	1.46	41	19	55	9.4	96	82	—	—	—	$$$
Polysulfone	Udel PSU	1.24	70	80	103	5.6	171	149	—	H-B	167	$$$
Polyetherlmide	Ultem 1000 (unfilled)	1.28	114	80	138	5.6	200	171	0.13	V-0	327	$$$
	Ultem 2300 (30% gf)	1.51	117	3	207	2.0	210	171	0.13	V-0	303	$$$$
Polyimide	Vespel SP-1 (unfilled)	1.43	86	8	110	5.4	360	316	0.35	V-0	221	$$$$$
	Vespel SP-21 (graphite)	1.51	65	5	110	4.9	360	316	0.87	V-0	91	$$$$$

SpG, Specific gravity (gs/cm^3).
Tens, Tensile strength (MPa).
TEB, Tensile Elongation at break %.
Flex, Flexural strength (MPa).
CTE, Coefficient of linear thermal expansion $\times 10^{-5}$ (cm/cm/K).
HDT, Heat deflection temperature (°C).
Tmax, Max. recommended use temperature (°C).
Tcon, Thermal conductivity (W/m K).
UL94, UL94 flammability rating, UL rating.
DiSt, Dielectric strength (kV/mm).
$, Relative cost factor (estimated).

Poor resistance to sunlight has been a characteristic of past polymers, but new alloys are said to have overcome this problem. The lost cost of ABS and PVC resins have made possible the wide distribution of water in many countries.

8.5.2.2 *Acetals*

Acetals are tough, strong polymers used for many pipe and plumbing parts. The good stability of the resins in cold and hot water recommends their use for many household and appliance parts. Gears, cams, and bushings are often made of the tough material for appliance and automotive parts. The material is not resistant to continued ultraviolet exposure. However, this polymer has an outstanding performance in steam and high humidity environments.

8.5.2.3 *Acrylics*

Methyl methacrylate (Lucite and Plexiglas) is the best known of this group. It is used for a wide variety of applications where its transparency rivals that of glass, without the brittleness. However, the acrylics are soft, easily scratched, and not very temperature resistant.

8.5.2.4 *Cellulosics*

Cellulosics are best represented by the cellulose acetate/butyrate product, which has long been used for piping to convey natural gas, oils, brines, and sour crudes. It is unaffected by exposure in the atmosphere over long periods of time. Care must be taken to assure that streams of crude products do not contain aromatic hydrocarbons or oxygenated solvents, which will readily attack the polymer.

8.5.2.5 *Fluorocarbons*

Fluorocarbons are composed of polymer chains of carbon and fluorine, although chlorine atoms are present in certain polymers. The first TFE was produced by Du Pont and designated Teflon. It is the "noble metal" of the polymers with corrosion resistance to practically all environments up to 285°C.

The very inertness of TFE was also the reason why this polymer was originally difficult to manufacture. However, Teflons have since then grown into a great variety of products, each with differing mechanical and corrosion-resistance properties. The appropriate fluorocarbon polymer must therefore be identified for a specific application. A number of companies now produce fluorocarbon polymers.

In addition to corrosion resistance, the low coefficient of friction of Teflon is the basis for a very successful application of the polymer illustrated in Figure 8.23. A confined Teflon sleeve separates the plug and body of this plug valve and acts as a lubricant, thereby preventing sticking or freezing of the metal parts. Ordinary petroleum greases that might be used to lubricate moving parts such as these are attacked by many

Figure 8.23 Nonlubricated, coated plug valve with tetrafluoroethylene (TFE) sleeve.

environments and consequently cannot be used in corrosive environments. Other uses are seals and gaskets, wire insulation, expansion joints, linings for pipe, tape, tubing, valve diaphragms, coatings, and even heat exchangers (thin-walled tubing).

8.5.2.6 Nylon

Nylon is a group of materials best known for its use in hosiery but which also finds much use for strictly corrosion-resistant applications. Its main applications depend on strength, low coefficient of friction, and wear resistance. These are gears, drawer and shelf rollers, fishline, sutures, tennis-racket strings, dentures, automobile-door catches, pipe fittings, and brush bristles. It is also utilized for electrical insulation and permits higher-temperature operation than rubber.

8.5.2.7 Polycarbonates

Like acetal and polysulfone polymers, the polycarbonates have good dimensional stability with temperature changes and a high modulus for a thermoplastic. The material is thus widely used in automobile and aircraft parts, lighting fixtures, tool housings, and electrical parts where properties and good impact strength are desired.

The polymer in the stressed condition (residual from molding stresses or applied) can stress crack in certain very mild chemical exposures such as ketone vapors. Consequently, the stress relief of fabricated parts may be desirable for many applications.

8.5.2.8 Polyethylenes

Polyethylene materials represent the greatest tonnage of all the polymers and are marketed by many producers. They are used for packaging film, sheet, squeeze bottles, pans, tumblers, ice trays, and many other household items. Piping for water and other chemicals is common. Certain solvents produce SCC of polyethylene. Considerable latitude in properties is possible, as indicated by Table 8.20.

8.5.2.9 Polypropylenes

Polypropylene was first produced in Italy in 1955. It exhibits better heat and corrosion resistance and is stiffer than polyethylene. It is made into valves and bottles to be sterilized by heat, pipe, rope, fittings, and scooter parts. Polypropylene's outstanding characteristics are its resistance to strong acids, even at elevated temperatures. A limitation of polypropylene is brittleness when exposed to freezing temperatures.

8.5.2.10 Polyimides

Polyimide materials have the highest heat stability of any of the thermoplastics. Continuous use of the materials up to temperatures of 285°C is possible, with shorter excursions to temperatures as high as 480°C permissible. Some 90% of the tensile strength is retained after 1000 h of exposure at air temperatures of 300°C.

This polymer has also an exceptional resistance to low-temperature embrittlement. Its use in cryogenic systems operating at temperatures as low as −250°C has been successful. No other thermoplastic has such a range of temperature stability.

The polymer also has an unusual resistance to nuclear-radiation sources where most thermoplastics become embrittled. Compressor parts, valve parts, bearings, and parts for electrical insulation have also been made from the material. The material is resistant to fungus attack but loses strength in sunlight exposures.

8.5.2.11 Polyphenylene Sulfide

Polyphenylene sulfide has unique corrosion resistance. Only strong oxidizing agents appear to attack the material. The outstanding corrosion resistance combines with good strength, high modulus, and good heat stability to provide a useful product for many chemical exposures. The material has been used for pumps, valves, piping, coatings, insulators, and other applications in severe industrial exposures.

8.5.2.12 Polystyrene

Polystyrene is used for wall tile, battery cases, flow meters, radio cabinets, bottle closures, and refrigeration equipment. It possesses good chemical resistance but is too brittle for many structural applications. Polystyrene shows good resistance to HF acid.

8.5.2.13 Polysulfones

The good heat stability, toughness at high and low temperatures, and good electrical insulating properties brought the polysulfones quickly into substantial use both in home and industry. Electric coffee pots, housing for many electric and electronic devices, eye glasses, microwave-oven internals, and chemical-processing equipment are but some of the past applications for this polymer.

The material can be continuously steamed without difficulty but weathering in sunlight will produce some change in properties. Like polyethylene, polycarbonate, and some other thermoplastics, the polysulfones can be stress cracked when in the stressed condition. Chlorinated or oxygenated chemicals are known cracking agents when sufficient stress is present.

8.5.2.14 Polyvinyl Chlorides (PVC)

The most extensive use of polymer piping, particularly for water service, involves the PVC materials. A "high-impact" grade is normally used. The lightness, strength, resistance to attack in soils and waters, good acid and alkali resistance, and ease of joining have made the material a standard for many utility operations and industrial applications.

The product is resistant to essentially all inorganic acid, alkali, and salt exposures, oils, and alcohols. Most aliphatic and aromatic chemicals will cause rapid attack. The material has a relatively low modulus of elasticity and weakens quickly with increasing temperatures. Vinylidene chlorides (Saran) are related to the PVC materials but contain a higher percentage of chlorine in the polymer chains.

8.5.3 Thermosets

Thermosetting polymers are composed of long chain molecules that are strongly cross linked to form three-dimensional network structures. These polymers are generally more rigid and stronger but more brittle than thermoplastics. Thermosets do not have a fixed melting temperature and cannot easily be reprocessed after the cross-linking reaction has occurred. Thermosets will harden when heated and retain hardness when cooled. They "set" into permanent shape by catalysis or when heated under pressure. Generally, they cannot be recycled.

8.5.3.1 Epoxies

Epoxies represent perhaps the best combination of corrosion resistance and mechanical properties. Epoxies are available as castings, extrusions, sheet, adhesives, and coatings. They are used as sinks, bench tops, pipe, valves, pumps, small tanks, potting compounds, adhesives, linings, protective coatings, dies for forming metal, printed circuits, insulation, and containers. Liquid epoxy resins can be cast into molds (as in a foundry) and then set by heating. Epoxies are generally resistant to weak acid and to weak alkalis, attacked by some strong acids and slightly attacked by strong alkalis.

8.5.3.2 Phenolics

Phenolic materials are among the earliest and best-known polymers. They are mostly based on phenol formaldehyde. Bakelite was developed by Leo Baekeland shortly after the turn of the century. Applications include radio cabinets, telephones, electrical sockets and plugs, piping, pumps, valves, trays, auto distributors, rollers, and coatings. Phenolics may resist or undergo slight attack by weak acids and are attacked by strong agents oxidizing agents and strong alkalis, undergoing decomposition.

8.5.3.3 *Polyesters*

Polyester polymers are best known to the general public for their use as polyester fabric. Corrosion resistance of the polyesters compares favorably with many other polymers. One of the main uses of polyesters involves reinforced material. Polyesters are resistant to some weak acids but are attacked by strong acids, alkalis, and by oxidizing agents.

8.5.3.4 *Silicones*

Silicones offer outstanding heat resistance. Mechanical properties change little with variations in temperature. These polymers differ from most polymers in that an important ingredient is inorganic silicon. Silicones are used for molding compounds, laminating resins, and insulation for electric motors and electronic equipment. Resistance to attack by chemicals is not outstanding.

8.5.3.5 *Ureas*

Ureas were the second important resins to be developed. They are based on urea and formaldehyde. Ureas are fairly resistant to weak acids but may undergo surface attack and decomposition in strong acids. Moderately attacked by weak alkalis, they decompose in strong alkalis.

8.5.3.6 *Composites*

These materials usually consist of thermosetting resins filled or laminated with cloth, mats, paper, chopped fabrics, or fibers such as fiberglass or graphite, which are commonly used. The main advantage is that tensile strength can be dramatically increased, providing a high strength–weight ratio for these light materials. Availability and uses include tanks, pipe, ducts, sheets, rod, car bodies, boats, and missile and satellite parts.

8.5.4 SCC of Polymeric Materials

Thermoplastic and thermosetting polymers will stress crack in a manner physically comparable to that of the metals. As in the case of the metals, a loss of cohesive bonding in the material allows the matrix to separate without any other evidence of attack. The type of bonding between atoms and chains of atoms may be different in metals and polymers, but the loss of cohesive strength along weak lines in the structure results in the same brittle cracking.

Cracking of stressed polymer structures can occur almost instantaneously when specific cracking environments are introduced. As an example, simply waving a stressed section of polycarbonate or polysulfone polymer over an open bottle of an oxygenated organic chemical, such as acetone, can produce immediate cracking. Other polymer-cracking-agent combinations may require a significant period of contact before cracking is obtained (e.g., polyethylene in detergents). It is important to understand that such a mechanism of attack can occur on polymeric materials and often must be evaluated before selecting the proper material for a specific service.

Many of the means for alleviating the cracking of metals are applicable for avoiding the cracking of polymers. These include stress relief, coating, alloying or compounding, redesign, or a change of material.

8.6 OTHER NONMETALS

8.6.1 Ceramics

Ceramic materials consist of compounds of metallic and nonmetallic elements. A simple example is MgO or magnesia. Other ceramics include alumina (Al_2O_3), thoria (ThO_2), glass, brick, stone, fused silica, stoneware, clay tile, porcelain, concrete, abrasives, mortars, and high-temperature refractories.

In general, compared with metals, ceramics resist higher temperatures, have better corrosion and abrasion resistance, including erosion-corrosion resistance, and are better insulators. However, ceramics are brittle, weaker in tension, and subject to thermal shock. Most ceramic materials exhibit good resistance to chemicals, with the main exceptions of HF acid and caustic. Parts are formed by pressing, extrusion, or slip casting.

8.6.2 Acid Brick

This material is made from fireclay with a silica content about 10% greater than ordinary firebrick. A common application is lining of tanks and other vessels to resist corrosion by hot acids or erosion–corrosion. A brick-lined steel tank usually contains an intermediate lining of lead, rubber, or a polymer. Acid-resistant cements and mortars join the brick. Floors subject to acid spillage are made of acid brick.

8.6.3 Stoneware and Porcelain

Both of these find many applications because of their good corrosion resistance. Porcelain parts are usually smaller in size than stoneware and the porcelain is less porous. Both can be glazed for easy cleaning. Stoneware and porcelain show tensile strengths of about 14 and 35 MPa (2000 and 5000 psi), respectively. Stoneware sinks, crocks, and other vessels, absorption towers, pipes, valves, and pumps are available. Porcelain can be made into similar equipment (e.g., acid nozzles) and is widely used as insulators and spark plugs.

8.6.4 Structural Clay

Structural clay products include building, fire, sewer, and paving bricks, terra cottas; pipe and roofing; and wall tiles. Hot acids sometimes attack these materials.

8.6.5 Glass

Glass is an amorphous inorganic oxide, mostly silica, cooled to a rigid condition without crystallization. Glass laboratory ware, such as Pyrex and containers, are well known. Piping and pumps are available. Transparency is utilized for equipment such as flow meters. Glass fibers are widely used for air filters, insulation, and reinforced polymers. HF acid and caustic can attack glass, as can hot water in a moderate manner.

8.6.6 Vitreous Silica

Vitreous silica, also called fused quartz, has better thermal properties than most ceramics and excellent corrosion resistance at high temperatures. It is used for furnace muffles, burners, reaction chambers, absorbers, piping, and so forth particularly where contamination of the product is undesirable.

8.6.7 Concrete

Vast quantities of concrete are used throughout the world because of the strength, durability, and low cost of the product. Concrete is used to construct large tanks, large-diameter pipe, vessel linings, pump bases, and flooring for industrial work areas. Concrete is also used as linings on the interior of steel pipe, a ballast coating on the exterior of steel pipe, and foundations for many steel and polymer tanks, in addition to the more common paving and structural forms observed each day.

The term concrete identifies a mass of inert aggregate surrounded by a binder. Most of the binders used in concrete are based on portland cement. There are five basic types of portland cement as defined by ASTM C 150:

Type I: Type I is a general-purpose portland cement suitable for all uses where the special properties of other types are not required. It is used where cement or concrete is not subject to specific exposures, such as sulfate attack from soil or water, or to an objectionable temperature rise caused by heat generated by hydration. Its uses include pavements and sidewalks, reinforced-concrete buildings, bridges, railway structures, tanks, reservoirs, culverts, sewers, water pipes, and masonry units.

Type II: Type II portland cement is used where precaution against moderate sulfate attack is important, as in drainage structures where sulfate concentrations in ground waters are higher than normal but not unusually severe (Table 8.21). Type II cement will usually generate less heat at a slower rate than Type I. With this moderate heat of hydration (an optional requirement), Type II cement can be used in structures of considerable mass, such as large piers, heavy abutments, and heavy retaining

Table 8.21 Attack on Concrete by Soils and Waters Containing Various Sulfate Concentrations

Relative Degree of Sulfate Attack	Percentage Water-Soluble Sulfate (as SO_4) in Soil Samples	Sulfate (as SO_4) in Water Samples, (ppm)	Cement Type
Negligible	0.00–0.10	0–150	I
Positive	0.10–0.20	150–1500	II
Severe	0.20–2.00	1500–10,000	V
Very severe	2.00 or more	10,000 or more	V plus pozzolan

walls. Its use will reduce temperature rise, an important quality when the concrete is placed in warm weather.

Type III: Type III is a high–early strength portland cement that provides high strengths at an early period, usually a week or less. It is used when forms are to be removed as soon as possible, or when the structure must be put into service quickly. In cold weather, its use permits a reduction in the controlled curing period. Although richer mixtures of Type I cement can be used to gain high–early strength, Type III, high–early-strength portland cement, may provide it more satisfactorily and more economically.

Type IV: Type IV is a low heat of hydration cement for use where the rate and amount of heat generated must be minimized. It develops strength at a slower rate than Type I cement. Type IV portland cement is intended for use in massive concrete structures, such as large gravity dams, where the temperature rise resulting from heat generated during curing is a critical factor.

Type V: Type V is a sulfate-resisting cement used only in concrete exposed to severe sulfate action, principally where soils or ground waters have a high sulfate content. Table 8.21 describes sulfate concentrations requiring the use of Type V portland cement.[11] Low tricalcium aluminate (C_3A) content, generally 5% or less, is required when high sulfate resistance is needed.

The first three types are also available as air-entraining cements (Types IA, IIA, and IIIA). The cements contain lime, gypsum, silica, alumina, magnesia, some iron, and a little free alkali. Upon reaction with water, interactions take place to produce various hydrated compounds, the most important of which are the silicates and aluminates (e.g., tricalcium aluminate). Varying the ratio of these components, the size of the void space, the rate at which the reactions take place, and the possible addition of other compounds all have an influence on the mechanical, physical, and corrosion-resistance properties of the cured cement.

An aggregate must also be chosen to resist the proposed environment. Naturally occurring rock, gravel, mill slags, and metal lumps have been used as aggregate for various reasons. With the aggregate constituting some three-quarters of the total concrete volume, it is important to select an inert, dense material compatible with the environment.

Unlike metals, the strength of concrete is normally described in terms of compressive strength. Although great variations in strength can be obtained, a value of 40 MPa (6000 psi) should be achieved in a fully cured concrete.

Concrete is readily attacked by soft, aggressive waters, which tend to leach calcium from the structure. Acidic waters corrode concrete by direct attack on calcium-carbonate constituents. A special problem is the direct attack on tricalcium-aluminate constituents by sulfate ions (Ettringite or formerly "cement bacillus"), requiring increasingly resistant grades of concrete as the sulfate-ion content increases. Ettringite is a naturally occurring mineral characterized as a high-sulfate calcium sulfoaluminate. It is formed as a cement-hydration product or by sulfate attack in mortar and concrete. It is also the product of the principal expansion-producing reaction in expansive cements.

Type V portland cement will tolerate up to 1500 ppm of sulfate ions. Above that concentration in fresh waters, special types of cement are prepared by adding a finely powdered material such as found in volcanic ashes (e.g., Pozzolans and Brown-millerite), following an old recipe used by the Romans. Oddly, the sulfate attack is not encountered in seawater, probably because of the ratio among chlorinity, alkalinity, and hardness. A very common problem with concrete structures is internal corrosion occasioned by rusting of steel rebars often used to strengthen the concrete. This topic is covered in much detail in Chapter 6.

The deterioration of concrete by freezing–thawing cycles, vibration, and attack by strong acids is well known. However, any environmental agent that can break the inorganic bonds of the cement binder will create corrosion of the concrete. Sulfates, ammonium salts, strong alkalies, weak acids, organic esters, carbon dioxide, magnesium salts, and many other agents can destroy the binder over a period of time.

8.6.8 Super Refractories

Table 8.22 lists some properties of a few very high-temperature refractories or super refractories.[12] They are used to resist molten metals, slags, and hot gases. Alumina has also found application for pump and valve seats because of its high hardness and good wear and corrosion resistance. Silicon carbide performs well as spray nozzles for hot sulfuric acid and for many other severe applications.

8.6.9 Carbon and Graphite

These are unique nonmetallics in that they are good conductors of heat and electricity. High thermal conductivity results in excellent thermal-shock resistance. They are used for heat exchangers, columns, pumps, and impressed-current anodes. Carbon and graphite are inert to many corrosive environments. They are weak and brittle compared to metals. Tensile strength varies between about 3 to 20 MPa (500 to 3000 psi) and impact resistance is nil. Abrasion resistance is poor. High-temperature stability is good, and they can be used at temperatures up to 2760°C if protected from oxidation (burning). Silicon-base coatings (silicides or silicon carbide) and iridium coatings are claimed to give protection up to around 1593°C in air.

Carbon exhibits good resistance to alkalies and most acids, but oxidizing acids such as nitric, concentrated sulfuric, and chromic acid attack it. Fluorine, iodine, bromine, chlorine, and chlorine dioxide are likely to attack carbon. Karbate, a resin-bonded graphite, has found wide application in the chemical-process industries.

Table 8.22 Some Physical and Mechanical Properties of Some Fired Refactory Bricks

	Magnesia (95% MgO)	Chrome Alumina (30% Cr_2O_3)	90% Alumina	70% Alumina	Zircon	Superduty Firebrick	Silicon Carbide	Silica (Superduty)
Bulk density, g/cm^3	2.80–2.95	3.06–3.14	2.90–2.96	2.53–2.60	3.60–3.72	2.31–2.37	2.56–2.66	1.78–1.87
Potosity, %	15–19	16–20	14–18	17.5–21.5	19–23	11–14	11–15	20–24
Cold crushing strength, MN	48–70	35–55	62–95	27–48	48–76	12–21	69–83	27–41
Modulus of rupture, MPa	17–24	14–21	17–21	7.6–11	15–23	4.8–6.9	21–24	4.1–6.9
Reheat test, % permanent linear change after heating to	—	—	—	—	—	—	—	—
1600°C	—	+3.5 to 6.0	—	0.0 to 0.9	—	—	—	—
1650°C	—	—	—	—	0.0	—	—	—
1725°C	-0.2–1.0	—	+0.1 to 1.0	—	—	—	–0.1 to 0.1	—
Load test at 170 KPa (25 ksi); withstands load to temperature, °C	1620	1400	1760	1450	1600	1450	1650	1680
Relative cost	2.8		3.1		10		2.1	

Pyrolytic graphite is a unique form of graphite manufactured by decomposition of a hydrocarbon gas at very high temperature in a vacuum furnace. The result is an ultrapure product that is near theoretical density and extremely anisotropic. As an example, pyrolytic graphite exhibits a thermal conductivity consistent with the best conductors in one orientation and lower than alumina brick in the other. Mechanical, thermal, and electrical properties are generally far superior to conventional graphites.

8.6.10 Wood

Although modern polymers are finding a place in cooling-tower construction, wood has been the traditional material of construction, and large amounts are still in service today. Chemically, the cellulose constituents in wood are susceptible to acid attack and the lignin by alkali, especially in the presence of oxidizing agents.

Deterioration in cooling towers is usually in the form of delignification, occasioned by the concentrated alkalinity in the water and aggravated by the requisite chlorination.

Biological attack may be in the form of internal decay or surface (soft) rot. In soft rot, the organisms consume cellulose and leave lignin behind. The wood turns dark, becomes brittle, soft, punky, and cross checked or fillibrated, and eventually loses its strength.

Cypress, pine, oak, and redwood are the main woods used for corrosion applications. There is no "better" wood. The choice depends on the exposure. Filter-press frames, structural members of buildings, barrels, and tanks are often made of wood. Containers must be kept wet or the staves will shrink, warp, and leak. Generally speaking, wood is limited to water and dilute chemicals. Strong acids and dilute alkalies attack woods. They are also subject to biological attack. Impregnation with waxes and polymer resins helps reduce chemical and biological attack.

References

1. H.L. Logan, The Stress Corrosion of Metals (New York, NY: Wiley, 1966).
2. Corrosion Data Survey Metals, 6th ed. (Houston, TX: NACE 1985).
3. A.J. Sedriks, Corrosion of Stainless Steels (New York, NY: Wiley, 1979).
4. Stainless Steel from Avesta Sheffield—Steel Grades, 2nd ed. (Avesta, Sweden: Avesta Sheffield, 1997).
5. C.P. Dillon, Corrosion Resistance of Stainless Steels (New York, NY: Marcel Dekker, 1995).
6. D.R. Askeland, P.P. Phulé, The Science and Engineering of Materials, 4th ed. (Pacific Grove, CA: Thomson, 2003).
7. A. Froats, T.K. Aune, D. Hawke, W. Unsworth, J. Hillis, "Corrosion of Magnesium and Magnesium Alloys," in Metals Handbook: Corrosion (Metals Park: ASM International, 1987), p. 740–754.
8. J.R. Davis, Metals Handbook, Desk Edition (Materials Park, OH: ASM International, 1998).
9. "Chemical Resistance of Plastics and Elastomers Used in Pipeline Construction" (Tustin, CA: George Fischer Inc., 2004).
10. "Materials Properties Tables" (Shiner, TX: Boedeker Plastics, Inc., 2004).
11. Concrete Manual, a Manual for the Control of Concrete Construction, 7th ed. (Denver, CO: U.S. Department of the Interior Bureau of Reclamation, 1963).
12. P.L. Mangonon, The Principles of Materials Selection for Engineering Design (Upper Saddle River, NJ: Prentice Hall, 1999).

9

Chapter

Forms of Corrosion

9.1 INTRODUCTION

General corrosion is usually the least threatening type of attack, allowing one to forecast with some degree of accuracy the probable life of equipment. Unfortunately, most corrosion encountered in engineering practice is more of a localized type.

Localized corrosion can be defined as selective attack by corrosion at small special areas or zones on a metal surface in contact with an environment. It usually occurs under conditions where the largest part of the original surface is either not attacked or is attacked to a much smaller degree than at the local sites.

The most common type of localized corrosion is pitting, in which small volumes of metal are removed by corrosion from certain areas on the surface to produce craters or pits. Pitting corrosion may occur on a metal surface in a stagnant or slow-moving liquid. It also may be caused by crevice corrosion, poultice corrosion, deposition corrosion, cavitation, impingement, and fretting corrosion. All of these forms of localized corrosion are discussed in this chapter.

Another common type of corrosion attack is intergranular or intercrystalline corrosion during which a small volume of metal is preferentially removed along paths that follow the grain boundaries to produce what might appear to be fissures or cracks. The same kind of subsurface fissures can be produced by transgranular or transcrystalline corrosion. In this, a small volume of metal is removed in preferential paths that proceed across or through the grains. This occurs only under certain conditions and with certain alloys, as is described later.

Intergranular and transgranular corrosion sometimes are accelerated by tensile stress. In extreme cases, the cracks proceed entirely through the metal, causing rupture or perforation. This condition is known as SCC. Intergranular and transgranular subsurface cracks also can be produced by hydrogen. Caustic embrittlement and corrosion fatigue are two other mechanisms of metal deterioration that form fissures at or beneath the surface.

In a completely different type of corrosion that may become localized, one of the metals in an alloy may be selectively leached out without producing visible pits or cracks and without changing the dimensions of the metal. At a casual glance the metal may appear to be intact. Under a microscope, however, it can be seen to be porous. The mechanical properties of the alloy are greatly reduced by the selective attack.

The most common example of this type is dezincification of brass in which the zinc is selectively dissolved out of the alloy. Another case is graphitic corrosion of cast iron, in which the iron is selectively dissolved or leached away, leaving a porous mass which appears intact, but in reality consists largely of graphite.

The several forms of corrosion (Figure 9.1) may be divided into three groups[1]:

1. those recognizable with the unaided eye;
2. those that are more easily discerned with specific aids (e.g., dye penetrants, magnetic particles, or low-power microscopy); and
3. those that can only be identified definitely by optical or electronic microscopy.

Much can be deduced from examination of materials that have failed in service. It is often possible by visual examination to decide which corrosion mechanisms have been

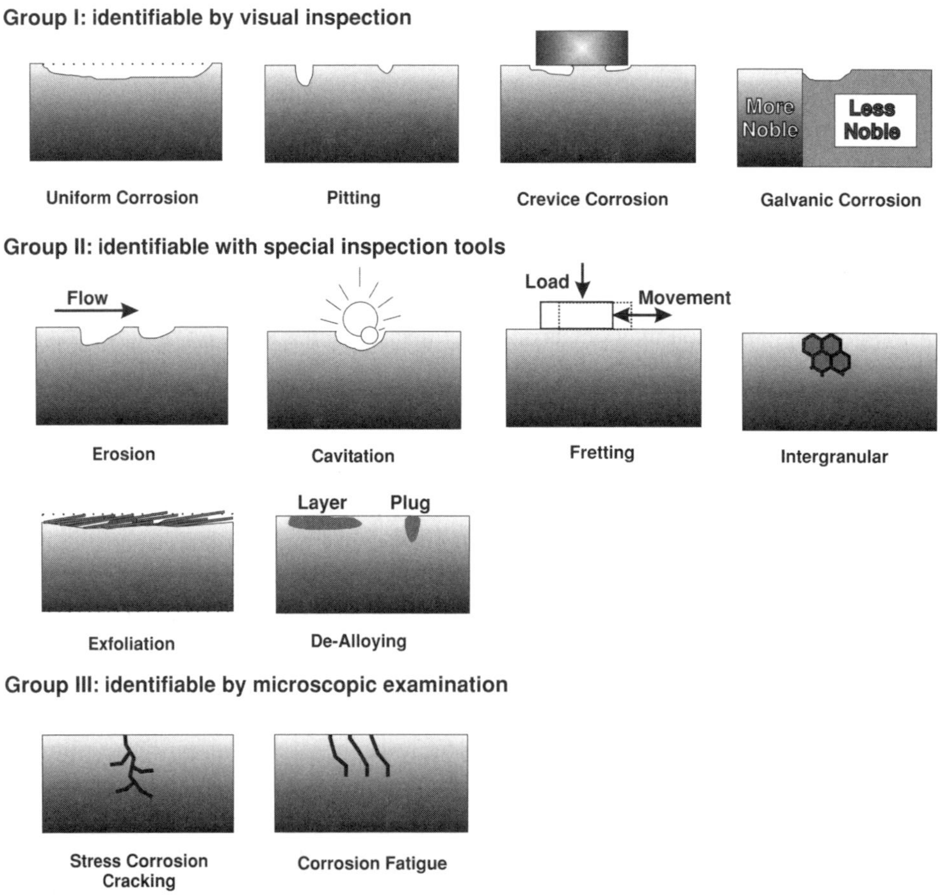

Figure 9.1 Main forms of corrosion attack are regrouped by their ease of recognition.[1]

Table 9.1 Causes of Corrosion Currents

Impurities in the metal
Orientation of grains
Grain boundaries
Differential grain size
Differential thermal treatment
Surface roughness
Local scratches or abrasions
Difference in shape
Differential strain
Differential pre-exposure to air or oxygen
Differential concentration or composition of solution
Differential aeration
Differential heating
Differential illumination
Differential agitation
Contact with dissimilar metals
Externally applied potentials
Complex cells

at work and what corrective measures are required to solve the problem. However, a more complete analysis of a problem should be made. Too many offhanded analyses have been made with costly results. Even metallurgical examinations may require sophisticated tools to differentiate between a corrosion-fatigue crack and a stress-corrosion crack. However, this is not meant to denigrate the value of visual inspection.

The important point is to understand that a variety of corrosion mechanisms can occur, that these are caused by different combinations of factors, and that corrective procedures are widely varied.

9.2 UNIFORM ATTACK

Although representing potentially the greatest metal weight loss, uniform attack is highly visible and its effects predictable; hence, it may be less troublesome than other forms of corrosion. Designing a corrosion allowance in a system is one of the simplest methods of dealing with uniform attack. Unfortunately, the literature does not always describe the character of corrosive attack when discussing corrosion behavior of materials. It would be dangerous to assume that corrosion will be uniform, unless direct evidence is available to verify this assumption.

9.3 LOCAL-CELL CORROSION

The sources of the electrochemical driving force for corrosion reactions are of considerable interest and merit special attention. In a landmark paper, Mears and Brown have summarized 18 mechanisms by which differences in potential may develop on metal surfaces[2]. These mechanisms are listed in Table 9.1.

A difference in potential will be observed if electrically connected specimens of the same metal are immersed in solutions that have different concentrations of ions of the metals. This is one common type of concentration cell. Another type is the oxygen concentration cell. This type and the metal-ion cell can oppose one another. For example, variations in oxygen concentration can establish oxygen concentration cells. In solutions containing metal ions, the solubility of oxygen may vary, depending on the concentration of metal ions.

For this reason, although it would be expected that a metallic material immersed in a solution with the lower concentration of metal ions would be anodic to a similar piece of metal immersed in a more concentrated solution, the greater solubility of oxygen in the dilute solution may mask the electrochemical effect of this difference in metal-ion concentration. It may even cause the metal in the more dilute solution to become the cathode. Therefore, in real situations, when differences in chemical or oxygen concentration exist, one cannot predict with certainty which part of the metal will be the anode or the magnitude of the current flow.

Local cells can be produced by differences among small nearby areas on the metal surface. They may result from differences in the metal, or the environment, or from impressed currents. Metal variations may be the result of composition differences. A second-phase constituent will have a different corrosion potential compared with that of an adjacent solid solution. The difference may be in the thickness of a surface film at adjacent sites, which in turn may reflect metal differences in the substrate.

9.3.1 Pitting Corrosion

Pitting corrosion is a localized form of corrosion by which cavities or "holes" are produced in the material. Pitting is considered to be more dangerous than uniform corrosion damage because it is more difficult to detect, predict, and design against. A small, narrow pit with minimal overall metal loss can lead to the failure of an entire engineering system. Only a small amount of metal is corroded, but perforations can lead to costly repair of expensive equipment.

Pitting cavities may or may not become filled with corrosion products. Corrosion products may form caps over pit cavities, which are described as nodules or tubercles. Although the shapes of pits vary widely (Figure 9.2), they usually are roughly saucer-shaped, conical, or hemispherical. Pit walls usually are irregular when viewed under a microscope. The following are some factors contributing to the onset and propagation of pitting corrosion:

- localized chemical or mechanical damage to a protective oxide film that is almost, but not completely, resistant to corrosion;
- water-chemistry factors that can cause breakdown of a passive film are acidity, low dissolved-oxygen concentrations (which tend to render a protective oxide film less stable), and high concentrations of chloride (as in seawater);
- localized damage to or poor application of a protective coating; and
- the presence of nonuniformities in the metal structure of the component, for example, nonmetallic inclusions.

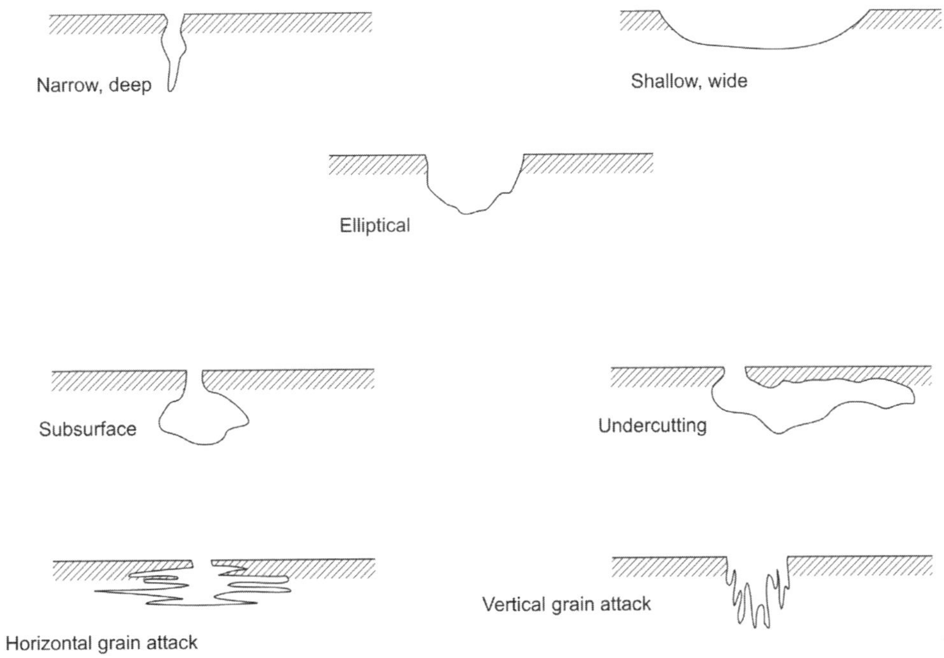

Figure 9.2 Cross-sectional shapes of various corrosion pits.

Corrosion-rate data appearing in the literature must be viewed with considerable suspicion, unless complete information regarding the geometrical character of attack is fully known. Some authors make use of a pitting factor to provide this information. The pitting factor is the ratio of the depth of the deepest pit resulting from corrosion divided by the average penetration as calculated from weight loss.

The practical importance of pitting depends on the thickness of the metal and on the penetration rate, which could be small. The rate usually decreases with time. Thus, on thin sections pitting may be serious, whereas on a thick section it may be almost unimportant. In general, the rate of penetration decreases if the number of pits increases. This is because adjacent pits have to share the available adjacent cathodic area, which controls the corrosion current that can flow. Movement of the solution over a metal surface often reduces and even may prevent pitting that otherwise would occur if the liquid was stagnant.

Pitting is often associated with other forms of corrosion. Intergranular cracks, for example, may progress from the main pit cavity further into the metal. Crevice corrosion, described later, can be considered to be an aggravated case of pitting corrosion. Stray-current corrosion that occurs when an electric current leaves a metal surface and flows into the environment can cause a very characteristic form of macroscopic pits (see Figure 1.9 in Chapter 1). Stray-alternating currents generally do not cause very much corrosion of metals, except in the case of aluminum. In this case, the metal is unable to plate back onto the surface during the reverse part of the cycle. Most severe stray-current corrosion is caused by direct current leakage, for example, from DC welding of reverse polarity or from DC power of electrified systems.

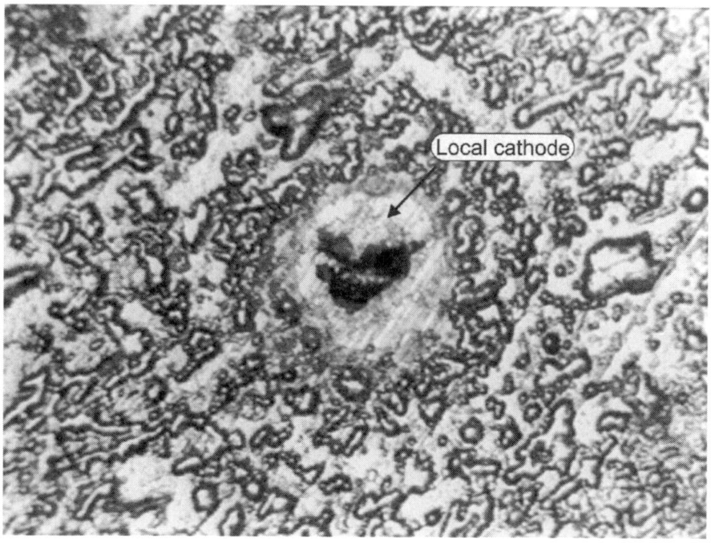

Figure 9.3 Pitted surface of aluminum A92519 exposed to a 3.5% NaCl solution for 7 days (×100 magnification).

Theoretically, a local cell that leads to the initiation of a pit can be caused by an abnormal anodic site surrounded by normal surface, which acts as a cathode, or by the presence of an abnormal cathodic site surrounded by a normal surface in which a pit will have disappeared because of corrosion. In the second case, post-examination should reveal the local cathode because it will remain unattacked (Figure 9.3). Most cases of pitting are believed to be caused by local cathodic sites in an otherwise normal surface.

A pit may go through four separate stages: (a) initiation, (b) propagation, (c) termination, and (d) reinitiation. The causes of initiation (i.e., local cell, action) have been discussed. In the propagation stage, the rate increases because of changes in the anodic and cathodic environment (these become more acidic and alkaline, respectively). A pit may terminate because of increased internal resistance of the local cell (caused by filling with corrosion products, filming of the cathode, etc.). If a pitted surface is dried out, the pits will, of course, terminate. When rewetted, some of the pits may reinitiate. This may be due to the reestablishment of the conditions or to differential aeration between the solution in the main pit cavity and solutions in some of the cracks that emanate from it deeper into the metal.

When evaluating the condition of a pitted metal, most weight should be given to the deepest pit, although this is not always done. Pit depths can be measured using a calibrated microscope focused first on the shoulder and then on the bottom of the pit, by using a dial gauge with a needle probe, by metallographic sectioning, or by using a metal-cutting tool in a carpenter's router and cutting down until no corrosion is evident. This last method is the most accurate because even metallographic sectioning requires taking a section at the proper location; which may not be evident. A variation of the routing method that can be used to measure the maximum pit depth on the outside of a pipe is to place a sample length in a lathe and machine it until no further corrosion is evident.

Efforts have been made to treat the rate of penetration statistically. For example, equations have been developed to predict the perforation of buried steel pipe lines. The depth of the deepest pit on a specimen is the upper tail of the distribution of depths of all pits on that specimen.[3] Engineers concerned with soil corrosion and underground steel piping are aware that the maximum pit depth found on a buried structure is somehow related to the percentage of the structure inspected. Finding the deepest actual pit requires a detailed inspection of the whole structure. As the area of the structure inspected decreases, so does the probability of finding the deepest actual pit. A number of statistical transformations have been proposed to quantify the distributions in pitting variables. Gumbel is given the credit for the original development of extreme value statistics (EVS) for the characterization of pit-depth distribution.[4]

The EVS procedure is to measure maximum pit depths on several replicate specimens that have pitted, and then arrange the pit-depth values in order of increasing rank. The Gumbel or extreme-value cumulative probability function [F(x)] is shown in Equation (9.1), where λ and α are the location and scale parameters respectively. This probability function can be used to characterize the data set and estimate the extreme pit depth that possibly can affect the system from which the data was initially produced:

$$\mathrm{F}(x) = \exp\left(-\exp\left[-\frac{x-\lambda}{\alpha}\right]\right) \tag{9.1}$$

Because pitting is an electrochemical process, it can be stopped by CP. It can also be prevented by the use of inhibitors, which alter the electrode reactions of the local cell and remove their driving force.

Penetration by pitting often can be prevented by coating the surface of a metal with a sacrificial layer of another alloy, such as zinc on steel or Alclad* on aluminum, or by protective coatings. A zinc-rich paint is sacrificially active and will prevent the pitting of either steel or aluminum. In some cases, agitation of the environment will prevent environmental differences from developing and will prevent pitting that otherwise would occur.

9.3.2 Crevice Corrosion

Crevice corrosion occurs in cracks or crevices formed between mating surfaces of metal assemblies, and usually takes the form of pitting or etched patches. Both surfaces may be of the same metal or of dissimilar metals, or one surface may be a nonmetal. It can also occur under scale and surface deposits and under loose-fitting washers and gaskets that do not prevent the entry of liquid between them and the metal surface. The crevice may proceed inward from a surface exposed to air or may exist in an immersed structure.

*A composite wrought product composed of an aluminum alloy core that has on one or both surfaces a metallurgically bonded aluminum or aluminum alloy coating that is anodic to the core electrochemically protects the core against corrosion.

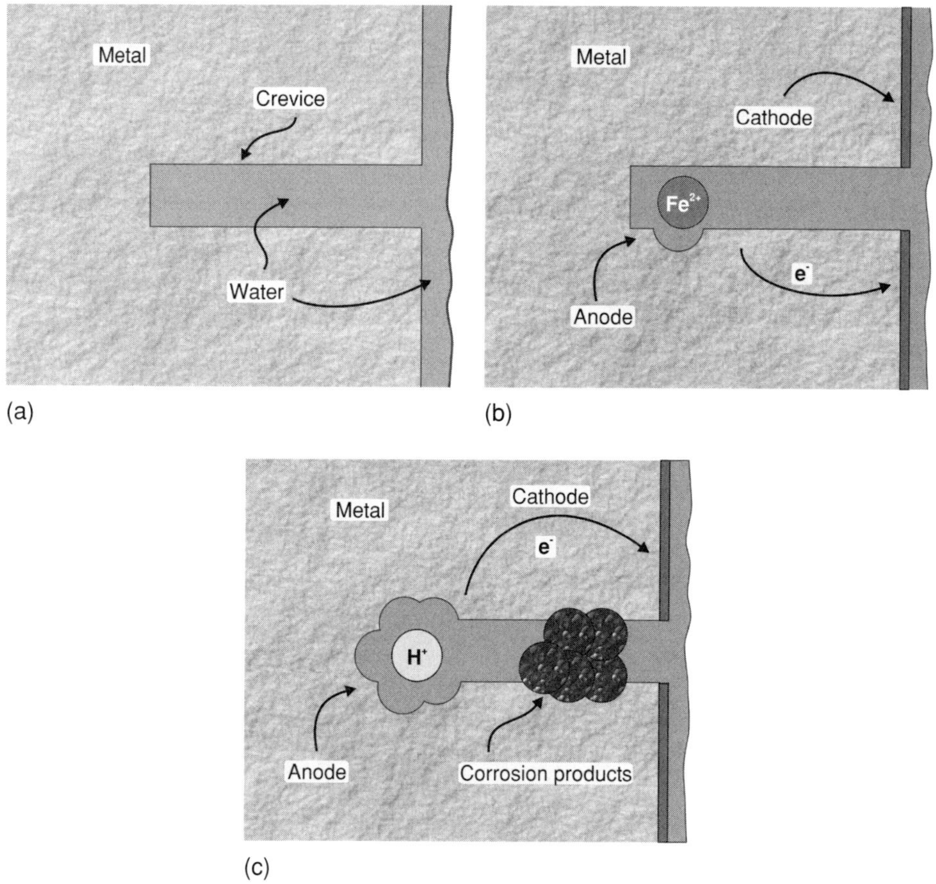

Figure 9.4 Schematic shows the stages of a crevice formation: a) first stage, b) second stage, c) third stage.

The series of events leading to the formation of a severe crevice can be summarized in the following 3 stages. First, crevice corrosion is believed to initiate as the result of the differential-aeration mechanism mentioned earlier. Dissolved oxygen in the liquid that is deep in the crevice is consumed by reaction with the metal (Figure 9.4a). Second, as oxygen diffusion into the crevice is restricted, a differential-aeration cell tends to be set up between crevice (microenvironment) and the external surface (bulk environment). The corrosion reactions now specialize in the crevice (anodic) and on the open surface (cathodic). The large cathodic surface (S_c) vs. anodic surface (S_a) ratio (S_c/S_a) that forms in these conditions is a definitive aggravating factor of the anodic (corrosion) reaction (Figure 9.4b).

Third, the cathodic oxygen-reduction reaction cannot be sustained in the crevice area, giving it an anodic character in the concentration cell. This anodic imbalance can lead to the creation of highly corrosive microenvironmental conditions in the crevice, conducive to further metal dissolution. It is also thought that subsequent pH changes at anodic and cathodic sites further stimulate local-cell action (Figure 9.4c).

The aggravating factors present in a fully developed crevice can be summarized in the following points:

- Metal ions produced by the anodic corrosion reaction readily hydrolyze, giving off protons (acid) and forming corrosion products. The pH in a crevice can reach very acidic values, sometimes equivalent to pure acids.
- The acidification of the local environment can produce a serious increase in the corrosion rate of most metals.
- The corrosion products seal even further the crevice environment.
- The accumulation of positive charge in the crevice becomes a strong attractor to negative ions in the environment, such as chlorides and sulfates that can be corrosive in their own right.

Crevice corrosion is known to be a serious derating factor when the load-bearing capacity of a bridge or of any other infrastructure component is evaluated during inspection.

9.3.2.1 Poultice Corrosion

Poultice corrosion is a special form of crevice corrosion, which usually takes the form of pitting when absorptive debris such as paper, wood, asbestos, sacking, cloth, and so forth are in contact with a metal surface that becomes wetted periodically. No action occurs while the entire assembly is wet, but during the drying period, adjacent wet and dry areas develop. Near the edges of the wet zones, differential aeration develops, which leads to pitting.

Areas on automobiles where poultice has occurred, for example, include the hem flange within doors, wheel wells, and inside frames. These areas remain wet almost continuously with a highly corrosive liquid because of the moisture-entrapment effect of the poultice. The aggravation caused by deicing salts can be quite serious in these areas because of wet–dry cycling and accumulation that can reach saturation. The stationary electrolyte can become increasingly acidic in this particular form of crevice corrosion.

A detailed analysis of poultices collected on 50 cars driven in four major North American cities has revealed the presence of large quantities of ions such as sodium, calcium, sulfate, and chloride Table 9.2. In these cities, sodium, calcium, and chloride

Table 9.2 Average Composition of Poultice (ppm) Found in 50 Cars

	Chloride	Sulfate	Calcium	Sodium
Boston	33	150	63	42
Dallas	23	133	348	82
Detroit	679	531	431	140
Montreal	63	3390	1404	77

(a)

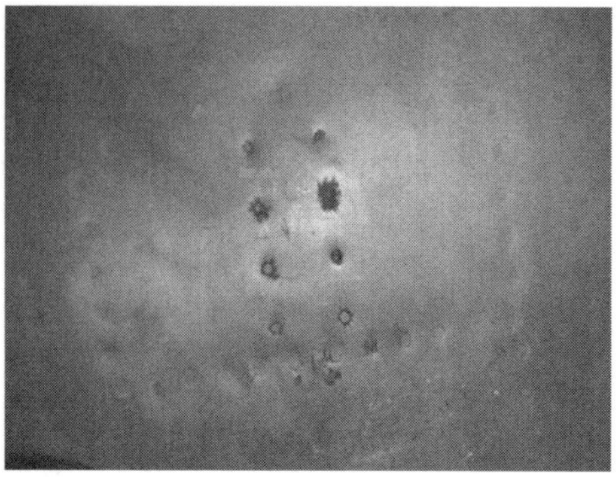

(b)

Figure 9.5 "Pillowing" on the fuselage of an aircraft: a) early stage (courtesy of Kingston Technical Software) and b) advanced stage on the belly of a Boeing 737, where the corrosion products have expanded to the point where a number of rivets have actually popped their heads (courtesy of Pacific Corrosion Control Corp).

deposit in poultices through road deicing and dust-control practices. Sulfate can be attributed to acid deposition commonly associated with pollution.[5]

9.3.2.2 *Pillowing*

Deformation caused by corrosion in lap joints of commercial airlines is accompanied by a bulging ("pillowing") between rivets, resulting from the increased volume of the corrosion product over the original material. Corrosion processes and the subsequent build-up of voluminous corrosion products inside the lap joints lead to so-called pillowing, whereby the faying surfaces are separated as illustrated in Figure 9.5a (early stage) and Figure 9.5b (advanced stage). A schematic description of the pillowing process is shown in Figure 9.6. The prevalent corrosion product identified in corroded fuselage joints is hydrated alumina [$Al(OH)_3$], with a particularly high volume expansion relative to aluminum as shown in (Figure 9.7). The build-up of voluminous corrosion products also leads to an undesirable increase in stress levels near critical fastener holes;

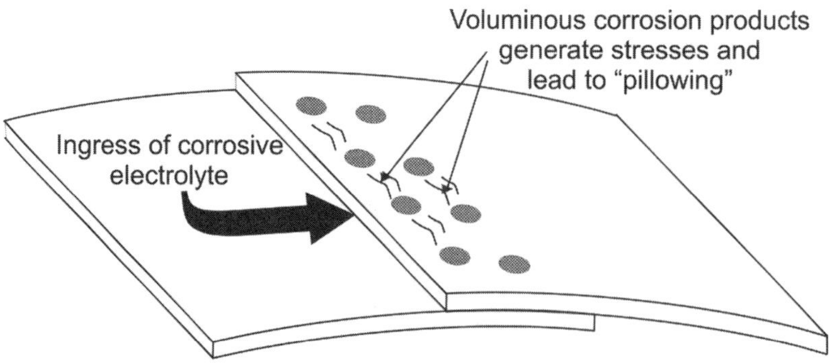

Figure 9.6 Schematic shows the "pillowing" of lap splices.

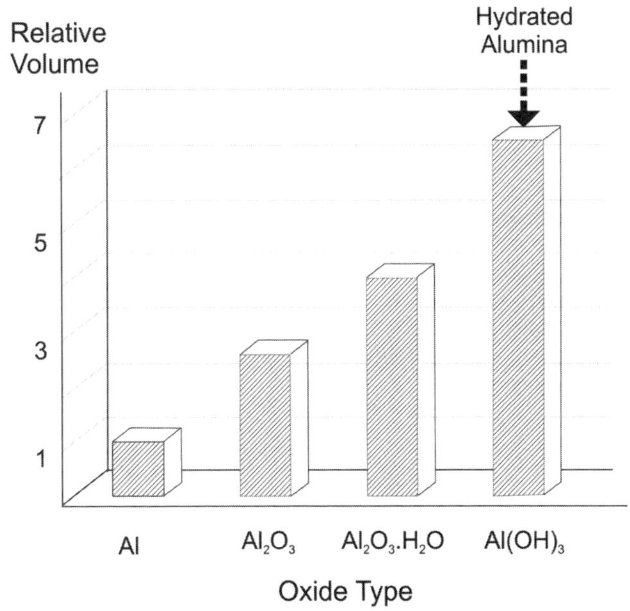

Figure 9.7 Relative volume of aluminum corrosion products.

rivets have been known to fracture because of the high tensile stresses resulting from pillowing.[6]

9.3.2.3 Filiform Corrosion

Filiform corrosion is a special form of oxygen-cell corrosion occurring beneath organic or metallic coatings on steel, zinc, aluminum, or magnesium. Filiform corrosion normally starts at small, sometimes microscopic, defects in the coating. Lacquers and "quick-dry" paints are most susceptible to the problem. The attack results in a fine network of random "threads" of corrosion product developed beneath the coating material with a shallow grooving of the metal surface (Figure 9.8). Such attack develops beneath semipermeable films in a high-humidity environment on such items as coated cans, office furniture, cameras, aircraft structures, auto interiors and exteriors, and a host of other common products.

In filiform corrosion, the head of the advancing filament (about 0.1 mm wide) becomes anodic, with a low pH and a lack of oxygen as compared with the cathodic area immediately behind the head where oxygen is available through the semipermeable film. Corrosion proceeds as the cathode follows behind the anodic head. The water and oxygen present in the cathodic area convert the anodic products to the usual oxides of the metal.

The microenvironments produced by filiform corrosion of steel are illustrated in Figure 9.9.[7] Essentially, a differential aeration cell is set up under the coating, with the lowest concentration of oxygen at the head of the filament because oxygen has to diffuse through the porous tail to the head region. A characteristic feature of such a differential aeration cell is the acidification of the environment with a low level of

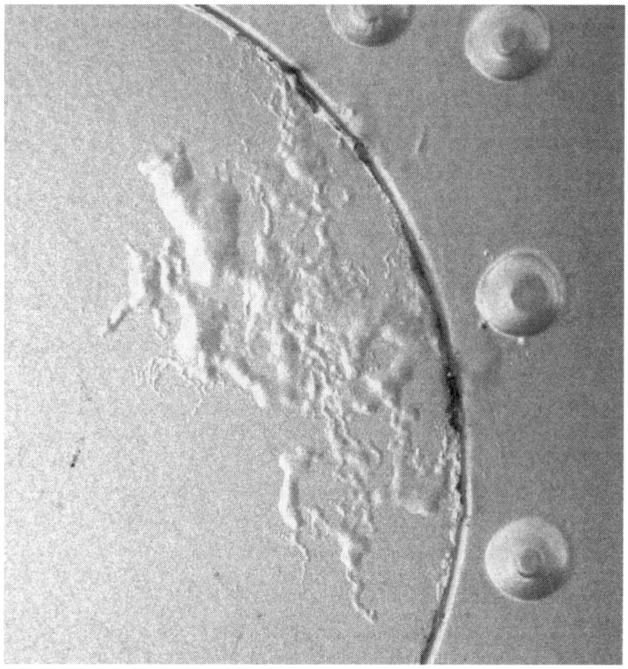

Figure 9.8 Filiform corrosion on the skin of an aircraft (courtesy of Kingston Technical Software).

dissolved oxygen. This leads to the formation of an anodic metal-dissolution site at the front of the head of the corrosion filament (Figure 9.9).

For iron, pH values at the front of the head of one to four have been reported with a corrosion potential around −0.44V (SHE). In contrast, at the back of the head where the cathodic reaction dominates, the prevailing pH is around 12. For aluminum, an

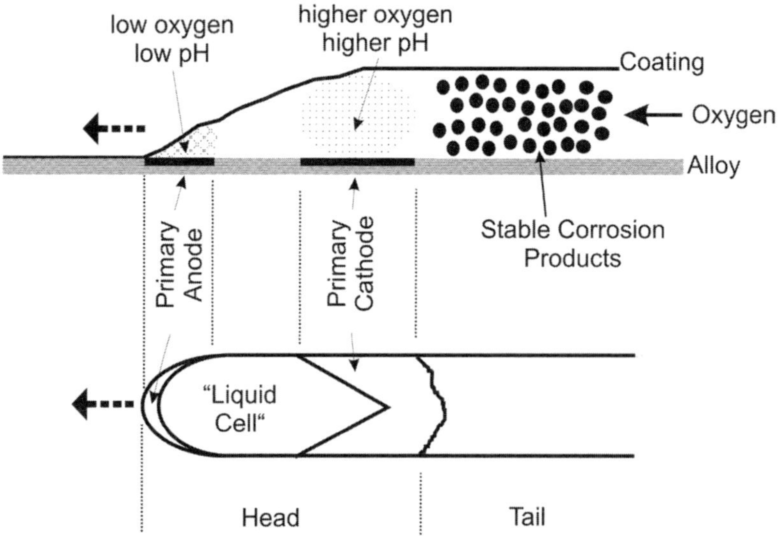

Figure 9.9 Graphical representation shows the microenvironments created by filiform corrosion.[7]

electrochemical potential at the front of the head of −0.73 V (SHE) has been reported, together with a 0.09-V difference between the front and the back of the head.[8] Reported acidic pH values close to 1 at the head and higher fluctuating values in excess of 3.5 associated with the tail.

9.3.2.4 *Crevice Corrosion Mitigation*

Crevice corrosion is ideally prevented at the design stage by avoiding crevices in the first place or during construction by filling them with a durable jointing compound that will exclude moisture and remain resilient. Several synthetic types are available with much longer lives than the vegetable-based compounds that have been used for decades.

Crevices are present in most types of equipment. Examples are gasketed flanges, rolled joints between tubes and tube sheets in heat exchangers, and faying surfaces between tanks and supporting structures. Moisture and chemical solutions may be trapped within crevices and held stagnant. The use of angle-iron construction, skip welding, high outlets in tank bottoms, "dead" flow areas in piping stubs, heat exchangers, and vessels must be avoided. A number of procedures may be used to avoid crevice corrosion (see Chapter 13, Section 4 for more details on this topic).

1. Use butt joints in preference to lap joints.
2. Eliminate or carefully seal lap joints so that they will not open up.
3. Provide complete drainage.
4. Inspect regularly; clean thoroughly.
5. Avoid the use of packing materials or thermal insulation, which can hold moisture in contact with metals.

Poultice corrosion is prevented by avoiding the contact of absorptive materials with a metal surface, by painting the surface that will contact such materials, or by designing to prevent such materials from becoming wet in service.

Filiform corrosion is minimized by careful surface preparation prior to coating, by the use of coatings that are resistant to this form of corrosion, and by careful inspection of coatings to insure that holidays, or holes, in the coating are minimized.

Pillowing is prevented by targeted inspection because it can be measured with various inexpensive, rapid, and nondestructive optical-imaging techniques. However, the relationship between percent material loss and pillowing surface deformation is complex.

9.3.3 Galvanic Corrosion

Galvanic corrosion (also called "dissimilar-metal corrosion" or, wrongly, "electrolysis") refers to corrosion damage induced when two dissimilar materials are coupled in a corrosive electrolyte. In a bimetallic couple, the less-noble material becomes the anode and tends to corrode at an accelerated rate compared with the uncoupled condition and the more noble material will act as the cathode in the corrosion cell.

The bimetallic driving force was discovered in the late part of the eighteenth century by Luigi Galvani in a series of experiments with the exposed muscles and nerves of a frog that contracted when connected to a bimetallic conductor. The principle was later put into a practical application by Alessandro Volta, who built, in 1800, the first electrical cell or battery: a series of metal disks of two kinds, separated by cardboard disks soaked with acid or salt solutions. The principle was also engineered into the useful protection of metallic structures by Sir Humphry Davy and Michael Faraday in the early part of the nineteenth century. The sacrificial corrosion of one metal such as zinc, magnesium, or aluminum has become a widespread method of cathodically protecting metallic structures.

In the galvanic couple example shown in Figure 2.8 (Chapter 2), the corrosion of one of the metals (steel) is accelerated and the corrosion of the other (copper) is reduced or stopped. Therefore, the first thing to establish is which of the metals will be affected one way or the other. The answer to this is provided by the direction in which an electric current will flow from the one metal (the anode) to the other (the cathode) as a result of the galvanic effect. This direction in turn will be determined by the difference in the potentials of the two metals in the solution in question. In the case of an aluminum/steel couple, the aluminum acts as the anode.

9.3.3.1 Galvanic Series

The potential of a metal in a solution is related to the energy that is released when the metal corrodes. Differences in corrosion potentials of dissimilar metals can be measured in specific environments by measuring the direction of the current that is generated by the galvanic action of these metals when immersed in a given environment. Such measurements could be repeated with all the possible combinations of metals in any corrosive solution.

Examination of the results would make it possible to arrange the metals in what is called a galvanic series. If the experiments were to be repeated using a different solution, a different concentration of sodium chloride, a different degree of aeration, a different velocity of movement, or a different temperature, the values recorded could be different, and the positions of some of the metals relative to each other in the new galvanic series may change.

However, the situation may not be as bad as the preceding statement implies. The relative tendencies of metals to corrode remain about the same in many of the environments in which they are likely to be used. Consequently, their relative positions in a galvanic series may be about the same in many environments. Because more observations of potentials and galvanic behavior have been made in seawater than in any other single environment, an arrangement of metals in a galvanic series based on observations in seawater, as shown in Table 9.3,[9] is frequently used as a first approximation of the probable direction of the galvanic effects in other environments. For sake of comparison Table 9.4 presents the galvanic series that was constructed for metals exposed to neutral soils and water.[10]

There has been some confusion in the literature regarding electromotive force (EMF) potentials and the galvanic series. Although there are similarities between the

Table 9.3 Galvanic Series of Selected Metals and Alloys in Stagnant Seawater (Active at Top and Noble, Less Active, and Cathodic at Bottom)

Magnesium and its alloys
Zinc (hot dip, die cast, or plated)
Beryllium (hot pressed)
Aluminum alloys: Al 7072 clad on 7075, Al 2014-T3, Al 1160-H14, Al 7079-T6
Cadmium (plated)
Uranium
Aluminum alloys: Al 218 (die cast), Al 5052-0, Al 5052-H12, Al 5456-0, H353, Al 5052-H32, Al 1100-0, Al 3003-H25, Al 6061-T6, Al A360 (die cast), Al 7075-T6, Al 6061-0
Indium
Aluminum alloys: Al 2014-0, Al 2024-T4, Al 5052-H16
Tin (plated)
Stainless steel 430 (active)
Lead
Steel 1010
Iron (cast)
Stainless steel 410 (active)
Copper (plated, cast, or wrought)
Nickel (plated)
Chromium (plated)
Tantalum
AM350 (active)
Stainless steels (active): 310, 301, 304, 430, 17-7PH
Tungsten
Niobium (columbium) 1% Zr
Yellow brass (268)
Uranium 8% Mo
Brass, naval brass (464), muntz metal (280), brass (plated)
Nickel/silver (18% Ni)
Bronze (220), copper (110), red brass
Stainless steel 347 (active)
Molybdenum, commercial pure
Copper/nickel (715), admiralty brass
Stainless steel 202 (active)
Bronze, phosphor 534 (B-1), Monel 400
Stainless steel 201 (active)
Alloy 20 (active)
Stainless steels (active): 321, 316, 309
Stainless steel 17-7PH (passive)
Silicone bronze 655
Stainless steels (passive): 304, 301, 321, 201, 286, 316
AM355 (active)
Stainless steels (passive): alloy 20, AM355, A286
Titanium alloys
AM350 (passive)
Silver
Gold
Graphite

Table 9.4 Practical Galvanic Series for Metals in Neutral Soils and Water

Metal	Potential V (CSE)
Commercially pure magnesium	−1.75
Magnesium alloy (6% Al, 3% Zn, 0.15% Mn)	−1.6
Zinc	−1.1
Aluminum Alloy (5% zinc)	−1.05
Commercially pure aluminum	−0.8
Mild steel (clean and shiny)	−0.5 to −0.8
Lead	−0.5
Cast iron (not graphitized)	−0.5
Mild steel (rusted)	−0.2 to −0.5
Mill scale on steel	−0.2
High silicon cast iron	−0.2
Copper, brass, bronze	−0.2
Mild steel in concrete	−0.2
Platinum	0 to −0.1
Carbon, graphite, coke C	+0.3

Typical potentials normally observed in neutral soils and water, measured in relation to copper-sulfate reference electrode.

galvanic series presented in Table 9.3 and the standard reduction potentials in Table 2.5 (Chapter 2), there are also some fundamental differences. Although standard potentials can provide an indication of the stability of a metal, as it is done with E-pH or Pourbaix diagrams, galvanic series are used to predict whether galvanic corrosion will occur, and if so, which of the two coupled metals will exhibit increased corrosion. Thus, these two tabulations have entirely different uses and should therefore not be confused.

In a galvanic couple involving any two metals in a galvanic series, corrosion of the metal higher in the list is likely to be accelerated, whereas corrosion of the metal lower in the list is likely to be reduced. Metals with more positive corrosion potentials are called noble or cathodic, and those with more negative corrosion potentials are referred to as active or anodic metals and alloys.

Note that several metals in Table 9.3 are grouped. The potential differences within a group are not likely to be great and the metals can be combined without substantial galvanic effects under many circumstances.

Values of potential can change from one solution to another or in any solution when influenced by such factors as temperature, aeration, and velocity of movement. Consequently, there is no way, other than by direct potential measurements in the exact environment of interest, to predict the potentials of the metals and the consequent direction of a galvanic effect in that environment. As an example, zinc is normally very negative or anodic to steel at ambient temperatures, as indicated in the galvanic series shown in Table 9.3. However, the potential difference decreases with an increase in temperature until the potential difference may be zero or actually be reversed at 60°C.

9.3.3.2 Recognizing Galvanic Corrosion

For galvanic corrosion to occur, three conditions must be present:

1. electrochemically dissimilar metals must be present,
2. these metals must be in electrical contact, and
3. the metals must be exposed to an electrolyte.

Consider, for instance, that Type 201 stainless steel (S20100) in electrical contact with Type 316 stainless steel (S31600) is observed to be rapidly corroding. Table 9.3 indicates that this should not be the result of galvanic corrosion because these alloys are grouped in the galvanic series. Therefore, separating these two metals would not improve the corrosion resistance of these stainless steels.

Consider also that a piece of aluminum connected to a cast-iron motor block immersed in hot motor oil was rapidly attacked. This is not a case of galvanic corrosion, because motor oil and most organic liquids are not electrolytes. Again, separating these two metals would not improve the resistance of the aluminum.

In addition to the three previously listed conditions necessary for galvanic corrosion, another way of recognizing this kind of attack is to look for localized corrosion near the junction between two dissimilar metals. Galvanic corrosion is usually most intense immediately adjacent to the cathodic material.

9.3.3.3 Magnitude of Galvanic Effects

Although the determination of the direction of galvanic action in dissimilar-metal situations is important, one should also be concerned with the intensity of the galvanic action. This intensity is determined by the amount of current or, more specifically, the current density (current per unit area).

When exposed to a given environment, the potential of a metallic material is determined by many factors, such as temperature, liquid flow rate, the level of aeration, and so on. However, the relative ranking of alloys in the galvanic series remains largely unaffected by such changes, even though their corrosion potential may shift by several hundred millivolts. Figure 9.10 shows the electrochemical series in a well-documented environment, flowing seawater.[11] The difference between the most electropositive and the most electronegative metals is nearly 2 V, and the coupling of these materials could generate serious currents and hence high corrosion rates on the most anodic material.[12]

However, the magnitude of the potential difference alone will not necessarily indicate the amount of galvanic corrosion. For instance, metals with a potential difference of only 50 mV have shown severe galvanic-corrosion problems, whereas metals with a potential difference of 800 mV have been successfully coupled together. This is because the potential difference gives no information about the kinetics of galvanic corrosion, which depends on the current flowing between the two metals in the couple.

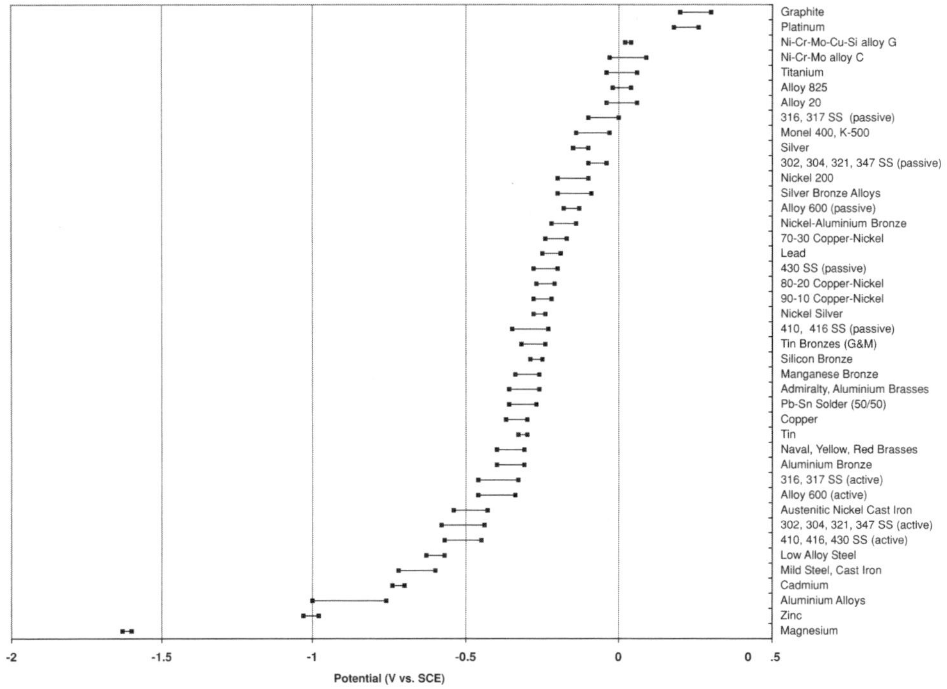

Figure 9.10 Galvanic series for slow-moving seawater.[11]

9.3.3.4 *Area Effects*

Another important factor in galvanic corrosion is the area effect or the ratio of cathodic to anodic area, which was discussed in Chapter 2. The larger the cathode compared with the anode, the more oxygen reduction, or other cathodic reaction, can occur, and hence the greater the galvanic current. Such a situation can result in more corrosion on the anode as shown schematically in Figure 2.8 (Chapter 2).

From the standpoint of practical corrosion resistance, the most unfavorable ratio is a very large cathode connected to a very small anode. This effect is illustrated in Figures 9.11 and 9.12. Table 9.3 indicates that iron is anodic with respect to copper and therefore is more rapidly corroded when placed in contact with it. This effect is greatly accelerated if the area of the iron is small in comparison with the area of the copper, as shown in Figure 9.11. However, under the reverse conditions when the area of the iron is very large compared with the copper, the corrosion of the iron is only slightly accelerated.

Under static or slow-flow conditions, where the galvanic-corrosion current is often dependent on the rate of diffusion of dissolved oxygen to the cathode, the total amount of galvanic corrosion is independent of the size of the anode and proportional to the area of the cathodic-metal surface. This is sometimes known as the catchment area principle and has important implications in designing to minimize the risk of galvanic corrosion. Thus, for a constant area of cathode metal the total amount of corrosion of the anode is constant, but the corrosion per unit area increases as the area of the anode

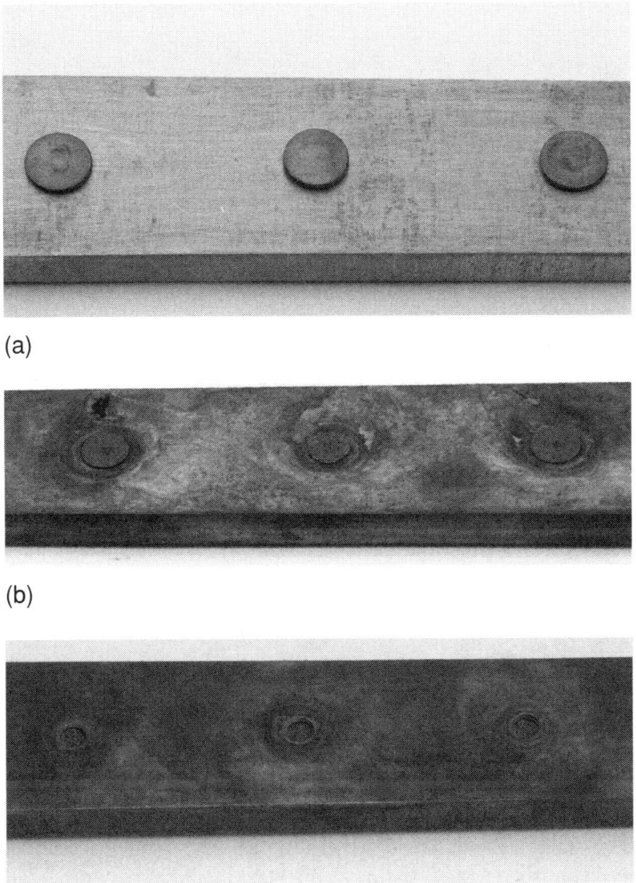

(a)

(b)

(c)

Figure 9.11 Steel rivets on a copper bar a) at the start of the experiment, b) 6 months after being submerged in 3% sodium chloride solution, and c) after 10 months in the same solution.

is decreased. That is, the corrosion current is constant but the current density increases as the anode area decreases.[12]

If the electrical conductance of the electrolyte bridging the galvanic contact is low, either because the bulk conductivity is low or because the electrolyte is present only as a thin film, the effective areas taking part in galvanic cell reactions are small and the total amount of corrosion is relatively small, although it may be severe immediately adjacent to the metal junction.

Under immersed conditions in many supply waters, which generally have a relatively low electrical conductivity, adverse effects are uncommon if the contacting metals are of similar area. Thus, galvanized steel pipes can be used with brass or gunmetal connectors, but serious corrosion to the pipe end is likely to result if the contact is directly to a large area of copper, such as a tank or cylinder. Similarly, stainless steel and copper tubes can usually be joined without problems, but accelerated corrosion of the copper tube is likely to occur if it is attached to a stainless-steel tank.

(a)

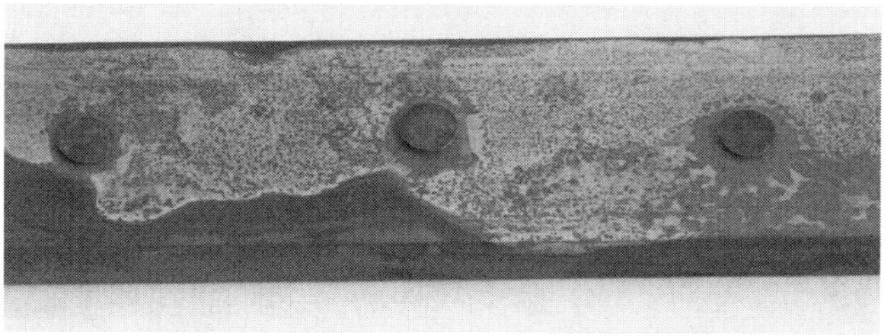

(b)

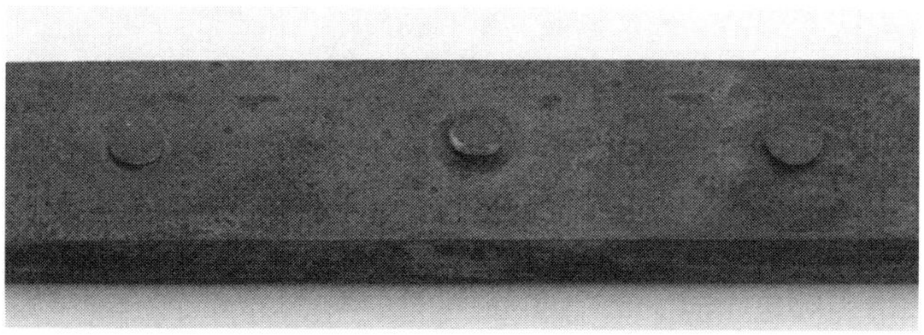

(c)

Figure 9.12 Copper rivets on a steel bar a) at the start of the experiment; b) 6 months after being submerged in 3% sodium chloride solution, and c) after 10 months in the same solution.

Under immersed conditions in a highly conducting electrolyte, such as seawater, effective areas will be greater and severe corrosion may be encountered on small anodic areas of many metals. Extremely small anodic areas exist at discontinuities, such as cracks or pinholes, in cathodic coatings such as magnetite (millscale on iron) and copper plating on steel.

By correct selection of area ratio, dissimilar metals have frequently been used successfully together. One example is the use of type 316 stainless-steel pump impellers in austenitic cast-iron pump casings in seawater. Another is the use of high-alloy valve trim (seats, stems, etc.) in such metals as alloy 625 (UNS N06625), Ni-Cu alloy K-500 (N05500), and so on with gunmetal or aluminum bronze valve bodies in seawater. In both cases the large area of the anode compared to the cathode means that the increased corrosion is small and is well within design limits for the valve or pump body.[12]

9.3.3.5 Deposition Corrosion

Deposition corrosion is a subtle form of galvanic cell that can produce pitting in a liquid environment when a more cathodic metal is plated out of solution onto a metal surface.

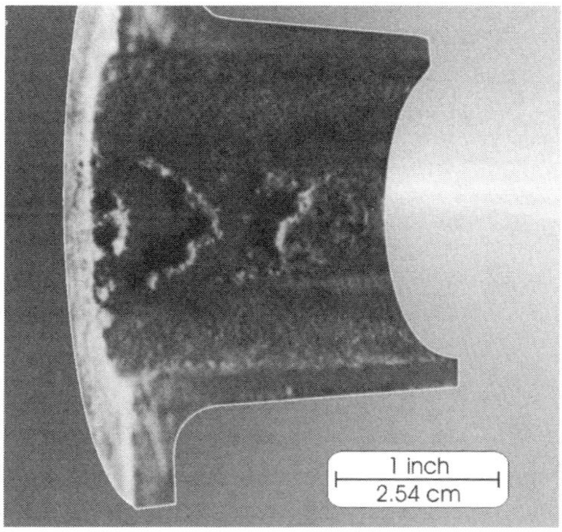

Figure 9.13 Deposition corrosion. Copper deposited on the interior of an aluminum pipe caused the severe pitting.[1]

For example, soft water passing through a copper water pipe will accumulate some copper ions. If water is then admitted to a galvanized or aluminum vessel, particles of metallic copper will plate out, that is, deposit on the surface and stimulate pitting by local cell action (Figure 9.13). Common cathodic "activators" are mercury and copper ions in solution. In the particular case shown in Figure 9.13, the failure of the aluminum (A93003) pipe occurred after 2 years of service in handling glacial acetic acid contaminated with less than 10 ppm of Cu^{2+} ions.[1]

The listing of metals in Tables 9.3 and 9.4 can provide an approximate tendency of one metal to plate out on another metal when one metal is placed in a solution containing ions of another metal. A common example of this is the plating out of silver on a piece of copper immersed in a solution of silver nitrate (Figure 9.14). The bubbling, or "plating out of hydrogen" on zinc (Figure 9.15) is analogous.

This plating-out action, or deposition corrosion, may be an important factor in the corrosion of the more reactive metals near the top of the series (e.g., magnesium, zinc, and aluminum) when these latter metals come into contact with solutions containing ions of metals (particularly copper) lower in the series. Copper ions in concentrations less than 1 ppm have been observed to have a significant effect on the corrosion of aluminum by water. Metals, such as copper, that can aggravate corrosion of aluminum are sometimes referred to as heavy metals in solution. The fact that they are heavier than aluminum is less significant than that they occupy positions lower than aluminum in the electromotive or galvanic series.

9.3.3.6 Galvanic-Corrosion Mitigation

There are a number of ways that galvanic corrosion can be prevented. These can be used singly or in combination. All of these preventive measures follow directly from the basic mechanism of galvanic corrosion.

1. Avoid the use of dissimilar metals wherever possible. If this is not practical, try to use metals that are close together in the galvanic series (Table 9.3).

Figure 9.14 A common example of the plating out of silver on a piece of copper immersed in a 5% solution of silver nitrate (courtesy of Kingston Technical Software).

Figure 9.15 Bubbling or "plating out of hydrogen" on zinc immersed in a 0.1-M sulfuric-acid solution (courtesy of Kingston Technical Software).

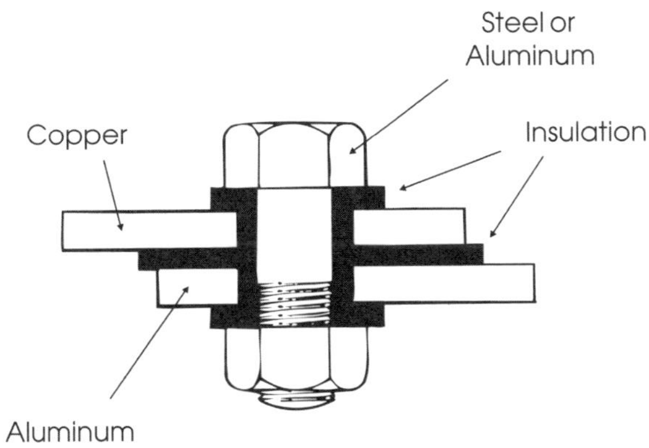

Figure 9.16 Two dissimilar metals are insulated for protection against galvanic corrosion.

2. Avoid an unfavorable area ratio whenever possible. Under no circumstances should a small anode be connected to a large cathode.
3. If dissimilar metals are used, insulate these electrically from one another. An example of this technique is shown in Figure 9.16, which illustrates the insulation of aluminum and copper plates joined by a steel nut-and-bolt assembly. It is very important to make sure that there is no electrical contact between metallic components apart in the galvanic series when they are exposed to a corrosive environment.
4. If it is necessary to use dissimilar metals, and these cannot be insulated, then the more anodic part should be designed for easy replacement or should be constructed of thick materials to longer absorb the effects of corrosion.
5. Coat the cathode (or both anode and cathode) near the junction to reduce the effective cathodic area. Never coat the anode alone.

Deposition corrosion can be avoided by preventing the pick-up of cathodic ions that will enter the equipment or by scavenging them by passing the contaminated product through a tower packed with more anodic-metal turnings on which the ions can deposit (Figure 9.17). Another approach to this problem is the use of glass-reinforced plastic pipe.

9.4 VELOCITY-RELATED CORROSION

9.4.1 Erosion–Corrosion

A metal surface over which a smooth, linear flow of liquid is maintained will corrode at essentially the same rate as the metal would in the stagnant solution. This is true because at the surface–liquid interface there is a static film of the liquid molecules. Unless the liquid is aggressive as a corrosive, the metal will form some semiprotective film over the surface and remain stable.

However, if the flow of liquid becomes turbulent, the random liquid motion impinges on the surface to remove the thin film. Additional oxidation then occurs by

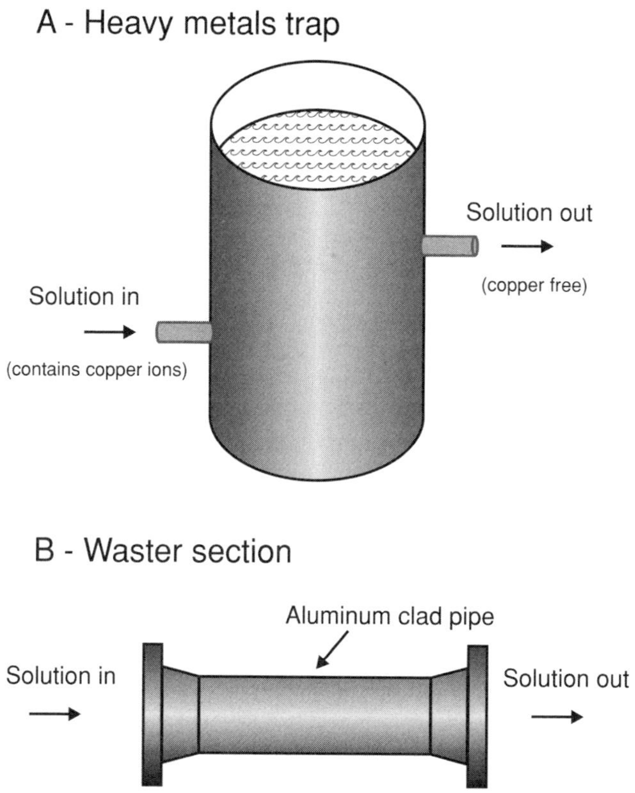

Figure 9.17 Method for removing troublesome ions from solution: a) heavy metal trap: solutions containing copper ions enter a barrel filled with aluminum shavings and b) waster section: aluminum clad pipe inserted in a system removes heavy-metal ions. The section is replaced once corroded.

reaction with the liquid. This alternate oxidation and removal of the film will accelerate the rate of corrosion. The resulting erosive attack may be uniform but quite often produces pitted areas over the surface that can result in full perforation (Figure 9.18).

Obviously, the presence of solid particles or gaseous bubbles in the liquid can accentuate the attack. Also, if the fluid dynamics are such that an impingement or cavitation attack is developed, even more severe corrosion can occur.

9.4.2 Cavitation

Cavitation damage (sometimes referred to as cavitation corrosion or cavitation erosion) is a form of localized corrosion combined with mechanical damage that occurs in turbulent or rapidly moving liquids and takes the form of areas or patches of pitted or roughened surface. Cavitation has been defined as the "deterioration of a surface caused by the sudden formation and collapse of bubbles in a liquid." It has been similarly defined as the "localized attack that results from the collapse of voids or cavities due to turbulence in a liquid at a metal surface." Cavitation also occurs in areas of high vibration such as on engine pistons and piston liners (Figure 9.19).

Figure 9.18 Erosion–corrosion of a brass tube carrying out seawater (courtesy of Kingston of Technical Software).

Cavitation is the formation of voids or cavities in a liquid caused by turbulence or temperature, which causes the pressure in local zones of the liquid to fall below the vapor pressure. This happens on the backside of ship propellers, water-turbine blades in pumps, high-velocity flow lines, and the submerged planes of hydrofoil craft, depending the design of equipment and degree of turbulence.

The voids are actually "holes" in the liquid that contain only water vapor. These usually exist for only a very short time because changes in liquid pressure in the turbulent local zones close the voids (in times as short as a microsecond [one millionth of a second]). As the liquid surfaces meet during bubble collapse, considerable kinetic

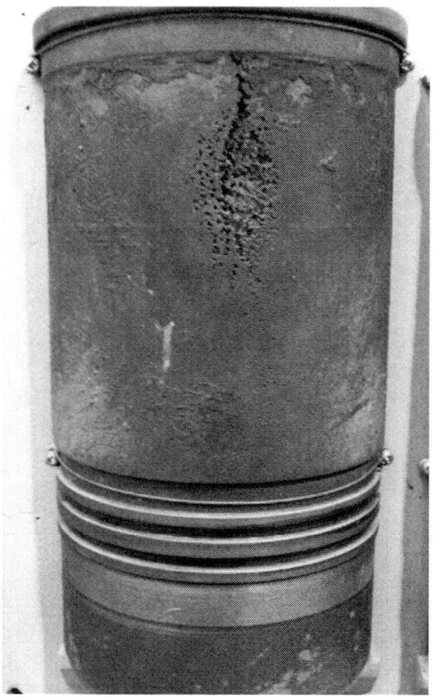

Figure 9.19 Cavitation damage of a diesel-engine piston liner on the return stroke (courtesy of Kingston Technical Software).

Figure 9.20 Pump impeller perforated by cavitation attack. Note that the same area on each vane is attacked.

energy is released. This had been estimated to be as high as 687,000 kPa (100,000 psi). Depending on the speed of collapse, the energy released can produce a clap, similar to the familiar water hammer that occurs in water pipes when a valve is closed suddenly. The claps of energy impinge on the metal surface and may break a protective surface film and lead to the initiation of corrosion.

A great many cavities may form and collapse in a small zone (perhaps more than a million) in the course of one second. As a result, cavitation damage can occur very rapidly. For example, a cast aluminum alloy with 10% Mg installed to drive a 4 meters boat at 57 km/h developed visible cavitation damage areas on the backsides of the blades in only 15 min. Cavitation of a pump impeller is also quite common (Figure 9.20).

In some instances, CP has been successful in reducing or preventing cavitation damage, but because cavitation damage usually involves physical as well as electrochemical processes, it cannot always be prevented by this means. In some cases, inhibitors have been used successfully to limit cavitation corrosion, as in the water side of diesel-engine cylinder liners.

It is believed that work hardening of the surface occurs, and ultimately bits of metal are knocked from the surface by mechanical cracking and fatigue failure. Ductile metals resist cavitation damage longer than more brittle materials. On the other hand, 11 to 13% of chromium martensite stainless steels are often superior to bronze in water service. As might be expected, there is good correlation between fatigue life and cavitation resistance.

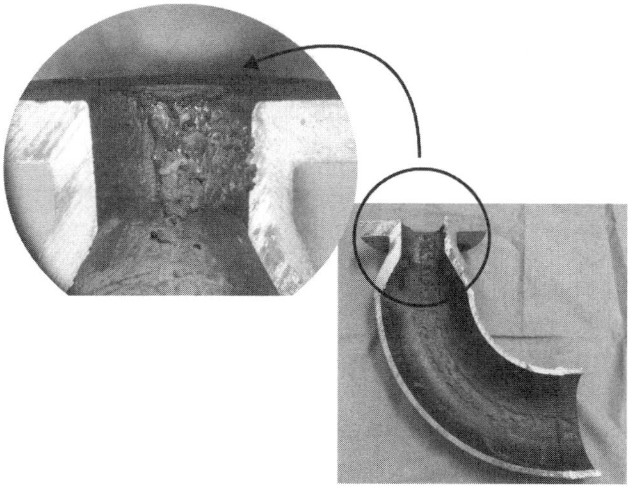

Figure 9.21 Cavitation on the downstream side of the service water system downstream of a valve (courtesy of Russ Green, TMI).

Cavitation resistance increases as the grain size of a metal becomes smaller. The nature of the grain boundaries also plays a role. Stainless steels are among alloys more resistant to cavitation damage. This is due to their ductility, toughness, high corrosion fatigue limit, homogeneity, fine grain size, and ability to work harden.

Cavitation can occur in pipes when liquid is moving at high velocity, especially at fittings where the smooth bore of the pipe is interrupted. The effect is usually to produce severe pitting on the downstream side of the turbulence as is shown in the case of the service water system downstream of a valve (Figure 9.21).

The prevention of cavitation damage requires the use of the most resistant alloys and, where possible, design of the system to avoid turbulence and cavitation. Design changes may aim to raise pressures or eliminate abrupt changes in cross section (of pipes). The shape of a propeller or impeller also may be altered.

9.4.3 Impingement

Impingement attack is related to cavitation damage and has been defined as "localized erosion–corrosion caused by turbulence or impinging flow." Entrained air bubbles tend to accelerate this action, as do suspended solids. This type of corrosion occurs in pumps, valves, orifices, and heat-exchanger tubes and at elbows and tees in pipe lines.

Impingement corrosion usually produces a pattern of localized attack with directional features. The pits or grooves tend to be undercut on the side away from the source of flow, in the same way that a sandy river bank at a bend in the river is undercut by the oncoming water (Figure 9.22).

When a liquid is flowing over a surface (e.g., in a pipe), there is usually a critical velocity below which impingement does not occur and above which it increases rapidly. Impingement attack first received attention because of the poor behavior of some copper alloys in seawater.

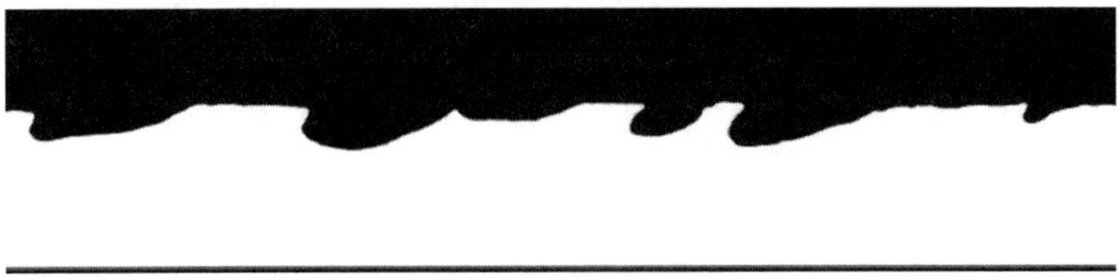

Figure 9.22 Cross section of copper tube attacked by turbulent erosion–corrosion.

In practice, impingement and cavitation may occur together, and the resulting damage can be the result of both. Impingement may damage a protective oxide film and cause corrosion or it may mechanically wear away the surface film to produce a deep groove.

The solution to impingement attack is the use of more resistant alloys. The 90–10 copper–nickel has better impingement resistance than aluminum brass, which in turn is better than copper.[13] The 70–30 alloy shows better resistance than the 90–10 alloy. This is not a complete representation, however. Rates of attack are not only dependent on seawater velocity but also on pipe work diameter. Table 9.5 compares results of tests on condenser-tube alloys using the jet-impingement technique in two independent laboratories. General experience has shown that 90–10 copper–nickel can successfully be used in condensers and heat exchangers with water velocities up to 2.5 m/s. When the 70–30 brasses began failing in seawater service (e.g., condenser tubes), it was found that 70–30 cupronickel containing 0.4 to 1.0% Fe was considerably more resistant, whereas 90–10 cupronickel is even more resistant.

A Ti-6A1-4V alloy has a high resistance to impingement attack. Titanium is fully resistant to natural seawater regardless of chemistry variations and pollution effects (i.e., sulfides). Twenty-year corrosion rates well below 0.0003 mm y^{-1} have been measured on titanium exposed beneath the sea and in splash or tidal zones. In the sea, titanium alloys are immune to all forms of localized corrosion and withstand seawater impingement and flow velocities in excess of 30 m s^{-1}. Table 9.6 compares the erosion-corrosion

Table 9.5 Comparison of Test Results Obtained on Condenser Tube Alloys by Two Independent Laboratories

	Average Depth of Attack (μm)	
Material	BNFMA[(a)]	LCCT[(b)]
Arsenical admiralty brass	340	270
Arsenical copper	300	—
70–30 copper–nickel, 0.04% Fe	110	220
Aluminum brass	40[(c)]	200
70–30 copper–nickel, 0.8% Fe	20	—
70–30 copper–nickel, 0.45% Fe	—	100
90–30 copper–nickel, 2% Fe	0	150

Testing conditions: velocity of jet, 4.58 m s^{-1}, air added, 3% by volume; duration, 28 days.

(a) British Non Ferrous Metal Research Association Laboratories, UK.

(b) LaQue Centre for Corrosion Technology, NC, USA.

(c) One specimen out of 20 pitted to a depth of 650 μm. No other specimen greater than 200 μm.

Table 9.6 Erosion of Unalloyed Titanium in Seawater Containing Suspended Solids

Flow Rate ($m\ s^{-1}$)	Suspended Matter	Duration (h)	Erosion Corrosion ($\mu m\ y^{-1}$) Ti Grade 2	Cu-Ni 70/30[a]	Al B
7.2	None	10,000	Nil	Pitted	Pitted
2	40 g/l 60 mesh sand	2,000	2.5	99.0	50.8
2	40 g/l 10 mesh sand	2,000	12.7	Severe erosion	Severe erosion

[a]High-iron, high-manganese 70–30 copper-nickel.

resistance of unalloyed titanium with two commonly used seawater materials.[14] In addition, the fatigue strength and toughness of most titanium alloys are unaffected in seawater and many titanium alloys are immune to seawater stress corrosion.

9.5 OTHER TYPES OF LOCAL ATTACK

9.5.1 Fretting Corrosion

Fretting corrosion refers to corrosion damage at the asperities of contact surfaces (Figure 9.23). This damage is induced under load and in the presence of repeated relative surface motion, as induced for example by vibration. Pits or grooves and oxide debris characterize this damage, typically found in machinery, bolted assemblies, and ball or roller bearings. Contact surfaces exposed to vibration during transportation are exposed to the risk of fretting corrosion.

Damage can occur at the interface of two highly loaded surfaces that are not designed to move against each other. The most common type of fretting is caused by vibration. The protective film on the metal surfaces is removed by the rubbing action and exposes fresh, active metal to the corrosive action of the atmosphere. Conditions

Figure 9.23 Fretting corrosion at the interface of two highly loaded surfaces that are not designed to move against each other (courtesy Pacific Corrosion Control Corp.).

necessary for the occurrence of fretting are (a) the interface must be under load and (b) vibratory or oscillatory motion of small amplitude must result in the surfaces striking or rubbing together. The results of fretting are as follows:

1. metal loss in the area of contact;
2. production of oxide debris;
3. galling, seizing, fatiguing, or cracking of the metal;
4. loss of dimensional tolerances;
5. loosening of bolted or riveted parts; and
6. destruction of bearing surfaces.

The mechanism of fretting corrosion involves both chemical or electrochemical and mechanical factors. The relative motion scrubs off tiny high points or surface irregularities on the metal surface, and these, in turn, are chemically converted to oxide. They also may be sites for galling, in which case the mechanical action results in tearing out small bits of metal from one or the other of the interfaces, and these oxidize to form debris.

Two different theories have been advanced to explain the mechanism of fretting corrosion. In the first, it is assumed that pressure causes small areas of the mating surfaces to weld together. Motion then tears a small volume of metal from one of the surfaces (galling). Subsequent motion detaches this, and then a small volume of metal from the other surface is removed to produce another discrete fine particle of metal. In the presence of oxygen and moisture, these metal particles oxidize or corrode to produce a metal-oxide corrosion product.

The term fretting is also used to describe the result when abrasion caused by the relative motion removes protective surface films, leading to oxidation or corrosion of the fresh metal surface to form corrosion products, which are detached from the surface by subsequent further abrasion.

Fretting corrosion can be prevented by eliminating any slipping movement between the two surfaces. Thus, it is sometimes possible to overcome fretting by increasing the load on the surfaces, if it prevents relative motion. Alternatively, decreasing the load can minimize the effect of vibratory motion. In other cases, roughening the surfaces can increase the friction between them and stop the movement. Fretting corrosion can be greatly retarded by lubricating the contacting surfaces with an oil or grease with sufficient load-bearing characteristics that it will separate the surface from the environment and thus from local welding even under the loads and vibrations encountered. Listed are four means of mitigating fretting damage:

1. Use mating surfaces of differing character, for example soft metal against harder metal.
2. Avoid small relative motion between surfaces, for example, roughen interface and/or increase load. Alternatively, greatly increase relative motion.
3. Exclude air, that is, use cements or sealants.
4. Lubricate the faying surfaces either by use of lubricants such as molybdenum sulfide or by making one of the contacting surfaces of a low-friction material such as Teflon.

In the case of metal sheets in packages, a rigid package and pressure packing of the contents will eliminate relative motion during transit and thus eliminate fretting corrosion. Interleaving with paper and oiling are also beneficial.

9.5.2 Intergranular Corrosion

Intergranular corrosion is a form of localized surface attack in which a narrow path is corroded out preferentially along the grain boundaries of a metal. It initiates on the surface and proceeds by local cell action in the immediate vicinity of a grain boundary. Although the detailed mechanism of intergranular corrosion varies with the metal system, its physical appearance at the microscopic level is quite similar for most systems. The effects of this form of attack on mechanical properties may be extremely harmful.

The driving force is a difference in corrosion potential that develops between a thin grain-boundary zone and the bulk of the immediately adjacent grains. The difference in potential may be caused by a difference in chemical composition between the two zones. This can develop as a result of migration of impurity or alloying elements in an alloy to the grain boundaries. If the concentration of alloying elements in the grain-boundary region becomes sufficient, a second phase or constituent may separate or precipitate. This may have a corrosion potential different from that of the grains (or matrix) and cause a local cell to form.

The constituent may be anodic, cathodic, or neutral to the base metal or adjacent denuded zone. Examples of anodic constituents are the intermetallic phases Mg_5Al_8 and $MgZn_2$ in aluminum alloys and Fe_4N in iron alloys. Examples of cathodic constituents are $FeAl_3$ and $CuAl_2$ in aluminum alloys and Fe_3C in iron alloys. Examples of neutral constituents are Mg_2Si and $MnAl_6$ in aluminum alloys and Mo_6C and W_6C in wrought Ni-Cr-Mo alloys.

If the precipitate is anodic to the denuded zone, it corrodes preferentially, and if it is present in a continuous network throughout the grain boundary, it is removed to produce a thin channel, fissure, or crack between adjacent grains as is discussed later in this chapter.

In mild cases, the penetration rate may be extremely slow and even after a considerable time may penetrate to a depth of only a dozen grains or so. In more severe cases, penetration may go quickly completely through sections that are much thicker. Shallow intergranular corrosion may escape visual inspection but is seen easily under a microscope. As attack proceeds, it can be recognized more readily by the naked eye.

Intergranular corrosion causes a drop in the ductile property of a metal, which can become appreciable before any significant loss of tensile stress or yield stress can be detected. In severe cases, there can be a marked loss of tensile properties, even though only a small volume of metal has corroded.

If the constituent is cathodic to the denuded zone, it remains intact and local-cell action develops in the adjacent denuded zone. Because most intergranular corrosion is the result of small differences in composition at grain boundaries, the metallurgical history of an alloy can be as important as its average chemical composition.

This is because heat treatments and cold work affect not only the size and shape of the grains, but also the composition, location, amount, and size of the constituents.

Heating an alloy to a high temperature just below the melting point tends to dissolve the alloying elements. This homogenization treatment is often carried out in practice and is called a solution heat treatment. If the alloy is then quenched rapidly, the elements are trapped in solid solution. The solution is now supersaturated and there is a tendency for the elements to precipitate out as constituents, providing additional strength to the alloy. However, depending on the composition and location of these precipitated constituents, they may become the cause of intergranular corrosion.

The amount and distribution of the precipitated constituents is important, because even anodic and cathodic constituents will not cause intergranular corrosion if they are present in small sizes and are uniformly dispersed throughout the grain structure. This condition can be achieved by overaging, which means heating for a time longer than that which develops maximum mechanical properties of the alloy.

9.5.2.1 Aluminum Alloys

The aluminum/copper alloys (e.g., A92024), containing about 4% copper, exhibit intergranular attack under certain circumstances. The occurrence of intergranular corrosion may be sensitive to precipitation and the prior metallurgical history of the alloys.

After fabrication of Al-Cu alloys, it is customary to solution heat treat at approximately 490°C and water quench. This class of alloys will age harden at room temperature. If, however, the quench is not severe (e.g., quenching in hot or boiling water), $CuAl_2$ precipitates as a "string of pearls" along grain boundaries, leaving a zone adjacent to the grain boundaries that is depleted of copper (Figure 9.24).

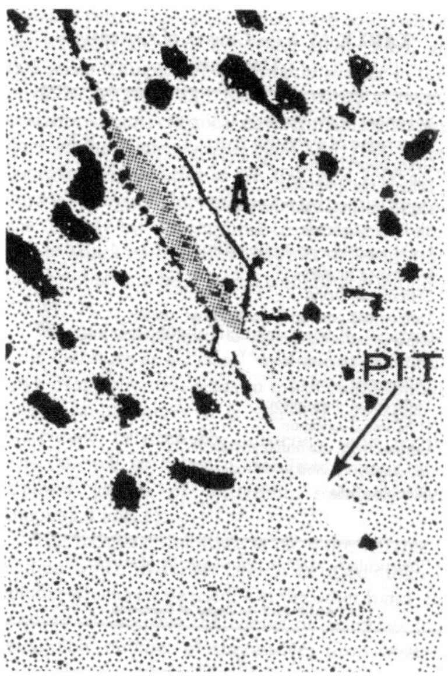

Figure 9.24 Consequences of precipitation at grain boundaries in Al-Cu alloys. In zone A, one alloy constituent is reduced in local concentration with respect to bulk alloy. The dark copper-rich phase is present at grain boundaries and in the grains. Specimen was aged 4 h at 190°C (×50,000 magnification).

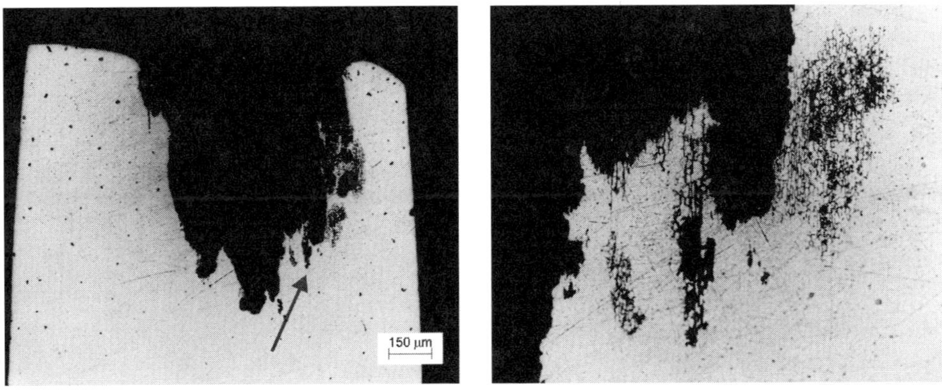

(a) (b)

Figure 9.25 Photomicrograph of a section through an edge of an aluminum lithium (A98090) panel immersed in seawater for 4 months at a) ×64 and b) ×320 magnification.

This depleted zone is anodic both to the main body of the grain and to the precipitated $CuAl_2$. Therefore, the depleted zone is attacked selectively, giving rise to intergranular attack. Such compositional variations usually result from a reaction between one alloy constituent and other elements in the alloy (either as impurities or alloy additions) to form precipitates. Although such precipitates may occur either at grain boundaries or within grains, or both, the grain boundary case is particularly important because intergranular corrosion may result.

A practical example of overaging to reduce the susceptibility to intergranular corrosion is the special T-73 aging treatment for the Al-Zn-Mg-Cu alloy (A97075), which almost eliminates susceptibility to intergranular corrosion. If the alloy is quenched slowly from a high temperature, or if there is a delay between removal from a heat-treatment furnace and immersion in the quenching medium, there is a marked tendency toward susceptibility to intergranular corrosion.

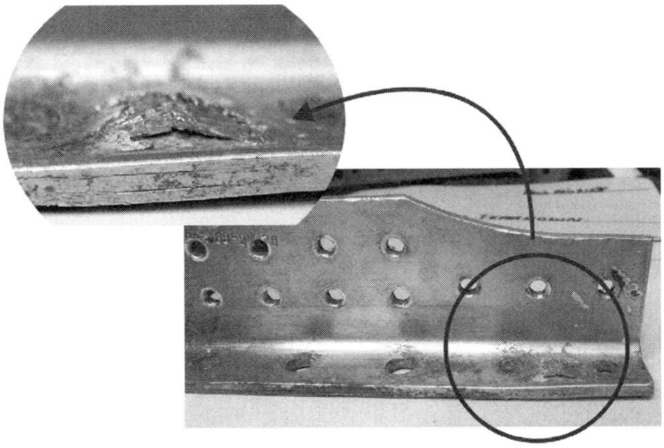

Figure 9.26 Lamellar exfoliation corrosion of an aluminum alloy (A97075) (courtesy of Kingston Technical Software).

Figure 9.27 Leaf-like exfoliation corrosion of corrugated A92024 aluminum alloy (courtesy of Kingston Technical Software).

Extensive cold work will also tend to cause precipitation in the grain boundaries and susceptibility to intergranular corrosion. An example is Al-5Mg alloy (A95056), which should not be used for rivets because the heavily deformed region just under the formed head suffers intergranular precipitation and intergranular corrosion. SCC also may occur in this case.

A second influence of cold work is to produce an elongated grain structure parallel to the surface. When grain-boundary precipitation and susceptibility to intergranular corrosion occur, intergranular penetration tends primarily to be parallel to the surface (Figure 9.25). The expansive force of insoluble corrosion products tends to force the grains apart and leads to exfoliation corrosion, sometimes known as lamellar or layer corrosion (Figure 9.26). In extreme cases, the edges of the affected area are leaf-like

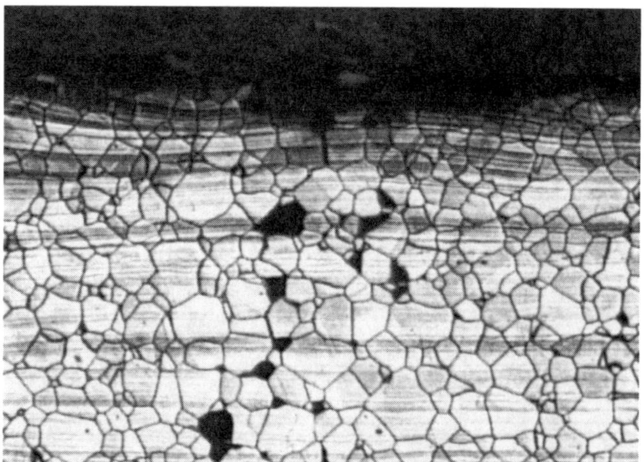

Figure 9.28 Microstructure is typical of sensitized austenitic stainless steel (×67 magnification). This particular problem occurred on Type 316H stainless steel after 7 years of service (courtesy of Corrosion Testing Laboratories).

Figure 9.29 End fitting of a welded support rod (S30300) after service in seawater. This particular stainless steel is not recommended for welding (courtesy of Kingston Technical Software).

and resemble the separated pages of a wetted book that has become swollen and begun to open up (Figure 9.27).

9.5.2.2 Stainless Steels

Attack on the austenitic stainless steels occurs in certain specific media after sensitization of the alloy has been developed by some heating process, normally welding. In this case, a precipitation of chromium carbide occurs around the grain boundaries of the alloy, which reduces the chromium content of the grain immediately adjacent to the boundary. The deficient alloy in this area can then be attacked so severely in certain environments that the entire grain is surrounded and will fall from the structure (Figure 9.28). Welding of the unstabilized stainless steels can be a major source of this problem (Figure 9.29).

The austenitic stainless steels may be "sensitized" to intergranular attack by heating in the temperature range 400 to 900°C (Figure 9.30). A short time (minutes) above

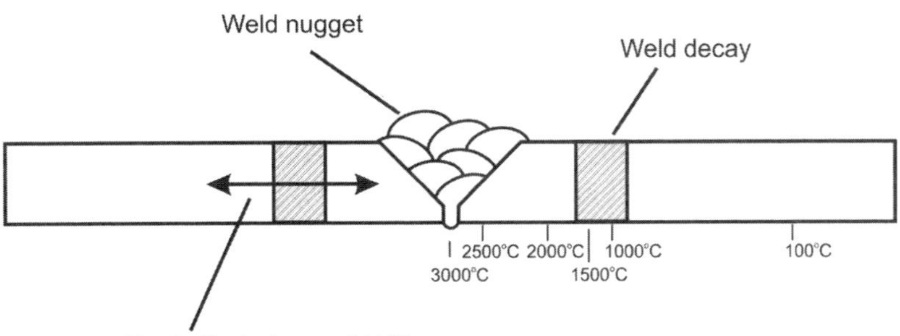

Figure 9.30 Weld decay zone as a function of the welding temperature of stainless steel.

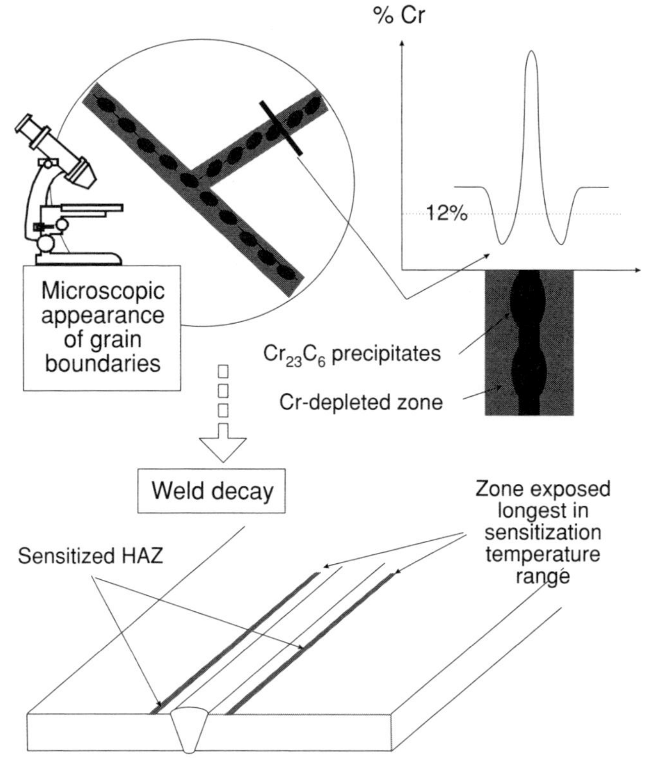

Figure 9.31 Sensitization of stainless steel in the heat adjacent zone.

750°C will produce sensitization; longer times (hours) are required to produce sensitization at lower temperatures. The commonly accepted mechanism attributes intergranular corrosion to a difference in potential between a chromium-depleted zone (at the grain boundaries) and the chromium-rich central zones of the grains (Figure 9.31). Weld decay is an example of intergranular attack adjacent to a stainless-steel weld.

9.5.3 Dealloying

Another type of localized corrosion involves the selective removal by corrosion of one of the elements of an alloy by either preferential attack or by complete dissolution of the matrix. The various kinds of selective dissolution have been named after the alloy family that has been affected, usually on the basis of the element dissolved (except in the case of graphitic corrosion). Table 9.7 gives several examples, including dezincification or the selective removal of zinc from brass. Because the inference in Table 9.7 that each of several designations indicates a separate mechanism may be misleading, this system of terminology is undesirable and should be replaced by the generally applicable terms dealloying, parting, leaching, or selective dissolution.

9.5.3.1 Dezincification

The term dezincification is usually associated with brasses containing more than 15% Zn, in which a porous copper surface zone develops as a result of the depletion of

Table 9.7 Nomenclature of Selective Dissolution Related to Alloy Constituents

Selective Attack Called	Removes	From Alloy System
Dealuminization	Aluminum	Copper–aluminum
Decobaltification	Cobalt	Stellite (Co-Cr-W-C)
Decuprification	Copper	Copper–silver copper–gold
Demanganization	Manganese	Copper–manganese
Denickelification	Nickel	Copper–nickel
Desiliconification	Silicon	Silicon–copper
(no name yet)	Silver	Silver–gold
(no name)	Tin	Lead-tin (solders)
Graphitic corrosion	Iron	Iron–Carbon (gray cast iron)

zinc. Often the gross appearance and size of the part that has suffered dezincification is unchanged except for the development of a copper hue. The part, however, will have become weak and embrittled, and therefore subject to failure without warning. To the trained observer, dezincification is readily recognized under the microscope and even with the unaided eye, because the red copper color is easily distinguished from the yellow of brass.

Conditions generally conducive to dezincification include the presence of an electrolyte (e.g., seawater), slightly acid conditions, presence of carbon dioxide, and appreciable oxygen. There are two general types of dezincification. The more common is the layer type in which dezincification proceeds uniformly, as shown in Figure 9.32. The second type is referred to as the plug type and occurs at localized areas. The site of plugs may be recognized by the presence above them of a deposit of brownish-white zinc-rich corrosion product.

The layer type is frequently observed in high brasses, whereas the plug type usually occurs in low brasses. (The terms high and low brasses refer to zinc content.) High temperatures and aggressive corrosion conditions seem to favor plug-type attack in either class of alloy.

Dezincification does not occur in red brass (below 15% Zn). It may be controlled in the yellow brasses by additions of tin, arsenic, antimony, or phosphorus.

Figure 9.32 Layer dezincification of a brass fitting.

Admiralty-metal heat-exchanger tubes are "inhibited" by use of such alloy additions. Dezincification can therefore be prevented by the proper choice of alloys.

9.5.3.2 *Graphitic Corrosion*

Graphitic corrosion is a form of selective leaching specific to the deterioration of metallic constituents in gray cast iron, which leaves the graphitic particles intact. (The term graphitization is commonly used to identify this form of corrosion but is not recommended because of its use in metallurgy for the decomposition of carbide to graphite). Cast-iron pipe may suffer graphitic corrosion as a result of the selective dissolution of the ferrite in of the alloy. Graphitic corrosion usually progresses uniformly inward from the surface, leaving a porous matrix of the remaining alloying element, that is, carbon, in the form of graphite. Graphitic corrosion occurs in salt waters, acidic mine waters, dilute acids, and soils, especially those containing chlorides from deicing salts or seawater and SRB.

There is no outward appearance of damage, but the affected metal loses weight and becomes porous and brittle. Depending on alloy composition, the porous residue may retain appreciable tensile strength and have moderate resistance to erosion. For example, a completely graphitized, buried, cast-iron pipe may continue to hold water under pressure until jarred by a workman's shovel.

The addition of several percent of nickel to cast iron greatly reduces its susceptibility to graphitic corrosion. It is possible to produce graphitic corrosion in the laboratory by immersing gray cast iron in very dilute sulfuric acid.

9.6 ENVIRONMENTAL CRACKING

SCC and other types of environmental cracking are among the most insidious forms of corrosion. Environmental cracking is defined as the brittle fracture of a normally ductile material in which the corrosive effect of the environment is a causative factor. Environmental cracking can occur with a wide variety of metals and alloys and includes all of the types of corrosion failures listed:

- corrosion fatigue;
- hydrogen embrittlement;
- hydrogen-induced cracking (HIC), a cathodic process;
- hydrogen stress cracking;
- liquid metal cracking(LMC), usually a physicochemical process;
- SCC an anodic process; and
- sulfide stress cracking (SSC).

The following terms have been used in the past in connection with environmental cracking but are now obsolete and should not be used: caustic embrittlement, delayed cracking, liquid-metal embrittlement, season cracking, static fatigue, sulfide-corrosion cracking, and sulfide SCC.

Table 9.8 Materials Susceptibility to Environmental Cracking

Material	SCC	HIC	LMC
Magnesium	$Cl^- + K_2CrO_4$	—	Na, Zn
Aluminum	Cl^- + Oxidants	Hg, Ga, Na, In, Sn, Zn	
Steels, soft	OH^-, NO_3^-, CN^-, CO_3^{2-}, NH_3, $CO/CO_2/H_2O$	—	Cu, Pb, Sb, Sn, Zn, As, P, Cd
Steels, hard	OH^-, Cl^-	H°, H_2S, H_2Sb	Cu, Pb, Sb, Sn, Zn, As, P, Cd
Stainless steels			
Martensitic	OH^-	H°, H_2S, Cl^-	?
Ferritic	OH^-, Cl^-	—	?
Austenitic	OH^-, Cl^-, $S_2O_6^{-}$(a), O_2/H_2O(a)	—	Zn, Cd, Al, Pb, Cu, In
Copper alloys	NH_3, NO_3^-, steam	H°	Hg, Li, In
Monel	OH^-, $HF + O_2$	H_2S	Hg, S
Nickel	OH^-	—	S
Inconel	OH^-	—	S
Alloy C276, alloy 625	OH^-	H_2S/Cl^-(b)	S
Titanium	HNO_3, Cl^-(300°C)	MeOH	Cd, Zn, Sn, P

(a) Sensitized.
(b) Severely cold worked and in galvanic couple.

Mechanical forces (e.g., tensile or compressive forces) will usually have little if any effect on the overall corrosion of metals. However, a combination of tensile stresses and a specific corrosive environment is one of the most important causes of sudden cracking-type failures of metal structures. All commercial metals and alloys will crack in a brittle manner in some environment.

Table 9.8 is a brief summary of the resistance of metals and alloys, indicating the species that may cause environmental cracking (i.e., SCC, HIC, and LMC, which is discussed further later in the book).

9.6.1 Fracture Mode

Environmental cracks are microscopic in their early stages of development. In many cases, they are not evident on the exposed surface by normal visual examination, and can be detected only by special techniques. Optical or scanning electron microscopy (SEM) of failed sections may be required to fully identify them. As the cracking penetrates farther into the material, it eventually reduces the effective supporting cross section to the point where the structure fails by overload or, in the case of vessels and piping, escape (seepage) of the contained liquid or gas occurs.

Cracking usually is designated as either intergranular (intercrystalline) as shown in Figures 8.19 (Chapter 8) and Figure 9.33 or transgranular (transcrystalline) as illustrated in Figure 9.34. Occasionally, both types of cracking are observed in a failure. Intergranular cracks follow the grain boundaries in the metal. Transgranular cracks cross the grains without regard for the grain boundaries. The morphology of the cracks may change with the same material in different environments.

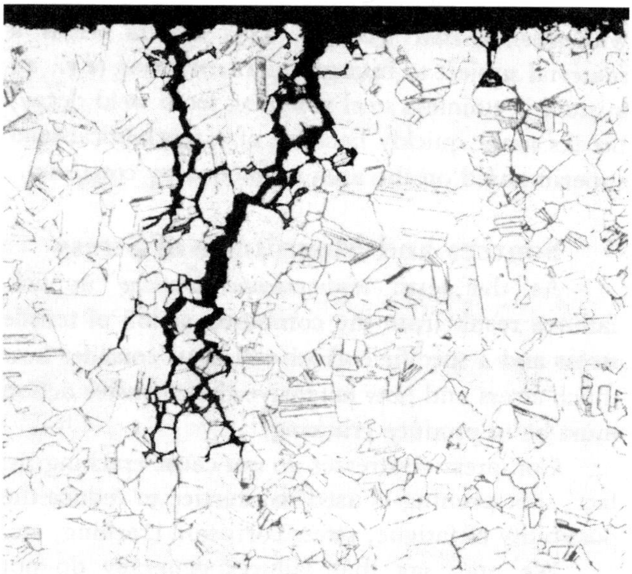

Figure 9.33 Typical intergranular SCC in cartridge brass (70 Cu–30 Zn), etched 30% H_2O_2, 30% NH_4OH, 40% H_2O (×75 magnification).

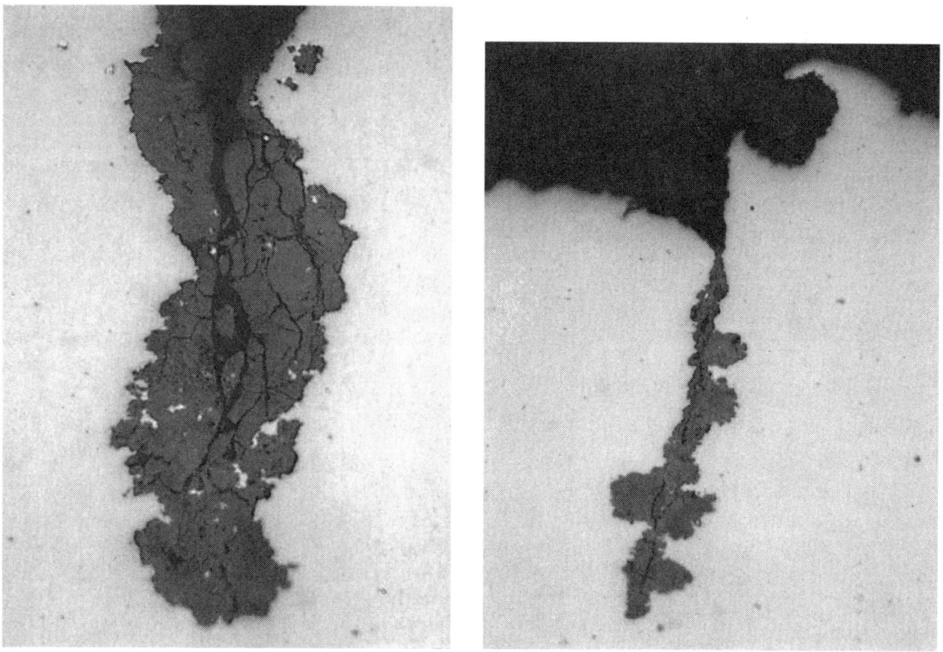

(a) (b)

Figure 9.34 Axial or transgranular cracks filled with oxide or scale: a) observed on the external surface of a pipe line. These cracks may occur in association with b) pits and general corrosion (courtesy of *MACAW's Pipeline Defects*, published by Yellow Pencil Marketing Co.).

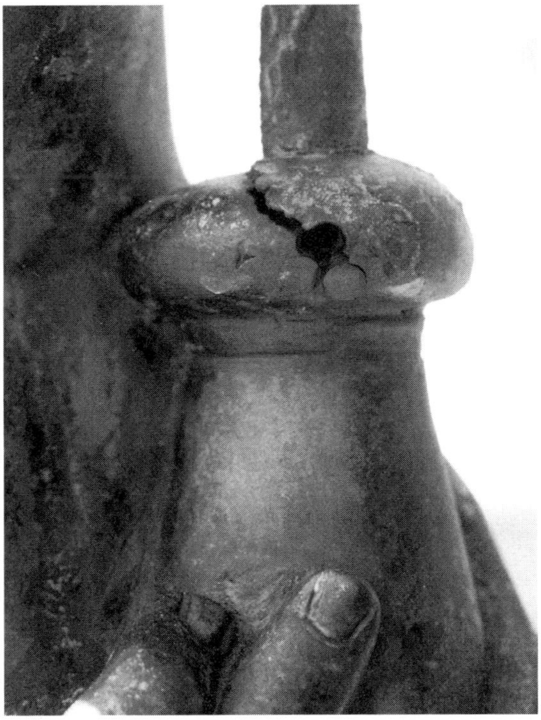

Figure 9.35 SCC in a bronze monument caused by build-up of rust around a decorative steel rod (courtesy of Kingston Technical Software).

9.6.2 Sources and Magnitude of Stress

Compressive stresses do not cause cracking. In fact, shot peening is used in practice to reduce the susceptibility of metallic materials to fatigue, SCC, and other forms of cracking. Failures generally do not result from the ordinarily applied engineering loads or stresses. Engineering loads, however, are often additive to the residual stresses already present in the structure. These residual stresses result from fabricating processes (e.g., deep drawing, punching, rolling of tubes into tubesheets, mismatch in riveting, spinning, welding, etc.).

Residual stresses will remain in the structure unless it is annealed or otherwise thermally stress relieved following fabrication, a practice that becomes increasingly impractical as a system gets larger or more complex. Cooling from the high temperatures required may also induce internal stresses because of nonuniform cooling (e.g., quenching stresses in the solution annealing and water quenching of austenitic stainless steels). In fact, very slow, controlled cooling is a prerequisite for effective stress relief. Cooling rates after tempering of hardened steel, as a palliative measure for HIC, may be less critical once the proper maximum hardness is attained.

Corrosion products from general corrosion or other forms of corrosion may build up between mating surfaces and, because they occupy so much more volume than the metal from which they are produced, generate sufficient stresses to cause SCC (Figure 9.35). In this example moisture working down the steel rod in combination with the galvanic corrosion resulting from the contact with the bronze support caused enough rust build-up to generate high stresses and induce SCC.

Even less is known about the magnitude of these stresses. To be effective, the stresses must be tensile in nature. As mentioned previously, and as is discussed further, compressive stresses actually militate against cracking.

Some alloy–environment combinations seem to require stresses at or above the yield point (e.g., steel in caustic). Others appear to have a definite threshold stress below which cracking will not occur. This is the case for SSC of alloy steels, in which a hardness below about Rockwell C 22 seems to result in practical immunity to this type of attack.

Other systems seem to have no apparent minimum stress. For example, chloride SCC of austenitic stainless steels has been reported at stresses as low as one-tenth the yield strength. It should be noted, however, that notches or stress raisers, even of a microscopic nature, can concentrate the apparent stresses locally to a value several times the known macrostress. Laboratory work suggests that, at least in the region in which the cracks originate, the stresses must be above the yield strength (i.e., the stress must produce some plastic deformation of the material in that region, sometimes very small, at which cracking occurs).

9.6.3 SCC

SCC is a mechanical–chemical process leading to the cracking of certain alloys at stresses below their tensile strengths. A susceptible alloy, the proper chemical environment, plus an enduring tensile stress are required. It is likely that there are no alloy systems that are completely immune to SCC in all environments. Usually, there is an induction period, during which cracking nucleates at a microscopic level. This is followed by actual propagation.

Table 9.9 lists some environments in which SCC has been observed for at least some alloys of the systems listed. This listing does not imply that all alloys of a given material will be equally susceptible or that there are none in the class that may be immune to the environments listed.

Table 9.9 Some SCC Environments for Metals

Aluminum alloys	NaCl-H_2O, NaCl solutions, seawater, mercury
Copper alloys	Ammonia vapors and solutions, mercury
Gold alloys	$FeCl_3$ solutions, acetic acid/salt solutions
Inconel	Caustic soda solutions
Lead	Lead acetate solutions
Magnesium Alloys	NaCl-$K_2Cr_2O_7$ solutions, rural and coastal atmospheres, distilled water
Nickel	Fused caustic soda
Carbon steels	NaOH solutions; NaOH-$NaSiO_3$ solutions; $CaNO_3$, NH_4NO_3, and $NaNO_3$ solutions; mixed acids, (H_2SO_4 HNO_3); HCN; H_2S; seawater; NaPb alloy
Stainless steels	$BaCl_2$, $MgCl_2$ solutions; NaCl-H_20_2 solutions; seawater; H_2S, NaOH-H_2S solutions
Titanium	Red fuming nitric acid

SCC is an anodic process, a fact that can be verified by the applicability of CP as an effective remedial measure. SCC may occasionally lead to fatigue or vice versa. Usually, the true nature of the cracking can be identified by the morphology of the cracks themselves. Typically, SCC occurs under only very mildly corrosive conditions, although there are exceptions, as in the cracking of Monel 400 (N04400) in aerated hydrogen-fluoride vapors.

In a failure by SCC there is usually little metal loss or general corrosion associated with it. If there is severe general corrosion, SCC usually will not occur. Thus, the failure of a stress bolt rusted away until it eventually cannot sustain the applied load is not classified as SCC. However, if products from general corrosion are trapped so as to exert stress in a structure, they can cause SCC, as in the pillowing examples shown earlier.

The idea, once prevalent, that only alloys and not pure metals are susceptible to SCC is quite possibly correct. The question is, "How pure is pure?" Copper containing 0.004% phosphorous or 0.01% antimony is reported to be susceptible to SCC in ammoniacal environments (i.e., those containing ammonia or ammonium ions). Cracking has been produced in a decarburized steel (containing less than 0.01% carbon, but, of course, small amounts of manganese, sulfur, and silicon) in a boiling ammonium-nitrate solution. SCC has been produced in commercial titanium containing, among other constituents, 600 ppm of oxygen and 100 ppm of hydrogen. Hence, the idea that a given material cannot fail by SCC because it is commercially pure is not correct.

The following sections describe some general aspects on the occurrence and control of SCC. More information on this particularly important form of catastrophic corrosion can be found under specific alloy systems in Chapter 8.

9.6.3.1 Electrochemical Aspects

In the electrochemical part of the mechanism, it is assumed that there is at least a thin film of electrolyte on the metal surface (in many cases the surface is completely immersed in the electrolyte) and that both anodic and cathodic areas exist on the surface covered by the liquid film. Almost without exception, very thin oxide films form almost instantaneously on the surface of all metals exposed to moisture. This oxide-covered surface is much less chemically active than a bare or unfilmed surface, and it will serve as the cathode in the corrosion processes.

The anodic sites may be located at discontinuities in the oxide film covering the metal surface or at a grain boundary where the mismatch of metal lattices is large, or they may result from inhomogeneous composition of the metal at an area on the surface.

It is postulated in one of the proposed mechanisms for SCC (i.e., the film-rupture mechanism) that stresses applied to susceptible alloys will rupture this oxide film on the metal surface. The film-free surfaces are normal to the applied stress and very narrow compared to the film-covered surfaces separating them (Figure 9.36).[15]

Because the anodic sites are very small compared with the cathodic area, the current density will be large at the anodes and trenches or grooves will be formed. These grooves concentrate the stresses at the groove tips, preventing the film from reforming there.

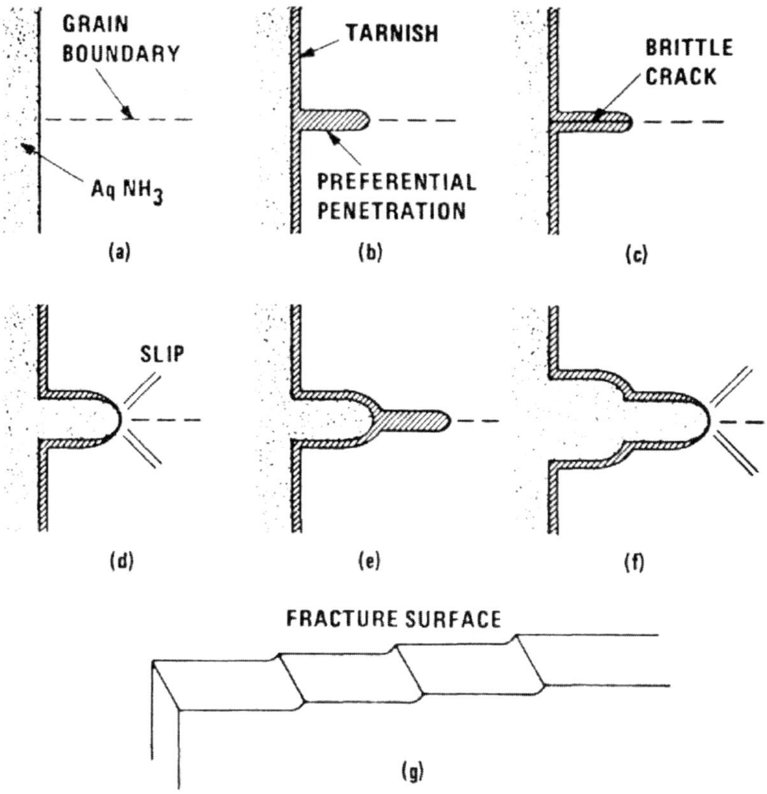

Figure 9.36 Schematic shows the film-rupture mechanism for a) brass exposed to a tarnishing ammonium solution. Brittle tarnish grows along b) grain boundary and c) cracks, and exposes more substrate to continue (d–f).

Because effective stresses soon disappear on the sides of the groove, a film will form there and thus produce cathodic areas very close to the anodes to accelerate chemical action.

In the case of intergranular SCC, the initial notch may well form at grain boundaries that may be more chemically active than the grain face because of a discontinuity in the metal lattice there. Once a notch is formed, stress concentrations at its tip will cause the film there to rupture and prevent it from reforming, as was discussed previously. It has also been suggested that stresses may produce areas of inhomogeneous metal composition at the metal surfaces, which could form anodic sites.

9.6.3.2 Mechanical Aspects

As the electrochemically propagated stress corrosion crack penetrates into the material, stress concentrations will build up at the crack tip until there is sufficient energy available to initiate mechanical cracking. This energy is required to form new surfaces and to deform material at the tip of the crack. (The energy to form new surfaces is very small compared to that needed to deform the metal at the tip of the crack.) The deformed (work-hardened) material becomes increasingly more difficult to fracture mechanically

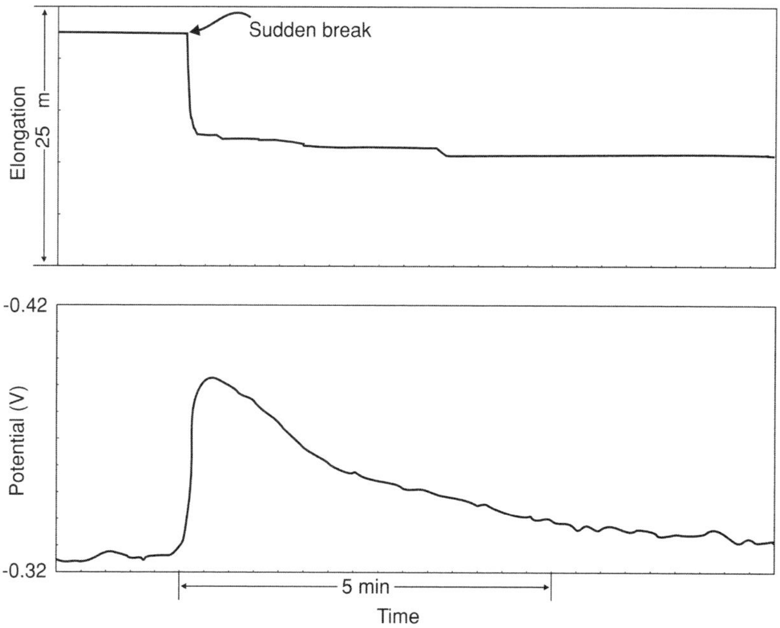

Figure 9.37 Automatically recorded elongation time and electrochemical-solution potential–time curves for notched low-carbon steel specimen exposed in boiling NH_4NO_3 solution. Sudden break in elongation curve indicates mechanical fracture that exposed film-free and anodic metal to corrosive agent (indicated by change in potential). Fracture was prevented from going to completion by energy required to form new surfaces and plastically deform material at tip of crack.

(the energy available as the result of the stress concentration being used up) and the mechanically propagated crack stops. The electrochemical process again takes over, building up stress concentrations at the tip of the crack, and the process is repeated.

Figure 9.37 shows evidence that crack propagation is in part mechanical. The upper curve is a plot of the elongation of a notched low-carbon steel specimen, exposed in a stressed condition to a boiling 20% NH_4NO_3 (ammonium nitrate) solution. The break in the curve indicates that a small mechanical fracture has occurred. If cracking were purely electrochemical, the curve should fall slowly and uniformly without sudden breaks.

The lower curve is a plot, recorded simultaneously with the extension curve, of the electrochemical solution potential of the specimen. The change in potential, approximately 0.08 V, occurring simultaneously with the sudden extension of the specimen, confirms the idea that there had been a small mechanical fracture. This fracture exposed bare metal to the corrodent. This bare metal was much more reactive than the film-covered metal, which was the "old surface," and hence, the sudden change in potential.

9.6.3.3 *Transgranular SCC*

Transgranular SCC is a form of localized subsurface attack in which a narrow path is corroded randomly across grains without any apparent effect on the crack direction by the presence of the grain boundary. Transgranular SCC initiates on the surfaces and proceeds inward, presumably by local-cell action. It can occur during the SCC of austenitic stainless steels and less commonly during the SCC of low-alloy steels. It can also occur in the SCC of copper alloys in certain media (e.g., ammonia). It seldom occurs in aluminum alloys.

9.6.4 Corrosion Fatigue

Fatigue is the failure of a metal by cracking when it is subjected to cyclic stress. The usual case involves rapidly fluctuating stresses that may be well below the tensile strength. As stress is increased, the number of cycles required to cause fracture decreases. There is usually a stress level below which no failure will occur, even with an infinite number of cycles, and this is called the endurance limit. In practice, the endurance limit is defined as the stress level below which no failure occurs in 1 million cycles. A fatigue curve, commonly known as an S-N curve, is obtained by plotting the number of cycles required to cause failure against the maximum applied cyclic stress. A typical S-N curve is shown by the dotted line in Figure 9.38.

When a metal is subjected to cyclic stress in a corrosive environment, the number of cycles required to cause failure at a given stress will be reduced below that shown in Figure 9.38. This acceleration of fatigue by corrosion is called corrosion fatigue. A corrosion-fatigue S-N curve can also be drawn, as shown by the solid line in Figure 9.38. It is lower than the normal fatigue curve obtained for the same metal in air. In both cases, the frequency of stressing should be reported because this factor influences the endurance limit.

The solid S-N curve obtained in this case is the result of both corrosion and fatigue. The solid curve indicates that metal life under such conditions can be much lower than the reference curve established in air. The S-N curve with corrosion tends to keep dropping, even at low stresses, and thus does not level off, as will the ordinary fatigue curve.

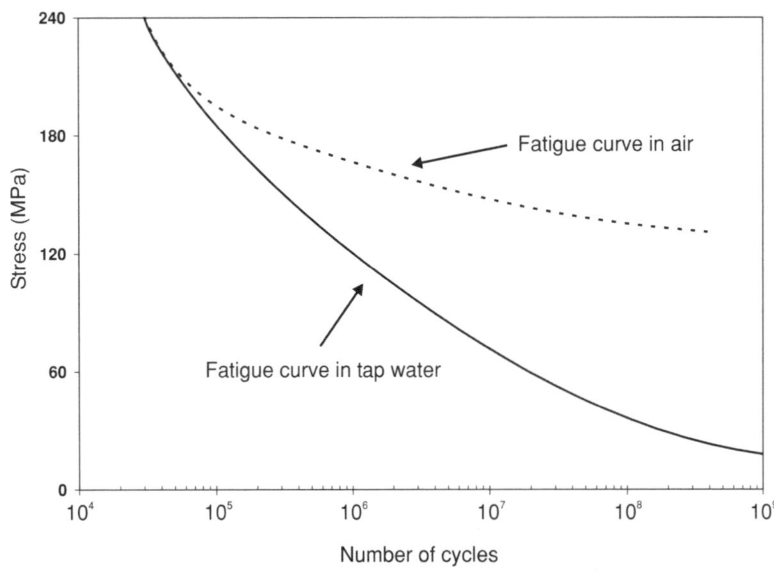

Figure 9.38 Fatigue and corrosion fatigue curves for an aluminum alloy.

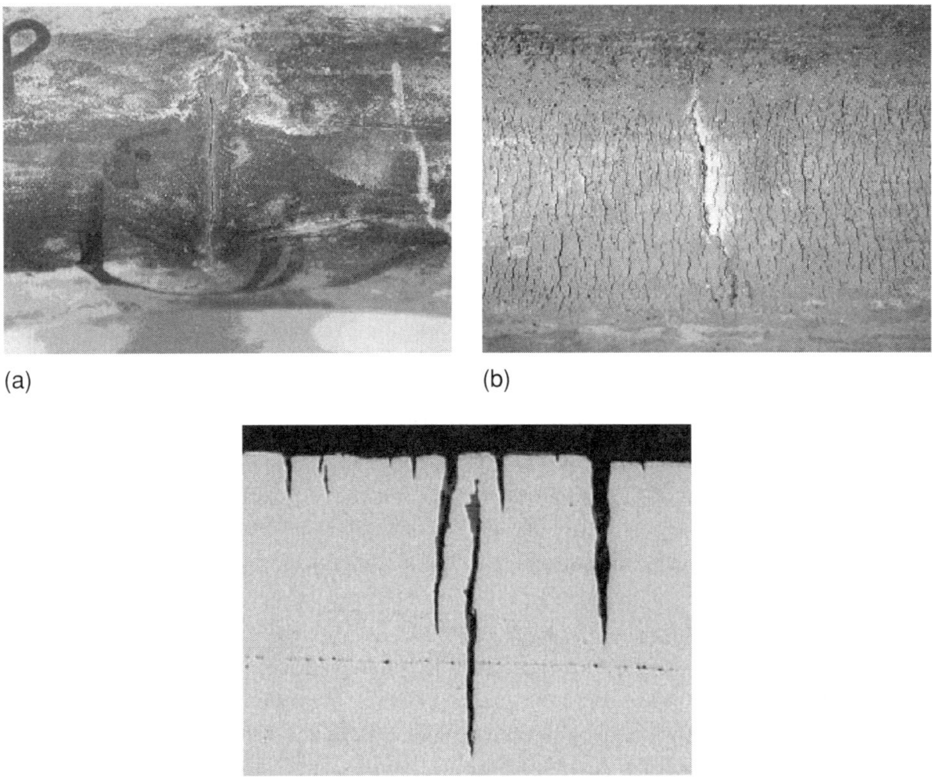
(a) (b) (c)

Figure 9.39 a) As-received appearance of boiler tube section with crack failure. Note remnants of tack weld to strap on the back of the tube and that the top of tube is marked establishing orientation. b) Close-up ID surface of failed tube section shows band of numerous parallel partial through-wall cracks in line with through-wall crack (arrow) (×2 magnification). c) Photomicrograph shows as-polished longitudinal metallographic specimen through cracked area on failed tube section. Cracks originate on ID surface. Note wedge- or needle-shaped, nonbranching cracks typical of corrosion fatigue. Faint black line near bottom of photo is an inclusion not related to the failure (×125 magnification) (courtesy of Corrosion Testing Laboratories).

A marked drop in or elimination of the endurance limit may occur even in a mildly corrosive environment, especially in the case of a film-protected alloy. For example, deionized water, which ordinarily will produce only film growth on an aluminum alloy immersed in it, will appreciably reduce the endurance limit of the same alloy subjected to cyclic stressing. This is because stress reversals cause repeated cracking of the otherwise protective surface film, and this allows access of the water to the unprotected metal below with resultant corrosion.

Failures that occur on vibrating structures (e.g., taut wires or stranded cables) exposed to the weather under stresses below the endurance limit are usually caused by corrosion fatigue. Corrosion fatigue also has been observed in steam boilers, because of alternating stresses caused by thermal cycling (Figure 9.39).

The petroleum industry also encounters major trouble with corrosion fatigue in the production of oil. The exposure of drill pipe and of sucker rods to brines and sour crudes encountered in many producing areas results in failures that are expensive both from the standpoint of replacing equipment and from loss of production during the time required for "fishing" and rerigging.

For uniaxial stress systems, there will be an array of parallel cracks that are perpendicular to the direction of principal stress. Torsion loadings tend to produce a system of crisscross cracks at roughly 45 degrees from the torsion axis. Corrosion-fatigue cracks found in pipes subjected to thermal cycling usually will show a pattern made up of both circumferential and longitudinal cracks.

There are several effective remedial procedures for reducing or preventing corrosion fatigue:

- CP useful in aqueous solutions, and under ideal conditions can increase the fatigue limit to that predicted from fatigue tests in vacuum.
- Shot peening, which induces compressive stresses in the surface, tends to reduce corrosion fatigue, at least until subsurface corrosion penetrates into metal under tensile stress.
- Electroplated sacrificial coatings (e.g., zinc or cadmium on steel) protect against corrosion fatigue. Zinc coating on steel by spray metallizing also is effective. The corrosion-fatigue life of high-strength aluminum alloys can be improved by spray metallizing (with an aluminum alloy that is anodic to the base metal).
- Properly applied inhibitors also can be effective in aqueous environments. The supply of inhibitor must be uninterrupted; otherwise it is possible to stimulate localized attack, which can form focal points for corrosion fatigue.

9.6.5 HIC and Hydrogen Embrittlement (HE)

There are several categories of hydrogen phenomena that are localized in nature. The little hydrogen atom (not the molecule) is about one-third the size of the crystal cube in a metal lattice and can diffuse easily through the structure. When the crystal lattice is in contact or is saturated with atomic hydrogen, the mechanical properties of many metals and alloys are diminished. Nascent atomic hydrogen can be produced as a cathodic reaction when certain chemical species are present that act as negative catalysts (i.e., poisons) for the recombination of atomic to molecular hydrogen shown in Equation (9.2).

$$2H^\circ \rightarrow H_{2(g)} \tag{9.2}$$

If the formation of molecular hydrogen is suppressed, the nascent atomic hydrogen can diffuse into the interstices of the metal instead of being harmlessly evolved as a gaseous reaction product. If these are in a highly stressed condition, they may be subject to HIC.

There are many chemical species that poison this recombination (e.g., cyanides, arsenic, antimony, or selenium compounds). However, the one most commonly encountered is hydrogen sulfide (H_2S), which is formed in many natural decompositions (e.g., rotten eggs), and in many petrochemical processes.

Processes or conditions involving wet hydrogen sulfide are called sour services, and the high incidence of sulfide-induced HIC has resulted in the term SSC. The SSC of medium-strength steels has been a continuing source of trouble in the oil fields, and from these troubles evolved the NACE Standard MR-01-75, "Materials Selection for Sour Service." However, similar problems are encountered wherever wet hydrogen sulfide is encountered (e.g., acid gas scrubbing systems, heavy-water plants, waste-water treatment).

Although structural, bolting, and pressure-vessel steels, as well as piping materials, are of primary concern, other materials are also susceptible. MR-01-75 indicates the heat treatments and hardness that should be specified for a variety of steels. HIC is a cathodic phenomenon that is suppressed by anodic protection and aggravated by CP.

Failures have occurred in the field when storage-tank roofs have become saturated with hydrogen by corrosion and then been subjected to a surge in pressure, resulting in a brittle failure of the circumferential welds. In rare instances, even copper and Monel 400 (N04400) have been subjected to HIC. More resistant materials, such as Inconels and Hastelloys often employed to combat HIC, can become susceptible under the combined influence of severe cold work, chemical poisons, and a direct current from the galvanic couple resulting from electrical contact with a more anodic member.

The mechanism of HIC has not been definitely established. Various factors are believed to contribute to unlocking the lattice of the metal, such as hydrogen pressure at the crack tip, the competition of hydrogen atoms for the lattice-bonding electrons, the easier plastic flow and dislocation formation in the metal at the crack tip in the presence of hydrogen, and the formation of certain metal hydrides in the alloy. What is known is that the maximum effect takes place at room temperature, that as little as 5 ppm of hydrogen in the lattice produces the full HIC, that the hydrogen concentrates at stressed areas in the metal matrix, and that carbon and manganese downgrade the resistance of a steel, among other things.

Three additional phenomena have been reported quite commonly in relation to hydrogen weakening of metallic components.

9.6.5.1 *Loss of Ductility and Strength*

All metals will lose some portion of their ductility and to a lesser extent their strength when exposed to hydrogen. Copper, aluminum, and the austenitic stainless steels are less affected than iron or nickel-base alloys. The effect can be noted by conducting a slow strain-rate tensile test of the materials in a hydrogen environment and comparing the results with the standard data. Removing the alloy from the source of the hydrogen and baking the metal at around 200°C to remove any absorbed hydrogen will restore the mechanical properties.

9.6.5.2 *Hydrogen Blistering*

When the atomic hydrogen enters the metal structures, nonmetallic inclusions can catalyze the formation of molecular hydrogen within the metal lattice, generating

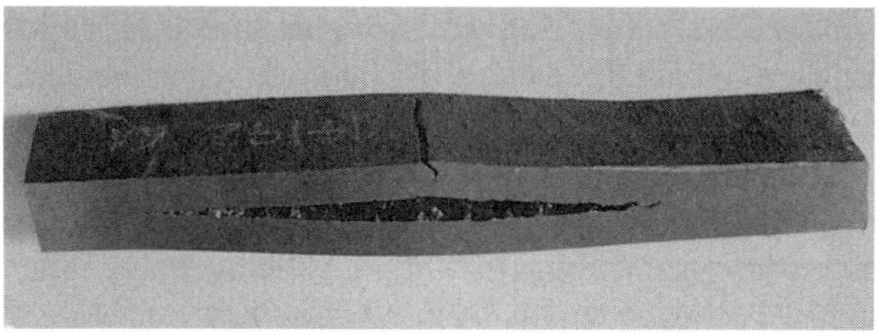

(a)

(b)

Figure 9.40 a) HIC is shown with mid-wall cracks running parallel to the pipe-line wall and b) surface blisters may also contain cracks (courtesy of *MACAW's Pipeline Defects*, published by Yellow Pencil Marketing Co.).

tremendous internal pressures and causing splits, fissures, and even blisters on the metal surface (Figure 9.40). Normally 3 cm or so in diameter, some blisters have been observed larger than 1 m in diameter in special cases. The tendency to blister can be combated to some extent by using steels of the same grain size and cleanliness as are specified for low-temperature service.

9.6.5.3 *High-Temperature Effects*

At elevated temperatures and high partial pressures of hydrogen, the tendency for molecular hydrogen to split into hydrogen atoms is greater and the atoms themselves are more active and mobile. Above about 400°C, cuprous oxide inclusions in metallic copper can be reduced, generating steam within the metal lattice and causing internal fissuring.

Of more importance from the engineering standpoint is the methanation of steel. At temperatures above about 220°C, atomic hydrogen will react with the iron carbides in steel, forming methane within the structure and causing localized decarburization and fissuring.

9.6.6 LMC

Certain metals subjected simultaneously to a high tensile stress and specific molten metals suffer a form of environmental cracking known as LMC (or embrittlement). The

Figure 9.41 Stressed Monel part from a distillation column cracked from exposure to mercury.

mechanism of failure is said by some to be closely related to the HIC process. The particular cracking of highly stressed yellow brass in mercurous-nitrate solution (the basis for a test method) is actually a combination of electrochemical attack and LMC. In this attack, mercurous ions seem to be cathodically reduced to metallic mercury at local sites and the susceptible material then suffers LMC by the metallic mercury.

Metallic mercury from broken thermometers, blown manometers, or from certain catalysts can move freely around a process (mercury has a relatively high vapor pressure) and raise havoc with any highly stressed copper alloy or Monel (Figure 9.41).

Mercury can cause LMC of high-strength aluminum alloys (as could sodium, gallium, indium, tin, and zinc). Molten copper (as from welding of copper-clad steel) will cause LMC of steels or stainless steels. Cadmium, aluminum, and lead can cause LMC of austenitic stainless steels, and cadmium can attack titanium. Steel can be cracked by exposure to tin, cadmium, zinc, lead, bismuth, copper, and antimony.

The biggest problem, however, in engineering practice is zinc, because it is widely employed in galvanizing and in the zinc-pigmented paint system. Fortunately, zinc as a molten metal is only encountered during welding or brazing, or in the unfortunate event of a fire or explosion.

Nevertheless, overspray of zinc-rich paint from adjacent structures will render stainless steel or high-nickel alloys unweldable. Although most people should know better, problems still arise from those trying to weld galvanized structural members to alloy equipment. Table 9.8 presents a summary of the liquid-metal embrittlement couples that have been observed with pure metals and alloys.

9.6.7 Corrosion Testing for Environmental Cracking

SCC test methods are the subject of an introductory book by NACE International.[16] However, a few remarks on testing as it pertains to environmental cracking are pertinent here.

In alloy-development work, numerous alloy compositions must be screened. This suggests the need for testing many specimens in a laboratory environment. The more promising materials will then be further exposed in the field. If the purpose of the test is

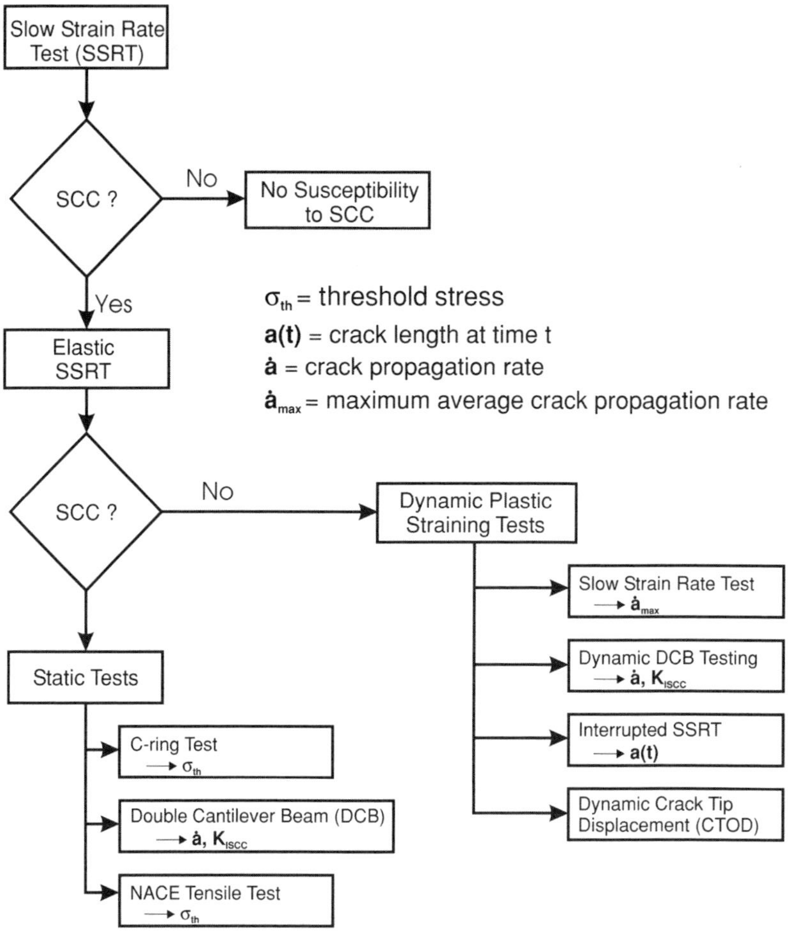

Figure 9.42 Decision tree for the selection of corrosion tests to verify the susceptibility of steels to SCC.

to determine whether the material in the warehouse is sufficiently resistant to justify its use in the final product, the test must be of relatively short duration and, at the same time, correlate well with results that would be obtained in several years of ordinary exposure.

One difficulty is that a negative proposition cannot be proven. If a test results in a cracking failure, it can be concluded that failures may occur in service. If no cracking results, it may simply be that the test was not run long enough, or at a suitable stress, or did not really reproduce the true service conditions. The development of the slow strain-rate test has done much to ameliorate this problem. When a tensile coupon is pulled very slowly in a corrosive environment, if it does not develop cracks it is unlikely to do so in service. If it does crack, it may not necessarily do so under less severe stress or strain in service in that environment.

There are a number of standardized tests available to study different modes of environmental cracking. With such a long list of questions and the continuously increasing number of testing methods, it is important to simplify the design of test plans by adopting a testing strategy relating requirements with the main test parameters. The decision tree presented in Figure 9.42 has been developed to facilitate the selection of tests designed to verify the susceptibility of steels to various forms of SCC. The strategy would be to start with the most severe and least expensive SCC test, that is, the slow strain-rate test in which a bar made from the relevant material is exposed to the environment of interest and slowly monotonically strained to fracture.[17] When cracks are found, the susceptibility of the material should then be further evaluated by performing a battery of other tests designed to differentiate between the various mechanisms leading to SCC and hydrogen embrittlement.

9.6.8 Mitigation of Environmental Cracking

It is something of a joke among practicing corrosion engineers to say, "It's easy to prevent environmental cracking. Simply eliminate the stress, isolate the metal from the environment, or change the environment!" This is, of course, much easier said than done. Nevertheless, the following five basic methods of corrosion control apply to the mitigation of environmental cracking.

9.6.8.1 *Change of Material*

A total or even partial change of material (e.g., "safe-ending" of heat-exchanger tubes) is a common approach. For complete reliability, use of a totally crack-resistant material is often the most cost economical approach.

9.6.8.2 *Change of Environment*

The removal of chlorides, caustic, or other major cracking-type species is an effective solution where possible. However, much less drastic changes are often effective as well, such as the following:

- removing oxygen or oxidizing agents,
- changing the pH, and
- adding inhibitors. Nitrates, for example, have been used as an inhibitor against caustic embrittlement of steel.

9.6.8.3 *Barrier Coatings*

Temporary rust-preventative greases have been found to be helpful against atmospheric cracking of high-strength steels, although good commercial organic coatings are much more reliable. Certain silicone-base paints are routinely used to prevent external chloride stress cracking of insulated stainless-steel vessels and piping.

9.6.8.4 *Electrochemical Techniques*

CP has been found effective against anodic SCC. Lead/tin solders and nickel plating have been used to protect stainless-steel tube ends against SCC by water. This technique must be carefully controlled to be effective. Austenitic stainless steel has been known to suffer chloride SCC while simultaneously causing galvanic corrosion of steel components.

9.6.8.5 Design

This is probably the most important consideration when use of suitable alternative materials of construction is not possible.

Residual stresses may be minimized by careful thermal-stress relief appropriate to the alloy. Compressive stresses may be introduced by controlled shot peening. Stress raisers may be eliminated or minimized. Crevices should be eliminated, as well as areas where deposits can accumulate.

Conditions conducive to evaporation and concentration of corrosive species can be minimized. Vertical condensers with water on the tube side should be pressed full, the tube sheets being adequately vented. Tube-to-tube sheet rolls can be seal welded or strength welded. All welds should be full penetration to prevent accumulation of species in the crevice that otherwise exists.

In short, every effort should be made to minimize stress as well as to reduce the concentration of specific species in the equipment to a level compatible with engineering performance.

References

1. C.P. Dillon, Forms of Corrosion: Recognition and Prevention (Houston, TX: NACE, 1982).
2. R.B. Mears, R.H. Brown, "Designing to Prevent Corrosion," Corrosion 3 (1947): pp. 97–120.
3. F.H. Haynie, "Statistical Treatment of Data, Data Interpretation, and Reliability," in Corrosion Tests and Standards, MNL 20, ed. R. Baboian (Philadelphia, PA: ASTM, 1995), pp. 62–67.
4. E.J. Gumbel, "Statistical Theory of Extreme Values and Some Practical Applications," National Bureau of Standards, Mathematics Series 33, Washington, DC, 1954.
5. R. Baboian, Automotive Corrosion and Proledion (Houston, TX: NACE, 1991).
6. J.P. Komorowski, S. Krishnakumar, R.W. Gould, N.C. Bellinger, F. Karpala, O.L. Hageniers, "Double Pass Retroreflection for Corrosion Detection in Aircraft Structures," Materials Evaluation 54 (1996): pp. 80–86.
7. R.T. Ruggeri, T.R. Beck, "An Analysis of Mass Transfer in Filiform Corrosion," Corrosion 39 (1983): 452–465.
8. W.H. Slabaugh, W. DeJager, S.E. Hoover, L.L. Hutchinson, "Filiform Corrosion of Aluminum," Journal of Paint Technology 44 (1972): pp. 76–83.
9. MIL-STD-889B(3), "Dissimilar Metals" (Aberdeen, MD: Army Research Laboratory, 1993).
10. A.W. Peabody, R.L. Bianchetti, Peabody's Control of Pipeline Corrosion, 2nd ed. (Houston, TX: NACE, 2001).
11. H.P. Hack, "Evaluation of Galvanic Corrosion," in Metals Handbook: Corrosion, vol. 13 (ASM International, 1987), pp. 234–238.
12. R. Francis, Galvanic Corrosion: A Practical Guide for Engineers (Houston, TX: NACE, 2001).
13. P.T. Gilbert, "Corrosion Resisting Properties of 90/10 Copper Nickel Iron Alloy with Particular Reference to Offshore Oil and Gas Applications," British Corrosion Journal (1979).
14. "Titanium and Its Alloys," retrieved October 24, 1996, from http://www.titanium.org/alloychartext.html.
15. R.W. Staehle, "Predictions and Experimental Verification of the Slip Dissolution Model for Stress Corrosion Cracking of Low Strength Alloys," in Stress Corrosion Cracking and Hydrogen

Embrittlement of Iron Base Alloys, eds. R.W. Staehle, J. Hochmann, R.D. McCright, J.E. Slater (Houston, TX: NACE, 1977), pp. 180–207.

16. A.J. Sedriks, Corrosion Testing Made Easy: Stress Corrosion Cracking Test Methods (Houston, TX: NACE, 1990).
17. J.G. Erlings, H.W. de Groot, J. Nauta, "The Effect of Slow Plastic and Elastic Straining on Sulphide Stress Cracking and Hydrogen Embrittlement of 3.5% Ni Steel and API 5L X60 Pipeline Steel," Corrosion Science 27 (1987): pp. 1153–1167.

Part IV

Methods of Control

10
Chapter

Protective Coatings

10.1 INTRODUCTION

Putting an organic or metallic barrier coating between a corrosive environment and the material to be protected is a fundamental method of corrosion control. It is, in fact, the most widely used method of protecting most metals. According to the U.S. Department of Commerce Census Bureau, the total amount of organic coating material sold in the United States in 1997 was 5.56 billion L, at a value of $16.56 billion. The total sales can be broken down into architectural coatings, product original equipment manufacturers' (OEM) coatings, special-purpose coatings, and miscellaneous paint products. A portion of each of these was classified as corrosion coatings for a total estimate of $6.7 billion.[1]

The most widely used metallic-coating method for corrosion protection is galvanizing, which involves the application of metallic zinc to carbon steel for corrosion-control purposes. Hot-dip galvanizing is the most common process, and as the name implies, it consists of dipping the steel member into a bath of molten zinc. Information released by the U.S. Commerce Department in 1998 stated that about 8.6 million metric tons of hot-dip-galvanized steel and 2.8 million metric tons of electrolytic-galvanized steel were produced in 1997. The total market for metallizing and galvanizing in the United States is estimated at $1.4 billion.

The concept of adding coatings to surfaces is so ancient that its beginnings are lost in the mists of history. Paleolithic cave drawings exist that are reputed to be tens of thousands of years old. They were applied by fingers, a splayed twig, or, as some believe, were sprayed by blowing pigments through a hollow reed. Until the Egyptians began using varnishes around 4000 B.C., there is no evidence that coatings were used for protection. Polychrome Greek statues became common by 300 B.C., and evidence exists that the Romans used coatings for both decorative and protective purposes. Coated Chinese artifacts are even more ancient than those in Western Europe.

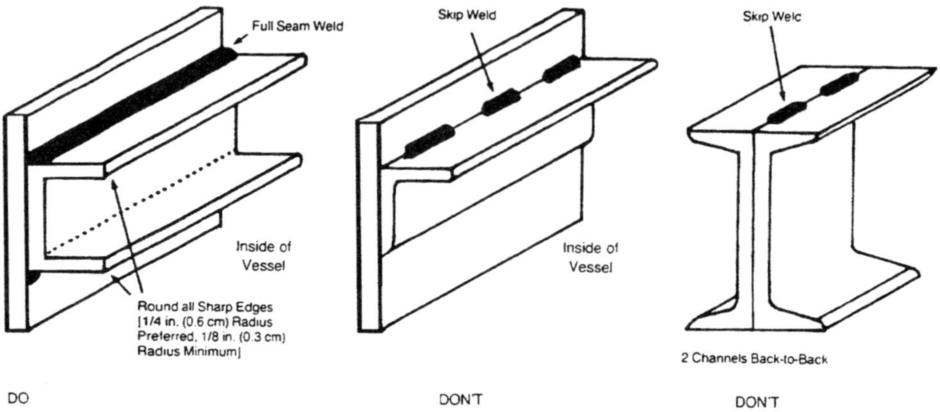

Figure 10.1 All construction involving pockets or crevices that will not drain or that cannot be properly sandblasted and lined should be avoided.

10.2 PROTECTIVE COATINGS: THE SYSTEMS APPROACH

Mechanisms by which coatings protect other substances can be summarized as follows:

1. They prevent contact between the environment and the substrate, for example, nickel electroplating.
2. They restrict or limit contact between the environment and the substrate, as with most organic coatings.
3. They release substances that are protective or that inhibit attack, for example, molybdate, phosphate, and other inhibitive primers that partially solubilize and passivate a surface.
4. They produce an electrical current that is protective, for example, galvanizing.

A common reason protective coatings do not perform well is that they have not been considered as systems[2]. Successful coating engineers approach a coatings project in much the same way that they approach any other engineering problem, beginning with the design of surfaces to be protected and ending with schedules for monitoring the

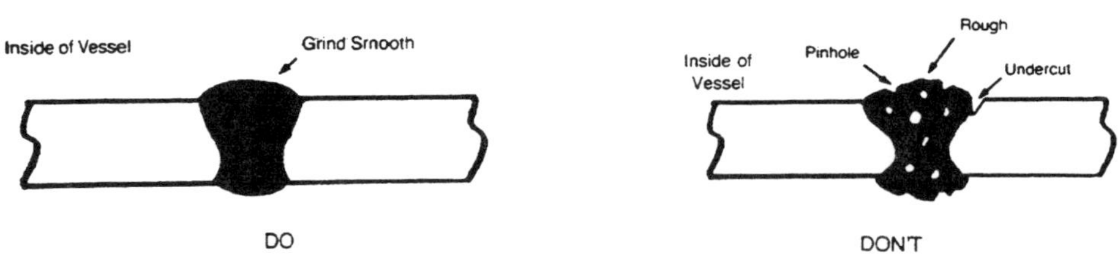

Figure 10.2 All joints should be continuous full-penetration welds. In tanks that require a 100% holiday-free lining, all welds must be smooth with no porosity, holes, high spots, lumps, or pockets. Grinding is required to eliminate porosity, sharp edges, and high spots. The use of weld metal to fill in undercuts and so forth is required for severe service conditions and/or high bake linings. Compatible caulking materials may be used for filling in mild service such as dry bulk storage.

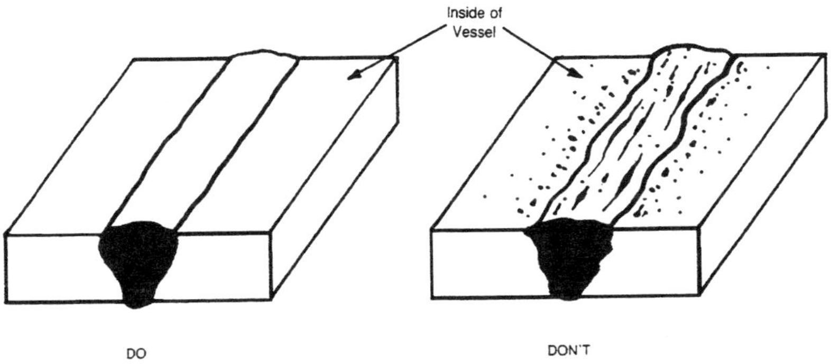

Figure 10.3 All weld spatters should be removed.

completed work. It is useful to consider protective coatings as systems in which the component parts or stages of development are related and mutually reactive. The various elements of good coating system are described in this chapter.

Prefabrication decisions and posterection corrections include elimination of sharp corners and hard-to-reach places; adoption of designs that will not collect or hold water or debris; provision for drainage of recessed zones, smoothing of rough surfaces, and rounding of corners; removal of weld splatter; avoidance of skip welding; and caulking of crevices. Figures 10.1 to 10.5 illustrates some of these considerations.[3]

Tubular or box shapes should be used whenever possible to minimize edges and inside corners. If possible, the metallic materials selected should have the best resistance to corrosive attack or be alloyed to enhance coating performance. Tests have shown that alloying specific elements to steel, for instance, will lengthen the effective life of coatings (e.g. HSLA steels).

Particular attention to design should be made to minimize the dwelling time of atmospheric condensation and provide special attention to surfaces shaded from the sun. Large enclosed spaces that have not been totally sealed, such as the insides of box girders, should be coated or ventilated, with fans if necessary.

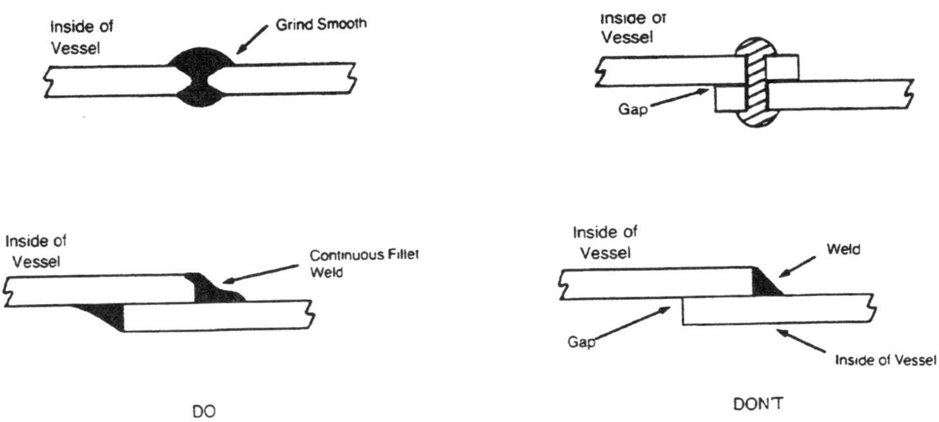

Figure 10.4 Butt welding should be utilized rather than lap welding or riveted construction.

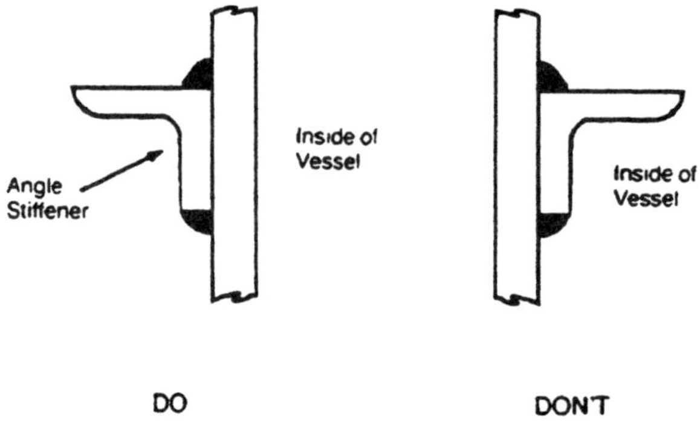

Figure 10.5 Stiffening members should be on the outside of the vessel or tank.

10.3 SURFACE PREPARATION

An essential preliminary to any coating operation is proper surface preparation. It is false economy to skimp on surface preparation in the belief that the coatings applied will make up for surface deficiencies. This is especially true of high-performance coating materials.

Surface preparation is a critical part of coating operations and must provide a surface that is compatible with the coating material to be applied. The major considerations concern the cleanliness of the surface required and the surface roughness or surface profile to be obtained. It is believed that, of the cost of a coating job, as much as one-half to two-thirds is for surface preparation and labor.

In some cases it is impossible to provide the best prepared surface because there is insufficient money or time to do the job and/or it is not permitted because of possible product contamination, fire hazard, or some other sufficient reason. It must be recognized that when there is a lesser surface preparation, coating-system longevity will likely be compromised.

10.3.1 Principles of Coating Adhesion

The cleaning of a metal surface is performed to achieve two basic goals:

- to obtain a hard continuous surface that will remain stable as a coating substrate and
- to provide a cleaner metallic base to which the coating material will adhere more strongly.

Heavy brushing of the surface with wire brushes, carborundum, or other materials can attain some part of these goals. A firm oxide surface is produced in this way, which is probably much less contaminated with atmospheric pollutants (chlorides, sulfates, and nitrates) than the original surface. If the remaining scale is firm, dry, and not excessive in thickness, a heavy coating can be applied to cover the rough substrate and keep active species from the atmosphere from diffusing to and through the scale. However,

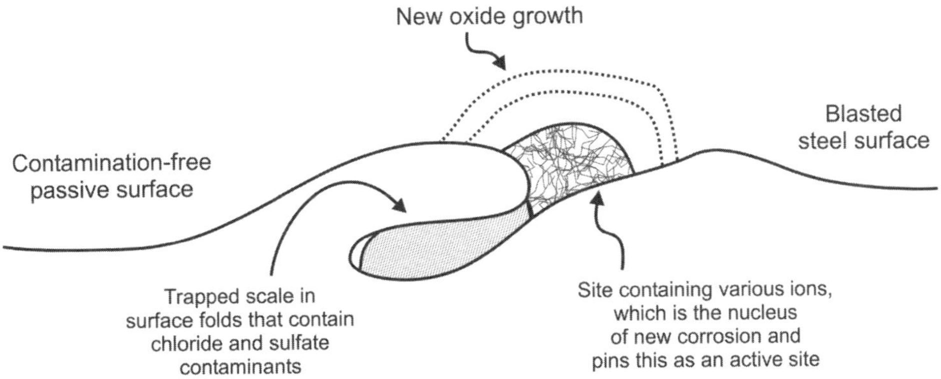

Figure 10.6 Schematic depicts the new surface-corrosion processes of an industrially corroded, blasted, and environmentally tested steel.

if all solid foreign material could be removed from the metal surface, a much improved adhesion would be achieved.

Unfortunately, when a metallic surface is cleaned even in a normal atmosphere, the surface can still be badly contaminated. Oxygen is immediately adsorbed and the cleaning procedure can leave some undesirable residue, and the original scale or other foreign material may not have been fully removed (Figure 10.6).

Of the cleaning methods available to provide the cleanest and largest effective surface per unit area, the use of abrasive blasting produces a surface closest to the ideal surface profile. However, the best of blasted surfaces can still be badly contaminated with residual metal oxides, extraneous dirt, particles of the abrasive materials, and adsorbed gases on the metal. As a consequence, it is necessary to construct coating formulations with the maximum capacity to compete for and lock into the available bonding sites. For this reason, coating materials containing polar functionalities such as OH, and others, particularly C=O functional groups, provide the best adhesion to a metal substrate.

Good wetting of the surface by liquid coatings is obviously a requisite for good adhesion. The coating material must remain stable to maintain adequate adhesion. If embrittlement occurs through oxidation, cross-linking, or volatilization of a portion of the coating (i.e., such as plasticizers), the resultant shrinkage stresses in the coating can pull it from the surface. Any loss of adhesion from such sources is obviously undesirable.

10.3.1.1 Abrasive Cleaning

The two principal methods of abrasive cleaning before application of coatings are centrifugal blasting and air-pressure blasting. Other abrasive cleaning methods, such as barrel cleaning, are discussed here.

Centrifugal blasting, as shown in Figure 10.7, is accomplished by machines that propel abrasives against the surface to be cleaned by imparting velocity to them by means of a rapidly rotating wheel. The abrasive material (commonly grit and/or shot, but also specialized abrasives such as cut wire, various hard oxides, or carbides) impinges

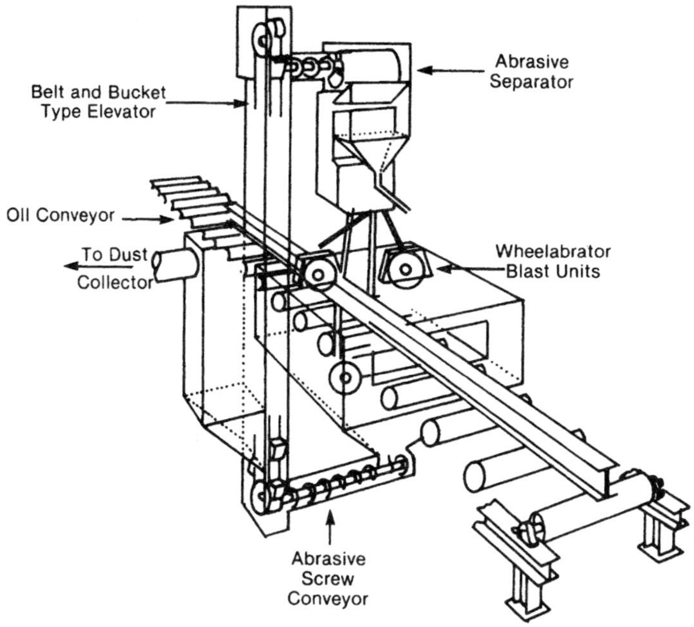

Figure 10.7 Schematic shows a centrifugal blasting machine. Abrasive is fed into the wheel from the center (a), from where it is impelled against the rapidly moving vanes (arrows show direction of rotation) and flung from the wheel perimeter (b) against the structural-steel shape.

against the surface, removing both the surface contamination and making a pattern of indentations, such as that shown in Figure 10.8. This pattern, consisting of peaks and valleys, serves as a mechanical assist to the chemical bonding forces of materials applied to the surface. The total surface area exposed in a profile of this type is obviously much greater than that exposed by a smooth surface.

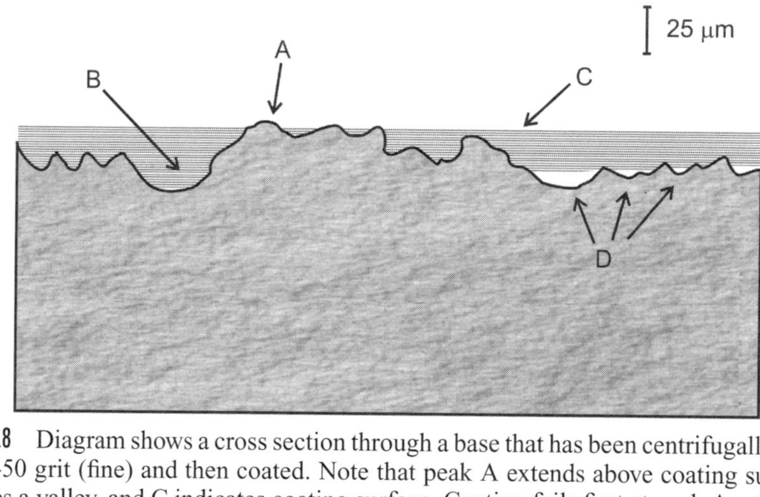

Figure 10.8 Diagram shows a cross section through a base that has been centrifugally blasted with G-50 grit (fine) and then coated. Note that peak A extends above coating surface, B indicates a valley, and C indicates coating surface. Coating fails first at peak A, whereas D indicates "bridging" of valleys where coating failure may start.

(a) (b)

Figure 10.9 Dry abrasive blasting of a) a ship hull and b) a pipe line (courtesy of Barton Mines Co.).

Abrasives used in centrifugal-blasting systems are collected and recycled. They are usually screened and cleaned during the recycling to remove dust, undersized particulate fines, and oversized debris removed during cleaning. Despite this, usually some particles of surface contamination recirculate with the abrasive.

Centrifugal-blasting machines are most economical when continuous or intermittent shipments of steel are cleaned by the wheel machine, such as in a steel-fabrication shop. All steel, as it is received, can be run through the machine and immediately primed with a preconstruction primer to provide an advantageous start on the coating system to be used. Without this initial treatment of the steel, a more tedious, expensive, and difficult blast-cleaning operation may be required after erection. Even then, overlapping surfaces and many recesses in the erected structure cannot be properly cleaned and primed and must be touched-up by air-pressure blast cleaning.

Cleaning of steel by blasting or pickling, followed immediately by a prime coat, or in many cases a complete coating system, can also be provided by many mills at an attractive price. The major problem then becomes one of handling the steel throughout transportation and handling from the mill through field erection without excessively damaging the shop-applied coating.

Field blasting involves injecting a supply of abrasive into a rapidly moving air stream and expelling it through a nozzle of the proper configuration so that the solid particles impinge against the surface to be cleaned (Figure 10.9). Important considerations in good and efficient air blasting include the following:

1. the skill of the workman,
2. good equipment producing the proper volume of abrasive at the proper velocity to do good work,
3. the quality of the abrasive, and
4. contamination problems.

A typical rig for efficient pressure blasting is shown in Figure 10.10. This illustrates a standard commercial setup. For smaller items, or for touch-up, venturi-abrasive blast-cleaning cup guns may be used. The abrasive in this equipment is stored in a small

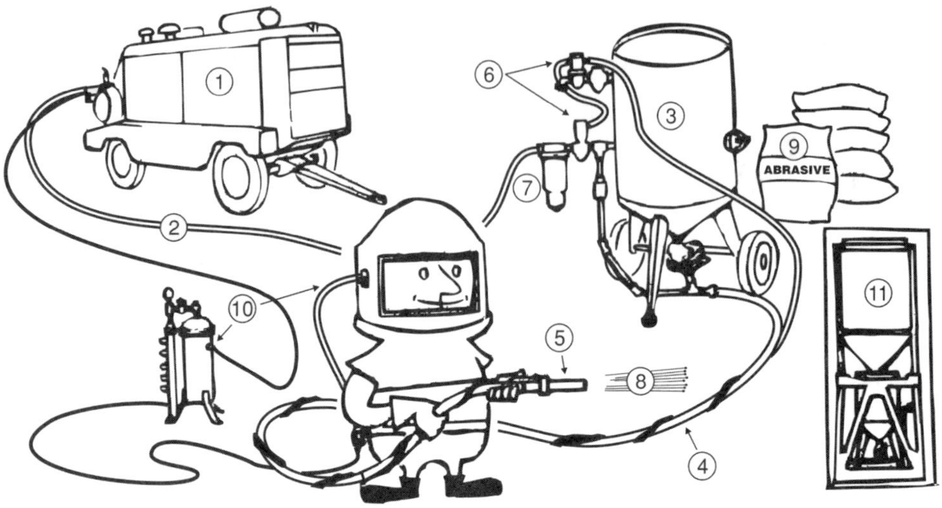

Figure 10.10 Typical rig for compressed-air abrasive blasting: (1) an adequate and efficient air supply (compressor); (2) air hose, couplings, and valves of ample size; (3) a portable, high-production blast machine; (4) the correct size of antistatic pressure hose with externally fitted quick couplings; (5) high-production venturi nozzle; (6) pneumatic-remote-control valves for safety and cost savings; (7) an effective moisture separator; (8) high nozzle air pressure; (9) the proper blasting abrasive; (10) a safe air-fed helmet and air purifier; and (11) inset showing bulk abrasive storage hopper over pressure blast vessel equipped with remote-control valves—when pop-up filling valve opens, abrasive from the hopper fills the machine.

container fastened beneath the blasting gun and educted into the air stream. This equipment is used by hobbyists, by automotive-repainting shops, and for small areas or inside shops for work on small pieces.

The skill of the workman in keeping the blast stream at the proper distance from and angle to the work, and his conscientiousness in being sure that all areas are properly cleaned, are a prerequisite to a good job. He must have the right kind of equipment, including the proper size and type of nozzle for the job being done, the correct size and lengths of hose, proper air pressure, and abrasive of the right quality and size to produce the desired surface. When skill and equipment are correct, achievement of a given grade of cleanliness is a function of time.

NACE International and the Society of Protective Coatings (SSPC) have jointly adopted a schedule of five degrees of cleaning steel by blasting operations. These are described briefly as follows.

1. NACE 1/SSPC 5: White metal blast cleaned—produces a gray/white, uniform, metallic color with no obvious foreign material remaining.
2. NACE 2/SSPC 10: Near-white blast cleaned—foreign matter is removed, but with differing shades of metallic gray surface allowable.
3. NACE 3/SSPC 6: Commercial blast cleaned—rust and foreign matter is removed, except for tight specks of oxide or paint uniformly distributed over a minor portion of the surface; some residues in pits.

(a) (b)

Figure 10.11 a) Water-jetting operation and b) pressurizing equipment (courtesy of Termarust Technologies).

4. NACE 8/SSPC 14: Industrial blast cleaned—similar to NACE 3/SSPC 6, except that evenly disbursed small islands of mill scale are also allowed to remain on the cleaned surface.
5. NACE 4/SSPC 7: Brush-off blast cleaned—light mill scale and tightly adhered rust is allowed if distributed over the surface.

Actual wording of the five grades of surface preparation and detailed information concerning methods that may be used to achieve proper surface preparation are available from documents published by NACE and SSPC. Blasting of the inside of tanks, piping, and other difficult surfaces can be achieved by ingenious available methods.

10.3.1.2 Water Jetting

Water jetting involves playing a stream of water moving at a high velocity (ca. 5000–40,000 psi) against a surface to be cleaned (Figure 10.11). The energy imparted to the water dislodges the scale or contamination from the surface to reveal the profile that is underneath existing coatings. This method is especially effective in removing resilient materials that may not be removed efficiently by abrasives and in cleaning surfaces with complex cross sections such as gratings or sieves. Water also can be used when blasting is a fire hazard. Water jetting excels at removing water-soluble salts from a surface. The coating industry uses water jetting primarily for recoating or relining projects where there is an adequate preexisting profile.

NACE and the SSPC have jointly adopted a schedule of four degrees of cleaning steel with water jetting to parallel the degrees of cleaning with abrasives. NACE 5/SSPC12 describes the degrees briefly as follows:

WJ-1: Clean to bare Substrate—surface is free of all visible rust, dirt, previous coatings, mill scale, and foreign matter.

WJ-2: Very thorough or substantial cleaning—allows for randomly dispersed stains of rust, tightly adherent thin coatings, and other tightly adherent foreign matter. The staining or tightly adherent matter is limited to a maximum of 5% of each unit area of the surface.

WJ-3: Thorough cleaning—allows staining or tightly adherent matter to a maximum of 33% of each unit area of the surface.

WJ-4: Light cleaning—provides a lesser degree of cleaning than thorough cleaning (WJ-3). The objective of light cleaning is to allow as much of an existing coating or foreign matter to remain as possible and to roughen the surface prior to coating application.

SSPC-VIS 4/NACE VIS 7, *Guide and Reference Photographs for steel Surfaces Prepared by Waterjetting*, shows a total of six different steel surfaces (two uncoated and four previously coated) before and after waterjetting. Photographs illustrate four separate degrees of cleaning (WJ-1, WJ-2, WJ-3, and WJ-4) for each initial condition, with additional photos that depict the appearance of light, moderate, and heavy flash rust after cleaning. It also contains a written guide to the use of reference photographs and additional explanatory notes. Cleanliness requirements for conditions depicted are defined by the joint SSPC/NACE surface-preparation specification for high- and ultrahigh-pressure water jetting (SP 12/NACE 5).

The use of remotely controlled robotic systems with vacuum suction has become very prevalent for large, fairly flat surfaces such as ships, storage tanks, and water towers.

10.3.1.3 Wet Abrasive Blasting

A variation of abrasive particle air blasting, useful in special cases, is abrasive wet blasting, which involves introducing abrasives into a rapidly moving stream of water and air. The combination obviously cuts down on the atmospheric pollution experienced when using the dry abrasive alone. Inhibitors are often added to the water to reduce rusting of the "clean" steel before the prime coat is applied. See Figure 10.12 for a sketch of a nozzle used for such purposes.

Wet abrasive blast-cleaning systems commonly encountered include air/water/abrasive blast cleaning, which uses compressed air to propel the abrasive; pressurized water/abrasive blast cleaning, which uses water to propel the abrasive; and a combination of pressurized water/pressurized air with abrasives. The abrasive defines the profile.

Generic terms to describe specific air/water/abrasive blast-cleaning methods are water shroud or wet-head blasting, wet blasting, low-volume water abrasive blasting, and slurry blasting. Other generic terms to describe specific water/abrasive blast-cleaning methods are slurry blasting, abrasive water jetting (AWJ), or abrasive injected water jetting/blasting (AIWJ), or ultrahigh-pressure abrasive blasting (UHP AB).

SSPC-VIS 5/NACE VIS 9, *Guide and Reference Photographs for Steel Surfaces Prepared by Wet Abrasive Blast Cleaning*, shows two uncoated, rusted steel surfaces (conditions C and D) before and after wet abrasive blast cleaning. Photographs illustrate two degrees of cleaning (WAB-6 and WAB-10) for each initial condition, with additional photos that depict the appearance of light, moderate, and heavy flash rust after cleaning. It also contains a written guide to the use of reference photographs

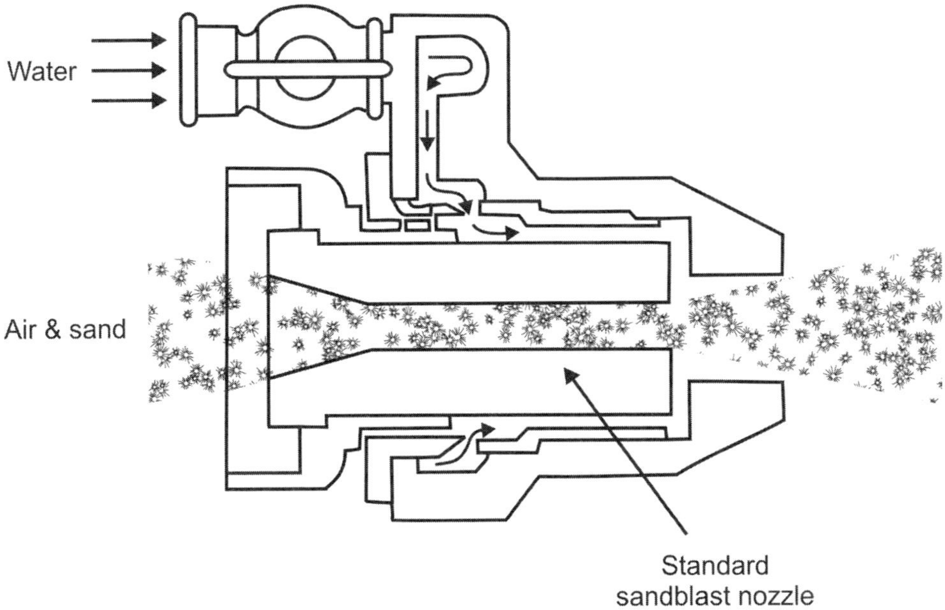

Figure 10.12 An attachment makes wet-abrasive blasting possible by adding water as the abrasive leaves the nozzle. This technique is also widely used to clean stone, brick, concrete, and stucco.

and additional explanatory notes. Cleanliness requirements for conditions depicted are defined by the joint SSPC/NACE surface-preparation specifications for commercial (SP 6/NACE 3) and near-white blast cleaning (SP 10/NACE 2). Photos of commercial and near-white surfaces achieved by dry abrasive blast cleaning can be found in new SSPC-VIS 1.

SSPC-TR 2/NACE 6G198, *Wet Abrasive Blast Cleaning*, covers procedures, equipment, and materials involved in a variety of air/water/abrasive, water/abrasive, and water-pressurized abrasive blast-cleaning systems. Equipment usage and safety are also discussed.

10.3.1.4 Other Surface-Preparation Methods

Although abrasive blasting and water jetting are widely used for plants, ships, bridges, and other large systems, the following procedures are also frequently used.

Solvent Wipe For certain surfaces, it is only necessary to wipe the surface with a selected solvent to remove oil, grease, and loose dirt. The use of detergent solutions, isopropanol, ketones, and aliphatic and aromatic hydrocarbons for this purpose can be effective.

The major requirement for the process is that the wipe shall not leave a residual contaminant, but rather a clean, dry metal surface. Such a cleaning procedure is used on previously cleaned steel, stainless-steel surfaces, and old paint films before overcoating, and on dirty steel prior to abrasive blasting.

Vapor or Solution Degreasing A surface that is dry and free from oily residues can be obtained by suspending the item in a closed booth in which the vapors of a solvent are condensing. The vapors condense on the part, coalesce, dissolve the contaminant, and drip from the surface. 1,1,1 Trichloroethane (Trichlor) and perchloroethylene are commonly used solvents for this purpose. Many production-line items are cleaned in this manner before coating.

Baths of cleaning solvents may also be prepared to remove oily or loose contaminants. Aqueous detergent systems, emulsion systems, and alkali baths are also used with good mechanical or ultrasonic agitation to ensure scouring of the surface by the liquid.

Steam or Hot-Water Degreasing Hot-water heaters or steam generators ("gennys" or "gensets") are available for use in the field to clean metals, coatings, and concrete surfaces. Strong detergents or alkalies are added to the water to emulsify oils and similar organic contaminants on contact. Impingement of the hot, strong emulsifying agent on the surface removes even heavy soil at a rapid rate. This method can often be used with good results prior to abrasive blasting.

Flame Cleaning The process of flame cleaning has found greater favor in Europe than in America. A wide oxy-acetylene flame is played on the surface of steel parts to produce two effects. First, the heavier portions of rust scale or mill scale will pop from the surface because of the thermal expansion difference between the scale and the substrate. Second, the surface is freed of moisture if sufficient heat is applied to the substrate. Removal of loose rust particles after the treatment, followed by prompt coating can produce an adequate coating procedure for certain applications. When using an open flame for cleaning, extreme caution must be used to prevent fires and explosions.

Mechanical Abrasion Manual or power-tool cleaning of an oxidized surface is sometimes the least thorough and slowest means of surface preparation. However, cost, location of the part, or the availability of tools often dictates the use of such a technique. Normally, only the top loose layers of rust on a piece of steel are removed through this method. Moisture and other contaminants remain in the residual scale. Chipping and grinding can remove the heavier scale but will impress oxide and salts into the metal surface. Rotary flapper wheels and wire wheels can clean a metal surface to a cleanliness approaching "white metal" and additionally can impart a profile suitable for painting.

When using brushes or grinding wheels, it is important to consider the compatibility of the tool with the substrate. For instance, the use of steel or bronze brushes to clean aluminum can leave sufficient heavy-metal (Cu or Fe) contamination on the surface to initiate severe pitting of the metal.

An air-actuated needle gun is also available to clean small areas by mechanical action. Barrel cleaning, where parts are placed in a rotating drum containing an abrasive, is also used for small parts.

Pickling Great quantities of sheet, plate, coil stock, and other forms of metal are pickled in production mills or by reprocessors of the metals. On steel products, the goal is the removal of the mill scale ($FeO\text{-}Fe_3O_4$) formed on the surface during hot processing (>575°C). The bath is an aqueous hot hydrochloric, sulfuric, or less frequently phosphoric acid solution containing an inhibitor. The inhibitor must allow uniform attack on the metal, as little metal loss as possible, and, when pickling steel, must leave a clean surface free of carbon smut.

Such an inexpensive cleaning procedure followed by thorough rinsing and drying provides one of the cleanest and most active surfaces for further processing. More steel is pickled, treated, and coated than by all other methods of preparation and protection.

The pickling process is conducted in large vats and does not lend itself readily to use in the field. Acid gels or washes based on phosphoric acid can be prepared (typified by MIL C-10578, Type C) that will lightly etch a steel surface when held on the surface once the heavy scale has been removed. This type of pickling is restricted to smaller articles where the acid can be managed properly.

10.3.1.5 Cementitious Surfaces

As with the application of coatings to metals, the important first step with concrete is that of surface preparation. When considering concrete, preparation problems are compounded because the character of the substrate may vary widely. Concrete block construction may present a fairly uniform, hard surface, whereas some hydraulic concrete surfaces must be actually repaired before attempting coating of the structure. Somewhere between these two cases arise the problems of coating polymeric concretes, precast slabs, and gunite surfaces.

When present, any laitance (thin, cement-rich, weak surface) must be removed, cavities filled, other surface imperfections leveled, all dust and loose material removed, the surface neutralized if there exists alkali from salts of the cement, and the surface dried. To accomplish this end, hand brushing, acid etching, water blasting, steam cleaning, and abrasive blast cleaning are often used.

Cracks in concrete can sometimes be prepared by routing and applying an elastic epoxy, urethane, or other material that will expand and contract without rupturing as the concrete moves. A large variety of coating materials are used on concrete, with the epoxies, acrylics, certain urethanes, asphalts, and polyureas heading the list.

If sealing of the concrete surface is desired, repeated treatment with a silicone wash may make the surface of concrete, granite, and so forth satisfactorily water resistant. Another treatment that can be said to produce a conversion coating is repeated spraying with a dilute solution of magnesium fluorosilicate. The chemical reacts with the outer portion of the concrete to produce a hard, water-resistant finish.

10.4 APPLICATION OF ORGANIC COATINGS

The coating material must have application characteristics that would allow its proper application under all conditions existing during the coating process. Proper identification of the physical or other properties of the film expected from the application must be specified. Inspection procedures to verify the properties and to maintain uniformity of the system must also be stated. The safety of personnel involved in the work or of others in the vicinity must be considered.

Proper application of a coating can mean the difference between a successful job or a severe economic loss. In some instances, the choice of the application method can be the most important decision. At other times, more subtle influences determine the outcome of the work.

Atmospheric conditions are an important consideration. In high-humidity environments, a coating must be applied very quickly after the surface is cleaned. Gross oxidation of the surface by moisture and oxygen in the air cannot be allowed.

Many coating resins are intolerant of moisture. Application of these materials on a dewy morning, under rainy conditions, or where rain is anticipated can be a waste of time and money. Also, the rheology of most coating materials is such that flow-out of the film and proper solvent release will not occur below a given temperature. For many materials, this point is around 5 to 10°C.

The most difficult influence to control is that of contamination of the surface by some airborne species before the coating is applied. This can be a whiff of hydrogen sulfide, sulfur trioxide, or other chemical from a stack. It may be an alkaline material, such as fly ash, ground seashells, or limestone dust. In a marine environment salt contamination is a problem. When a coating is applied over such contaminants, a more rapid failure is certain to occur once moisture permeates the film.

Materials to be applied must be properly mixed. This process can be difficult for some materials, and adequate equipment to do the job should be available. When catalysts are to be added to the mixture, close attention to the proportions required must be given, and thorough postmixing must be accomplished. The pot life of the mixture must be known and understood.

Many jobs have been downgraded because excessive or inadequate thinning of the coating material was made before the application. Overthinning to make the mixture more workable is the more common error.

Many methods of applying a coating are available. The familiar techniques involve brushing, rolling, dipping, palming, and spraying. Other procedures have been developed for more specific types of organic coatings and economy of operation. Some of the more important techniques are as follows.

10.4.1 Brushing and Rolling

Brushing and rolling rank first as the easiest and most commonly used application techniques.

10.4.2 Palming

Palming is sometimes referred to as a mitten application.

10.4.3 Dipping

Dipping is a technique that should be considered for intricate shapes such as grating.

10.4.4 Air Spray

Air spray is an older spraying technique where a separate air stream was used to break up the coating stream, resulting in a fine atomization and a smooth, glossy coating.

10.4.5 Airless Spray

Airless spray is a more popular spraying technique where more pressure is used on the pot, and the tip of the gun is designed to disperse the coating. Less overspray results as well as a higher production rate and a greater coating thickness. Airless spraying is the most common means of applying most corrosion-protective coatings.

10.4.6 Electrostatic Spray

With electrostatic spray, electrically charged particles of coating are directed through the air toward a surface whose charge is opposite that of the particles. The coating is thus attracted to the surface and will tend to deposit with little overspray. Good buildup on edges is also a positive factor. It is often used on production lines.

10.4.7 Electrophoretic Dipping

With the electrophoretic-dipping technique, electrically charged particles suspended in a conductive liquid within a dip tank are impelled toward a surface when a DC current is applied through the liquid. The particles tend to collect on surfaces of opposite polarity and will go into crevices and occluded areas in a relatively uniform thickness. Edges are particularly well coated. Automobile bodies are coated in this manner.

10.4.8 Fluidized Bed

When using the fluidized-bed technique, thermoplastic polymers are powdered and suspended (levitated) in a tank through which air is pumped in such a way that the particles behave like a liquid. Surfaces to be coated are preheated and run through the bed. The hot surface melts the thermoplastic, fuses the coating, and subsequent cooling produces a uniform plastic sheathing. Thickness of the coating depends on the temperature of the part and the time of exposure.

10.4.9 Powder Spray

Parts to be coated by the powder-spray method are heated to the appropriate temperature and then moved to an area where a fusible coating powder is sprayed onto the surface. Powder spray is a production-line technique that can produce very uniform coatings.

10.4.10 Flame Spray

During the flame-spray method, thermoplastic powders or rods are fed into an oxy-acetylene or other flame and impelled toward the surface to be coated. The particles

(a) (b)

Figure 10.13 Prefabricated rubber linings are glued to a) a water treatment unit and b) a large open top tank (courtesy of Blair Rubber Company).

coalesce and solidify to form the coating. Metals are applied in the same manner at much higher temperatures.

10.4.11 Troweling

Concrete, mortars, epoxy, polyester, furane, and certain other polymers may be applied as heavy coatings by hand troweling. The organic materials are normally reinforced with glass cloth, whereas wire mesh might be used for the inorganic products.

10.4.12 Vacuum Deposition

With vacuum deposition, the coating material is placed on a heating source in the lower part of a container, which is then put under vacuum. The item to be coated is suspended above the coating material. The coating material (organic or inorganic) is then heated until vaporization occurs. The vaporized coating molecules condense on the cold surface of the target object to produce the coating. Glass, plastics, and other nonmetallic bases may be coated in this manner.

10.4.13 Calendered or Sheet Lining

Prefabricated linings may be glued to surfaces, as shown in Figure 10.13. Applications are made inside tanks and large-diameter pipes. Polymeric linings can be designed to be inserted loose inside tanks (Figure 10.14). Large indoor or outdoor ponds and sumps designed to hold corrosives or ecologically restricted fluids may also be lined with polymeric materials (Figure 10.15).

10.5 FACTORS IN COATING SELECTION

The number of coats to be applied, the compatibility of the various coats, and the requirements for their maintenance must all be considered to provide proper stability

(a) (b)

Figure 10.14 Flexible liners protect a) the ceilings and b) walls of anaerobic digester tanks from corrosive agents (courtesy of Flexi-Liner Corporation).

Figure 10.15 Secondary containment made with flexible lining material (courtesy of Flexi-Liner Corporation).

Table 10.1 Common Coating Types and Areas of Use

Area of Use	Coating Type	Environment
Exterior iron and steel	Oil	Rural, mild industrial
	Alkyd	
	Urethane	
	Oil-based phenolic	High-humidity mild industrial
	Inorganic zinc	Corrosive
	Vinyl	Industrial/marine
	Epoxy	Severe weather
Interior iron and steelmetal	Alkyd	
	Urethane	Mild industrial
	Epoxy ester	
	Vinyl	Acid or other corrosive fumes or dust fallout
	Epoxy	
	Chlorinated rubber	
Interior storage tanks chemical solutions	Vinyl	Salt solutions
	Epoxy phenolic	Dilute acids
	Epoxy	Alkalies
Interior water tanks	Vinyl	
	Epoxy	
	Chlorinated rubber	Water immersion
	Coal tar epoxy	High humidity
	Coal tar asphalt	
Interior concrete tanks or reservoirs	Vinyl	Water immersion
	Epoxy	High humidity
	Chlorinated rubber	
Interior solvent storage tanks	Inorganic zinc	Solvent immersion
	Epoxy phenolic	Fumes/condensation
	Epoxy	
Exterior concrete plaster/masonry	Cement	Mild industrial
	Acrylic	
	Vinyl acetate	
	Vinyl	Corrosive
	Epoxy	Humidity
	Chlorinated rubber	
Exterior wood	Oil	Rural, mild industrial
	Alkyd	
Interior plaster or wall board	Acrylic	Mild industrial
	Alkyd	
	Chlorinated rubber	
Wooden floors	Alkyd	Mild industrial
	Polyurethane	
Concrete floors	Epoxy ester	Mild industrial
	Polyurethane	
	Chlorinated rubber	Spillage water
	Vinyl	Dilute chemicals
	Epoxy	

Table 10.2 Maintenance Procedure for Common Coating Types

Class of Coating	Type of Vehicle	Maintenance
Air-reactive coating	Oil, varnish, enamel, alkyds, oil-modified phenolic, epoxy ester	(1) Remove all oil or grease. (2) Wash/scrub with clean water. (3) Power sand or light brush blast. (4) Brush first coat on repair areas. (NOTE) If general surface has heavy or medium chalk, sanding may be eliminated.
Internally reactive coatings (catalyzed)	Epoxy amine, epoxy polyamide, epoxy phenolic, coal tar epoxy, urethanes, polyesters	(1) Remove all oil or grease. (2) Wash/scrub with clean water. (3) Power sand or light brush blast. (4) Brush first coat on repair areas. (NOTE) Surface of coating must be abraded for proper topcoat adhesion.
Moisture-reactive coating	Moisture cure urethane	Same as for internally reactive coatings.
Solvent drying lacquers	Vinyl, vinyl acrylic, coal tar, nitrocellulose, asphalt	(1) Remove all oil and grease. (2) Wash/scrub with clean water. (3) Reapply solvent dry coatings.
Water-based coatings	Vinyl acetate P.V.A., vinyl acrylic, acrylic, coal tar emulsion, asphalt emulsion	(1) Remove all dirt, dust, and grease. (2) Wash/scrub with clean water. (3) Reapply water-base coating.
Inorganic zinc coatings	Water-based silicate, solvent-based silicate	(1) Remove all oil and grease. (2) Wash/scrub with clean water. (3) Light power sand or light sweep blast. (4) Reapply coating. (NOTE) Light abrasion of surface is necessary to open up new zinc. (NOTE) Inorganics will not adhere to any organic or/oil/grease coating.

during the life of a project. Tables 10.1 and 10.2 provide some general information where common coating types have been used (Table 10.1) and the maintenance procedure that is required (Table 10.2).[2]

10.5.1 Substrate Properties Affecting Coatings' Performance

Properties of the substrate to which coatings are applied will influence performance. Tests have shown advantages of certain alloying elements in steel, which are conducive to improved service. It usually is conceded that most, if not all, metals have oxides on their surfaces to which coatings are applied, the only exception being in certain exotic vacuum systems used for electronic equipment. Atmospheric oxide layers are usually thin on nonferrous metals, unless special treatments are used to purposefully thicken these layers, for example, anodization of aluminum and magnesium alloys.

Surfaces in compression often increase coating performance because they tend to be attacked less than surfaces that are not so stressed; on the other hand, surfaces

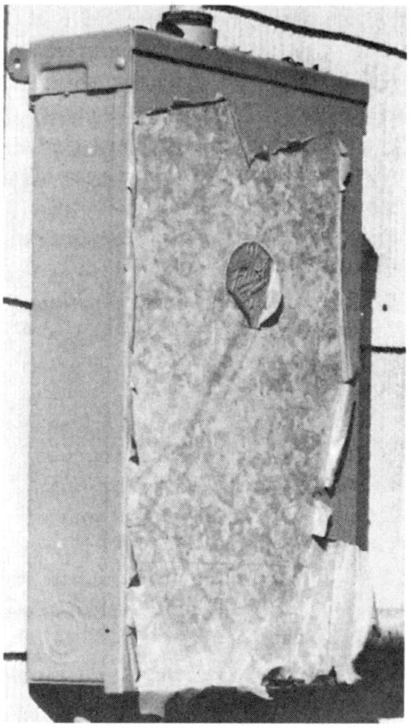

Figure 10.16 Because the galvanized surface of this circuit-breaker box was not properly prepared, there was substantial loss of adhesion in less than a year.

in tension may cause coatings to fail earlier than they would otherwise. Differences in expansion rates between a coating and a substrate impose stresses on the coating, which may lead to cracking and failure. Both inadequate and excessive coating thickness may lead to failure.

Metals high in the electrochemical series, such as magnesium and zinc, require special attention. Because organic coatings are permeable, corrosive solutions that reach the base metal will attack, causing generation of hydrogen which can cause blistering and ultimate failure. Galvanized surfaces are prone to reject organic coatings (Figure 10.16).

Wood substrates are porous and require sealers. Coatings applied to wood may suffer attack from water penetrating from an uncoated side. Some coatings will permit reverse transmission of vapor better than others. Examples are certain linseed-oil formulations, some alkyds, and latex water-based materials.

10.5.2 Characterization of Coatings

A convenient way to characterize coatings is as follows:

1. primers (first coatings applied to a surface),
2. secondary or intermediate (if required by some specifications), and
3. topcoats (selected for specific resistance to the expected environment and compatible with second or primer coats).

10.5.2.1 Primers

The original term "primer" implied a coating first applied to a substrate because it had a singular adhesive affinity for it and/or because it provided better adhesion for a subsequent coat than the subsequent coat could achieve if applied directly. A primer also functioned as a carrier for an inhibitor or for an anodic metal loading such as zinc, both of which are discussed further in later sections of this chapter.

In modern usage and for many applications, primers usually are thin films (75 μm; 3 mils or less) and may be used on both metal and wood surfaces. On wood, their function is to seal the grain or to provide a smooth base for topcoats. On concrete, they should be compatible with the alkaline surface and help improve both adhesion and the life of subsequent coats.

10.5.2.2 Intermediate Coats

A second coat may be either the top or final coat, or an intermediate coat added when multiple thin films are required. The intermediate coat must serve as a tie coat between the primer and the subsequent coats and may be of a different composition than either of them. When the three are of different formulations, it is the intermediate coat that is often used to provide the major film thickness for the coating.

10.5.2.3 Topcoat

A topcoat may be required to extend the life of the preceding coats. When designed as a topcoat, the film is normally more dense and hydrophobic than the remainder of the system to reduce the rate of moisture permeation to the underlying coats.

A topcoat also may be selected to confer reflectivity, reduce photodegradation, or to affect color, among other reasons.

Glasses and other silica-based materials require coatings that can be fired (melted) on, as a rule. Limited adhesion of organic coatings can be achieved on etched surfaces.

Polymeric materials are seldom protected with organic coatings, although the coating of epoxy and polyester piping and vessels to prevent deterioration by ultraviolet rays is common.

10.5.3 Coating Materials

Coating materials have increased in complexity in parallel with developments in surface science and engineering, coating technologies, and environmental requirements. More can be achieved by the use of a proper coating material today than by simply covering a surface to reduce contact of an environment with the substrate. Coatings of various types are applied to produce the following effects:

1. to provide a dense, inert barrier to shut off the environment;
2. to provide good adhesion to a metal surface for the application of other materials;
3. to serve as a better substrate for topcoats than a bare metal surface would afford;

4. to sacrificially protect the metal surface by CP;
5. to resist abrasion;
6. to cover a substrate while allowing the escape of gases from beneath;
7. for aesthetic reasons;
8. to provide a smoother, low-friction surface;
9. to prevent catalytic effect of a surface on a product in contact with the surface;
10. to hold an inhibitor at a metal surface for use when moisture permeates the film;
11. to provide surfaces that are more easily cleaned;
12. to prevent contamination of a contained product; and
13. to protect the surface from high-temperature oxidation.

Coating materials consist fundamentally of a resin, pigments, additives, and a solvent or dispersant (except in the case of 100% solid coatings, where the latter two items are not used). The resinous material, plus the pigments, nonvolatile additives, and fillers, comprise the solids of the coating. These are carried in a vehicle (liquid solvent or dispersing medium), which may be varied in viscosity by the addition of a thinner. The vehicle also may ultimately be the binder of the coating (binding the pigment particles together and to the substrate).

Only the solids remain on a surface after the final coating has formed. However, a portion of or the total vehicle may be reacted with the resin (100% solid coating) to produce the final film. Pigments are added to provide color or as inhibitors in a prime (first) coat. Fillers are selected to increase the bulk of the coating, improve the density, increase abrasion resistance, and increase the opacity of the film. Selection of the wrong filler or addition of excess filler can downgrade a coating by increasing the permeability or decreasing the cohesion in the film.

Lacquers are those coatings formed by the evaporation of the vehicle from a solvent-dissolvable (thermoplastic) polymer (e.g., coal-tar resin dissolved in a hydrocarbon leaves a lacquer finish), whereas an enamel is normally described as a pigmented coating of greater thickness and hardness, which may be formed from a resin that completes its "cure" by baking or oxidation in the atmosphere. Curing indicates the ultimate reaction of the film, whether effected by the atmosphere, catalysts, or baking. The process is normally associated with some cross-linking of the polymer chains of the coating resin, unless simple drying of the film occurs.

Obviously, the solid content of a liquid coating determines the final film thickness of the coating, depending on the amount applied over a specific area. Thus,

$$\text{Film thickness} = \frac{\text{Vol. liquid coating} \times \%\,\text{Solids} \times \text{Factor}}{\text{Surface area covered}} \tag{10.1}$$

If the volume of liquid is in liters and the surface area in square meters, the factor in this equation is 1000 and the film thickness is in μm. If the volume of liquid is in U.S. gallons and the surface covered in square feet, the factor becomes 1604 for thickness in mils.

This theoretical coverage will not be precisely obtained because of the roughness of the surface and losses occurring during application of the wet film. Some 15% loss is to be anticipated for spray applications.

Resins used in the coating formulations can be natural or synthetic. Asphalts, copal, and tung are examples of the naturally occurring molecules, whereas urethanes, epoxy, and acrylics are examples of synthetic polymers prepared by the process industry.

If the resin product is not designed to provide a lacquer coating after evaporation of the vehicle and thinner, some reaction must take place to produce the desired polymeric film on the surface. When using linseed oil and certain other molecules, curing is accomplished by reaction with oxygen from the air (oxidation). When a coreactant (accelerator, hardener) is added to a specific formulation, it reacts by cross-linking with the resin to produce the hard, inert final film. Baking of certain types of coatings can also cause cross-linking, and additionally will help volatilize unwanted solvents.

Some of the more widely used coating materials are described in the following paragraphs. It is most important to realize that each of the basic polymers listed does not produce a single coating. They serve as a base upon which it is possible to construct innumerable coating combinations. Although the base resin may impart some property of the coating that is fairly consistent (e.g., epoxies resist alkalies), it is possible to alter the final polymer in such a manner that entirely new properties are obtained.

10.5.3.1 Bituminous

Natural asphalts are the oldest of all known coating materials. These products were used alone or mixed with other materials before the time of written history. Today, the asphalts and the coal tars derived from petroleum or coal are important coatings. Unless exposed to direct sunlight, the materials are stable, inexpensive, heavy, protective coatings.

As with other coating materials, no single product exists. The asphalt coatings are formulated from a number of naturally occurring beds of bitumen or from petroleum residues. The coal-tar product is recovered from a mix of molecules that allows production of coatings with varying properties. These can be applied as hot melts or as cut backs (diluted with hydrocarbon solvents) and have greater moisture resistance per unit thickness than do the asphalts.

The products are widely used for the coating of steel and concrete structures below ground, as roofing coatings, and in certain severely acidic atmospheric exposures. Heavy bituminous coatings are often reinforced with glass cloth, burlap, or other materials to improve lateral strength in the completed covering.

10.5.3.2 Oils

For many years, the formulation of oleoresinous protective paints was based on naturally occurring "oils" (long-chain, unsaturated acid and ester molecules) and rosins such as linseed from flax, tung, pine, castor, fish, and other sources, which would provide a resinous film when the oxygen from the air reacted with the molecule to harden the film and produce some cross-linking. This atmospheric-drying process was slow, with obvious disadvantages. Also, once the film was cured, the chemical resistance was not adequate to withstand many of the more severe exposures of an industrialized environment.

However, each material has its useful place, and the proper formulation of a long oil can produce a cured coating that is tough, ductile, dense, continuous, and very protective of steel exposed to rural and milder urban environments.

10.5.3.3 *Phenolics*

It is possible to synthesize a great diversity of phenolic resins. As a consequence, it is best to describe how the various resins are used rather than discuss specific chemical combinations.

First, there are a large group of substituted phenols that are added to oil, alkyd, and other coatings to upgrade their water resistance. Such a coating is phenolic fortified but is not a phenolic coating.

Second, preparation of oil-soluble phenolic resins allowed the preparation of a group of true phenolic coatings to be used for atmospheric and fresh-water immersion. These are phenol esters combined with oils, rosins, or newer synthetic resins. Certain of these phenolic resins are formulated into an inexpensive low-baking finish.

Last, there is a class of phenolic resins applied as a lacquer and baked at high temperature (150°C or greater) to produce unique chemical-resistance properties, such as excellent acid and solvent resistance. Alkali and strong oxidants are essentially the only environments in which they are not stable. The coatings are used extensively on the interior of cans, drums, tank cars, and storage vessels to prevent contamination of the contents by iron.

10.5.3.4 *Alkyds*

The alkyd resins (polyesters) replaced the oils to a major extent as the easiest material to apply at the lowest cost, beginning in the 1920s. These can be formulated for application under most any condition, including baking cycles. The enamel produced has good color stability with little chalking in the atmosphere. The excellent films obtained can be used in the milder industrial environments and most other atmospheric exposures but should not be exposed to strong acid, alkali, or solvent conditions.

This category of coating should always be considered when assessing a job because of the ease of application.

10.5.3.5 *Acrylics*

As a category of coating resin, the same comments may be made for the acrylic resins as for the phenolic products; that is, the various resins available are used in three different manners.

Some of the acrylic resins are designed to be used as additions to upgrade other coating bases. Others are used in formulations that are applied on home appliances, office furniture, and so forth and baked on production-line systems. The product has high gloss, toughness, and aging stability.

Today, the greatest use of acrylic resins is in the preparation of latex paints. The original water dispersions were based on styrene/butadiene resins but quickly yielded the field to the acrylic acid/vinyl polymers. Little need be said about the utility of the latex formulations available today.

The environmental restrictions on the use of organic solvents in paint formulations makes the use of the water-based materials most attractive. Great strides have been made in producing nonrusting water-based primers and water-based electrophoretic coatings.

The materials do not have good solvent or alkali resistance. However, their durability in the atmosphere is excellent, although color retention can be a problem. They are normally compatible with previously applied oil, phenolic, or alkyd coatings. Acrylic latexes can be particularly beneficial in eliminating blistering problems of other coatings on wood where the more dense coatings prevented breathing (vapor transmission) through the paint.

10.5.3.6 Polyureas

Polyureas are becoming a commonly used coating on concrete and metals because of the diversity of properties and resistances that can be formulated using the materials. Although technologies regarding their formulation and application are recent, the term "polyurea" is an old term denoting the chemical cross-linking bond that results when, for example, an isocyanate reacts with moisture. All moisture-cured polyurethanes are actually polyureas. A modern polyurea can be defined as a reaction product of an aliphatic or aromatic isocyanate component and a resin-blend component that is composed of amine-terminated polymer resins and/or amine-chain extenders.

The new polyureas, are often formulated as 100% solid elastomeric thick-film coatings (applied at thickness of 1 mm or more). They are used to coat concrete containments around tanks in refineries and as roofing coatings. Polyureas are perhaps most widely known as sprayable truck-bed liners. They have a very short pot life (from 3–30 minutes) and therefore must be applied using heated, plural-component spray equipment. However, modification with polyaspartic esters extend the pot life so application with conventional airless-spray equipment is permissible. Polyaspartic ureas can be formulated as hard thin-film coatings (up to 400 μm) with properties similar to polyurethanes. Both polyureas, with and without aspartic modification, dry, cross link and cure within hours, and thus are suitable for service almost immediately after application.

Another unique group of coatings are the vinyl plastisols and organosols. These are heavy, tough coatings applied on such items as dishwasher trays and other production line items.

10.5.3.7 Epoxy

The epoxy molecule was prepared in Germany in the 1800s, but it was in the 1940s that use of the resins based on the molecule became commercially important. Their popularity increased rapidly because of the excellent adhesion attained by the resin and the broad spectrum of chemical resistance inherent in most of the formulations.

The epoxy molecule was combined with oils to produce the epoxy esters, a less costly, easily applied, less chemical-resistant coating that should not be confused with the catalyzed epoxies.

The two-package (catalyzed, activated) epoxies offer some of the best properties of any coating material available, once properly applied. However, care must be exercised in their application. Heavy films with fine adhesion to clean metal can be achieved in two coats. The film is resistant to acids, alkalies, and most solvents but not to strong oxidants. Rather heavy chalking of the surface can occur in sunlight, depending on the formulation chosen, and the surface can be difficult to overcoat at a later time because of continued curing and chemical resistance of the film.

The amine-cured epoxies are used extensively for the internal lining of storage vessels, hopper cars, and concrete structures. A hand lay-up of epoxy coating reinforced with glass cloth provides an exceptionally heavy coating for immersion service.

The combination of epoxy resin with coal-tar polymer produces a less expensive coating with properties appropriate for use on the interior and exterior of piping and many other severe-exposure conditions.

10.5.3.8 Urethane

The urethane polymer is most familiar in the form of plastics, elastomers, foams, and coatings. Obviously, the different urethane resins must be capable of combining with many other types of molecules to produce this variety of end products. As a result, the category is divided into six types by ASTM. In each of these various types, a range of properties can be obtained for the coating, depending on the chemistry of the system.

Type I: One-package urethane coating in which the polyisocyanate has been reacted.
Type II: One-package moisture-cured urethane coating.
Type III: One-package heat-curable coating with blocked polyisocyanate groups.
Type IV: Two-package-catalyzed urethane coatings: one package contains the polyisocyanate and the other contains the catalyst.
Type V: Two-package polyol-urethane coating: one package contains the polyisocyanate and the other contains polyols, with or without a catalyst.
Type VI: One-package urethane coating in which the polyisocyanate has already reacted.

Elastic coatings, hard coatings, coatings with poor ultraviolet-light stability, coatings with excellent ultraviolet-light stability, coatings difficult to pigment, coatings readily pigmented with the usual additives, coatings with a chemical resistance no better than an alkyd, or coatings with very fine chemical resistance can all be obtained in this grouping. There is no such thing as a "urethane coating," so it is important to be specific.

The most unique property of this category of coating material is the superb abrasion resistance of the moisture-cured varieties. No other organic material with a hard, glossy finish can compare with a Type II urethane in abrasion resistance.

The materials are used widely for floor and other wood finishes, hopper-car linings, and as maintenance coatings. Adhesion of the basic urethane resin to metal is not good, and the coatings are applied over primers of other resins such as epoxies.

10.5.3.9 Temporary Preservatives

One large area of coating use is that of the temporary preservatives. The term temporary can mean intervals from hours to months to years. Protection of metal during a period of days or months, depending on whether the exposure is indoors or outdoors, is normally expected. The coating materials may be divided into the following groups.

Oils, Greases, and Microcrystalline Waxes Materials such as oils, greases, and microcrystalline waxes have been identified and divided into various types by MIL (military) specifications. These range from thin oils applied to steel, aluminum sheets, or coil stock to the heavy, cohesive greases or waxes that provide relatively permanent coverings over the metal substrate. The materials are often termed slushing compounds because they are applied by dipping, rough brushing, or spraying. Many are formulated with inhibitors to stabilize the surface of the metal.

Such materials may be used to preserve metals between periods of processing, to protect of precision surfaces during shipment or storage, to maintain the interior surfaces of equipment during storage, or to prevent corrosion of steel surfaces when in contact with stagnant waters (e.g., fire water tanks).

Grease Cloth impregnated with heavy, inhibited grease is also available for atmospheric exposures. Although too tacky for use in areas where there is personnel

Figure 10.17 Highly polished silver surfaces are subject to rapid oxidation or sulfidation, so a temporary, strippable vinyl coating is used. Sprayed-on silicones also are used for protection of surfaces such as these.

traffic, the materials provide an exceptionally heavy, stable covering over metal where undisturbed.

Strippable Coatings Synthetic polymers can be used to "cocoon" equipment to prevent interchange of moisture with the atmosphere (Figure 10.17).

10.5.3.10 Antifouling Coatings

Because the life processes of living organisms often induce or accelerate deterioration of coatings, there is a wide range of additives designed to kill or reduce the activity of bacteria and other micro- and macroscopic living things.

Because steel is anodic to copper oxides in salt water, copper-containing boat or ship hull coatings require an electrically insulating coating on the steel. Copper compounds should not be used on aluminum, because the galvanic difference is even greater than it is for steel.

Antifouling coatings have been in use for centuries on hulls of ships, and some copper formulations developed long ago are still in use. Old antifouling compounds include variations of organic tin esters. In these materials the metal salts are bound into the organic molecule control and extend their dissolution rates into water, thus improving surface concentration levels vs. time and thereby extending effective life of the coating. However, recently, organotins have been banned for all U.S. flag carriers and from most waters in North America and Europe.

Copper-containing antifouling coatings are still being used, but because the copper is toxic to marine life (thus its benefit for an antifouling coating) it also likely will be restricted or eliminated from use in the near future. Research and experimentation is being conducted regarding the use of silicone or polyfluoride nonstick coatings as antifoulants. The marine foulants have difficulty sticking to the coating, and the friction of water on the hull of a moving ship (above 10 knots) is enough to remove much of the fouling. However, in practice, a practical antifoulant to replace the toxic organotins has not been perfected as yet.

Fungicides are widely used in water-based organic coatings to control bacterial attack in containers and after the coating has been applied. Biocides and/or biostats include various salts of metals, arsenic, tin, lead, and zinc, as well as organic poisons. Fungicides include many of those listed previously, and such compounds as chloromethyl oxetane, chlorophenyl compounds, sulfuryl/sulfonyl pyridine, and many others. They prevent attachment or growth of organisms.

10.5.3.11 Composite or Multiphase Coatings

Various combinations of primers, synthetic polymer coatings, inorganic fabrics, and thick-film coatings are used for protection against corrosive fluids and gases, usually in tanks or vessels. A membrane of polymeric material will be laid over the inside surfaces of a steel tank to be lined with acid-resistant brick, such as that used for resistance to heat and impact in steel-mill pickling tanks.

Composite coatings are typified by systems used on pipe lines. As one example, a coating may consist of inhibited primers, followed by cross-linked butyl rubber-based adhesives and polyethylene tape. They also may consist of a hot-sprayed thermoplastic, along with cut glass fibers or other additives.

10.5.3.12 Zinc-Rich Materials

Because of their singular merits in corrosion control in many environments and their worldwide use, special attention is necessary for zinc-loaded organic and inorganic materials. The materials have exhibited excellent protection of steel when used as primers, and in many instances, as a total coating system.

The zinc functions as an anodic material to provide cathodic protection of steel at those areas where moisture has penetrated the coating (pores, scratches, etc.). To achieve this end, the zinc particles must be in electrical contact with one another and the substrate to provide the metallic path of the cell (Chapter 2). Thus, the loading or metallic content of the coating must be very high. Coating thickness should be about 70 μm.

The binder for the small zinc spheroids can be one of three kinds:

1. inorganic silicate produced by evaporation of water,
2. inorganic silicate produced by reaction with water, or
3. organic resin resulting from loss of solvent, and/or cross-linking.

The original coating of this type used an aqueous sodium-silicate solution as the vehicle (type 1) with a final rinse of phosphoric acid to set the coating. Such a procedure is obviously not as simple to use as the straightforward mixing, application, and drying of the standard two-package (zinc and vehicle) coating that succeeded it. However, the coating had certain characteristics that have not been duplicated by the later products.

The two-package system of the Type 1 dries by the evaporation of water from the vehicle to leave the silicate binder holding the zinc particles in place. A one-package version is also available.

The Type 2, two-package product has ethyl silicate and ethanol as the vehicle. The ethyl silicate then reacts with moisture from the air once the coating is applied. Ethanol is formed by the reaction and evaporates rapidly to leave the silicate binder. This product is more tolerant of moisture than is the Type 1 material, but should not be used where the relative humidity is very low.

The third category of zinc-rich coating utilizes customary organic polymers as vehicles that dry by solvent evaporation. Catalyzed epoxy resins are most common, although alkyds, acrylics, polyurethanes, and other resins are also used.

As indicated, these materials serve well as primers beneath organic topcoats. Underfilm creepage of corrosion is reduced where the substrate becomes exposed to the elements. Care must be taken in the selection and application of the topcoat to ensure that the resin is compatible with the zinc when moisture is present and that adequate adhesion with good continuity is achieved.

All zinc-rich coatings require good surface preparation. The materials are the best primers available for use on galvanized metal.

The zinc-rich coatings are particularly good in chloride environments (marine areas, etc.) where the corrosion products form a hard, dense film over the zinc. The coatings should not be used in sulfurous atmospheres with or without topcoats.

In many instances, the zinc-rich coating has sufficient resistance to be used without a topcoat, or the application of a topcoat can be deferred for an appreciable period of time. Such a deferred system can save considerable money because the investment is delayed.

10.5.4 Inorganic (Nonmetallic) Coatings

Inorganic coatings include numerous classes of materials, among them the hydraulic cements (hydraulic cements are waterproof and can even set underwater), ceramics and clays, glass, carbon, silicates, and others. On a volumetric basis, hydraulic cements are probably the most widely used inorganic coating, although they are not always considered to be coatings per se.

10.5.4.1 Hydraulic Cement

Hydraulic cements are used to coat pipe inside and outside, especially pipes that are to be buried or submerged, for example, water or sewer lines. In the case of underwater piping to transport gas or liquid hydrocarbons, for example, it may be mixed with barites or other heavy materials to confer negative buoyancy. It also may be used in mixtures of organic materials where its ability to maintain a noncorrosive pH at the surface of steel is exploited.

Cast iron, ductile iron, or steel pipe may be coated at the mill by a process during which the pipe is spun on the center of its longitudinal axis while a mortar mixture is sprayed onto the inside surface in a uniform, dense layer. After proper curing, provided the pipe is handled carefully, this coating can protect the pipe interior against attack by water and many other liquid and gaseous corrosive environments. Solid concrete pipes reinforced by heavy wire mesh or by prestressed wires are also available.

A concrete coating on steel of any configuration can be protective to the steel as long as it is intact because the alkaline reaction of hydraulic cement maintains a high pH at the steel surface, which effectively prevents corrosion. It does this when encasing reinforcing steel, for example, where it functions both as strength and protective component.

10.5.4.2 Ceramics

Ceramics may also be used in much the same way as hydraulic cements, but they usually are applied to confer resistance to heat and/or attack by hot, high-velocity gases. Thus, they are used in exhaust passages emitting hot gases such as rocket nozzles, furnace linings, and other similar applications.

Also to be considered are the specialty mortars derived from the acid-resistant cements. These mortars can be applied to steel or concrete to provide remarkable acid, alkali, and heat resistance. They may be reinforced with wire mesh to sustain the heavy thicknesses normally applied. Coefficients of expansion close to that of steel have been attained to resolve one of the major problems when applying inorganic materials on steel. These can be used as floors, tank interiors, in scrubbers, and other process equipment of clean interior design.

10.5.4.3 Glass

Glass, a material now being produced in a variety of new and interesting formulations, usually is applied as a slurry, usually in a production line, but seldom for maintenance purposes. Once applied and as long as protected from mechanical abuse, glass surfaces are highly resistant to acidic and many mildly alkaline corrosives and are easily cleaned but difficult to repair. Glass-lined and coated pipe, valves, pumps, and vessels of all kinds are widely used in the chemical, pharmaceutical, and food industries.

Glass coatings are available in two basic types and in various grades of integrity. As with the basic borosilicate glass from which the coatings are derived, the materials may be used up to about 175°C in most acid mixtures other than those containing fluorides. Porcelain coatings are used extensively for their excellent atmospheric stability but are not normally used for immersion service of any type.

10.5.4.4 Specialty Coatings

The preparation of a coating system can be made from many other sources. As an example, coatings are used to "stop off" areas of a metal when it is to be plated, etched, or treated at high temperatures.

10.5.5 Conversion Coatings

As the name implies, conversion coatings are produced by converting the surface of the base metal into a new, uniform, and normally more corrosion-resistant form. Some are used as prepared, but many others serve as a base for subsequent chemical treatments or overcoating. The more common conversion treatments are as follows.

10.5.5.1 Anodizing

Aluminum is always encased in a thin oxide film, which can provide protection of the metal itself. This film can be fortified or increased in thickness by making the metal the anode in an acid bath. This anodizing condition produces a much heavier oxide film on the metal. Oxide-film thicknesses of 1 to 25 μm (0.04 to 1 mil) are nominal. The films may be colored during formation or treated after formation to increase the stability. Greater resistance to permeation of contaminating ions on the surface is produced by this heavy film. Many aluminum building materials, automotive parts, and other structural components are treated in this manner.

Other metals, such as magnesium, titanium, zirconium, and stainless steels can also be "anodized," but the term is normally reserved for the treatment of aluminum and magnesium.

10.5.5.2 *Phosphatizing*

One very successful treatment of steel is phosphatizing, also called phosphating or phosphate-conversion coating. The steel is pickled or otherwise well cleaned and immediately held in a bath of hot phosphoric acid containing zinc and perhaps manganese salts, plus other selected additives. A coating of iron/zinc phosphates is formed. The steel is removed, washed, and dried. The resulting film of some 1 to 10 μm (0.04 to 0.5 mil) thickness and is then often given a chromate treatment to upgrade the corrosion resistance, because the film as formed has little resistance to attack. For many years, car bodies have been given this treatment prior to coating.

The requirements of a large vat to hold the hot acid/salt solution limits the process to mills and other limited areas where large tonnages of steel are to be processed. Phosphatizing of zinc, cadmium, and aluminum can be conducted.

Cold phosphating of metal surfaces by spraying with a warm acid/salt solution should not be compared with the phosphatizing treatment described previously. Pickling of a metal surface can be obtained in this manner, and some residual phosphate formed, but not a full phosphate-conversion coating.

10.5.5.3 *Chromating*

The oxidation of a clean metal surface greatly increases the corrosion resistance of the surface if a thin, continuous, dual-metal oxide can be formed. The most satisfactory method of producing this film has been by treatment with chromic-acid washes of one type or another. Aluminum, zinc, and cadmium are often treated in this manner, and the golden color left by the treatment can be seen on commercial items.

As with phosphatizing, the film produced serves as an excellent coating base. Indeed, it can be difficult to make most polymeric coatings adhere to the nonferrous alloys without such a treatment. The thin films of chromium and substrate metal oxides are only some 0.01 to 0.2 μm in thickness and can be readily removed by abrasion.

Wash primers are a special type of chromate solution used to produce a conversion coating. A phosphoric-acid solution containing chromate salts and a small amount of vinyl-butyral resin is sprayed on the clean metal surface (steel, aluminum, etc.) and reacts with the alloy in the manner described earlier. Although the treatment does not serve as a primer, the film is an excellent base for most coatings. Recent concerns with the toxicity of the hexavalent chromium used in chromating have limited its usage.

10.5.5.4 *Metal Oxides*

The bluing of steel has long been used as a process for producing a protective oxide on steel. When oiled and maintained, the thin oxide film prevents the rusting of steel in the atmosphere. Such oxide coatings can be produced by exposing a clean steel surface to dry steam, strong alkalies, cyanides, or a great number of other chemicals. The films must be very thin and continuous to have the mechanical and physical properties desired.

Oxide coatings have been produced on copper, nickel, tin, and gold for specific applications.

10.5.5.5 *Pack Cementation*

Certain materials can be diffused into the surface of a metal to produce a surface product entirely different from the metal. Powders of carbonaceous materials, aluminides, chromium salts, borides, or silicides are packed around the object in a container. The mass is taken to an appropriate high temperature for sufficient time for the carbon, aluminum, chromium, or boron, to diffuse into the metal surface. A hard, abrasion-resistant surface of some 20 to 50 μm (1 to 2 mils) in thickness is usually produced. These coatings are used to impart abrasion and/or high-temperature resistance to gas-turbine blades and vanes, among other similar uses. Aluminum coatings are frequently used on heat-exchanger surfaces subject to attack by sulfur gases (H_2S, SO_2, and SO_3).

The sherardizing process involves diffusion of a zinc coating into steel by tumbling steel parts with zinc dust at high temperatures.

10.5.6 Metal Coatings

For many rugged services, the choice of a metal coating is preferred. Where severe impact, abrasion, or high temperatures are a part of the materials-selection consideration, the use of metallic coatings should be considered. These are provided by a variety of techniques, as described in the following paragraphs.

Be aware that metallic coatings on a metal substrate may be electrochemically anodic, cathodic, or neutral. If there is a breach in the protective coating and a contact established between the two different metals by the presence of some aqueous electrolyte, corrosion of either may be accelerated or accentuated, depending on the existing conditions (Figure 10.18).

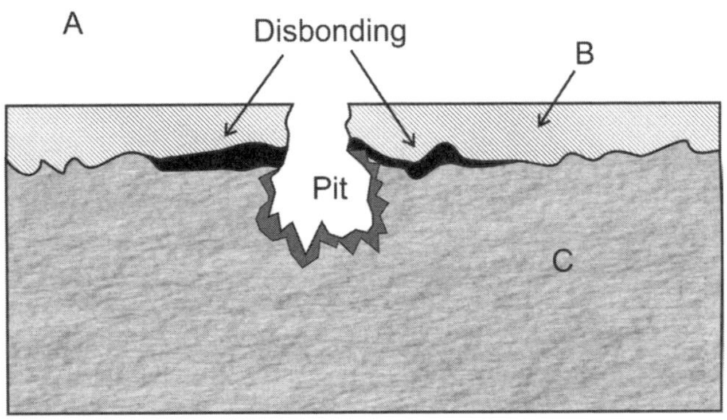

Figure 10.18 Schematic shows how a holiday (hole) in a coating cathodic to a base may cause a pit in the base metal. Arrows show approximate paths of ions leaving base metal: (a) electrolyte, (b) cathodic layer, and (c) base metal. The indicated disbonding may or may not take place.

10.5.6.1 Electroplating (Electrodeposition)

Metal plating on a cathodic surface can be achieved on a commercial basis to provide aesthetic and protective coatings. The part to be coated is immersed as a cathode in a solution of the metal ions to be plated. A power supply drives the current from an appropriate anode into the electrolyte and the soluble metal plates out on the cathode. The current densities used are usually high, and the plating is done rapidly. The more common plated metals are chromium, nickel, copper, cadmium, and zinc.

Chrome plate, as used on automotive parts and other items requiring protection under difficult conditions, is actually a three-ply coating. A flash of copper is first laid down, followed by a nickel coating, which constitutes most of the finished thickness. A thin, hard coating of bright chromium is then applied as the topcoat. The total thickness is normally 25 to 50 μm (1 to 2 mils).

A plated coating should be pore-free. When the coating is porous or perforated, accelerated corrosion of the steel substrate can occur because the steel is anodic to the topcoats. Severe attack can occur because of the adverse electrode area effects of the corrosion cell (Chapter 2).

Electroplated coatings normally fail by cracking from repeated expansion and contraction or impact. Another problem associated with plating is the presence of large quantities of hydrogen on the cathodic surface. Some of this hydrogen can enter the substrate metal (normally steel). For low-strength steels (hardness $< R_C$ 22), hydrogen permeation seldom poses a problem. However, the higher-strength steels can be badly embrittled and fail prematurely. When poisons for the recombination into molecular hydrogen (e.g., S, CN, As, Cd, etc.), are present in the bath, the amount of atomic hydrogen absorbed can be quite high. If immediate damage is not done, the parts may be baked at temperatures around 200°C to drive most of the hydrogen from the metal.

For small, intricate shapes such as bolts and screws, plating metal may be applied by barrel plating. A rotating barrel containing the parts to be plated is turned slowly in the electrolyte, and current is discharged to the tumbling parts inside.

Brush plating is also performed, particularly for touch-up of small surfaces. In this case, an absorbent pad holding the electrolyte is wrapped around a metal anode and the brush is moved over the cathodic surface.

Today, chrome plating is also applied over plastic materials (primarily ABS) for many uses where a bright finish is desired.

Thin electroplated films of gold, platinum, and silver are used on electrical contacts and in electronic equipment. Because gold and platinum resist oxidation, they are important in contacts where small currents are being switched or passed through connections, because any oxides on the contact surfaces impose unwanted resistances in the circuits.

Platinum-group metals and their oxide coatings on CP impressed-current anodes permit imposition of higher voltages and current densities that other metals cannot sustain. Silver reflectance coatings must be protected by transparent topcoats against oxidation and sulfidation. Gold, rhodium, and osmium are widely used as coatings for small parts of operating equipment in industry, such as on meter and valve parts.

Gold leaf, not electroplated, has also been used for advertising and industrial signs and as coatings for roofs, especially those on churches and government buildings, because of excellent resistance to atmospheric attack. For such uses, it was often the most economical coating.

10.5.6.2 Molten Bath

Galvanizing, tinning, and aluminizing are accomplished by cleaning the steel and immersing it in a controlled melt of the coating metal. The zinc coating produced is actually alloyed to the steel.

Aluminum is applied to thin sheet and coil stock only in the mill. The coating is very continuous and very stable in marine atmospheres and high-temperature exposures. The aluminum is not to be considered as an anodic (sacrificial) coating on the steel.

10.5.6.3 Cladding

Metal layers of varying thicknesses can be applied to other metals by the following methods. These procedures are designed to obtain a more corrosion-resistant interior, while using the less expensive steel as a backing for strength.

Arc or Gas Welding In arc or gas welding, relatively thick layers of weld metal are deposited either by manual or machine methods on surfaces. The interior of pulp digesters or other pressure vessels requiring an alloy composition to resist the chemical conditions can be constructed in this manner.

Roll Bonding In roll bonding, layers of two metals are mated by heavy rolling in a mill after the surfaces have been thoroughly cleaned and treated. Clad thicknesses of 5 to 10% of the base-steel thickness are common. Some small areas of unbonded metal will be present. Two metals may also be coextruded through a die.

Explosion Bonding In explosion bonding, a base metal and a covering metal coating are contacted and placed in an appropriate enclosure with the coating material on top. A layer of explosive is placed over the coating metal. When the explosion is activated, the resulting shock wave merges the two materials, as shown in Figure 10.19.

Melting or Brazing Lead, copper, and other low-melting alloys can be applied by the technique of melting or brazing.

Soldering A metal coating can be soldered to a base by flowing a solder layer on the mating surfaces, putting the metals in contact, and applying heat sufficient to melt the solder.

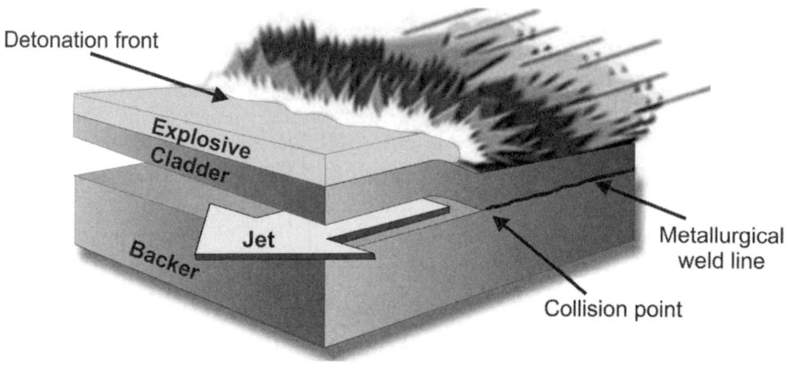

Figure 10.19 Manner in which explosives bond a layer of cladding metal to a base. The arrow shows how a "jet" tends to secure cladding layer to base.

Metallizing (Thermal Spray) The metallizing technique for applying metal coatings consists of thermally spraying one type of metal onto another metal (often steel) surface. The technique is accomplished in several ways:

1. feeding a wire or metal powder into oxyacetylene or oxyhydrogen flames (Figure 10.20),
2. feeding a wire into the heated zone created by electrodes arcing alternating current from which molten metal is impelled by a stream of compressed air toward the surface to be coated (Figure 10.21), and
3. blowing metallic particles suspended in a gas through a blow-pipe flame, twin arc, (Figure 10.22) or plasma (Figure 10.23), a process that can be used to coat inside pipes and tubes.

Thorough surface preparation usually is required because the molten-metal droplets must achieve adhesion before solidification. The coating achieved is somewhat

Figure 10.20 High-velocity oygen-fuel thermal-spray process (courtesy of M. Gariepy, WEIR Engineering Services, Montreal, Canada).

Figure 10.21 Arc spraying process (courtesy of M. Gariepy, WEIR Engineering Services, Montreal, Canada).

oxidized and porous. Common applications are on steel surfaces exposed on off-shore petroleum-production platforms, chemical-tanker compartments, stack breechings, and other structural steel-locations. Metallized surfaces frequently are topcoated (Figure 10. 24).

10.5.6.4 High Temperature

Ceramics, cermets (ceramic/metal mixtures), and various oxides are applied by a variety of methods, including plasma-arc spraying, for high-temperature applications as in the

Figure 10.22 Twin arc blowing torch for metallizing inside pipes and tubes (courtesy of M. Gariepy, WEIR Engineering Services, Montreal, Canada).

Figure 10.23 Plasma blowing torch for metallizing inside pipes and tubes (courtesy of M. Gariepy, WEIR Engineering Services, Montreal, Canada).

liquid-metal vessel shown in (Figure 10.25) and for cutting edges. This classification overlaps with the diffusion category and with coatings designed to provide hard, abrasion-resistant surfaces, such as those available with tungsten carbide, aluminum oxide, and nitrides.

10.5.6.5 Peen Plating

In peen plating, powders of soft metals, such as aluminum or zinc, are deposited on surfaces when objects are tumbled in a rotating barrel at high temperatures.

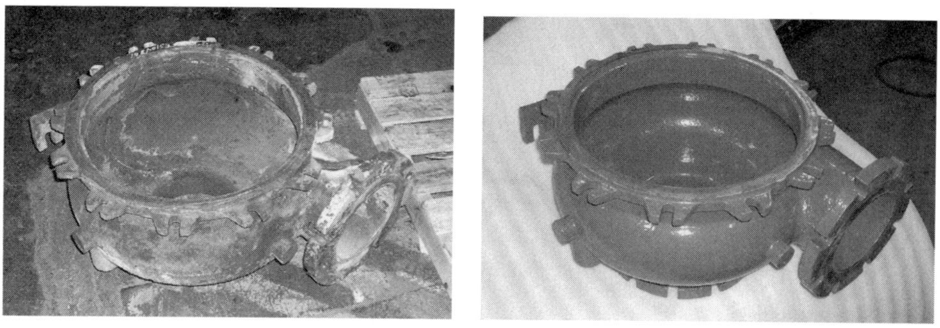

(a) (b)

Figure 10.24 Pump casing a) before and b) after being arc sprayed with stainless steel S43000 and top-coated with epoxy (courtesy of M. Gariepy, WEIR Engineering Services, Montreal, Canada).

Figure 10.25 Ceramic plasma-coated high-temperature vessel with an antiadhesive top coat (courtesy of M. Gariepy, WEIR Engineering Services, Montreal, Canada).

10.5.6.6 Cathodic Sputtering

The metal to be deposited through cathodic sputtering is vaporized in an atmosphere of gas or in a vacuum from which it deposits on the object to be coated.

10.5.6.7 Chemical Reduction or Deposition

Particularly used for nickel, for example, electroless nickel, chemical reduction or deposition is conducted at 85 to 95°C and involves immersing the surface to be coated in a solution of nickel chloride and phosphoric acid, or borohydrides or alkyl amine boranes. A smooth, nonporous layer is produced, which may contain up to 14% of phosphorus. The method covers surfaces in recesses and other irregularities and is particularly suited for coating the insides of fabricated tanks, piping, valves, and so forth.

Salts can be obtained for the deposition of copper, tin, and certain other metals by the reduction process.

10.6 TESTING

Coating operations of most any size should have some type of meaningful test program. The program should evaluate materials, surface preparation, application, and inspection procedures. Such programs are the only objective basis on which to make decisions relating to future work. These may be done in-house or contracted to the many consulting firms available to help those without the necessary personnel.

It is important that all parties having an interest in this work subscribe to the program to be instituted and agree on the basis by which the decisions are to be made.

Whenever possible, standard test procedures and means of evaluation provided by ASTM, NACE, SSPC, and others should be consulted and followed. Only in this manner can correlation with the results of other workers be established. The following sections describe procedures in which coating evaluation is conducted in laboratory and field conditions.

10.6.1 Laboratory Testing

The number and types of tests that can be conducted in the laboratory are almost limitless. Because field testing of coatings is so costly in time, labor, and facilities, it has been the persistent aim of formulators, raw-materials producers, and users of coatings to establish meaningful accelerated tests. These may involve a wide variety of comparatively simple as well as highly sophisticated procedures. Chapter 14 contains additional information on some of these procedures.

The tests can define certain characteristics of a coating system. This discrimination usually is specific to the type of accelerated test and frequently does not indicate in advance how a given coating will perform in service. However, the tests are invaluable when developing coating or appraising new concepts in coating application or use.

10.6.1.1 General Immersion Tests

General immersion tests can be as simple as the standard salt-water immersion test for coatings to be used in the atmosphere or as complicated as "cold wall" effect immersion tests for heavy coatings to be used constantly in immersion service. Reproduce the anticipated conditions of the field exposure as closely as possible when evaluating coatings for immersion service.

10.6.1.2 Salt-Spray Testing

Exposure of coatings to various formulations of salt spray or fog, with and without cyclic immersion and exposure to highly actinic light, heat, and various wetting-and-drying cycles is a common practice.

10.6.1.3 CP Exposures

One accelerated test involves subjecting the material to which the coating has been applied to various voltages of impressed current, often in excess of that normally required to achieve CP, with the aim of demonstrating the ability of the coatings to resist the disbonding effect of the cathodic current or to resist electroendosmosis (Figure 10.26).

10.6.1.4 Impedance Measurements

Various approaches to the measurement of changes in the electrical resistance of a coating during exposure are available.

10.6.1.5 Condensation Apparatus

Relying on the known susceptibility of coatings to permeation by condensed water, a screening test is available in a laboratory-type cabinet that results in condensation of water on the painted surface. Disbonding of poorer coatings occurs.

Figure 10.26 Laboratory setup to test for disbanding during the application (courtesy of Corrpro).

10.6.1.6 *Environmental Test Rooms*

Environmental test rooms that permit various combinations of temperature, fogging, humidity levels, and shower effects to simulate rain can reproduce in an accelerated mode many of the factors present in an actual exposure.

10.6.2 Field or Service Testing

Although the delays involved in field or service testing may seem interminably long in many cases (years may be required for completion of a full-scale field test), when coating expenditures run into the hundreds of thousands, and in some cases into millions of dollars, it pays to be sure about performance. Most field testing is done on-site under the supervision of the customer's testing staff. Ordinarily, coupons coated with the materials to be tested are exposed in multiples at several test sites in a plant as a first step.

After varying periods of exposure, depending on the design of the test program, coupons are removed from the exposure sites to laboratories where their residual properties are measured. At the end of such a program, it usually is possible from among a group of coatings similarly exposed. This coating can be applied with some confidence that good service will be obtained.

As for any other tests, a field test panel and application procedure should be selected to represent common situations or other conditions of interest.

Once the panel testing is completed, the usual procedure is to apply the most promising two or three candidate materials on tanks or other large metal areas in the plant site, fully realizing that, when proceeding with this step, all surfaces in the field have far from the same exposure. The four quadrants of a tank are different in sunshine exposure, and the top and bottom halves of the tank are not the same because of the internal changes in liquid level.

In any large chemical or industrial plant, corrosivity of one plant location may be more or less severe by several orders of magnitude than the corrosivity at another location. Thus, exposure at a single location may not give results that can be applied with confidence everywhere in the plant. Furthermore, because some locations are less corrosive than others, they may not require systems as expensive as those used in the worst locations.

10.7 SPECIFICATIONS AND INSPECTION

10.7.1 General Concepts of Specifications

It is a good engineering practice to have a clear goal in mind in the application of a coating. Much waste can be avoided by providing proper specifications applicable to a coating project and planning in advance an inspection program that will ensure conformity with the specifications. Specific properties of the coating to be measured should be identified, and the method of measuring the quantity should be stipulated.

There is no substitute for a conference among the owner or the owner's representative, the applicator, the material supplier, and the inspector or inspection agency before the job to establish what is to be done, how it is to be done, what inspection procedures are to be used, and the acceptable limitations imposed by the specifications. At this meeting, all parties concerned with the job safety and conduct should be present. Agreements by people having no oversight of the work normally have little meaning. All parties should be properly informed as to any special requirements or limitations on work activities.

Some items that may be contained in any specification relate to such basics as surface preparation. Before a specification for surface preparation can be made, the type of coating system to be used must be established. If a thin-film system is selected, the surface-preparation method or surface-profile characteristics should be known. Once this is known, the acceptable inspection standard should be identified.

Routines for surface preparation, acceptable intervals between preparation and coating application (relevant to the environment present), intervals between coats, drying and curing conditions and durations, and other pertinent details may be given.

Careful identification of the components of the system may be necessary. Characteristics of the primer and second, third, or final coats, their respective thicknesses and drying-time allocations, and the range of permissible final coating thickness may be stated. Limits on addition of diluents, thinners, accelerators, or other changes in basic formulation can be established.

10.7.2 Inspection

The point at which the job is started is no time to decide on the techniques and thoroughness of the inspection to be made. As indicated, these factors should be succinctly

Figure 10.27 Keane–Tator surface-profile comparator disc (courtesy of KTA-Tator, Inc.).

stated in the specifications. There are many tests that may be specified that are essentially standardized. Consult these standard procedures. Some of the tests which may be used relate to the following subjects.

10.7.2.1 Surface Preparation

The degree of cleanliness of a steel surface has been discussed early in the chapter. Characteristics of a properly prepared surface may include specification of a surface profile in terms of depth as well as other criteria. This is important in some instances, especially when a thin-film coating is to be applied. If the surface profile is such that an inadequate thickness of coating is deposited on the peaks or the coating bridges the valley (Figure 10.8), a coating failure may follow, or, at best, a greater volume of coating will be required for adequate coverage of the peaks than would be the case if the profile were of proper depth and spacing.

Gauges and equipment should be available for checking surface profiles (Figures 10.27 to 10.32). In the hands of an experienced person, a good statistical estimate can be made in a short time of a properly prepared surface. Profilometers (useful

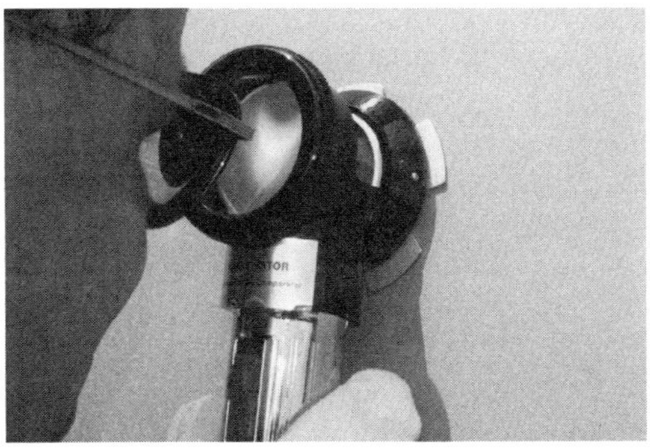

Figure 10.28 Use of Keane–Tator surface-profile comparator method for measurement of surface-profile depth (courtesy of KTA-Tator, Inc.).

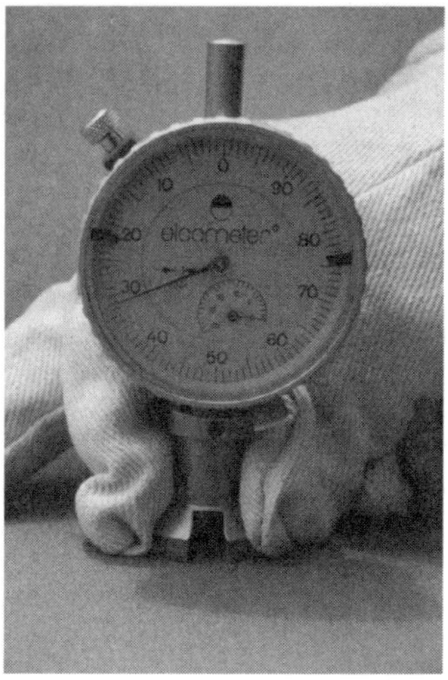

Figure 10.29 Measuring surface-profile depth using depth micrometer (courtesy of KTA-Tator, Inc.).

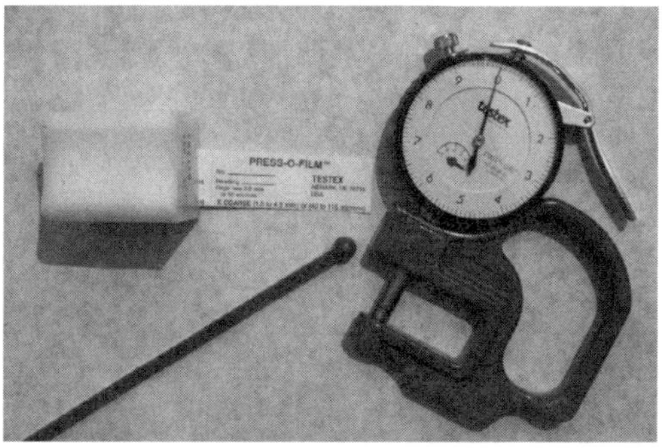

Figure 10.30 Testex Press-O-Film replica tape surface-profile measurement equipment (courtesy of KTA-Tator, Inc.).

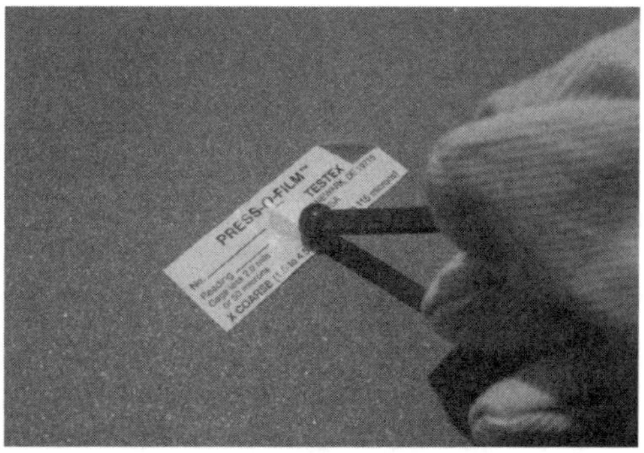

Figure 10.31 Burnishing Testex Press-O-Film replica tape (courtesy of KTA-Tator, Inc.).

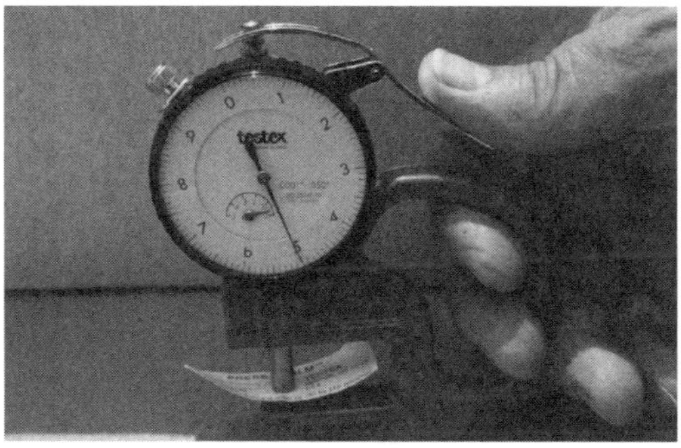

Figure 10.32 Measuring surface-profile depth using Testex micrometer (courtesy of KTA-Tator, Inc.).

laboratory instruments) give exact information on the profile, but it is difficult and usually impossible to use them in the field.

The following sections are brief descriptions of some of these inspection test methods for organic coatings.

10.7.2.2 Wet-Film Gauging

Thickness of a wet film can be determined by any of several gauges that consist of a series of calibrated indentations along the edge of a handheld ruler (Figure 10.33). Probable thickness of the dried film can be calculated if solid content by volume is known.

10.7.2.3 Dry-Film Gauging

Electrical and magnetic nondestructive and visual destructive equipment for determining thickness are described in several places (Figures 10.34 and 10.35).

10.7.2.4 Coverage Completeness

Coverage completeness can be checked by using different colors of multiple coats or conductive and nonconductive coats.

10.7.2.5 Visual Inspection

Visual inspection can be conducted with or without magnification. The Tooke gauge, which makes a diagonal cut through a coating system to the substrate, permits measurement and identification of the coating layers under magnification (Figures 10.36 and 10.37). These gauges are less commonly used because they are slower to operate, and create a cut through the film.

10.8 FAILURES OF ORGANIC COATINGS

A coating may fail as a result of a large number of potentially adverse conditions. Some of these can be defined as mechanical, as when abrasion or impact removes the coating

Figure 10.33 Notch-type wet-film-thickness gauges (top: stainless steel, bottom: aluminum) (courtesy of KTA-Tator, Inc.).

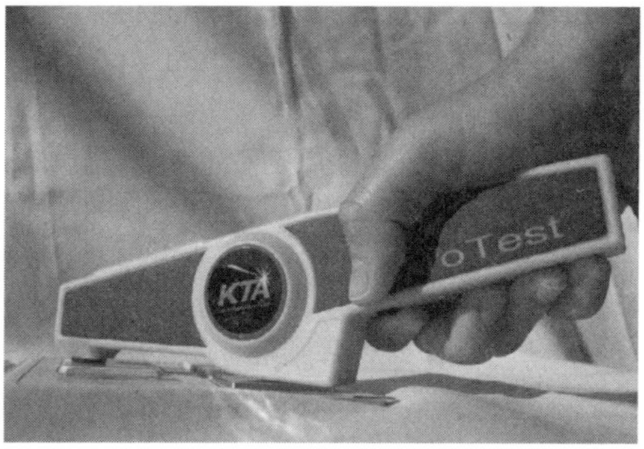

Figure 10.34 Magnetic pull-off dry-film-thickness measurement gauge (courtesy of KTA-Tator, Inc.).

Figure 10.35 Electronic dry-film-thickness measurement gauge (courtesy of KTA-Tator, Inc.).

Figure 10.36 Tooke gauge (courtesy of KTA-Tator, Inc.).

or more subtly locally damages the coating and creates a corrosion site that will show in a few months (Figure 10.38). Thermal expansion of organic materials is high. Continued expansion and contraction of the coating over steel, and particularly concrete, requires a material that can resist significant internal stresses before failure occurs.

The rate at which diffusion is occurring through a coating is greatly controlled by the thickness and formulation of the coating. Based on empirical data, economics, and some theoretical considerations, the thickness of a coating necessary to resist moisture permeation from the atmosphere and otherwise resist deterioration is about 125 μm (5 mils).

What happens when moisture and oxygen penetrate the coating? Perhaps nothing, if a truly good bond has been achieved, but a number of reactions may occur. Where

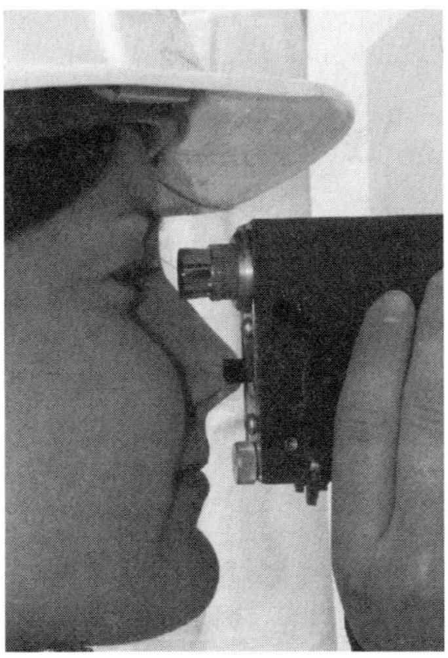

Figure 10.37 Use of Tooke gauge for measurement of paint layers (courtesy of KTA-Tator, Inc.).

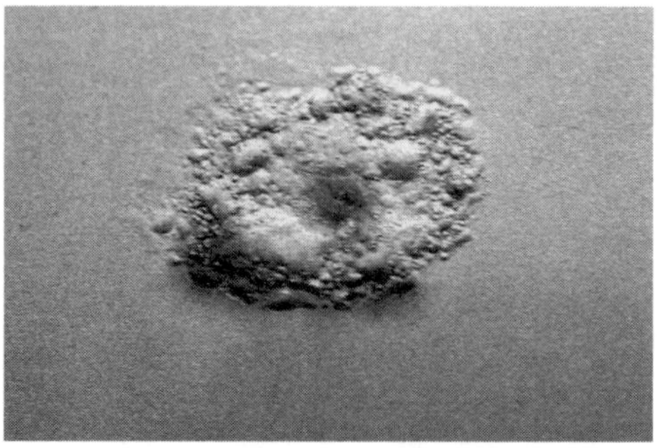

Figure 10.38 This corrosion blister formed on a car door months after an impact caused a break in the coating (courtesy of Kingston Technical Software).

the moisture contacts the steel, for example, the initiation of further oxidation of the iron can occur. A corrosion cell comparable to that in a pit on freely exposed iron will develop. If ionic contaminants are present at the interface between the coating and the steel, the electrolyte would be more conductive to implement the cell action.

Many contaminants, such as chlorides and other soluble salts, will actually enter into the chemistry of the cell to accentuate the attack. This is why many contractors are specifically requested to measure the level of these salts and continue washing a surface until the salt level is below an acceptable limit (Figure 10.39). If mill scale exists in the area, a galvanic cell of 230 mV can be established, with the iron as the anodic electrode.

In given conditions of humidity and oxygen permeation, the initial corrosion site may begin to move in a random manner as the corrosion product reduces the oxygen content at the surface and the area becomes highly anodic to the surrounding cathode area of oxygen saturation. The worm track of corrosion that then occurs is termed filiform corrosion (see Chapter 9).

If the corrosion reaction is allowed to proceed, two of the cathodic reactions may disbond the coating beyond the minute point of corrosion. Hydrogen is produced at the cathode and will compete for the bonding forces that hold the coating to the surface. Molecular hydrogen (H_2) is a highly active gas that can simply pry the coating loose.

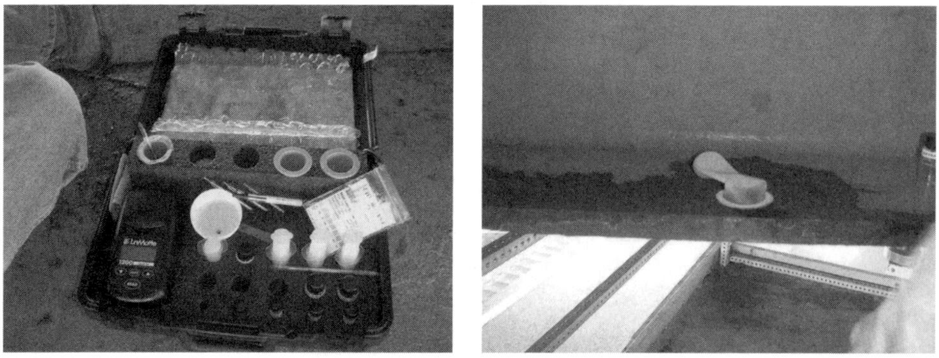

(a) (b)

Figure 10.39 a) Soluble salts detector kit and b) sample being taken on a surface in preparation (courtesy of Termarust Technologies).

In addition, hydroxyl ions (OH^-) are formed at the cathode. Coating materials poorly resistant to alkaline conditions can be attacked by the hydroxyl ion and lose adhesion. External cathodic currents increase the possibility of failure by the hydrogen and hydroxyl formation because much greater quantities of cathode products are formed.

In addition, external impressed cathodic currents on the steel can force more water through the coating than would be the case without the current. This forced diffusion of water through the coating is termed electroendosmosis (not to be confused with osmosis that occurs when water is drawn at a higher than normal rate through the film by a soluble salt lying beneath the coating). Thus, cathodic currents of sufficient magnitude can strip coatings from a steel surface.

The importance of adequate coating thickness cannot be overemphasized. If good flowout is not obtained during application, thin spots (particularly at the edges) or actual holidays (holes) may exist in the film. At times, the profile pattern may be too great for the coating, resulting in thin areas over the peaks of metal constituting the profile roughness. All of these obviously produce a more rapid failure of the coating.

A holiday detector will help to find coating defects by impressing an electrical voltage across the coating (Figures 10.40 and 10.41). An electrode passes over the entire coating surface and as it crosses a defect, it gives off an electrical discharge, or spark, which signals to the operator that a holiday has been detected.[4] The operator can then mark the holiday for subsequent repair prior to acceptance. Holiday detectors must be selected for the thickness and type of coating being tested. Excessive voltage can stress or damage thin coatings if the detector has been set too high. Adherence to the manufacturer's instructions is recommended, including both the manufacturer of the coating and the manufacturer of the holiday detector. NACE International also publishes a recommended practice (RP0274) for high-voltage holiday detection.

Coatings overloaded with fillers or the wrong filler can be "cheesy" and porous. Rapid penetration of moisture to the surface will occur, with subsequent corrosion and rupturing of the weakly cohesive film. Such failures are often observed as a "star-burst" rupture in the film, rather than as a ductile tear observed in the better coatings.

Another source of failure may be from continued polymerization or oxidation of the coating once it is applied. The film becomes harder, less ductile, and changes in volume when these mechanisms prevail. Cracks in the coating then develop.

Esterification of the coating by foreign metallic cations to produce "soaps" of the coating can occur under the proper conditions when a coating containing ester groups is used. Zinc ions will produce these soaps with some alkyds and oil-based coating materials.

Compatibility of the various coats must be considered. Even if two coats are known to be compatible, application of the second coat too early or too late can be disastrous. Figure 10.42 shows an example of the former case where an alkyd topcoat was applied and dried over an oil primer coat that had not completely dried or cured. An example of the latter case can be the overcoating of a fully cured, catalyzed epoxy. The topcoat has little chance to "bite" into such a solvent-resistant surface.

The importance of cleanliness during application, uniform coverage, and good density and thickness in the coating should be evident when some of the more common modes of failure are understood.

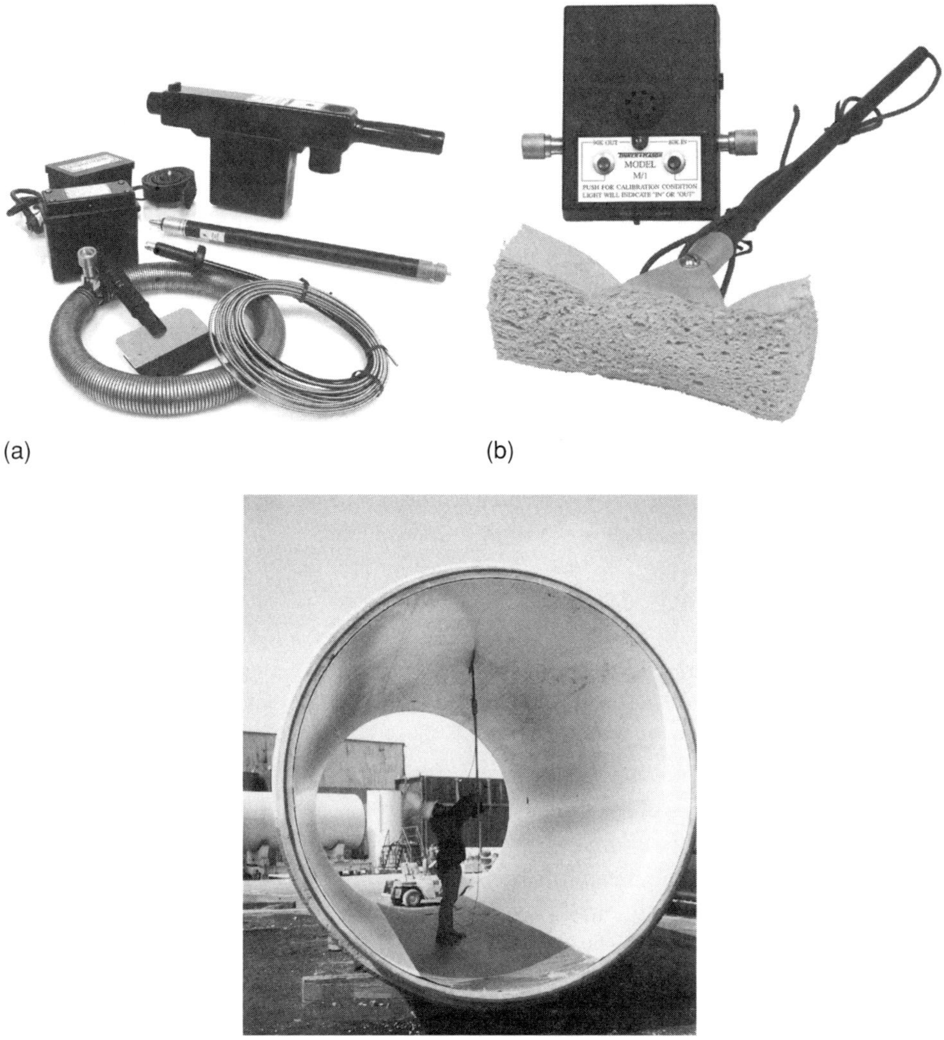

(a) (b) (c)

Figure 10.40 a) High-voltage holiday-detector kit, b) low-voltage holiday-detector kit, and c) holiday testing with high-voltage detector (courtesy of Tinker & Rasor).

A handy, pocket-sized manual that illustrates a range of coating defects is commercially available.[5] This guide provides the user with a greater understanding of coating defects. Each coating is illustrated and described along with its probable causes, prevention measures, and repair advice. Additionally this guide contains sections on welding faults, surface preparation, and marine fouling.

10.9 ECONOMIC ASPECTS OF COATING USE

When considering the true cost of a coating job, the application of economic principles must be the same as for any other investment or maintenance money. The life of system A must be compared with system B, the type of money being spent (depreciated capital

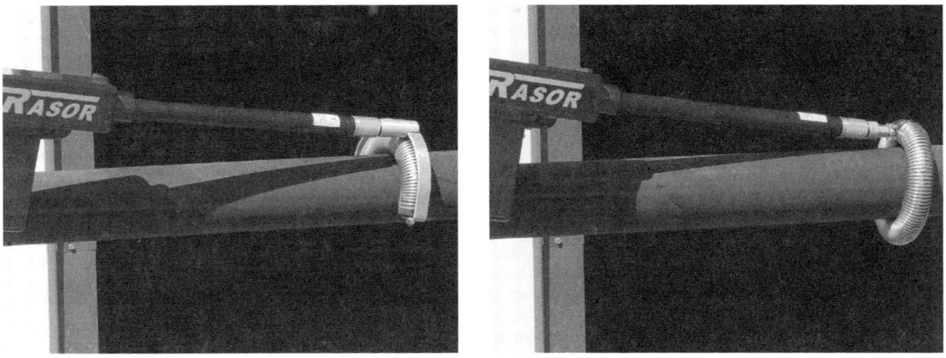

(a) (b)

Figure 10.41 Holiday testing with a) half-circle spring electrode and b) full-circle spring electrode (courtesy of Tinker & Rasor).

vs. maintenance) must be established, the tax rate known, and the value of money to the spender must be considered. System A can then be compared accurately with System B as an investment.

Expressions such as cost/m^2, cost/m^2/y (better), percent of maintenance costs, percent of capital costs, and so forth, are often manipulated as economic terms but are unrelated to true economic comparisons necessary for a business decision.

The cost of coating materials on metals can range from 5 to 21% of the total costs, whereas surface preparation is about 45% of the total. High-performance epoxy polyamide, urethane, or zinc-rich systems over 10 years can represent savings of nearly 40% over "less expensive" systems in demanding environments.

Figure 10.42 Cracking of a topcoat applied over a primer that had not completely dried or cured.

10.10 SAFETY

In every industrial environment, the safety factor is an important consideration. In some environments, it is of overriding importance. Some of the safety factors that need to be considered in coating application are as follows.

10.10.1 Safety in Surface Preparation

Workmen using abrasive blasting equipment should be furnished with good equipment in proper working order. They should be equipped with air-filter masks to prevent their breathing dust and scale. Clothing should be adequate and safe. Safety shoes should be worn. Necessary goggles or safety glasses should be mandatory. Ear plugs or protective ear coverings should be worn.

No equipment should be operated in areas where it will create sparks that might ignite explosive or flammable materials. It should not be operated where abrasives, scale, or overspray will damage or interfere with the operation of other equipment.

10.10.2 Safety in Materials

Coating materials are frequently flammable, explosive, or poisonous, and sometimes all three. The characteristics of any material used should be known in advance, and any precautions required should be taken and rigidly maintained during the progress of the job. In summary:

1. Coatings incorporating flammable or explosive materials should not be used in the vicinity of open flames, sparks, or electrical equipment. Every precaution should be taken to prevent accidental fire or explosion by prohibiting smoking, requiring the use of non sparking tools, or whatever other safety requirements are appropriate.
2. When used in enclosed places, solvent concentration should be kept both below the explosive limit and below the acceptable toxicity level. Both limits vary among materials, so safety rules should be a function of materials used. Ventilation of enclosed places should be continuous during the operation and for 3h afterward when explosive or flammable solvents are used. Safety-approved electrical equipment is mandatory.

Solvent vapors should be removed from tanks by suction because many vapors are heavier than air. Thus, the remotest and lowest ends of tanks should receive special attention. Workmen should wear approved compressed-air masks. Shoes should have rubber soles and heels and no exposed steel nails.

10.10.3 Safety in Equipment

Whenever ventilation is a factor, it should be planned carefully and checked frequently. When necessary, automatic equipment should be used to make constant checks of air for poisons or explosive concentrations.

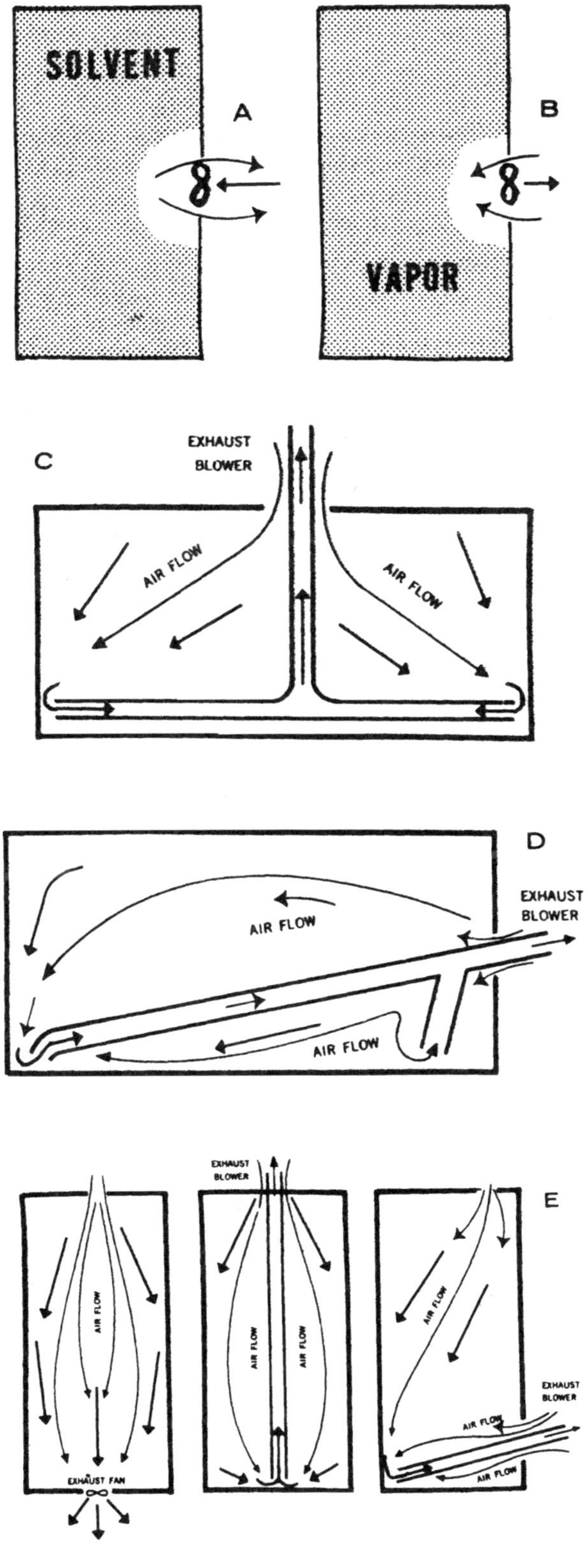

Figure 10.43 Schematic shows incorrect (a and b) and correct (c through e) methods of ventilating tanks where coatings are being applied. (a) and (b) show little or no ventilation of the enclosed volume. (c) through (e) show how gases displaced from remote or low areas are aspirated into the open and how fresh air comes in to replace them.

Diagrams showing the incorrect and correct ways to ventilate tanks are shown in Figure 10.43. Failure to ventilate according to the principles illustrated may lead to injury or death of workmen in enclosed places.

All rigging, lifts, platforms, hoses, or any other equipment used on the job should be inspected and maintained in safe order. Rigging should be done by experienced operators.

10.10.4 Sources of Safety Information

Plant safety practices should be known and understood by plant paint crews or by outside crews doing contract work. In addition, the advice of plant safety engineers should be solicited and recommendations followed. Published information is available from trade associations, testing bureaus, and most important, is usually printed on labels and instruction sheets received with materials and equipment.

10.11 SUMMARY

Coating systems for the protection of materials need and probably demand more organized study than they have been given in many instances, and the benefits attainable from proper preliminary planning, job execution, on-site inspection, and postcompletion surveillance are well worth the effort and expense.

If anything is demonstrated by the facts set forth in this chapter concerning the problems associated with coatings, it is that decisions on coatings should not be left to the casual attention of inexperienced personnel. It is unmistakably clear that

1. decisions concerning effective protective coatings can be made properly only by experienced personnel;
2. the economic consequences of poor coating practices can be disastrous;
3. the number of coating materials available necessitates careful study if a proper material is to be selected;
4. proper surface preparation is a prime economic factor for coatings in any environment;
5. miscellaneous factors may determine the success or failure of an expensive coating job, including the skill of the workmen, proper inspection of the job in progress and at its end, and provision of equipment to apply the coatings properly and promptly, among numerous others; and
6. continuous testing and evaluation of candidate systems and systems actually in service more than pays by way by providing a basis for proper selection.

References

1. G.H. Koch, M.P.H. Brongers, N.G. Thompson, Y.P. Virmani, J.H. Payer. "Corrosion Costs and Preventive Strategies in the United States," National Technical Information Service, FHWA-RD-01-156, Springfield, VA, 2001.

2. C.G. Munger, L.D. Vincent, Corrosion Prevention by Protective Coatings, 2nd ed. (Houston, TX: NACE, 1999).
3. Coatings and Lininigs for Immersion Service, rev. ed. (Houston, TX: NACE, 1998).
4. M. Lewis, "Instrumentation," in Peabody's Control of Pipiline Corrosion (Houston, TX: NACE, 2001), pp. 101–130.
5. B. Fitzsimons, R. Weatherhead, P. Mogran, Fitz's Atlas of Coating Defects (Hartley Wintney, UK: MPI Group, 2004).

11

Chapter

Cathodic Protection

11.1 HOW CP WORKS

The purpose of this chapter is to introduce the concept of CP, a widely used method for the control of corrosion throughout industry. After a brief description of the meaning of CP, this chapter discusses practical aspects of CP systems and various factors that may influence CP designs.

DC electricity is associated with the corrosion process on a buried or submerged metallic structure. Concentration cells, such as shown in Figure 11.1, are responsible for much of the corrosion that takes place in soils. Pipe lines, for example, pass through different soils on their way across the land. Oil- and gas-well casings penetrate various strata of different compositions. In every case where two different soils are in contact with a single piece of metal, a concentration cell is possible.

In most cases occurring in naturally occurring soils, that portion of pipe lying in the more conductive soil is the anode and the portion in the less conductive soil is the cathode. Moist soils themselves act as the electrolyte (in this case, a compound electrolyte), and the pipe provides the connecting circuit. As shown in Figure 11.1, current flows from the anodic area to the soil and then through the soil toward the cathodic area, where it comes back through the pipe itself to complete the circuit.

For a given driving voltage (the galvanic potential between anode and cathode), the amount of current flowing is limited by such factors as the resistivity of the environment (expressed normally as ohm-centimeters) and the degree of polarization at anodic and cathodic areas (see Chapter 2 for more details on general polarization methods).

Corrosion occurs where the current discharges from the metallic structure into the soil at anodic areas. Where current flows from the environment onto the pipe (cathodic areas), there is no corrosion. In applying CP to a structure, then, the objective is to force the entire structure surface exposed to the environment to collect current from the environment. When this condition has been attained, the exposed surface becomes a cathode, and corrosion is then successfully mitigated.

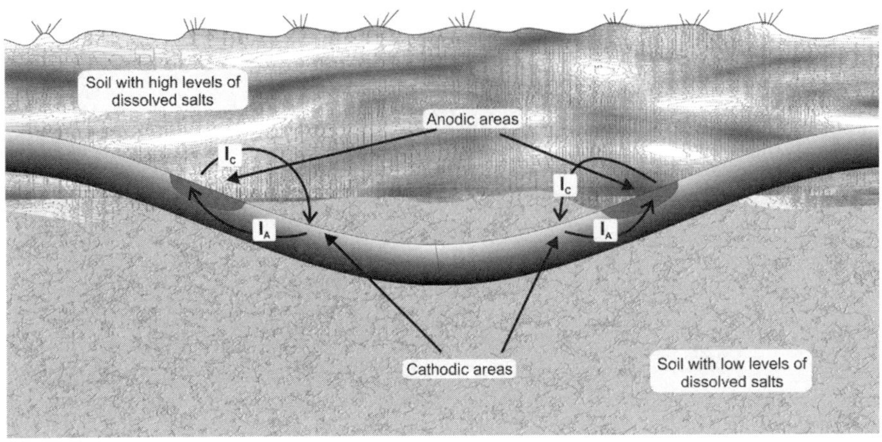

Figure 11.1 Direct current (I_C) and (I_A) flow respectively through earth from anodic area to cathodic area and through the pipe steel from cathodic area back to anodic area to complete the circuit. The anodic area, where current leaves the steel to enter the surrounding earth, is where corrosion occurs.

The basic manner in which CP is accomplished is illustrated in Figure 11.2, which shows how the originally corroding section of pipe used in Figure 11.1 becomes a cathode with cancellation of all current-discharging areas on the pipe surface.

From inspection of Figure 11.2, it can be seen that the CP current must flow into the environment from a special ground connection (usually called a groundbed) established for the purpose. By definition, the materials used in the groundbeds are anodes, and material consumption (corrosion) must occur. It follows, then, that corrosion has not been eliminated by the application of CP, but has been transferred from the structure to be protected to known locations (the groundbeds), which can be designed to discharge the CP current for a reasonably long time and which, when consumed, may be replaced without interruption of the normal function of the structure under protection.

Although the basic principle of applying CP is quite simple, some practical questions have to be resolved, such as these:

- How much minimum current is needed to protect a given structure?
- What source of DC should be used?
- How should the installation actually be designed?
- How can it be ascertained, on a completely buried structure, whether the entire structure surface has, in fact, been made a cathode?

There are limitations to what CP can do. The CP current must be able to flow through a conducting environment onto the metallic surface being protected. Taking a pipe line as a convenient example, a basic CP installation, as shown by Figure 11.2, can protect only those external surfaces of the pipe that are in contact with the conducting environment. The internal surfaces of the pipe will receive no protection. This will be

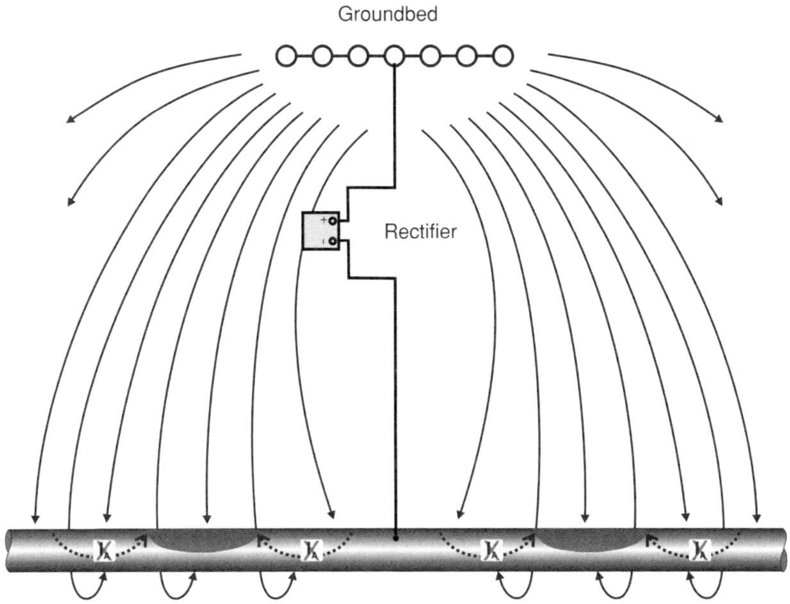

Figure 11.2 Basic ICCP installation. Shaded area on the pipe line indicates anodic areas prior to application of CP and dotted lines represent current discharge from anodic areas that have been eliminated.

true even though the pipe may contain a conductive material, because the current is intercepted by the pipe and carried back to its source.

Any portions of the pipe-line system in air (such as aboveground portions, valves, aboveground header manifolds, etc.) cannot receive any of the CP current because air will not carry current to the unburied surfaces. If a closely spaced bank of structures such as a number of lead-sheathed power or communication cables is to be protected, current from the simple CP system shown in Figure 11.2 will flow mostly to the outermost surfaces of the cables in the bank. Relatively little current will flow to the innermost structures in the bank.

Because the current flow to the innermost structures in the bank is limited, the amount reaching these structures may not be sufficient to adequately protect them, even though the outermost surfaces in the bank are receiving full cathodic protection. This is referred to as electrical shielding because the metallic structure surfaces on the outermost portions of the bank intercept the flow of protective current and shield the innermost structures.

Piping in congested areas, such as pumping stations and tank farms, may encounter this form of shielding. In Figure 11.3, a condition is represented wherein a network of piping in a restricted area is protected by a remote-type groundbed. The remote groundbed and suitably sized rectifier may change the potential of the entire structure sufficiently to give an indication of full protection when measured with respect to a remote reference electrode, such as the −1.5 V indicated in Figure 11.3. However, if a measurement is made between remote earth and earth in the midst of the congested area,

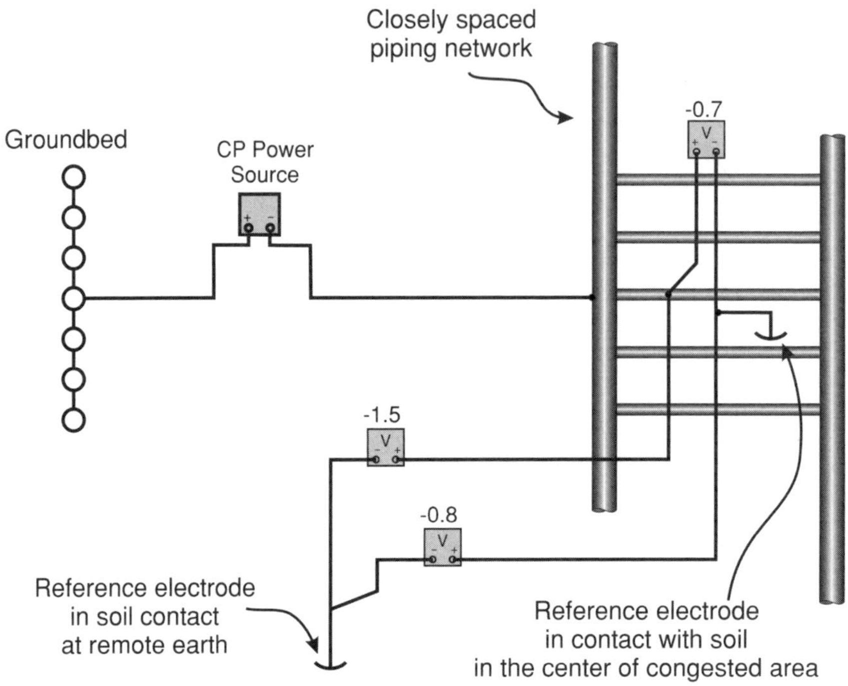

Figure 11.3 Electrical shielding in congested areas.

the potential of the whole earth mass in the area may have been changed as indicated by the −0.8-V reading.[1]

When this occurs, there may be relatively low potentials between the pipe and adjacent earth. This is indicated by the −0.7-V reading in Figure 11.3, which is less than full protection criterion (i.e., −0.85 V vs. a copper-sulfate reference electrode). The shielding effect will tend to be greatest near the center of the congested area.

Where congested-area shielding is a problem, it may not be practical to rely solely on a remote CP system. However, CP can still be attained by the use of the close anodes (either galvanic or impressed current) distributed throughout the congested area in such a way that the areas of influence surrounding the anodes overlap sufficiently to permit development of protective potentials on the entire interconnected structure.

11.1.1 CP Current Sources

This section outlines the various sources of electrical energy needed to overcome corrosion currents flowing on a structure and thus cause it to become a cathode.

11.1.2 Galvanic Anodes

An anode is, by definition, the corroding member in a dissimilar-metal combination. A galvanic or sacrificial anode can be described as a metal that will develop a negative voltage with respect to the corroding structure and will discharge current that will

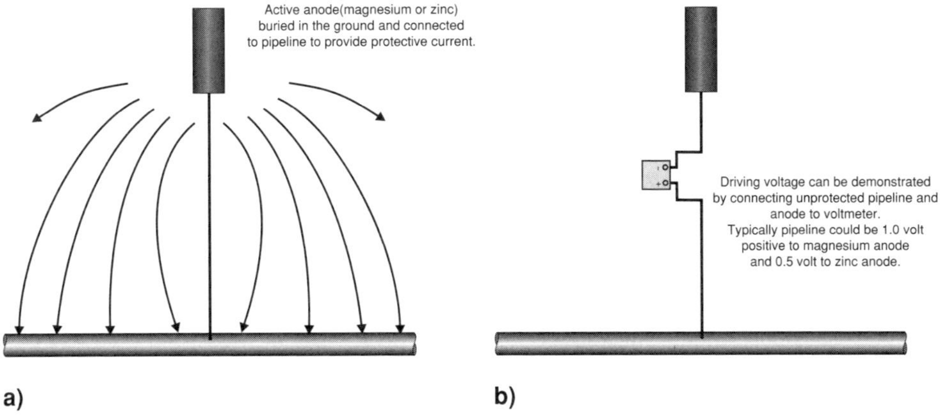

Figure 11.4 CP with galvanic anodes: a) active galvanic anode buried in earth and connected to pipe line to provide protective current and b) driving voltage measured between unprotected pipe line and anode with voltmeter.

flow through the environment to the structure. To do this the galvanic anode must be electrically connected to the structure to be cathodically protected and must also be in contact with the conducting environment containing that structure.

This is not as complicated as it seems. Figure 11.4 contains the essential elements. As shown in the figure, a voltage can be measured between the corroding structure and the material to be used as an anode. The structure must be positive (+) with respect to the anode before the anode can discharge current. There are certain requirements for a metal to be a practical galvanic anode material:

1. The potential between the anode and corroding structure must be large enough to overcome the anode–cathode cells on the corroding structure.
2. The anode material must have sufficient electrical energy current to permit reasonably long life with a practical amount of anode material.
3. Anodes must have good efficiency, meaning that a high percentage of the electrical energy content of the anode should be available for useful CP current output. The balance of the energy that is consumed in self-corrosion of the anode itself should be very small.

Materials suitable for use as galvanic anodes include aluminum, magnesium, and zinc. According to Faraday's law, the amount of electrical energy that can be obtained from a galvanic anode depends on the electrochemical equivalent of the metal used (in terms of ampere-hours per unit weight) and the efficiency of the working anode. An ampere-hour (Ah) is 1 A flowing for 1 h, or any combination that will be equal to this, such as 0.5 A flowing for 2 h, 2 A flowing for 0.5 h, and so forth.

The efficiency of a galvanic anode is the ratio of an anode weight sacrificed for CP purposes divided by the total theoretical ampere-hours or capacity of the material used. Galvanic anode materials are subject to self-corrosion, which uses some of its energy. This is why the efficiency is less than 100%.

For example, pure zinc has a theoretical maximum capacity of 820 Ah per kilogram. This means that if a zinc anode were to discharge 1 A continuously, 1 kg would be consumed in 820 h. If this kilogram was discharging one-tenth of an ampere, it would be totally consumed in 8200 h or 48 weeks. Actually, zinc anodes operate, typically, at about 95% efficiency. This means that the energy content available for useful current output would be 820 × 0.95, or 779 Ah per kg.

Another way of expressing this is in terms of kilograms per ampere-year. At 779 Ah per kg useful output for zinc, the conversion would be

$$\frac{8760\,(\mathrm{h\,y^{-1}})}{779\,(\mathrm{Ah\,kg^{-1}})} = 11.2\,(\mathrm{kg\,A^{-1}y^{-1}}) \tag{11.1}$$

This equation indicates that 11.2 kg of zinc would be consumed from an anode discharging 1 A for 1 year. The "kg per ampere-year (Ay)" value is useful in calculating the expected life of a galvanic anode installation.

Zinc anodes are made of high-purity zinc (99.99% purity or better) for soil use. For marine use, small amounts of aluminum and cadmium may be added to ensure maximum efficiency. If zinc of lesser purity is used, efficiency may suffer and the anodes may tend to "go passive," that is, cease to discharge useful amounts of protective current. Zinc anode working potential with respect to a copper-sulfate electrode is on the order of −1.10 V.

Aluminum anodes have a theoretical capacity of 2960 Ah per kg. In recent years, aluminum anodes for seawater use have been developed that can operate at 95% efficiency with a useful output of approximately 2800 Ah per kg. Working potential is about the same as that of zinc, although there are variations depending on the alloy used. Aluminum anodes may have small amounts of alloying elements, and in some cases also are heat treated to obtain desired characteristics.

Magnesium anodes have a theoretical capacity of 2200 Ah per kg. Efficiency varies with current output density in terms of amperes per square meter of anode surface, but typically can be about 50% (useful output of 1100 Ah per kg) at a current output density of 0.3 A per m^2. Anode working potentials to a copper-sulfate electrode can range from approximately −1.45 V for standard alloy magnesium to approximately −1.70 V with proprietary alloys, also called "high-potential anodes."

Standard magnesium anodes have appreciable amounts of alloying materials. Typically, these may be 6% aluminum, 3% zinc, and 0.2% manganese. Anode materials are cast in numerous weights and shapes to meet CP design requirements. Data on available anodes can be obtained from suppliers of CP materials. The appearance of typical magnesium anodes is shown in Figure 11.5.

11.1.3 Impressed-Current Anodes

Figure 11.2, which was used to show how CP is applied, is basically the schematic of an impressed-current CP system. With such a system, the groundbed anodes are not depended upon as a source of electrical energy. Instead, some external source of

(a)

(b)

(c)

Figure 11.5 Magnesium anodes are being prepared for use in soil: a) electrical connection being soldered, b) insulating tar added to protect the electrical connection, and c) sulfate-rich powder added to the porous container to provide good anode current efficiency (courtesy of Drinking Water Services, City of Ottawa).

DC power is connected (or impressed) between the structure to be protected and the groundbed anodes.

The positive terminal of the power source must always be connected to the groundbed, which is then forced to discharge as much CP current as is desirable. This is important. If a mistake is made and the positive terminal is connected erroneously to the structure to be protected, the structure will become an anode instead of a cathode and will corrode actively, which is the opposite of the desired results.

Groundbed anodes forced to discharge current will corrode. It is important to provide anode materials that are consumed at relatively low rates and thus permit designing groundbeds that can discharge large amounts of current and still have a long life expectancy.

Although scrap steel pipe, rail, rod, or other similar iron or steel material may be used, such materials are consumed at the rate of about 44 kg per Ay (1 A flowing for 1 year). This means a relatively large amount of material is needed to achieve the desired operating life for an installation and may be justified only when it is available as scrap at low cost.

Table 11.1 Chemical Composition of High-Silicon Cast-Iron Anodes

Element	Content (%)
Carbon	0.70–1.10
Magnesium	1.50 Max
Silicon	14.20–14.75
Chromium	3.25–5.00
Molybdenum	0.20 max.
Copper	0.50 max.
Iron	Remainder

Materials currently popular for use as impressed current anode material in soils include graphite, high-silicon cast iron, mixed-metal oxide, platinum, and steel.[2]

Graphite anodes may be obtained in various sizes, although 8-cm diameters by 150-cm rods are most commonly used for pipe-line groundbeds. These anodes are typically supplied by the manufacturers with insulated copper leads, usually high-molecular-weight polyethylene. Graphite is brittle so anodes of this material should be handled accordingly.

Graphite anodes are consumed at no more than one kg per ampere per year when discharging current into an electrolyte. When used with carbonaceous backfill by direct electrical contact, most of the material consumed is backfill material rather than the anode itself.

High-silicon cast-iron anodes normally contain between 14 and 15% silicon with lesser quantities of other alloying elements (Table 11.1) . Silicon alloy behaves differently from ordinary cast iron when discharging current. Ordinary cast iron loses approximately 9 kilos of its iron content per ampere per year. High-silicon cast iron, on the other hand, loses material at a much lower rate.

High-silicon cast-iron anodes also have insulated leads, as does graphite. The material is somewhat brittle and must be handled with care. They are available in various sizes and available from suppliers of CP material. A common size used in impressed-current groundbeds is 5 cm in diameter by 150 cm in length. They are used commonly with carbonaceous backfill, which absorbs most of the consumption resulting from current discharge.

Mixed-metal oxide anodes are available in a variety of sizes and shapes. Mixed-metal oxide coated titanium anodes are based on electrode technology developed in the early 1960s for production of chlorine and caustic soda. Usually the mixed metal-oxide films are thermally applied to a noble metal such as titanium or niobium cores. These oxide coatings have excellent conductivity, are resistant to acidic environments, are chemically stable, and have relatively low consumption rates. Groundbed installation in soils usually specifies that the anode be prepackaged in a canister with carbonaceous backfill material.

Platinum and platinized-niobium anodes are available in a variety of sizes and shapes. Platinum anodes are available in wire and rod configurations and in different platinum coating thickness. Table 11.2 shows typical chemical composition of platinum anodes.

Table 11.2 Chemical Composition of Platinum Anodes

Element	Content (%) of Diameter
Copper	79.5
Niobium	19.5
Platinum	Less than 1 %

Typically the anode is composed of a copper core surrounded by a niobium substrate with the platinum metallurgically bonded to the niobium substrate. The consumption of the platinum coating is extremely low (40 to 80 mg/A). To achieve the desired design life of the anode the platinum coating thickness can be varied. Typical platinum coating thicknesses are 0.65, 1.3, and 2.5 μm.

For marine work, considerable success has been achieved in the past with lead/silver alloys. Although lead was consumed at around 32 kg/Ay, suitable lead/silver alloys were forming a lead peroxide film on this outer surface, which discharges current at a low consumption rate for the base anode material.

Mixed-metal oxide anodes have replaced most other types of impressed-current anodes in seawater applications. Platinum metals and their oxides work very well as an impressed-current anode and are consumed at an extremely low rate. Because of its cost, it is usually plated onto a less expensive material such as titanium or niobium. Platinized titanium and niobium are used in marine work on ships, hulls, and for applications involving the interior parts of structures such as condenser water boxes, internal parts of pumps, certain pipeline-interiors, and so forth.

11.1.4 Rectifier Current Sources

When impressed current CP systems are specified, the most common power source used is a rectifier (Figure 11.6). This is a device that is provided with power from electric utility-system lines and which converts the alternating current to a lower-voltage DC by means of a step-down transformer and a rectifying device utilizing, commonly, selenium or silicon elements. These elements have a low resistance to current flow in one direction and a high resistance in the opposite direction. Figure 11.7 shows a simplified circuit of a rectifier utilizing a "bridge-connected" rectifying element, which is used in most CP rectifier systems.

The arrowhead symbols on the rectifying elements show the direction that current can flow readily. For the usual 60-cycle AC power source, the direction of current flow reverses 120 times per second. Refer to Figure 11.7. At some instant in time, current may be originating at connection (1) on the transformer secondary. The only path that this current can take will be through leg (c) of the bridge-connected rectifier, through the external circuit (groundbed to protected structure), and through leg (b) of the rectifier to return to the secondary winding at (2).

A one-hundred-and-twentieth of a second later, the AC flow will have reversed and it will originate at connection (2). Under this condition, the only path that it can take to reach connection (1) will be through rectifier leg (d), the external circuit, and

Figure 11.6 Five 300-A marine oil-cooled rectifiers are supported by an automatic dewatering oil-spill containment pan (courtesy of Dean Rookes, West Coast Corrosion Prevention Ltd).

rectifier (a). For either direction of AC flow, the current flow through the external circuit is in one direction only—DC.

Rectifiers operate at less than 100% efficiency, meaning that the DC output power is less than the AC input power from the supply line. Rectifiers operate at maximum

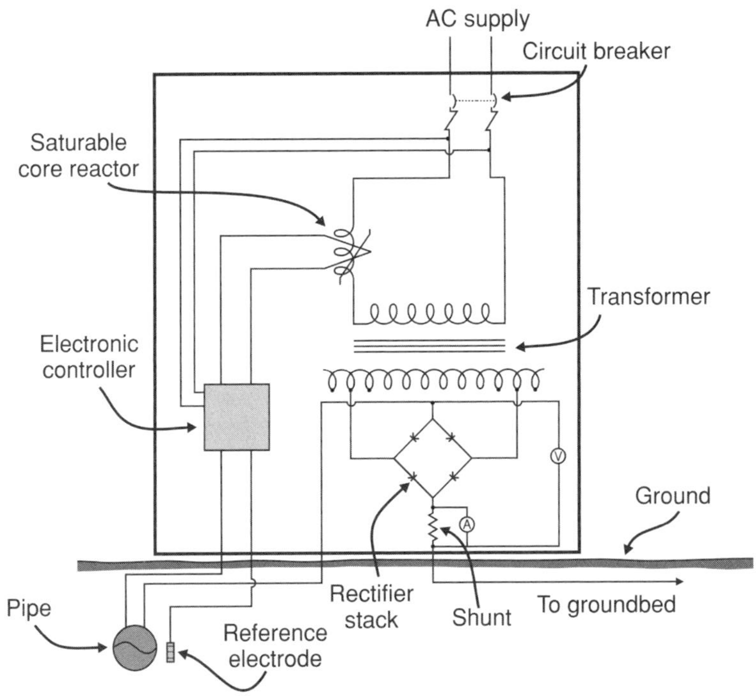

Figure 11.7 Diagram of a typical constant potential rectifier.

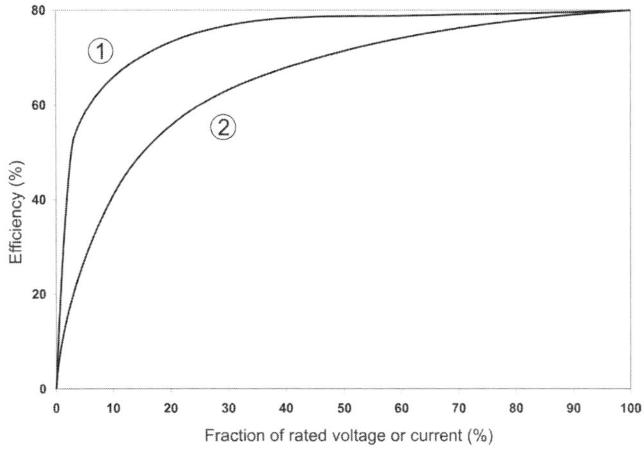

(a)

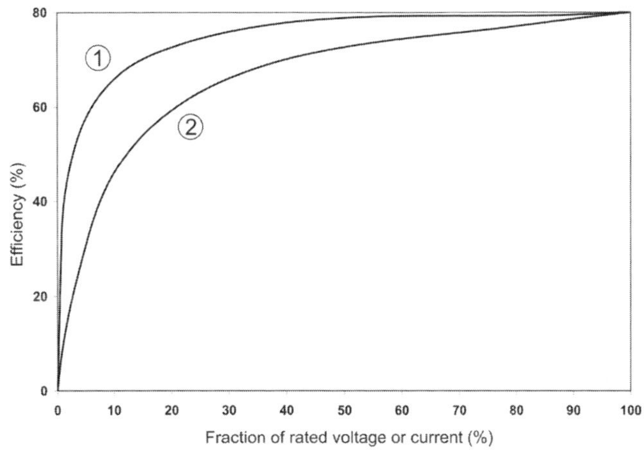

(b)

Figure 11.8 Rectifier efficiencies: a) Selenium typical overall-efficiency curves for custom line air-cooled selenium bridge-connected rectifier unit rated at 60 V, 60-A full-load DC output and 230 V, single-phase AC input; b) Silicon typical overall-efficiency curves for custom line air-cooled silicon bridge-connected rectifier unit rated at 40 V, 34-A full-load DC output and 230 V, single-phase AC input; (1) efficiency vs. percent rated current at full rated voltage; (2) efficiency vs. percent rated voltage at full rated current.

efficiency when operated at full-rated load. Where large rectifiers are used (as for large, bare structures), the rectifier rating should be selected so that it will operate at close to the full-load rating. The effect of reducing either current output or voltage is illustrated by Figure 11.8.

The efficiency of a CP rectifier may be determined by measuring the AC power input with a watt meter, measuring the DC volts and amperes output, and calculating the efficiency as follows:

$$\%\,\text{Efficiency} = \frac{\text{DC voltage (V)} \times \text{DC current (A)} \times 100}{\text{AC power (W)}} \tag{11.2}$$

For example, a rectifier operating at 20 A and 32 volts DC output is found to have an AC input of 940 W. The efficiency will then be $(20 \times 32 \times 100)/940 = 68\%$.

The difference between AC input power and DC output power represents the power loss in the rectification process. This power loss becomes heat. Cooling is sometimes necessary to prevent the rectifying components from becoming too hot and failing. One cooling method utilizes an air-cooled assembly with the rectifier cabinet arranged so that there will be a natural draft of air through the rectifier components. In another method, the rectifier components are immersed in electrical insulating oil in a large steel case. The rectifying elements give up their excess heat to the oil. The oil is in turn cooled by radiation from finned outer surfaces of the case.

One precaution bears repeating: Never connect the positive (+) terminal of an impressed current power source to the structure to be protected. If this is done, it will rapidly corrode instead of being protected.

11.1.5 Other Current Sources

In some instances where an impressed current CP system is desired, a commercial AC power line may not be available. In such instances, there are other DC power sources that may be used instead of rectifiers.[3]

11.1.5.1 Engine Generators

Engine–generator sets may be used to provide the electrical energy for CP rectifiers if a large power source is needed and AC power lines are not available. Gas taken directly from the pipe line can be used to power the engine for a CP system on a natural-gas pipe line. If the line carries a petroleum product suitable for engine fuel, this may be taken directly from the line as well. Otherwise, fuel must be brought to the generator station periodically. Present practice is to use an engine coupled with an AC alternator.

Engine–generator installations must be designed with reliable equipment that will operate unattended for several weeks. The cost of operating such an installation will be relatively high, particularly if engine fuel must be brought in.

11.1.5.2 Turbogenerators

Closed-cycle vapor turbogenerators (CCVTs) are commercially available as power sources for remote CP systems. The available CCVT systems can supply up to 5 kW and 100 V. The system consists of a Rankine cycle turbine and an alternator. A burner heats an organic liquid that vaporizes and expands. The vapor is directed through the rotating turbine wheel, providing power to the alternator.

CCVT systems can operate on a variety of fuels, including natural gas, liquified petroleum gas, kerosene, jet fuel, and diesel fuel. With natural gas, the burner should be inspected and cleaned, if necessary, annually. More frequent cleaning may be required if less clean fuels are used.

Figure 11.9 Thermoelectric field generator (courtesy of Global Thermoelectric).

11.1.5.3 Thermoelectric Generators

It has long been known that heating a junction of certain dissimilar metals (thermocouples) could generate electricity. In recent years, there has been rapid development of higher-capacity semiconductor thermoelectric materials designed specifically for power-generation use (Figure 11.9).

Standard thermoelectric generator units are available at power outputs up to 600 W and voltages up to 48 V. Because thermoelectric generators contain no moving parts, maintenance is minimal. Annually, it is necessary to replace the fuel filter and clean the fuel orifice.

11.1.5.4 Solar Electric-Power Systems

In areas where sunlight can be expected for relatively large percentages of the time, a combination of solar cells and storage batteries can be used to provide a continuous flow of current to a CP installation.

The initial cost of the solar electric-powered CP systems has dropped dramatically over the past three decades as the technology has advanced. Systems are now available that operate at power outputs up to 1 kW, voltages up to 20 V, and currents up to 50 A. Battery storage capacities up to 3,200 Ah (at 12 V) are available. Such a battery backup could supply a 10-A rectifier for almost two weeks with no recharging.

11.1.5.5 Batteries

The cost of electrical energy from batteries is high. Occasionally, however, they have been used to supply CP current to isolated sections of well-coated pipe where power lines are not available and where galvanic anodes will not supply the necessary protective current at less cost. Batteries have also been used to power temporary CP installations to evaluate the current requirements at different locations of a system.

11.1.5.6 Wind-Powered Generators

Wind-powered generators may be used as a source of power in areas where prevailing winds are of sufficient intensity and duration. Such units were used fairly extensively in the early days of pipe line CP. However, they are expensive and require extensive maintenance. Their use for this application has declined with the development of more cost-effective, reliable power sources, such as solar cells, CCVTs, and thermoelectric generators. Because the power output from a wind-powered generator will be neither steady nor continuous, some means must be used to assure a steady supply of current to the CP groundbed. This can be done by using storage batteries. Wind-powered generator installations require significant maintenance.

11.1.5.7 Gas Turbines

Gas turbines can be used to drive DC generators on natural-gas pipe-lines if there is an adequate pressure drop available. The gas is diverted to the turbine through a by-pass line and returned to the system without loss. Similar systems can be established at other locations on gas systems where reasonably constant pressure drops are available, such as in producing gas fields, at well heads, and on gas transmission pipe lines. The principal disadvantage of gas turbine power sources is the restricted number of locations where they can be used. For this reason, they are not widely installed.

11.1.5.8 Fuel Cells

Fuel cells are another source of DC energy that is subject to intensive development as a result of space exploration and military applications and, most recently, automotive applications. A fuel cell consists of two electrodes sandwiched around an electrolyte. Oxygen passes over one electrode and hydrogen over the other, generating electricity, water, and heat. Development is active also in the technology of fuel cells that use fuels other than hydrogen and oxygen. The pipe-line corrosion engineer will have another useful DC power source for impressed current CP installations when fuel cells are developed commercially at practical cost that can use natural gas or propane as fuels.

11.2 CRITERIA FOR CP

The criteria used for the control of a CP system also provide a direct method to monitor the efficiency of that system and determine how well a structure supposedly under CP is actually protected against corrosion.

11.2.1 Potential to Environment

Potential measurements are used most commonly as a criterion of protection. The basis for this is that if current is flowing onto a protected structure, there must be a change in the potential of the structure with respect to the environment. This is because the current flow causes a potential change, which is a combination of the voltage drop across the resistance between the protected structure and the environment and the polarization potential developed at the structure surface. (If necessary, review Chapter 2 for a discussion of polarization.)

The resistance between the protected structure and the environment includes the resistance of any electrically insulating paint or coating on the structure. The net result is that the structure being protected will become more negative with respect to its environment. This is illustrated by Figure 11.10.

The desirable place to measure potential would be across the interface between the pipe and the environment, as is represented by the terminals marked "polarization potential" on the equivalent circuit of Figure 11.10.

Actually, in practice, this is rarely feasible (particularly on buried structures), and it becomes necessary to resort to potentials measured between the structure and the environment surface directly above (or otherwise nearest to) the structure. The measured potential now includes the polarization potential plus a portion of the voltage drop across the structure-to-remote-earth resistance, as shown by the potential measured between the structure and point A on Figure 11.10. Under some conditions, measurement of the potential between structure and remote earth, point C, is desirable.

The expression "remote earth" used previously and in Figure 11.10 means the infinite conductor (practically no resistance) that is the earth's mass. With some idea as to where to measure the potential, the next questions are how to do it and how much of a voltage reading is needed to indicate protection.

11.2.2 Reference Electrodes

Measuring the potential of a structure is done by contacting the earth (or other conducting environment) with a reference electrode (see Chapter 2 for details). This is a stable device, suitable for field use, which will permit reproducible results. Figure 11.11 shows how the measurement may be taken.

A common reference electrode (sometimes referred to as a half-cell), shown in the detail of Figure 11.11, is a copper/copper sulfate reference electrode widely used in CP work. Any voltage reading taken between the structure and the copper sulfate electrode will consist of two parts (half-cell potentials). One of these will be the half-cell potential between the electrode itself and earth through the porous plug contact, which is relatively constant under most field conditions. The other will be the half-cell potential between structure and earth, the main variable of interest.

In practice, it is not necessary to separate the two half-cell potentials because we are interested primarily in changes in the structure-to-earth portion, and this change shows up well enough in the total value registered by the voltmeter. If we are talking about a steel structure, a value of −0.85 V or more on the voltmeter of Figure 11.10

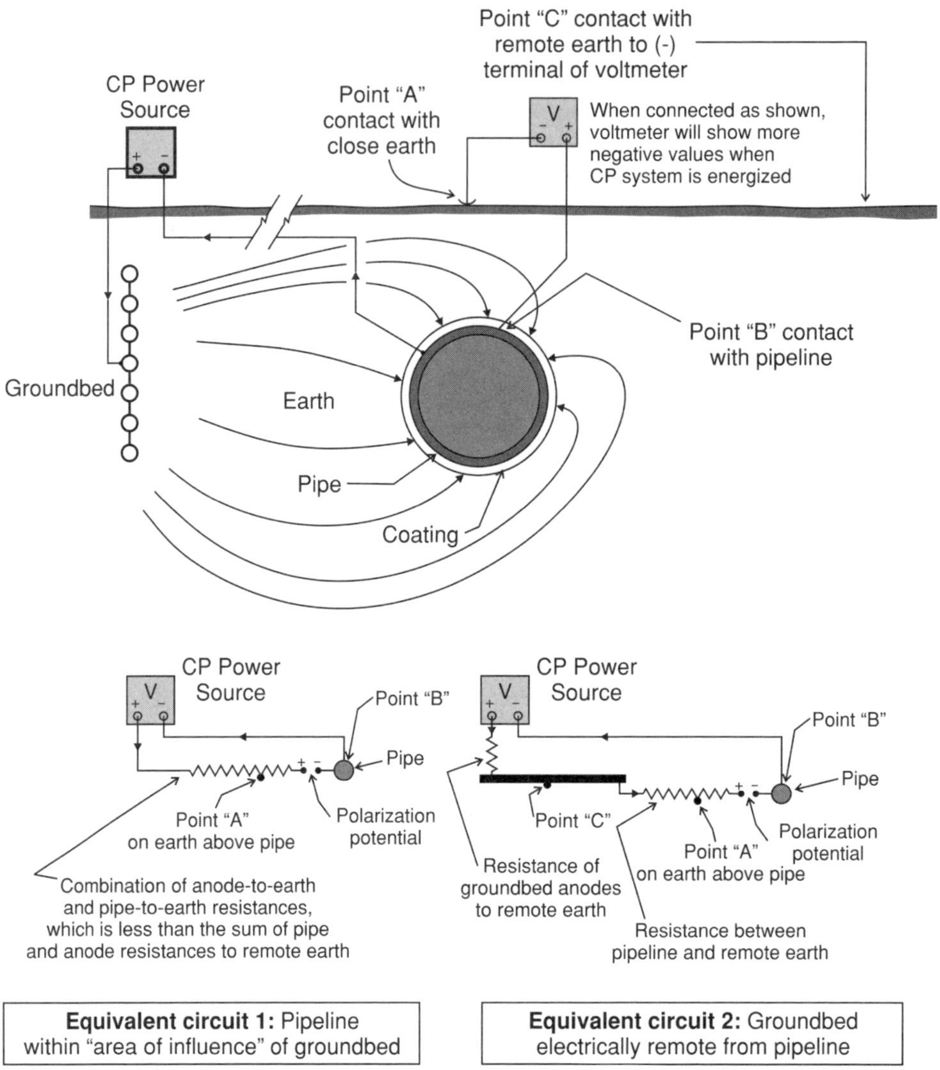

Figure 11.10 Pipe-to-environment potential change with flow of cathodic current.

would indicate full CP. This is considered to mean that the structure potential is −0.85 V with respect to a copper-sulfate electrode and is based on the value −0.80 V assigned to the most highly anodic steel found in practical situations. Because in many cases the copper-sulfate electrode cannot be placed close to the structure being checked (such as when it has to be placed on the earth surface above a buried structure), and to allow for some latitude in the potential of the most highly anodic areas, the practical value of −0.85 V has been adopted.

When metals other than steel are to be cathodically protected, different potentials to a copper sulfate reference may be used as a criterion of protection. For lead-sheathed cables, a customary value is −0.70 V. For aluminum, the potential should be held between the limits of −1.00 and −1.20 V.

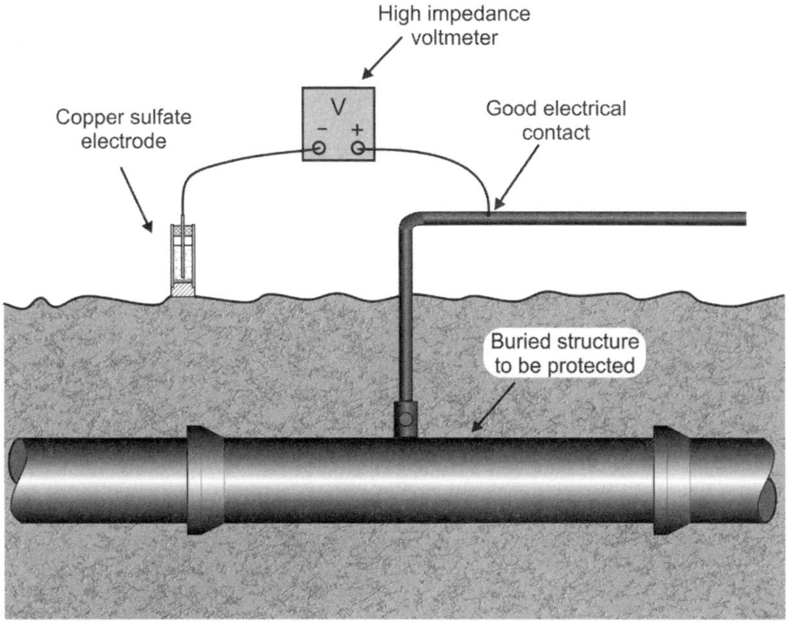

Figure 11.11 Pipe-to-earth potential measurement setup.

Notice the maximum negative limit given for aluminum. This is because the alkalinity built up at the aluminum surface by the CP process is corrosive to aluminum. This metal, if subjected to excessive CP may actually corrode more rapidly than if not protected at all. Steel is not subject to this effect. Lead may be to some extent, but usually is not a problem unless the CP current source is interrupted for an extended period. In this case, the alkalinity developed at the lead surface will have an opportunity to corrode the lead.

The degree of CP attained also may be evaluated using other stable reference electrodes for which the potentials indicative of protection will be different when their half-cell potentials differ from that of the copper/copper sulfate electrode. The calomel electrode is used by some workers for field investigations, although it is much more commonly used as a reference electrode in laboratory studies. The silver/silver chloride electrode is frequently used in work in seawater where the electrode must be submerged, because it is not subject to contamination by seawater as a copper/copper sulfate electrode can be (pressure at moderate depth forces seawater into the cell, causing the contamination).

Pure zinc in packaged backfill material makes a good permanently installed reference electrode for periodic checking of protective potentials at key points. Pure zinc without backfill may be used in seawater but should be checked against another type of reference prior to each use. Contacting the earth with a steel pin is not a suitable reference because the steel reacts with the soil and does not have a stable potential.

A practical guide to the potential differences among the reference electrodes mentioned is included in Table 11.3.

Table 11.3 Comparison of Other Reference Electrode Potentials with That of the Copper/Copper-Sulfate Reference Electrode at 25°C

Type of Comparative Reference Electrode	Structure-to-Comparative Reference Electrode Reading Equivalent to −0.85 V with Respect to CSE	To Correct Readings Between Structure and Comparative Reference Electrode to Equivalent Readings with Respect to CSE
Calomel (saturated)	−0.776 V	Add −0.074 V
Silver/silver chloride (0.1 N KC 0.1 Potassium chloride Solution)	−0.822 V	Add −0.028 V
Silver/silver chloride (silver screen with deposited silver chloride)	−0.78 V	Add −0.07 V
Pure zinc (special high grade)	+0.25[a]	Add −1.10 V

[a] Based on zinc having an open circuit potential of −1.10 V with respect to copper-sulfate reference electrode.

11.2.3 Test Coupons

Better proof that CP is working can be obtained by using coupons of the same metal as that in the protected structure. These are weighed carefully beforehand and are electrically connected to the protected structure. The coupons should be placed where they can receive the same exposure to CP current as does the structure; then, after a known exposure time, they may be removed and weighed. Any weight loss indicates incomplete CP.

Coupons may be placed where protective potentials are lowest (as indicated by measurements to a suitable reference electrode) or at locations where it is suspected that the degree of protection may not provide complete freedom from corrosion.

11.2.4 Potential Change

Some workers, when dealing with large, bare structures, use a potential change of 0.2 to 0.3 V when protective current is applied as an indication of a reasonable degree of CP. This involves measuring the structure-to-reference electrode potential with the current off, and again with the current on. A potential change in the negative direction of 0.2 to 0.3 V indicates that the structure is collecting current.

11.3 DESIGN OF A CP SYSTEM

In many instances, prior to designing a CP installation, it is necessary to conduct a field survey. Such a survey is made to assemble the data necessary to permit a workable design.

11.3.1 Current-Requirement Tests

The current needed to protect a given buried metallic structure may vary over wide limits, depending on the nature of the environment and whether it has a protective coating, and, if coated, on the quality and effectiveness of the coating as applied.

For example, if it is assumed that a steel structure to be protected is buried in corrosive soil and has an exposed surface area of 90 m^2, the current required (assuming reasonably uniform distribution of that current) could range from about 3 A if the structure is bare, to as low as 30 μA or less if the structure has a superior coating. This means that a bare structure may require 100,000 times as much current as the same structure would if it were well coated.

It cannot be assumed, however, that just because a structure is coated, it will take a small amount of current to cathodically protect it. A poor coating material (or even an excellent material poorly applied) can take much more current than the low figure given previously.

For the same 90 m^2 structure, a relatively poor coating could result in a current requirement for CP on the order of 15 mA or more. A current of 15 mA may not sound like much, but it is 500 times as great as the 30 μA figure given for a superior coating. The difference is considerable when working with large-structures (such as large-diameter cross-country pipe lines).

11.3.1.1 Tests for a Coated Line

When surveying a coated pipe line, data on the electrical strength of the coating and on current requirements for CP can be taken concurrently. If the coating is in reasonably good condition, current requirements are much smaller than on bare lines as discussed later. This makes it possible to test large distances (kilometers or miles) of pipe line with one test set up and a modest power supply. This may be done by actually applying current using a temporary test setup and adjusting the current from the power source until suitable protective potentials are obtained. Automotive rechargeable batteries are usually sufficient.[4]

Although the same coating specification may be used throughout the length of a given pipe line, the effective electrical strength of that coating may vary considerably along the route because of the type of terrain, construction difficulties, changes in average soil resistivity, and different degrees of quality of pipe line construction work and inspection.

When making the combined survey on a coated line, the test arrangement shown in Figure 11.12 may be used. At the test-battery location, a current interrupter is used to automatically switch the current source on and off at a convenient time interval, for example, 10 s on and 5 s off, to gather the data needed for coating-resistance calculations.

Testing section by section can be continued in each direction from the temporary CP location until the changes in the observed currents and potentials as the current interrupter switches on and off are no longer large enough to result in accurate data.

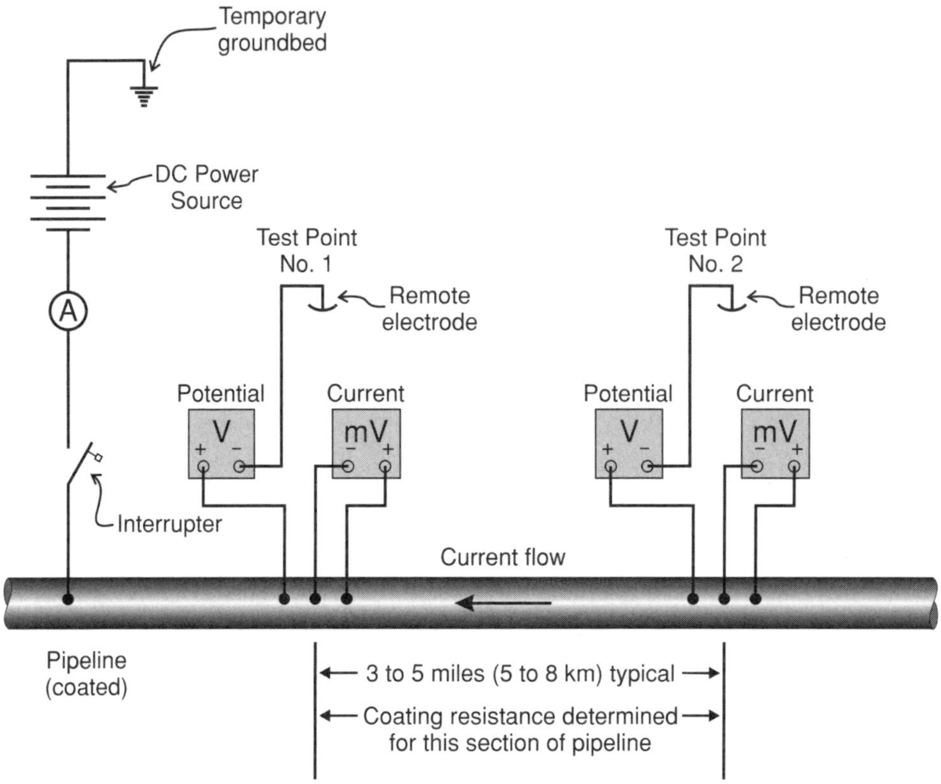

Figure 11.12 Coating resistance and CP current-requirement tests.

The limits of the area that can be maintained above the protected criterion of −0.85 V or better can be established at this same time.

Coated pipe lines polarize very rapidly and the better the coating, the faster the polarization. This means that conditions stabilize within the first few minutes and sometimes in a matter of seconds after the test current is applied.

On coated pipe-line systems provided with test points for potential and line current measurement, a survey will proceed rapidly. Data may be taken with reasonable accuracy by a single test engineer. For maximum accuracy, however, two engineers in radio communication can observe data simultaneously at each end of each section tested. This becomes essential if the pipe line under test is affected by stray current, in which case it may be necessary to take a series of simultaneous readings and average them to obtain usable data.

11.3.1.2 Tests for a Bare Structure

The preceding discussion dealt with a coated structure. When current-requirement tests are made on an all-bare structure, results may be quite different. Assume, as shown by the illustration in Figure 11.13, that DC is forced to flow from a temporary ground

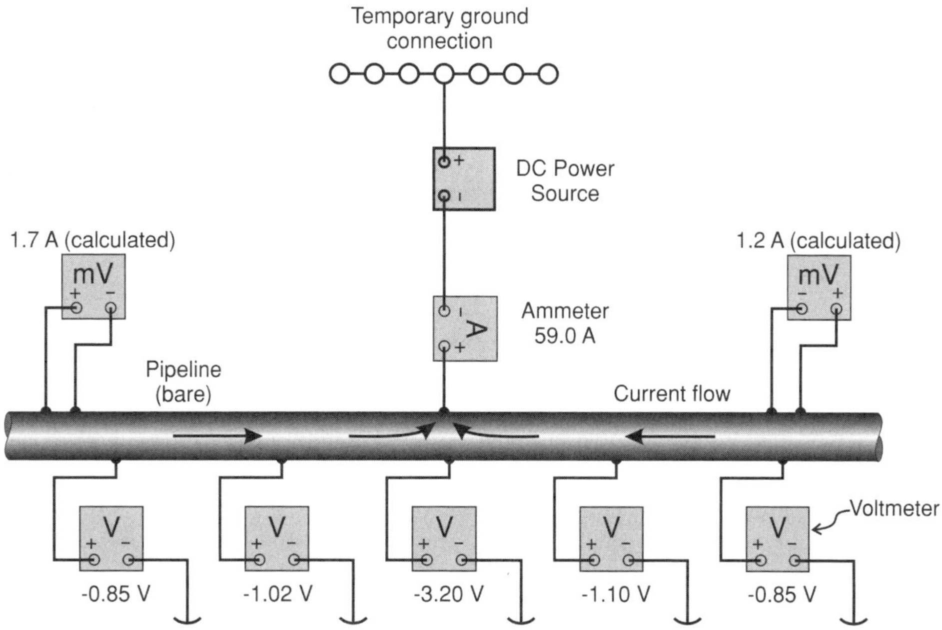

Figure 11.13 Evaluation of current requirement for CP.

connection to the pipeline section being studied, and determine how much current will be needed to protect that section.

The output of the current source shown in this figure may be adjusted until protective potentials are attained at the ends of the section to be protected. When working with bare line, the current should be allowed to flow steadily, which will permit the line to polarize to some degree, depending on the duration of the test. Full polarization on a bare line may take weeks to achieve. However, if the increase in protective potential during the test is plotted versus time at a fixed current output, the curve can be extended to give a rough approximation of the potentials that would be obtained with complete polarization.

Current flowing into the protected section from the pipe line on either side may be measured by the voltage drop across a known pipe span. These two values may then be subtracted from the total current to obtain the net flow into the protected section. Getting the net current in this manner is significant if, when making permanent CP installations, there will be CP units in adjacent areas that would reduce or eliminate the current flow from remote areas into the area being tested as described.

Current-requirement tests of the type described do not give more than a reasonable close approximation of current needs. Variations will be caused by such things as groundbed resistance and location with respect to the pipe line, polarization effects, the amount of line that is protected by each installation, and whether or not adjacent installations will affect the amount of current flowing to the pipe line beyond the area to be protected. Such variables can cause appreciable differences between survey results and the ultimate performance of permanent installations. Interpretation of survey test results on bare lines is seldom simple.

As a general rule, longer pipe lengths can be tested when working with larger-diameter lines because their longitudinal resistance is lower. When planning a current requirement test on a bare line, the current source needed should have sufficient capacity to supply, as a guide, at least 10–30 mA/m^2 of pipe surface plus 25% more. The voltage of the power source must be sufficient to force the needed current through the temporary ground connection used. Storage batteries may be adequate for testing short lengths of small-diameter pipe, whereas a source such as a DC welding generator may be needed for longer lengths of larger-diameter pipe.

11.3.2 Coating-Conductance Tests

Coatings are subject to gradual deterioration with time, with the deterioration rate depending on quality, coating type, and environmental conditions. Coating conductance tests are helpful on newly coated structures and at intervals during their operating life to evaluate coating performance. This information is particularly useful when selecting coatings for subsequent projects.

These tests are performed by determining the total current passing from the soil to the pipe surface in a specially designated pipe section that has been prepared for this test. The net average change in pipe-to-soil potential is measured and coating conductance is calculated. Periodic tests will indicate the long-term ability of a coating to maintain the low conductance that is desirable from a CP current-demand viewpoint over long periods of time.

11.3.3 Soil-Resistivity Measurements

Soil resistivity is an electrical characteristic of the soil/groundwater that affects the ability of corrosion currents to flow through the electrolyte (soil/groundwater). Resistivity is a function of soil moisture and the concentrations of ionic soluble salts and is considered to be most comprehensive indicator of a soil's corrosivity. Typically, the lower the resistivity, the higher will be the corrosivity.

Because ionic current flow is associated with soil-corrosion reactions, high soil resistivity will arguable slow down corrosion reactions. Soil resistivity generally decreases with increasing water content and the concentration of ionic species. Sandy soils are high up on the resistivity scale and therefore considered the least corrosive. Clay soils, especially those contaminated with saline water, are on the opposite end of the spectrum. The generally adopted corrosion-severity ratings are shown in Table 5.1 (Chapter 5).

Measurements of the electrical resistivity of the soil or water environment must be made as part of the corrosion survey conducted prior to designing a CP system. As a minimum, resistivity data, usually reported in ohm-centimeters, are necessary at locations where groundbeds (or single anodes) are to be installed. Such data permit the designer to determine the type, size, and number of sacrificial galvanic anodes, if used, or to select the most favorable combination of groundbed, size, and power-source DC-output rating if an impressed-current system is to be used.

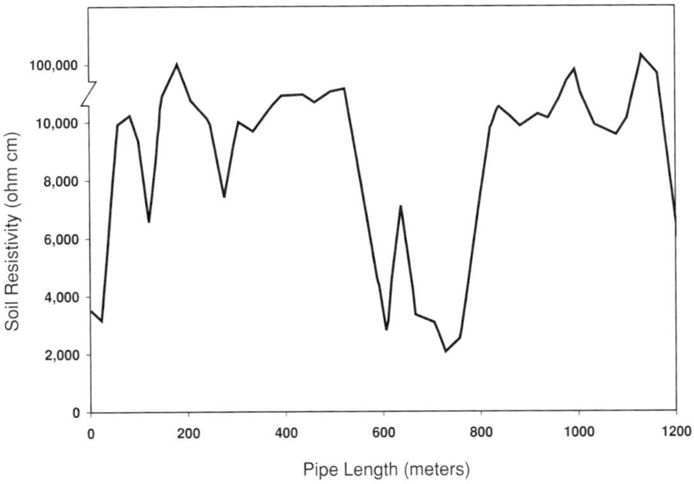

Figure 11.14 Soil-resistivity survey.

On very large structures, such as long, bare pipe lines, soil-resistivity measurements made at intervals along the structure may be used to help locate areas where corrosion can be most severe. In general, a structure that is subject to wide variations of soil or water resistivity from point to point along its surface will be particularly subject to corrosion damage at the low-resistivity areas, which tend to be the anodic portions of the structure.

Soils are particularly apt to have widely varying electrical resistivity within very short distances. For this reason, taking a single soil sample from the surface at a proposed groundbed location and measuring its electrical resistivity can be extremely misleading.

Figure 11.14 is an example of a plot of the soil-resistivity measurements taken along the same section of pipe line. Data plotted in this figure represent average soil resistivity to approximate pipe depth. In this example, a very wide range of soil resistivity is represented. Other cases would not necessarily have such a large difference between maximum and minimum resistivities along a similar length of line.[4]

To obtain meaningful data, it is also necessary to know the average soil resistivity at various depths below the surface. For example, at the usual groundbed installation involving anodes (impressed current or galvanic) buried in the first 3 m below the surface, it is helpful to have figures on average resistivity by increments of 0.8 m or so, down to depths of 5 to 8 m.

If wide changes in resistivity are suspected, it may be desirable to determine the average resistivity to depths of 15 m, 30 m, or even more, particularly for large groundbeds. Fortunately, it is possible to measure soil resistivity to various depths without actually digging down to the depths in question. This is done, basically, as shown in Figure 11.15.

In the example shown in Figure 11.14, resistivity of the soil was measured at 30-m intervals along the proposed alignment. Soil-resistivity measurements were conducted by the Wenner four-pin method,[5] utilizing a soil-resistance meter. The Wenner method requires the use of four metal probes or electrodes, driven into the ground along a

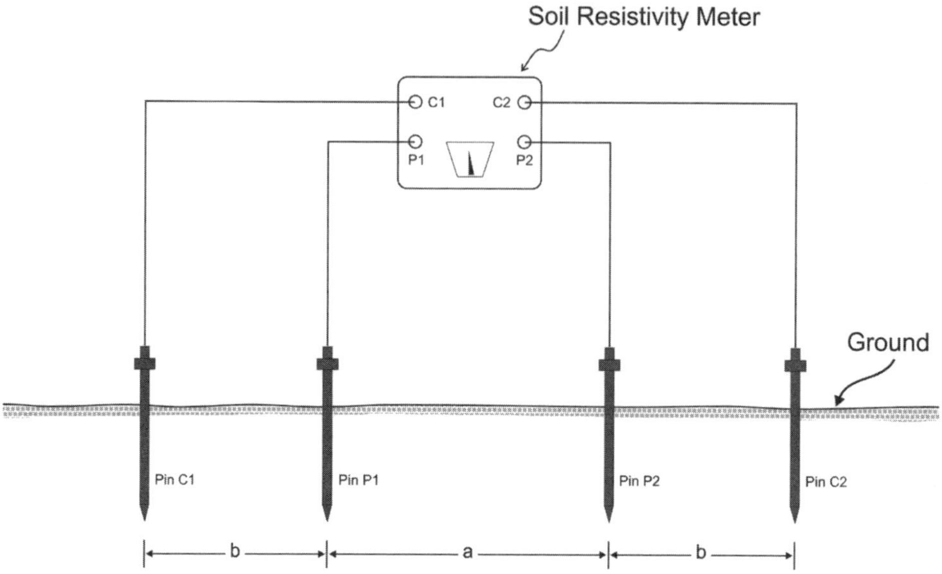

Figure 11.15 Wenner four-pin soil-resistivity test setup.

straight line, equidistant from each other, as shown in Figure 11.15. Soil resistivity is a simple function of the voltage drop between the center pair of pins, with current flowing between the two outside pins. An AC from the soil-resistance meter causes current to flow through the soil between pins C1 and C2. The voltage or potential is then measured between pins P1 and P2. The meter then registers a resistance reading. Resistivity of the soil is then computed from the instrument reading, according to the following formula.[4]

$$\rho = 2\pi \ AR \tag{11.3}$$

where: ρ = soil resistivity (ohm-centimeters); A = distance between probes (centimeters); R = soil resistance (ohms), instrument reading; and $\pi = 3.1416$.

The resistivity values obtained represent the average resistivity of the soil to a depth equal to the pin spacing. Resistance measurements are typically performed to a depth equal to that of the pipe line being evaluated. Typical probe spacing is in increments of 0.8 m.

If the line of soil pins used when making four-pin resistivity measurements is closely parallel to a bare, underground pipe line or other metallic structure, the presence of the bare metal may cause the indicated soil resistivity values to be lower than it actually is. Because a portion of the test current will flow along the metallic structure rather than through the soil, measurements along a line closely parallel to pipe lines should be avoided. When making soil-resistivity measurements along a pipe line, it is good practice to place the line of the pins perpendicular to the pipe line with the nearest pin at least 4.5 m from the pipe or even further, if space permits.

Soil-resistivity data taken by the four-pin method should be recorded in tabular form for convenience in calculating resistivities and evaluating obtained results. The tabular

Table 11.4 Format for Recording Soil-Resistivity Measurements

Test Location	Nominal Pin Spacing, Ft	Resistance, Ohms	Factor	Resistivity in Ohm-cm
No. 1				
Pipe-line station	2.5	4.40	500	2,200
1000 + 00. Nearest	5	2.05	1,000	2,050
pin 100 ft west of pipe.	7.5	1.26	1,500	1,890
Line of tests	10	0.96	2,000	1,920
perpendicular to	12.5	0.78	2,500	1,950
pipe. Clay moist.	15	0.62	3,000	1,860
	50	0.17	10,000	1,700
No. 2				
Station 1001 + 00.	2.5	1.30	500	650
Nearest pin 200 ft	5	0.60	1,000	600
east of pipe. Line of	7.5	0.45	1,500	675
tests parallel to pipe.	10	0.36	2,000	720
Edge of swamp. Wet.	12.5	0.34	2,500	850
	15	0.33	3,000	990
	25	0.34	5,000	1,700

arrangement may be as shown in Table 11.4. Where many soil-resistivity measurements are to be made, field time will be saved by using printed forms arranged for entering the necessary data.

With experience, much can be learned about the soil structure by inspecting series of readings to increasing depths. The recorded values from four-pin resistivity measurements can be misleading unless it is remembered that the soil resistivity encountered with each additional depth increment is averaged, in the test, with that of all the soil in the layers above. The indicated resistivity to a depth equal to any given pipe spacing is a weighted average of the soils from the surface to that depth. Trends can be illustrated best by inspecting the sets of soil resistivity readings in Table 11.5.

The first set of data in Table 11.5, set A, represents a uniform soil conditions. The average of the readings shown (~960 ohm-cm) represents the effective resistivity that may be used for design purposes for impressed current groundbeds or galvanic anodes.

Table 11.5 Typical Soil-Resistivity Readings Using Four-Pin Method

Pin Spacing	Pin Spacing	Soil Resistivity (Ohm-cm)			
(M)	(Ft)	Set A	Set B	Set C	Set D
0.76	2.5	960	1,100	3,300	760
1.5	5	965	1,000	2,200	810
2.3	7.5	950	1,250	1,150	1,900
3.0	10	955	1,500	980	3,800
3.8	12.5	960	1,610	840	6,900
4.6	15	955	1,710	780	12,500

Data set B represents low-resistivity soils in the first few feet. There may be a layer of somewhat less than 1000 ohm-cm around the 1.5-m depth level. Below 1.5 m, however, higher-resistivity soils are encountered. Because of the averaging effect mentioned earlier, the actual resistivity at 2.3 m deep would be higher than the indicated 1250 ohm-cm and might be on the order of 2500 ohm-cm or more. Even if anodes were placed in the lower-resistivity soils, there would be resistance to the flow of current downward into the mass of the earth. If designs are based on the resistivity of the soil in which the anodes are placed, the resistance of the completed installation will be higher than expected. The anodes will perform best if placed in the lower-resistance soil. The effective resistivity used for design purposes should reflect the higher resistivity of the underlying areas. In this instance, where increase is gradual, using horizontal anodes in the low-resistivity area and a figure of effective resistivity of ~2500 ohm-cm should result in a conservative design.

Data set C represents an excellent location for anode location even though the surface soils have relatively high resistivity. It would appear from this set of data that anodes located >1.5 m deep, would be in low-resistivity soil of ~800 ohm-cm, such a figure being conservative for design purposes. A lowering resistivity trend with depth, as illustrated by this set of data, can be relied upon to give excellent groundbed performance.

Data set D is the least favorable of these sample sets of data. Low-resistivity soil is present at the surface but the upward trend of resistivity with depth is immediate and rapid. At the 2.3-m depth, for example, the resistivity could be tens of thousands of ohm-centimeters. One such situation could occur where a shallow swampy area overlies solid rock. Current discharged from anodes installed at such a location will be forced to flow for relatively long distances close to the surface before electrically remote earth is reached. As a result, potential gradients forming the area of influence around an impressed-current groundbed can extend much farther than those surrounding a similarly sized groundbed operating at the same voltage in more favorable locations such as those represented by data sets A and C.

In some areas, experience will show that soil resistivities may change markedly within short distances. A sufficient number of four-pin tests should be made in a groundbed construction area, for example, to be sure that the best soil conditions have been located. For groundbeds of considerable length (as may be the case with impressed-current beds), four-pin tests should be taken at intervals along the route of the proposed line of groundbed anodes. If driven rod tests or borings are made to assist in arriving at an effective soil resistivity for design purposes, such tests should be made in enough locations to ascertain the variation in effective soil resistivity along the proposed line of anodes.

11.3.4 Local Conditions Affecting Design

During CP surveys, observations must be made regarding local conditions that may have a bearing on the CP design finally selected. These include the following:

1. availability of electric power for impressed current CP systems;
2. locations suitable for CP installations (subject to satisfactory soil resistivity) that will be accessible for routine inspection and maintenance and that are not

subject to construction or other activities that may make the site untenable within the expected life of the installation;
3. presence of metallic structures, other than the one to be protected, that might be affected by the planned system;
4. presence of CP systems on other structures that may have an effect on the structure to be protected;
5. presence of sources of stray DC such as DC railway systems;
6. presence of AC systems, which can have an effect on the readings taken and on the criteria for full protection of the metal; and
7. unusual environmental conditions such as acid-waste discharge from local manufacturing operations.

A later section ("Factors Affecting CP Design") covers some of these points in greater detail.

11.3.5 Instrumentation Requirements

To carry out corrosion survey work before the application of CP and performance tests on completed protection systems, specialized equipment is required. A complete discussion of this subject requires more space than is available in this chapter. However a more complete discussion is available in a NACE International publication.[6] This publication also covers much more complete information on CP design details than is possible in this chapter and is suggested as a useful supporting reference. Basic, necessary equipment are described in the following sections condensed from Ref. 7.

11.3.5.1 Voltmeters

Because of improvements in measurement technology, electronic voltmeters are manufactured with an enormous variety of capabilities. Many commonly available voltmeters are referred to as multimeters because they have the ability to measure multiple electrical values in addition to voltage. The features offered on a given instrument may be tailored to the needs of a particular industry or discipline, or the device may offer a combination of test capabilities common to many applications.

It is important to understand the versatility of these devices as well as their limitations. Certain multimeter functions and features are valuable for corrosion-engineering work, whereas others can be insufficient or unnecessary. When measuring, for example, the voltage difference between a pipe line and a reference electrode, two important resistance values must be kept in mind. To determine a pipe-to-soil potential value, a voltmeter must measure across an external-circuit resistance, which may vary widely from one environment to another. For instance, the resistance of a reference electrode in contact with moist or wet soil will be considerably less than one in contact with dry sand or frozen or oily soil. Therefore, the contact resistance of a reference electrode represents a large portion of the overall resistance of the circuit across which a voltage is to be measured.

To compensate for these variations in the measuring circuit, voltmeters must be equipped with high impedance or input resistance. A high external resistance requires a high input resistance to maintain accuracy during measurements. Most conventional multimeters currently produced have an internal impedance of 10 million ohms (10^7 ohms, or 10 megohms) or more. However, some inexpensive voltmeters are available that are not intended for use in the measurement of high-resistance circuits, such as those encountered in corrosion testing. The use of a meter with, for example, an internal resistance of only 2,000 ohms per volt can cause significant error when used to measure pipe-to-soil potentials.

11.3.5.2 Resistivity-Test Instruments

As mentioned earlier, soil-resistivity field measurements are often made using the four-pin Wenner method.[5] Four-point voltmeters can provide the necessary current supplied to the outer pins while simultaneously measuring the voltage drop between the center pair of pins. The resistance, in ohms, is then the reading on the dial multiplied by the range switch position. The instrument can measure resistance values between 0.01 ohm and 1.1 megohms depending on which multiplier (scale) is selected. The meter also uses AC, which overcomes the problem of polarization of the pins, an inherent problem with DC tests.

These meters are also useful for other measurements, such as during anode installation or for measuring the resistance of anode beds and grounding systems. The resistance-to-earth of ground rods can be measured when used in the three-pin configuration. It is important to remember not to measure across energized circuits to avoid damaging the meter. In addition, the resistance of the test leads might be included in the measurement, depending on the test-lead configuration.

11.3.5.3 Recording Instruments

Some CP measurements are best made in relation to specific periods of time, or distance, to quantify values that change or fluctuate. A number of recording voltmeters are available that can measure and record readings while left unattended, such as during interference testing.

Other recording instruments are used to measure and store pipe-to-soil potentials while traversing a pipe line during an over-the-line-type survey. Figure 11.16 shows a data recorder that can store many thousands of data entries, including text comments.

11.3.5.4 Current Interrupters

The ability to interrupt CP current is an important and frequently used technique in CP testing. A variety of current interrupters are available, most of which offer the benefits of precision timing and the ability to synchronize multiple interrupters.

When using any interrupter it is important to follow the manufacturer's recommendations on the maximum allowable current (either AC or DC) that the instrument

Figure 11.16 Battery-operated compact data logger. The small diameter of this unit (3.8 cm) allows it to be concealed in the riser of most test stations (courtesy of Tinker & Rasor).

can safely handle. It is also important to recognize the amount of drift in the timing of multiple current interrupters.

11.3.5.5 Pulse Generators and Analyzers

Pulse generators are gaining acceptance as an alternative to traditional current interruption. The pulse generator is installed in the CP power supply and momentarily interrupts the output in a precisely timed pattern. The interruptions are extremely brief and occur in a repetitive cycle. Pulse generators can be installed in multiple rectifiers (or other CP power sources, including galvanic systems) and an "on–off" survey conducted with a portable analyzer.

The analyzer recognizes these multiple, momentary current interruptions and, using an algorithmic analysis, calculates the overall "off" potential of the protected structure. It is critical to ensure that all pulse generators are operating, and that they are each set to a timing cycle that exactly corresponds to the algorithm analyzer. These analyzers also have a finite limit to the number of pulse generators that they can recognize, so this must also be taken into consideration when surveying or interpreting survey data. When used correctly they offer the advantage of measuring "off" potentials at every test site, without the installation of numerous specialized test stations. A disadvantage to permanent pulse-generator installation is that the pulsing signal may be confusing to the operators of neighboring utilities who happen to see this pattern during surveys of their own structures. To avoid this, permanently installed generators can each be set to steady-state operation.

11.3.5.6 Pipe and Cable Locators

During the course of corrosion testing, it is sometimes necessary to determine the location of buried items such as the pipe line being worked on, the location of foreign utilities, groundbed cables, and other concealed metallic structures. The pipe and cable, locator can be a great timesaver in such a case. Pipe and cable locators impress an electrical signal onto the target line, and this permits a hand-held receiver to detect their location. Electrical contact with the metallic structure being located need not be made, although conductive locating is generally much more accurate than inductive.

The type of signal that is transmitted to the pipe (or cable) can vary from manufacturer to manufacturer, and can even be variable within a single locator. Alteration of the signal frequency allows the operator to select the most appropriate signal for the application. For example, poorly coated pipes, well-coated pipes, shorted pipe lines, and insulated cables might each be best located using different frequencies. It is imperative to remember to set both the transmitter and the receiver to matching frequencies.

Advances in locator technology have also enabled a variety of attributes to be incorporated into pipe and cable locators. Depth-measurement capability has greatly improved and allows the operator to measure, with increasing confidence, the depth of the target.

11.3.5.7 Ammeter Clamps

The measurement of AC or DC current flowing through a pipe line or cable is obviously a useful kind of measurement in CP work. However, shunts or "current spans" may not exist at a location where a measurement is necessary. If the pipe or cable is made accessible, a portable noncontact ammeter clamp can measure the current magnitude and direction of flow. These devices rely on the relationship between current flow in a conductor and the electromagnetic field that it generates. They use amplifying coils positioned around the pipe or cable. These clamps are quite specialized in that they must fit around the pipe being tested. They are also a tool for which there is no substitution in certain applications. They can be used in interference testing, short locating, fault detection, and to identify bad insulating joints. The proper use of ammeter clamps requires careful attention to coil orientation, magnetic interference, nearby current flow, and polarity interpretation.

11.3.5.8 Insulator Testers

Portable, battery-operated instruments that check the integrity of insulators find a very valuable place in CP and corrosion control.

Devices such as that shown in Figure 11.17a cannot only identify shorted flanges, but when tested individually can detect shorted bolt–sleeve–washer assemblies. They can also test other insulating assemblies, such as dielectric insulating unions. These instruments are affordable and can rapidly test critical insulating devices before they are backfilled or after if a provision as been made to reach the points to measure Figure 11.17b.

(a)

(b)

Figure 11.17 Insulator tester used a) above ground and b) below ground (courtesy of Tinker & Rasor).

11.3.5.9 Reference Electrodes

The type of reference electrode the corrosion technician uses is dictated by the environment in which the electrode is used. Predominantly, copper copper-sulfate (CSE) electrodes are used, but other types are common and often necessary. Silver/silver chloride and zinc/zinc sulfate are also used in corrosion-measurement applications,

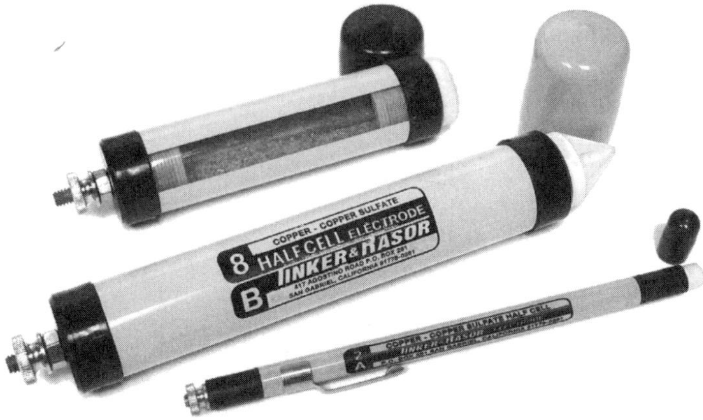

Figure 11.18 Copper sulfate electrodes of different configurations (courtesy of Tinker & Rasor).

particularly in high-chloride environments such as seawater. All reference electrodes are available in transportable models as well as models manufactured for permanent burial or immersion.

Portable reference electrodes have a variety of configurations as seen in Figure 11.18. Note that some have a large sensing area for use where maximum soil contact is desirable. Other slim models can be used where only a narrow soil access is available, such as in small test holes drilled through pavement. These can also be easily transported in a shirt pocket or briefcase.

The antimony electrode is an accessory that can measure the pH (acidity) of soil or water. It does not measure a CP potential but can help conduct soil evaluations and is helpful during forensic work.

Reference electrodes should always be selected for compatibility with the environment in which they will be used. Copper-sulfate electrodes are easily contaminated by saline or brackish environments that have high levels of ionic chlorides. If a portable reference electrode is suspected of contamination, it should be cleaned and refilled with fresh solution. Reliance on a single electrode for field work is never advisable, because it precludes both cell-to-cell comparison and the redundancy of having a spare cell on hand.

11.4 FACTORS AFFECTING CP DESIGN

A number of factors must be considered before a CP system is selected for any specific project. The following serves as a checklist of some of the items that should be considered, together with a brief description of their significance. Some of the more important factors have already been discussed in greater detail earlier in this chapter.

11.4.1 Total Current Requirement

The total amount of current needed for CP of a project must be known. This may be determined by a current-requirement test using a temporary setup. If the current

Table 11.6 Rough Current Requirements for CP of Steel[(a)]

Conditions	Current Required (mA/m^2)
Bare steel in moving seawater	100–160
Bare steel in quiet seawater	55–85
Bare steel in earth	10–30
Poorly coated steel in earth or water	1
Well-coated steel in earth or water	0.03
Very well-coated steel in earth or water	$\leq$0.003

[(a)] Approximate only—not to be used for design.

requirement per CP installation is low (up to 1.5 or 2 A), galvanic anodes may be the best choice, whereas large amounts of current may be more economically provided with impressed current systems using rectifiers or other sources of DC power. This is only a rough indication, and other factors explained later may govern the final decision.

Although an actual current-requirement test will be needed to determine the amount of current that should be provided for a specific application, Table 11.6 can be used to get a rough idea of the magnitude of the test current that may be needed. Take particular note that the table is not to be used for design purposes. The data in Table 11.6 also illustrate the variability of the current requirements that strongly depend on conditions in the local environment.

11.4.2 Variations in Environment

For soils, the variations include heavy, poorly aerated soils (little oxygen access to protected surface) as well as those that are loose, sandy, and well aerated. Metal in soil that is poorly aerated tends to be relatively easy to polarize. More current may be needed to polarize a structure in soil with ready oxygen access to its surface. Acidic soil conditions also may make it more difficult to attain polarization. Both oxygen and the acidic condition tend to have a depolarizing effect by removing hydrogen in the polarization films as it is formed by the CP process.

When immersed in water, the amount of water movement can have a pronounced effect. Structures in quiet waters tend to take the least current for protection. Highly turbulent waters produce a strong mechanical depolarizing effect from scrubbing by the moving water. Furthermore, the turbulent water is apt to be highly aerated and thus have an added depolarizing effect.

Variations in electrical resistivity of the environment should be evaluated. Where other factors permit, the lowest available soil resistivity is usually the most suitable for anode locations for either galvanic anode or impressed current rectifier systems.

11.4.3 Protective Coatings

If the structure to be cathodically protected is coated, the type of coating and in-place condition should be known. The stability of the coating is an important factor. Stability refers to the coating's ability to maintain a high degree of electrical resistance with minimum deterioration over time.

Measurement of the in-place resistance to earth of a coated structure made immediately after installation can be misleading, particularly in soils. Measurements made before the backfill has settled around the structure and has become thoroughly water saturated can indicate a much higher resistance than after conditions have stabilized.

11.4.4 Electrical Shielding

Where an underground system to be protected involves groups of closely spaced metallic components (such as banks of underground cable or conduits), a shielding effect can occur. Current from a remote CP source can collect readily on outermost layers in the bank but relatively little current and perhaps an amount insufficient for good protection may reach interior components. The outer layer of structures, then, comprises an electrical shield. Such conditions may require placement of CP anodes within the shielded area to force protective current to flow where it is needed.

11.4.5 Economic Considerations

The first consideration is whether the expense of a CP system is justified at all. Structure replacement could be the better choice in some instances. In other cases, the structure may not be continued in service long enough to justify the cost of CP unless hazardous conditions are a factor.

If CP is an economic solution to the corrosion problem, the type of CP system selected should be that having the least cost. In addition to the initial cost of design and installation of the system, the cost of power (if required) and the cost of maintaining the system (inspections, repairs, and replacements) should be considered. Depreciation schedules, the value of money, and the salvage value also must be recognized.

11.4.6 Metal to Protect

Although steel is not adversely affected by a reasonable degree of overprotection, certain metals that may be cathodically protected can be damaged if protective potentials are too high.

Aluminum, for example, is sensitive to alkali attack. The CP process creates an alkaline environment at the surface of the protected structure. If the protective potential maintained on aluminum is too high, the environment can become so alkaline that the aluminum will corrode faster than if it is not cathodically protected at all. This type of attack is called cathodic corrosion. If CP is used on aluminum, it is suggested that protective potentials be maintained within the range of −1.0 to −1.1 V, as measured to a copper sulfate reference electrode, but not allowed to exceed −1.2 V in any case.

11.4.7 Life Requirements

The expected useful life of the structure to be protected should be known. Where practical to do so, the design life of the CP system should be coordinated with that of the structure to be protected. Specifically, there is no point in designing a CP system that will last for a longer time than needed. The additional cost for the longer life is money wasted.

11.4.8 Maintenance Capabilities

The type of CP system selected can be influenced by the availability and reliability of maintenance facilities and personnel. If there is reason to believe that they will not be adequate, the simplest, most foolproof CP design may be the better choice, even though the initial cost of installation is higher. This, under some circumstances, tends to favor galvanic anodes over impressed current systems that have stopped supplying protective current for a reason such as blown fuses, groundbed failure, or accidental cutting of a cable and have been allowed to remain in this condition for extended periods, sometimes years. During these periods, no benefit is derived from the installation, and corrosion damage to the supposedly protected structure can and generally does result.

Galvanic-anode systems can, of course, be damaged as well. However, if the design provides for this it can be arranged so that damage suffered will affect only a small part of the protection system without necessarily a loss of protection potential elsewhere. Damage to impressed current systems is much more apt to cause complete loss of protection for relatively large structure areas.

11.4.9 Stray-Current Effects

Before preparing a CP design, it must be known whether or not stray currents are present. Currents that will most severely affect the structure to be protected are apt to be those of substantial magnitude that continually vary in quantity. These include stray currents from DC railway systems, mining operations using DC power, DC welding operations, and similar sources.

Depending on the severity of the stray-current effect, the magnitude of the anodic potentials at areas where stray current is discharging from the affected structure and causing corrosion may be such that they cannot be counteracted readily with the usual CP system. This is particularly true where, at areas of current discharge, the structure may be a matter of several volts positive with respect to its environment. Such conditions necessitate special stray-current control techniques involving metallic bonds from the affected structure to the source of the damaging current or other means of removing the stray current without harm to the affected structure.

The stray-current sources described earlier can result in very rapid corrosion, which is usually much more severe than the corrosion caused by other environmental factors (Figure 11.19). Another type of stray current that is variable in nature may be observed during periods of "magnetic storm" activity. Long structures such as pipe lines or cables are most apt to be affected. During magnetic storms, the intensity of the earth's magnetic field can vary. When these variations occur, potentials are induced in the pipe or cable in much the same manner as potentials are induced in an electric generator.

Man-made variable stray currents usually demonstrate some patterns, as revealed by recording-instrument records that show rush-hour peaks on transit systems, shift

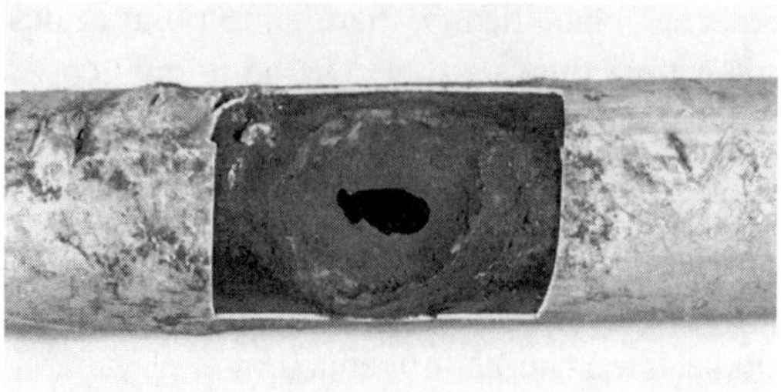

Figure 11.19 Pipe line perforated by a large fault current (courtesy of *MACAW's Pipeline Defects*, published by Yellow Pencil Marketing Co.).

changes in mining operations, and so forth. Variable stray currents resulting from magnetic disturbances, however, rarely, if ever, demonstrate any pattern. Furthermore, they may be active in a given area for a time and then not appear again for a long time thereafter. Although sometimes intense for short periods, stray telluric currents resulting from magnetic induction will seldom cause as much corrosion as uncontrolled man-made stray current because the currents are of relatively short duration and usually are not concentrated in a given area for an appreciable length of time.

In addition to the variable type of stray currents described, steady-state stray currents may be encountered. These may be caused by an ICCP system on an adjacent but electrically separate structure if the groundbed is too close to the unprotected structure. Under such a condition, the affected structure passes through the potential gradient field around the groundbed in which the soil is positive with respect to the pipe line. This positive potential difference causes current to flow onto that part of the affected structure in the gradient field. Because there is no metallic path by which this current can flow back to the groundbed where the stray current originates, it must discharge into the earth from those parts of the affected structure that are outside the groundbed potential gradient field. This situation may involve installation of metallic bonds between the affected structure and the structure to which the offending impressed current system is connected. In severe cases, moving the groundbed that is causing the stray-current condition may be necessary.

When planning a CP system, the anode systems or groundbeds should be located so that stray current from them will not be picked up by other structures and damage them. This is particularly important when using impressed current systems. Figure 11.20 illustrates the general features of typical stray-current conditions that may be encountered in association with an electrified DC transit system. Pipe lines are used, in this illustration, as a convenient means of illustration.

11.4.10 Temperature

The effect of temperature on CP design is primarily to change environment resistivity. Resistivities of soils and waters normally decrease as their temperature increases. The greatest change occurs with freezing. Resistivities of frozen soils or waters can be extremely high compared with those in the unfrozen condition. For this reason,

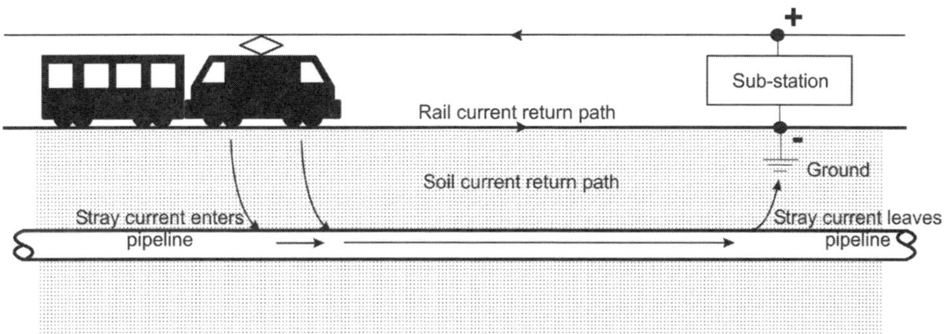

Figure 11.20 Stray-current conditions may be encountered in association with an electrified DC transit system.

measurements of the resistivity of frozen material would be meaningless for CP groundbed-design purposes.

In marine work, the resistivity of seawater in tropical locations can be significantly lower than the same water in colder areas.

Groundbeds should be installed below freezing depth in soils or waters where this may be a problem.

When measuring the resistivity of water samples from, for example, heat-exchanger water boxes that are to be protected, the measurement should be made at the normal water-box operating temperature.

11.4.11 Anode Materials

Materials available to serve as galvanic or impressed-current anodes have been described previously, as well as the sources of power for the impressed current system. The selection of these basic elements in the CP design must be made on a sound economic basis, with proper attention to maintenance cost during the life of the system.

11.4.12 Wire and Cable

In impressed-current systems, all underground or submerged cable from the positive DC terminal of the rectifier (or other power source) to the groundbed is at a positive voltage with respect to earth. For this reason, it must be very well insulated.

It is important to remember that all buried parts of the groundbed assembly connected to the positive terminal of the power source may discharge current and corrode at any point where metal contacts the conducting environment in which the groundbed assembly is placed. This includes the cable from the rectifier to groundbed anodes and cable interconnecting anodes within the groundbed.

To prevent current discharge from the cables, wire that has high-quality electrical insulation suitable for underground use must be provided, and all splices and connections must be insulated perfectly. At any defects in the cable-insulation system, there will be current discharged that will corrode the wire until it separates, thus breaking the electrical connection between the rectifier and all or part of the groundbed installation.

By contrast, all buried wire connecting self-powered galvanic anodes to a protected structure are subject to current collection from the environment and are free of corrosion. Insulation on such wires is used to prevent their picking up unnecessary current and associated loss of anode efficiency. This is also true of the insulated cable extending from the negative terminal of the impressed-current power source to the protected structure.

Cables from the negative terminal of impressed current systems to the protected structure, wires from galvanic anodes to the protected structure, and wires connected to the protected structure for testing purposes are all under protection and not subject to current discharge and corrosion if there are breaks in their insulation. However, it is a good practice to use well-insulated cable and wire for such applications to prevent unnecessary consumption of CP current. The junction between a bare copper wire and a steel structure would create, unnecessarily, a strong dissimilar-metal corrosion cell. Where such junctions are necessary, careful and durable insulation from the environment is required.

11.4.13 Anode Backfill

Chemical backfill is used to surround galvanic anodes, in most cases, when they are buried in earth. There are several advantages to this. The special backfill provides a uniform environment surrounding the anode to promote uniform anode consumption and maximum efficiency. It isolates the anode material from direct contact with the earth, which is desirable to prevent action by soil chemicals that might otherwise cause high-resistance "passive" films to form at the anode surface and prevent discharge of useful amounts of current. It has a low electrical resistivity, so where the surrounding soil is of higher resistivity, this permits a lower anode-to-earth resistance and greater current output.

A typical backfill material used with magnesium anodes consists of 75% hydrated gypsum, 20% bentonite clay, and 5% sodium sulfate. A mixture of 50% molding plaster (plaster of Paris) and 50% bentonite clay works well with zinc anodes. These backfill mixtures are available from suppliers or are provided in prepackaged containers as shown in Figure 11.5. Figures 11.21 and 11.22 illustrate how these anodes are installed in their prepackaged backfill containers.

For impressed current anode systems in earth, it is common practice to use a coke or graphite backfill to surround anodes (Figure 11.23). There are two advantages to this. One is that graphite or suitable coke has a very low electrical resistivity (lower than most soils) and helps to produce low resistance between anode and earth.

The second advantage is that if the coke or graphite is solidly packed around the impressed-current anode, the greater part of the current discharged by the anode passes

(a) (b) (c)

Figure 11.21 Holes are being drilled for sacrificial-anode installation: (a) thermit welding of anode connection on water main and (b) anode cluster ready to be buried and connected to fire hydrant tap (c) (courtesy of Drinking Water Services, City of Ottawa).

Figure 11.22 Ductile iron municipal water main is protected with tape wrap and sacrificial magnesium anodes (courtesy of Dean Rookes, West Coast Corrosion Prevention Ltd.).

Figure 11.23 Conventional groundbed using petroleum-coke backfill and 5-cm cast-iron anodes (courtesy of Dean Rookes, West Coast Corrosion Prevention Ltd.).

from the anode to the backfill through direct contact between the anode material and the backfill particles. This reduces the consumption rate of the anode. The current discharging from the outer surfaces of the backfill column will, however, consume the backfill material at a slow rate. The net result is a lower-resistance, longer-life anode installation than would exist if backfill were not used.

Suitable coke or graphite backfill materials are available from suppliers. Coke should be either coal coke or calcined (heat-treated) petroleum coke. Non calcined petroleum coke can have high electrical resistivity and be unsuitable for groundbed backfill use.

Backfills customarily are not used with either galvanic or impressed current anodes in water-immersion applications.

11.4.14 Miscellaneous

Electrical connections, particularly in galvanic-anode systems, must be of very low electrical resistance to permit maximum current output. All underground or submerged connections should be welded, brazed, or soldered for permanent low resistance. This is also a good practice for impressed current systems and for test-wire connections to protected structures.

Exposed metal at underground or submerged connections between impressed current anodes and cables to the power source must be perfectly insulated to prevent current discharge and accompanying corrosion.

Enclosed blocks suitable for terminating test wires from protected structures are available to facilitate periodic tests of the degree of protection attained.

Catalogs from suppliers of CP equipment and material give detailed information on the various components of CP installations.

11.5 MONITORING PIPE-LINE CP SYSTEMS

Monitoring pipe line CP systems is a technically complex field. The two main aspects of this type of work are monitoring the condition and performance of the CP system hardware and monitoring the condition of the pipe line itself.

11.5.1 Monitoring CP-System Hardware

Because CP systems are expected to operate in demanding environmental conditions over long time periods, the reliability requirements of the associated hardware are high. Selected test stations should be visited on a regular basis. Rectifier units and selected sacrificial anode locations should be visually inspected and their output measured. These comprehensive steps also apply to annual monitoring, which should additionally focus on the detailed inspection of system components (bonds, shunts, fuses, surge, divertors, distribution boxes, cable connections, mounting systems, etc.).

An increasing trend toward selective, remote rectifier monitoring, using modern communication systems and computer networks, is emerging to accomplish these tasks with reduced resources. Wireless cell phone and satellite communication systems are available for interrogating rectifiers in remote locations.

11.5.2 Structure-Condition Monitoring

As mentioned previously, monitoring the condition of structures protected by CP is a highly specialized subject. In many cases, condition-monitoring requirements are specified by regulatory authorities. The following sections briefly describe some of the techniques used to carry out this work.

11.5.2.1 Close-Interval Potential Surveys

The principle of a close-interval potentials surveys (CIPS) is to record the potential profile of a pipe line over its entire length by taking potential readings at intervals of approximately 1 m.

A reference electrode is connected to the pipe line at a test post and positioned in the ground over the pipe line at regular intervals (around 1 m) for the measurement of the potential difference between the reference electrode and the pipe line (Figures 11.24 and 11.25) .

A three-person crew is typically required to perform these measurements. One person walking ahead locates the pipe line with a pipe locator to ensure that the potential measurements are performed directly overhead the pipe line. This person also carries a tape measure and inserts a distance marker (a small flag) at regular intervals over the pipe line. A second person carries a pair of electrodes that are connected to the test post by means of a trailing, thin copper wire and the potential-measuring instrumentation. This person is also responsible for entering specific features as a function of the measuring

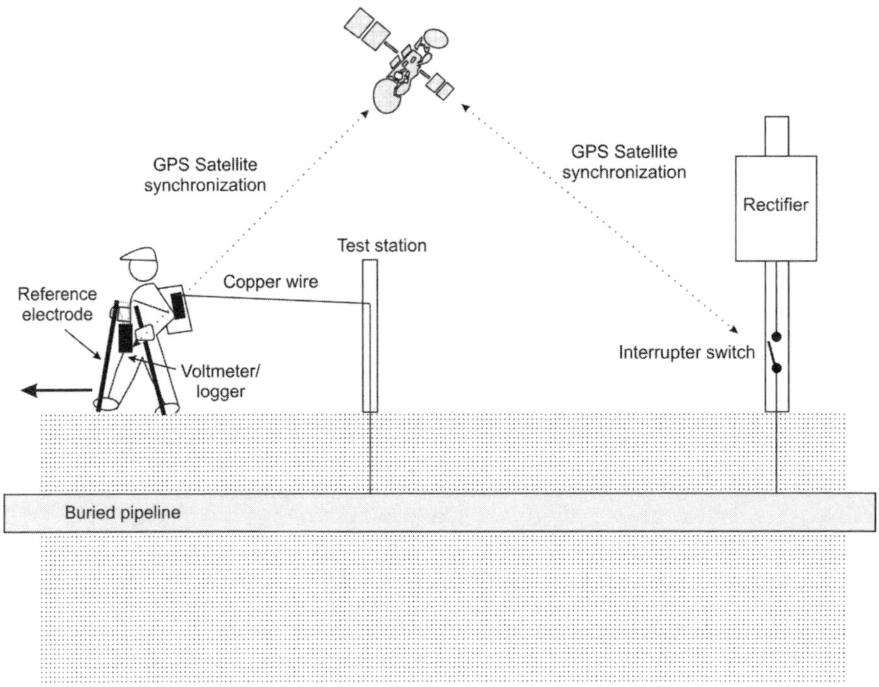

Figure 11.24 Schematic shows the CIPS methodology.

distance. The third person collects the trailing wire, after individual survey sections have been completed.

CIPS measurements are quite demanding on the field crews and require extensive logistical support from both the pipe-line operator and the CIPS contractor. Field crews are typically required to move over/around fences, roads, highways, and other obstacles and difficult terrain. Breakage of the trailing copper wire is not uncommon and special strengthened wire has to be taped down onto road surfaces that are crossed.

An important consideration in the potential readings is the "IR" or ohmic-drop error that is included in the potential measurements when a CP system is operational. The "on" potential measurements taken at the surface do not represent this interfacial potential and include, among other effects, an "IR" drop from the cathodic polarization of the pipe line. In the presence of the IR drop, the measured "on" potentials tend to be more negative than the actual interfacial potential and a false sense of security may be obtained.

To accomplish the on–off switching for these potential measurements, a current interrupter has to be installed on the rectifiers. Modern interrupters are based on solid-state switches and are programmable to perform switching only when the survey is performed during the daytime. This feature minimizes the depolarization of the pipe line that may occur gradually as a result of the cumulative effects of the "off" periods. Importantly, a particular section of a pipe line will typically be under the influence of several rectifiers and thus a number of rectifiers will have to be interrupted in a synchronized manner to perform the "on" and "off" potential readings. Pipe-line operators

Figure 11.25 Over-the-line potential survey being performed using GPS synchronized data recorder and interrupters (courtesy of Dean Rookes, West Coast Corrosion Prevention Ltd.).

usually specify that at least two rectifiers ahead of the survey team and two rectifiers behind the survey team have to be interrupted in a fully synchronized manner.

An example of graphical CIPS data is presented in Figure 11.26.[4] In the simplest format, the "on" and "off" potentials are plotted as a function of distance. The usual sign convention is for potentials to be plotted as positive values. The difference between the "on" and "off" potential values should be noted. As is usually the case, the "off" potentials are less negative than the "on" values. If the relative position is reversed, some unusual condition such as stray-current interference is likely to have arisen.

The CIPS technique provides a complete pipe-to-soil potential profile, indicating the status of CP levels. The interpretation of results, including the identification of defects, is relatively straightforward.

11.5.2.2 Pearson Survey

The Pearson survey, named after its inventor, is used to locate coating defects in buried pipe lines. Once these defects have been identified, the protection levels afforded by the CP system can be investigated at these critical locations in more detail.

An AC signal of around 1000 Hz is imposed onto the pipe line by means of a transmitter (Figure 11.27), which is connected to the pipe line and an earth spike as illustrated in Figure 11.28. Two survey operators make earth contact either through aluminum poles or metal-studded boots (Figure 11.29). A distance of several meters typically separates the operators. Essentially, the signal measured by the receiver is the potential gradient over the distance between the two operators. Defects are located by a change in the potential gradient, which translates into a change in signal intensity.

As in the CIPS technique, the measurements are usually recorded by walking directly over the pipe line. As the front operator approaches a defect, increasing signal

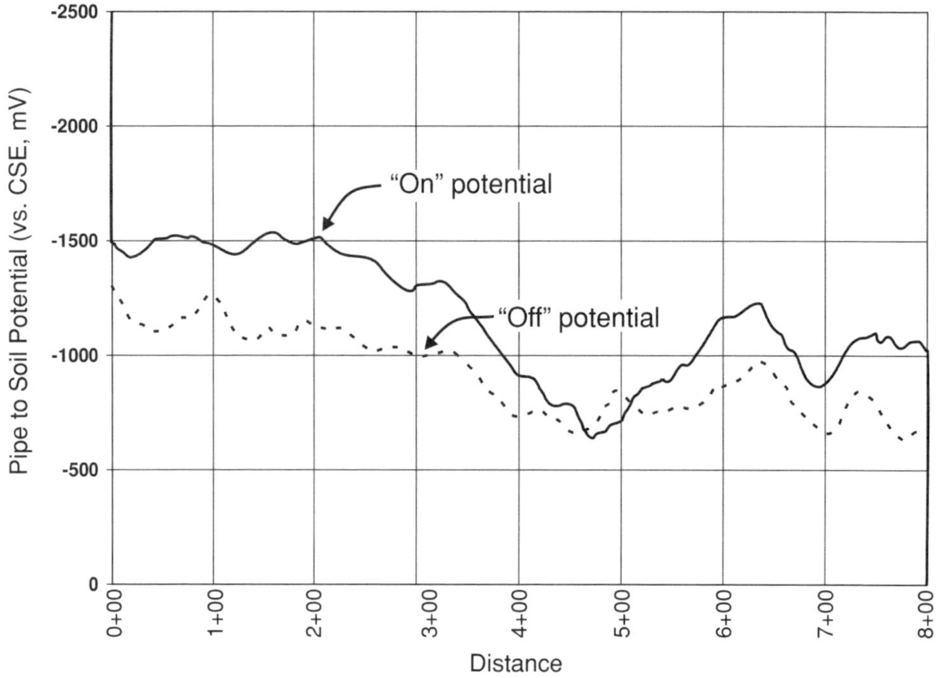

Figure 11.26 Over-the-line survey with CP.

intensity is recorded. As the front person moves away from the defect, the signal intensity drops and later picks up again as the rear operator approaches the defect. The interpretation of signals can obviously become confusing when several defects are located between the two operators.

In principle, a Pearson survey can be performed with an impressed CP system remaining energized. Sacrificial anodes should be disconnected, because the signal

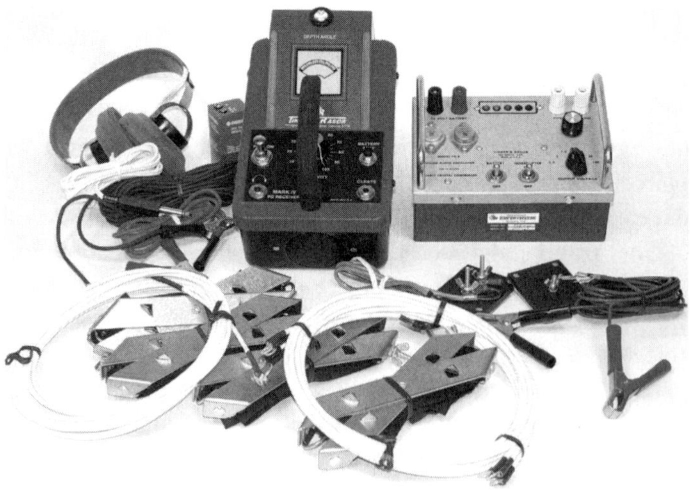

Figure 11.27 Instrumentation to carry out a Pearson survey (courtesy of Tinker & Rasor).

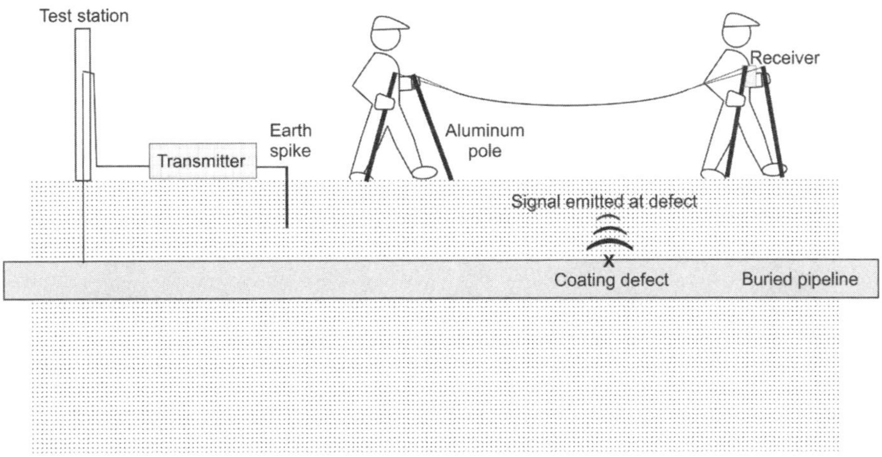

Figure 11.28 Schematic shows the Pearson survey technique.

from these may otherwise mask actual coating defects. A three-person team is usually required to locate the pipe line, perform the survey measurements, place defect markers into the ground and move the transmitters periodically.

By walking the entire length of the pipe line, an overall inspection of the right-of-way can be made together with the measurements. In principle, all significant defects and metallic conductors causing a potential gradient will be detected. There are no trailing wires and the impressed CP current does not have to be pulsed.

The disadvantages associated with Pearson surveys are similar to those of CIPS, as the entire pipe line has to be walked and contact established with ground. The technique is therefore unsuitable to roads, paved areas, rivers, and so forth.

Figure 11.29 Pipe-line CP fault testing with Pearson-type detector (courtesy of Tinker & Rasor).

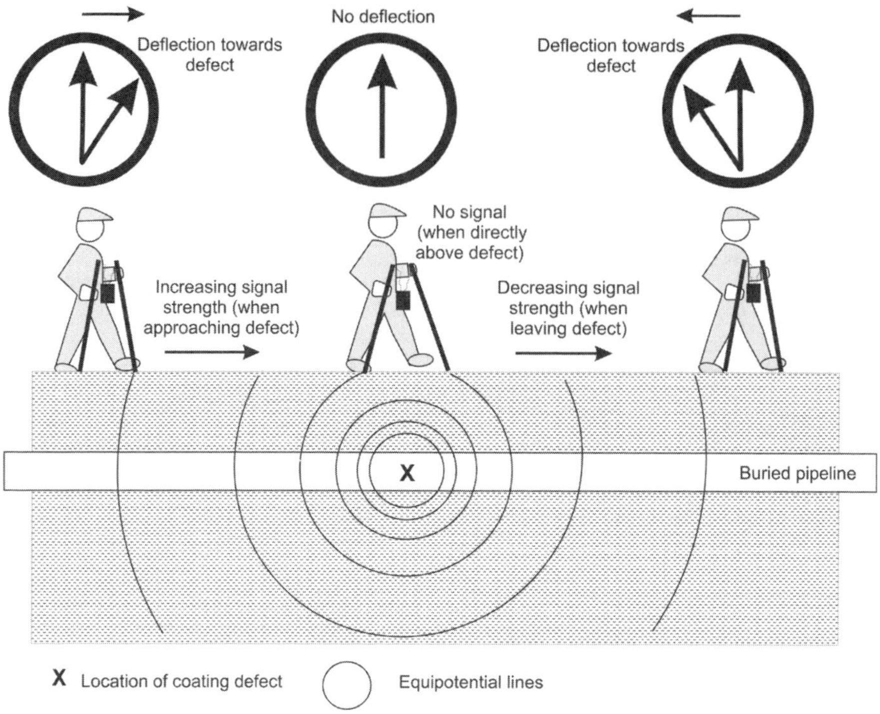

Figure 11.30 Schematic shows the DCVG methodology.

11.5.2.3 Direct-Current Voltage-Gradient (DCVG) Surveys

DCVG surveys are a more recent method of locating defects on coated, buried pipe lines and making assessments of their severity. The technique again relies on the fundamental effect of a potential gradient being established in the soil at coating defects under the application of CP current. In general, the greater the size of the defect, the greater the potential gradient.

The potential gradient is measured by an operator between two reference electrodes, for example, copper sulfate electrodes, separated by a distance of approximately half a meter. A pulsed DC signal is imposed on the pipe line for DCVG measurements. The pulsed input signal minimizes interference from other current sources, for example, CP systems, electrified rail transit lines, and telluric effects. This signal can be obtained with an interrupter on an existing rectifier or through a secondary current pulse superimposed on the existing "steady" CP current.

The operator walking the pipe line observes voltage deflections on a precision voltmeter to identify defect locations. The presence of a defect is indicated by an increased deflection as the defect is approached, no deflection when the operator is immediately above the defect, and a decreasing deflection as the operator walks away from the defect (Figure 11.30). It is claimed that defects can be located with an accuracy of 0.1–0.2 m, which represents a major advantage in minimizing the work of subsequent digs when corrective action has to be taken.

The DCVG technique is particularly suited to complex CP systems, for example, areas with a relatively high density of buried structures. These are generally the most difficult survey conditions. The DCVG equipment is relatively simple and involves no trailing wires.

11.5.2.4 Corrosion Coupons

Corrosion coupons connected to cathodically protected structures are finding increasing application for performance monitoring of the CP system. Essentially these coupons, installed uncoated, represent a defect simulation on the pipe line under controlled conditions. These coupons can be connected to the pipe line via a test post outlet, facilitating a number of measurements such as potential and current flow.

Perhaps the most important consideration in the installation of corrosion coupons is that a coupon must be representative of the actual pipe-line surface and defect. The exact metallurgical detail and surface finish as found on the actual pipe line are therefore required on the coupon. The influence of corrosion product build-up may also be important. Furthermore the environmental conditions of the coupon and the pipe should also be matched, for example, temperature, soil conditions, soil compaction, oxygen concentration, and so forth. Current shielding effects on the bonded coupon should be avoided.

Several measurements can be made, after a coupon-type corrosion sensor has been attached to a cathodically protected pipe line. "On" potentials measured on the coupon are in principle more accurate than those measured on a buried pipe, if a suitable reference electrode is installed in close proximity to the coupon. The potentials recorded with a coupon sensor may still contain a significant "IR" drop error, but this error is lower than that of surface "on" potential measurements. Instant-off potentials can be measured conveniently by interrupting the coupon bond wire at a test post. Similarly, longer-term depolarization measurements can be performed on the coupon without depolarizing the entire buried structure. Current flow to and from the coupon can be measured and its direction can also be determined, for example by using a shunt resistor in the bond wire. It is also possible to determine corrosion rates from the coupon. Electrical-resistance sensors provide an option for in situ corrosion-rate measurements as an alternative to weight-loss coupons.

11.6 SPECIAL REQUIREMENTS FOR SPECIFIC APPLICATIONS

11.6.1 Pipe-Line Applications

Most pipe-line CP applications involve either galvanic anode or impressed current systems installed in earth for protection of external surfaces. Of the galvanic anode installations, most use magnesium as the anode material. Rectifiers are the most common source of DC power for impressed current systems.

For pipe lines installed in the ocean bed (as for long harbor crossings and lines to offshore drilling operations), considerable use has been made of galvanic-anode bracelets. These are essentially a ring of specially cast anodes encircling the pipe and attached directly to it. This permits having the anodes already attached to the pipe as it is laid. By so doing, the pipe line will be cathodically protected as soon as it becomes submerged. Used in conjunction with a good coating, sufficient anode material may be provided for long useful life. Zinc has been used most frequently for this type of installation.

Where surface soil conditions for pipe lines on land are not suitable for groundbeds installed near the surface, deep groundbeds (vertical) may be installed if underlying earth resistivity is more favorable. Such groundbeds usually are installed in a single hole, which may be a few hundred meters deep in some instances. Deep groundbeds are used with impressed current systems. Particular care must be exercised during installation to avoid premature failures of anodes or anode leads that may not be repairable and may necessitate the installation of a complete new groundbed.

Other instances where deep groundbeds are necessary include sites where right-of-way for surface groundbeds cannot be obtained. For example, a deep bed can be installed on a pipe-line right-of-way. They also are used in congested distribution systems where remote groundbeds are needed, but where available sites for surface groundbeds are not sufficiently remote from the pipes to be protected or from structures of other ownership.

In some congested areas, anodes (galvanic or impressed current) are distributed along the length of pipe to be protected. This permits placing the anodes close to the pipe, with each anode protecting a short length. The effect on other structures also may be controlled more readily. This type of installation may be more expensive than remote groundbeds placed at much longer intervals but may, nevertheless, be the best solution in some instances.

Where pipe lines are banked, giving rise to severe shielding (as was illustrated by Figure 11.3), a continuous ribbon anode may be used within the bank and parallel to the pipe lines to provide protective current within the bank. Such material is available in zinc or magnesium. For impressed current systems, platinum-coated wire or rod anodes are available if required.

The interiors of large pipe lines carrying corrosive liquids (such as seawater or industrial waste) may be lined with a suitable coating and protected with strip-type galvanic-anode material. If the pipe interior is bare, relatively large amounts of current may be needed. In this case, an impressed current system may be used with platinum-coated anodes penetrating the pipe walls at intervals.

11.6.2 Fixed Structures in Seawater

Structures such as steel bulkheads, steel piles supporting piers or wharfs, offshore drilling platforms, and other similar structures may be protected with either sacrificial galvanic anode systems or impressed current systems (Figure 11.31).

Galvanic-anode systems in seawater, for the most part, use much heavier anodes than those used in soil. This is because it is possible to obtain high current from them in a low-resistivity seawater environment and because the corrodible mass is needed to provide reasonably long life.

Figure 11.31 Sled with six cast-iron anodes, integrated cable, and protective hose for an operation capacity of 75 A for 15 years (courtesy of Dean Rookes, West Coast Corrosion Prevention Ltd.).

For structures that can be polarized, a low potential galvanic-anode material such as zinc or aluminum is generally preferable to a high-potential material such as magnesium. Magnesium will work perfectly well, but may discharge more current than is needed.

Zinc or suitable aluminum alloy anodes can polarize a steel structure in seawater to within a few millivolts of the potential of the anode itself. Assume that potential is 50 mV, although it can be less. This means that the anode current will stabilize at a value sufficient to maintain polarization.

If magnesium anodes are used, however, the structure does not tend to polarize beyond about −1.1 V to a silver chloride reference electrode because the hydrogen overvoltage potential is reached and this results in the evolution of free hydrogen. This means that with a magnesium working voltage of about −1.4 V to a silver chloride electrode, there will be a driving voltage of approximately 0.3 V (300 mV). Thus, on a comparable basis, the magnesium will discharge about five times as much current as is actually required to achieve the required polarization.

Because less powerful anodes can provide an efficient CP protection, the surplus current from a more powerful anode is, in effect, wasted. However, there is one advantage. The more cathodic voltage provided by magnesium anodes tends to force the deposition, in seawater, of a thicker protective calcareous coating on the structure surface than would be obtained with less powerful anodes.

Special chemical backfills are not needed in the uniform seawater environment because galvanic anodes work satisfactorily without them.

Impressed current systems for fixed seawater structures may use suitable anode materials, without backfill, suspended in the seawater from the structure being protected or placed on the ocean floor. In the past, various materials such as treated graphite, high-silicon cast iron, platinized titanium, or lead/silver have been used as anodes. However, the introduction of highly efficient dimensionally stable anodes (DSAs), which are basically mixed-metal oxide coatings on titanium, has rendered these other anodes almost obsolete.

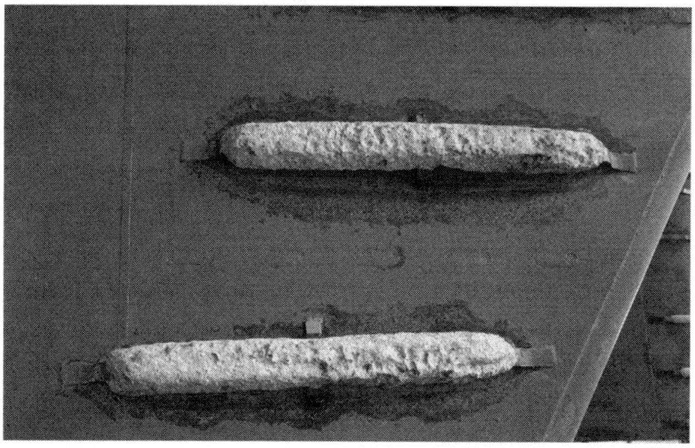

Figure 11.32 Partially depleted zinc anodes on a ship hull (courtesy of Dean Rookes, West Coast Corrosion Prevention Ltd.).

Particular attention must be given to the design of the rectifier placement, header cable distribution system, and anode suspension or placement details. Above-water components are subject to severe marine atmospheric attack, whereas other portions must be protected from or designed to withstand the mechanical forces exerted by moving seawater as well as by water-carried debris or shipping traffic.

11.6.3 Marine Craft

CP systems on hull exteriors must be capable of a wide range of current output. As a rule, a craft tied up at dockside will require much less current to keep the hull in a polarized condition than will be needed when the same craft is underway. This is because of the mechanical depolarizing effect of the rapidly moving seawater.

Furthermore, a craft that is fresh from dry docking after application of a new paint system initially will require relatively little current. As the paint deteriorates in service, current requirements may increase substantially. This is particularly true with a craft that may encounter sandbars, which remove paint rapidly through abrasive action.

Galvanic anodes (zinc and aluminum) are used to a considerable extent on ship hulls and are satisfactory if sufficient anode material is provided (Figure 11.32). Current output of the anodes automatically increases as the need for it increases. Design problems involve providing sufficient anodes to provide maximum expected current demand. Enough anode weight must be provided to give reasonable life. This is usually matched to the normal interval between dry-docking or some multiple thereof.

Impressed current systems use hull-mounted anodes and some means of automatically adjusting the power-source DC output to meet varying requirements. Present practice favors the use of potential-controlled automatic rectifiers. These devices are so arranged that the rectifier current output to hull anodes is automatically raised or lowered through control circuitry as the need changes. The control circuitry is guided by the potential between the hull and permanently installed through-hull reference electrodes placed at critical locations.

In tankers, tanks that are filled with salt water for ballast after discharging cargo are susceptible to severe corrosion. Galvanic anodes are used widely to prevent corrosion during the ballast period. Where there is an explosion hazard (as with tankers in petroleum-products service), the use of impressed current CP systems would impose critical safety problems such as accidental short circuits and sparking. Although there is no probable spark hazard in galvanic-anode circuits, an anode that broke away from its mounting and fell might, as it struck the steel tank bottom, create a spark with sufficient energy to ignite an explosive mixture. Zinc would appear to be the least likely to cause such sparking.

11.6.4 Marine Growth

The fouling of surfaces exposed to seawater can affect both corrosion and the application of CP. Very high applied-current densities (much higher than needed for protection) will kill marine growth, whereas lower values seem to have only slight effects. The current can pass through the bodies of the organisms and protect the metal surface; it is even possible to deposit cathode scale beneath attached barnacles by this process.

11.6.5 Tanks

Tank exteriors normally may be handled by conventional CP systems. Tank interiors may be protected when they contain waters or other conductive fluids that are not strongly acidic. If tank interiors are well coated, galvanic anodes may be a good choice. If they are bare, impressed current systems may be used with anodes mounted inside the tank. Anode leads are normally brought through the tank wall to a rectifier or other source of DC power.

11.6.6 Industrial-Plant Underground Networks

Many manufacturing plants, refineries, electrical generating stations and so forth have complex networks of underground structures. Dissimilar metals are frequently involved, and where soil resistivity is low and corrosive, CP can provide substantial savings.

In complex networks, there usually is a considerable amount of shielding, so that a simple CP system using a remote groundbed may not permit sufficient current to reach structures within the shielded area. In such instances, CP systems (either galvanic anode or impressed current) using anodes distributed throughout the shielded area may be required.

Where impressed current systems are used in areas subject to explosion hazards (such as petroleum refineries), explosion-proof wiring systems and rectifiers must be used.

11.6.7 Piles and Well Casings

Corrosion on steel piles driven into earth tends, in most cases, to be concentrated near the upper end. This is particularly true where the piles are driven through fill or otherwise previously disturbed earth. Lower areas are generally much less susceptible to corrosion unless it can be shown that there are unusual subsurface conditions or a stray-current situation that can justify CP.

If conditions are such that the upper ends require CP and/or if many piles are involved, a distributed galvanic or impressed current anode system may be required. Such a system will concentrate the applied current at the upper pile ends and, if designed properly, will avoid shielding effects.

Piles used in marine structures are most susceptible to corrosion by the effects of the salt water to which they are exposed. Under normal conditions, areas where worst corrosion on such piles can be expected are as follows:

1. just below the mudline (the point where the pile enters earth below the salt water),
2. just below mean low-water level (in the continuously immersed portion of the pile), and
3. in the splash zone (where wind and wave action wets the pile surface in the area above the mean high-water level).

Usual practice is to concentrate CP current on those areas of the pilings continuously immersed in salt water. This will take care of conditions 1 and 2. Condition 3 is not helped much by CP because there is an insufficient amount of seawater electrolyte to carry protective current to the affected areas. Suitable protective-coating systems are the best approach to corrosion control in splash-zone areas.

11.7 OTHER USES OF CP

11.7.1 Limiting Damage by Cavitation

Numerous experiments have proven, and widespread practical application has confirmed, that CP can help reduce the damage caused by cavitation. CP of surfaces exposed to the cavitation effect caused by the creation and collapse of microscopic vacancies in rapidly moving volumes of water (e.g., on the surface of ship propellers) has been beneficial over a wide range of materials, velocities, and corrosive waters.

It is a common practice, for example, to affix anodes to the hulls of ships in the vicinity of propellers and to apply CP to the surfaces of the propellers themselves. It is believed that under cavitation conditions CP helps by limiting the thickness of corrosion films on the affected surfaces. These films are dislodged by the impact of water from collapsing voids, exposing fresh metal to corrode anew. When corrosion is reduced, less film is exposed to be eroded by the cavitation effect.

11.7.2 Crevice Corrosion-Damage Reduction

Some materials susceptible to crevice corrosion, including Monel 400 (N04400), certain stainless steels, mild steel, and others, display marked benefits from CP's effectiveness in limiting the extent and severity of this form of attack. This could be important in such installations as desalting plants, where large volumes of water are handled.

CP also has proven beneficial in preventing or reducing the incidence of SCC in some materials. Some tests show that CP applied after SCC has started will bring the cracking to a halt.

11.7.3 Domestic Hot-Water Heaters

Magnesium anodes can be used to protect the inside surfaces of glass-lined domestic water heaters. The anodes provide current to help polarize the defects in the glass coatings of the heaters and greatly extend their life. Impressed currents can also be applied to these and to commercial hot-water storage tanks to cathodically protect them.

11.7.4 Other Applications

Among other CP installations are systems on the steel reinforcing wires of prestressed concrete pipe, reinforcing steel in reinforced-concrete structures (described in Chapter 6), and such things as brewery bottle-washing machines and the exposed metal underwater parts of wooden-hulled boats. In each case, careful study of the corrosive environment, the areas and types of metals to be protected, and possible side effects needs to be properly undertaken.

Care needs to be taken, for instance, in applying CP to stressed steel under some conditions because it may cause hydrogen embrittlement (Chapter 9) and catastrophic failure. In the case of wooden-hulled boats, for example, it is necessary to avoid applying excessive current because the hydrogen generated by the excess current (as explained earlier) will create a highly alkaline condition near the metal parts that can potentially destroy the wood.

References

1. J.A. Beavers, "Cathodic Protection—How It Works," in Peabody's Control of Pipeline Corrosion, eds. A.W. Peabody, R.L. Bianchetti (Houston, TX: NACE 2001), pp. 21–47.
2. R.L. Bianchetti, "Impressed Current Cathodic Protection," in Peabody's Control of Pipeline Corrosion, eds. A.W. Peabody, R.L. Bianchetti (Houston, TX: NACE, 2001), pp. 157–176.
3. J.A. Beavers, "Cathodic Protection with Other Power Sources," in Peabody's Control of Pipeline Corrosion, eds. A.W. Peabody, R.L. Bianchetti (Houston, TX: NACE, 2001), pp. 201–210.
4. R.L. Bianchetti, "Survey Methods and Evaluation Techniques," in Peabody's Control of Pipeline Corrosion, eds. A.W. Peabody, R.L. Bianchetti (Houston, TX: NACE, 2001), pp. 65–100.
5. ASTM G 57-95a, "Standard Test Method for Field Measurement of Soil Resistivity Using the Wenner Four-Electrode Method" (Philadelphia, PA: ASTM, 2001).
6. A.W. Peabody, R.L. Bianchetti, eds., Peabody's Control of Pipeline Corrosion, 2nd ed. (Houston, TX: NACE, 2001).
7. M. Lewis, "Instrumentation," in Peabody's Control of Pipeline Corrosion, eds. A.W. Peabody, R.L. Bianchetti (Houston, TX: NACE, 2001), pp. 101–130.

12
Chapter

Corrosion Inhibitors

12.1 INTRODUCTION

An inhibitor is a substance that retards or slows down a chemical reaction (in the present context, a corrosion reaction). Corrosion inhibitors are commonly added in small amounts to acids, cooling waters, steam, and many other environments, either continuously or intermittently, to prevent serious corrosion.

The use of chemical inhibitors to decrease the rate of corrosion processes is quite varied. In the oil extraction and processing industries, corrosion inhibitors have always been considered to be the first line of defense against corrosion. A great number of scientific studies have been devoted to the subject of corrosion inhibitors. However, most of what is known has grown from trial-and-error experiments, both in the laboratories and in the field. Rules, equations, and theories to guide inhibitor development and use are very limited.

Some inhibitors retard corrosion by adsorption to form a thin, invisible film only a few molecules thick; others form visible bulky precipitates that coat the metal and protect it from attack. Another common mechanism consists of causing the metal to corrode in such a way that a combination of adsorption and corrosion product forms a passive layer.

Another type of inhibitor either causes conditions in the environment to be more favorable for the formation of protective precipitates or removes an aggressive constituent from the environment. Corrosion inhibitors have grown to be one of the most universal methods of combating corrosion.

Prevention of corrosion by inhibition may be desirable for several reasons:

1. to extend the life of equipment,
2. to prevent shutdowns,
3. to prevent accidents resulting from brittle (or catastrophic) failures,
4. to avoid product contamination,

5. to prevent loss of heat transfer, and
6. to preserve an attractive appearance.

Potential savings for each of these goals must be evaluated to determine if a program of corrosion inhibition will be economical. Because costs are sometimes difficult to estimate, the best method is to obtain data on maintenance, replacements, and so forth from the past history of the system that is to be protected or a similar system.

There are several costs associated with the use of inhibitors. In fact, the cost of one or more of the following must be factored into any economic evaluation of corrosion inhibition:

1. installation of injection equipment,
2. maintenance of injection equipment,
3. purchase of inhibitor chemical(s),
4. monitoring inhibitor concentration(s),
5. system changes to accommodate the inhibitor,
6. operational changes to accommodate the inhibitor,
7. system cleaning,
8. waste disposal, and
9. personnel safety equipment.

In cases where major shutdowns can be avoided through the use of inhibitors, the economic advantages of inhibition undoubtedly will be clear. Other cases will require detailed economic evaluations.

12.2 BASIC TYPES OF INHIBITORS AND HOW THEY WORK

By definition, a corrosion inhibitor is a chemical substance that, when added in a small concentration to an environment, effectively decreases the corrosion rate of an exposed metal. The efficiency of an inhibitor can be expressed by a measure of this improvement:

$$\text{Inhibitor Efficiency (\%)} = 100 \times \frac{(CR_{\text{uninhibited}} - CR_{\text{inhibited}})}{CR_{\text{uninhibited}}} \qquad (12.1)$$

where $CR_{\text{uninhibited}}$ = corrosion rate of the uninhibited system and $CR_{\text{inhibited}}$ = corrosion rate of the inhibited system.

Example: Mild steel corroded in a cooling water at a rate of 1650 μm/y (65 mpy). When 10 parts per million (ppm) of an inhibitor was added, the corrosion rate dropped to 380 μm/y (15 mpy). What is the inhibitor efficiency?

Substituting $CR_{\text{uninhibited}} = 1650$ and $CR_{\text{inhibited}} = 380$ in Equation (12.1), we get

$$\text{Inhibitor Efficiency (\%)} = 100 \times \frac{1650 - 380}{1650} = 77\%$$

In general, the efficiency of an inhibitor increases with an increase in inhibitor concentration; for example, a typically good inhibitor would give 95% inhibition at a concentration of 0.008% and 90% at a concentration of 0.004%. A synergism, or cooperation, is often present between different inhibitors and the environment being controlled, and mixtures are the usual choice in commercial formulations. The scientific and technical corrosion literature has descriptions and lists of numerous chemical compounds that exhibit inhibitive properties. Of these, only very few are actually used in practice. This is partly due to considerations of cost, toxicity, availability, and environmental friendliness, which are normally of considerable importance.

Nearly 40 years ago, Hugel tested a variety of inhibitors for steel in 6 M of HCl at 60°C and found that alkenyl and aromatic aldehydes were very effective.[1] Cinnamaldehyde was one of the best, providing almost 99% protection. Numerous patents have been issued since then on the use of aldhydes, and cinnamaldehyde in particular, as steel-corrosion inhibitors in acid media.

The following example illustrates how the corrosion efficiency of an inhibitor can be evaluated with a relatively simple corrosion test. Trans-cinnamaldehyde (TCA) corrosion-inhibiting efficiency was evaluated with an electrochemical technique called linear polarization resistance (LPR), a testing method described in Chapter 14. TCA can be used to reduce the corrosion of steel during pickling or oil-field acidizing treatments. The LPR polarization resistance (R_p) is typically calculated from the slope of a polarization curve where

$$R_p = \frac{(\Delta E)}{(\Delta i)_{\Delta E \to 0}} \tag{12.2}$$

where R_p is the polarization resistance.

$$i_{corr} = \frac{B}{R_p} \tag{12.3}$$

where i_{corr} is the corrosion current and B an empirical polarization resistance constant that can be related to the anodic (b_a) and cathodic (b_c) Tafel slopes according to the following equation:

$$B = \frac{b_a \times b_c}{2.3(b_a + b_c)} \tag{12.4}$$

The polarization curves presented in Figure 12.1 were obtained with carbon steel exposed to a solution containing respectively a) no inhibitor, b) 250, c) 1000, and d) 2000 of TCA in a 6-M HCl solution. One can first estimate the R_p from the slopes of the polarization curves. Assuming that b_a and b_c have the same value (i.e., 0.1 mV per decade of current), each R_p can then be converted in a corrosion current with the help of Equations (12.3) and (12.4). The inhibitor efficiency of each solution can subsequently be calculated using Eq. (12.1) presented earlier. These results are presented in

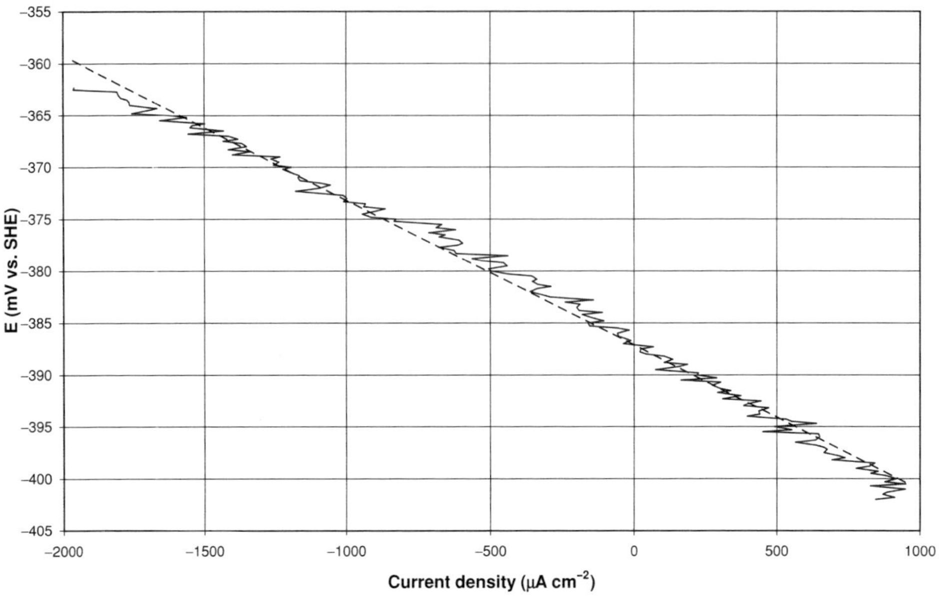

(a)

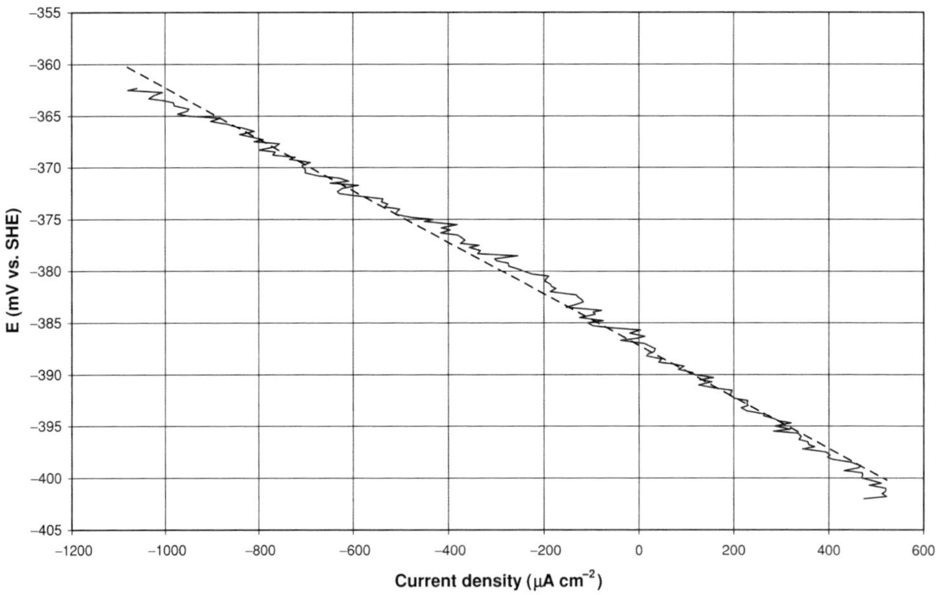

(b)

Figure 12.1 Corrosion of AISI 1018 carbon steel in 6 M of HCl containing a) no inhibitor, b) 250 ppm of trans-cinnamaldehyde (TCA), c) 1000 ppm of TCA, and d) 5000 ppm of TCA. The dotted line in these figures has been added to estimate the slope of the curves that can be related to the polarization resistance (R_p).

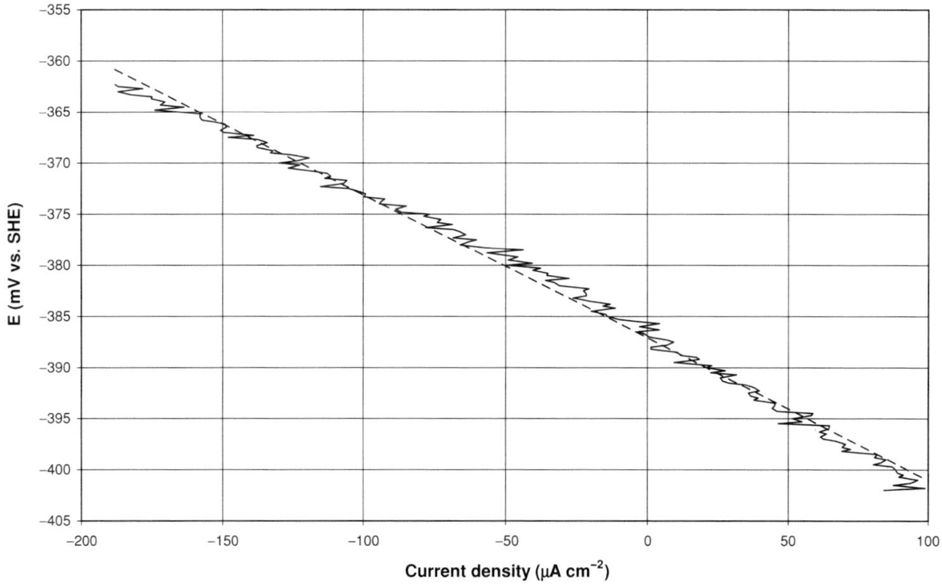

(c)

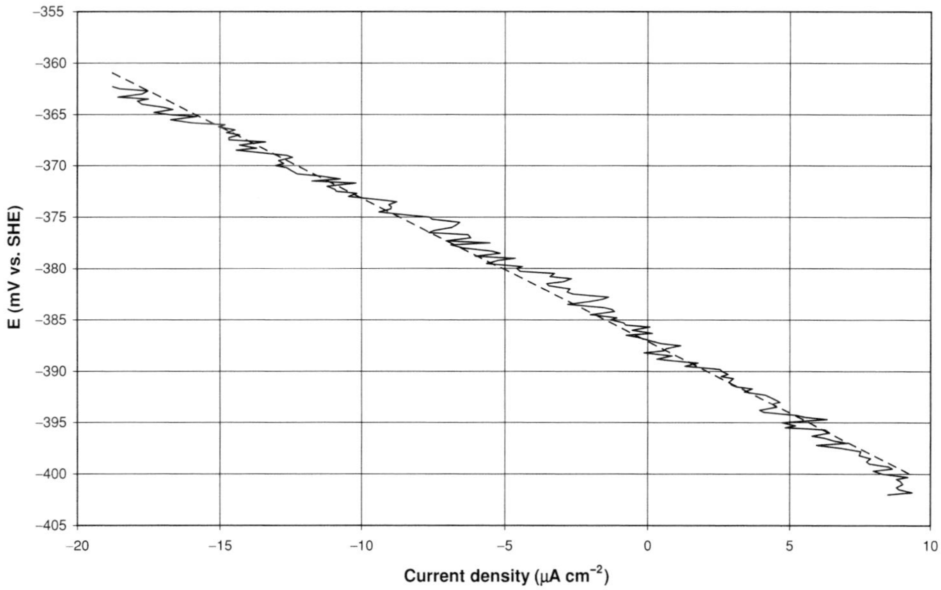

(d)

Figure 12.1 (Continued)

Table 12.1 Inhibitor Efficiency of Trans-Cinnamaldehyde (TCA) to the Corrosion of Carbon Steel Exposed to a 6-M HCl Solution

TCA (ppm)	Rp (Ω cm^2)	Corrosion Current (mA cm^{-2})	Corrosion Rate (mm y^{-1})	Efficiency (%)
0	14	1.55	18	0
250	25	0.87	10.1	44
1000	140	0.155	1.80	90
2000	1400	0.0155	0.18	99

Table 12.1 along with corrosion rates converted in millimeters per year with the help of Table 2.3 in Chapter 2.

12.2.1 Polarization Diagrams

The electrochemical nature of corrosion and the concept of mixed potentials were explained in Chapter 2. In the present chapter, the relationship of corrosion inhibitors to anodic and cathodic polarization will be explained.

All three components of a corrosion cell (anode, cathode, and electrolyte) are affected by a corrosion inhibitor to retard corrosion. The inhibitor may cause

1. increased polarization of the anode (anodic inhibition) or
2. increased polarization of the cathode (cathodic inhibition), or it may
3. increase the electrical resistance of the total corrosion cell circuit by forming a thick deposit on the surface of the metal.

Of course, the bulk film-forming inhibitors also restrict diffusion of depolarizers (such as dissolved oxygen) to the surface of the metal; hence, they may play a dual role. The effects of a corrosion inhibitor on a corrosion cell are conveniently determined by polarizing the corroding metal in a suitable electrolyte with varying amounts of current from an external source. A simple laboratory apparatus for polarization measurements has been shown in Figure 2.23 of Chapter 2.

In an experiment to compare the effects of several inhibitors on the polarization of steel in a corrosive environment, such as dilute acid, the electrochemical cell would be assembled using steel as the working electrode, dilute acid as the electrolyte, and an inert material such as carbon or platinum as an auxiliary electrode. A plot of the current versus potential in the absence and presence of an inhibitor shows the effects of the inhibitor on the polarization characteristics of the steel. Typical polarization characteristics of steel are illustrated without an inhibitor, in the presence of an anodic inhibitor, and with a cathodic inhibitor in respectively Figure 12.2 a-c.

When no current is applied and the working electrode has achieved a steady state, the measured potential is then the corrosion potential, E_{corr}, of the electrode material. The corrosion potential is the mixed potential to which the anode and cathode of the system under study are polarized by the corrosion reaction (see Chapter 2 for more details). It is also an indication of the effects of corrosion inhibitors.

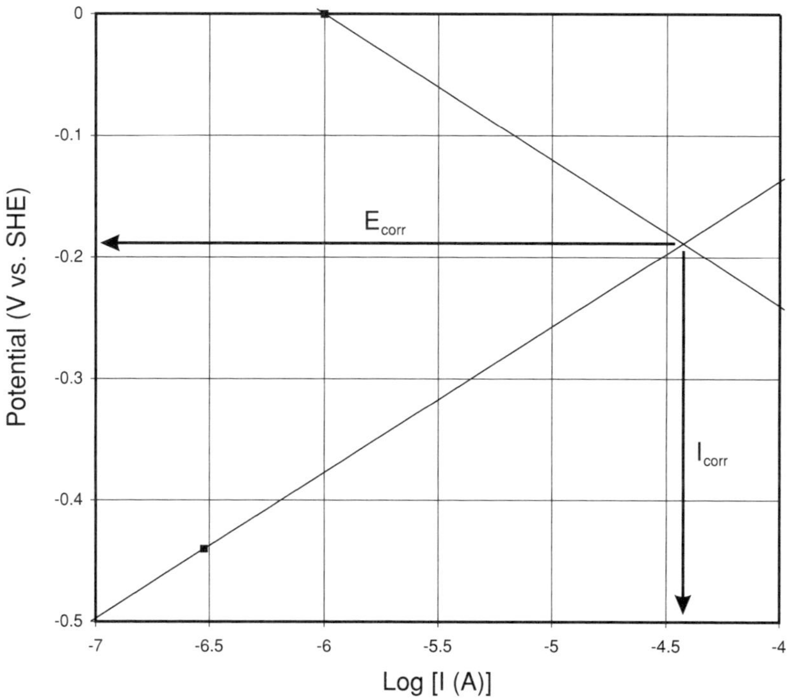

(a)

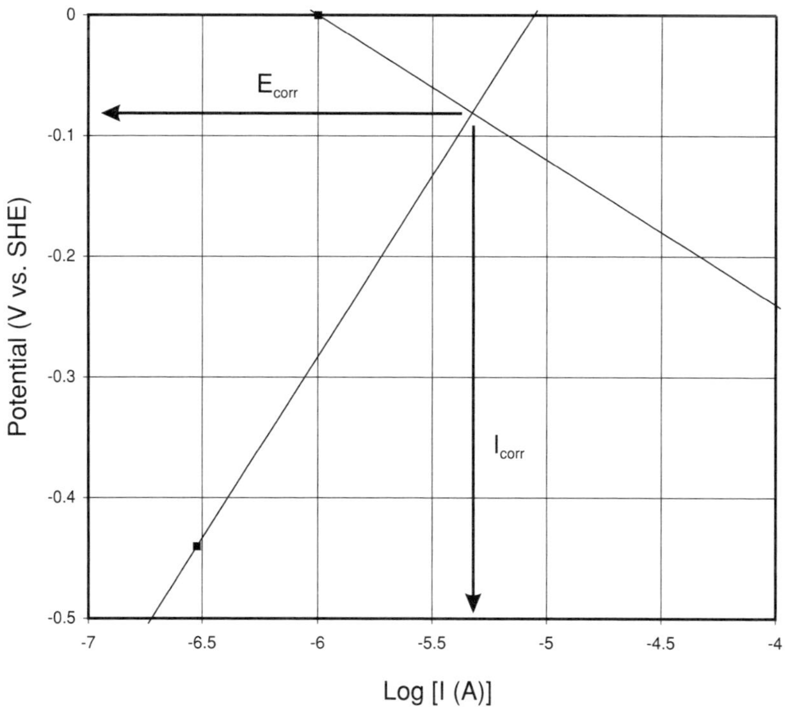

(b)

Figure 12.2 The effects of the inhibitor on the polarization characteristics of the steel: a) no inhibitor, b) anodic inhibitor, and c) cathodic inhibitor. (Continued)

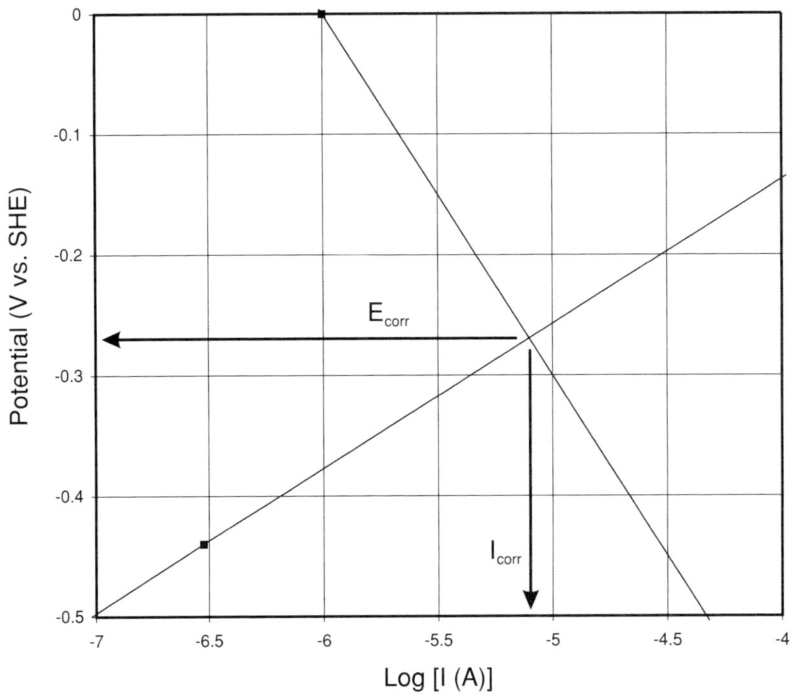

(c)

Figure 12.2 (Continued)

12.2.2 Types of Inhibitors

Six classes of inhibitors are discussed:

1. passivating (anodic),
2. cathodic,
3. ohmic,
4. organic,
5. precipitation inducing, and
6. vapor-phase inhibitors.

12.2.2.1 *Anodic Passivating Inhibitors*

Anodic inhibition is illustrated in Figure 12.2b, which shows an increase in the polarization (a large potential change results from a small current flow) of the anode in the presence of an anodic inhibitor. Addition of the inhibitor causes the corrosion potential to shift to more positive values.

Anodic inhibitors that cause a large positive shift in the corrosion potential are called passivating inhibitors. There are two types of passivating inhibitors: oxidizing anions such as chromate, nitrite, and nitrate, which can passivate steel in the absence of oxygen, and nonoxidizing ions such as phosphate, tungstate, and molybdate, which require the presence of oxygen to passivate steel. With careful control, however, passivating inhibitors are used because they are very effective in sufficient quantities.

The mechanism by which chromate passivates steel has been studied extensively and remains an excellent corrosion-inhibition model even if the practicality of using chromate inhibitors is severely limited by environmental constraints. These inhibitors can provide corrosion protection by a combination of adsorption and oxide formation on the metallic surface. Adsorption helps to polarize the anode to sufficient potentials to form very thin hydrated ferric oxides, which protect the steel. Because the oxide film is invisible on steel, articles protected by chromate remain bright in otherwise aggressive environments. The oxide film is a mixture of ferric and chromic oxides and is kept in good repair by adsorption and oxidation with very little loss of metal as long as sufficient chromate remains in solution.

Nonoxidizing passivators such as sodium benzoate, polyphosphate, and sodium cinnamate require the presence of oxygen to cause passivation and do not inhibit corrosion in the absence of oxygen. They apparently function by promoting the adsorption of oxygen on the anodes, thereby causing polarization into the passive region. Nonoxidizing passivators are also dangerous when used in insufficient amounts because the oxygen that is required for passivation is a good cathodic depolarizer.

12.2.2.2 Cathodic Inhibitors

Cathodic inhibitors either slow the cathodic reaction itself or they selectively precipitate on cathodic areas to increase circuit resistance and restrict diffusion of reducible species to the cathodes. The effects of cathodic inhibitors on cathodic polarization are shown in Figure 12.2c. In this case, where the anodic polarization is unaffected, the corrosion potential is shifted to more negative values.

The cathodic reaction is often the reduction of hydrogen ions to hydrogen gas. Some cathodic inhibitors make the discharge of hydrogen gas more difficult, and they are said to increase the hydrogen overvoltage. Compounds of arsenic and antimony are examples of this type of inhibitor and are often used in acids or in systems where oxygen is excluded. Another possible cathodic reaction is the reduction of oxygen. The inhibitors for this cathodic reaction are different from those mentioned for the more acidic systems.

Other cathodic inhibitors utilize the increase in alkalinity at cathodic sites to precipitate insoluble compounds on the metal surface. The cathodic reaction, hydrogen ion, and/or oxygen reduction causes the environment immediately adjacent to the cathodes to become alkaline; therefore, ions such as calcium, zinc, or magnesium may be precipitated as oxides to form a protective layer on the metal. Many natural waters are self-inhibiting because they deposit a scale on metals by precipitation of naturally occurring ions.

Inhibition by polarization of the cathodic reaction can be achieved in several ways, and several examples have already been given. The three main categories of inhibitors that affect cathodic reaction are cathodic poisons, cathodic precipitates, and oxygen scavengers.

Cathodic Poisons Cathodic poisons are substances that interfere with the cathodic-reduction reactions, for example, hydrogen-atom formation and hydrogen-gas evolution. The rate of the cathodic reaction is slowed, and because anodic and cathodic reactions must proceed at the same rate, the whole corrosion process is slowed.

Although some cathodic poisons such as sulfides and selenides are adsorbed on the metal surface, compounds of arsenic, bismuth, and antimony are reduced at the cathode to deposit a layer of the respective metals. Sulfides and selenides generally are not useful inhibitors because they are not very soluble in acidic solutions, precipitate many metal ions, and are toxic. Arsenates are used to inhibit corrosion in strong acids, but in recent years the trend has been to rely more on organic inhibitors (discussed later in this chapter) because of the toxicity of arsenic.

A serious drawback of the use of cathodic poisons is that they sometimes cause hydrogen blistering of steel and increase its susceptibility to hydrogen embrittlement. Because the recombination of hydrogen atoms is inhibited, surface concentration of atomic hydrogen is increased and a greater fraction of the hydrogen produced by the corrosion reaction is absorbed into the steel. Note that hydrogen is adsorbed on the surface, but some of it is absorbed into the steel.

Hydrogen atoms that penetrate steel may pass through and diffuse out the other side. Blisters are formed when hydrogen atoms combine to form hydrogen molecules (H_2) inside the steel (see Chapter 9 for more details). Molecular hydrogen does not diffuse through steel; therefore, it collects at defects or voids to create pressures that may reach tremendously high pressures (thousands of MPa or even more).

Cathodic Precipitates The most widely used cathodic precipitation–type inhibitors are the carbonates of calcium and magnesium because they occur in natural waters and their use as an inhibitor usually requires only an adjustment of pH. Zinc sulfate ($ZnSO_4$) precipitates as zinc hydroxide [$Zn(OH)_2$] on cathodic areas and is considered an inhibitor of this type. Phosphates and silicates are not distinctively cathodic or anodic inhibitors but appear to be a combination of both types, so they are considered later in the section on precipitation inhibitors.

Many natural waters and municipal water supplies contain calcium carbonate (limestone, $CaCO_3$) in solution. Limestone is dissolved in water to form soluble calcium bicarbonate [$Ca(HCO_3)_2$]. Limestone can be caused to precipitate again, forming a milky-white suspension by making the calcium bicarbonate solution more alkaline or by adding more calcium. Usually, lime is added to accomplish both objectives.

The objective of corrosion-inhibiting water treatment is to increase the alkalinity of the water to a pH at which precipitation of $CaCO_3$ is just about to occur. If the appropriate pH is exceeded, $CaCO_3$ will precipitate to form a slimy, porous deposit that does not provide corrosion protection and may increase corrosion by creating concentration cells involving oxygen. At the correct pH, the deposit will be fairly hard and smooth and similar to an eggshell. Once a protective deposit is formed, the pH of the water must be maintained at the equilibrium level, because if it is allowed to become acidic, it will redissolve the protective deposit. A convenient way to express the condition of a water with respect to its tendency to deposit $CaCO_3$ is the scaling indices discussed in Chapter 4.

Oxygen Scavengers Corrosion of steel in waters above about pH 6.0 is due to the presence of dissolved oxygen which depolarizes the cathodic reaction and increases corrosion. Neutral water of low salt content in equilibrium with air at 21°C will contain about 8 ppm of dissolved oxygen. The concentration of oxygen decreases with increasing salt concentration and increasing temperature. Only 0.1 ppm of oxygen is required to seriously increase corrosion rates in a dynamic system. In static systems, a higher concentration of oxygen is required to seriously increase the corrosion rate because the corrosion reaction soon depletes the oxygen supply in the immediate vicinity of the metal.

Oxygen scavengers help inhibit corrosion by preventing the cathodic depolarization caused by oxygen. Oxygen scavengers are added to water, either alone or with a corrosion inhibitor, to retard corrosion. Organic corrosion inhibitors alone in aerated brine water will slow general corrosion but will not always prevent pitting attack. The most common oxygen scavengers used in water at ambient temperatures are sodium sulfite (Na_2SO_3), sodium metabisulfite ($Na_2S_2O_5$), sodium bisulfite ($NaHSO_3$), and ammonium bisulfite (NH_4HSO_3). All these sulfites react with oxygen to form sulfates as shown in Equation (12.5) for sodium sulfite. At elevated temperatures, hydrazine (N_2H_4) is used to react oxygen to form water and nitrogen (Equation [12.6]).

$$2Na_2SO_3 + O_2 \rightarrow 2Na_2SO_4 \tag{12.5}$$

$$N_2H_4 + O_2 \rightarrow 2H_2O + N^{\uparrow}_{2\,(gas)} \tag{12.6}$$

The reaction rate of sulfites with oxygen at low temperature is slow, so a catalyst is usually added. Cobalt, manganese, and copper salts are the best catalysts. Cobalt gives the greatest increase in reaction rate. Copper should not be added to water that contacts steel or aluminum because it lowers the hydrogen overvoltage and increases the corrosion rate. Consequently, cobalt is preferred and manganese is a close second choice.

Hydrazine reacts very slowly with oxygen in water at low temperatures in the absence of a catalyst and thus is not often used at low temperatures. The hazards of hydrazine are significant, especially in the hands of untrained personnel. However, in high-pressure boilers, hydrazine is the preferred oxygen scavenger.

12.2.2.3 Ohmic Inhibitors

Inhibitors that increase the ohmic resistance of the electrolyte circuit already have been considered to some extent in discussions of anodic and cathodic film-forming inhibitors. Because it usually is impractical to increase the resistance of the bulk electrolyte, increased resistance is practically achieved by the formation of a film of microscopic dimension on the metal surface. If the film is deposited selectively on anodic areas, the corrosion potential shifts to more positive values; if it is deposited on cathodic areas, the shift is to more negative values; and if the film covers both anodic and cathodic areas, there may be only a slight shift in either direction.

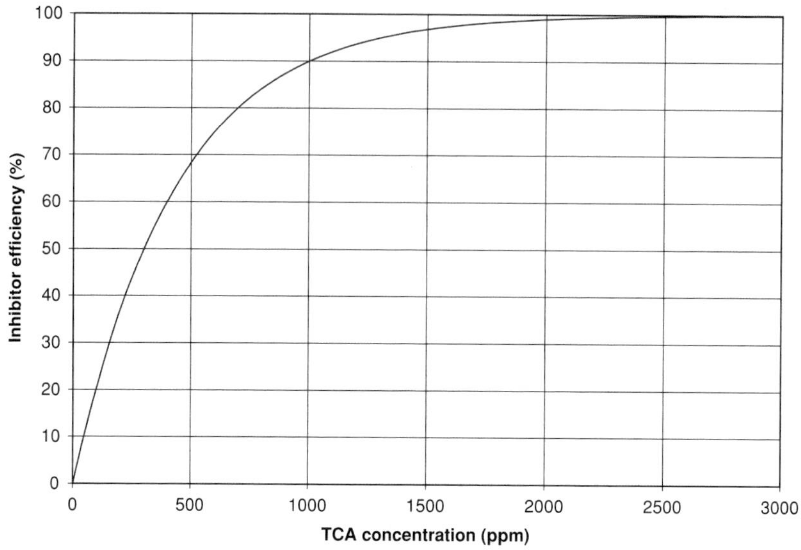

Figure 12.3 Corrosion-inhibitor efficiency as a function of inhibitor concentration for the results in Figure 12.1 and Table 12.1.

12.2.2.4 Organic Inhibitors

Organic compounds constitute a broad class of corrosion inhibitors that cannot be strictly classified as anodic, cathodic, or ohmic inhibitors. Anodic or cathodic effects alone are sometimes observed in the presence of organic inhibitors, but, as a general rule, organic inhibitors affect the entire surface of a corroding metal when present in sufficient concentration. Both anodic and cathodic areas probably are inhibited but to varying degrees, depending on the potential of the metal, chemical structure of the inhibitor molecule, and size of the molecule.

The increase in corrosion inhibition with the concentration of inhibitor shown in Figure 12.3, drawn from the results discussed earlier and presented in Table 12.1, suggests that this type of corrosion inhibition is the result of adsorption of the inhibitor on the metal surface. The film formed by adsorption of soluble organic inhibitors, trans-cinnamaldehyde in the present example, is only a few molecules thick and is invisible.

Organic inhibitors will be adsorbed according to the ionic charge of the inhibitor and the charge on the metal surface. Cationic inhibitors (positively charged, +), such as amines, or anionic inhibitors (negatively charged, −), such as sulfonates, will be adsorbed preferentially, depending on whether the metal is charged negatively or positively (opposite-sign charges attract). The in-between potential at which neither cationic nor anionic molecules are preferred is known as the zero point of charge or ZPC. Thus, a combination of CP and an inhibitor that is adsorbed more strongly at negative potentials gives greater inhibition than either CP or an inhibitor when used alone.

Synergism with Halogen Ions The efficiency of organic amines as corrosion inhibitors is improved when certain halogen ions are present. Halogen ions alone inhibit

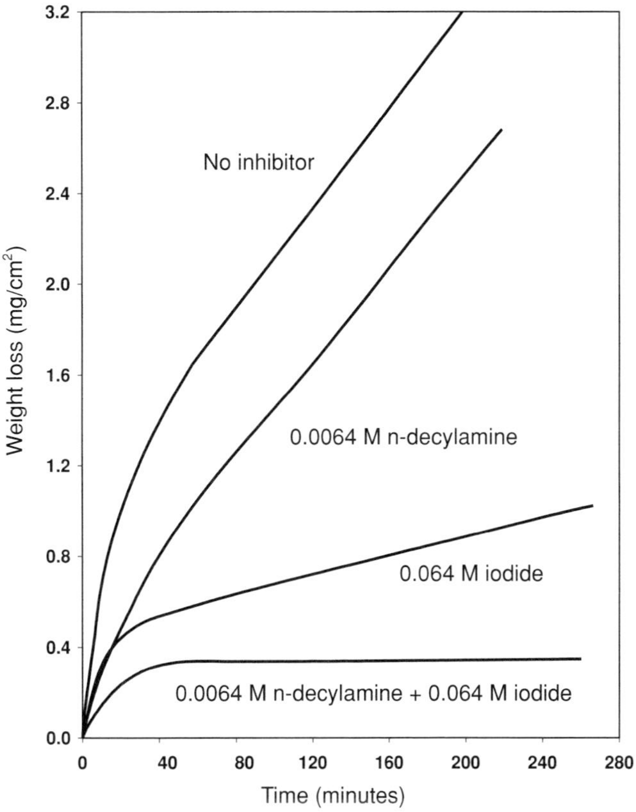

Figure 12.4 Corrosion of mild steel in 3 M of phosphoric acid (30%) shows synergistic effect of n-decylamine and iodide ion.

corrosion to some extent in acid solutions. The iodide (I^-) ion is the most effective, followed by bromide (Br^-) and chloride (Cl^-). Fluoride (F^-) does not have significant inhibitive properties. Chloride ions, for example, lower the rate of attack on steel by sulfuric acid. A combination of amine and iodide may be more inhibitive than either additive alone, that is, the two additives are synergistic.

Figure 12.4 shows a comparison of corrosion rates of mild steel in 3.0 M of phosphoric acid (H_3PO_4) in the presence of n-decylamine, iodide ion, and a combination of n-decylamine and iodide ion.[2] One explanation for synergism is that steel adsorbs iodide ions whose charge shifts the surface potential in a negative direction, thereby increasing adsorption of the cationic amine.

Effects of Molecular Structure How the size of organic molecules influences their effectiveness as corrosion inhibitors has been thoroughly investigated. However, the results are not consistent enough to permit formulation of a general rule regarding the effect of increasing molecular weight.

Primary amines such as n-decylamine become more efficient inhibitors as the chain length is increased, but, in contrast, primary aliphatic mercaptans such as n-butyl

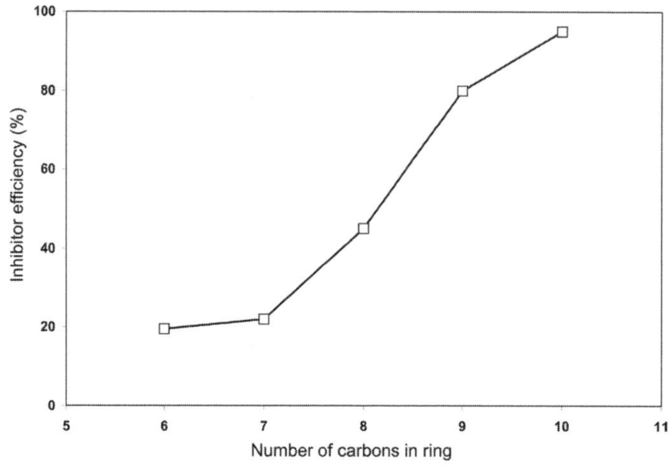

Figure 12.5 Inhibitor efficiencies of saturated cyclic imines for corrosion of steel in 6 M of hydrochloric acid (22%).

mercaptan and some aldehydes decrease in efficiency as the chain length is increased. These results probably are due to the interaction of various factors that influence the strength of the adsorption bond, compactness of the adsorbed layer, and tendency of the adsorbed molecules to cross link or otherwise interact with neighboring molecules. There is little doubt that the bonding of amines to a metal surface is through the nitrogen atom. For example, in the series of saturated cyclic imines, inhibitor efficiency increases, as shown in Figure 12.5, as the number of carbon atoms is increased in the ring up to at least 10.[3]

Adsorption The observations given previously indicate that for soluble organic inhibitors, the strength of the adsorption bond is the dominant factor. Adsorption of an inhibitor from solution establishes the following equilibrium:

$$I_{\text{solution}} \rightleftarrows I_{\text{surface}} \tag{12.7}$$

where I is the concentration of a soluble organic inhibitor.

It is characteristic of an equilibrium that if the concentration of one species is changed, then the concentration of the species with which it is in balance will change in the same direction to preserve equilibrium. From Equation (12.7), it is seen that the amount of inhibitor on the surface is increased by increasing I_{solution}; that is, the concentration of inhibitor used in the environment to be inhibited. Inhibitor efficiency increases with concentration until the surface is saturated; that is, it has adsorbed inhibitor molecules on all available sites. Therefore, the stronger the adsorption bond, the lower the concentration of inhibitor in solution required to achieve a given coverage of the surface.

Soluble organic inhibitors form a protective layer only a few molecules thick, but if an insoluble organic inhibitor is added by dispersion as fine droplets, the film may continue to build to a thickness of 100 μm and more. Such films show good persistence; that is, they continue to inhibit corrosion for a time when an inhibitor is no longer injected into the environment. Persistency is an important property when inhibitors can be injected into a system only in slugs.

12.2.2.5 *Precipitation Inhibitors*

Precipitate-inducing inhibitors are film-forming compounds that have a general action over the metal surface and that, therefore, indirectly interfere with both anodes and cathodes. The most common inhibitors of this class are the silicates and phosphates.

In waters with a pH near seven a low concentration of chlorides, silicates, and phosphates cause passivation of steel when oxygen is present; hence, they behave as anodic inhibitors. Another anodic characteristic is that corrosion is localized in the form of pitting when insufficient amounts of phosphate or silicate are added to saline water. However, both silicates and phosphates form deposits on steel that increase cathodic polarization. Thus, their action appears to be mixed (i.e., a combination of both anodic and cathodic effects).

Silicate is used most often in low-salinity waters that contain oxygen. It has the rare property of inhibiting corrosion of steel that is already scaled with rust. Although the concentration of silicate required for protection depends on the salinity of the water, for most city water supplies, 5 to 10 ppm are required initially, followed by a gradual reduction to 2 to 3 ppm after a protective deposit is established. High concentrations of calcium and magnesium interfere with inhibition by silicates, but this problem is often overcome by adding 2 to 3 ppm of polyphosphates in addition to the silicates.

Sodium silicate is used in many private water softeners to prevent occurrence of red water or rust water, which is caused by suspended ferric hydroxide. The silicates remove iron by precipitation, so they should not be used where formation of scales cannot be tolerated.

In aerated hot-water systems, sodium silicate protects steel, copper, and brass. However, protection is not always reliable and depends on pH and composition of the water. The best procedure is to adjust the saturation index as already described to facilitate formation of a protective silicate-containing film.

Phosphates, like silicates, require oxygen for effective inhibition. A concentration of about 10 ppm sodium hexametaphosphate, a typical polyphosphate, provides corrosion inhibition in aerated water if the water is in motion. In stagnant areas, corrosion might be increased because of the establishment of oxygen concentration cells.

In addition to motion, the presence of calcium ions is essential to inhibition. Phosphates inhibit the deposition of $CaCO_3$; therefore, the concentration of phosphate and calcium must be in proper balance to obtain effective inhibition. A rule of thumb is that the phosphate level must not exceed twice the concentration of calcium carbonate. If a water contains 10 ppm of calcium carbonate, for example, then up to 20 ppm of phosphate can be used for inhibition. If the water is exceedingly soft, calcium can be increased by the addition of lime.

Industrial cooling towers can be inhibited with about 50 ppm of sodium hexametaphosphate $(NaPO_3)_6$. Addition of a soluble zinc salt often improves inhibition by polyphosphates. Pitting and excessive scale formation are prevented by maintaining the pH in a range of six to seven.

Silicates and phosphates do not provide the degree of protection that was available by using chromates and nitrites, which are now considered environmentally toxic. The main drawbacks of silicates and phosphates are their dependence on water composition and the careful control required to achieve maximum inhibition.

12.2.2.6 Vapor-Phase Inhibitors

Vapor-phase inhibitors (VPIs), also called volatile corrosion inhibitors (VCIs), are compounds that are transported in a closed system to the site of corrosion by volatilization from a source. In boilers, volatile basic compounds such as morpholine or octadecylamine are transported with steam to prevent corrosion in condenser tubes by neutralizing acidic carbon dioxide. Compounds of this type inhibit corrosion by making the environment alkaline. In closed vapor spaces, such as shipping containers, volatile solids such as the nitrite, carbonate, and benzoate salts of dicyclohexylamine, cyclohexylamine, and hexamethylene-imine are used. The mechanism of inhibition by these compounds is not entirely clear, but it appears certain that the organic portion of the molecules merely provides volatility.

On contact with a metal surface, the inhibitor vapor condenses and is hydrolyzed by any moisture present to liberate nitrite, benzoate, or bicarbonate ions. Because ample oxygen is present, nitrite, and benzoate ions are capable of passivating steel as they do in aqueous solution. The mechanism for carbonate may not be the same, and here the organic amine portion of the VPI may serve to aid inhibition by adsorption and by providing alkalinity.

It is desirable for a VPI to provide inhibition rapidly and to have a lasting effect. Therefore, the compound should have a high volatility to saturate all of the accessible vapor space as quickly as possible, but at the same time it should not be too volatile, because it would be lost rapidly through any leaks in the package or container in which it is used. The optimum vapor pressure of VPI then would be just sufficient to maintain an inhibiting concentration on all exposed metal surfaces.

Vapor pressures and other properties of some VPIs are given in Table 12.2. Note that the vapor pressure of cyclohexylamine carbonate is 2000 times higher than that of dicyclohexylamine nitrite, thus making it a better choice for containers that are opened occasionally because a more volatile inhibitor will resaturate the vapor space rapidly.

Dicyclohexylamine nitrite is advantageous for once-opened containers that may be stored for extended periods. The amount of VPI required depends on conditions, but 2.2 kg per 100 m^2 of surface has been suggested for dicyclohexylamine nitrite and 2.2 kg per 30 m^2 space for cyclohexylamine carbonate. VPIs attack nonferrous metals to varying degrees, so it is suggested that a potential user test several of the commercially available VPIs for the particular application. Compatibility of the amines and nitrites with the copper alloys should especially be considered.

Table 12.2 Saturated Vapor Pressures of Common VCIs

Substance	Temperature (°C)	Vapor Pressure (mm Hg)	Melting Point (°C)
Morpholine	20	8.0	—
Benzylamine	29	1.0	—
Cyclohexylamine carbonate	25.3	0.397	—
Diisopropylamine nitrite	21	4.84×10^{-3}	139
Morpholine nitrite	21	3×10^{-3}	—
Dicyclohexylamine nitrite	21	1.3×10^{-4}	179
Cyclohexylamine benzoate	21	8×10^{-5}	—
Dicyclohexylamine caprylate	21	5.5×10^{-4}	—
Guanadine chromate	21	1×10^{-5}	—
Hexamethyleneimine benzoate	41	8×10^{-4}	64
Hexamethyleneamine nitrobenzoate	41	1×10^{-6}	136
Dicyclohexylamine benzoate	41	1.2×10^{-6}	210

12.3 ENVIRONMENTAL FACTORS

12.3.1 Aqueous Systems

Aqueous systems are by far the most common corrosive environments to which corrosion inhibitors are applied. Water is a powerful solvent capable of carrying many different ions at the same time, thus requirements for corrosion inhibition may vary greatly, depending on the type and amount of dissolved species present. Because there is no universal inhibitor for water systems, an inhibitor that may be satisfactory for one system may be ineffective or even harmful in another. The main factors that must be considered in the application of corrosion inhibitors to aqueous systems are salt concentration, pH, dissolved-oxygen concentration, and interfering-species concentration.

12.3.1.1 Effects of Various Dissolved Species

Industrial and domestic waters contain dissolved substances, which affect their aggressiveness and corrosion-inhibitor requirements in various ways, depending on the nature of the substances. The most common dissolved substances and their effects on corrosion inhibition are as follows.

Oxygen (O_2) In neutral water, oxygen causes corrosion; therefore, if it is reduced to less than 0.1 ppm by scavenging compounds or by stripping, sufficient control is thereby provided for some systems, for example, in boilers and hot-water supplies. Oxygen can be utilized in passivating steel by adding a passivating inhibitor. Organic inhibitors are seldom effective against oxygen-caused attack unless they contain passivating groups such as benzoate or nitrite.

Chloride (Cl^-) Steel, as well as many other metals, is more difficult to passivate in the presence of chloride ions. A higher concentration of passivating inhibitor is therefore required if chlorides are present. Nonpassivating inhibitors must also be used in higher concentrations because chloride ions are strongly adsorbed by steel.

Sulfate (SO_4^{2-}) The effects of sulfate on passivity are similar to those of chloride but to a lesser degree. Sulfates or chlorides must not be allowed to build up by evaporation in a system because depassivation may occur.

Bicarbonate (HCO_3^-) Bicarbonate in hard waters can be utilized for natural inhibition by forming precipitates. In soft water, corrosion inhibitors must be used if excess carbon dioxide is present because of the acidic condition it produces.

Sulfides (S^{2-}) Sulfides precipitate many metal ions; for example, inhibitors that contain zinc cannot be used. Oxidizing inhibitors are reduced by sulfides, which in turn form free sulfur. They are effective only if an excess greater than the amount required to react with sulfide is used and the colloidal precipitate of free sulfur can be tolerated.

Metal Cations Sodium (Na^+) and potassium (K^+) ions have no particular effects on inhibitors; calcium (Ca^{2+}) and magnesium (Mg^{2+}) may be used to form protective precipitates, but at high concentrations they interfere with inhibitors by precipitating nonprotective deposits and also by precipitating inhibitors such as phosphate (PO_4^{3-}) and silicate (SiO_3^{2-}). Very small concentrations of heavy-metal ions, such as copper and mercury, can cause severe interference with inhibitors.

Acid (H^+) Hydrogen ions increase corrosion rates and increase the difficulty of passivating steel. Passivation is used in sulfuric (H_2SO_4) and phosphoric acids (H_3PO_4), but not in hydrochloric acid (HCl). Nonpassivating organic or cathodic inhibitors (e.g., guanidine or sodium arsenate) are preferred in pickling acids to avoid the disastrous consequences of depassivation.

Alkali (OH^-) In alkaline solutions, corrosion of steel is controlled by the rate of oxygen diffusion through the precipitated corrosion product, usually ferrous hydroxide [$Fe(OH)_2$], and thus corrosion rates are low. Steel is easily passivated in alkaline solutions. Amphoteric metals such as aluminum, zinc, and lead corrode slowly at low alkali concentrations, but above a pH of nine their rates are very high and inhibitors are required.

12.3.1.2 *Waters of Low to Moderate Salt Concentrations*

Waters of low to moderate salt concentrations are encountered in municipal water systems, cooling waters, marine and offshore activities, and oil-field water injection systems. Because metals adsorb ions of dissolved salts in water, an inhibitor has more difficulty reaching the metal surface and displacing adsorbed ions than it has in demineralized water; hence, a higher concentration of inhibitor is required. Furthermore, chloride ions have a depassivating effect; that is, they make it more difficult to control corrosion by passivating inhibitors. It is therefore important to maintain inhibitor concentrations at a safe level in waters containing dissolved salts, particularly if these include chlorides.

Municipal drinking water cannot be treated with most inhibitors because of their toxicity. Fortunately, treatment with lime to raise the pH usually affords sufficient protection to steel or cast-iron water pipes. If the water is high in chlorides or sulfates, then polyphosphates may be used for added inhibition. Silicates may also be used in municipal waters, but they have the disadvantage of forming precipitates with iron and calcium, which scale pipes and heat-transfer surfaces.

Cooling water systems may be of either recirculating or once-through types. In closed recirculating systems, oxygen can be excluded and corrosion often can be controlled by adjusting the pH to an alkaline value. Recirculating systems are more easily controlled by inhibitors because higher concentrations can be applied.

Glycol/water mixtures, such as those used to cool engines and to transfer solar heat, are usually inhibited with a mixture of borax for maintaining an alkaline pH and mercaptobenzothiazole, which inhibits the corrosion of brass and copper. Borax alone is satisfactory for steel in contact with glycol/water mixtures, but borax and glycol attack zinc, galvanizing rapidly, and attack the zinc in brass because of the formation of complex zinc compounds at low temperatures.

Thus, the addition of mercaptobenzothiazole is necessary in mixed-metal cooling systems. A soluble oil also is often added to increase protection and to lubricate moving parts in the cooling system. In some mixed-metal systems, silicates and nitrates are now used. Amine phosphates have also long been used in such systems.

Once-through cooling systems require inexpensive corrosion inhibitors. In open systems, corrosion is more severe and good inhibition is imperative. The situation is similar to that in municipal water supplies, so comparable remedial measures, namely addition of lime or polyphosphates, are used.

Waters that may contain appreciable quantities of organic matter, such as seawater and oil-field injection brines, usually are not inhibited with oxidizing inhibitors because of the high consumption of inhibitor caused by oxidation of the organics. Nonoxidizing inorganic inhibitors such as sodium silicate must also be used in high concentrations because of the high chloride content of brines. Generally, organic inhibitors offer the best means for protection in organic-contaminated brines. Inhibitor concentrations as low as 10 to 20 ppm in fatty amines often effectively control corrosion in oil-field brines.

12.3.1.3 High Salt Concentrations

Extremely high salt concentrations are used in aqueous solutions for heat transfer in refrigeration systems. The temperatures encountered are always low, and because the brines are recirculated, a high concentration of inhibitor can be maintained economically. Sodium chromate is effective in refrigeration brines, provided there is no limitation because of its toxicity. If physiological effects are a factor, then disodium phosphate can be used, although it is not as effective as sodium chromate in controlling corrosion.

12.3.1.4 *Effects of pH*

The pH of aqueous solutions is extremely important in determining the type of corrosion inhibitor that is most effective and most economical. Natural hard waters retain compounds of calcium, including calcium carbonate ($CaCO_3$) and calcium bicarbonate [$Ca(HCO_3)_2$], along with carbon dioxide, in solution. There is an equilibrium that occurs among these species as shown by Equation (12.8):

$$CaCO_3(\text{insoluble}) + CO_2 + H_2O \rightarrow Ca(HCO_3)_2(\text{soluble}) \qquad (12.8)$$

At high temperatures the reverse reaction occurs, and heated surfaces become coated with $CaCO_3$. A protective scale is produced also when $Ca(HCO_3)_2$ becomes alkaline in the region of a cathodic area. The scale thus deposited inhibits corrosion by reducing the cathodic area, restricting diffusion of cathodic depolarizers and increasing ohmic resistance. This scale is often developed on cathodically protected steel surfaces exposed to seawater in what is often called a calcareous deposit. For this reason, some natural hard waters are less corrosive than softened waters. The addition of zinc sulfate ($ZnSO_4$) in alkaline solutions also inhibits corrosion by precipitating insoluble zinc hydroxide [$Zn(OH)_2$] on the cathodic area.

Hydrogen sulfide is a particularly troublesome problem. The dissolved gas attacks steel only slowly when first exposed because of the formation of a protective layer of iron sulfide. However, the iron sulfide film affords only a temporary protection because it becomes permeable to hydrogen sulfide, and the corrosion rate increases with time, producing blistering, high metal loss, and possibly hydrogen embrittlement. Organic corrosion inhibitors prolong the interval preceding higher corrosion rates, but the iron sulfide film eventually prevents access of the inhibitor to the steel surface and, as a result, the corrosion process can proceed uninhibited.

The most effective chemical control measures against hydrogen sulfide (sour) corrosion are removal of the hydrogen sulfide from the water by countercurrent gas stripping or by cleaning the steel periodically with acid to allow access of the inhibitor to the metal surface. Steel sometimes can be cleaned sufficiently for inhibition to be effective by the use of a powerful wetting agent.

12.3.2 Strong Acids

High acid concentrations are encountered in pickling processes, in oil-well acidizing, and during the transportation of acids for use in chemical processes. Hydrochloric acid of all concentrations requires an inhibitor if steel is to be used. The use of an inhibitor in pickling processes also allows the acid to dissolve scale from steel without appreciable attack on the metal. Pickling acids are effectively inhibited by adding about 0.2% of an organic compound such as anilines, pyridines, thiourea, or sulfonated castor oil. Cathodic inhibitors such as arsenates (e.g., sodium arsenate [Na_3AsO_4]) are also good inhibitors for pickling acids but are less popular than organics because they cause blistering and hydrogen embrittlement of some steels. Arsenic compounds should never be used in fluids that are to be catalytically processed because arsenic is a poison to most catalysts.

Sulfuric and phosphoric acid concentrations up to 70% can be inhibited by methods similar to those used for hydrochloric acid. Concentrations of sulfuric acid higher than 70% are strongly oxidizing, attack steel slowly, and do not require inhibition. Fertilizer—grade phosphoric acid (73% black acid) attacks steel readily and usually is inhibited with potassium iodide. Organic inhibitors are not effective in concentrated phosphoric acid when used alone, but it has been reported that a lower concentration of potassium iodide is required for inhibition if a fatty amine is also added.

12.3.3 Nonaqueous Systems

Corrosion in nonaqueous liquids such as fuels, lubricants, and edible oils is usually caused by the small amounts of water often present. Water is slightly soluble in petroleum products, and its solubility increases with temperature. If a nonaqueous solvent is saturated with water and the temperature is lowered, then some of the water will separate to attack the steel that it contacts. Oils that have been subjected to high temperatures in air will contain organic acid, which will be extracted by any water present to increase the rate of attack on steel.

Corrosion in steel systems handling wet oils can be inhibited with both organic and inorganic compounds. Effective organic compounds include various amines, lecithin, and mercaptobenzothiazole. The inorganic inhibitors include sodium nitrite and sodium nitrate. Chromates were never a good option in nonaqueous environments because they can readily react with organics.

Small amounts of water inhibit corrosion in some nonaqueous solvents. Halogenated (containing chlorides, fluorides, bromides, or iodides), nonaqueous solvents can be particularly troublesome. Organic amines are effective inhibitors for steel-degreasing vessels that contain hot, chlorinated solvents.

Inhibition of corrosion of aluminum when in contact with halogenated solvents is more difficult. Aluminum in contact with many of the one- and two-carbon chlorocarbons can explode after a varying incubation period if the solvent is dry. A few parts per million of water will inhibit the reaction. Conversely, if water is added exceeding the limited solubility in the solvent, the water layer will become highly acidic from hydrolysis of the organic compound. Thus, inhibition in systems containing these materials must be approached with caution.

Water also inhibits SCC of steel in ammonia and titanium in methanol, as well as the attack on titanium by "dry" chlorine. A trace of water (0.001%) in liquid HF behaves as a passivating inhibitor for nickel. This is an extreme example of the importance of solvent-inhibitor interactions. The exact mechanism of inhibition by water in HF is unknown, but the passivating effect is similar to that observed on steel in the presence of chromates in aqueous solutions.

Solubility is an important factor to be considered in evaluating corrosion inhibitors for nonaqueous fluids, because they do not have the tremendous solvent-power range of water. Because an inhibitor must be transported through the environment to sites where corrosion occurs, it must be either soluble in the environment or sufficiently dispersed in fine droplets so that settling does not occur. Also, the inhibitor must not form filter-plugging insoluble products by reaction with metals or components of the nonaqueous fluid. Some corrosion inhibitors formerly used in gasoline were found to react with zinc in galvanized fuel tanks to form a precipitate that clogged fuel filters.

Testing inhibitors in nonaqueous media is more difficult than in aqueous solutions, especially if corrosion is due to water separating to form a two-phase system. This condition is difficult to duplicate in the laboratory, and polarization curves cannot be used effectively because most nonaqueous solvents are poor electrolytic conductors. Furthermore, corrosion coupons placed in pipes or tanks carrying fuel or similar products may give misleading results because they may not be contacted or wetted by the water phase.

A widely accepted laboratory test for two-phase systems (called the wheel test) consists of alternately wetting a corrosion specimen by the organic and water phases by rotating a bottle containing the two phases. This test is very useful but is not always reliable because of the restricted volume of the solvents and the difficulty in duplicating velocity and stagnation effects in real systems.

12.3.4 Gaseous Environments

Gaseous environments include the open atmosphere, the vapor phase in tanks, natural gas in wells, and the empty space in packaging containers. Here again, water and oxygen are the principal corrosive agents, but the main problem in providing inhibition is to transport the inhibitor from a source to the sites where corrosion may occur.

12.3.4.1 Open Atmosphere

Inhibitors for corrosion in the open atmosphere are applied directly to the metal surfaces to be protected. Zinc chromate and red lead have been used for many years in primer coats. In another application, rivet heads were coated with a slurry of microencapsulated zinc chromate. When the rivet was driven against another surface, the capsules would rupture to provide lasting inhibition to the crevice under the rivet head. Volatile inhibitors are never used in the open atmosphere because they are impractical and cannot saturate the vapor space.

12.3.4.2 Closed Vapor Spaces

The walls of tanks above a water line are subject to extensive corrosion because the relative humidity is always high and oxygen is plentiful if the tank is vented to the atmosphere. Where water contamination is not a factor, a layer of oil on the surface helps to maintain a low humidity and, as the level is raised and lowered, the walls are coated with a layer of oil. The oil may contain an organic inhibitor and an agent (usually an amine) to cause the oil to spread on the metal surface. An oil layer containing about 15% lanolin has been used in ship ballast tanks to control corrosion.

Gas wells corrode mostly in the reflux zone, which is an area of the well somewhere between the bottom and the wellhead where condensation occurs. As the gas flows up the well, its temperature drops because of expansion, and this causes condensation when the temperature reaches the dew point of the gas. Volatile inhibitors, such as formaldehyde and ammonia, injected into gas wells have been used successfully to inhibit corrosion. Many gas wells today are protected by injecting amine inhibitors continuously, in slugs, or by squeeze, which means they coat the well when injected and also enter the gas stream partially by vaporization and partially by entrainment.

Packaged materials may be protected from corrosion in several ways. Packages that can be sealed and that contain parts that cannot be coated with an inhibitor or exposed to volatile inhibitors (such as electronic parts) are protected by placing a desiccant, such as silica gel, in the package to maintain the humidity at a low level.

VPIs can be placed in a package in bulk or by wrapping an article in paper impregnated with a VPI. These compounds are volatile organics, so the package in which they are used must be fairly well sealed. The most common VPIs in use are dicyclohexylamine nitrite (DHN) and cyclohexylamine carbonate (CHC). These inhibitors are very effective for steel, but they should be tested if metals other than steel are present because they attack some nonferrous metals.

Inhibited coatings provide a cheap, effective method for controlling corrosion of packaged materials. Easily strippable coatings that do not harden are called soft coatings or slushing compounds. Oils and greases containing amines may be used. Steel and zinc articles can be protected with a thickened aqueous solution of sodium benzoate or sodium nitrite.

Metals that are very sensitive to hydrogen sulfide, such as copper and silver, are protected by enclosing them in paper impregnated with copper or zinc compounds. These materials are not corrosion inhibitors in a strict sense because they adsorb gaseous sulfur compounds to prevent reaction with the silver or copper.

12.3.5 Effect of Elevated Temperatures

Most effects of elevated temperatures are adverse to corrosion inhibition. High temperatures increase corrosion rates, and they decrease the tendency of inhibitors to adsorb on metal surfaces. Precipitate-forming inhibitors are less effective at elevated temperatures because of the greater solubility of the protective deposit. Thermal stability of corrosion inhibitors is an important consideration at high temperatures. Polyphosphates, for example, are hydrolyzed by hot water to form orthophosphates, which have little inhibitive value. Most organic compounds are unstable above about 200°C; hence, they may provide only temporary inhibition at best.

In neutral or slightly alkaline, oxygen-free aqueous systems, corrosion of fairly clean steel occurs at a very low rate at elevated temperatures. This principle is the basis for most boiler-water treatment to prevent corrosion (i.e., treatment is designed to provide alkalinity, to remove oxygen, and to prevent scale deposition). Other additives are also used to prevent foaming, but these will not be considered here.

In oxygen-free hot water, steel is protected by a natural coating of magnetite (Fe_3O_4 or black rust) formed when iron reacts with water in the following reaction:

$$3Fe + 4H_2O \rightarrow Fe_3O_4 + 4H_2\uparrow \quad (12.9)$$

If oxygen is present, then nonprotective Fe_2O_3 (red rust) is formed. At elevated temperatures, oxygen is removed readily from boiler waters by reaction with sodium sulfite or hydrazine. Hydrazine is preferred because it does not increase the salt content of the boiler water; moreover, it reacts faster than sodium sulfite at elevated temperatures, the dosage required to react with a given amount of oxygen is lower, and it is easier to apply because it is a liquid.

Boiler waters are maintained at an alkaline pH to facilitate formation and maintenance of the Fe_3O_4 protective film (Chapter 4). It is desirable to use an additive that will be carried into steam condensate lines to maintain an alkaline condition in these areas also. Volatile amines such as ammonia, morpholine, cyclohexylamine, and octadecylamine are used.

Deposition of scales in boilers reduces heat transfer and produces pitting-type corrosion. Water softeners often are used to treat boiler feed to remove objectionable ions such as calcium, magnesium, and iron, but usually a scale inhibitor such as sodium phosphate also is added. The phosphates prevent scaling by increasing the supersaturation of $CaCO_3$ and $CaSO_4$ in water. In high-pressure, high-temperature boilers, demineralized water is often used. In these cases, only oxygen removal with hydrazine is required.

High temperatures always increase the rate of attack of metals in acids because the driving force for the anodic and cathodic reactions is greater and hydrogen overvoltage is decreased. Other factors, such as the greater solubility of corrosion products and the higher rate of solution of metal oxides, also increase the rate of attack.

Although hot acids are handled best by resistant alloys or coated steel, an exception is the acidizing of oil-wells. It is often impractical to coat or line oil-well casing because the coating is soon destroyed by the insertion of tools and so forth. Bottomhole temperatures of oil wells are often 93 to 150°C and sometimes as high as 230°C. Inhibitors for acids used to increase the permeability of oil-producing formations (usually HCl) are effective but have temperature limitations. Amine-type pickling-acid inhibitors are often used for oil-well acidizing.

12.4 GREEN INHIBITORS

Green products are produced following a set of principles designed to reduce or eliminate the use or generation of hazardous substances in the manufacture and application of chemical products. Although this short definition appears straightforward, it marks a significant departure from the manner in which environmental issues have been considered or ignored in the upfront design of the molecules and molecular transformations that are at the heart of the many corrosion-prevention compounds.

12.4.1 Green Scale Inhibitors

Scale inhibitors are typically used in water treatment (cooling towers and boilers) and industrial water treatment, such as mining and oil-field production applications. Scale inhibitors function either by binding to the insoluble scale-forming mineral, keeping them in solution, or by modifying the crystal structure of the precipitate to prevent the formation of scaling on surfaces. Organic phosphates and phosphonates are widely used as scale inhibitors. Many of the conventional organic phosphates and phosphonates used as scale inhibitors are quite toxic to the environment. These have been replaced with less-toxic organic phosphorous compounds such as carboxyhydroxymethylphosphonic acid and dibutylphosphorodithoic acid.[4]

New classes of compounds that are not based on phosphorous chemistry and that are less toxic have been introduced into the marketplace. These are acrylate-based polymers such as polyacrylic acid, polyacrylamide, and various polyacrylates. Of these, compounds from phosphoric acid also function as corrosion inhibitors in waters. Some new chemistries are being proposed including natural compounds and carboxylated plant polysaccharides such as carboxylated inulins. The applications of these materials have ranged from water treatment, boiler treatment, and mining to oil-field applications. These newer green compounds also seem to have corrosion-inhibition properties.

12.4.2 Green Corrosion Inhibitors

In the area of corrosion inhibitors, it is interesting to note that old discoveries are coming back. The first patent in corrosion inhibition was to Baldwin, British patent 2327, which specified the use of molasses and vegetable oils for pickling sheet steel in acids. U.S. Patent 640491 in 1900 to Robinson and Sutherland used starch, a readily biodegradable material.[4]

12.4.2.1 Inorganic Corrosion Inhibitors

Inorganic inhibitors typically function by passivating the metallic surface without requiring oxygen. These include chromate and nitrate ions, which can be reduced while they oxidize the metal surfaces to form a passive oxide film. Other inorganic compounds require oxygen to work. These include phosphates, silicates, borates, tungstates, and molybdates. Eliminating the use of more environmentally toxic compounds, such as the chromates and dichromates, and replacing them with more environmentally friendly chemicals has been a major directional change. The development of alternate inorganic corrosion inhibitors has been slow. They are mostly being replaced with organic corrosion inhibitors.

Zinc chromate is one of the anticorrosive pigments most frequently used in the formulation of primers. Different green alternatives have been proposed to replace the chromate. Zinc phosphate was extensively researched as a replacement in the past decade. Second-generation phosphate pigments, incorporating elements such as molybdenum, aluminum, or iron, have been developed. These phosphate-based pigments can be proposed as realistic alternatives to the chromates in the formulation of protective paints for use.[5]

The replacement of chromate-based pretreatments and anticorrosive pigments with environmentally compatible coating systems, particularly in the oil industry, has been successfully achieved. These use a relatively new class of inhibitive pigments called ion-exchange pigments. Commercially available pigments based on calcium-exchanged silica gel are incorporated into polyester primer and show effectiveness as cathodic inhibitors.[6]

In chloride solutions, tinplate exhibits localized corrosion resulting from the defects or imperfections of the outer tin layer. The addition of cerium salts to the aggressive media decreases the pitting susceptibility of tinplate by means of cerium precipitation on the cathodic sites.[7]

12.4.2.2 *Organic Corrosion Inhibitors*

For several years, the research on green inhibitors has concentrated on moving from the toxic inorganic inhibitors to organic inhibitors. However, all organic inhibitors are not necessarily green.[8] Additionally, older synthetic organic inhibitors were often not very efficient for high-temperature service.

Organic amines are probably the most important class of organic corrosion inhibitors. Aliphatic and aromatic amines; mono-, di-, or polyamines, and their salts are all used as corrosion inhibitors. Aliphatic amines adsorb by the surface-active amine ($-NH_2$) groups, which forms a chemisorptive bond with the metal surface. The hydrocarbon tails orient away from the metal surface toward the solution. Further protection is provided by the formation of a hydrophobic network that excludes water and aggressive ions from the metal surface.

In acidic conditions, organic amines' acidic component forms an amine salt, which then forms a coating on the metal, thereby preventing further corrosion of the metal. Newer and greener inhibitors have resulted in the reduction of use of many of the more toxic amine compounds or salts and the introduction of new, less toxic, organic amine compounds or salts as replacements. An example of these changes is the use of aliphatic amines instead of aromatic amines.

Several amines and amine salts used as VPIs are being replaced by greener products. Dicyclohexylamine nitrate has been replaced by phytic acid. Soy-based solvents have been used in the place of the oil in conventional products to yield environmentally friendly, biodegradable products.[4]

The use of natural products as green inhibitors has been growing in popularity. Green algae has been shown to be an effective natural additive for a paint formulation based on vinyl chloride copolymer. The interest and activity in the development and utilization of green corrosion inhibitors has spanned a wide range of industries including cooling water, boiler, oil field, pipe line, and industrial cleaning, among others.[9]

12.5 APPLICATION TECHNIQUES

12.5.1 Continuous Injection

Continuous injection of corrosion inhibitors is practiced in once-through systems where slugs or batch treatment cannot be distributed evenly through the fluid. This method is used for water supplies, oil-field injection water, once-through cooling water, open annulus oil or gas wells, and gas lift wells. Liquid inhibitors are injected with a chemical injection pump. These pumps are extremely reliable and require little maintenance. Most chemical injection pumps can be adjusted to deliver at the desired injection rate.

Another form of continuous application is by the use of slightly soluble forms of solid inhibitors. The inhibitor (such as a glassy phosphate or silicate in the form of a cartridge) is installed in a flow line where the inhibitor is continuously leached out by passage of fluid through the cartridge. Inhibitors in the form of sticks or pellets are used in oil and gas wells to continuously supply inhibitor by their natural slow dissolution.

Boilers, closed cooling-water systems, and other closed circulating-fluid systems can be treated with inhibitors with continuous injection. When such systems are started up after construction or major maintenance, the inhibitor is often injected at higher-than-normal concentration to permit rapid development of protective films.

12.5.2 Batch Treatment

The most familiar example of batch treatment is the automobile cooling system. A quantity of inhibitor is added at one time to provide protection for an extended period. Additional inhibitor may be added periodically, or the fluid may be drained and replaced with a new supply. In most aerated, closed-loop cooling systems, it is important that the inhibitor concentration be measured occasionally to ensure that a safe level is maintained.

Batch treatment is also used in treating oil and gas wells. An inhibitor is diluted with an appropriate solvent and injected into the annulus of open-hole wells or into the tubing of gas wells that have a packer. In this application, it is important that the inhibitor contact all surfaces and that it has good persistence. Most wells require treatment about every two weeks.

12.5.3 Squeeze Treatment

The squeeze treatment is a method of continuously feeding an inhibitor into an oil well. A quantity of inhibitor is pumped into a well and is followed by sufficient solvent to force the inhibitor into the formation. The inhibitor is absorbed by the formation from which it slowly escapes to inhibit the produced fluids. Protection applied in this manner has been known to last for a year.

12.5.4 Volatilization

Volatilization has already been discussed under VPIs in connection with boilers and closed containers. Another application is the inhibition of gas-condensate corrosion. However, the treatment here is essentially the same as used in batch or squeeze treatments.

12.5.5 Coatings

Inhibitors are used in coatings exposed to the open atmosphere. When moisture contacts the paint, some inhibitor is leached out to protect the metal. Thus, the inhibitor must be soluble enough to be leached out in sufficient amounts to protect the metal but not so soluble that it will be lost rapidly. The most common coating inhibitors used for many years were zinc chromate and plumbous orthoplumbate (red lead), which could readily passivate steel by providing chromate and plumbate ions, respectively, as well as the zinc and lead cathodic inhibitors. However, red lead and zinc chromate are highly toxic and can produce a range of adverse health effects. Lead compounds are bioaccumulative and ingestion, inhalation, or skin absorption of red lead can lead to skin disorders, neuromuscular dysfunction, possible paralysis, and even death. Likewise, hexavalent chrome in zinc chromate pigments is a known carcinogen and is likely to cause dermatitis, skin sensitization, and reproductive disorders.

The move to ban the use of lead- and hexavalent chromate-based pigments in paints in developed countries, notably in the United States, has triggered active research in the development of new, environmentally friendly, and nontoxic anticorrosive pigments. Many new pigments have since been introduced. Substitutes such as zinc phosphate, calcium phosphate, aluminum phosphate, barium metaborate, phosphosilicates, borosilicates, and some proprietary compositions have been widely studied to determine their performance as primers vs. that of the red lead and zinc chromate primers.

Heavy coatings that act as sealants for crevices have also been developed for the aircraft and aerospace industries. These coatings contain proprietary inhibitor formulations that are especially effective in minimizing corrosion associated with dissimilar-metal fasteners.

12.5.6 System Condition

A system must be carefully examined before a program of corrosion inhibition can be planned effectively. The examination must include a survey of any adverse effects an inhibitor may have on the process in which it is to be used and an analysis to detect the presence of interfering substances.

Another possible adverse effect of inhibition is an increased rate of corrosion of a metal in the system other than the one for which the inhibitor was selected to protect. For example, some amines protect steel admirably but will severely attack copper and brass. Nitrites may attack lead and lead alloys such as solder. In some cases, the inhibitor may react in the system to produce a harmful product. An illustration of this is the reduction of nitrite inhibitors to form ammonia, which causes SCC of copper and brass. The only way to avoid these problems is to know the metallic components of a system and be thoroughly familiar with the properties of the inhibitor to be used.

The examination must include preparation of a complete list of materials, both metallic and nonmetallic, that will be in contact with the fluid that will be inhibited. Such small items as gaskets, instrument probes, and control devices may be made from materials that will not be compatible with some inhibitors. The results of this examination may suggest that certain parts of a system should be changed to permit the use of a particular inhibitor.

The examination must also include a determination of the cleanliness of the surfaces of the system that will be in contact with the inhibited fluid. A system can be plugged as the result of an inhibitor loosening scale and suspending it in the fluid. This problem is best avoided by planning ahead. The best preventive measure is to clean the system as thoroughly as possible before an inhibitor is applied.

Cleaning may be accomplished with chemical cleaners, mechanical cleaners, ultrasonic energy, or thermal shock. Inhibitors can reach cleaned metal surfaces much more easily than they can reach heavily fouled or scaled surfaces.

12.5.7 Inhibitor Selection

There are several strategies for the selection of a proper inhibitor for a given system. An outside consultant can be asked for advice. Various suppliers can be asked for advice. Table 12.3 presents some inhibitors that have been used with success, in typical corrosive environments, to protect the metallic elements of industrial systems. Commercial inhibitors are available under various trade names and labels that usually provide little or no information on their chemical composition. It is sometimes very difficult to distinguish between products from different sources because they may contain the same basic anticorrosion agent. Commercial formulations generally consist of one or more inhibitor compounds with other additives such as surfactants, film enhancers, demulsifiers, oxygen scavengers, and so forth. The inhibitor–solvent package used can be critical with respect to the solubility and dispersibility characteristics and hence application and performance of the products.

A frequent cause of ineffective inhibition is loss of the inhibitor before it has a chance to contact the metal surfaces or effect the desired changes in the environment. An inhibitor might be lost by precipitation, adsorption, reaction with a component of the system, or insufficient solubility or slow dissolvability.

Typical examples of losses of an inhibitor resulting from these factors are precipitation of phosphates by calcium ions, reaction of chromates with sulfides or organics, adsorption of inhibitors on suspended solids, and injection of a poorly soluble inhibitor without an adequate dispersing agent. To avoid these problems, inhibitors should be tested in the actual fluids to be treated rather than in simulated environments. If possible, testing should be done in the process stream or in a small side stream.

Table 12.3 Some Corrosive Systems and the Inhibitors Used to Protect Them

System	Inhibitor	Metals	Concentration
Acids			
HCl	Ethylaniline	Fe	0.5%
..	MBT*	..	1%
..	Pyridine + phenylhydrazine	..	0.5% + 0.5%
..	Rosin amine + ethylene oxide	..	0.2%
H_2SO_4	Phenylacridine	..	0.5%
H_3PO_4	NaI	..	200 ppm
Others	Thiourea	..	1%
..	Sulfonated castor oil	..	0.5–1.0%
..	As_2O_3	..	0.5%
..	Na_3AsO_4	..	0.5%
Water			
Potable	$Ca(HCO_3)_2$	Steel, cast iron	10 ppm
..	Polyphosphate	Fe, Zn, Cu, Al	5–10 ppm
..	$Ca(OH)_2$	Fe, Zn, Cu	10 ppm
..	Na_2SiO_3	..	10–20 ppm
Cooling	$Ca(HCO_3)_2$	Steel, cast iron	10 ppm
..	Na_2CrO_4	Fe, Zn, Cu	0.1%
..	$NaNO_2$	Fe	0.05%
..	NaH_2PO_4	..	1%
..	Morpholine	..	0.2%
Boilers	NaH_2PO_4	Fe, Zn, Cu	10 ppm
..	Polyphosphate	..	10 ppm
..	Morpholine	Fe	variable
..	Hydrazine	..	O_2 scavenger
..	Ammonia	..	neutralizer
..	Octadecylamine	..	variable
Engine coolants	Na_2CrO_4	Fe, Pb, Cu, Zn	0.1–1%
..	$NaNO_2$	Fe	0.1–1%
..	Borax	..	1%
Glycol/water	Borax + MBT	All	1% + 0.1%
Oil-field brines	Na_2SiO_3	Fe	0.01%
..	Quaternaries	..	10–25 ppm
..	Imidazoline	..	10–25 ppm
Seawater	Na_2SiO_3	Zn	10 ppm
..	$NaNO_2$	Fe	0.5%
..	$Ca(HCO_3)_2$	All	pH dependent
..	$NaH_2PO_4 + NaNO_2$	Fe	10 ppm + 0.5%

*MBT = mercaptobenzotriazole.

12.5.8 Concentration and Performance Technology

Corrosion inhibitors are sold in solid or liquid form. Most solids are relatively pure, but sometimes a solid inhibitor is fused with another ingredient or encapsulated where a controlled rate of solubility is required. Liquids are usually preferred because of the ease with which they can be transported, measured, and dispersed.

Liquid inhibitors are rarely pure for several reasons. Organic inhibitors seldom have optimum characteristics of viscosity or freezing or boiling points; therefore, they are dissolved in an appropriate solvent to achieve the properties desired. Furthermore, it is often desirable to blend the inhibitor with a demulsifier, dispersant, surfactant, antifoaming agent, or synergistic agent.

Liquid inhibitors are sold by the liters, part of which is solvent. The amount of inhibitor present is expressed as percent active, that is, a liter of inhibitor that is 20% active contains 20% by weight of inhibitor. In cold climates where inhibitors are likely to be stored or used in subfreezing temperatures, it may be impossible to use as concentrated a solution as in warmer climates without resorting to more expensive solvents.

Inhibitor concentrations are expressed as parts per million; for solids the units are on a weight basis (e.g., kilograms of inhibitor per million kilograms of fluid) and for liquid inhibitors, volumes are used, (e.g., liters of inhibitor per million liters of fluid). To obtain the amount of inhibitor required for a given system, simply divide the amount of fluid to be inhibited by 1,000,000 and multiply by the number of ppm desired:

$$Q_{\text{inhibitor}} = \frac{V}{1,000,000} \cdot \text{ppm} \tag{12.10}$$

where Q is quantity of inhibitor required, V is the amount of fluid to be inhibited, and ppm is the inhibitor concentration in parts per million. Note that the quantity of inhibitor must be in the same units as those for the amount of fluid.

Example: What is the dosage of sodium silicate (a solid) required to add 10 ppm to 620,000 liters of water? The volume of water is first converted to weight.

$$\text{Kilograms of water} = 620,000 \text{ liters} \times 1 \text{ or}$$

$$(\text{Pounds}) = (165,000 \text{ gallons} \times 8.316/\text{gal}), \text{ then}$$

$$V = 620,000\text{kg}(1,369,500 \text{ lbs.})$$

Because the required inhibitor concentration is 10 ppm, by substituting in Equation (12.10), we get:

$$Q_{\text{inhibitor}} = \frac{620,000}{1,000,000} \cdot 10 = 6.2\text{kg}$$

$$\text{or, in pounds, } \left(Q_{\text{inhibitor}} = \frac{1,369,500}{1,000,000} \cdot 10 = 13.7 \text{ lbs}\right)$$

12.6 SAFETY PRECAUTIONS

Safety precautions cover personnel safety, environmental protection, and equipment operational protection.

12.6.1 Handling

Toxic effects of inhibitors must be considered in processes where the compounds may be inhaled or contacted. Extreme care must be taken to avoid the use of toxic compounds in or near food- or beverage-processing equipment and potable-water supplies.

When inhibitor solutions are prepared for injection, care should be taken to follow label instructions regarding skin contact, eye contact, ingestion, and inhalation. All safety information labeling should be carefully studied before opening containers of inhibitors.

12.6.2 Disposal

Because a number of inhibitors contain ions and compounds that may be toxic, it is often difficult to dispose of inhibited fluids that are drained, dumped, or leaked from systems. Chromates are a prime example of this problem. In fact, the use of chromate-containing inhibitors has been severely reduced in recent years by this disposal problem. Disposal must be considered during the inhibitor-selection process.

12.6.3 Heat Transfer

Heat transfer is an important consideration in applying corrosion inhibitors. Scaling of heat-transfer surfaces, which can reduce heat-transfer, should be avoided or held to a minimum. Excessive deposits of phosphates, silicates, or sulfates should be avoided; the latter in particular because they are difficult to remove by chemical means.

12.6.4 Foaming

The most appropriate action to take in avoiding difficulty from foaming is to determine where foam-forming conditions exist in the system. These will consist of places where the inhibitor-containing fluid is agitated with a gas such as in a gas separator, in a countercurrent stripper, or in an aerator. The next step is to obtain a sample of the fluid and gas from the process, add the inhibitor in question, adjust the temperature to that corresponding to the process step, and shake vigorously. If this test produces a stable foam, a potential problem exists. There are three alternative remedies:

1. an antifoaming agent may be added (this must be tested also),
2. tests can be made to select an inhibitor that does not cause foaming, and
3. the system can be shut down periodically and treated with a slug of persistent inhibitor.

The last two remedies are the least palatable because the need for an inhibitor is at hand, and there are few processes that can be shut down with sufficient frequency to maintain effective inhibition by slug treatment.

12.6.5 Emulsions

Emulsions are similar to foams in that they consist of one phase dispersed in another. Emulsions consist of two liquid phases that will not mix, whereas foams are composed of a gas and a liquid phase. Conditions favorable for emulsion formation are the presence of two liquid phases, agitation, and an emulsion stabilizer. In this case, the corrosion inhibitor may behave as an emulsion stabilizer, so the procedure is to shake the two liquid phases with inhibitor and measure the time required for them to separate. If the time required for the separation is longer in the presence of the inhibitor than in its absence, the inhibitor is an emulsion stabilizer. The remedy is the same as for foams: add a demulsifier, use another inhibitor, or inhibit with slugs during shutdown.

References

1. G. Hugel, "Corrosion Inhibitors—Study of their Activity Mechanism," 1st European Symposium on Corrosion Inhibitors (Ferrara, Italy: University of Ferrara, 1960).
2. N. Hackerman, E.S. Snaveley, "Inhibitors," in Corrosion Basics, eds. L.S. Van Delinder, A.deS. Brasunas (Houston, TX: NACE, 1984), 127–146.
3. N. Hackerman, R.M. Hurd, R.R. Annand, "Some Structural Effects of Organic N-Containing Compounds on Corrosion Inhibition," Corrosion 18 (1962): pp. 37–42.
4. D.V. Satyanarayana Gupta, "Green Inhibitors—Where Are We?" CORROSION 2004, paper no. 406 (Houston, TX: NACE, 2004).
5. M. Bethencourt, F.J. Botana, M. Marcos, R.M. Osuna, J.M. Sanchez-Amaya, "Inhibitor Properties of 'Green' Pigments for Paints," Progress in Organic Coatings 46 (2003): pp. 280–207.
6. F. Deflorian, I. Felhosi, "Electrochemical Impedance Study of Environmentally Friendly Pigments in Organic Coatings," Corrosion 59 (2003): pp. 112–120.
7. M.A. Arenas, A. Conde, J.J. de Damborenea, "Cerium: A Suitable Green Corrosion Inhibitor for Tinplate," Corrosion Science 44 (2002): pp. 511–520.
8. W.P. Singh, J. O'M. Bockris, "Toxicity Issues of Organic Corrosion Inhibitors: Applications of QSAR Model," CORROSION/96, paper no. 225 (Houston, TX: NACE, 1996).
9. M.A. Quraishi, I.H. Farooqi, P.A. Saini, "Investigation of Some Green Compounds as Corrosion and Scale Inhibitors for Cooling Systems," Corrosion 55 (1999): pp. 493–497.

13
Chapter

Failure Analysis and Design Considerations

13.1 PREVENTING CORROSION FAILURES BY DESIGN

System failures and subsequent failure investigations have become increasingly important in our modern societies. Besides liability issues, an important reason for conducting a failure investigation is to identify the mechanisms and causes of a problem to prevent its recurrence.

Corrosion engineers are often confronted with a corroded sample taken from an apparatus that has failed. Usually, some stopgap action must be immediately taken so that the apparatus can again be placed in service. What further action should be taken to avoid a recurrence of the failure must then be decided on. To determine which factors were most responsible for the failure often involves planning and conducting a test program under simulated conditions.

Engineered structures should be designed to provide the desired functional qualities for the required period of service. This implies that a structure should not be underdesigned (too risky) or overdesigned (too costly). Thoughtful design requires more than the provision of adequate strength. The part must also last for a given period of time.

13.2 RUDIMENTS OF CORROSION FAILURE ANALYSIS

13.2.1 What Constitutes a Failure?

Neglecting to get to the underlying causes of corrosion failures and to take corrective action can expose an organization to litigation, liability, and loss of customer and public confidence in its product(s) or services. Such risks are often unacceptable in the modern, global, competitive business environment.

The term failure is literally defined as "a falling short, a deficiency or lack, an inability to perform." For the corrosion engineer, the term failure is defined in terms of how well or badly a material fulfills all aspects of the functional requirements of the application for which it was selected. The use of correct and consistent terminology in failure analyses is vital. The value of information in reports is greatly diminished by inattention to this detail.

Conducting a failure analysis is not an easy or straightforward task. Early recognition of corrosion as a factor in a failure is critical, because much important corrosion information can be lost if a failure scene is altered or changed before appropriate observations and tests are made. To avoid these pitfalls, certain systematic procedures have been proposed to guide an investigator through the failure-analysis process. However, failure analysis is ultimately best learned by experience and a failure analyst must earn proper credentials by living through actual investigations and having successfully solved a variety of problems.

The depth of the analysis into the roots of a failure is the key to accurately unearthing all of the factors that caused that failure in the first place. Looking at machinery breakdowns, one finds that there are[1]

1. physical roots: the physical reasons why the parts failed,
2. human roots: the human errors of omission or commission that resulted in the physical roots, and
3. management system (latent) roots: the deficiencies in the management systems or the management approaches that allow the human errors to continue unchecked.

The more detailed an analysis is, the easier it is to understand all the events and mechanisms that contribute to the roots of the problem. These analyses are usually divided into three categories in order of complexity and depth of investigation:

1. Component failure analysis (CFA): CFA is conducted in a laboratory and looks at the piece of the machine that has failed, a bearing or a gear for example, to determine that it resulted from a specific mechanism such as fatigue, overload, or corrosion and that there were these x, y, and z significant influences.
2. Root-cause investigation (RCI): RCI includes a field investigation in addition to the CFA and is conducted in much greater depth than the CFA. The RCI goes substantially beyond the physical roots of a problem, but it usually stops at the major human causes and does not involve the latent system deficiencies.
3. Root-cause analyses (RCA): RCA includes everything the RCI covers plus the minor human-error causes and, more important, the management system that allows the human errors and other system weaknesses to exist.

13.2.2 Conducting a Failure Analysis

The first step in a failure analysis is to learn as much as possible from the existing sample. Care must be taken in handling the sample to avoid destroying valuable evidence. Much can be deduced from the physical appearance of the corrosion as viewed with the unaided eye or a low-power magnifying glass or microscope. The geometry of attack, color, and form of corrosion products, all provide valuable clues.

Other important information can be obtained only by altering the shape of or partially destroying the sample. For example, metallographic observation of type and depth of attack usually will require cutting and polishing the sample. In planning the sequence of investigation, all visual examinations, collection of corrosion products for chemical and X-ray analysis, and so forth should be completed before cutting the sample.

The following sections discuss the various options to consider during a corrosion failure investigation.[2]

13.2.2.1 Planning the Analysis

Early recognition of corrosion as a factor in a failure analysis is critical to any such investigation. Therefore, it is generally desirable to conduct the analysis as soon as possible after the apparent failure. It is always desirable to protect the physical evidence until the analysis can begin. Much important corrosion information can be lost if a failure scene is altered or changed before appropriate observations can be made.

A written plan for the detailed analysis should be prepared. The plan may include methods of documentation (photographs before and during analysis, sketches, statements), responsibilities of parties, reporting needs, and scheduling.

If the capability (corrosion knowledge and experience) of in-house personnel and availability of resources are inadequate to make the analysis in a timely manner, it may be expedient to seek third-party services.

13.2.2.2 Conditions at the Failure Site

An overall examination of the conditions at a failure site prior to cleaning, moving, or, sampling debris should be conducted. Impressions of physical arrangements, odors, colors, textures, and conditions of adjacent structures can provide important clues as to active corrosion processes and should be reported appropriately.

Annotated photographs, videotapes, and sketches and drawings with notes can serve as documentation of the observations and become truly beneficial during later evaluations.

Interviews with those who were present or nearby when the failure occurred would be appropriate. Information on time, materials, and supplies may be significant and should be evaluated.

13.2.2.3 Operating Conditions at Time of Failure

Special attention should be given to the stability of the operating conditions at the time of failure. Conditions of corrosion concern could be temperature, pressure, flow rate, velocity, pH and chemical composition, time, and weather.

Special attention should be given to out-of-specifications or any other unusual upset conditions. It may be necessary to plot or track operating conditions for a long period of time prior to the detection of failure to reveal possible unusual contributory operating conditions.

The operating conditions of any similar equipment should also be noted and use as a reference. Any present corrosion-monitoring instruments and coupons should, of course, be examined to help document operating conditions at the time of failure.

13.2.2.4 Historical Information

Historical information, when available, is extremely useful in understanding some situations. Often in the cases of older equipment, historical information may be nearly impossible to find because of lost files or retired personnel. Additionally not all types of information will turn out to be useful. Based on cost, time, and anticipated benefit, a judgment must be made as to the effort one should expend in retrieving historical information.

Useful details regarding original constructions may include, for example, design drawings and specifications, material specifications, joining, and surface treatments. Details regarding modifications made subsequent to original fabrication may be extremely important because they often reveal less-than-optimum field work. Modifications may have been made for one or more reasons, including problems with the original design, changes in service requirements, corrected earlier failures, and correction of safety and environmental concerns.

13.2.2.5 Sampling

Careful sampling is critical to the successful investigation of corrosion-related failures. Sampling in corrosion investigations is similar to that used in forensic investigations by criminologists. ASTM Guide E 1459 and Practice E 1492 address issues of labeling and documenting field evidence. These standards may provide useful guidance during sampling for corrosion investigations.

Contamination should be avoided during sampling by using clean tools and by wearing gloves. Sample containers should be clean and sealable to protect samples from contamination and damage. The material of sample containers should be selected carefully to avoid undesirable interaction with samples. Each container should be dated and identified according to the sampling plan.

Samples of corroded and uncorroded materials can be quite useful in the identification of causative factors. Samples should be as large as practical to provide sufficient material to work with during subsequent analysis. Critical corroded areas should be protected from damage during cutting and transporting. If the location of failure initiation is apparent, it should be sampled. Samples should be cut with care to prevent heating the samples and to avoid the introduction of cutting and cooling fluids that could alter the surface and metallurgical conditions. Because of the solubility in water of many corrosion products, samples must be protected from extraneous moisture.

Corrosion products and deposits should be given special sampling treatment because they are often key elements in understanding the failure. Care should be used in the selection of tools for collecting these samples. Nonmetallic tools are often preferred because they present less chance for contamination of the sample or for damaging critical corroded surfaces.

It may be desirable to obtain samples of the local environment, for example, process stream, soils, concrete, and so forth. The most useful samples are those taken from the failure location as soon after the event as possible. Samples obtained later are less useful because there is no assurance that they represent the conditions at failure. Special sampling procedures can also be required when microbiological factors are suspected of being involved.

13.2.2.6 Evaluation of Samples

Metallic samples should be evaluated for metallurgical condition and structure. In some cases it may be necessary to remove corrosion products from a sample to permit further evaluation. The evaluation itself may involve mechanical and physical property tests, metallographic examination of cross sections, and corrosion tests:

- mechanical and physical property tests to assess the influence of service temperature and time on the properties being evaluated,
- metallographic examination techniques to verify the influence of service temperature and time on the metallurgical structures being examined, and
- corrosion testing of affected material to identify metallurgical and environmental factors associated with the failure.

The location and orientation of each specimen must be documented by photographs, drawings, or written descriptions. Each specimen should be labeled to aid in identifying its original location within the sample. Failure locations, such as pits, fracture surfaces, crevices, and generally attacked surfaces, should be examined, and measurements should be made to document surface chemistry, pit depths, crack dimensions, metal losses, and other modes of attack. These examinations often require the use of optical microscopes, SEM microscopes, X-ray powder diffraction, and other instruments.

13.2.2.7 Assessment of Corrosion-Related Failure

The type and extent of corrosion should be noted. The extent of corrosion may be determined by measurements and calculations of general corrosion rate, pitting penetration, or crack growth rate. It may also be useful to compare these rates with expected rates reported in the literature or from experience. Rate discrepancies should be investigated by laboratory simulations, for example.

From these observations and findings, the investigator should be able to identify the causes and factors involved in the failure. In many cases, more than one factor will have to be suggested as having played a role in the failure. The investigator may provide explanations and rationales for suggested corrective actions.

13.3 INFORMATION TO LOOK FOR

It should now be obvious that the corrosion behavior of materials under conditions of service should be an important consideration during the design of a system. However, although it is relatively straightforward to obtain quantitative information on physical and mechanical properties such as tensile strength, yield strength, impact values, fatigue limit, effect of temperature on properties, and so forth, truly representative corrosion data are much more difficult to obtain.

It is not always possible to anticipate the actual environment in which a metal will operate. Even if initial conditions were known completely, often there is no assurance that operating temperatures, pressures, or even chemical compositions will remain constant over the expected life span of equipment. Under these circumstances, one is tempted to ask, "Why bother?"

Experience has shown, however, that even when impossible circumstances apparently are present, substantial cost savings are possible by intelligent use of corrosion data and good design practices.

13.3.1 Cautionary Use of Test Data

Most published data have limited usefulness because the character of attack is often omitted. Another factor to keep in mind is that published data usually are obtained under carefully controlled laboratory conditions using high-purity, reagent-grade materials. This is very different than plant conditions using commercial-grade materials, which are less pure. Minor elements or impurities can play an important role in the stability of metals and alloys when they are exposed to service conditions. A classic example is the catastrophic influence on water-handling aluminum equipment even when only a minute amount of copper ions is present in the water.

If laboratory tests indicate that a given chemical is seriously corrosive to a particular material of construction, a decision can usually be made to eliminate this material from further consideration. Favorable laboratory or field-test data, on the other hand, are not positive assurances of good performance of a material in service, unless the data were developed under precisely the same circumstances that will be encountered in use. Even two different petroleum refineries belonging to the same oil company and using the same basic processes find differences in performance of materials traceable to differences in local conditions, for example, different source of crude, different mixture of crudes, and different pretreatment because of different end products (lube oil, asphalt, fuel oil, etc.).

In a soda-ash plant, for example, consideration was being given to the use of aluminum heat-exchanger tubes and shells for the ammonia-recovery condensers. This is a standard aluminum application in many plants of this type. Sheet specimens of aluminum alloys were carefully mounted in the existing cast-iron shells using brackets coated with electrical insulating material to avoid dissimilar-metal attack.

After six months of exposure, examination of the specimens showed no discernible attack. Therefore, because this evidence seemed to confirm experience in other soda-ash plants, a decision was made to build these larger, water-cooled condensers of aluminum alloys.

One week after the plant went in service with the new condensers, it was necessary to shut the plant down because the new condensers had failed. Although the walls of the condensers were unaffected by corrosion, the tubes were completely perforated from the condensate side (exterior).

The evaluation tests using aluminum coupons had not taken into account the influence of heat transfer on corrosion behavior. Actually, the specimens accurately simulated only the sidewall conditions. In this particular plant, the ratio of carbon dioxide (CO_2) to ammonia (NH_3) in the moist gaseous stream was lower than was customary in other soda-ash plants; consequently, the inhibiting effect of CO_2 was incomplete. This illustrates the importance of local situations that can complicate the task of material selection.

13.3.2 Identifying the Form of Corrosion

The character and distribution of attack is important in evaluating the potential danger of corrosion damage. Therefore, in contemplating a specific design, it is useful to consider the likely character of attack under the prospective operating conditions. For example, pitting attack could result in premature failure of heat-exchanger tubing or chemical storage tanks or drums. For these services, materials that suffer uniform attack are much more desirable. By contrast, for many structural applications, a material subject to localized pitting may be at least as suitable as one that suffers general attack (assuming stress raisers are not important in the particular use). (See Chapter 9 for more details of the various forms of corrosion.)

13.3.3 Product Degradation

When designing equipment for handling chemicals, synthetic fibers, or paper, it may be desirable to select a material whose corrosion products are not highly colored. Food-handling equipment or apparatus for microbiological processes will require materials whose corrosion products are nontoxic.

The textile and paper industries, for example, are concerned with the color of their finished products. Because the corrosion products of many metals are highly colored, care must be exercised in the selection of materials to be used in contact with the finished products. Mills of this type generally are highly automated, so accidental product contamination may go unnoticed until the finished product reaches a customer's plant.

In a particular case, undesirable rust spots were discovered on paper stock being processed at a paper-specialties plant. Careful investigation revealed that a steel electrical conduit passed above the paper machine in the paper-mill. Corrosion of the steel in the humid paper-mill environment permitted iron-corrosion products to fall and contaminate the paper.

Edible products also must be protected from degradation. Vegetable oils tend to become rancid in contact with some materials. Copper-base alloys have significant undesirable effects on edible oils. Accordingly, such products normally are handled in organic coating, nickel, stainless steel, plastic, or aluminum vessels.

Corrosion of certain metals can introduce toxic reaction products into process streams. For this reason, in the concentration of sap in the maple sugar–making process, use of lead equipment or lead-base solder for fabrication of steel equipment is forbidden. Similarly, lead alloys are forbidden in the processing of edible gelatin.

13.3.4 Contact with Nonmetals

Severe attack of metal can occur if it is in contact with a nonmetal such as cork, hairfelt, wood, cloth, or paper (especially severe when impregnated with certain flame proofing or biocides). The cause of such attack is usually the formation of differential aeration cells; the term poultice corrosion was used to describe this in Chapter 9. Such attack can be prevented by one or more of the following:

1. using nonmetallic materials known to be without corrosive action,
2. preventing nonmetallic materials from becoming wet,
3. providing adequate drainage and air circulation,
4. coating the metal surfaces,
5. using water-impervious mastics for attachments of nonmetallics,
6. impregnating nonmetals with inhibitors, and
7. "Building in" CP by use of metal coatings (or clad aluminum products).

13.3.5 Residual and Fabrication Stresses

Often, even when the proper materials are selected for the anticipated service conditions, some will fail because of improper fabrication. The assembly of parts for transportation and airframe structures, for example, must be done under closely controlled conditions to avoid introduction of high assembly stresses that might lead to SCC.

For this reason, special care is exercised in fit-up. Fasteners are selected that do not introduce localized regions of residual tensile stress (e.g., avoid tapered drive pins). Inasmuch as SCC requires an enduring tensile component of stress, sometimes it is desirable to shot peen critical structures to ensure the presence of residual compressive stresses at the surface.

Residual stresses can be introduced by service conditions. Differences in thermal expansion between dissimilar metals forming parts of the same structure may result in either tensile or compressive failure, or in corrosion fatigue.

Corrosion products usually occupy more volume than the metal from which they were formed. Corrosion product build-up between the shanks of fasteners and fastener holes or between the faying surfaces of lap joints can cause pressure and eventual failure of the joint in a form of corrosion called "pillowing" (see Chapter 9).

13.3.6 Surface Contamination

Roll forming of plates is a common source of surface contamination caused by chips, shavings, and other foreign particles that are embedded in the surface of the metal. Special procedures are used to avoid surface contamination. These include cleaning the rolls, filtering rolling lubricants, coating rolls with heavy paper, and coating plates with protective tape. High-purity aluminum plates used for fabrication of hydrogen peroxide storage tanks usually are coated with protective tape at producer mills to avoid surface contamination during shipment and fabrication. In this example, such surface contamination, in addition to being a potential corrosion hazard, can contribute to catalytic breakdown of the peroxide.

13.3.7 Temperature Effects

In most chemical reactions, an increase in temperature is accompanied by an increase in reaction rate. A rough rule-of-thumb suggests that the reaction rate doubles for each 10°C rise in temperature. Although there are numerous exceptions to this "rule," it is important to take into consideration the influence of temperature when analyzing why materials fail and when designing to prevent corrosion.

Figure 13.1 illustrates the influence of temperature and composition on the corrosion behavior of materials exposed to hydrochloric acid solutions. This isocorrosion chart is drawn for a corrosion rate of 500 μm/y (20 mpy) or less. The upper (high-temperature) boundary of each area represents a corrosion rate of 500 μm/y. The materials listed in each area are metals having a corrosion rate of 500 μm/y or less under the indicated conditions of temperature and acid concentration. The boiling-point curve for atmospheric pressure is shown because at temperatures greater than this curve, the metal is no longer in a liquid environment at 1 atm pressure.

Changing the temperature of a solution can influence the corrosion tendency. Many household hot-water-heater tanks, for example, have historically been made of galvanized steel. The zinc coating on the mild steel base offered a certain amount of CP to the underlying steel, and the service life (usually judged by how long it took to produce red water, i.e., rusty water) was considered adequate. Water tanks seldom were operated above 60°C.

With the development of automatic dishwashers and automatic laundry equipment, the average water temperature was increased so that temperatures of about 80°C are not unusual in household hot-water tanks. Coinciding with the widespread use of automatic dishwashers and laundry equipment was a sudden upsurge of complaints of short life of galvanized-steel water-heater tanks. Electrochemical measurements showed that in many cases, iron was anodic to zinc at 77°C, whereas zinc was anodic to iron at temperatures below 60 C°. This explained why zinc offered no CP at 77°C, and why red water and premature perforation of galvanized water tanks occurred so readily at higher temperatures. The problem was partly solved by using magnesium sacrificial anodes or protective coatings and by new alloys.

13.3.8 Fluid-Velocity Effects

Metals generally owe their corrosion resistance to a tightly adherent, protective film on the metal surface. This film may consist of reaction products, adsorbed gases, or both. Any mechanical disturbance of this protective film can stimulate attack of the underlying metals until either the protective film is reestablished or the metal has been corroded away. The mechanical factor may result from abrasion, impingement, turbulence, or cavitation.

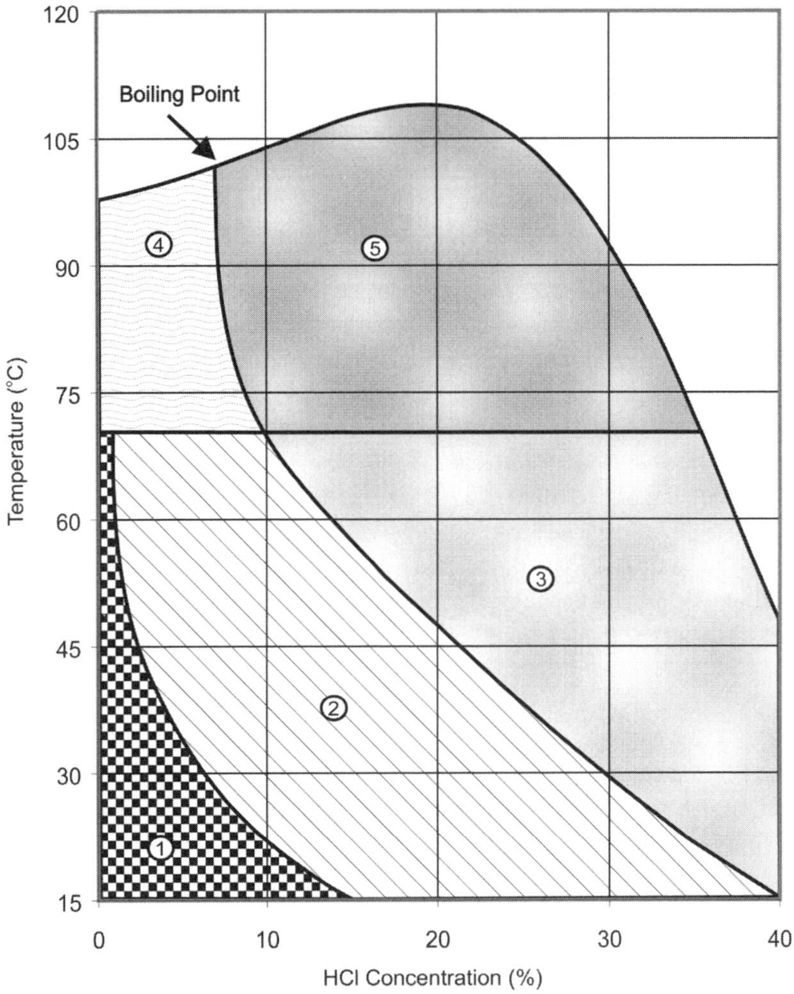

Materials in shaded zones have reported corrosion rates of < 0.5 mm/year

Zone 1	Zone 2	Zone 3	Zone 4	Zone 5
20Cr 30Ni[a]	Zirconium	62Ni 28Mo[e]	66Ni 32Cu[b, f]	62Ni 28Mo[e]
66Ni 32Cu[b]	62Ni 32Cu	Molybdenum	62Ni 28Mo[e]	Platinum
62Ni 28Mo	Molybdenum	Platinum	Platinum	Silver
Copper[b]	Platinum	Silver	Silver	Tantalum
Nickel[b]	Silicon bronze[b]	Tantalum		Zirconium
Platinum	Silicon cast iron[c]	Zirconium		
Silicon bronze[b]	Silver			
Silicon cast iron[c]	Tantalum			
Silver	Zirconium			
Tantalum				
Titanium[d]				
Tungsten				
Zirconium				
Tantalum				
Tungsten				

[a] < 2% at 25°C
[b] No air
[c] No $FeCl_3$
[d] < 10% at 25°C
[e] No chlorine
[f] < 0.05% concentration

Figure 13.1 Corrosion rate map of metallic materials in hydrochloric acid.

Erosion–corrosion is encountered most frequently in pumps, valves, centrifuges, elbows, impellers, inlet ends of heat-exchanger tubes, agitated tanks, and so forth. Experience with lead valves is an illustration. Hard lead specimens exposed to hot, stagnant, dilute sulfuric acid have suffered negligible attack. However, throttled valves of the same material exposed to high-velocity effects have failed in less than a week.

Locations in flowing systems where there are sudden changes in direction or flow cross section, as in heat exchangers where water flows from the water boxes into the tubes, are likely places for erosion–corrosion. Under these conditions, which stimulate some corrosion of the metal surface, the effects of flow velocity may be to displace the corrosion product, thereby exposing fresh metal to the corrosion action of the solution. This action leads to a much increased corrosion rate (see Chapter 9). The corrosion products themselves can subsequently precipitate downstream and cause the plugging of tubes, screens, and valves.

When the fluid velocity is higher than a certain threshold value, cavitation or the formation and collapse of voids in fluids (e.g., water) can occur because of the momentary occurrence of low pressure. It occurs whenever the absolute pressure at a point in the liquid stream is reduced to the vapor pressure of the fluid. Cavitation damage is the wearing away of metal by mechanical damage resulting from repeated impact blows produced by collapse of voids within a fluid. Pump impellers (Figure 13.2) and ship propellers often show cavitation-damage effects.

Wet steam traveling at high velocities (900 to 1200 m/s) can cause considerable damage to power-plant condenser tubes, turbine blades, valve seats and disks, piston rings in engines, and so forth. Removal of moisture droplets from the steam is helpful but not always possible. Consequently, the practical solution of steam-impingement problems usually involves a combination of several procedures: improved flow design, reduction of water-droplet content of the steam, and the use of materials having higher

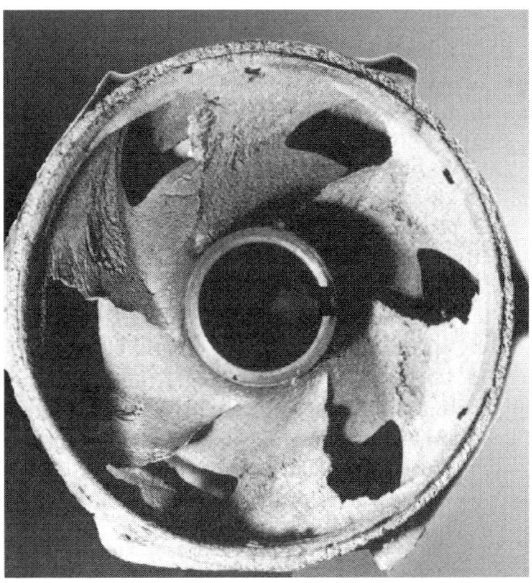

Figure 13.2 A pump impeller is subjected to cavitation-enhanced corrosion caused by the collapse of cavities in the liquid at the metal-liquid interface.

resistance to erosion–corrosion damage (e.g., stainless steels) in areas where impingement is anticipated.

13.3.9 Impurities

Impurities present in minute amounts often have more influence on the corrosion behavior of materials than do substances present in much greater quantities. Sometimes trace quantities of impurities serve to accelerate attack, whereas at other times they may act as inhibitors. Figure 13.3 shows the behavior of commercially pure aluminum in acetic acid at various temperatures and concentrations. In the concentration range between 90 to 100% acetic acid, the water content of the acid controls the rate of attack. Violent attack occurs at elevated temperatures if the water content is reduced below about 0.2%. Apparently, so long as at least 0.2% moisture is present, the protective film that gives aluminum its corrosion resistance is maintained. Therefore, a trace of moisture, in this particular case, acts as an inhibitor.

The introduction of small amounts of ions of metals such as copper, lead, or mercury can cause severe corrosion of aluminum equipment. For example, corrosion of upstream copper-alloy equipment can result in contamination of cooling water. Under

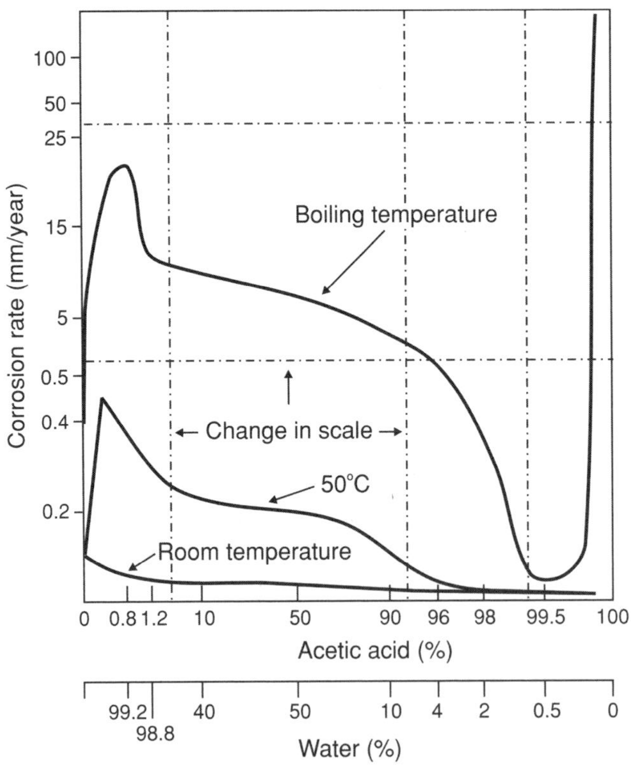

Figure 13.3 Behavior of 1100-H-14 aluminum in acetic acid solutions. The corrosion rate in the 90 to 100% range is controlled by water content. The curve can be applied generally to aluminum alloys.

Figure 13.4 Effect of mercury contamination cracks a brass tube. Very small concentrations of mercury can produce this effect.

these circumstances, copper can plate out (cathodic reaction) on downstream equipment and pipe, setting up local galvanic cells, which can result in severe pitting and perhaps perforation.

Mercury is a corrosion hazard for most metallic materials of construction. Mercury causes severe pitting of aluminum and lead and SCC of copper-base alloys (Figure 13.4). Common sources of mercury contamination have been attributed to blown manometers, broken thermometers and seals, or the use of mercury-containing biocides or wood preservatives. The reduced use of mercury in these applications, for obvious environmental reasons, has coincidentally benefited the overall corrosion performance of metallic materials.

Titanium alloys resist an extensive range of acidic conditions. Many industrial acid streams contain contaminants that are oxidizing in nature, thereby passivating titanium alloys in normally aggressive acid media. Metal-ion concentration levels as low as 20–100 ppm can inhibit corrosion extremely effectively. Potent inhibitors for titanium in reducing acid media are common in typical process operations. It is this potent metal-ion inhibition that permits titanium to be successfully used for equipment handling hot HCl and H_2SO_4 acid solutions in metallic-ore leaching processes. Figure 13.5 shows the inhibiting effect of ferric chloride on Grade 2 titanium exposed to hydrochloric acid at various concentrations and temperatures.

13.3.10 Presence of Microbes

Buried steel pipes are known to have suffered severe corrosion as the result of bacterial action. In unaerated (anaerobic) soils, this attack is attributed to the SRB. The mechanism is believed to involve both direct attack of the steel by hydrogen sulfide and cathodic depolarization aided by the presence of bacteria. Even in aerated (aerobic) soils, there are sufficiently large variations in aeration that the action of SRB cannot be neglected. For example, within active corrosion pits, the oxygen content becomes exceedingly low.

Bacteria, fungi and other microorganisms can play a major part in soil corrosion. Spectacularly rapid corrosion failures have been observed in soil because of microbial action and it is becoming increasingly apparent that most metallic alloys are susceptible

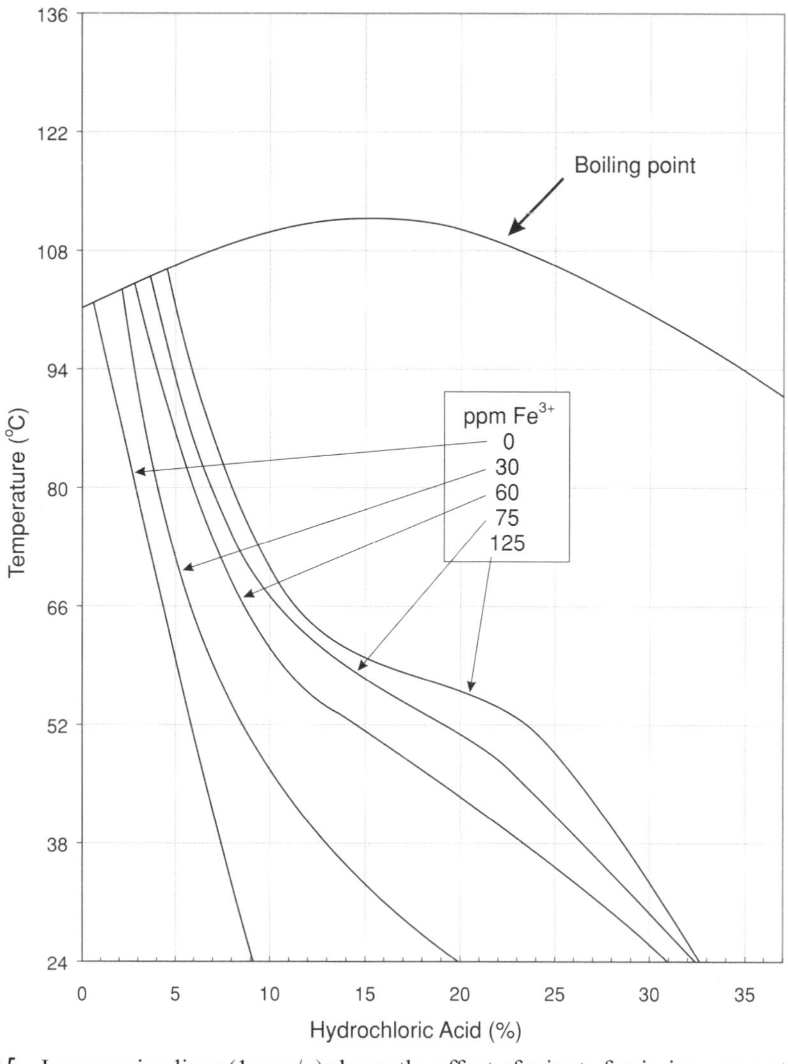

Figure 13.5 Isocorrosion lines (1 mm/y) shows the effect of minute ferric-ion concentrations on the corrosion resistance of Grade 2 titanium in naturally aerated HCl solutions.

to some form of MIC. The mechanisms potentially involved in Microbiologically Influenced Corrosion (MIC) are summarized as follows:

- Cathodic depolarization, whereby the cathodic rate limiting step is accelerated by micro biological action.
- Formation of occluded surface cells, whereby microorganisms form "patchy" surface colonies. Sticky polymers attract and aggregate biological and nonbiological species to produce crevices and concentration cells, the basis for accelerated attack.
- Fixing of anodic reaction sites, whereby microbiological surface colonies lead to the formation of corrosion pits, driven by microbial activity and associated with the location of these colonies.

- Underdeposit acid attack, whereby corrosive attack is accelerated by acidic final products of the MIC "community metabolism," principally short-chain fatty acids.

Certain microorganisms thrive under aerobic conditions, whereas others thrive in anaerobic conditions. The pH conditions and availability of nutrients also play a role in determining what type of microorganisms can thrive in a particular soil environment. Microorganisms associated with corrosion damage are classified as follows:

- anaerobic bacteria that produce highly corrosive species as part of their metabolism;
- aerobic bacteria that produce corrosive mineral acids;
- fungi that may produce corrosive by products in their metabolism, such as organic acids. Apart from metals and alloys these can also degrade organic coatings and wood; and
- slime formers that may produce concentration corrosion cells on surfaces.

These microorganisms are present almost everywhere in soils, fresh water, seawater, and air. Therefore, the mere detection of microorganisms in an environment does not necessarily indicate a corrosion problem. What is important is the number of microorganisms of the relevant types.

MIC is responsible for the degradation of a wide range of materials. A useful representation of materials degradation by microbes was made in the form of a pipe cross section (Figure 13.6).[3] Most metals and their alloys (including stainless steel, aluminum, and copper alloys), polymers, ceramic materials, and concrete can be attacked by microorganisms. The synergistic effect of different microbes and degradation mechanisms is noted in Figure 13.6.

The muddy bottom of any relatively stagnant body of water with a high biological oxygen demand often supports massive growth of SRB, as may waterlogged soils. Any metallic installations buried or immersed in such environments can be expected to suffer badly from microbiological corrosion. The most serious economic problem is to pipe lines, although sheet piles, hulls of ships, piers, and so on are frequently attacked. In some instances, cast-iron pipes of 6.3 mm in thickness have become perforated within one year under such conditions, while perforation in three years is common.

13.3.10.1 Acid-Producing Fungi

Certain fungi are capable of producing organic acids and have been blamed for corrosion of steel and aluminum, as in the corrosion failures of aluminum aircraft fuel tanks. In addition, fungi may produce anaerobic sites for SRB and can produce metabolic byproducts that help the growth of various bacteria.

13.3.10.2 Aerobic Slime Formers

Aerobic slime formers are a diverse group of aerobic bacteria. They are important to corrosion mainly because they produce extracellular polymers that make up what is

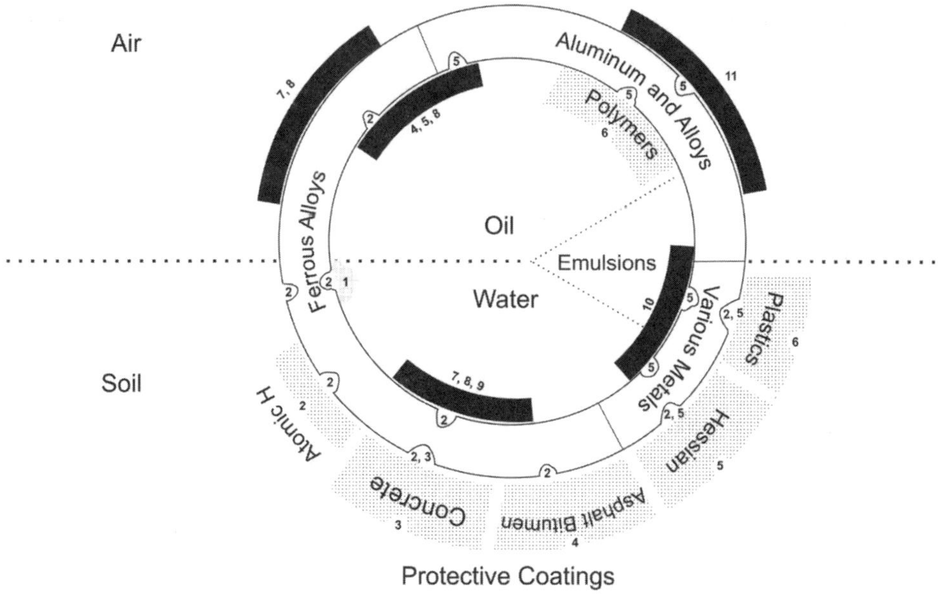

Figure 13.6 Schematic illustrates the principal methods of microbial degradation of metallic alloys and protective coatings: 1) tubercle leading to differential aeration corrosion cell and providing environment for "2"; 2) anaerobic SRB; 3) sulfur-oxidizing bacteria, producing sulfates and sulfuric acid; 4) hydrocarbon utilizers, breaking down aliphatic and bitumen coatings and allowing access of "2" to underlying metallic structure; 5) various microbes producing organic acids as end products of growth, attacking mainly nonferrous metals and alloys and coatings; 6) bacteria and molds breaking down polymers; 7) algae forming slimes on aboveground damp surfaces; 8) slime forming molds and bacteria (which may produce organic acids or utilize hydrocarbons), providing differential aeration cells and growth conditions for "2"; 9) mud on river bottoms providing matrix for heavy growth of microbes (including anaerobic conditions for "2"); 10) sludge (inorganic debris, scale, corrosion products, etc.) providing matrix for heavy growth and differential aeration cells, and organic debris providing nutrients for growth; 11) debris (mainly organic) on metal above ground, providing growth conditions for organic acid-producing microbes.

commonly referred to as "slime." This polymer is actually a sophisticated network of sticky strands that bind the cells to the surface and control what permeates through the deposit.

The stickiness traps all sorts of particulates that might be floating by, which, in dirty water, can result in the impression that the deposit or mound is an inorganic collection of mud and debris. Slime formers can be efficient "scrubbers" of oxygen, thus preventing oxygen from reaching the underlying surface. This creates an ideal site for SRB growth.

13.3.10.3 Algae

Algae are photosynthetic organisms with relatively simple nutritional requirements of light, water, air, and certain inorganic compounds. They can be found in wide ranges of salinity (seawater to distilled water) and survive in different degrees of light intensity.

Algae have been associated with fouling and related corrosion problems. Corrosion reactions can be stimulated by the metabolic production of gaseous oxygen (under the photosynthetic principle, algae produce oxygen when exposed to light) and also corrosive species such as organic acids.

13.3.10.4 Iron/Manganese-Oxidizing Bacteria

Bacteria that derive energy from the oxidation of Fe^{2+} to Fe^{3+} are commonly reported in deposits associated with MIC. They are almost always observed in tubercles (discrete hemispherical mounds) over pits on steel surfaces. The most common iron oxidizers are found in the environment in long protein sheaths or filaments. Although the cells themselves are rather indistinctive in appearance, these long filaments are readily seen under the microscope and are not likely to be confused with other life forms.

The observation that filamentous iron bacteria are "omnipresent" in tubercles might be, therefore, more a matter of their easy detection than of their relative abundance. An intriguing type of iron oxidizers is the Gallionella bacterium, which has been blamed for numerous cases of corrosion of stainless steels.

Besides the iron/manganese oxidizers, there are organisms that simply accumulate iron or manganese. Such organisms are believed to be responsible for the manganese nodules found on the ocean floor. The accumulation of manganese in biofilms is blamed for several cases of corrosion of stainless steels and other ferrous alloys in water systems treated with chlorine or chlorine/bromine compounds.

13.3.10.5 Methane Producers

Only in recent years have methane-producing bacteria (methanogens) been added to the list of organisms believed responsible for corrosion. Like many SRB, methanogens consume hydrogen and thus are capable of performing cathodic depolarization. Although they normally consume hydrogen and carbon dioxide to produce methane, in low-nutrient situations these strict anaerobes will become fermenters and consume acetate instead.

13.3.10.6 Organic Acid–Producing Bacteria

Various anaerobic bacteria such as *Clostridium* are capable of producing organic acids. Unlike SRB, these bacteria are not usually found in aerated macroenvironments such as open, recirculating water systems. However, they are a problem in gas-transmission lines and could be a problem in closed water systems that become anaerobic.

13.3.10.7 SRB

SRB are anaerobes that are sustained by organic nutrients. Generally they require a complete absence of oxygen and a highly reduced environment to function efficiently. Nonetheless, they circulate (probably in a resting state) in aerated waters, including those treated with chlorine and other oxidizers, until they find a "ideal" environment supporting their metabolism and multiplication.

SRB are usually lumped into two nutrient categories: those that can use lactate and those that cannot. The latter generally use acetate and are difficult to grow in the laboratory on any medium. Lactate, acetate, and other short-chain fatty acids usable by SRB do not occur naturally in the environment. Therefore, these organisms depend on other organisms to produce such compounds.

SRB reduce sulfate to sulfide, which usually shows up as hydrogen sulfide or, if iron is available, as black ferrous sulfide. In the absence of sulfate, some strains can function as fermenters and use organic compounds such as pyruvate to produce acetate, hydrogen, and carbon dioxide. Many SRB strains also contain hydrogenase enzymes, which allow them to consume hydrogen. Most common strains of SRB grow best at temperatures from 25° to 35°C. A few thermophilic strains capable of functioning efficiently at more than 60°C have been reported.

Tests for the presence of SRB have traditionally involved growing the organisms on laboratory media, quite unlike the natural environment in which they were found. These laboratory media will only grow certain strains of SRB, and even then some samples require a long lag time before the organisms will adapt to the new growth conditions. As a result, misleading information has been obtained regarding the presence or absence of SRB in field samples. SRB have been implicated in the corrosion of cast iron and steel, ferritic stainless steels, 300-series stainless steels (also very highly alloyed stainless steels), copper/nickel alloys, and high nickel/molybdenum alloys. They are almost always present at corrosion sites because they are in soils, surface-water streams, and waterside deposits in general. Their mere presence, however, does not mean they are causing corrosion. The key symptom that usually indicates their involvement in the corrosion process of ferrous alloys is localized corrosion filled with black sulfide corrosion products.

13.3.10.8 Sulfur/Sulfide-Oxidizing Bacteria

This broad family of aerobic bacteria derives energy from the oxidation of sulfide or elemental sulfur to sulfate. Some types of aerobes can oxidize the sulfur to sulfuric acid, with pH values as low as one reported. These *Thiobacillus* strains are most commonly found in mineral deposits, and are largely responsible for acid mine drainage, which has become an environmental concern. They proliferate inside sewer lines and can cause rapid deterioration of concrete mains and the reinforcing steel therein. They are also found on surfaces of stone buildings and statues and probably account for much of the accelerated damage commonly attributed to acid rain. Where *Thiobacillus* bacteria are associated with corrosion, they are almost always accompanied by SRB. Thus, both types of organisms are able to draw energy from a synergistic sulfur cycle.

13.3.11 Presence of Stray Currents

Stray currents causing corrosion may originate from DC distribution lines, substations, or street railway systems and flow into an adjacent metallic structure. Alternating currents very rarely cause corrosion. The corrosion resulting from stray currents coming from external sources is similar to that from galvanic cells that generate their own current.

However, stray-current strengths may be much higher than those produced by galvanic cells and, as a consequence, corrosion may be much more rapid. Another difference between galvanic currents and stray currents is that the latter are more likely to operate over long distances because the anode and cathode are more likely to be remotely separated from one another. Seeking the path of least resistance, stray currents from a foreign installation may travel along a pipe line, causing severe corrosion where they leave the line. Knowing when stray currents are present becomes very important when remedial measures are undertaken because a simple sacrificial anode system is likely to be ineffective in preventing corrosion under such circumstances.

Where several buried or submerged structures are in close proximity (e.g., a network of buried pipe lines), special attention should be given to the possibility that leakage of current between them could cause localized corrosion. If nonferrous structures of aluminum, zinc, or lead are involved, there is danger of CP overprotection, even at locations where the structure receives current. Similarly, coated-steel structures can have their protective coatings disbonded ("popped off") by hydrogen evolution if the CP currents are too great.

Each situation involving stray currents requires individual study. If an entire buried structure can be brought to the same potential, or if provision can be made to discharge leakage currents through either inert ground rods or replaceable structures such as galvanic anodes, damage to the structure will be minimized.

13.4 DESIGN FOR CORROSION SERVICE

No single text or academic course can give an engineer or technologist the answers to all corrosion problems. A checklist of some potential problems facing them when designing a new installation is provided in Table 13.1. A reference text on this subject has also been published by NACE International.[4]

The avoidance of corrosion should be the long-range goal of all engineers and designers. The best way to avoid corrosion is the imaginative use of knowledge. Proper analysis of failures, intelligent use of technical information, and creative use of engineering principles, combined with a deep appreciation of the economic aspects involved, will result in improved materials selection and eliminate many needless and expensive corrosion failures.

Much can be done in the detailed design to improve corrosion resistance and obtain better service from economical materials and affordable maintenance procedures. There are three important things to remember when designing a system for demanding service[5]:

1. design for complete and free drainage,
2. eliminate or seal weld service, and
3. make it easy to inspect.

Table 13.1 Some Design Considerations to Control Corrosion

Plant Construction
Prevailing wind direction OK?
Method of using cooling water defined (open recirculated, salt water, double exchange, ponds, air cooling)?
Cooling tower downwind?
Closed construction used (pipe or box stanchions, etc.)?
Exterior designs free of pockets?
Tank bottoms protected (cathodically or sealed, depending on type)?
Metal-to-metal contacts protected?
Equipment designed properly for coating?
Will hydrotesting corrode the steel or leave chlorides on stainless steels?
Is a system of material verification needed?
Will all steel be cleaned and primed before erection?
Are temporary preservatives needed for some items?
Should CP be installed now?
Is electrical equipment properly grounded to prevent stray currents?

Equipment Design:
No 'dead' spaces in lines, tanks, vessels?
All possible crevices sealed or flushed?
Inlets and outlets properly contoured? (If not, is protection provided?)
Offsets on mechanical equipment properly filleted?
Other potential fatigue sites?
Can equipment be cleaned adequately if needed?
Will end-grain attack be a problem?
Homogenizing anneal needed?
Stress relief required to forestall SCC?
Are weld deposits compatible with parent metal in chemistry and hardness?
Corrosion allowances added to metal thickness?
Possibility of liquid-metal embrittlement (e.g., zinc on stainless)?
Desiccating vents needed?

Temperature
Evaporation area—vaporization or ebullition cause problem?
Hot spots—from stream injection point?
Hot spots—due to impinging flame?
Cold spots—due to attachment of external metal?
Cold spots—condensation zone a problem? (If so, keep +10°C above dew point.)
Breakaway point exceeded (e.g., organic acids on stainless steels)?
Freeze–thaw zones properly protected?

Velocity
Excessive linear flow?
Turbulent flow (from excess velocity; from design, such as raised tube ends)?
Insufficient flow? (Passive metals need oxygen [O_2]. Vapor voids created?)
Adequate radii in bends?
Stream mixing tangential or in a tank?

Impingement and Cavitation
Restrictions out of lines?
Adequate radii in bends?
Varying diameters in a line?
Pump problems analyzed?
Stream inlets baffled?

Table 13.1 (Continued)

Bimetal Problems
Galvanic couples in an aerated electrolyte?
Heavy metal (Fe, Cu) upstream of aluminum or steel?
Differential expansion problems?
Environmental Control
Neutralization, dehydration, or deaeration of streams considered?
Parameters of operation to reduce corrosion registered with operators?
Two-phase liquid systems adequately evaluated?
Gas–liquid interfaces evaluated?
Solids in any gas or liquid stream?
Adequate alarms for breakthrough of corrosive agent (ppm can be disastrous)?
Will atomic hydrogen be present? What effect?
Could bacteria create a corrosive situation?
Testing
Initial testing defined required materials?
Extra nozzles available on equipment for later test-rack insertion?
Resistance wire probes of help?
Linear polarization instrument of help?
Can side stream be built for critical test area?
Corrosion test program planned firmly?
Ultrasonic bench-mark readings made on critical equipment?

One major oil refinery installed approximately 65 km of aluminum instrument airline tubing serving a catalytic cracking unit. The tubing was clustered where possible and was supported in steel channels. Unfortunately, these supporting channels were, in many cases, installed with the flanges pointing upward. This allowed rain water, insulating material, test solutions, cleaning solutions, and so forth to collect in the channels containing the aluminum tubes during the construction process. When final tests were made, many of the aluminum tubing runs would not hold pressure. Examination of the tubing revealed extensive pitting and perforation of a number of the tubes. If adequate drainage had been provided in the design, it is likely that no such difficulty would have been encountered.

13.4.1 Designing Adequate Drainage

It is good engineering practice to provide adequate drainage. It cannot be overemphasized that corrosion control must start at the drawing board and that design details are critical for ensuring adequate long-term corrosion protection. It is generally good practice to provide adequate ventilation and drainage to minimize the accumulation of condensation (Figures 13.7 and 13.8) and avoid depressed areas where drainage is inadequate (Figure 13.8).

In the design of tanks, the drain should be at the lowest point (the bottom should slope toward the drain). The design of tank supports also is of vital importance. Water left standing in the bottom of stainless-steel tanks has been a source of tank-bottom failures in both fresh and saline waters.

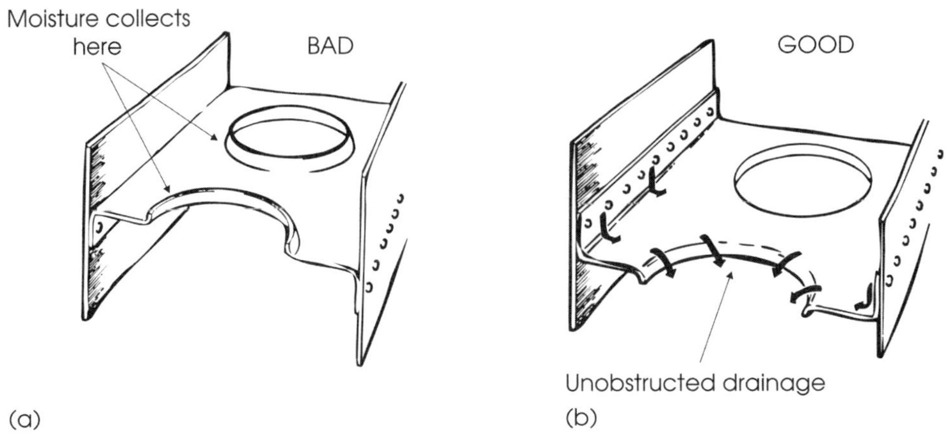

Figure 13.7 Lightening holes in horizontal diaphragms.

Poor side outlets and center outlets designs, shown in Figures 13.9a and b, allow for convenient construction but invite early failure of tank bottoms. Not only is a layer of stagnant water held on the tank bottom but sediment cannot be easily flushed out.

Good designs of a flush side outlet and a recessed bottom outlet as shown in Figures 13.9c and d allow the bottom to be completely drained and all debris and sediment to be flushed out, leaving the bottom clean and dry. However, the sloped arrangements shown in Figures 13.9e and f make it easier to flush out and clean and are therefore the best designs.

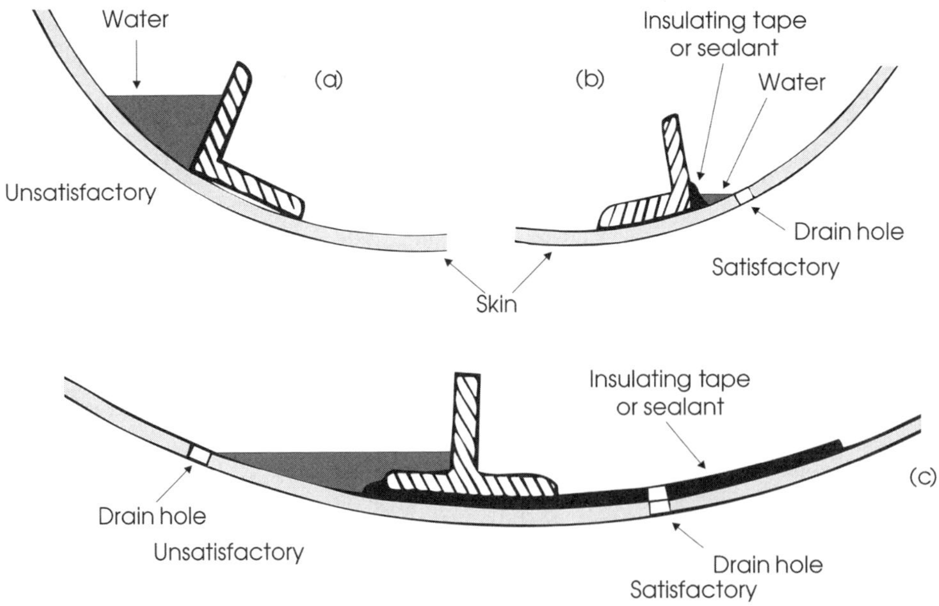

Figure 13.8 Water traps and faying surfaces.

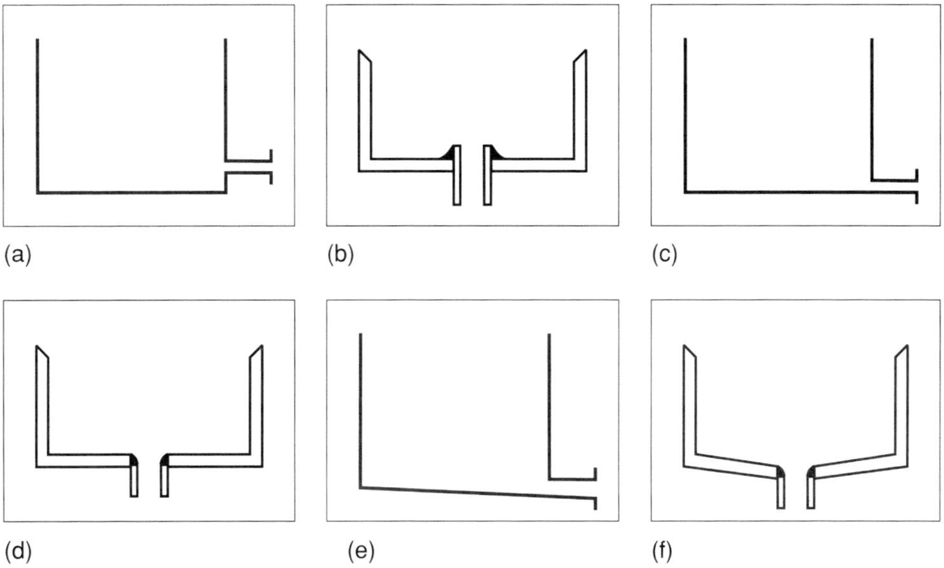

Figure 13.9 Tank bottom outlets: a) and b) poor designs; c) and d) good designs; e) and f) best designs.

Crowned pads for flat-bottomed tanks will encourage free drainage of contents, particularly where a sump is provided. Crowning also encourages free drainage of liquids from under the tank floor under variations of hydrostatic head, which can pull corrosive liquids under the tank. Use of pads slightly smaller in diameter than the tank base will reduce the danger of moisture and chemicals being drawn under tanks by surface-tension effects.

Corrosion of a large brewery fermenter had resulted in considerable loss of brew. Close examination revealed that corrosion started from the outside of the aluminum fermenter under vertical, hat-shaped stiffeners used to stay the sidewalls of this rectangular, open-top tank against hydrostatic loadings. Apparently, for aesthetic reasons, the tank fabricator had plugged the bottom of each of the sidewall stiffeners with a welded plate. The tops of the stiffeners were above eye level and so were left open. Moisture and dirt accumulated and corrosion resulted, causing failure. The cost of repairs was almost as much as the original cost of the tank.

Six common tank-bottom arrangements are illustrated here with comments from a corrosion-engineering point of view.

1. **Worst**(Figure 13.10a): The square-corner flat-bottom arrangement invites early failure from the inside at the corner weld where sediment will collect, increasing the probability of crevice attack. Moisture penetrating the flat bottom to pad support invites rapid crevice corrosion from the underside.
2. **Good corners, poor outside** (Figure 13.10b): The rounded bottom shown here is much more resistant from the inside, but is actually worse from the outside because condensation is funneled directly into the crevice between the tank bottom and pad support.

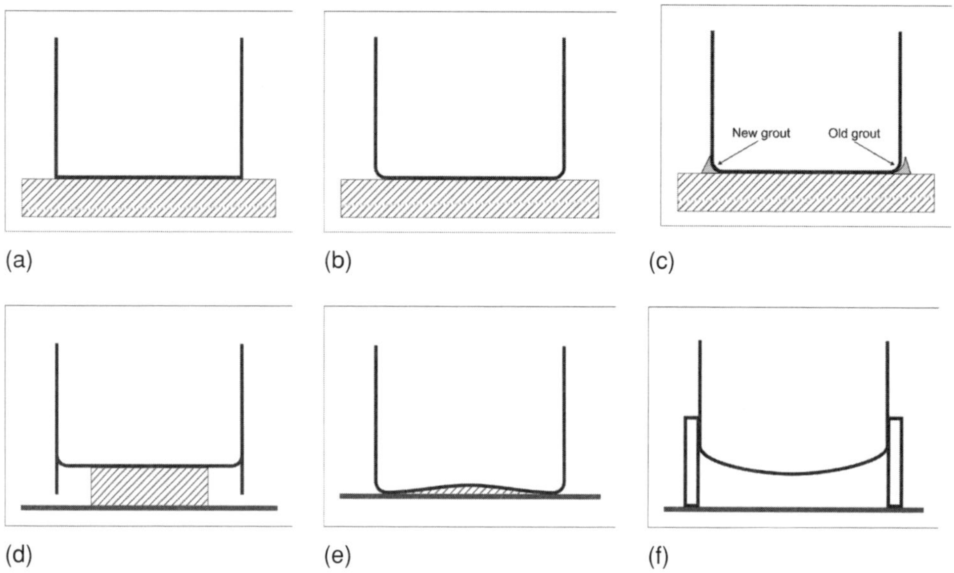

Figure 13.10 Tank bottoms: a) worst; b) good corners, poor outside; c) poor inside, poor outside; d) good inside, good outside; e) good inside, good outside, fatigue resistant; f) best inside, best outside, fatigue resistant.

3. **Poor inside, poor outside** (Figure 13.10c): The grout used here to divert such condensation does help initially but soon shrinks back and becomes a maintenance demand itself.
4. **Good inside, good outside** (Figure 13.10d): The drip skirt shown here is the best arrangement for flat-bottom tanks.
5. **Good inside, good outside, fatigue resistant** (Figure 13.10e): The concave bottom shown here and the dished head bottom on supports below are very good and superior to all flat-bottom tanks not only in corrosion resistance but also in fatigue. Fatigue stresses from filling and emptying are seldom considered in design but can be significant and have led to failures in flat-bottom tanks. The concave and dished-head arrangements can withstand much greater fatigue loadings than can flat bottoms.
6. **Best inside, best outside, fatigue resistant** (Figure 13.10f): Dished-head bottom.

13.4.2 Adequate Joining and Attachments

Nonstabilized austenitic stainless steels (e.g., S30400, S31600, etc.) must sometimes be quench-annealed following fusion welding to avoid sensitization to intergranular attack caused by chromium carbide precipitation and chromium depletion. Therefore, fabricated parts that cannot be heat treated readily after welding should be specified in low-carbon or stabilized grades such as S32100 and S34700. However, even the stabilized grades can suffer a form of intergranular attack (knife-line attack) under specific circumstances.

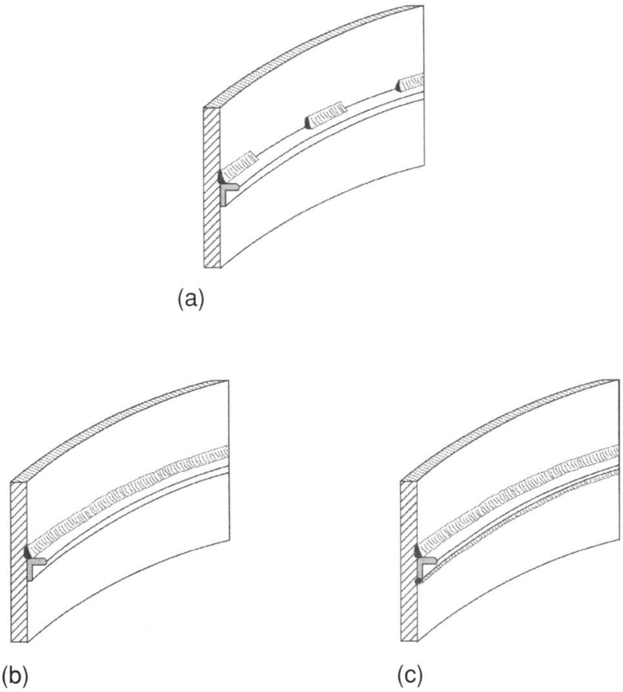

Figure 13.11 Structural elements: a) adequate support, severe crevice; b) good crevice resistance; c) best crevice resistance.

All attachments create potential crevice sites. Figure 13.11a shows a tray support angle with intermittent welds adequate for strength. There is, however, a severe crevice between the angle and the inside wall of the vessel, which will become filled with debris and invite premature failure from crevice corrosion.

Figure 13.11b shows the same tray support, with a continuous seal weld at the top preventing unwanted material from finding its way down the wall and into the crevice, is still open from the bottom but this is a much less severe crevice into which severe crevice vapors but not material can still enter. Figure 13.11c is a full seal weld at the top and bottom of the tray support angle. Here the crevice is fully sealed.

When the side wall of bottom corner welds forms a right angle with the bottom, the fillet weld is seldom as smooth as shown in Figure 13.12a. It is usually rough

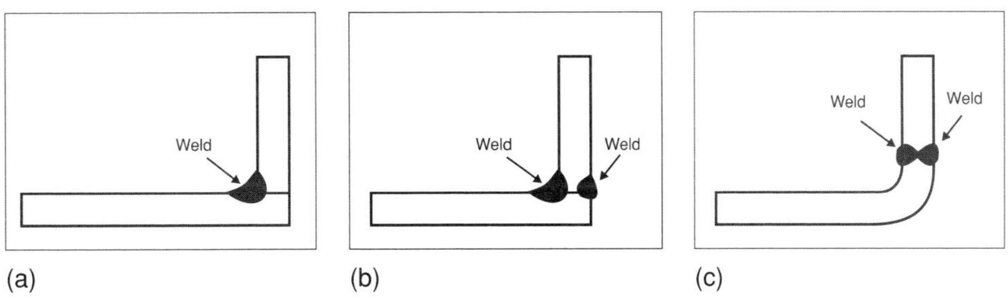

Figure 13.12 Bottom corner welds: a) poor inside, worst outside; b) poor inside, good outside; c) best inside, good outside, fatigue resistant.

and frequently varies in width, compensating for variations in fit-up. Because of the location, it is very difficult to grind and blend the weld into the adjacent sides. Debris tends to collect and is difficult to remove, leading to under-sediment-type crevice attack. Unless welded from the outside as shown in Figure 13.12b, the crevice is vulnerable to crevice attack. However, the inside weld area still is a cause for concern.

Rounding the corner and moving the weld to the side wall overcomes both shortcomings as shown in Figure 13.12c. This construction has much improved corrosion resistance and also has better fatigue resistance.

In handling chemical products such as fatty acids containing minute traces of sulfuric acid in aluminum piping systems, alloy selections for pipe and weld wire should be made with care. The 6000-series aluminum alloys have often been welded with 4043 (Al–5% Si) weld wire. This combination can suffer knife-line attack adjacent to the welds in the presence of dilute mineral acids. A better combination under these conditions is 3003 (A93003)-alloy pipe with 1100 (A91100)-alloy weld wire.

Care should be taken to make certain that all brazing flux residues are removed from brazed assemblies. These residues are often hygroscopic and can draw moisture from the air, thereby establishing their own corrosive solutions.

References

1. N.W. Sachs, "Understanding the Multiple Roots of Machinery Failures," Reliability Magazine 8 (2002): pp. 18–21.
2. ASTM G 161-00, "Standard Guide for Corrosion-Related Failure Analysis" (Philadelphia, PA: ASTM, 2000).
3. E.C. Hill, Microbial Aspects of Metallurgy (New York, NY: American Elsevier, 1970).
4. R.J. Landrum, Designing for Corrosion Control (Houston, TX: NACE, 1989).
5. R.E. Avery, A.H. Tuthill, "Guidelines for the Welded Fabrication of Nickel-Containing Stainless Steels for Corrosion Resistant Services," Nickel Development Institute (NiDI), NiDI Report 11007, Toronto, Canada, 1992.

Part V

Testing and Monitoring

14

Chapter

Corrosion Testing and Monitoring

14.1 INTRODUCTION

Metals and their alloys are used in a multitude of changing conditions and corrosion failures are not uncommon. The limited predictability of metal performance is the main reason why corrosion tests and corrosion-monitoring programs are so important. Properly conducted, these tests can mean serious savings. In many cases they are built-in requirements.

Corrosion can lead to failures in plant infrastructure and machines that can be costly to repair, costly in terms of lost or contaminated product, costly in terms of environmental damage, and ultimately costly in terms of human safety. Decisions regarding the future integrity of a structure or its components depend entirely upon an accurate assessment of the conditions affecting its corrosion and rate of deterioration.

Corrosion tests are probably the simplest means to avoid using materials under unsuitable conditions or using a more expensive material than is required. Corrosion tests also help in the development of new materials that perform more inexpensively, efficiently, longer, or more safely than materials presently used. Quality-control corrosion tests are also a means of ensuring that the materials have the capabilities expected of them.

Corrosion-testing programs can be relatively simple and be completed in a few minutes or hours, or they can be complex and require the combined work of a number of investigators over a period of weeks, months, or years.

Conversely, modern corrosion-monitoring technologies can reveal the highly time-dependent nature of corrosion damage. The integration of corrosion-monitoring technology in existing systems can also provide early warning of costly corrosion damage and provide information on the location where damage possibly could occur.

14.2 CORROSION TESTING

Corrosion tests are divided into two broad categories: (a) tests made in the laboratory under controlled conditions and (b) tests made in the field under natural or service conditions.

14.2.1 Laboratory Controlled Tests

Tests in the laboratory are made with pure chemicals under closely controlled conditions. Each test should be somewhat reproducible in repeat runs of fixed duration. An assessment of this reproducibility is especially important when a new test method is used or when a new alloy or fabricated item is evaluated. If reproducible conditions can be achieved, then differences in data truly reflect a difference in the resistance to corrosion of the materials being tested and the investigator knows if an improvement has been made.

Laboratory tests provide an early indication of what can happen in actual practice. However, the time required for an "indication" depends on the purpose and nature of the tests. Laboratory corrosion tests generally are conducted over a period of time ranging from a few minutes to as long as a year or more.

14.2.1.1 Tests of the Metal

Laboratory tests usually are made with small metal specimens of specified shape and size. Some typical test specimens are shown in Figure 14.1 Such specimens have the advantages of being relatively inexpensive, small, and reproducible. They are intended to evaluate the corrosion-response characteristics of a given metal. For example, they should show how "steel A" responds differently than "steel B" or how some other metal would perform under these same test conditions.

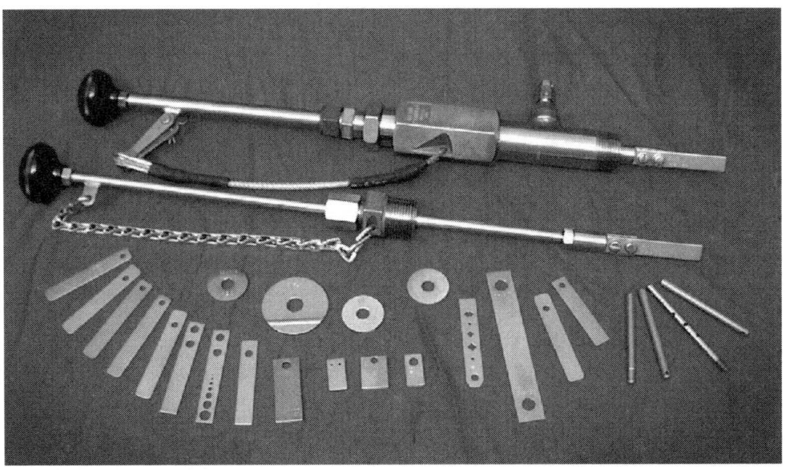

Figure 14.1 Various corrosion coupon shapes and forms for laboratory and field testing (courtesy of Metal Samples).

Even for the relatively simple types shown (disk, panels, and tensile specimens), a variety of sizes and shapes are employed. Certain specimens contain the full thickness of the product, such as the flat tensile specimen and the cast and sheet panels. Others, such as the round tensile specimens and the stepped panel, are machined from a particular region of thicker products.

In protective-coating evaluations, panels generally have well-rounded edges to facilitate uniform coating formation. Sometimes coated or clad specimens are intentionally damaged; one form of coating damage is various scribe marks through the film. Other forms of damage might be impact blows to cause crazing, or countersunk holes.

Many other types of specimens are used to evaluate the effect of applied stress on corrosion.[1] These types of specimens, however, are designed for evaluation of the resistance of the alloy, without much regard to its end use. Certain types, such as the two tensile bars and the sheet tensile specimen, are exposed unstressed as well as stressed. This permits the determination of the effect of stress on reduction of strength by corrosion. Numerous other specimens are also used to evaluate specific products or methods of stressing.

14.2.1.2 *Tests of Fabricated Items*

Fabricated items also are evaluated in laboratory corrosion tests. Usually, this is done on a limited basis to establish a correlation between tests on cut or machined specimens and the fabricated item to see if results with small test specimens are giving valid answers. This may be necessary when the final product is so complex that the investigators cannot be sure that the test data obtained for each small component give a true evaluation of the behavior of the final assembly.

14.2.2 Field Tests

Field tests can be much closer to end use than the usual laboratory tests and normally range in duration from several months to many years. Some tests have been conducted for more than 50 years at well-established outdoor facilities. Field tests, for example, may consist of exposure to these corrosive conditions:

1. **Natural Environments**
   a) Atmospheres: seacoast, industrial, rural (see Chapter 3).
   b) Waters: distilled, fresh, salt, flowing (Figure 14.2), stagnant, tidal (Figure 14. 3).
   c) Soils: dry, moist, clay, sandy.
2. **Particular Industrial Applications**
   a) Architectural.
   b) Processing plants.
   c) Transportation vehicles (Figure 14.4).
   d) Storage tanks.

Figure 14.2 Seawater erosion–corrosion test rig (courtesy of Corrpro Companies).

Figure 14.3 Tidal-wave test rack (courtesy of Corrpro Companies).

(a) (b)

Figure 14.4 Salt corrosion/temperature/humidity chamber is designed to provide accelerated corrosion life testing of body, frame, and undercarriage components of a) passenger cars and trucks as well as b) full vehicles (courtesy of Despatch Industries).

Figure 14.5 Oil-recovery microbial test lines evaluate biocide programs (courtesy Kingston Technical Software).

3. **Particular Substances**
   a) Concrete
   b) Fertilizers
   c) Biocides (Figure 14.5)
   d) Petroleum fuels (Figure 14.6)
   e) Flue gas
   f) Acids
   g) Bases
   h) Solvents

For the most part, these tests use small cut or machined specimens, but assemblies of simulated components and complete systems are also tested as shown in Figure 14.4.

Figure 14.6 Oil flow line evaluates inhibitor efficiency (courtesy of Kingston Technical Software).

Field tests are often conducted for many of the same reasons that laboratory tests are made. New and promising materials must be evaluated, changes in the environment and the effect on materials of construction must be monitored, and a constant appraisal of the corrosion occurring in many environments is necessary. Such testing is obviously of great value when making plans for maintenance of equipment and can improve the economics and safety of a system.

14.2.3 Test Objectives

Corrosion tests can also be subdivided according to their basic objectives, which in turn depend on the types of environments and test durations that interest an investigator.

14.2.3.1 Fundamental Investigation

In fundamental corrosion research, the investigator seeks to determine how or why a particular form of corrosion occurs. The aims of the investigation are not necessarily tied to any particular product or use, and seldom is there a set time for completion of the objective. Tests in this area are almost exclusively conducted in a laboratory setting using small specimens and highly specialized techniques suited to the investigation.

14.2.3.2 Alloy and Product Development and Improvement

Researchers in the field of alloy and product development and improvement can seek to improve for tomorrow the materials and products of today. They might want the bridge, automobile, or home of tomorrow to be more resistant to deterioration by corrosion than those presently in use. The aims of such efforts are tied directly to related applications with the intent of achieving improvements in the near future.

The researcher will start with small specimens and laboratory tests, and, as the material shows promise, will proceed through all phases of testing.

14.2.3.3 Material Selection

The engineer who designs a bridge, automobile, or building must decide which of many available materials will be used. Often, data developed by suppliers must be relied upon to provide this information. If this is insufficient, comparative corrosion tests must be conducted among candidate materials for the specific application.

One area that must be considered is the compatibility of different alloys with contacting metallic or nonmetallic materials. Actual parts testing may be involved, but test durations usually are short term and are geared toward answering questions about a particular application or environment. As such, the tests conducted must be directly correlated with the end use of the material.

14.2.3.4 *Quality Control*

After a choice of the material for an application has been made, the fabricator and purchaser then need to know if the quality of incoming material is as specified. Quality-control corrosion tests, such as the Huey test (5 to 48 h in boiling 65% nitric acid) and Strauss test (72 h in boiling 3% cupric sulfate, 10% sulfuric acid) for stainless steels or the intergranular corrosion test [6 h in 1% hydrogen peroxide, 5% sodium chloride at 30°C for heat-treatable aluminum alloys,[2] ensure that the material meets specifications. Such tests must be sufficiently rapid to avoid undue delay of metal shipments. Corrosion tests of this type do not require any particular correlation to end use. Alone, these tests simply ensure that the metal shipped has the same corrosion resistance as the metal previously tested.

14.2.3.5 *Plant Maintenance*

As indicated previously, corrosion tests are important in maintaining the reliability of buildings and equipment already in use. Periodic tests determine whether a material fulfills design requirements. Such tests can involve cut specimens, coated and uncoated, exposed under operating conditions. They might also involve evaluations of parts removed from the actual building or operating unit. Tests of this type also yield practical information for the selection of materials for future buildings.

14.2.4 Procurement of Test Materials

The first step in corrosion testing is procurement of the material to be tested. In some cases, such as quality-control tests or failure analyses, the type and lot of material to be tested are predetermined. In most other cases, there is considerable freedom of choice.

14.2.4.1 *Commercial Alloys*

If commercial alloys are to be tested, the best procedure is to obtain mill-fabricated material representative of current production. If possible, a fabrication history should be obtained listing the major fabricating steps, together with a precise analysis of the metal composition. Table 14.1 shows an example of one method used to report the fabrication history. A standard form is prepared and the pertinent data is then merely filled in. Although the data required may vary with the type of metal and product being made, the more important data, such as ingot size, composition, times and temperatures of heat treatments, and tensile properties, are required for all metal systems and products.

At the very least, the material should be certified as being within the composition limits specified for the alloy and meeting the strength or hardness specifications. Microscopic examination may also be necessary to ensure that the material is in proper metallurgical condition. Such basic information may prevent misleading conclusions as a result of off composition or improper fabrication.

Table 14.1 Fabrication History for Standard Evaluations of Aluminum-Alloy Extrusions

EXTRUDED SECTION: DATE:
ALLOY: LOT NO:

A. **Melting and Casting Operations**

1. Charge Components:
2. Alloying Procedure:
3. Fluxing Procedure:
4. Casting Data:

Casting Method	Ingot Diameter	Pouring Temperature	Metal Head	Lowering Rate	Water Volume	Skimmer

5. Cast Analysis: (percent of alloying elements, remainder aluminum)
   Cast No. Cu Fe Si Mn Mg Zn Ni Cr Ti Br Pb
6. Gas Determinations:
7. Etch Results:

B. **Preheating and Extrusion Operations**

1. Preheat: Soaking Temperature: Time at Temperature:
   Total Cycle Time:
2. Reheat: Furnace Seeing: Cycle Time:
   Exit Time:
3. Extrusion: Type of Die: Ingot Six:
   Maximum Effort: Maximum Exit Speed:
   Runout Length: End Scrap, Front Rear:

C. **Finishing Operations**

1. Solution Heat Treatment: Furnace Type:
   Soak Temperature: Maximum: Minimum:
   Time at Temperature: Cycle Time:
   Type of Quench: Rate of Quench:
2. Stretcher Straightening Operation:
   No. of inches per foot: inches per feet.
3. Additional Straightening Operations (Give Sequence):
4. Precipitation Heat Treatment (Aging): Furnace Type:
   Soak Temperature: Maximum: Minimum:
   Time at Temperature: Cycle Time:
5. Etch Result:

D. **Mechanical Tensile Properties**

Test Direction:
Elongation:

Test Location	Tensile Strength	Yield Strength	(Elongation)
FRONT			
MIDDLE			
REAR			

14.2.4.2 *Nonstandard Alloys*

When nonstandard alloys or tempers are to be evaluated, it may be economically impractical to fabricate them using large production equipment. In such cases, small laboratory casts or heats are made, sometimes weighing only a few kilograms. Although this is a valid procedure for initial tests, before a promising alloy candidate may "go commercial," it should be evaluated several times using specimens from quantities of material large enough to be representative of production variations. Evaluation of several production lots is necessary because, all too often, promising results from a single lot of material are not reproduced on subsequent fabrication runs.

14.2.4.3 *Metal Form*

Another preliminary decision is the form of the metal to be tested. Metals are available in two basic forms: cast and wrought (hot worked). These two forms usually cannot be interchanged in testing. The various methods of casting (die, permanent mold, sand mold, etc.) and of working (drawing, extruding, forging, rolling, etc.) will affect grain structure and homogeneity that, in turn, affect the corrosion resistance. The metal procured should be as nearly as possible of the type that will be used in the final product.

If it is anticipated that the final use of a metal will be in a cast form, then cast specimens should be evaluated. The same applies for wrought-metal specimens. Data obtained from cast- and wrought-metal forms cannot be interchanged reliably.

14.2.5 Specimen Preparation

14.2.5.1 *Size and Shape*

Having selected and procured the test material, the next step is to prepare test specimens. The size and shape of the specimen should be limited by the material being tested and the test environment. The type of specimen used must permit ready evaluation. If several characteristics are to be evaluated, more than one type of specimen may be required. In addition, the following considerations can be important.

14.2.5.1.1 Visual Examination. Visual examinations of the specimens should be made in all corrosion tests. When the appearance of the end product is important, such as for decorative trim or architectural applications, a fairly large surface area should be used, typically 0.25 m^2 or larger, to permit a reliable appraisal in the event that corrosion is not uniform. Specimens of only a few square centimeters can give misleading results.

14.2.5.1.2 Depth of Attack. Specimens used to evaluate corrosion by measurement of the depth of penetration should be thick enough so that they are not perforated by the corrosion attack. Other than this thickness consideration, no particular size or shape is required, but the size or area of the specimen will determine the amount of corrosive agent needed. Also, the specimen should be large enough, or a sufficient number of small specimens should be exposed, to include all important metallurgical or manufacturing variables. Specimens for this purpose usually are kept as small as possible for convenience, and because there are no strict requirements, data often can be obtained from specimens designed for some other test.

14.2.5.1.3 Weight Loss or Gain. Weight-change measurement also does not require any definite size or shape, but a large area-to-volume ratio is used for better sensitivity. Usually, a flat square or rectangle is used to simplify measurement of surface area. Specimens should be kept relatively small to permit simple and accurate weight measurements. It is accepted practice to use a standardized size and shape for all specimens in a given test series so that the same surface area is exposed with the same degree of accuracy in measurement and calculations.

14.2.5.1.4 Loss in Tensile Properties. If loss in tensile properties is to be evaluated, the best procedure is to use one of the standard ASTM tensile specimens, such as those shown in Figure 14.1.

Round tensile specimens are machined prior to exposure. When sheet or thin plate, for example, up to 1.3 cm in thickness, is to be tested, a full-thickness, flat tensile specimens of the type shown in Figure 14.1 can be used. In this case, tensile specimens can either be fully machined before exposure or they can be machined from a corroded panel after completion of the corrosion test.

Full-thickness specimens machined before exposure will give an earlier indication of the effect of corrosion, but the indicated degree of loss in tensile properties, particularly elongation, usually is unrealistically high because of notch effects produced by corrosion of the edges of the specimen. When edge corrosion is not critical in actual applications, for example, roofing sheet, a more realistic appraisal of the loss in strength and elongation can be obtained by machining full-thickness specimens from a corroded test panel, and in this way edge corrosion effects in the appraisal are avoided.

14.2.5.1.5 SCC Tests. The selection of a specimen for SCC tests is complex but is primarily dependent upon the ability to apply stress uniformly in a specific metallurgical direction. (A more detailed discussion of these factors is given in Chapter 9).

14.2.5.1.6 Corrosivity of the Test Environment. Another important factor to be considered in conducting a corrosion test is the corrosivity of the test environment. Shorter exposure times and thicker specimens are required when the test conditions are highly corrosive. If too small a specimen is used and it is completely dissolved or perforated, or fragments from it are lost, only limited information can be obtained. Conversely, it is also a mistake to use a specimen that is so thick that the overall effect of corrosion is too slight to be accurately determined.

14.2.5.1.7 Suitability to Other Test Purposes. Finally, if factors other than the metal and the corrosive environment are to be evaluated, the investigator must be sure that the specimen is suited to the particular test purpose. For example, if protective paints or metallic coatings are to be evaluated, the corners of the specimen should be well rounded before coating. Films are always thin or otherwise disturbed at sharp corners, and this presents unrealistically weak points for corrosion to begin. If galvanic corrosion is to be evaluated, the cathode–anode size ratios and geometry must be known and controlled.

14.2.5.2 *Specimen Identification*

An investigator never relies on memory to ensure identification. This is especially important when many specimens are tested, when tests are conducted over several months, or when several persons process the same specimens.

A suitable identification, usually a specimen number, should be given to each specimen. The first part of the number should be the identification given to the particular lot of metal from which the specimen was obtained. The second part of the number, sometimes called a dash number, is the identification of the particular specimen. When several specimens are taken from the same lot of metal, the various dash numbers may be assigned in numerical order. Frequently, a code letter is used to indicate a particular type of specimen or environment. In such a case, the various specimens of the same type or for the same purpose are then numbered consecutively. The important feature is that the investigator must be sure each specimen can be fully identified.

Whenever possible, the identification number should be marked on the specimen. The mark on the specimen should be such that it can withstand the entire corrosion test. The most common and best method is an impression stencil, except for cases where it is known that the performance of the specimen would be affected by the stresses or surface damage introduced. SCC has often been detected around these numbers.

When only minor corrosion is expected, scribing with an engraving tool, indelible ink, or paint markers may be adequate. If extensive corrosion is expected, the specimen number should be protected with a suitable inert masking material such as wax, tape, or paint. Another method is to attach a nonmetallic marker that is not affected by the corrosive medium. Strips of plastic sheet, scribed with an engraving tool and fastened with plastic cord, are excellent for this purpose.

If test conditions are such that paint or other "foreign" substances might affect the results, they should not be used. In such cases, specimen identity is retained by marking the rack or container and recording the location of the specimen. The common "lead" pencil should never be used for marking specimens, as the clay/graphite mixture is electrochemically reactive and can initiate sites of galvanic corrosion on most metals.

14.2.5.3 *Replicate Specimens*

A certain amount of data scatter is inevitable in any testing procedure. The amount of unavoidable scatter depends upon

1. the uniformity of the material tested,
2. the accuracy in preparing the test specimens, and
3. the stability of test conditions.

All these factors influence the accuracy of corrosion testing. Therefore, a good testing procedure will provide some method of cross-comparison or double-checking to eliminate the possibility of an incorrect conclusion based on a single atypical result.

The use of a single specimen may be a valid procedure, provided more than one measurement can be made on the specimen. For example, loss in strength can be

Figure 14.7 Bank of 50 NACE TM0177 method "A" constant-strain test facility with uniaxially loaded tensile specimens (courtesy of Bodycote Materials Testing).

determined at several places in a single, corroded, 10 × 23 cm panel. Using a single specimen may be a poor procedure, however, if it provides only one value, such as loss in strength on a premachined tensile specimen or failure of a single SCC specimen. This becomes especially important if it is known that the test procedure can vary. In such tests, a cross-check is provided by the use of multiple specimens.

Generally, three or more specimens are used to establish a trend (Figures 14.7 and 14.8). Duplicate specimens are not suitable because they can result in two widely different values, which do not clarify the results. The number of specimens used in a test must be a compromise among

1. the accuracy needed (statistical evaluation),
2. the cost of a test,
3. the known reproducibility of the test, and
4. the number of variables investigated.

14.2.5.4 Machining of Specimens

Many products, such as sheet, certain structural shapes, tube, and pipe are used in service just as they are received from the supplier. Consequently, if their surface condition is fairly consistent, it may be possible to test the corrosion resistance of these products after simple degreasing. When the surface is inconsistent because of dirt, scale, or oxide, steps must be taken to reach a more uniform metal surface. Generally, an acid pickle or a caustic etch is used, depending on the metal.

Other products such as plate, rod, bar, or forgings are generally machined into the final configuration before they are used. Such products are best tested with a machined surface, employing good shop procedure to minimize roughness variations.

Close dimensional tolerances are required for specimens used in tests to evaluate characteristics such as loss in strength or resistance to SCC. Accordingly, such

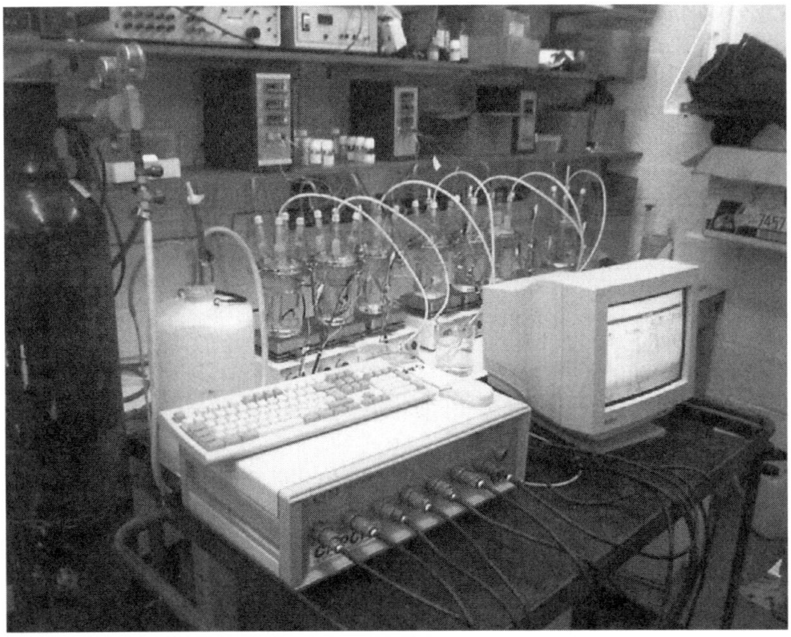

Figure 14.8 Parallel inhibitor testing in the presence of hydrogen sulfide with polarization measurements made on many similar systems at the same time (courtesy of CAPCIS Ltd.).

specimens should be made by experienced machinists. A few of the more important considerations during machining are

1. properly sharpened tools and appropriate feed rates to avoid heating or mechanically damaging the specimen;
2. noncorrosive cutting lubricants (especially important if water-soluble oils are used—in certain cases, dry machining may be necessary);
3. dissimilar metals such as steel, brass, and aluminum should be finished on separate pieces of equipment to prevent contamination of one metal with particles or chips of another metal. This is especially important when using grinding equipment.

A sawed edge may be used if the specimen is evaluated merely for depth attack on the surface. Edges should be filed, machined, or ground smooth if edge attack is to be determined or if weight loss will be used in the final determination. Sheared or blanked edges must be machined, ground, or filed to remove the layer of severely cold-worked metal resulting from this operation. The amount of metal removed from the edges should be at least half the thickness of the piece. Drilled or reamed holes can be left as is, but punched holes should be treated as sheared edges, unless the purpose of the investigation is to evaluate the effect of cold work.

The usual procedure in preparing specimens is to keep the area of cut edges small in comparison to the surface area because this is generally the case in actual parts. In any case, the edge-to-surface-area ratio should be known and kept uniform from test to test.

14.2.5.5 *Simulated Machining of Specimens*

If machine-shop facilities are not available, only the simpler type of specimens such as coupons or panels can be prepared. These can be evaluated by depth of attack or weight loss. The usual method of simulating a machined surface is to polish the metal on a suitable abrasive cloth or paper, or to scrub it with a fine abrasive such as wet pumice. Such specimens are most often cut from sheet or thin plate.

Other methods of producing specimens of very hard metals are electrical discharge (EDM) or electrochemical machining (ECM) techniques. In these somewhat complicated methods, special precautions are required.

14.2.5.6 *Degreasing and Final Measurements*

After machining, if weight loss or loss in strength after corrosion is to be determined, the specimens should be degreased just prior to initial measurements of weight and dimensions.

The easiest method of removing oils and greases (except silicone oils) is to use a solvent vapor degreaser. If such equipment is not available, the specimens should be wiped with a fast-drying, low-residue solvent such as alcohol, acetone, or methyl ethyl ketone (MEK). Solvents that are highly flammable (gasoline), are toxic (carbon tetrachloride), or leave a film residue (kerosene) should not be used.

The specimens should be weighed and measured as carefully as possible because subsequent data can be no more accurate than the original data. When the accuracy needed is dependent upon the expected weight loss, measurements should be carried to enough significant figures to keep the possible error in calculations less than 1%. In some cases, particularly for large or heavy specimens, measurement to the first decimal place is sufficient. However, measurements generally are made to as many decimal places as is practicable, for example, to the nearest 0.01 mm and 0.0001 g for small specimens measured with a micrometer and an analytical balance, respectively.

To remove contamination caused by handling, the specimen should be degreased again just prior to exposure to the corrosive medium. After this final degreasing, specimens should be given a "white glove" treatment and handled only at edges to prevent fingerprinting or other contamination that can cause accelerated corrosion. If, for some reason, the specimens cannot be exposed in a reasonably short time, they should either be stored in a desiccator or coated with a protective oil and stored in a dry, dust-free place. If the latter method is used, the specimen will, of course, have to be degreased again just prior to exposure.

14.2.6 Controls

A vital consideration in the corrosion testing of metals is the use of suitable controls, both for the metal and for the test environment. This aspect of corrosion testing is perhaps the greatest single source for error, even by experienced investigators.

14.2.6.1 *Control of the Metal*

Control data for the uncorroded metal consists of measurement of those properties that can be affected by corrosion so that they may be used to assess the effects of corrosion. Typical metal properties used as controls are strength, electrical conductivity or resistance, and reflectivity. Specimens used to obtain control data should be from the same vicinity in the metal stock and of the same type as those used for corrosion tests. Also, the same method of measurement and degree of accuracy should be used for both sets of specimens. This ensures that any differences in data before and after exposure represent only the effects of corrosion.

When corrosion tests are of a short duration, or when the metal is such that the characteristic being evaluated is not changed during long-term storage, the original control data may be obtained just once prior to any corrosion tests and used for reference when needed. Sometimes, certain characteristics will change in time, even in the absence of corrosion. In such cases, a set of control specimens must be provided for each period of corrosion testing. This set is then tested at the same time as the corroded specimens. Some metals, for example, gradually change in strength over long periods of time at room temperature, and the strength of most metals is affected by time at elevated temperatures.

Specimens to be used for periodic controls should also be protected from corrosion and stored at the same temperature at which the corrosion tests are conducted. In long-term field tests, for which temperature obviously cannot be controlled (such as tests in natural atmospheres), it may be desirable to store the controls on site in dry, air-tight containers. This ensures that both the corrosion and control specimens are exposed to the same time and temperature conditions.

In tests at elevated temperatures, special precautions may be necessary to ensure that the control specimens themselves do not undergo high-temperature corrosion.

14.2.6.2 *Control of the Test Environment*

The other required control is a method of determining whether the corrosive medium maintains the intended degree of corrosivity. When an alloy is repeatedly tested by the same method, as in quality-control tests, the investigator may be sufficiently experienced to know whether the visual appearance and resultant data are of the order expected. Often this is not so, especially with new alloys or products, when corrosion performance is unpredictable.

The usual procedure is to expose a well-established, previously tested alloy as a control to determine whether the results on it are consistent with past experience. If they are, the new alloy is considered to have received a valid exposure. However, if the effects of corrosion on the well-known alloy are not typical, the investigator is alerted to the possibility of an error in the test environment and the need for a rerun.

Another control method is to expose specimens for several periods of increasing duration. At least three periods of exposure usually are used (Figure 14.9). Changes in the corrosion rates of metals can be caused either by a change in the condition

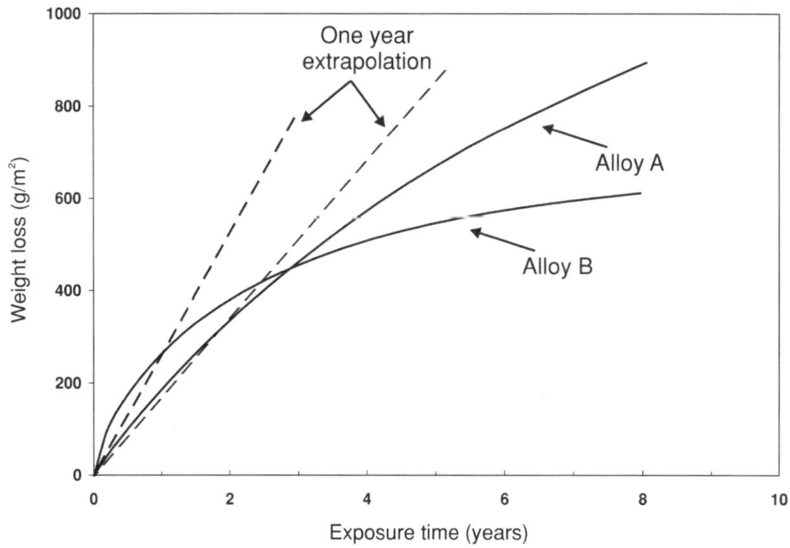

Figure 14.9 Schematic shows why data from several exposure periods are essential. Better resistance of Alloy A was not evident until fourth year of testing. Extrapolation rates based on one-year data are in error for both alloys.

of the metal itself or by a change in the corrosive environment. A properly designed corrosion test should enable the investigator to determine which of these two factors was operative, or whether both had an effect. Comparison of the data on the different groups of specimens then permits evaluation of whether the corrosion rate of the metal was constant (linear rate) or varied with time (either an accelerating or a self-limiting rate).

14.2.7 Specimen Arrangement for Corrosion Tests

The final point to consider, before selecting the corrosion tests themselves, is the manner in which specimens should be arranged or framed for exposure.

In laboratory beaker tests, usually one specimen per container is desirable. If more than one specimen of the same lot of metal is used, multiple specimens may be run in the same container (Figure 14.10), provided the volume of corrosive medium is sufficient to maintain its original properties throughout the exposure.

For plant and field tests, some sort of framing sequence is needed, depending upon the type of specimen and test method. An example of specimens racked for exposure is given in Figure 14.11. The important considerations are as follows:

1. All the specimens must be electrically isolated from one another and from direct contact with the racks. Inert, nonabsorptive racks or spacers of polymeric material or porcelain are usually used for this purpose.
2. The specimens should be arranged so that corrosion products from one specimen will not contaminate other specimens.
3. The corrosive environment should be equally accessible to all specimens.

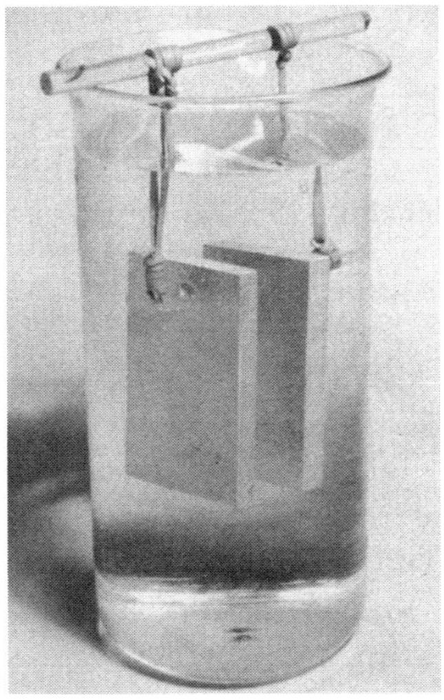

Figure 14.10 Beaker test exposes two specimens. Note specimens are electrically insulated from each other and that supporting rod is suspended from a nonconductor.

4. In liquid immersion tests, specimens of different metals should not be exposed in the same container because one metal can affect the corrosion of another, even though they are not in contact. In fact, even in the same metal system, it may be desirable to separate one alloy type from another. For example, aluminum alloys containing copper are often separated from copper-free aluminum alloys.

In cases where metal containers are used to hold the test media, special precautions are required to prevent metal-to-metal contact between the container and the specimens.

Figure 14.11 Typical specimen exposure rack. Note specimens are electrically separated from each other and from rack. Specimens with bolted-on straps may register galvanic, edge, or crevice effects and the influence of materials such as gaskets. Sequence of specimens should be carefully recorded.

14.2.8 Effects of Variables

The results of a corrosion test on metals can be affected significantly by variations in environmental factors. The single or combined effects of these factors are not always known, but it is essential that the corrosion investigator appreciates their complex interactions.

14.2.8.1 *Temperature*

The temperature of the test solution should be controlled to within ±1.0°C in any laboratory corrosion test. If the temperature of the solution differs from the temperature of the metal it is important to know the actual metal temperature. This can be measured by use of a "dummy control" specimen, equipped with a peened-in thermocouple. In such a case, the thermocouple contact point and leads should be masked by a sleeve or suitable coating. The difference between ambient and metal temperatures becomes especially important in a vapor or humidity test.

Vapors condense on surfaces that are colder than the environment. Consequently, near 100% humidity at 38°C, condensation may have little or no effect on metals warmer than 38°C. In the case of aluminum alloys, condensation produces water staining, a surface discoloration. If the aluminum-alloy temperature is greater than the humidity test temperature, condensation and possibly staining do not occur.

When appraising the effect of heat transfer through a metal, the simple exposure of a specimen in the appropriate solution at a given temperature may not provide the desired information. The metal surface should be heated on one side and cooled on the other to obtain proper results. This is particularly true when determining the corrosion or pitting rate of a metal surface to be used to boil a liquid.

Higher temperatures, greater than 100°C generally produce a more aggressive corrosion attack. Although the condensation in these cases is not a problem, the actual temperature can determine the thickness of the corrosion product oxide layer and extent of attack.

14.2.8.2 *Solution Concentration*

The effect of solution concentration may be an even more important variable than small variations in temperature. Often, metals are corroded by dilute solutions and not corroded by more concentrated ones. Solutions react with metals in a variety of ways, including the following:

1. Some plate out metallic components (cations) on cathodic areas and concentrate anionic components (negatively charged ions) at anodic (corroding) areas.
2. Some solutions corrode metals vigorously for a time and then cease action abruptly.
3. Some organic solutions do not react for a time (induction time) and then react vigorously with the metal.

Hydrogen-ion (H^+) concentrations and hydroxyl-ion (OH^-) concentrations reflect the acidity or alkalinity of a solution (pH). These values can be quite variable on local areas of the metal surface and in the solution, depending upon initial solution concentration, agitation, and corrosion activity. The important considerations are as follows.

1. The bulk solution concentrations must be known and controlled before, during, and after the test.
2. The volume of the solution should be sufficient to avoid expending it before the test is completed.
3. The bulk solution concentration can be quite different from the solution concentrations on the metal surface.

Parts per million of contaminants in a solution may be responsible for essentially all of the corrosion occurring, and their effect must be recognized.

14.2.8.3 Agitation

To assess the comparative corrosion effects on metals of quiescent and moving solutions, they are mildly stirred or otherwise circulated by pumps or jets. Agitation changes the concentrations of the solution on the local anodes and cathodes and provides a more homogeneous test environment.

Simple stirring of a solution is not adequate to determine the actual effect of velocity where liquid flow rate across the specimen face must be under control. An appropriate standard test should be used when evaluating the effect of agitation. A recent publication by NACE International provides a good introduction to this subject.[3]

In gaseous test environments, it usually is necessary to provide some agitation to be certain that vapor concentrations are uniform in the test chamber. In humidity tests, where vapor pressures can vary markedly in a quiet test chamber, air flow ensures equilibrium.

14.2.8.4 Aeration

Aeration, or oxygenation, has a marked effect on the corrosion of metals. Consequently, oxygen is often added by bubbling it through the test solution. Fine gas bubbles are better than large ones because there is more opportunity for the gas to dissolve in the solution during the time it rises to the solution surface.

The counterpart of aeration is deaeration, which can be controlled in a number of ways. Oxygen can be discouraged from entering the test solution by use of a closed system equipped with a trap (Figure 14.12). Another method is to deaerate by introducing an excess of an inert gas, such as nitrogen, into and above the test solution. Oxygen scavengers, such as hydrazine, are also available for use in minimizing the free-oxygen content of solutions.

Under special test conditions, other gases such as carbon monoxide, carbon dioxide, hydrogen sulfide, ozone, nitrous oxide, and sulfur dioxide are metered into liquid or gaseous test chambers to produce special effects on metal corrosion reactions.

Figure 14.12 Photograph shows how a liquid trap keeps oxygen out of test-vessel vapor space.

14.2.8.5 *Ultraviolet Light*

Some chemical reactions are altered photochemically; that is, light of certain wavelengths can alter the chemistry of solutions, gases, or corrosion reactions. Of special interest is the photochemical oxidation of nitrous oxide to peroxyacetal nitrate in the atmosphere, which contributes to air pollution (e.g., smog). In this case, the sunlight acts on the combined nitrous oxide and hydrocarbons from automobile exhausts.

14.2.8.6 *Bacteriological Effects*

In any one set of environmental conditions, several species of fungi or bacteria often may grow and reproduce in solutions, in nonmetallics, and on metal surfaces. Their metabolic (life-cycle) process can produce concentrations of organic or inorganic acids. General examples are corrosion by SRB under anaerobic (oxygen-free) conditions, crude-oil tank bottoms, and aircraft jet-fuel-tank corrosion.

Other organisms produce slime films that can screen oxygen or concentrate metal ions. Each of the innumerable species has its own peculiar requirement for food, oxygen, and temperature. Variations in corrosion test results are caused by changing growth rates and balance of bacterial species.

Although detailed testing for biological corrosion is beyond the scope of this chapter, it is important to recognize that corrosion tests, particularly long-term ones where organic nutrients are present, may be affected by biochemical influences in moist air or aqueous environments.

14.2.9 Conducting the Test

14.2.9.1 *Records*

It is essential that a detailed record be kept for each specimen to help in the eventual analysis of the test. The laboratory notebook and/or other permanent record forms are used for this purpose. When possible, these records should be in a tabulated form that permits easy insertion of test results as they are obtained. Meaningful data should be minimized and recorded promptly to avoid accumulation of unnecessary or confusing paperwork.

14.2.9.2 *Interim Inspections*

Usually, an exposure time is used that experience has shown will yield meaningful results. Because each test probes the unknown, however, the experimenter must be alert for unexpected trends. During the test, frequent examinations are made and notes recorded of any unusual effects on the specimens or in the test environment. Interim inspections often will permit corrections of problems with the specimens or environment, in the case of laboratory tests.

A regular schedule of examinations usually is established at the beginning of the test, with more frequent inspections in the beginning. Appropriate records of specimen condition and environment stability are often necessary to explain unexpected results. Personal judgment must be used to decide whether the original plans for interim inspections were correct. Often, progress reports are advisable to summarize segments of the work separately.

14.2.9.3 *The Unexpected*

When a test contains appropriate comparison specimens, it is unlikely that small environmental variations will be cause for alarm. Sometimes, when tests "accidents" occur, the shrewd investigator can turn the mishap into a "lucky" accident. Many valuable inventions have been the result of such accidents that were observed, evaluated, and reproduced by intelligent investigators who did not hastily discard unexpected results. After all, if the answers are already known, why conduct the test?

14.2.9.4 *Concluding the Test*

Obviously, a decision to terminate an investigation is easiest when there is a fail–no fail criterion. If an earlier investigator has established the proper test durations, the work can be greatly simplified. Innovative tests, on the other hand, rarely signal that the program is completed. The investigator must make such a judgment after weighing results and interpreting trends.

The danger of ending a test too soon is obvious when one realizes how many corrosion processes progress through nonlinear trends. The use of duplicate or multiple specimens helps to predict the long-range results by giving periodic indications of progress. When the graph of results is anything but a straight line with respect to time, periodic reviews are necessary to develop an understanding of what is happening. When satisfied that there is little to gain by continuation of the test, the investigator terminates it. Often, the decision is hastened because of

1. shortage of test space,
2. backlog of other programs,
3. requests by those who need information, and
4. curiosity.

The investigator must weigh these pressures and arrive at a decision. Once it has been decided to end the test, posttest appraisals of the specimen will often prevent its reexposure. A new test would then be required if the data obtained were insufficient.

14.2.10 Posttest Appraisals

14.2.10.1 Before Cleaning

When laboratory tests are concluded, it is advisable to perform certain tasks rather quickly to avoid changes in specimen appearance or additional corrosion. Depending upon the goals, initial steps may include

1. visual examination before cleaning,
2. photographs before cleaning,
3. careful collection of corrosion products for chemical or other analyses, and
4. saving test solutions for chemical study.

Visual examination before cleaning should include the apparent distribution of corrosion products (the relative severity and extent of corrosion will be recorded after cleaning). Also, it should be noted whether the identification number of each specimen has survived the exposure. If not, one should be able to determine the number from the location of the specimen in the frame or other cross-check methods included in the test procedure for such a contingency.

Before discarding any test solutions, the investigator must decide on the possible need for a chemical analysis to evaluate how the solutions may have changed during the test.

14.2.10.2 Analyses of Products and Solutions

Selection of techniques for analysis of corrosion products and solutions is a rather complex procedure requiring background experience. Some typical methods include the following:

1. Analyses for pH and certain ion concentrations using a technique such as ion chromatography for anions (e.g., sulfates, nitrates, etc.) and atomic absorption spectroscopy for cations (e.g., potassium, sodium, and most other ions of metals).
2. X-ray powder diffraction analysis of dry corrosion products. This technique is unique because it can reveal the crystallographic nature of minerals and some corrosion products.
3. Measurement of solution conductivities (the ease of passing an electrical current through a known volume).

14.2.10.3 *Cleaning the Specimens*

Before the extent of corrosion on metal specimens can be completely evaluated, it is usually necessary to clean them according to approved methods.[4] It is essential that special care be exercised in this procedure so that the cleaning will

1. not remove any metal,
2. remove all adherent corrosion products, and
3. facilitate the program of evaluation that will follow.

In some instances, it may be desirable to weigh the specimen before removing the corrosion product to obtain a measure of the weight gained during the test. This is done when the entire corrosion product is intact, as is often the case in high-temperature corrosion experiments on some metals.

14.2.10.4 *Weighing*

After rinsing and thoroughly drying the specimens, they should be weighed carefully, using the same equipment and care as used before the corrosion test.

A laboratory notebook should be used to tabulate final weights, as well as any changes in weight produced by the test. Usually, the level of significant numbers has been predetermined. For example, if a 30-g sample of metal is tested, the weights may be recorded to the sixth significant number. That is, the record may show these results:

	Weight (g)
Original weight:	30.0186
Final weight:	27.3243
Weight loss:	**2.6943**

In the case of a 500-g sample, it may be decided that accuracy to 0.1 g is sufficient:

	Weight (g)
Original weight:	531.6
Final weight:	486.2
Weight loss:	**45.4**

After the weight-change data have been obtained, it is necessary to convert the raw information into suitable units. Obviously, some units may vary between countries or for special criteria. Where possible, however, unique methods of reporting should be avoided. Unusual data forms distract the reader and thereby diminish the effectiveness of communication.

Corrosion data are either reported as relative percent changes in the measured property of a corroded specimen after cleaning (e.g., loss in weight, strength, elongation, conductivity, etc.) or as corrosion rates. There are basically two types of corrosion-rate expressions.

Engineering expressions: microns per year (μm/y), mils per year (mpy), or inches per year (ipy).
Laboratory expressions: milligrams/square decimeter/day (mdd) or mm per year (mm/y).
Sample calculation: If a specimen having a surface area of 6.4516 dm^2 lost 7.3216 g (7321.6 mg) in 50 days,

1. the corrosion rate in mdd would be

$$C.R. = \frac{7321.6 \text{ mg}}{(6.4516 \text{ dm}^2)?(50 \text{ days})}$$

$$C.R. = 22.698 \text{ mdd.}$$

2. For steel (density 7.87 g cm^{-3}) the corrosion rate also may be expressed in mils per year as

$$C.R. = \text{mdd} \times \frac{1.437}{\text{density}}$$

$$C.R. = 22.697 \times \frac{1.437}{7.87}$$

$$C.R. = 4.14 \text{ mpy.}$$

The conversion scheme provided in Table 2.2 for all metals or Table 2.3 adapted to iron or steel provides a simple way to make the conversion between the most common corrosion-rate units, including the corrosion current measurements associated with most electrochemical techniques.

Table 14.2 Mechanical Test Data for Corrosion Test Panels Exposed 12 Weeks to 3.5% Sea Salt by Alternate Immersion

Steel	Tensile Strength (MPa)	Yield Strength (MPa)	Elongation (%)
Before			
Mild steel	283	128	34
Alloy A	530	347	22.5
Alloy B	545	344	19.3
After			
Mild steel	211	91	23
Alloy A	356	242	11.3
Alloy B	405	262	11.8
Difference (%)			
Mild steel	25.3	28.4	32.4
Alloy A	32.8	30.2	49.8
Alloy B	25.7	23.9	38.9

The exposed specimens are those which were corrosion tested. They were cut from the same material for which the "unexposed" data are cited. The calculations therefore represent the losses as a percentage of original values.

14.2.10.5 Visual Examination

Visual examination of cleaned specimens is not made until weighing has been completed. This avoids unnecessary handling, which could affect weight.

Generally, the appearance of the test specimens should be recorded in the laboratory notebook in a form that will facilitate a stepwise appraisal of each specimen. Typical examples of frequent observations on cleaned specimens are listed here.

1. An SCC specimen represents a fail–no fail investigation, ordinarily requiring only a record of the time to failure. However, any unusual failures, such as failure outside of the most highly stressed region, should be noted. A general examination may be desirable on specimens that complete a predetermined exposure period without failing.
2. A panel, bar, or coupon specimen should be examined with some of the following questions in mind.
   a) Was corrosion uniform or was it located at pits, edges, identification marks, or mounting areas?
   b) Is there any pattern to the corrosion? Does it follow work lines, edges, and so forth?
   c) What is the shape and texture of the corrosion sites (blisters, exfoliation, minute pits, broad smooth pits, and brightness)?

   When the sites of attack are numerous, the tedious job of counting can be eliminated by using rating charts, as shown in Figure 14.13.[5] A weighting factor, to account for the form of the corrosion, might also be used. Caution should be exercised in conversion to numerical values, so as not to imply greater accuracy than warranted.
3. An assembly of components should be examined in a few additional ways.
   a) What is the galvanic effect between parts (accelerated or decelerated corrosion)?
   b) How far does the effect extend?
   c) Are crevices (faying surfaces) attacked?

If the investigator is sufficiently experienced, a more detailed tabulation of visual appearance may be desirable before selecting the column headings that will be permanently recorded in the notebook and ultimately appear in the test report.

14.2.10.6 Photographs

Photographs, micrographs, and X-ray radiographs can be used to supplement the visual-examination report. Entire specimen groups, representative ones, or even a magnified view of a typical area of one panel can be quite valuable in reducing the extent to which the visual examination must be recorded. Radiographs, particularly useful for flat sheet specimens, help to show severity (depth) of the corrosion sites, as well as their locations.

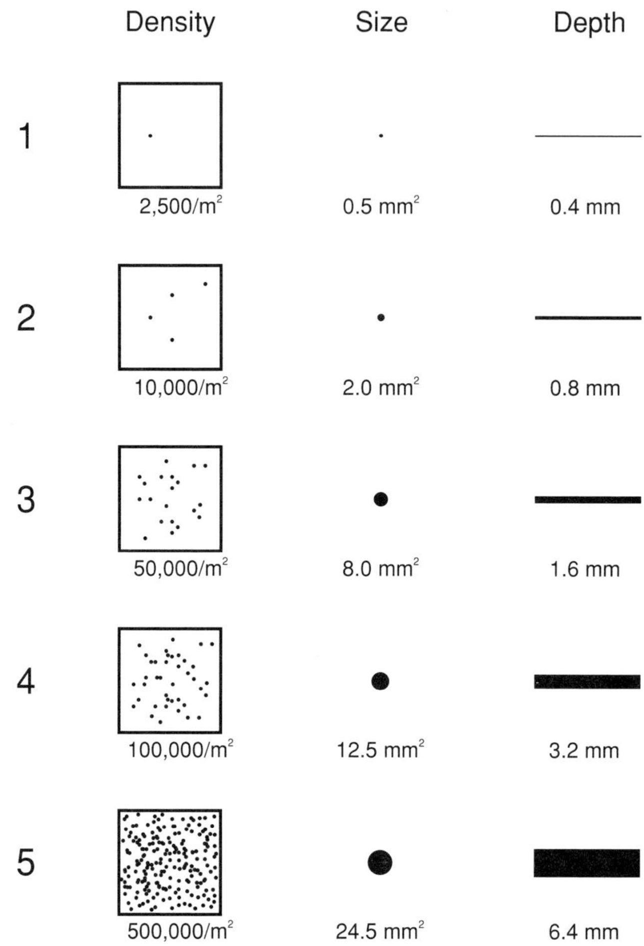

Figure 14.13 Rating charts eliminate tedious counting of pitting sites.

All photographs should be accompanied by descriptive identifications of the specimen(s) and magnifications used.

14.2.10.7 Mechanical and Physical Tests

The next step in the appraisal may be to obtain data required to establish effects of the test on mechanical properties. Usually, ultimate strength (tensile strength) and elongation (stretching caused by tensile testing) tests are compared on specimens that have and have not been corrosion tested. The differences are attributed to corrosion. Generally, the final data are recorded as percent losses. An example is given in Table 14.2.

Special note is taken of any unusual failure observed during testing (failure outside gauge length, at a scratch, etc.) to decide whether these data should be ignored in the final appraisal.

14.2.10.8 Microscopic Studies

One of the best corrosion-test evaluation tools is the metallurgical microscope. A careful selection must be made of metal areas to be studied. Special samples may be corroded specifically for this purpose, or remnants of test specimens, previously evaluated by other methods, may be used. A skilled metallographer then sections through the corrosion sites, polishes the cut surfaces, and examines them for

1. type of corrosion (relationship of corrosion with metallurgical structure),
2. corrosion depth,
3. path of corrosion, and
4. other changes caused by corrosion.

Metallographers use a variety of procedures, including cutting, polishing, chemical or electrochemical etching, and illumination techniques, to permit high-magnification views of the corrosion damage. Generally, the metallographer works closely with the investigator to prepare appropriate photomicrographs for use in the investigator's report.

Experience is necessary to select and interpret micrographs. From proper high-magnification photographs of surfaces and cross sections, the trained investigator often is able to identify factors relating to the mechanism of corrosion, as well as the type, depth, and metallurgical condition of the metal, and the presence of any unusual conditions.

Figure 14.14 (composite) shows a few of the types of corrosion that may be encountered.

14.2.10.9 The Notebook

All entries and calculations should be shown carefully in a permanent notebook. Although each company has its own requirements, everyone agrees that a notebook is a very important document and often serves to establish initial records for patent protection. Most patent attorneys will advise an investigator to write clearly, make a carbon copy of each entry, and have each page of the notebook witnessed and dated by someone who understands the entry. The carbon-copy pages usually are stored at another location to minimize the possibility of loss of records.

14.2.10.10 Communication—The Report

The written report serves to communicate results of any corrosion test. It should be as brief as possible but must include the "why, how, what, and where" of the test program. It should clearly state conclusions and make appropriate recommendations. If the report is written according to a plan, it will help the investigator as well, leading him through a process of self-interrogation to minimize omissions.

Because specific rules or procedures for report writing vary from company to company, this section will present only basic considerations for reporting. These reports

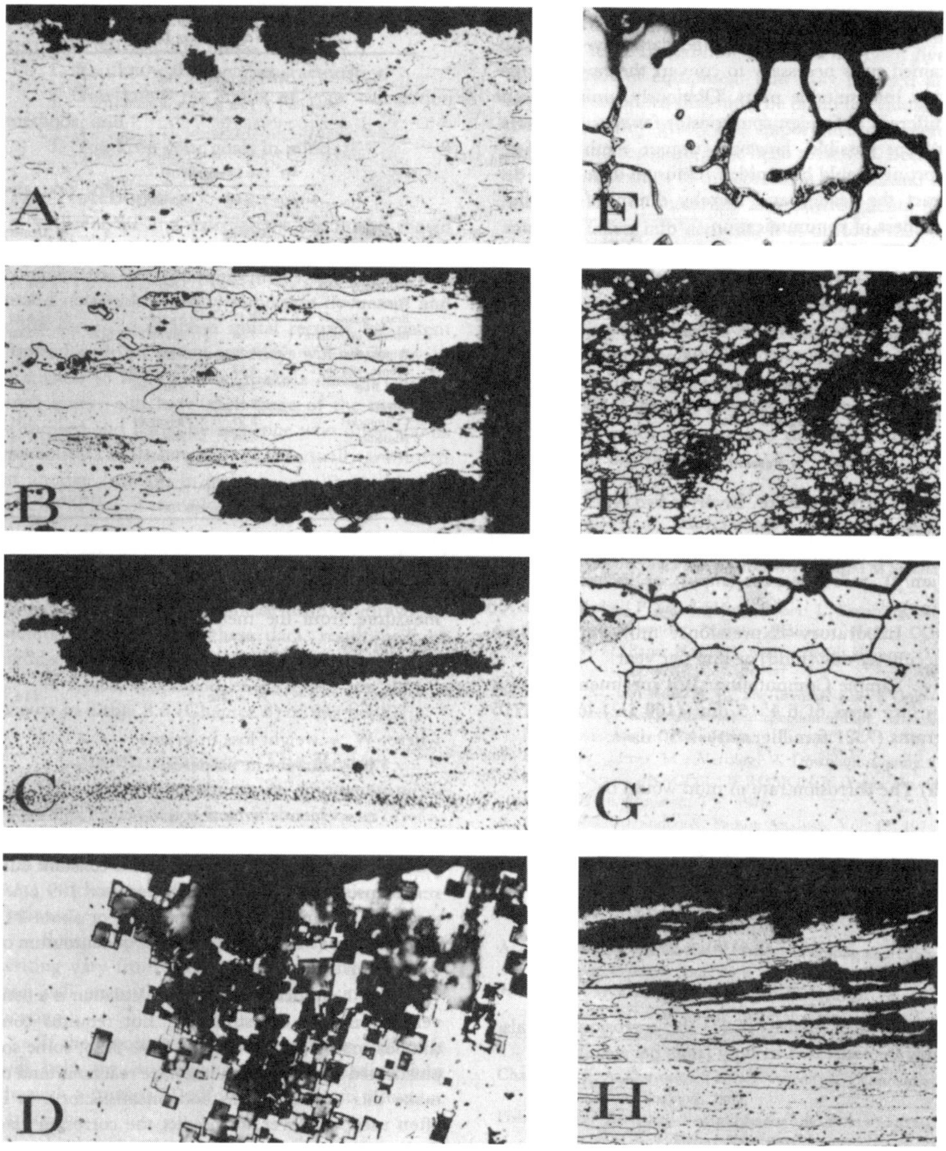

Figure 14.14 Examples of several types of corrosive attack on aluminum alloys. Pitting corrosion: a) surface pitting, b) edge pitting, c) undermining pitting of surface, and d) "cubic" pitting attack. Examples of intercrystalline corrosion are shown: e) interdendritic attack in cast structure, f) interfragmentary attack in wrought, nonrecrystallized structure, g) intergranular attack in wrought, recrystallized structure, and h) exfoliation attack (a, b, and h, ×100 magnification; others ×500).

are usually broken into parts or sections. Reports typically contain one or more of the following sections:

1. synopsis or abstract,
2. introduction,

3. object,
4. materials,
5. procedure,
6. results,
7. summary,
8. conclusions, status, or recommendations, and
9. appendix.

14.2.11 Accelerated Test Environments

Laboratory corrosion tests have been proven to be a reliable and simple means to generate information for the selection of materials by exaggerating the corrosion factors of a specific corrosion attack. These tests fall into three categories; cabinet tests, immersion tests, and high-temperature/high-pressure (HT/HP) tests.

14.2.11.1 Cabinet Tests

Corrosion cabinet tests are conducted in closed cabinets in which the conditions of exposure are controlled to emulate specific corrosion mechanisms. These tests are generally used to determine the corrosion performance of materials intended for use in natural atmospheres. To correlate test results with service performance, it is necessary to establish acceleration factors and to verify that the corrosion mechanisms are indeed following the same paths. Modern surface-analysis techniques can be quite useful to ascertain that the corrosion products have the same morphologies and crystallographic structures that are typically found on equipment used in service. There are basically four types of cabinet tests.

14.2.11.1.1 Controlled Humidity Test. Tests of this sort are used to evaluate staining characteristics or temporary protective coatings. Commercial test equipment is available that can be set for any combination of percent relative humidity and temperature. When only a few small specimens are to be exposed, an inexpensive method is to store them above water in a sealed glass container (Figure 14.15). The vessel is then stored in a controlled-temperature location. When lower humidity is desired, it can be obtained by using a mixture of water and sulfuric acid (Table 14.3) rather than pure water.[6] Another method is to use salt solutions which produce specific relative humidity (Table 14.4).

There are numerous ASTM standards relating different variations of creating and controlling fog and humidity in cabinets for corrosion testing of a broad spectrum of products, from decorative electrodeposited coatings to the evaluation of the corrosivity of solder fluxes for copper-tubing systems. The basic humidity test is most commonly used to evaluate the corrosion behavior of materials or the effects of residual contaminants. Cyclic humidity tests are conducted to simulate exposure to high humidity and heat typical of tropical environments.

Table 14.3 Constant Humidity with Sulfuric Acid Solution[a]

Desired Percent Relative Humidity	Specific Gravity of Required Acid/Water Solution	mL of Acid to Be Added[b]
100	1.00	none
95	1.09	69
90	1.14	135
85	1.17	190
80	1.20	229
70	1.25	309
60	1.29	379
50	1.34	469
40	1.37	586
30	1.44	766
20	1.49	945
10	1.59	1422

[a]Listed are concentrations of sulfuric acid/water solutions that provide different percent relative humidity at 20°C.
[b]To a liter of water to give a solution of the required specific gravity.

The cabinet performing such tests should be equipped with a solid-state humidity sensor reading the current humidity condition and a feedback controller. The mechanism used to control the humidity moves chamber air via a blower motor and passes it over a heater coil in the bottom of the chamber with an atomizer nozzle fogging into this air stream Figure 14.16.

Figure 14.15 Specimens are exposed to high humidity in a sealed container.

Table 14.4 Constant Humidity with Salt Solutions[a]

Percent Relative Humidity	Solid Phase
100	None
95	$Na_2SO_3 \cdot 7H_2O$
90	$ZnSO_4 \cdot 7H_2O$
86	$KHSO_4$
80	NH_4Cl
75	$NaClO_3$
66	$NaNO_2$
58	$NaBr \cdot 2H_2O$
52	$NaHSO_4 \cdot H_2O$
42	$Zn(NO_3)_2 \cdot 6H_2O$
32	$CaCl_2 \cdot 6H_2O$
20	$KC_2H_3O_2$
10	$H_3PO_4 \cdot 1/2H_2O$

[a] The table shows the percent humidity at 20°C within a closed space above a supersaturated aqueous solution of the given solid phase. Solutions of these types can be used to provide a constant relative humidity in tests of the type illustrated in Figure 14.15.

14.2.11.1.2 Corrosive Gas Test. Of considerable interest in recent years is the laboratory simulation and acceleration of atmospheric pollution effects used to evaluate degradation of materials. These tests involve injection of a variety of gases, vapors, and particulate matter into a controlled-pressure chamber with controlled condensation on metal specimens. Sometimes, ultraviolet light is used to produce photochemical changes in the gases or vapors, such as are known to occur under present-day smog conditions.

In these tests, controlled amounts of corrosive gases are added to humidity to replicate more severe environments. The ASTM B 775 (Test Method for Porosity in Gold Coatings on Metal Substrates by Nitric Acid Vapor) and B 799 (Test Method for Porosity in Gold or Palladium Coatings by Sulfurous Acid/Sulfur Dioxide Vapor)

Figure 14.16 Controlled-humidity test chamber.

employ very high concentrations of corrosive gases to amplify the presence of pores in gold or palladium coatings. The moist SO_2 test (ASTM G 87) is intended to produce corrosion in a form resembling that in industrial environments. A very sophisticated variation of these tests is the flowing of a mixed-gas test (ASTM B 827) that consists in introducing parts per billion levels of pollutants such as chlorine, hydrogen sulfide and nitrogen dioxide in a chamber at controlled temperature and humidity. This test is particularly adapted to the needs of the electronics industry.

14.2.11.1.3 Salt-Spray Testing. The oldest and most wildly used cabinet test is ASTM B 117 (Test Method of Salt Spray [Fog] Testing), a test that introduces a spray in a closed chamber where some specimens are exposed at specific locations and angles. The concentration of the sodium chloride (NaCl) solution has ranged from 3.5 to 20%. There is a wide range of chamber designs and sizes including walk-in rooms that are capable of performing this test. Although used extensively for specification purposes, results from salt-spray testing seldom correlate with service performance.

Hot, humid air is created by bubbling compressed air through a bubble (humidifying) tower containing hot, deionized water. Salt solution is typically moved from a reservoir through a filter to the nozzle by a gravity-feed system (Figure 14.17). When the hot, humid air and the salt solution mix at the nozzle, it is atomized into a corrosive fog. This creates a 100%-relative-humidity condition in the exposure zone. For a low-humidity state in the exposure zone of the chamber, air is forced into the exposure zone via a blower motor that directs air over the energized chamber heaters (Figure 14.18).

14.2.11.2 Immersion Testing

Probably the most common corrosion-test method is immersion in a liquid. Obvious differences in test procedures are the solutions used, agitation rates, and temperature. The environmental conditions that must be simulated and the degree of acceleration that is required often determine the choice of a laboratory test. In immersion testing, acceleration is achieved principally by these methods:

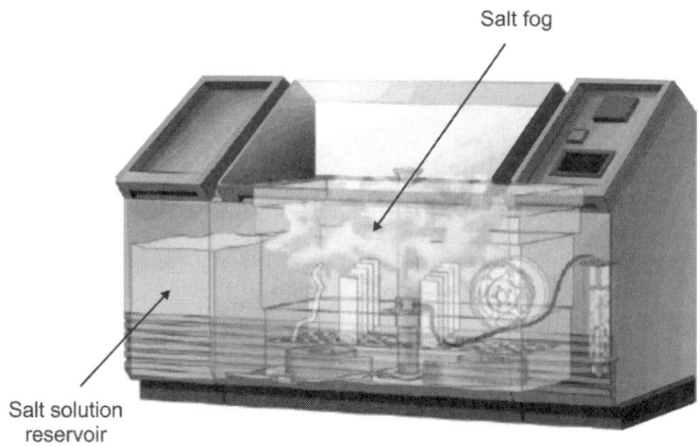

Figure 14.17 Controlled-salt-fog test chamber during a humid cycle.

Figure 14.18 Controlled-salt-fog test chamber during a dry cycle.

- Lengthening the exposure to the critical conditions that are suspected to cause corrosion damage. For example, if a vessel is to be batch processed with a chemical for 24 h, then laboratory corrosion exposure of 240 h should be considered.
- By intensifying the conditions to increase corrosion rates, that is, increasing solution acidity, salt concentration, temperature, or pressure.
- By provoking localized environmental changes that can result in the formation of acidic or high-chloride conditions, furthering the initiation and propagation of crevice corrosion of susceptible alloys, for example, ASTM G 78.[7] (Figure 14.19).

Once the environmental conditions have been determined and the test designed, then it should be repeated a sufficient number of times to determine whether it meets the desired standard for reproducibility. Immersion tests can be divided into two categories.

(a) (b)

Figure 14.19 a) Crevice-forming assembly in a beaker immersion test, and b) components of the assembly and S30400 washer coupons after 30 days in a solution containing 4% NaCl and 8% $FeCl_3$.

14.2.11.2.1 Simple Immersion Tests. In these tests, small sections of the candidate material are exposed to the test medium and the loss of weight of the material is measured for a period of time. Immersion testing is a powerful method of screening and eliminating from further consideration those materials that should not be considered for specific applications. Although these tests are quick and economical means for providing a preliminary selection of best-suited materials, there is no simple way to extrapolate the results obtained from these simple tests to the prediction of system lifetime.

14.2.11.2.2 Alternate Immersion Tests. Another variation of the immersion test are the cyclic tests in which a test specimen is immersed for a period of time in a test environment, and then removed and dried before being reimmersed to continue the cycle. Normally hundreds of these cycles have to be completed during the course of a test program (Figure 14.20).

14.2.11.2.3 Immersion Tests Under Load. The study of environmentally assisted cracking in its most basic sense involves the consideration and evaluation of the inherent compatibility between a material and the environment under conditions of either applied or residual stress. This is a very broad topic encompassing many possible combinations of materials and environments. Over the past 20 years a number of laboratory test methods to predict the behavior of steels in sour service have been developed, for example, that shown in Figure 14.7.

Figure 14.20 This large alternate immersion-testing machine is timed to rotate the racks 60 degrees every 10 min, thus immersing specimens for 10 min and exposing them to air for 50 min every hour.

Figure 14.21 Full ring test in preparation for a 150-cm pipe (courtesy of CAPCIS Ltd.).

One of the best ways of assessing pipe-line steels is to stress a full-ring specimen of the line pipe in a sour environment.[8] This can be achieved by welding end caps onto the sample and pressurizing it with a suitable gas or liquid medium (Figures 14.21 and 14.22) or by applying external loading as illustrated in Figure 14.23. The advantage of the full ring test is that it is not necessary to pressurize the line pipe full-ring specimen to achieve the required stress loading. Equivalent stresses can be produced using mechanical means to deform the pipe by ovalization.

14.2.11.3 HT/HP Testing

Autoclave corrosion tests are a convenient way to simulate service environments where high temperatures and pressures are common. There are many HT/HP environments of commercial interest, which include those in industries such as petroleum, nuclear, chemicals, aerospace, and transportation where reliability, serviceability, and concerns for corrosion are paramount.[9]

Corrosion coupons can be placed in the aqueous phase, vapor space, or at phase interfaces, depending on the specific interest involved. Additionally, it is also possible to conduct electrochemical tests in HT/HP vessels. If multiple liquid phases are present

Figure 14.22 Full ring test (60-cm diameter) as per OTI 95-635 containing 300 L of NACE TM0177-96 solution saturated with H_2S (courtesy of Bodycote Materials Testing).

it can be necessary to stir or agitate the media or test vessel to produce mixing and conditions whereby the corrosion test specimens are contacted by all of the phases present.

Special magnetic and mechanical stirrers are available that can be used to produce movement of the fluid and a mixing of the phases. In some cases where contact of the specimens with both liquid and gaseous phases is important in the corrosion process, it may be necessary to slowly rotate or rock the test vessel to produce the intended results.

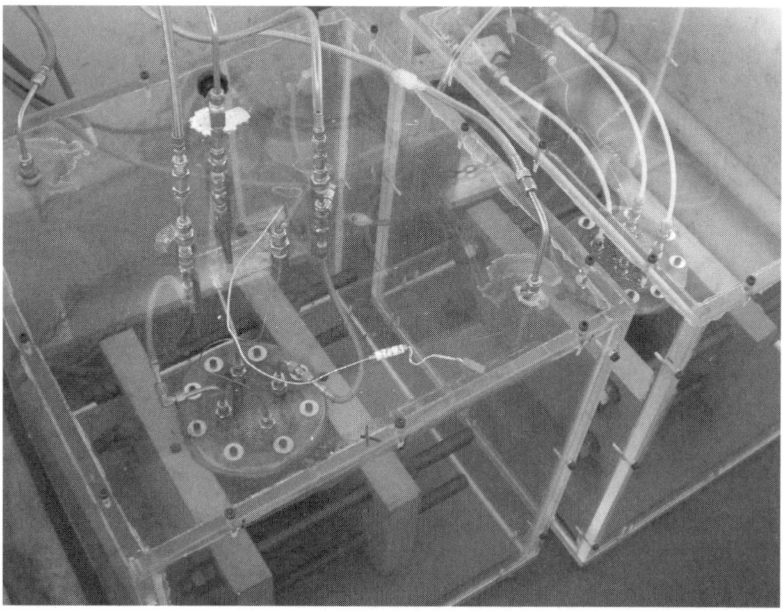

Figure 14.23 Super 13% Cr weldable pipe with external load in secondary containment tanks totally exclude oxygen ingress (courtesy of Bodycote Materials Testing).

Figure 14.24 Double-vessel system with two 5-L pressure-vessel bodies, two end flanges, and three seal rings for evaluation of high temperature (340°C at 35 MPa or 540°C at 25 MPa) hydrogen attack or hydrogen-induced disbonding (courtesy of InterCorr International, Inc.).

14.2.11.3.1 Static Tests. The simplest type of HT/HP corrosion test is conducted in a sealed and static pressurized test vessel (Figure 14.24). The test vessel would typically contain a solution and a vapor space above the solution. In static corrosion tests, the only form of agitation of the test environment is convection produced by heating of the solution. The solution itself can be anything from a single liquid to water-based solutions containing various dissolved salts such as chlorides, carbonates, bicarbonates, alkali salts, and other constituents or mixtures. The liquid and gas phases will be determined by the amount and vapor pressures of the constituents in the test vessel as well as by the test temperature. In general, the degree of difficulty and amount of expense required for these tests increase with increasing test pressure and temperature.

14.2.11.3.2 Refreshed and Recirculating Tests. The depletion of volume of corrosive environment in HT/HP tests is a serious limitation that can be overcome by the introduction of a fresh environment either continuously or by periodic replenishment of the gaseous and liquid phases being depleted by the corrosion processes. The ratio of volume of corrosive fluid to specimen surface area is a critical factor that should be maintained to no less than 30 cm^3 per square centimeter of tested surface area.

The need for agitation is particularly required when multiple liquid phases are present. Special magnetic and mechanical stirrers are available that can be used to produce fluid movement (Figure 14.25). Magnetic or mechanical stirring can also be employed to spin the specimens in the test environment or alternatively a high-velocity flow system can be employed to induce cavitation or erosion damage on the specimens.

Figure 14.25 Custom-designed 150-L vessel with C276 (N10276) construction for exposure of large-scale plate specimens to wet H_2S, magnetic stirring attachment with motor and control, and pressure-balanced Ag/AgCl reference electrodes (courtesy of InterCorr International, Inc.).

14.2.11.3.3 Factors Affecting HT/HP Test Environments. For simple HT/HP exposure tests involving either aqueous or nonaqueous phases, the total pressure is usually determined by the sum of the pressures of the constituents of the test environment that will vary with temperature. Where liquid constituents are being used for the test environment, the partial pressure is usually taken to be the vapor pressure of the liquid at the intended test temperature. In some cases, higher test pressures can be obtained by pumping additional gas into the test vessel using a special gas pump. Alternately, hydrostatic pressurization may be employed whereby there is no gas phase in the test vessel and the pressure is increased by pumping additional liquid into the test vessel in a controlled manner. The importance of partial pressure in HT/HP corrosion testing is that the solubility of the gaseous constituents in the liquid phase is usually determined by its partial pressure, which explains why the effect of some gaseous corrosives is often magnified by high pressures.

14.2.11.3.4 Special HT/HP Corrosion Test Conditions. It has been known for decades that atomic hydrogen can produce embrittlement in many metallic materials. Under high-hydrogen-environment pressure or electrochemical reaction, or both, atomic hydrogen can penetrate structural materials, where it can react destructively with the metal (Chapter 9).

There has been much interest in conducting HIC tests in aqueous media that can produce atomic hydrogen on the surface of materials as a result of corrosion or cathodic charging. When aqueous hydrogen charging is involved, pressure is usually not a major factor. However, as in the case of steels exposed to aqueous environments containing hydrogen sulfide, the atomic hydrogen is produced as a result of sulfide corrosion. The severity of the mass-loss corrosion and hydrogen charging is directly dependent on the amount of hydrogen sulfide dissolved in the aqueous environment. Special multiphase autoclave assemblies have been constructed to allow in situ corrosion monitoring under high temperate flowing conditions (Figure 14.26).

14.2.12 Electrochemical Test Methods

Given the fundamental electrochemical nature of corrosion processes (see Chapter 2 for more details) it is not surprising to see that electrochemical measurements are extensively used across the whole spectrum of corrosion science and engineering, from fundamental studies to monitoring and control in service.

14.2.12.1 DC Polarization Test Methods

DC polarization methods involve changing the potential of the working electrode and monitoring the current that is produced as a function of time or potential. For anodic polarization, the potential is changed in the anodic (or more positive direction), causing the working electrode to become the anode and forcing electrons to be withdrawn from the metal being tested. For cathodic polarization, the working electrode becomes more negative and electrons are added to the metal, sometime causing electrodeposition. For cyclic polarization, both anodic and cathodic polarizations are performed in a cyclic manner.[10] The instrumentation for carrying polarization testing consists of these items (cf. Figure 2.23 in Chapter 2):

- A potentiostat is needed that will maintain the potential of the working electrode close to a preset value.
- A current-measuring device for monitoring the current produced by the applied potential is needed. The ability of the current-measuring device to autorange or to change the scale automatically is also important.
- The ability of the data to be stored directly in a computer or plotted out directly is also important.
- Several polarization test cells are available commercially for making polarization measurements. Polarization cells can have various configurations specific to the testing requirements from testing small coupons to testing sheet materials to testing inside autoclaves. In a plant environment the electrodes may be inserted directly into a process stream. Some of the features of a cell include these[10]:

(a)

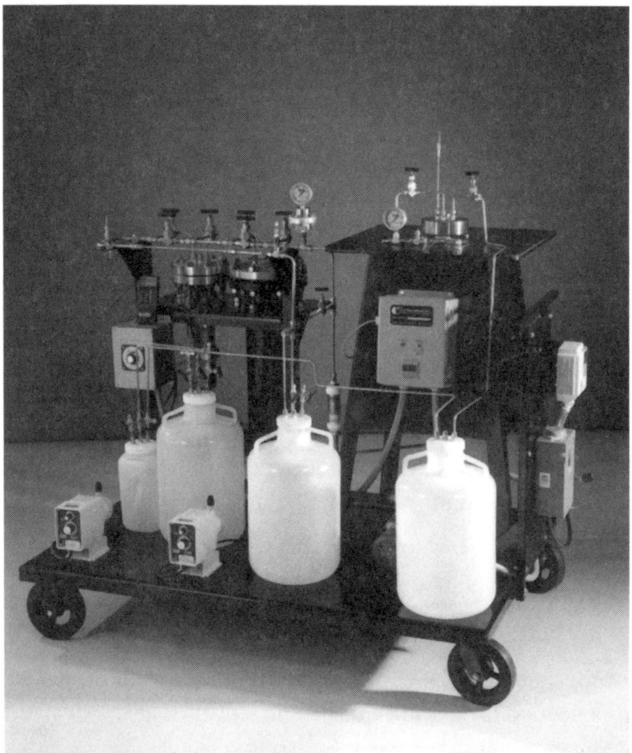

(b)

Figure 14.26 a) HP/HT flow loop for evaluation of velocity-accelerated corrosion and chemical treatment. N10276-lined, 5-L reservoir, three independently controlled N10276 flow-loop sections, and magnetic stirring attachment with motor and speed control, and b) higher-pressure version of a) that can go up to 14 MPa and 200°C for evaluation of velocity-accelerated corrosion and chemical treatment (courtesy of InterCorr International, Inc.).

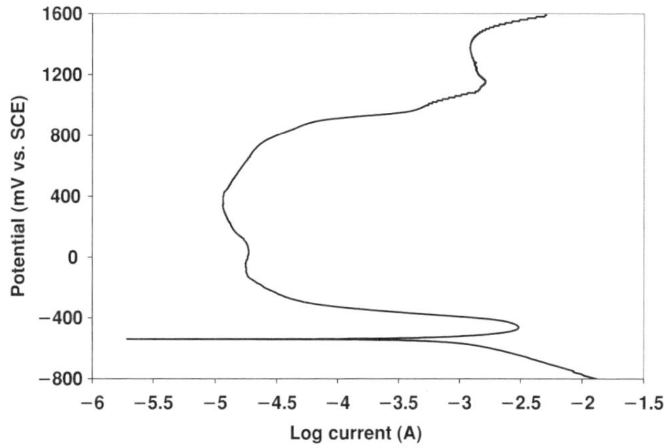

Figure 14.27 Typical anodic-polarization plot for S43000 steel in a 0.05-M H_2SO_4 solution.

a) The working electrode (i.e., the test sample), which may be accompanied by one or more auxiliary or counter electrodes.
b) The reference electrode, which is often separated from the solution by a solution bridge or a capillary Luggin probe (Figure 2.15). This combination eliminates solution interchange with the reference electrode but allows it to be moved very close to the surface of the working electrode to minimize the effect of the solution resistance.
c) A temperature-monitoring device.
d) An inlet and outlet for gas to allow deaeration, aeration, or the introduction of specific gases into the solution.
e) An electrical connection can be made directly with the working electrode, which will not be affected by the solution.
f) The working electrode must be introduced into the solution completely so as to eliminate any crevice at the solution interface, unless this is a desired effect.
g) The test cell itself must be composed of a material that will not corrode or deteriorate during the test, and which will not contaminate the test solution. The volume of the cell must be large enough to allow removal of the corroding ions from the surface of the working electrode without affecting the solution potential.
h) It may be necessary to include a mechanism for stirring the solution such as a stirring bar or bubbling gas to insure uniformity of the solution chemistry.

There several accepted methods to carry out DC polarization of specimens for corrosion testing.

14.2.12.1.1 Potentiodynamic Polarization. Potentiodynamic polarization refers to a polarization technique in which the potential of the electrode is varied over a relatively large voltage domain at a selected rate by the application of a current through the electrolyte. Figure 14.27 is an example of a polarization plot obtained with an S43000-steel specimen in a 0.05-M H_2SO_4 solution.

A potentiodynamic polarization variant is cyclic voltammetry, which involves sweeping the potential in a positive direction until a predetermined value of current or potential is reached; then, the scan is reversed toward more negative values until the original value of potential is reached. In some cases, this scan is done repeatedly to determine changes in the current–potential curve produced with scanning.

Another variation of potentiodynamic polarization is the potentiostaircase method. This refers to a technique for polarizing an electrode in a series of potential steps where the time spent at each potential is constant, while the current is often allowed to stabilize prior to changing the potential to the next step. The step increase may be small, in which case the technique resembles a potentiodynamic curve.[10]

Electrochemical potentiodynamic reactivation (EPR) is another polarization method that evaluates the degree of sensitization of stainless steels such as S30400 and S30403 steels. This method uses a potentiodynamic sweep over a range of potentials from passive to active (called reactivation).

However, probably the most popular variant is the cyclic-polarization test. This test is often used to evaluate the pitting susceptibility of a material. The potential is swept in a single cycle or slightly less than one cycle, usually starting at the corrosion potential. The voltage is first increased in the anodic or noble direction (forward scan). The voltage-scan direction is reversed at some chosen current or voltage toward the cathodic or active direction (backward or reverse scan) and terminated at another chosen voltage. The presence of the hysteresis between the currents measured in the forward and backward scans is believed to indicate pitting, whereas the size of the hysteresis loop itself is often related to the amount of pitting.

This technique has been especially useful to assess localized corrosion for passivating alloys such as S31600 stainless steel, nickel-based alloys containing chromium, and other alloys such as titanium and zirconiumal though the generation of the polarization scan is simple, its interpretation can be difficult.[11]

In the following example, the polarization scans were generated after one and four days of exposure to a chemical product maintained at 49°C. The goal of these tests was to examine if S31600 steel could be used for short-term storage of a 50% commercial organic-acid solution (aminotrimethylene phosphonic acid) in water. A small amount of chloride ion (1%) was also potentially present in this acidic chemical.

In this example, the potential scan rate was 0.5 mV s^{-1} and the scan direction was reversed at 0.1 mA cm^{-2}. Coupon-immersion tests were run in the same environment for 840 h. The S31600-steel specimens were exposed to the liquid, at the vapor–liquid interface, and in the vapor. The reason for the three exposures was that in most storage situations, the containment vessel would be exposed to a vapor–liquid interface and a vapor phase at least part of the time. Corrosion in these regions can be very different from liquid exposures. The specimens were also fitted with artificial crevice formers.

Figure 14.28 shows the polarization scan generated after one day and Figure 14.29 shows the polarization scan generated after four days of exposure. The important parameters considered were the position of the "anodic-to-cathodic" transition relative to the corrosion potential, the existence of the repassivation potential and its value relative to the corrosion potential, the existence of the pitting potential and its value relative to

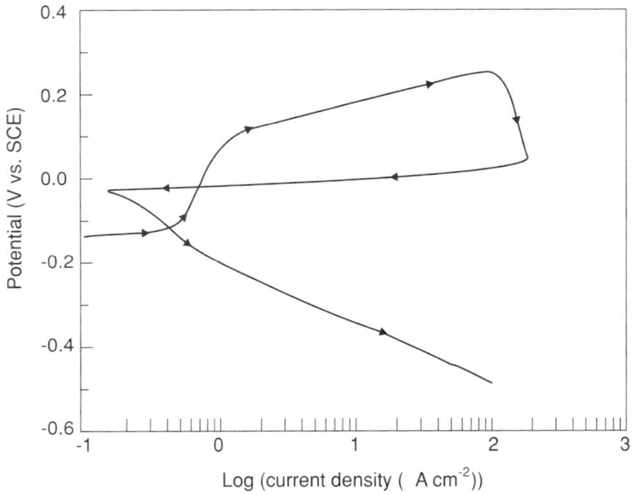

Figure 14.28 Polarization scan for S31600 steel in 50% aminotrimethylene phosphonic acid after one day of exposure (the arrow indicates scanning direction).

the corrosion potential, and the hysteresis (positive or negative). The interpretation of the results is summarized in Table 14.5.

The presence of the negative hysteresis would typically suggest that localized corrosion is possible depending on the value of the corrosion potential relative to the characteristic potentials present in these polarization plots. After the first day of exposure, pitting was not expected to be a problem because the pitting potential was far away from the corrosion potential. The currents generated were much higher than those normally associated with S31600 steel in a passive state. These observations suggested that there was a risk of initiation of corrosion, particularly in localized areas where the pH can decrease drastically.[11]

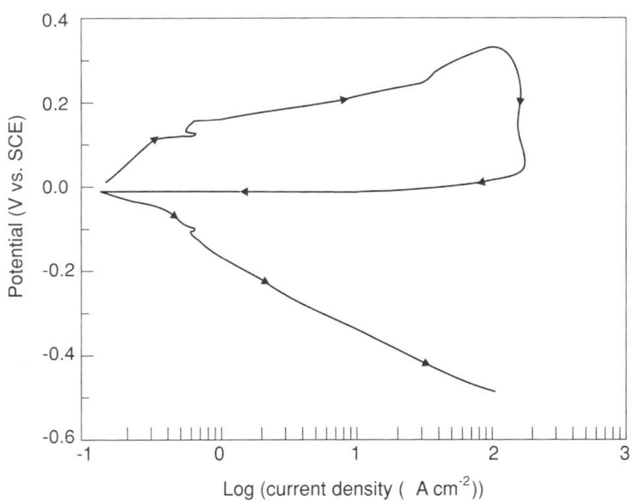

Figure 14.29 Polarization scan for S31600 steel in 50% aminotrimethylene phosphonic acid after four days of exposure (the arrow indicates scanning direction).

Table 14.5 Features and Values Used to Interpret Figures 14.28 and 14.29

Feature	Value in Figure 14.23	Value in Figure 14.24
Repassivation potential – corrosion potential	0.12 V	0.0 V
Pitting potential – corrosion potential	0.22 V	0.12 V
Potential of anoadic-to-cathodic rantsition – corrosion potential	0.12 V	0.0 V
Hysteresis	Negative	Negative
Active-to-passive transition	No	No

After four days, the risk of localized corrosion increased. At this time, the repassivation potential and the potential of the change from anodic to cathodic current were equal to the corrosion potential. The pitting potential was only about 0.1 V more noble than the corrosion potential and the hysteresis still negative. The risk of pitting had increased to become a concern.

Coupon-immersion tests confirmed the long-term predictions. Slight attack was found under the artificial crevice formers in the complete liquid exposure. The practical conclusion of this in-service study was that, because localized corrosion often takes time to develop, a few days of exposure to this chemical product could be acceptable. However, it was recommended to avoid long-term exposure because both pitting and crevice corrosion would be expected for longer exposure periods.

14.2.12.1.2 LPR. Another widely used polarization method is LPR. The polarization resistance of a material is defined as the slope of the potential–current density ($\Delta E/\Delta i$) curve at the free corrosion potential (Figure 14.30), yielding the polarization

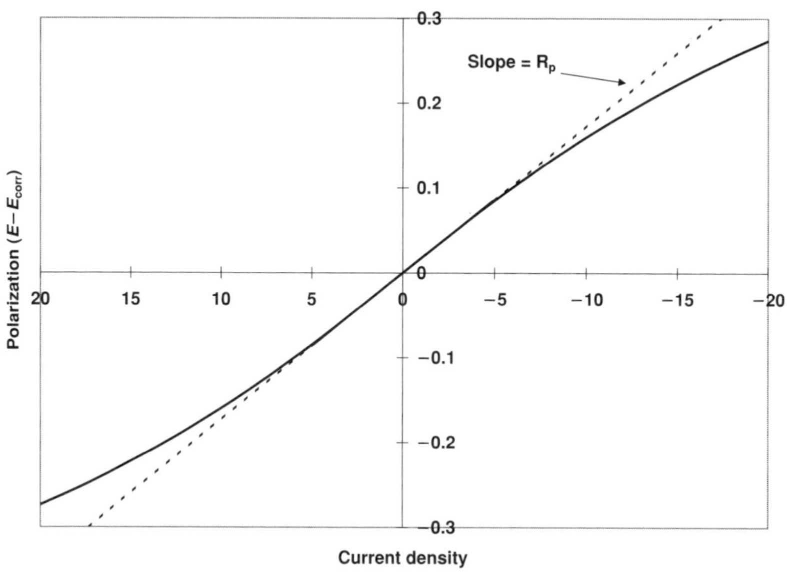

Figure 14.30 Hypothetical linear polarization plot.

resistance R_p that can be itself related to the corrosion current (i_{corr}) with the help of the following Stern–Geary equation[12]:

$$R_p = \frac{B}{i_{corr}} = \frac{(\Delta E)}{(\Delta i)}_{\Delta E \to 0} \tag{14.1}$$

where R_p is the polarization resistance, i_{corr} the corrosion current, and B an empirical polarization-resistance constant that can be related to the anodic (b_a) and cathodic (b_c) Tafel slopes with Equation (14.2):

$$B = \frac{b_a \cdot b_c}{2.3(b_a + b_c)} \tag{14.2}$$

The Tafel slopes in this equation can be evaluated experimentally using real polarization plots in the fashion described Figure 2.24 (Chapter 2) or obtained from the literature.[12] As explained in Chapter 2, the corrosion currents can then be converted into other corrosion-rate units using Faraday's law or, more simply, by using a conversion scheme provided in Table 2.2 for all metals or Table 2.3 adapted to iron or steel. An example of the use of LPR for the determination of the efficiency of an inhibitor system is shown in Chapter 12.

14.2.12.1.3 Complications with Polarization Methods. The widespread use of these polarization techniques does not mean that they are without complications. Both linear polarization and Tafel extrapolation need special precautions for their results to be valid. The main complications or obstacles in performing polarization measurements can be summarized in the following categories:

- Effect of Scan Rate: The rate at which the potential is scanned may have a significant effect on the amount of current produced at all values of potential.[10] The rate at which the potential is changed is an experimental parameter over which the user has control. The goal is for the polarization scan rate to be slow enough to minimize surface-capacitance charging. If not, some of the current being generated serves to charge the surface capacitance in addition to the corrosion process with the result that the measured current can be greater than the current actually generated by the corrosion reactions alone.
- Effect of Solution Resistance: The distance between the Luggin probe (Figure 2.15 in Chapter 2) and the working electrode is purposely minimized in most measurements to limit the effect of the solution resistance. In solutions that have extremely high resistivity (e.g., concrete, soils, and organic solutions), this can be an extremely significant effect.
- Changing Surface Conditions: Because corrosion reactions take place at the surface of materials, when the surface is changed as a result of processing conditions, active corrosion, or any other reason the potential is usually also changed. This can have a strong effect on the polarization curves.[10]

- Determination of Pitting Potential: In analyzing polarization curves the presence of a hysteresis loop between the forward and reverse scans often indicates that localized corrosion (e.g., pitting or crevice corrosion) is in progress.

14.2.12.2 EIS

EIS has been successfully applied to the study of corrosion systems and proven to be a powerful and accurate method for measuring corrosion rates for a few decades. In this approach for determining the polarization resistance of a metal, a measure of the electrochemical impedance (sometimes called AC impedance) is made at a series of predetermined frequencies. ASTM G 106 contains an appendix reviewing this technique that typically uses a small-amplitude sinusoidal potential to probe the working electrode at a number of discrete frequencies.[13]

An important advantage of EIS over other electrochemical techniques is the possibility of using very small amplitude signals without significantly disturbing the properties being measured. However, to estimate the polarization resistance (R_p) that is proportional to the corrosion rate at the monitored interface according to Equation (14.1), EIS results have to be interpreted with the help of a model of the interface.

Among the numerous equivalent circuits that have been proposed to model electrochemical interfaces, only a few really apply in the context of a freely corroding system. The first circuit is the simplest equivalent circuit that can describe a metal/electrolyte interface (Figure 14.31). Its behavior is described by Equation (14.3):

$$Z(\omega) = R_s + \frac{R_p}{1 + (j\omega R_p C_{dl})^{\beta}} \tag{14.3}$$

where R_s is the solution resistance, R_p the polarization resistance, ω the frequency, and C_{dl} the double-layer capacitance.

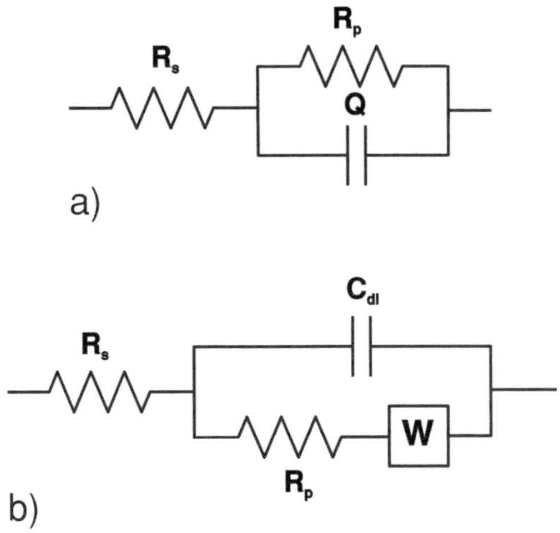

Figure 14.31 The equivalent circuit models proposed for the interpretation of EIS results measured in corroding systems: a) simplest representation of an electrochemical interface and b) one relaxation time constant with extended diffusion.

The term Q, in Figure 14.31, describes the "leaky capacitor" behavior corresponding to the presence of a constant-phase element (CPE).[14] Figure 14.32a illustrates the complex plane presentation of the EIS model circuit where $R_s = 10\ \Omega$, $R_p = 100\ \text{k}\ \Omega$, and Q decomposes into $C_{dl} = 40\ \mu$ F and $n = 0.8$. Figure 14.32b shows how the same data would appear in a Bode-plot format.

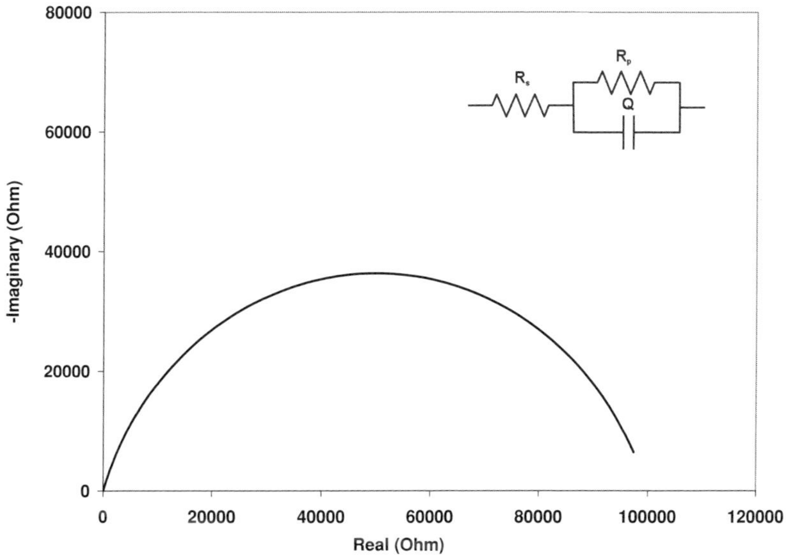

(a)

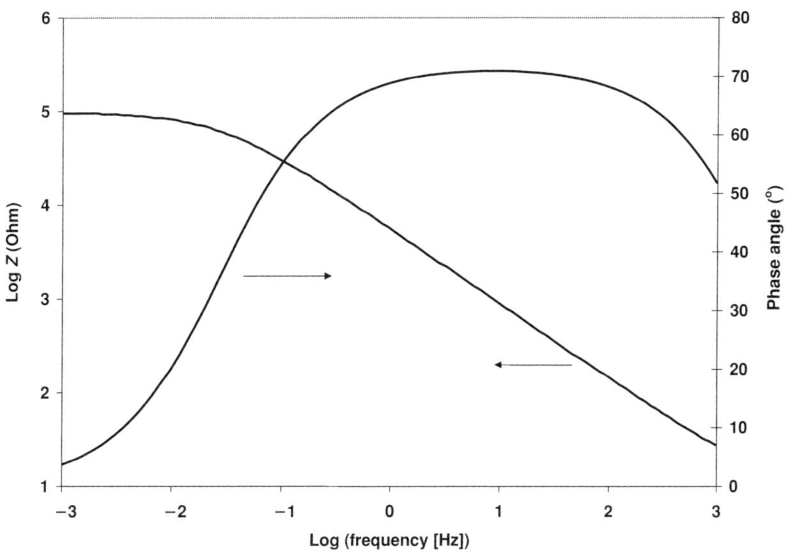

(b)

Figure 14.32 a) Complex plane and b) Bode plots of simulated data correspond to the model circuit a) in Figure 14.31, where $R_s = 10\ \Omega$, $R_p = 100\,\text{k}\Omega$, and Q decomposes into $C_{dl} = 40\,\mu\text{F}$, and $n = 0.8$.

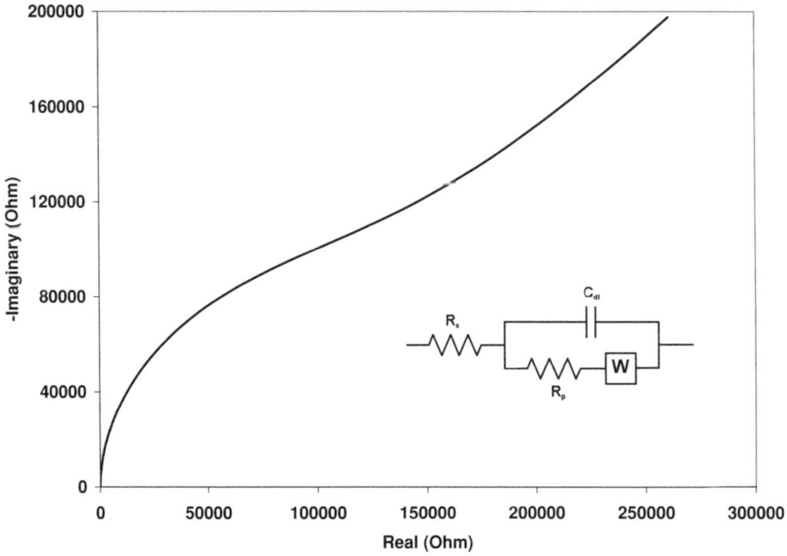

(a)

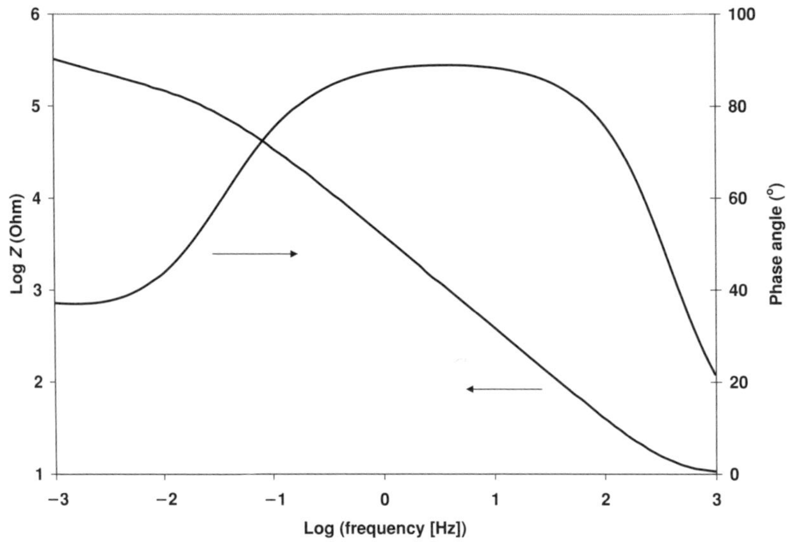

(b)

Figure 14.33 a) Complex plane and b) Bode plots of simulated data correspond to model circuit b) in Figure 14.31, where $R_s = 10\,\Omega$, $R_p = 100\,\text{k}\,\Omega$, $C_{dl} = 40\,\mu\text{F}$, and the exponent n of the Warburg component = 0.4.

The second circuit (Figure 14.31b) was proposed by Hladky et al.[15] to take into account a diffusion-limited behavior corresponding to a Warburg component, which can be described by Equation (14.4). The exponent n in Equation (14.4) can vary between 0.5 and 0.25 depending on the smoothness of the metallic surface (i.e., 0.5 for highly polished surfaces and 0.25 for porous or very rough materials).[16] R and C in Equation

(14.4) are the resistance and capacitance associated with the distributed *R-C* line of infinite length.

$$Z(\omega) = (0.5\ R/C)^{0.5}\ \omega^{-n} \tag{14.4}$$

Figure 14.33a illustrates the complex plane presentation of simulated data corresponding to the model circuit b) in Figure 14.31 where $R_s = 10\,\Omega$, $R_p = 100\,\text{k}\,\Omega$, $C_{dl} = 40\ \mu$F, and the exponent n of the Warburg component = 0.4. Figure 14.33b shows the same data in a Bode representation.

A critical problem in EIS, as well as any other scientific measurements, is the validation of the experimental data. To address the validation of impedance data, the use of Kramers–Kronig (KK) transforms has been proposed. Currently, the KK algorithm is commercially available and is used routinely in some laboratories to assess the quality of the measured impedance data.[17] These integral transforms were derived assuming four basic conditions, which are discussed prior to stating the transforms themselves.

- **Linearity:** A system is said to be linear if the response to a sum of individual inputs is equal to the sum of the individual responses. Practically, this implies that the impedance (or admittance) of a system should not be a function of the magnitude of the probing signal.
- **Causality:** The response of a system should be solely due to the input. Physically, this means that the system being tested should not generate noise independent of the applied signal. This is an important consideration in electrochemical systems, because charge-transfer interfaces are often active and do in fact generate noise in the absence of any external stimulus.[17]
- **Stability:** A system is said to be stable if it comes back to its original state after a perturbation is removed. This condition ensures that there is no negative resistance in the system. This apparent restriction on the presence of negative resistance has often been associated the presence of active-to-passive transitions in the system being investigated.[17]
- **Finite Value:** The real and imaginary components of a complex impedance must be finite over the entire frequency range sampled.

If a system satisfies the conditions of linearity, stability, and causality, it will a priori satisfy the KK transforms, provided that the frequency range is sufficiently broad for the integrals to be evaluated. Modern equipment often contains, in its control software, a KK transform algorithm to simplify this otherwise complicated validation process.

However, a simple validity test is more than often sufficient to ascertain that the EIS measurements are indeed meaningful. For measurements made in real time, that is, by changing the frequency for each frequency point, the validity of a forward frequency scan can be verified by scanning in the opposite direction. Repeated measurements during two successive scans in opposite directions should generally satisfy KK transforms.

14.2.12.3 Electrochemical Noise (EN)

Fluctuations of potential or current of a corroding metallic specimen are a well-known and easily observable phenomenon. The evaluation of EN as a corrosion tool has increased steadily since Iverson's paper in 1968.[8] The study of EN has been found to be uniquely appropriate for monitoring the onset of events leading to localized corrosion and understanding the chronology of the initial events typical of this type of corrosion.

During localized corrosion EN is believed to be generated by a combination of stochastic processes, such as passivation breakdown and repassivation events, and deterministic processes, which can be caused by film formation or pit-propagation processes. One electrode configuration that enables EN measurements involves connecting a zero-resistance ammeter (ZRA) between two nominally identical corroding electrodes immersed in the same solution. A third nominally identical electrode can be immersed in solution and connected to the first two using a high-impedance voltmeter. This electrode serves as a "noisy" pseudoreference electrode. This approach is attractive in field applications because of the rugged nature of the metallic electrode compared to laboratory electrodes.

Many believe that EN analysis can provide a parameter called the EN resistance (R_n) This parameter should be used in a manner analogous to polarization resistance (R_p).

A polarization resistance can be estimated from EN either statistically, spectrally, or through analysis of individual transients. The ratio of the standard deviation of the potential to that for the current fluctuations is a statistical measure of polarization resistance. The square root of the ratio of spectral power density of voltage to that for current is a frequency-domain measure of polarization resistance. The ratio of voltage to current for an individual transient is another measure of polarization resistance. Attempts to quantify localized corrosion include the pitting ratio, the slope and break point of the spectral power distribution, the average amount of charge and the frequency of charges from shot-noise analogy, and the intensity and frequency of individual transients.

14.2.13 Evaluation of Polymeric Materials

Structural polymeric materials (or, as they are often called, plastics) cannot be evaluated by the same criteria applied to metals. These materials often degrade from the effect of absorbed species from a solution at rates defined by Fick's laws of diffusion. A number of interactions between a polymer matrix and the absorbed compound are possible, such as breaking of the polymer chain, breaking of cross-links in the chains, oxidation of the polymer, reaction with terminal groups of the chains, or simply lubricating the area between chains to weaken the strength of the material.

One of the most important aspects of polymeric materials' behavior is that polymers can selectively absorb one or more constituents from an environment. This means that a continuous supply of only parts per million of an aggressive chemical in a stream can produce similar results as exposing the polymeric material to 100% of that chemical.

Many of these polymeric materials are used in composites (i.e., combinations with glass fibers or other strengthening material), and the variables introduced, in addition to resistance of the polymer matrix itself, may include the following:

1. extent to which the fibers are wetted by the polymer,
2. strength of the fibers,
3. mode of manufacture (i.e., whether fibers are single filaments wound about a mandrel or fabrics impregnated with the resins),
4. characteristics of the layers of the fabrication (i.e., whether glass fibers are exposed to the corrosive agents),
5. internal or external strengtheners or stiffeners,
6. percentage of resin to fiber, and many other variables.

Among the many types of tests that can be applied to polymeric materials are the following. Obviously, not all of these tests are necessary in evaluating polymeric materials for every use.

1. visual (changes in appearance vs. unexposed samples),
2. tensile strength,
3. flexural strength,
4. flexural cycling,
5. impact resistance,
6. notch sensitivity,
7. heat stability,
8. abrasion resistance,
9. static build-up,
10. coefficient of expansion,
11. chemical stability,
12. weight changes,
13. hardness,
14. dimensional change,
15. volume change,
16. flame-propagation rate,
17. biological effects,
18. creep rates, and
19. stress cracking resistance.

For many purposes, the change of hardness of a polymer during an exposure can be as good an indicator of the resistance of the polymeric material to the environment as any other mechanical test. However, working with these materials at ambient temperatures is similar to working with metals at high temperatures. Thus, creep rates at given stress levels while exposed to the environment are probably the best of all corrosion tests if the cost of the test has to be justified.

Whatever the test procedure employed, a test period of at least 100 days should be used. Changes in the properties of a polymeric material do not change linearly with time. Thus, appraisal of the properties at various times during the test period is advisable if possible.

14.2.14 Testing of Other Nonmetals

Other nonmetals, such as glass, ceramics, concrete, wood, stone, graphite, or mica, can be tested for resistance to corrosives by more or less the same means as can metals and polymeric materials, with similar reservations. Glass, for example, which is used for piping and as a lining to handle corrosive solutions, can be tested with comparative ease with results that have a high degree of precision.

Ceramics, because of the wide range of applications to which they are put, ranging from high-temperature, high-velocity gas environments to ambient-temperature aqueous environments, may require tests selected to evaluate them precisely for their intended use. Concrete, also a mixture, can be tested by many of the same methods used for other materials. For example, reinforced concrete exposed to salt water with severe cycling of temperatures, as in dock piling in arctic ports, must be tested for its ability to protect the structural steel it contains.

Tests are available for and are being used to evaluate nearly all inorganic materials, quite often with highly reproducible results. Usually, however, they involve long exposure times and do not always provide answers that permit extrapolation of performance data to periods greatly in excess of test periods. The tester must select the available parameters that will give results useful for the specific application.

Various tests are available for all these materials, but true evaluation of their properties often can be determined only from long-term exposure in the environments of interest (refer to ASTM standards).

14.3 CORROSION MONITORING

In a modern business environment, successful enterprises cannot tolerate major corrosion failures, especially when these involve personal injuries, fatalities, unscheduled shutdowns, and environmental contamination. Typically, once a system, a plant, or any piece of equipment is put into service, maintenance is required to keep it operating safely and efficiently. This is particularly true for aging systems and structures that will be required to operate beyond their original design life.

Correct and effective corrosion-monitoring strategies need to be used as a proactive tool to assist with operating a system more effectively, thereby prolonging its life and gaining optimum throughput. It also enables continuous monitoring of actual corrosion rates, allowing for timely preventative action if a variance is observed.

Various means of introducing coupons of materials of construction into operating equipment have been devised. For equipment operating under considerable pressure, special attachments are available from corrosion-equipment suppliers. A single high-pressure access fitting for insertion of a retrievable corrosion probe (Figure 14.34) can be used to accommodate most types of retrievable probes (Figure 14.35). With specialized tools (and brave specialist operating crews!) sensor insertion and withdrawal can be possible under pressurized operating conditions (Figure 14.36).

It is imperative that the corrosion sensors are representative of the actual component being monitored. If this requirement is not met, all subsequent signal processing and data analysis is negatively affected and the value of information greatly diminished or even rendered worthless. For example, if turbulence is induced locally around a protruding corrosion sensor, mounted in a pipe line, the sensor will in all likelihood give a very poor indication of the risk of localized corrosion damage to the pipe line wall. A flush-mounted sensor should be used instead (Figure 14.37).

An important aspect of a corrosion-monitoring program is to ensure that the coupon materials are fully and freely exposed in the environment. A sufficient number of sites should also be selected to ensure that variations in the exposure periods will be revealed. Dean has presented an example of identifying critical sensor locations in a distillation column.[19,20] The feed point, overhead product receiver, and bottom product line represent locations of temperature extremes and also points where products with different degrees of volatility concentrate.

In many cases, however, the highest corrosivity is encountered at an intermediate height in the column, where the most corrosive species concentrate. Initially, therefore, several monitoring points would be required in such a column, as shown in Figure 14.38. As monitoring progresses and data from these points become available, the number of monitoring points could be narrowed down further.

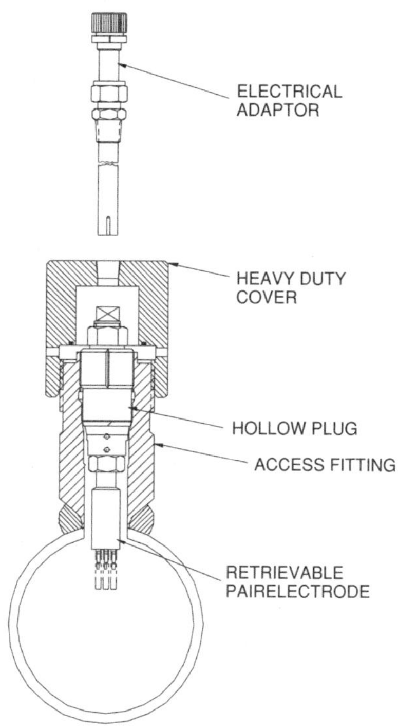

Figure 14.34 High-pressure access fitting for insertion of a retrievable corrosion probe (courtesy of Metal Samples).

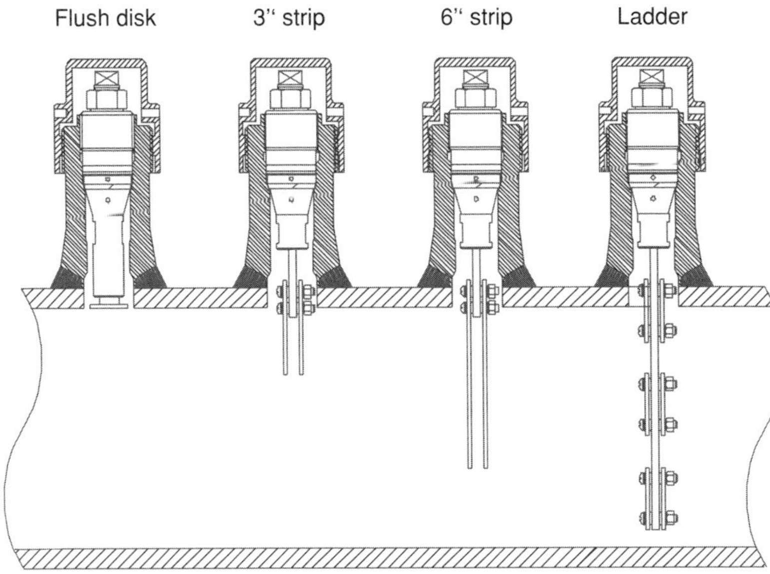

Figure 14.35 A single high-pressure access fitting can be fitted with different types of retrievable corrosion probes (courtesy of Metal Samples).

Sometimes the installation of a side stream on operating equipment is an economical, convenient method for field testing. This may be particularly true when attempting to appraise the effect of inhibitor additions or other attempts to alter the environment.

Assessment of corrosion in the field is complex because of the wide variety of applications, process conditions, and fluid phases that exist in industrial plants where corrosion occurs. A wide range of direct and indirect measurement techniques is available, but each technique has its strengths and weaknesses. In some applications certain techniques cannot be used at all. Some techniques can be used online, whereas others can only be used offline. Commonly more than one technique is used so that the weaknesses of one are compensated for by the strengths of another.

The categories of various available corrosion-monitoring techniques have been described in a recent NACE International report as being direct or indirect and as either intrusive or nonintrusive.[21]

Direct techniques are those that measure parameters directly affected by the corrosion process. Indirect techniques are those that provide data on parameters that either affect, or are affected by, the corrosiveness of the environment or the products of the corrosion process.

Intrusive techniques are any of those that require access through the pipe or vessel wall for measurements to be made. Intrusive or nonintrusive is not a definition of whether the probe projects into the process flow or not. Most commonly intrusive techniques make use of some form of probe or test specimen, and these techniques include flush probe designs.

The more sensitive monitoring techniques can detect corrosion-related changes as they occur and permit remedial actions to be taken before significant damage occurs. Such techniques are defined as real-time measurements.

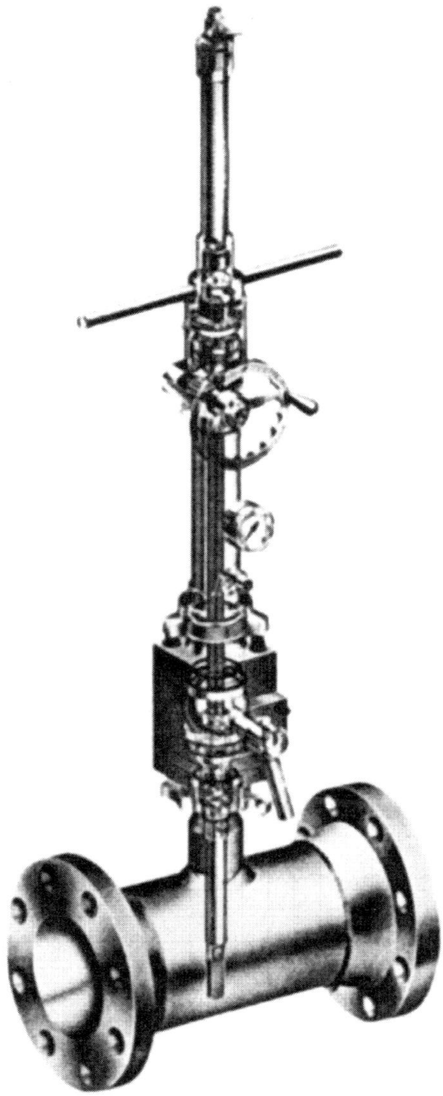

Figure 14.36 Retrieval tool for removing corrosion probes under pressure (courtesy of Metal Samples).

14.3.1 Direct Techniques

14.3.1.1 Corrosion Coupons (Intrusive)

In perhaps the simplest form of corrosion monitoring, small specimens are exposed to an environment for a specific period of time and subsequently removed for weight loss and more detailed examination. Even though the principle is very simple, numerous potential pitfalls exist, which can be avoided by following the recommendations of a comprehensive guide such as the ASTM G4 Standard.[22]

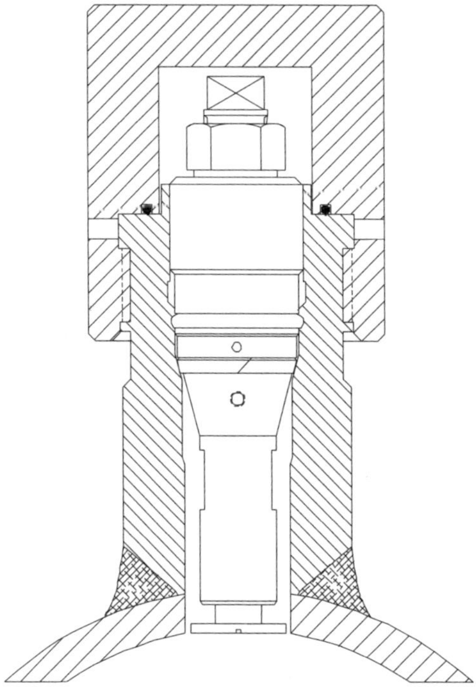

Figure 14.37 Flush-mounted corrosion sensor in an access fitting (courtesy of Metal Samples).

14.3.1.2 Electrical Resistance (ER) (Intrusive)

The underlying principle of the widely used ER probes is a simple concept, namely an increase in electrical resistance as the cross-sectional area of a sensing element is reduced by corrosion damage. Because temperature has a strong influence on electrical resistance, ER sensors usually measure the resistance of a corroding sensor element relative to that of an identical shielded element. Commercial sensor elements are in the form of plates, tubes, plates, or wires (Figure 14.39). Reducing the thickness of the sensor elements can increase the sensitivity of these sensors. However, improved sensitivity involves a trade-off against reduced sensor lifetime. ER probe manufacturers provide guidelines showing this trade-off for different sensor geometries. ER probes usually have a useful life up to the point where their original thickness has been halved (except wire sensors).

14.3.1.3 Inductive Resistance Probes (Intrusive)

This recently developed technology is a derivative of ER corrosion sensing.[23] The thickness reduction of a sensing element is measured by changes in the inductive resistance of a coil embedded in the sensor (Figure 14.40). Sensing elements with high magnetic permeability intensify the magnetic field around the coil; therefore, thickness changes affect the inductive resistance of the coil. Sensitivity or the decrease in response time has been claimed to be several orders of magnitude higher than with comparable ER probes.

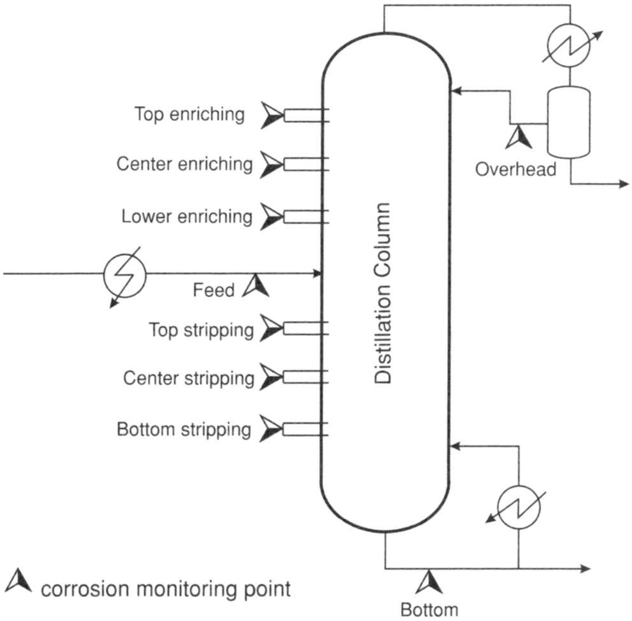

Figure 14.38 Corrosion-monitoring points in a distillation column.

14.3.1.4 LPR (Intrusive)

The LPR technique, which was described in the corrosion testing section of this chapter, is an electrochemical method using either two or three sensor electrodes (Figure 14.41). The small potential perturbation (typically of the order of 20 mV) used with LPR is applied to the sensor electrode of interest and the resulting current is measured. The ratio of the potential to current perturbations is known as the polarization resistance, which is inversely proportional to the uniform corrosion rate. The accuracy of the technique

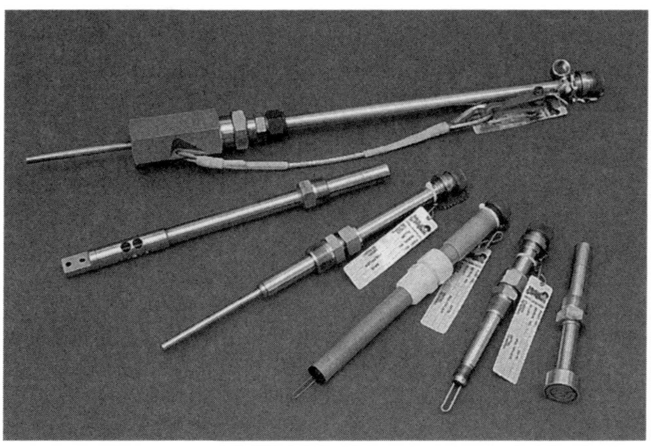

Figure 14.39 Commercial sensor elements to carry out electrical-resistance (ER) measurements (courtesy of Metal Samples).

can be improved by measuring the solution resistance independently and subtracting it from the apparent polarization-resistance value.

14.3.1.5 EIS (Intrusive)

As described in a previous section, the application of an alternating potential will produce an AC response, which is out of phase with the potential. The frequency range

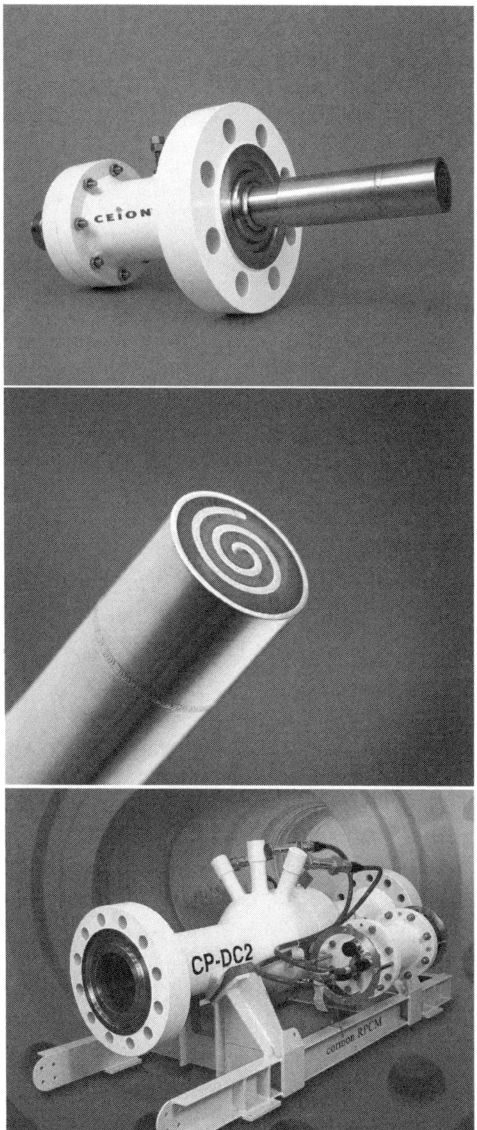

Figure 14.40 a) Subsea inductive-resistance probe, b) probe element, and c) spool instrument package for subsea (courtesy of JCE Operations & Marketing, Cormon Ltd.).

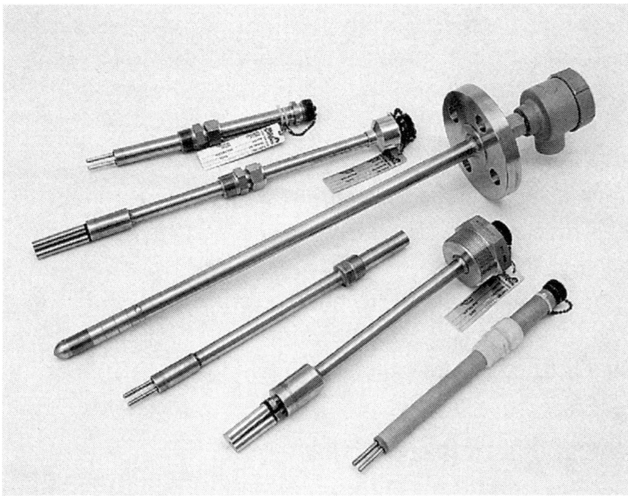

Figure 14.41 Commercial sensor elements to carry out linear polarization resistance (LPR) measurements (courtesy of Metal Samples).

of the applied perturbation signal for corrosion monitoring is typically between 0.1 Hz and 100 kHz. For characterizing the corrosion behavior in detail, measurements through the entire frequency range are required. Some simplified commercial systems rely on measurements at only two frequencies to extract kinetic information. Full-frequency scans provide phase-shift information that can be utilized in equivalent circuit models. With such models, relatively complex corrosion phenomena can be described and kinetic information derived.

14.3.1.6 EN (Intrusive)

Contrary to common perceptions, the underlying measurement principles of this electrochemical technique are extremely simple. The technique is not related to acoustic noise in any way; rather, fluctuations in potential and current between freely corroding electrodes are measured. However, the small scale of these fluctuations of interest, which is in many cases les than 1 μV and 1 nA, require sensitive instrumentation. Many high-precision digital multimeters facilitate these measurements.

A measurement frequency of 1 Hz usually suffices. For simultaneous measurement of electrochemical potential and current noise, a three-electrode sensor is required. The current noise is measured between two of these sensor elements. The potential noise is measured between the third element and the two coupled elements used for current measurements. For industrial corrosion-monitoring purposes, all three sensor elements are usually constructed from the same material.

Higher levels of electrochemical corrosion activity are generally associated with higher noise levels. Certain electrochemical phenomena, such as the breakdown of passivity during pit initiation, show up distinct noise "signatures," which can be exploited for corrosion-monitoring purposes. Pit initiation and growth can be detected with electrochemical-noise measurements long before it becomes evident by visual examination.

14.3.1.7 ZRA (Intrusive)

In this electrochemical technique, galvanic currents are measured between two electrodes or sensor elements. These currents would be due to the presence of dissimilarities between the sensor elements, for example, differences in composition, heat treatment, stress level, or surface condition. The technique may also be applied to nominally identical electrodes to indicate changes in the corrosive environment and serve as a broad indicator of changes in corrosion rate.

14.3.1.8 Potentiodynamic Polarization (Intrusive)

This method is best known for its fundamental role in electrochemistry for the measurement of mixed-potential or Evans diagrams. A three-electrode corrosion probe is utilized to polarize the electrode of interest. The current response is measured as the potential is shifted away from the free corrosion potential (E_{corr}). The basic difference as compared with the LPR technique is that the applied polarization potentials are normally stepped up to levels of several hundred millivolts. These polarization levels facilitate the determination of kinetic parameters such as the general corrosion rate and the Tafel constants. The formation of passive films and the onset of pitting corrosion can also be identified at characteristic potentials, which can assist in assessing the overall corrosion risk.

14.3.1.9 Thin-Layer Activation (TLA) and Gamma Radiography (Intrusive or Nonintrusive)

This technique was developed from principles of nuclear science. A small section of material is exposed to a high-energy beam of charged particles to produce a radioactive surface layer. For example, a proton beam may be used to produce the radioactive isotope Co-56 within a steel surface. This isotope decays to Fe-56, with the emission of gamma radiation. The concentration of radioactive species is sufficiently low that metallurgical properties of the monitored component are essentially unchanged.

The radioactive effects utilized are at very low levels and should not be compared to those of conventional radiography. The change in gamma radiation emitted from the surface layer is measured with a separate detector to study the rate of material removed from the surface. The radioactive surfaces can be produced directly on components (nonintrusive) or on separate sensors (intrusive).

14.3.1.10 Electrical-Field Signature Method (EFSM) (Nonintrusive)

The original development of this technique was largely directed at oil and gas production. The technique measures corrosion damage over several meters of an actual structure, clearly distinguishing it from other smaller sensor systems. An induced current would be fed into the monitored section and the resulting voltage distribution measured to detect corrosion damage. An array of pins would be attached strategically over the structure for measuring purposes. Increased pin spacing would then mean lower resolution for localized corrosion. Typical applications involve pin attachment to the external surface of a pipe line to monitor corrosion damage to the inside of the pipe walls.

14.3.1.11 Acoustic Emission (AE) (Nonintrusive)

This technique is based on measuring acoustic sound waves that are emitted during the growth of microscopic defects, such as stress corrosion cracks. The sensors, which are strategically positioned on a structure, can thus essentially be viewed as microphones. The sound waves are generated during the release of mechanical stresses produced by pressure or temperature changes. Background noise effects have to be taken into consideration and can be particularly troublesome in online measurements.

14.3.2 Indirect Techniques

A plethora of indirect corrosion-monitoring techniques is available for different types of corrosion damage. A multidisciplinary science and engineering team approach is often required for the organized implementation of these techniques. For example, the indirect monitoring of microbiologically induced corrosion in water systems typically requires expertise and laboratory infrastructure from the fields of corrosion engineering, microbiology, and chemistry.

14.3.2.1 Corrosion Potential (Nonintrusive)

Corrosion-potential measurements are usually classified as an intrusive indirect method. Either a reference electrode or possibly a separate sensor of the material to be monitored has to be introduced into the corrosive medium for these measurements, or an electrical connection has to be established to a structure in conjunction with an external reference electrode. The corrosion potential is usually measured relative to a reference electrode, which is characterized by a stable half-cell potential (see Chapter 2 for more details).

Monitoring corrosion potentials is a relatively simple method that is widely used in industry for a wide range of applications such as reinforcing-steel corrosion in concrete and buried pipe lines with or without CP. Monitoring potential changes resulting from the active–passive behavior of stainless steel in anodic-protection systems is another application area.

14.3.2.2 Hydrogen Monitoring (Nonintrusive)

The generation of atomic hydrogen, as part of the cathodic half-cell reaction in acidic environments, can be used for corrosion-monitoring purposes in both intrusive and nonintrusive forms. In the latter, hydrogen-monitoring sensors are often attached to the outside walls of vessels and piping. It is the diffusion of atomic hydrogen into the metallic substrate that is of most concern, because it can lead to problems such as HIC.

This "uptake" of hydrogen occurs when the recombination of hydrogen atoms and subsequent release as molecular hydrogen into the environment is impeded. Hydrogen monitoring is highly applicable to the oil-refining and petrochemical industries with hydrocarbon process streams. The presence of hydrogen sulfide in these industries promotes the uptake of hydrogen into plant items. Hydrogen-monitoring probes are based on any of the following three principles:

- pressure increase with time in a controlled chamber, as hydrogen passes through the material into the probe chamber;
- an electrochemical current resulting from the oxidation of hydrogen under an applied potential; and
- current flow in an external circuit, based on a fuel-cell principle whereby hydrogen entering the miniature fuel cell causes the current flow.

14.3.2.3 *Chemical Analyses*

Different types of chemical analyses can provide valuable information in corrosion-monitoring programs. The measurement of pH, conductivity, dissolved oxygen, metallic and other ion concentrations, water alkalinity, concentration of suspended solids, inhibitor concentrations, and scaling indices all fall within this domain. Several of these measurements can be made online using appropriate sensors.

References

1. W.B. Lisago, "Environmental Cracking—Stress Corrosion," in Corrosion Tests and Standards, ed. R. Baboian (West Conshohocken, PA: ASTM, 1995), pp. 240–252.
2. ASTM G110-92, "Standard Practice for Evaluating Intergranular Corrosion Resistance of Heat Treatable Aluminum Alloys by Immersion in Sodium Chloride + Hydrogen Peroxide Solution" (West Conshohocken, PA: ASTM, 1997).
3. P.R. Roberge, Corrosion Testing Made Easy: Erosion–Corrosion Testing (Houston, TX: NACE, 2004).
4. ASTM G1-03, "Standard Practice for Preparing, Cleaning, and Evaluating Corrosion Test Specimens" (West Conshohocken, PA: ASTM, 2003).
5. ASTM G46-94, "Standard Guide for Examination and Evaluation of Pitting Corrosion" (West Conshohocken, PA: ASTM, 1999).
6. CRC Handbook of Chemistry and Physics, 59th ed. (Boca Raton, FL: Chemical Rubber Publishing Company, 1979).
7. ASTM G78-01, "Standard Guide for Crevice Corrosion Testing of Iron-Base and Nickel-Base Stainless Alloys in Seawater and Other Chloride-Containing Aqueous Environments" (West Conshohocken, PA: ASTM, 2001).
8. "A Test Method to Determine the Susceptibility of Cracking of Linepipe Steels in Sour Service," Health and Safety Executive (HSE), OTI 95 635, Sudbury, UK, 1996.
9. R.D. Kane, "High-Temperature and High Pressure," in Corrosion Tests and Standards, ed. R. Baboian (West Conshohocken, PA: ASTM, 1995), pp. 106–115.

10. A.C. Van Orden, "Applications and Problem Solving Using the Polarization Technique," CORROSION/98, paper no. 301 (Houston, TX: NACE, 1998).
11. D.C. Silverman, "Tutorial on Cyclic Potentiodynamic Polarization Technique," CORROSION/98, paper no. 299 (Houston, TX: NACE, 1998).
12. R. Grauer, P.J., Moreland, G. Pini. A Literature Review of Polarisation Resistance constant (B) Values for the Measurement of Corrosion Rate (Houston, TX: NACE, 1982).
13. ASTM G106-89, "Standard Practice for Verification of Algorithm and Equipment for Electrochemical Impedance Measurements" (West Conshohocken, PA: ASTM, 1999.)
14. B.A. Boukamp, "Equivalent Circuit (Equivcrt.PAS) Users Manual," University of Twente, Report CT89/214/128, The Netherlands, 1989.
15. K. Hladky, L.M. Callow, J.L. Dawson, British Corrosion Journal 15 (1980): pp. 20.
16. R. de Levie, "Electrochemical Response of Porous and Rough Electrodes," in Advances in Electrochemistry and Electrochemical Engineering (1967), pp. 330–397.
17. M, Urquidi-MacDonald, P.C. Egan, "Validation and Extrapolation of Electrochemical Impedance Spectroscopy Data Analysis," Corrosion Reviews 15 (1997): pp. 169–194.
18. W.P. Iverson, "Transient Voltage Changes Produced in Corroding Metals and Alloys," Journal of the Electrochemical Society 115 (1968): pp. 617–618.
19. S.W. Dean, "Overview of Corrosion Monitoring in Modern Industrial Plants," in Corrosion Monitoring in Industrial Plants Using Nondestructive Testing and Electrochemical methods, ASTM STP 908, eds. G.C. Moran, P. Labine (West Conshohocken, PA: ASTM, 1986), pp. 197–220.
20. S.W. Dean, "Corrosion Monitoring for Industrial Processes," in Corrosion: Fundamentals, Testing, and Protection, vol. 13A, eds. D.S. Cramer, B.S. Coviner (Metals Park, OH: ASM International, 2003), pp. 533–541.
21. NACE 3T199, "Techniques for Monitoring Corrosion and Related Parameters in Field Applications" (Houston, TX: NACE, 1999).
22. ASTM G4-01, "Standard Guide for Conducting Corrosion Tests in Field Applications" (West Conshohocken, PA: ASTM, 2001).
23. A.F. Denzine, M.S. Reading, "An Improved, Rapid Corrosion Rate Measurement Technique for All Process Environments," Materials Performance 37 (1998): pp. 35–41.

A

Appendix

NACE Glossary of Corrosion-Related Terms

The following NACE International Glossary of Corrosion-Related Terms is published by the Technical Coordination Committee (TCC) Reference Publications Committee (RPC), and is reprinted with permission. This publication is revised periodically.

Abrasive Small particles of material that are propelled at high velocity to impact a surface during abrasive blast cleaning.

Abrasive Blast Cleaning Cleaning and roughening of a surface produced by the highvelocity impact of an abrasive that is propelled by the discharge of pressurized fluid from a blast nozzle or by a mechanical device such as a centrifugal blasting wheel. (Also referred to as *Abrasive Blasting*.)

Accelerator A chemical substance that increases the rate at which a chemical reaction (e.g., curing) would otherwise occur.

Acrylic Type of resin polymerized from acrylic acid, methacrylic acid, esters of these acids, or acrylonitrile.

Activator A chemical substance that initiates and accelerates a chemical reaction (e.g., curing). Heat and radiation may also serve as activators for some chemical reactions.

Active (1) The negative direction of electrode potential. (2) A state of a metal that is corroding without significant influence of reaction product.

Aeration Cell [See *Differential AerationCell.*]

Air Drying Process by which an applied wet coat converts to a drycoating film by evaporation of solventor reaction with oxygen as a result of simple exposure to air without intentional addition of heat or a curing agent.

Airless Spraying Process of spraying coating liquids using hydraulic pressure, not air pressure, to atomize.

Alkyd Type of resin formed by the reaction of polyhydric alcohols and polybasic acids, part of which is derived from saturated or unsaturated oils or fats.

Alligatoring Pronounced wide cracking over the surface of a coating, which has the appearance of alligator hide.

Amphoteric Metal A metal that is susceptible to corrosion in both acid and alkaline environments.

Anaerobic Free of air or uncombined oxygen.

Anion A negatively charged ion that migrates through the electrolyte toward the anode under the influence of a potential gradient.

Anode The electrode of an electrochemical cell at which oxidation occurs. Electrons flow away from the anode in the external circuit corrosion usually occurs and metal ions enter the solution at the anode.

Anode Cap An electrical insulating material placed over the end of the anode at the lead wire connection.

Anode Corrosion Efficiency The ratio of the actual corrosion (mass loss)of an anode to the theoretical corrosion (mass loss) calculated from the quantity of electricity that has passed between the anode and cathode using Faraday's law.

Anodic Inhibitor A chemical substance that prevents or reduces the rate of the anodic or oxidation reaction.

Anodic Polarization The change of the electrode potential in the noble (positive) direction caused by current across the electrode/electrolyte interface.[See *Polarization*.]

Anodic Protection Polarization to a more oxidizing potential to achieve a reduced corrosion rate by the promotion of passivity.

Anodizing Oxide coating formed on a metal surface (generally aluminum)by an electrolytic process.

Anolyte The electrolyte adjacent to the anode of an electrochemical cell.

Antifouling Preventing fouling. [See *Fouling*.]

Attenuation Electrical losses in a conductor caused bycurrent flow in the conductor.

Auger Electron Spectroscopy Analytical technique in which the sample surface is irradiated with low-energy electrons and the energy spectrum of electrons emitted from the surface is measured.

Austenitic Steel A steel whose microstructure at room temperature consists predominantly of austenite.

Auxiliary Electrode An electrode, usually made from a noncorroding material, which is commonly used in polarization studies to pass current to or from a test electrode.

Backfill Material placed in a hole to fill the space around the anodes, vent pipe, and buried components of a cathodic protection system.

Barrier Coating (1) A coating that has a high resistance to permeation of liquids and/or gases. (2) A coating that is applied over a previously coated surface to prevent damage to the underlying coating during subsequent handling

Beach Marks The characteristic markings on the fracture surfaces produced by fatigue crack propagation (also known as *clamshell marks, conchoidal marks, and arrest marks*).

Beta Curve A plot of dynamic (fluctuating) interference current or related proportional voltage (ordinate) versus the corresponding structure-to-electrolyte potentials at a selected location on the affected structure (abscissa).

Binder The nonvolatile portion of the vehicle of a formulated coating material.

Bituminous Coating An asphalt or coal-tar compound used to provide a protective coating for a surface.

Blast Angle (1) The angle of the blast nozzle with reference to the surface during abrasive blast cleaning. (2) The angle of the abrasive particles propelled from a centrifugal blasting wheel with reference to the surface being abrasive blast cleaned.

Blowdown (1) Injection of air or water under high pressure through a tube to the anode area for the purpose of purging the annular space and possibly correcting high resistance caused by gas blockage. (2) In conjunction with boilers or cooling towers, the process of discharging a significant portion of the aqueous solution in order to remove accumulated salts, deposits, and other impurities.

Blushing Whitening and loss of gloss of a coating, usually organic, caused by moisture (also known as *blooming*).

Bracelet Anodes Galvanic anodes with geometry suitable for direct attachment around the circumference of a pipeline. These may be half-shell bracelets consisting of two semi-circular sections or segmented bracelets consisting of a large number of individual anodes.

Brittle Fracture Fracture with little or no plastic deformation.

Brush-Off Blast Cleaned Surface A brush-off blast cleaned surface, when viewed without magnification, shall be free of all visible oil, grease, dirt, dust, loose mill scale,loose rust, and loose coating. Tightly adherent mill scale, rust, and coating may remain on the surface. Mill scale, rust, and coating are considered tightly adherent if they cannot be removed by lifting with a dull putty knife. [See NACE No. 4/SSPC-SP 7.]

Calcareous Coating A layer consisting of calcium carbonate and other salts deposited on the surface. When the surface is cathodically polarized as in cathodic protection, this layer is the result of the increased pH adjacent to the protected surface.

Calcareous Deposit [See *Calcareous Coating.*]

Case Hardening Hardening a ferrous alloy so that the outer portion, or case, is made substantially harder than the inner portion, or core. Typical processes are carburizing, cyaniding, carbonitriding, nitriding, induction hardening, and flame hardening.

Casein Paint Water-thinned paint with vehicle derived from milk.

Catalyst A chemical substance, usually present in small amounts relative to the reactants, that increases the rate at which a chemical reaction (e.g., curing) would otherwise occur, but is not consumed in the reaction.

Cathode The electrode of an electrochemical cell at which reduction is the principal reaction. Electrons flow toward the cathode in the external circuit.

Cathodic Corrosion Corrosion resulting from a cathodic condition of a structure, usually caused by the reaction of an amphoteric metal with the alkaline products of electrolysis.

Cathodic Disbondment The destruction of adhesion between a coating and the coated surface caused by products of a cathodic reaction.

Cathodic Inhibitor A chemical substance that prevents or reduces the rate of the cathodic or reduction reaction.

Cathodic Polarization The change of the electrode potential in the active (negative) direction caused by current across the electrode/electrolyte interface. [See *Polarization.*]

Cathodic Protection A technique to reduce the corrosion of a metal surface by making that surface the cathode of an electrochemical cell.

Catholyte The electrolyte adjacent to the cathode of an electrochemical cell.

Cation A positively charged ion that migrates through the electrolyte toward the cathode under the influence of a potential gradient.

Cavitation The formation and rapid collapse of cavities or bubbles within a liquid which often results in damage to a material at the solid/liquid interface under conditions of severe turbulent flow.

Cell [See *Electrochemical Cell.*]

Cementation The introduction of one or more elements into the surface layer of a metal by diffusion at high temperature. (Examples of cementation include carburizing [introduction of carbon], nitriding [introduction of nitrogen], and chromizing [introduction of chromium].)

Chalking The development of loose, removable powder (pigment) at the surface of an organic coating, usually caused by weathering.

Checking The development of slight breaks in a coating which do not penetrate to the underlying surface.

Chemical Conversion Coating An adherent reaction product layer on a metal surface formed by reaction with a suitable chemical to provide greater corrosion resistance to the metal and increase adhesion of coatings applied to the metal. (Example is an iron phosphate coating on steel, developed by reaction with phosphoric acid.)

Chevron Pattern A V-shaped pattern on a fatigue or brittle-fracture surface. The pattern can also be one of straight radial lines on cylindrical specimens.

Chloride Stress Corrosion Cracking Cracking of a metal under the combined action of tensile stress and corrosion in the presence of chlorides and an electrolyte (usually water).

Coat One layer of a coating applied to a surface in a single continuous application to form a uniform film when dry.

Coating A liquid, liquefiable, or mastic composition that, after application to a surface, is converted into a solid protective, decorative, or functional adherent film.

Coating System The complete number and types of coats applied to a substrate in a predetermined order. (When used in a broader sense, surface preparation, pretreatments, dry film thickness, and manner of application are included.)

Cold Lap (1) Discontinuity caused by solidification of the meniscus of a partially cast anode as a result of interrupted flow of the casting stream. The solidified meniscus is covered with metal when the flow resumes. Cold laps can occur along the length of an anode. (2) A protective film consisting of one or more coats, applied in a predetermined order by prescribed methods to an asspecified dry film thickness, including any reinforcing material that may be specified.

Cold Shut Horizontal surface discontinuity caused by solidification of a portion of a meniscus during the progressive filling of a mold, which is later covered with more solidifying metal as the molten metal level rises. Cold shuts generally occur at corners remote from the point of pour.

Commercial Blast Cleaned Surface A commercial blast cleaned surface, when viewed without magnification, shall be free of all visible oil, grease, dust, dirt, mill scale, rust, coating, oxides, corrosion products, and other foreign matter. Random staining shall be limited to no more than 33 percent of each unit area (approximately 58 cm^2 [9.0 $in.^2$]) of surface and may consist of light shadows, slight streaks, or minor discolorations caused by stains of rust, stains of mill scale, or stains of previously applied coating. [See NACE No. 3/SSPC-SP 6.]

Concentration Cell An electrochemical cell, the electromotive force of which is caused by a difference in concentration of some component in the electrolyte. (This difference leads to the formation of discrete cathodic and anodic regions.)

Concentration Polarization That portion of polarization of a cell produced by concentration changes resulting from passage of current though the electrolyte.

Conductive Coating (1) A coating that conducts electricity. (2) An electrically conductive, mastic-like material used as an impressed current anode on reinforced concrete surfaces.

Conductive Concrete A highly conductive cement-based mixture containing coarse and fine coke and other material used as an impressed current anode on reinforced concrete surfaces.

Conductivity (1) A measure of the ability of a material to conduct an electric charge. It is the reciprocal of resistivity. (2) The current transferred across a material (e.g., coating) per unit potential gradient.

Contact Corrosion [See *Galvanic Corrosion*.]

Continuity Bond A connection, usually metallic, that provides electrical continuity between structures that can conduct electricity.

Continuous Anode A single anode with no electrical discontinuities.

Conversion Coating [See *Chemical Conversion Coating*.]

Corrosion The deterioration of a material, usually a metal, that results from a reaction with its environment.

Corrosion Fatigue Fatigue-type cracking of metal caused by repeated or fluctuating stresses in a corrosive environment characterized by shorter life than would be encountered as a result of either the repeated or fluctuating stress alone or the corrosive environment alone.

Corrosion Inhibitor A chemical substance or combination of substances that, when present in the environment, prevents or reduces corrosion.

Corrosion Potential (E_{corr}) The potential of a corroding surface in an electrolyte relative to a reference electrode under open-circuit conditions (also known as *rest potential, open-circuit potential, or freely corroding potential*).

Corrosion Rate The rate at which corrosion proceeds.

Corrosion Resistance Ability of a material, usually a metal, to withstand corrosion in a given system.

Corrosiveness The tendency of an environment to cause corrosion.

Counter Electrode [See *Auxiliary Electrode*.]

Counterpoise A conductor or system of conductors arranged beneath a power line, located on, above, or most frequently, below the surface of the earth and connected to the footings of the towers or poles supporting the power line.

Couple [See *Galvanic Couple*.]

Cracking (of Coating) Breaks in a coating that extend through to the substrate.

Crazing A network of checks or cracks appearing on the surface of a coating.

Creep Time-dependent strain occurring under stress.

Crevice Corrosion Localized corrosion of a metal surface at, or immediately adjacent to, an area that is shielded from full exposure to the environment because of close proximity of the metal to the surface of another material.

Critical Humidity The relative humidity above which the atmospheric corrosion rate of some metals increases sharply.

Critical Pitting Potential (E_p, E_{pp}) The lowest value of oxidizing potential (voltage) at which pits nucleate and grow. The value depends on the test method used.

Curing Chemical process of developing the intended properties of a coating or other material (e.g., resin) over a period of time.

Curing Agent A chemical substance used for curing a coating or other material (e.g., resin). [Also referred to as *Hardener*.]

Current (1) A flow of electric charge. (2) The amount of electric charge flowing past a specified circuit point per unit time, measured in the direction of net transport of positive charges. (In a metallic conductor, this is the opposite direction of the electron flow.)

Current Density The current to or from a unit area of an electrode surface.

Current Efficiency The ratio of the electrochemical equivalent current density for a specific reaction to the total applied current density.

Dc Decoupling Device A device used in electrical circuits that allows the flow of alternating current (AC) in both directions and stops or substantially reduces the flow of direct current (DC).

Dealloying The selective corrosion of one or more components of a solid solution alloy (also known as *parting or selective dissolution*).

Decomposition Potential The potential (voltage) on a metal surface necessary to decompose the electrolyte of an electrochemical cell or a component thereof.

Decomposition Voltage [See *Decomposition Potential.*]

Deep Groundbed One or more anodes installed vertically at a nominal depth of 15 m (50 ft) or more below the earth's surface in a drilled hole for the purpose of supplying cathodic protection.

Depolarization The removal of factors resisting the current in an electrochemical cell.

Deposit Attack Corrosion occurring under or around a discontinuous deposit on a metallic surface (also known as *poultice corrosion*).

Dezincification A corrosion phenomenon resulting in the selective removal of zinc from copper-zinc alloys. (This phenomenon is one of the more common forms of dealloying.)

Dielectric Coating A coating that does not conduct electricity.

Dielectric Shield An electrically nonconductive material, such as a coating, sheet or pipe, that is placed between an anode and an adjacent cathode, usually on the cathode, to improve current distribution in a cathodic protection system.

Differential Aeration Cell An electrochemical cell, the electromotive force of which is due to a difference in air (oxygen) concentration at one electrode as compared with that at another electrode of the same material.

Diffusion-Limited Current Density The current density that corresponds to the maximum transfer rate that a particular species can sustain because of the limitation of diffusion (often referred to as *limiting current density*).

Disbondment The loss of adhesion between a coating and the substrate.

Dissimilar Metals Different metals that could form an anode-cathode relationship in an electrolyte when connected by a metallic path.

Double Layer The interface between an electrode or a suspended particle and an electrolyte created by charge-charge interaction leading to an alignment of oppositely charged ions at the surface of the electrode or particle. The simplest model is represented by a parallel plate condenser.

Doubler Plate An additional plate or thickness of steel used to provide extra strength at the point of anode attachment to an offshore platform.

Drainage Conduction of electric current from an underground or submerged metallic structure by means of a metallic conductor.

Driving Potential Difference in potential between the anode and the steel structure.

Drying Oil An oil capable of conversion from a liquid to a solid by slow reaction with oxygen in the air.

Elastic Deformation Changes of dimensions of a material upon the application of a stress within the elastic range. Following the release of an elastic stress, the material returns to its original dimensions without any permanent deformation.

Elastic Limit The maximum stress to which a material may be subjected without retention of any permanent deformation after the stress is removed.

Elasticity The property of a material that allows it to recover its original dimensions following deformation by a stress below its elastic limit.

Electrical Interference Any electrical disturbance on a metallic structure in contact with an electrolyte caused by stray current(s).

Electrical Isolation The condition of being electrically separated from other metallic structures or the environment.

Electro-Osmosis The migration of water through a semipermeable membrane as a result of a potential difference caused by the flow of electric charge through the membrane.

Electrochemical Cell A system consisting of an anode and a cathode immersed in an electrolyte so as to create an electrical circuit. The anode and cathode may be different metals or dissimilar areas on the same metal surface.

Electrochemical Equivalent The mass of an element or group of elements oxidized or reduced at 100% efficiency by the passage of a unit quantity of electricity.

Electrochemical Potential The partial derivative of the total electrochemical free energy of a constituent with respect to the number of moles of this constituent where all other factors are kept constant. It is analogous to the chemical potential of a constituent except that it includes the electrical as well as chemical contributions to the free energy.

Electrode A conductor used to establish contact with an electrolyte and through which current is transferred to or from an electrolyte.

Electrode Potential The potential of an electrode in an electrolyte as measured against a reference electrode. (The electrode potential does not include any resistance losses in potential in either the electrolyte or the external circuit. It represents the reversible work to move a unit of charge from the electrode surface through the electrolyte to the reference electrode.)

Electrokinetic Potential A potential difference in a solution caused by residual, unbalanced charge distribution in the adjoining solution, producing a double layer. The electrokinetic potential is different from the electrode potential in that it occurs exclusively in the solution phase. This potential represents the reversible work necessary to bring a unit charge from infinity in the solution up to the interface in question but not through the interface (also known *as zeta potential*).

Electrolyte A chemical substance containing ions that migrate in an electric field.

Electrolytic Cleaning A Process for removing soil, scale, or corrosion products from a metal surface by subjecting the metal as an electrode to an electric current in an electrolytic bath.

Electromotive Force Series A list of elements arranged according to their standard electrode potentials, the sign being positive for elements whose potentials are cathodic to hydrogen and negative for those anodic to hydrogen.

Ellipsometry An optical analytical technique employing plane-polarized light to study films.

Embrittlement Loss of ductility of a material resulting from a chemical or physical change.

Emf Series [See *Electromotive Force Series*.]

Enamel (1) A paint that dries to a hard, glossy surface. (2) A coating that is characterized by an ability to form a smooth, durable film.

End Effect The more rapid loss of anode material at the end of an anode, compared with other surfaces of the anode, resulting from higher current density.

Endurance Limit The maximum stress that a material can withstand for an infinitely large number of fatigue cycles.

Environment The surroundings or conditions (physical, chemical, mechanical) in which a material exists.

Environmental Cracking Brittle fracture of a normally ductile material in which the corrosive effect of the environment is a causative factor. Environmental cracking is a general term that includes all of the terms listed below. The definitions of these terms are listed elsewhere in the *Glossary*: Corrosion Fatigue Hydrogen Embrittlement Hydrogen-Induced Cracking—(Stepwise Cracking) Hydrogen Stress Cracking Liquid Metal Cracking Stress Corrosion Cracking Sulfide Stress Cracking

The following terms have been used in the past in connection with environmental cracking but are now obsolete and should not be used: Caustic Embrittlement Delayed Cracking Liquid Metal Embrittlement Season Cracking Static Fatigue Sulfide Corrosion Cracking Sulfide Stress Corrosion Cracking

Epoxy Type of resin formed by the reaction of aliphatic or aromatic polyols (like bisphenol) with epichlorohydrin and characterized by the presence of reactive oxirane end groups.

Equilibrium Potential The potential of an electrode in an electrolyte at which the forward rate of a given reactionis exactly equal to the reverse rate; the electrode potential with reference to a standard equilibrium, as defined by the Nernst equation.

Erosion The progressive loss of material from a solid surface due to mechanical interaction between that surface and a fluid, a multicomponent fluid, or solid particles carried with the fluid.

Erosion-Corrosion A conjoint action involving corrosion and erosion in the presence of a moving corrosive fluid or a material moving through the fluid, leading to accelerated loss of material.

Exchange Current The rate at which either positive or negative charges are entering or leaving the surface when an electrode reaches dynamic equilibrium in an electrolyte.

Exfoliation Corrosion Localized subsurface corrosion in zones parallel to the surface that result in thin layers of uncorroded metal resembling the pages of a book.

External Circuit The wires, connectors, measuring devices, current sources, etc., that are used to bring about or measure the desired electrical conditions within an electrochemical cell. It is this portion of the cell through which electrons travel.

Fatigue The phenomenon leading to fracture of a material under repeated or fluctuating stresses having a maximum value less than the tensile strength of the material.

Fatigue Strength The maximum stress that can be sustained for a specified number of cycles without failure.

Fault Current A current that flows from one conductor to ground or to another conductor due to an abnormal connection (including an arc) between the two. A fault current flowing to ground may be called a ground fault current.

Ferrite The body-centered cubic crystalline phase of iron-based alloys.

Ferritic Steel A steel whose microstructure at room temperature consists predominantly of ferrite.

Filiform Corrosion Corrosion that occurs under a coating in the form of randomly distributed thread-like filaments.

Film A thin, not necessarily visible layer of material.

Finish Coat [See *Topcoat.*]

Forced Drainage Drainage applied to underground or submerged metallic structures by means of an applied electromotive force or sacrificial anode.

Foreign Structure Any metallic structure that is not intended as a part of a system under cathodic protection.

Fouling An accumulation of deposits. This includes accumulation and growth of marine organisms on a submerged metal surface and the accumulation of deposits (usually inorganic) on heat exchanger tubing.

Fractography Descriptive treatment of fracture, especially in metals, with specific reference to photographs of the fracture surface.

Fracture Mechanics A quantitative analysis for evaluating structural reliability in terms of applied stress, crack length, and specimen geometry.

Free Machining The machining characteristics of an alloy to which an ingredient has been introduced to give small broken chips, lower power consumption, better surface finish, and longer tool life.

Fretting Corrosion Deterioration at the interface of two contacting surfaces under load which is accelerated by their relative motion.

Furan Type of resin formed by the polymerization or polycondensation of furfuryl, furfuryl alcohol, or other compounds containing a furan ring.

Galvanic Anode A metal that provides sacrificial protection to another metal that is more noble when electrically coupled in an electrolyte. This type of anode is the electron source in one type of cathodic protection.

Galvanic Corrosion Accelerated corrosion of a metal because of an electrical contact with a more noble metal or nonmetallic conductor in a corrosive electrolyte.

Galvanic Couple A pair of dissimilar conductors, commonly metals, in electrical contact in an electrolyte.

Galvanic Current The electric current between metals or conductive nonmetals in a galvanic couple.

Galvanic Series A list of metals and alloys arranged according to their corrosion potentials in a given environment.

Galvanostatic Refers to an experimental technique whereby an electrode is maintained at a constant current in an electrolyte.

General Corrosion Corrosion that is distributed more or less uniformly over the surface of a material.

Graphitic Corrosion Deterioration of gray cast iron in which the metallic constituents are selectively leached or converted to corrosion products, leaving the graphite intact.

Graphitization The formation of graphite in iron or steel, usually from decomposition of iron carbide at elevated temperatures. [Should not be used as a term to describe graphitic corrosion.]

Grit Small particles of hard material (e.g., iron, steel, or mineral) with irregular shapes that are commonly used as an abrasive in abrasive blast cleaning.

Grit Blasting Abrasive blast cleaning using grit as the abrasive.

Groundbed One or more anodes installed below the earth's surface for the purpose of supplying cathodic protection.

Half-Cell A pure metal in contact with a solution of known concentration of its own ion, at a specific temperature, develops a potential that is characteristic and reproducible; when coupled with another half-cell, an overall potential that is the sum of both half-cells develops.

Half-Cell Potential The potential in a given electrolyte of one electrode of a pair relative to a standard state or a reference state. Potentials can only be measured and expressed as the difference between the half-cell potentials of a pair of electrodes.

Hand Tool Cleaning Removal of loose rust, loose mill scale, and loose paint to degree specified, by hand chipping, scraping, sanding, and wire brushing. [See SSPC-SP 2.]

Hardener [See *Curing Agent.*]

Heat-Affected Zone That portion of the base metal that is not melted during brazing, cutting, or welding, but whose microstructure and properties are altered by the heat of these processes.

Heat Treatment Heating and cooling a solid metal or alloy in such a way as to obtain desired properties. Heating for the sole purpose of hot working is not considered heat treatment.

High-Pressure Water Cleaning Water cleaning performed at pressures from 34 to 70 MPa (5,000 to 10,000 psig).

High-Pressure Water Jetting Water jetting performed at pressures from 70 to 170 MPa (10,000 to 25,000 psig).

High-Temperature Hydrogen Attack A loss of strength and ductility of steel by high-temperature reaction of absorbed hydrogen with carbides in the steel, resulting in decarburization and internal fissuring.

Holiday A discontinuity in a protective coating that exposes unprotected surface to the environment.

Hydrogen Blistering The formation of subsurface planar cavities, called hydrogen blisters, in a metal resulting from excessive internal hydrogen pressure. Growth of near-surface blisters in low-strength metals usually results in surface bulges.

Hydrogen Embrittlement A loss of ductility of a metal resulting from absorption of hydrogen.

Hydrogen-Induced Cracking Stepwise internal cracks that connect adjacent hydrogen blisters on different planes in the metal, or to the metal surface (also known as *stepwise cracking*).

Hydrogen Overvoltage Overvoltage associated with the liberation of hydrogen gas.

Hydrogen Stress Cracking Cracking that results from the presence of hydrogen in a metal in combination with tensile stress. It occurs most frequently with high-strength alloys.

Impingement Corrosion A form of erosion-corrosion generally associated with the local impingement of a high-velocity, flowing fluid against a solid surface.

Impressed Current An electric current supplied by a device employing a power source that is external to the electrode system. (An example is direct current for cathodic protection.)

Impressed Current Anode An electrode, suitable for use as an anode when connected to a source of impressed current, which is generally composed of a substantially inert material that conducts by oxidation of the electrolyte and, for this reason, is not corroded appreciably.

Impulse Dielectric Test A method of applying voltage to an insulated wire through the use of electric pulses (usually 170 to 250 pulses per second) to determine the integrity of the wire's insulation.

Inclusion A nonmetallic phase such as an oxide, sulfide, or silicate particle in a metal.

Inorganic Zinc-Rich Coating Coating containing a metallic zinc pigment (typically 75 wt% zinc or more in the dry film) in an inorganic vehicle.

Instant-off Potential The polarized half-cell potential of an electrode taken immediately after the cathodic protection current is stopped, which closely approximates the potential without IR drop (i.e., the polarized potential) when the current was on.

Intercrystalline Corrosion [See *Intergranular Corrosion.*]

Interdendritic Corrosion Corrosive attack of cast metals that progresses preferentially along paths between dendrites.

Interference Bond An intentional metallic connection, between metallic systems in contact with a common electrolyte, designed to control electrical current interchange between the systems.

Interference Current [See *Stray Current.*]

Intergranular Corrosion Preferential corrosion at or along the grain boundaries of a metal (also known as *intercrystalline corrosion*).

Intergranular Stress Corrosion Cracking Stress corrosion cracking in which the cracking occurs along grain boundaries.

Internal Oxidation The formation of isolated particles of oxidation products beneath the metal surface.

Intumescence The swelling or bubbling of a coating usually caused by heating. [The term is commonly used in aerospace and fire-protection applications.]

Ion An electrically charged atom or group of atoms.

Ir Drop The voltage across a resistance in accordance with Ohm's Law.

Iron Rot Deterioration of wood in contact with iron-based alloys.

Knife-Line Attack Intergranular corrosion of an alloy along a line adjoining or in contact with a weld after heating into the sensitization temperature range.

Lamellar Corrosion [See *Exfoliation Corrosion.*]

Langelier Index A calculated saturation index for calcium carbonate that is useful in predicting scaling behavior of natural water.

Line Current The direct current flowing on a pipeline.

Lining A coating or layer of sheet material adhered to or in intimate contact with the interior surface of a container used to protect the container against corrosion by its contents and/or to protect the contents of the container from contamination by the container material.

Liquid Metal Cracking Cracking of a metal caused by contact with a liquid metal.

Long-Line Current Current though the earth between an anodic and a cathodic area that returns along an underground metallic structure.

Low-Carbon Steel Steel having less than 0.30% carbon and no intentional alloying additions.

Low-Pressure Water Cleaning Water cleaning performed at pressures less than 34 MPa (5,000 psig).

Luggin Probe A small tube or capillary filled with electrolyte, terminating close to the metal surface of an electrode under study, which is used to provide an ion-conducting path without diffusion between the electrode under study and a reference electrode.

Martensite A hard supersaturated solid solution of carbon in iron characterized by an acicular (needle-like) microstructure.

Metal Dusting The catastrophic deterioration of a metal exposed to a carbonaceous gas at elevated temperature.

Metallizing The coating of a surface with a thin metal layer by spraying, hot dipping, or vacuum deposition.

Mill Scale The oxide layer formed during hot fabrication or heat treatment of metals.

Mixed Potential A potential resulting from two or more electrochemical reactions occurring simultaneously on one metal surface.

Modulus of Elasticity A measure of the stiffness or rigidity of a material. It is actually the ratio of stress to strain in the elastic region of a material. If determined by a tension or compression test, it is also called Young's Modulus or the coefficient of elasticity.

Natural Drainage Drainage from an underground or submerged metallic structure to a more negative (more anodic) structure, such as the negative bus of a trolley substation.

Near-white Blast Cleaned Surface A near-white blast cleaned surface, when viewed without magnification, shall be free of all visible oil, grease, dust, dirt, mill scale, rust, coating, oxides, corrosion products, and other foreign matter. Random staining shall be limited to not more than 5% of each unit area of surface (approximately 58 cm^2 [9.0 $in.^2$]), and may consist of light shadows, slight streaks, or minor discolorations caused by stains of rust, stains of mill scale, or stains of previously applied coating. [See NACE No. 2/SSPC-SP 10.]

Negative Return A point of connection between the cathodic protection negative cable and the protected structure.

Nernst Equation An equation that expresses the exact electromotive force of an electrochemical cell in terms of the activities of products and reactants of the cell.

Nernst Layer The diffusion layer at the surface of an electrode in which the concentration of a chemical species is assumed to vary linearly from the value in the bulk solution to the value at the electrode surface.

Noble The positive direction of electrode potential, thus resembling noble metals such as gold and platinum.

Noble Metal (1) A metal that occurs commonly in nature in the free state. (2) A metal or alloy whose corrosion products are formed with a small negative or a positive free-energy change.

Noble Potential A potential more cathodic (positive) than the standard hydrogen potential.

Normalizing Heating a ferrous alloy to a suitable temperature above the transformation range (austenitizing), holding at temperature for a suitable time, and then cooling in still air to a temperature substantially below the transformation range.

Open-Circuit Potential The potential of an electrode measured with respect to a reference electrode or another electrode in the absence of current.

Organic Zinc-Rich Coating Coating containing a metallic zinc pigment (typically 75 wt% zinc or more in the dry film) in an organic resin.

Overvoltage The change in potential of an electrode from its equilibrium or steady-state value when current is applied.

Oxidation (1) Loss of electrons by a constituent of a chemical reaction. (2) Corrosion of a metal that is exposed to an oxidizing gas at elevated temperatures.

Oxidation-Reduction Potential The potential of a reversible oxidation-reduction electrode measured with respect to a reference electrode, corrected to the hydrogen electrode, in a given electrolyte.

Oxygen Concentration Cell [See *Differential Aeration Cell.*]

Paint A pigmented liquid or resin applied to a substrate as a thin layer that is converted to an opaque solid film after application. It is commonly used as a decorative or protective coating.

Paint System [See *Coating System.*]

Parting [See *Dealloying.*]

Passivation A reduction of the anodic reaction rate of an electrode involved in corrosion.

Passivation Potential [See *Primary Passive Potential.*]

Passive (1) The positive direction of electrode potential. (2) A state of a metal in which a surface reaction product causes a marked decrease in the corrosion rate relative to that in the absence of the product.

Passive-Active Cell An electrochemical cell, the electromotive force of which is caused by the potential difference between a metal in an active state and the same metal in a passive state.

Passivity The state of being passive.

Patina A thin layer of corrosion product, usually green, that forms on the surface of metals such as copper and copper-based alloys exposed to the atmosphere.

pH The negative logarithm of the hydrogen ion activity written as: $pH = -\log_{10}(a_H^+)$ where a_H^+ = hydrogen ion activity = the molar concentration of hydrogen ions multiplied by the mean ion-activity coefficient.

Pickling (1) Treating a metal in a chemical bath to remove scale and oxides (e.g., rust) from the surface. (2) Complete removal of rust and mill scale by acid pickling, duplex pickling, or electrolytic pickling. [See SSPC-SP 8.]

Pickling Solution A chemical bath, usually an acid solution, used for pickling.

Pigment A solid substance, generally in fine powder form, that is insoluble in the vehicle of a formulated coating material. It is used to impart color or other specific physical or chemical properties to the coating.

Pipe-to-Electrolyte Potential [See *Structure-to-Electrolyte Potential.*]

Pipe-to-Soil Potential [See *Structure-to-Electrolyte Potential.*]

Pitting Localized corrosion of a metal surface that is confined to a small area and takes the form of cavities called pits.

Pitting Factor The ratio of the depth of the deepest pit resulting from corrosion divided by the average penetration as calculated from mass loss.

Plastic Deformation Permanent deformation caused by stressing beyond the elastic limit.

Plasticity The ability of a material to deform permanently (nonelastically) without fracturing.

Polarization The change from the open-circuit potential as a result of current across the electrode/electrolyte interface.

Polarization Admittance The reciprocal of polarization resistance.

Polarization Cell A DC decoupling device consisting of two or more pairs of inert metallic plates immersed in an aqueous electrolyte. The electrical characteristics of the polarization cell are high resistance to DC potentials and low impedance of AC.

Polarization Curve A plot of current density versus electrode potential for a specific electrode/electrolyte combination.

Polarization Decay The decrease in electrode potential with time resulting from the interruption of applied current.

Polarization Resistance The slope (dE/di) at the corrosion potential of a potential (E)-current density (i) curve. (The measured slope is usually in good agreement with the true value of the polarization resistance when the scan rate is low and any uncompensated resistance is small relative to the polarization resistance.)

Polarized Potential The potential across the structure/electrolyte interface that is the sum of the corrosion potential and the cathodic polarization.

Polyester Type of resin formed by the condensation of polybasic and monobasic acids with polyhydric alcohols.

Postweld Heat Treatment Heating and cooling a weldment in such a way as to obtain desired properties.

Potential-pH Diagram A graphical method of representing the regions of thermodynamic stability of species for metal/electrolyte systems (also known as *Pourbaix diagram*).

Potentiodynamic Refers to a technique wherein the potential of an electrode with respect to a reference electrode is varied at a selected rate by application of a current through the electrolyte.

Potentiokinetic [See *Potentiodynamic.*]

Potentiostat An instrument for automatically maintaining a constant electrode potential.

Potentiostatic Refers to a technique for maintaining a constant electrode potential.

Pot Life The elapsed time within which a coating can be effectively applied after all components of the coating have been thoroughly mixed.

Poultice Corrosion [See *Deposit Attack.*]

Pourbaix Diagram [See *Potential-pH Diagram.*]

Power Tool Cleaning Removal of loose rust, loose mill scale, and loose paint to degree specified by power tool chipping, descaling, sanding, wire brushing, and grinding. [See SSPC-SP 3.]

Precipitation Hardening Hardening caused by the precipitation of a constituent from a supersaturated solid solution.

Primary Passive Potential The potential corresponding to the maximum active current density (critical anodic current density) of an electrode that exhibits active-passive corrosion behavior.

Prime Coat [See *Primer.*]

Primer A coating material intended to be applied as the first coat on an uncoated surface. The coating is specifically formulated to adhere to and protect the surface as well as to produce a suitable surface for subsequent coats. [Also referred to *as Prime Coat.*]

Profile Anchor pattern on a surface produced by abrasive blasting or acid treatment.

Protective Coating A coating applied to a surface to protect the substrate from corrosion.

Reduction Gain of electrons by a constituent of a chemical reaction.

Reference Electrode An electrode whose open-circuit potential is constant under similar conditions of measurement, which is used for measuring the relative potentials of other electrodes.

Reference Half-Cell [See *Reference Electrode.*]

Relative Humidity The ratio, expressed as a percentage, of the amount of water vapor present in a given volume of air at a given temperature to the amount required to saturate the air at that temperature.

Remote Earth A location on the earth far enough from the affected structure that the soil potential gradients associated with currents entering the earth from the affected structure are insignificant.

Resistivity (1) The resistance per unit length of a substance with uniform cross section. (2) A measure of the ability of an electrolyte (e.g., soil) to resist the flow of electric charge (e.g., cathodic protection current). Resistivity data are used to design a groundbed for a cathodic protection system.

Rest Potential [See *Corrosion Potential.*]

Reversible Potential [See *Equilibrium Potential.*]

Rimmed Steel An incompletely deoxidized steel. [Also called *Rimming Steel.*]

Riser (1) That section of pipeline extending from the ocean floor up to an offshore platform. (2) The vertical tube in a steam generator convection bank that circulates water and steam upward.

Rust Corrosion product consisting of various iron oxides and hydrated iron oxides. (This term properly applies only to iron and ferrous alloys.)

Rust Bloom Discoloration indicating the beginning of rusting.

Sacking Scrubbing a mixture of a cement mortar over the concrete surface using a cement sack, gunny sack, or sponge rubber float.

Sacrificial Anode [See *Galvanic Anode.*]

Sacrificial Protection Reduction of corrosion of a metal in an electrolyte by galvanically coupling it to a more anodic metal (a form of cathodic protection).

Scaling (1) The formation at high temperatures of thick corrosion-product layers on a metal surface. (2) The deposition of water-insoluble constituents on a metal surface.

Scanning Electron Microscope An electron optical device that images topographical details with maximum contrast and depth of field by the detection, amplification, and display of secondary electrons.

Sensitizing Heat Treatment A heat treatment, whether accidental, intentional, or incidental (as during welding), that causes precipitation of constituents (usually carbides) at grain boundaries, often causing the alloy to become susceptible to intergranular corrosion or intergranular stress corrosion cracking.

Shallow Groundbed One or more anodes installed either vertically or horizontally at a nominal depth of less than 15 m (50 ft) for the purpose of supplying cathodic protection.

Shielding (1) Protecting; protective cover against mechanical damage. (2) Preventing or diverting cathodic protection current from its natural path.

Shop Coat One or more coats applied in a shop or plant prior to shipment to the site of erection or fabrication.

Shot Blasting Abrasive blast cleaning using metallic (usually steel) shot as the abrasive.

Shot Peening Inducing compressive stresses in the surface layer of a material by bombarding it with a selected medium (usually steel shot) under controlled conditions.

Sigma Phase An extremely brittle Fe-Cr phase that can form at elevated temperatures in Fe-Cr-Ni and Ni-Cr- Fe alloys.

Slip A deformation process involving shear motion of a specific set of crystallographic planes.

Slow Strain Rate Technique An experimental technique for evaluating susceptibility to environmental cracking. It involves pulling the specimen to failure in uniaxial tension at a controlled slow strain rate while the specimen is in the test environment and examining the specimen for evidence of environmental cracking.

Slushing Compound Oil or grease coatings used to provide temporary protection against atmospheric corrosion.

Solution Heat Treatment Heating a metal to a suitable temperature and holding at that temperature long enough for one or more constituents to enter into solid solution, then cooling rapidly enough to retain the constituents in solution.

Solvent Cleaning Removal of oil, grease, dirt, soil, salts, and contaminants by cleaning with solvent, vapor, alkali, emulsion, or steam. [See SSPC-SP 1.]

Spalling The spontaneous chipping, fragmentation, or separation of a surface or surface coating.

Standard Electrode Potential The reversible potential for an electrode process when all products and reactions are at unit activity on a scale in which the potential for the standard hydrogen reference electrode is zero.

Standard Jetting Water Water of sufficient purity and quality that it does not impose additional contaminants on the surface being cleaned and does not contain sediments or other impurities that are destructive to the proper functioning of water jetting equipment.

Steel Shot Small particles of steel with spherical shape that are commonly used as an abrasive in abrasive blast cleaning or as a selected medium for shot peening.

Step Potential The potential difference between two points on the earth's surface separated by a distance of one human step, which is defined as one meter, determined in the direction of maximum potential gradient.

Stepwise Cracking [See *Hydrogen-Induced Cracking*.]

Stray Current Current through paths other than the intended circuit.

Stray-Current Corrosion Corrosion resulting from current through paths other than the intended circuit, e.g., by any extraneous current in the earth.

Stress Corrosion Cracking Cracking of a material produced by the combined action of corrosion and tensile stress (residual or applied).

Stress Relieving (Thermal) Heating a metal to a suitable temperature, holding at that temperature long enough to reduce residual stresses, and then cooling slowly enough to minimize the development of new residual stresses.

Structure-to-Electrolyte Potential The potential difference between the surface of a buried or submerged metallic structure and the electrolyte that is measured with reference to an electrode in contact with the electrolyte.

Structure-to-Soil Potential [See *Structure-to-Electrolyte Potential*.]

Structure-to-Structure Potential The potential difference between metallic structures, or sections of the same structure, in a common electrolyte.

Subsurface Corrosion [See *Internal Oxidation*.]

Sulfidation The reaction of a metal or alloy with a sulfur-containing species to produce a sulfur compound that forms on or beneath the surface of the metal or alloy.

Sulfide Stress Cracking Cracking of a metal under the combined action of tensile stress and corrosion in the presence of water and hydrogen sulfide (a form of hydrogen stress cracking).

Surface Potential Gradient Change in the potential on the surface of the ground with respect to distance.

Tack Coat A thin wet coat applied to the surface that is allowed to dry just until it is tacky before application of a thicker wet coat. (Use of a tack coat allows application of thicker coats without sagging or runs.)

Tafel Plot A plot of the relationship between the change in potential (E) and the logarithm of the current density (log *i*) of an electrode when it is polarized in both the anodic and cathodic directions from its open-circuit potential.

Tafel Slope The slope of the straight-line portion of the E log *i* curve on a Tafel plot. (The straight-line portion usually occurs at more than 50 mV from the open-circuit potential.)

Tarnish Surface discoloration of a metal resulting from formation of a film of corrosion product.

Thermal Spraying A group of processes by which finely divided metallic or nonmetallic materials are deposited in a molten or semimolten condition to form a coating.

Thermogalvanic Corrosion Corrosion resulting from an electrochemical cell caused by a thermal gradient.

Throwing Power The relationship between the current density at a point on a surface and its distance from the counterelectrode. The greater the ratio of the surface resistivity shown by the electrode reaction to the volume resistivity of the electrolyte, the better is the throwing power of the process.

Topcoat The final coat of a coating system. [Also referred to as *Finish Coat.*]

Touch Potential The potential difference between a metallic structure and a point on the earth's surface separated by a distance equal to the normal maximum horizontal reach of a human (approximately 1.0 m [3.3 ft]).

Transpassive The noble region of potential where an electrode exhibits a higher-than-passive current density.

Tuberculation The formation of localized corrosion products scattered over the surface in the form of knob-like mounds called tubercles.

Ultimate Strength The maximum stress that a material can sustain.

Ultrahigh-Pressure Water Jetting Water jetting performed at pressures above 170 MPa (25,000 psig.)

Underfilm Corrosion [See *Filiform Corrosion.*]

Vehicle The liquid portion of a formulated coating material.

Void (1) A holiday, hole, or skip in a coating. (2) A hole in a casting or weld deposit usually resulting from shrinkage during cooling.

Wash Primer A thin, inhibiting primer, usually chromate pigmented, with a polyvinyl butyral binder.

Water Cleaning Use of pressurized water discharged from a nozzle to remove unwanted matter (e.g., dirt, scale, rust, coatings) from a surface.

Water Jetting Use of standard jetting water discharged from a nozzle at pressures of 70 MPa (10,000 psig) or greater to prepare a surface for coating or inspection.

Weight Coating An external coating applied to a pipeline to counteract buoyancy.

White Metal Blast Cleaned Surface A white metal blast cleaned surface, when viewed without magnification, shall be free of all visible oil, grease, dust, dirt, mill scale, rust, coating, oxides, corrosion products, and other foreign matter. [See NACE No. 1/SSPC-SP 5.]

Weld Decay Intergranular corrosion, usually of stainless steel or certain nickel-base alloys, that occurs as the result of sensitization in the heat-affected zone during the welding operation. [This is not a preferred term.]

Wet Film Gauge Device for measuring wet film thickness of a coating.

Working Electrode The test or specimen electrode in an electrochemical cell.

Wrought Metal in the solid condition that is formed to a desired shape by working (rolling, extruding, forging, etc.), usually at an elevated temperature.

Yield Point The stress on a material at which the first significant permanent or plastic deformation occurs without an increase in stress. In some materials, particularly annealed low-carbon steels, there is a well-defined yield point from the straight line defining the modulus of elasticity.

Yield Strength The stress at which a material exhibits a specified deviation from the proportionality of stress to strain. The deviation is expressed in terms of strain by either the offset method (usually at a strain of 0.2%) or the total-extension- under-load method (usually at a strain of 0.5%.)

B
Appendix

SI Units Conversion Table

How to read this table

The table provides conversion factors to SI units.

These factors can be considered as unity multipliers. For example:

Length: m/X

0.0254 in 0.3048 ft

Means that

$$1 = 0.0254\ (\text{m/in}) \text{ or } 1\ \text{in} = 0.0254\ \text{m}$$

$$1 = 0.3048\ \text{m/ft} \quad \text{or } 1\ \text{ft} = 0.3048\ \text{m}$$

The SI units are listed immediately after the quantity; in this case: **Length: m/X**. The **m** stands for **meter**, and the "X" designates the non-SI units for the same quantity. These non-SI units follow the numerical conversion factors.

Note: In the following table at all locations, **ton** refers to U.S. rather than metric ton.

Area: m^2/X	1.0×10^{-4}	cm^2
	1.0×10^{-12}	μm^2
	0.0929	ft^2
	6.452×10^{-4}	in^2
	0.8361	yd^2
	4,047	acre
	$2.59 \times 10^{+6}$	mi^2
Density: (kg/m^3)/X	1,000.0	g/cm^3
	16.02	lbm/ft^3
	119.8	lbm/gal

Quantity	Multiply by	X
	27,700	lbm/in^3
	2.289×10^{-3}	$grain/ft^3$
Diffusion Coefficient: $(m^2/s)/X$	1.0×10^{-4}	cm^2/s
	2.78×10^{-4}	m^2/h
	0.0929	ft^2/s
	2.58×10^{-5}	ft^2/h
Electrical Charge: C/X	1	A-s
	10	abcoulomb
	3.336×10^{-10}	statcoulomb
Electrical conductance: (Siemens)/X	1	ohm^{-1}
Electrical Field Strength: (V/m)/X	1	$kg\text{-}m/A\text{-}s^3$
	100	V/cm
	39.4	V/in
Electrical Resistivity: (ohm-m)/X	1	$kg\text{-}m^5/A^2\text{-}s^3$
	1.0×10^{-9}	abohm-m
	$8.988 \times 10^{+11}$	statohm-m
Energy: J/X	$3.6 \times 10^{+6}$	kWh
	4.187	cal
	4187	kcal
	1.0×10^{-7}	erg
	1.356	ft-lbf
	1055	Btu
	0.04214	ft-pdl
	$2.685 \times 10^{+6}$	hp-h
	$1.055 \times 10^{+8}$	therm
	0.113	in-lbf
	$4.48 \times 10^{+4}$	hp-min
	745.8	hp-s
Energy Density: $(J/m^3)/X$	$3.6 \times 10^{+6}$	kWh/m^3
	$4.187 \times 10^{+6}$	cal/cm^3
	$4.187 \times 10^{+9}$	$kcal/cm^3$
	0.1	erg/cm^3
	47.9	$ft\text{-}lbf/ft^3$
	$3.73 \times 10^{+4}$	Btu/ft^3
	$1.271 \times 10^{+8}$	kWh/ft^3
	$9.48 \times 10^{+7}$	$hp\text{-}h/ft^3$
Energy, Linear: (J/m)/X	418.7	cal/cm
	$4.187 \times 10^{+5}$	kcal/cm
	1.0×10^{-5}	erg/cm
	4.449	ft-lbf/ft
	3461	Btu/ft
	$8.81 \times 10^{+6}$	hp-h/ft
	$1.18 \times 10^{+7}$	kWh/ft
Energy per Area: $(J/m^2)/X$	41,868	cal/cm^2
	$4.187 \times 10^{+7}$	$kcal/cm^2$
	0.001	erg/cm^2
	14.60	$ft\text{-}lbf/ft^2$
	11,360	Btu/ft^2
	$2.89 \times 10^{+7}$	$hp\text{-}h/ft^2$
	$3.87 \times 10^{+7}$	kWh/ft^2
Flow Rate, Mass: (kg/s)/X	1.0×10^{-3}	g/s
	2.78×10^{-4}	kg/h

	0.4536	lbm/s
	7.56×10^{-3}	lbm/min
	1.26×10^{-4}	lbm/hr
Flow Rate, Volume: (m^3/s)/X	1.0×10^{-6}	cm^3/s
	0.02832	cfs
	1.639×10^{-5}	in^3/s
	4.72×10^{-4}	cfm
	7.87×10^{-6}	cfh
	3.785×10^{-3}	gal/s
	6.308×10^{-5}	gpm
	1.051×10^{-6}	gph
Force: N/X	1.0×10^{-5}	dyn
	1	kg-m/s
	9.8067	kg(force)
	9.807×10^{-3}	g(force)
	0.1383	pdl
	4.448	lbf
	4,448	kip
	8,896	ton(force)
Heat Transfer Coefficient: (W/m-K)/X	41,868	cal/s-cm^2-°C
	1.163	kcal/h-m^2-°C
	1.0×10^{-3}	erg/s-cm^2-°C
	5.679	Btu/h-ft^2-°F
	12.52	kcal/h-ft^2-°C
Henry's Constant: (N/m2)/X	$1.01326 \times 10^{+5}$	atm
	133.3	mmHg
	6893	lbf/in^2
	47.89	lbf/ft^2
Length: m/X	0.01	cm
	1.0×10^{-6}	μm
	1.0×10^{-10}	Å
	0.3048	ft
	0.0254	in
	0.9144	yd
	1609.3	mi
Mass: kg/X	1.0×10^{-3}	g
	0.4536	lbm
	6.48×10^{-5}	grain
	0.2835	oz(avdp)
	907.2	ton (U.S.)
	14.59	slug
Mass per Area: (kg/m^2)/X	10	g/cm^2
	4.883	lbm/ft^2
	703.0	lbm/in^2
	3.5×10^{-4}	ton/mi^2
Power: W/X	4.187	cal/s
	4187	kcal/s
	1.0×10^{-7}	erg/s
	1.356	ft-lbf/s
	0.293	Btu/h
	1055	Btu/s
	745.8	hp

	0.04214	ft-pdl/s
	0.1130	in-lbf/s
	3517	ton refrigeration
	17.6	Btu/min
Power Density: $(W/m^3)/X$	$4.187 \times 10^{+6}$	cal/s-cm^3
	$4.187 \times 10^{+9}$	kcal/s-cm^3
	0.1	erg/s-cm^3
	47.9	ft-lbf/s-ft^3
	$3.73 \times 10^{+4}$	Btu/s-ft^3
	10.36	Btu/h-ft^3
	$3.53 \times 10^{+4}$	kW/ft^3
	$2.63 \times 10^{+4}$	hp/ft^3
Pressure, Stress: (Pa)/X	0.1	dyn/cm^2
	1	N/m^2
	9.8067	$kg(f)/m^2$
	$1.0 \times 10^{+5}$	bar
	$1.0133 \times 10^{+5}$	std. atm
	1.489	pdl/ft^2
	47.88	lbf/ft^2
	6,894	lbf/in^2 (psi)
	$1.38 \times 10^{+7}$	$ton(f)/in^2$
	249.1	in H_2O
	2,989	ft H_2O
	133.3	torr, mm Hg
	3,386	in Hg
Resistance: Ohm/X	1	kg-m^2/A^2-s^3
	1	V/A
	1.0×10^{-9}	abohm
	$8.988 \times 10^{+11}$	statohm
Specific Heat, Gas Constant: (J/kg-K)/X	1	m^2/s^2-K
	4187	cal/g-°C
	1.0×10^{-4}	erg/g-°C
	4187	Btu/lbm-°F
	5.38	ft-lbf/lbm-°F
Surface Tension: (N/m)/X	1.0×10^{-3}	dyn/cm
	14.6	lbf/ft
	175.0	lbf/in
Temperature: K/X (difference)	0.5555	°R
	0.5555	°F
	1.0	°C
Thermal Conductivity: (W/m-K)/X	418.7	cal/s-cm-°C
	1.163	kcal/h-m-°C
	1.0×10^{-5}	erg/s-cm-°C
	1.731	Btu/h-ft-°F
	0.1442	Btu-in/h-ft^2-°F
	2.22×10^{-3}	ft-lbf/h-ft-°F
Time: s/X	60.0	min
	3,600.	h
	86,400	day
	$3.156 \times 10^{+7}$	yr
Torque: N-m/X	1.0×10^{-7}	dyn-cm
	1.356	lbf-ft

	0.0421	pdl-ft
	2.989	kg(f)-ft
Velocity: (m/s)/X	0.01	cm/s
	2.78×10^{-4}	m/h
	0.278	km/h
	0.3048	ft/s
	5.08×10^{-3}	ft/min
	0.477	mi/h
Velocity, Angular: (rad/s)/X	0.01667	rad/min
	2.78×10^{-4}	rad/h
	0.1047	rev/min
Viscosity, Dynamic: (Kg/m-s)/X	1	N-s/m^2
	0.1	P
	0.001	cP
	2.78×10^{-4}	kg/m-h
	1.488	lbm/ft-s
	4.134×10^{-4}	lbm/ft-h
	47.91	lbf-s/ft^2
(g/cm-s)/X	1	P
Viscosity, Kinematic: (m^2/s)/X	1.0×10^{-4}	St
	2.778×10^{-4}	m^2/h
	0.0929	ft^2/s
	2.581×10^{-5}	ft^2/h
(cm^2/s)/X	1	St
Volume: m^3/X	1.0×10^{-6}	cm^3
	1.0×10^{-3}	l
	1.0×10^{-18}	μm^3
	0.02832	ft^3
	1.639×10^{-5}	in^3
	3.785×10^{-3}	gal (U.S.)
Voltage, Electrical Potential: V/X	1.0	kg-m^2/A-s^3
	1	W/A

Using the Table: The quantity in braces { } is selected from the table.

Example 1: To calculate how many **meters** are in **10 ft**, the table provides the conversion factor as **0.3048 m/ft**. Hence multiply **10 ft** × **{0.3048 m/ft}** = **3.048 m.**

Example 2: Convert Thermal Conductivity of 10 kcal/h-m-°C to SI units. Select the appropriate conversion factor for these units, **10 (kcal/h-m-°C)** × **{1.163 (w/m-K)/(kcal/h-m-°C)}** = **11.63 w/m-K.**

Index